A LEGACY OF SERVICE TO GEOGRAPHY

WILEY
Publishers Since 1807

John Wiley & Sons is the worldwide leader in geography publishing. We began our partnership with geography education in 1911 and continue to publish college texts, professional books, journals, and technology products that help teachers teach and students learn. We are committed to making it easier for students to visualize spatial relationships, think critically about their interactions with the environment, and appreciate the earth's dynamic landscapes and diverse cultures.

To serve our customers we have partnered with the AAG, the AGS, and the NCGE while building extraordinary relationships with Microsoft and Rand McNally that allow us to bundle discounted copies of Encarta and the Goode's Atlas with all of our textbooks.

Wiley Geography continues this legacy of service during academic year 2001-2002 with the publication of nine first editions and revisions for the undergraduate market.

Regional Geography

- de Blij/Muller — ***Geography: Realms, Regions, and Concepts 10e*** (0-471-31424-2)
- Blouet/Blouet — ***Latin America and the Caribbean 4e*** (0-471-85103-5)
- Weightman — ***Dragons and Tigers: Geography of South, East and Southeast Asia 1e*** (0-471-25358-8)

Physical Geography

- Strahler/Strahler — ***Physical Geography: Science and Systems of the Human Environment 2e*** (0-471-23800-7)
- Marsh/Grossa — ***Environmental Geography 2e*** (0-471-34552-9)
- MacDonald — ***Biogeography: Introduction to Space, Time and Life 1e*** (0-471-24193-8)

Human Geography

- Kuby/Harner/Gober — ***Human Geography in Action 2e*** (0-471-40093-9)

GIS

- Chrisman — ***Exploring GIS 2e*** (0-471-31425-0)
- DeMers — ***GIS Modeling in Raster 1e*** (0-471-31965-1)

ENVIRONMENTAL GEOGRAPHY

SCIENCE, LAND USE, AND EARTH SYSTEMS

Second Edition

WILLIAM M. MARSH
University of British Columbia

JOHN GROSSA, JR.
Central Michigan University

JOHN WILEY & SONS, INC.

New York • Chichester • Weinheim • Brisbane • Toronto • Singapore

For our parents

William R. and Audrey J. Marsh
John Q. and Margaret E. Grossa

ACQUISITIONS EDITOR Ryan Flahive
PROJECT MANAGER Joan Petrokofsky
SUPPLEMENTS EDITOR Mark Gerber
EDITORIAL ASSISTANT Denise Powell
MARKETING MANAGER Clay Stone
SENIOR PRODUCTION EDITOR Norine M. Pigliucci
SENIOR DESIGNER Kevin Murphy
PHOTO EDITOR Jennifer MacMillan
PRODUCTION MANAGEMENT SERVICES Suzanne Ingrao
COVER DESIGN Alison M. Mewett and William M. Marsh

This book was set in Minion by UG / GGS Information Services, Inc. and printed and bound by Courier Westford. The cover was printed by Pheonix Color Corp..

This book is printed on acid-free paper. ∞

Library of Congress Cataloging in Publication Data:
Marsh, William M.
Environmental geography : science, land use, and earth systems / William M. Marsh, John Grossa, Jr.—2nd ed.
p. cm.
Includes bibliographical references and index.
ISBN 0-471-34522-9 (cloth : alk. paper)
1. Environmental geography. I. Grossa, John. II. Title

GE105 .M37 2001
363.7—dc21

2001024553

Printed in the United States of America

10 9 8 7 6 5 4 3

PREFACE

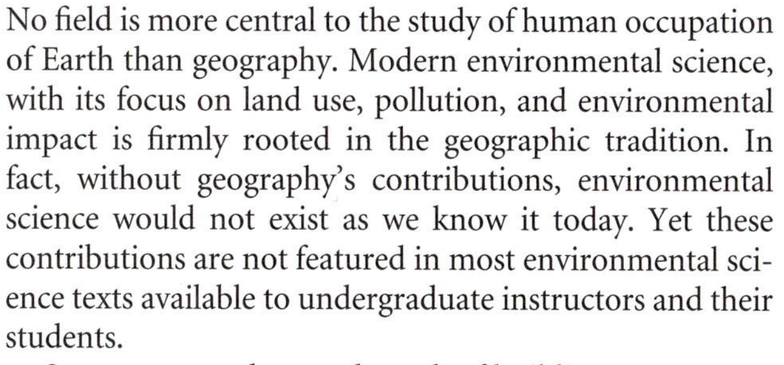

No field is more central to the study of human occupation of Earth than geography. Modern environmental science, with its focus on land use, pollution, and environmental impact is firmly rooted in the geographic tradition. In fact, without geography's contributions, environmental science would not exist as we know it today. Yet these contributions are not featured in most environmental science texts available to undergraduate instructors and their students.

So we set ourselves to the task of building an environmental text with geographic perspective, coverage, and presentation style. It draws on the traditional themes of geography including the concept of an integrated planet; the principles of space, scale, and distributions; the idea of landscape as the product of human–environment interplay; and the perspective of Earth as a dynamic planet where geographic knowledge enables us to see our environment, its opportunities, constraints, and risks with the insight of an educated person.

This edition of *Environmental Geography* builds on the 1996 edition and includes significant revisions and additions. Data and discussions have been updated, most graphics have been revised, and many new drawings and photographs have been added. Two largely new chapters have been added: Chapters 1 and 3. These chapters address the concepts of development, carrying capacity, limits to growth, and sustainability. Land use sustainability is featured in Chapter 3 and again in Chapter 18, the last chapter. Chapter 18 takes a broad perspective on the issue and also examines sources of geographic information essential to monitoring and managing the global environment.

The book is organized into 18 chapters of 20–25 pages each. Although the chapters form a logical sequence, individual chapters provide fairly complete treatments if read independently. Therefore, reading assignments can be arranged to suit different course organizational schemes. In addition, with a little over 400 pages of text, it is possible to cover the book in one semester. Instructors will also find a variety of supplemental materials available through a web site prepared by Randall L. Repic.

The first edition of *Environmental Geography* was several years in writing and production. Many people contributed to that effort, and we thank them all once again. A new group helped us with this edition, and we owe each person our sincere thanks. It was led by students and faculty users of the first edition who provided helpful suggestions for improving the text. Among them Randall L. Repic and his students at University of Michigan-Flint provided especially valuable advice. We benefitted equally from reviewers: Robert Brinkman of The University of South Florida at Tampa, Geoffrey L. Buckley of Ohio University, James Dyer of Ohio University, Melinda Laitun of Colorado State University, Robert Mason of Temple University, Gene J. Paull of University of Texas at Brownsville, and William Riebsame at University of Colorado. Credit must also go to the Wiley editorial and production team: Ryan Flahive, Jennifer MacMillan, Denise Powell, Suzanne Ingrao, Norinne Pigliucci, Kevin Murphy, and Nanette Kauffman. Finally, thanks to Carol Grossa, Alison Mewett, Martin Kaufman, Richard Hill-Rowley, Keith King, and Robert Chipman for support and direction.

W. M. M.
Courtenay, B.C.

J. G., Jr.
Mt. Pleasant, Michigan

INTRODUCTION

If the space program has given us nothing else, it has given us an invaluable perspective on Earth. It has shown that Earth is unique among the nine planets that orbit the sun. Unlike Jupiter, Saturn, Uranus, and Neptune, it has a stable, hard surface that is able to support the weight of oceans, glaciers, and living matter. Unlike Mercury and Venus, it is not blistering hot, scorched dry of all but rock. Unlike Mars, the Earth is insulated by a thick oxygen-rich atmosphere, with a modest surface temperature and abundant liquid water.

Earth is a glorious planet, unimaginably diverse and constantly changing. Its history is punctuated with dramatic events that have altered the course of environmental development—some advancing life, some setting it back. These events have been driven both by external factors, such as asteroid collisions, and internal ones, such as massive volcanic eruptions and, most recently, the explosion and spread of human population armed with technology.

Humans have a short tenure on Earth as a species, and a considerably shorter one as a significant environmental change agent. Earth is at least 4.5 billion years old. Life has been here at least 3 billion years, and advanced terrestrial life, such as vascular plants, has been around for about 500 million years. Humans equipped with agriculture and living in permanent settlements date back only about 12,000 years. The Industrial Revolution and the growth of cities with millions of people are less than 250 years old. Although humans have modified the environment since the origin of agriculture, the types and scales of environmental change we see today—the changes that cause us serious concern—are only one to two centuries old.

Set in the scale of Earth time, humans are a sudden flash on the planet, like the great asteroid collision 65 million years ago that so dramatically altered the environment that it ended the dinosaurs and thousands of other species. Could we be living in the midst of a similarly dramatic event driven by our own species? In 1900, world population was 1.5 billion people; it is now more than 6 billion and growing by more than 80 million people a year. By the end of this century, Earth can expect a human population between 9 and 11 billion. The pace of change accompanying this growth is dizzying as development lurches ahead and the planet's environment is retrofitted with human land use. It is an exciting time to be on Earth, but it is also a scary time, because we may be caught in the whirlwind of one of Earth's great environmental events.

No matter how we look at our place in Earth history, some monumental questions loom before us as we charge into the twenty-first century. Are we a species out of control, staged to overrun the planet and consume ourselves out of house and home? Are we unique because of our ability to use technology to solve the survival problem so that our species will ultimately govern the planet and ensure its own survival? Are we insightful and wise stewards with a capacity to foresee the perils, make adjustments to curb our growth, and design a sustainable future for ourselves and the Earth as a whole? The answers to these questions will likely unfold in your lifetime.

This book is about Earth, its environment, our use, misuse, and abuse of the environment, and where we seem to be headed as a species on the planet. This edition opens with some observations about the state of the global environment and the trends of change that can be expected in this century. Geographic perspective is important because there is no longer a handful of environmental crises scattered around the world, relatively isolated from one another. Local and regional problems have grown in both magnitude and scale, merged together, and taken on global proportions. We are indeed a global community in the way we use the environment, carry out development, and interact geographically. America's preoccupation with the activities of the rest of the world is more than an expression of might; it is a deeply held sense, driven by uncertainty (and sometimes fear), that like it or not we are all in one boat, but we are not rowing together and may not have a compass.

Chapter 2 sets the geographic stage by introducing some facts about Earth and the perspective that geographers bring to the study of our planet and its people. Chapter 3 discusses the issue of sustainability. Can we find a balance among the human population, a reasonable quality of life, and the use of Earth's resources that does not degrade the global environment and ultimately deprive future generations of a hopeful life? Can economic development continue to expand via global networks that appear to be insensitive to geographic differences in resources, environmental conditions, and human culture?

Chapter 4 examines the world's vast frontier regions such as the deserts and tropical rainforests where human occupation and environmental impact have traditionally been light. Pressure to use these lands is mounting rapidly and extensive tracts are now being settled and developed. The frontier environments may be the test cases signaling the next phase in the human planetary event. The next two chapters set the ecological framework, examining the Earth's systems of materials and energy and then exploring the nature of ecosystems, the building blocks of the biogeographical environment.

Chapters 7, 8, and 9 examine the principal human elements of the global environment: population, agriculture, and energy. No human force is more prominent in the modern environment than our population, and no land use is more widespread and environmentally demanding than agriculture. The world grows a population equal to that of the United States and Canada combined every 3 to 4 years. Feeding the existing population is a monumental task; feeding a population which may double Earth's present numbers by the end of the century seems unimaginable.

Chapters 10, 11, 12, and 13 explore the Earth's air and water systems, including their use, physical modification, and pollution by humans. These two systems are essential to life and are highly susceptible to abuses by human activities, especially where those activities exceed nature's self-renewing capabilities. Chapter 14 addresses hazardous waste, a serious problem related to rising energy use, increased consumption of manufactured goods, and the introduction of new technologies involving exotic chemical substances. Chapter 15 takes up one of the most valuable but least appreciated components of the environment, soil. There is no secret about where Earth's most favorable soils are located—farmers have already found them, and under current farming pressure, most of these soil reserves are declining rapidly.

The push for new land to house people, grow crops, and provide wood and mineral resources has taken a serious toll on Earth's rich reservoir of biological species. The tropical forests, which hold half or more of Earth's species, are being cleared at alarming rates with serious ecological consequences. This is covered in Chapter 16. Chapter 17 follows with a review of the world's open lands and the effort to preserve and manage them as parks, national forests, and other public spaces.

The final chapter comes back to the issues of development, environment impacts, and sustainability raised in the opening chapters. After a less-than-cheery prognosis, it examines some of the global efforts being made to find an environmentally sustainable system for the planet. The chapter ends with a brief overview of two environmental surveillance and mapping technologies: remote sensing surveillance and geographic information systems. These systems will be absolutely essential if we expect to build effective management programs for the global environment, and the field of geography will play a central role in the collection, analysis, and interpretation of the data they produce.

CONTENTS

1

A WORLD IN CRISIS: ENVIRONMENT AND HUMANITY IN THE TWENTY-FIRST CENTURY

1.1 INTRODUCTION

The world is more crowded, more polluted, more urban, more biologically stressed, and warmer than ever before in recorded history. During the twentieth century, human population increased more than threefold, from less than 2 billion to over 6 billion. The number of cities with more than a million people has grown from less than 20 to more than 300, and in the last 75 years, many cities have grown by 25 times or more. The largest cities in the world now contain a startling 30 million people.

During the same time, the number of automotive vehicles in the world has grown from a few tens of thousands to more than half a billion. The consumption of resources such as oil, water, and metals, has increased more than 10 times, while air pollution has increased even more. Human activities worldwide now add as much as 7 billion tons of carbon dioxide to the atmosphere every year as well as untold amounts of synthetic substances such as DDT.

It is also a world of stark contrasts. While the quality of the global environment declined in the twentieth century, the world economy grew enormously. But the economic growth was unevenly distributed over the planet. In fact, today the world is divided into three great camps: the *haves*, represented by the developed countries, the *have-nots*, represented by the less developed or developing countries, and the *deprived*, represented by the very poor, undeveloped countries (Figure 1.1). The first camp is led by Americans, Canadians, Europeans, and Japanese, who consume more resources, have higher standards of living, enjoy better health, have fewer children, live longer, and produce more waste than members of the second and third camps. The second camp includes populous countries such as China, India, Indonesia, Brazil, and Egypt, which are more or less in the process of developing their economies. The third camp includes environmentally stressed and economically destitute countries such as Bangladesh, Somalia, Sudan, Eritrea, Haiti, and Yemen. But there are five times more peo-

FIGURE 1.1 *Photographs representing the three main classes of countries in the world (a) developed, (b) developing, and (c) deprived. A suburban community in South Florida; a peasant village in Yunnan Province, China; and a residential neighborhood in Haiti.*

ple in developing and undeveloped countries than in developed countries, and the number of have-nots and deprived people is increasing at a rate more than 15 times that of haves. All the while, the economic gap between the haves and the rest of the world widens.

We know that all three groups of countries exact a heavy toll on the environment. We also know that human impact on the planet varies geographically with population density, land use, consumption rates, and the character of the local and regional environment in both more and less developed countries. But the global environment is more or less blind to the human differences among countries and regions. It interacts with all of us, irrespective of political borders, standards of living, and national economic power. We are all members of a global environmental community that links life, both human and nonhuman, together into one great, integrated whole. In short, we share the same resources and the same risks, and in the end our failures at managing the environment in one part of the world becomes the burden of the whole world. Although the solutions to the Earth's environmental crisis may depend on geographically specific solutions, it is the sum total effort of the global community that really counts.

The goal of this and the next two chapters is to introduce this book by examining the nature of the environmental dilemma in which we find ourselves. In this chapter we first look at some key environmental trends of the twentieth century that have carried into this century. We then examine the nature of development and the concepts of carrying capacity and limits to growth. The chapter ends with some thoughts on the role you can play in the great struggle of environment and survival.

1.2 PROSPECTS FOR THE NEW CENTURY

The twenty-first century may be the threshold century for humans, their quality of life, and the quality of life for many other organisms on the planet. The signs of wholesale environmental change—and, in the eyes of some observers, impending disaster—are all around us. Can we make reliable forecasts about population, land use, pollution, and environmental change for the decades ahead? Strictly speaking, of course not, especially forecasts aimed at the second half of the century. But forecasts aimed at the next several decades are certainly more probable.

Population

The annual rate of growth in world population reached a peak of around 90 million persons between 1985 and 1990. By the year 2000, this rate had fallen to about 80 million persons per year. Because of a number of factors, the populations in developed countries such as Germany, Italy, and Russia have stopped growing or even declined somewhat. This is not the case for less developed countries such as Brazil, Mexico, and India, where population growth rates are still substantial and will remain so for many years to come.

Because developing countries contain almost five times more people than developed ones, world population will continue to grow well into the twenty-first century. However, experts predict that growth will level off and world population will stabilize by the end of this century. The critical question is, what will the global population be when stability is reached?

The United Nations has calculated three different projections of world population based on high, medium, and low fertility rates. The human fertility rate is defined as average number of children born to a woman, and when it reaches a global average of 2.1, world population will stop growing within a few decades. The projections use lower fertility rates for developed countries and higher fertility rates for developing countries. At the highest rate, world population will reach 11.2 billion by 2050. At a medium rate it will reach 9.2 billion by 2050. At a low rate, it will reach 7.7 billion by 2050. For any of the three projections, more than 98 percent of all the growth will take place in the world's developing and deprived countries (Figure 1.2).

Economic Development

The world experienced unprecedented economic growth in the twentieth century. Since 1950 the global economic output has increased about five times. The vast majority of this growth took place in the developed world, where standards of living and wealth have increased substantially. This development was founded on increased rates of resource extraction, industrialization, consumption, and waste production. The resource extraction systems not only grew larger in terms of the amounts of fuel and raw materials used, but expanded geographically by reaching into developing countries as domestic supplies began to dwindle.

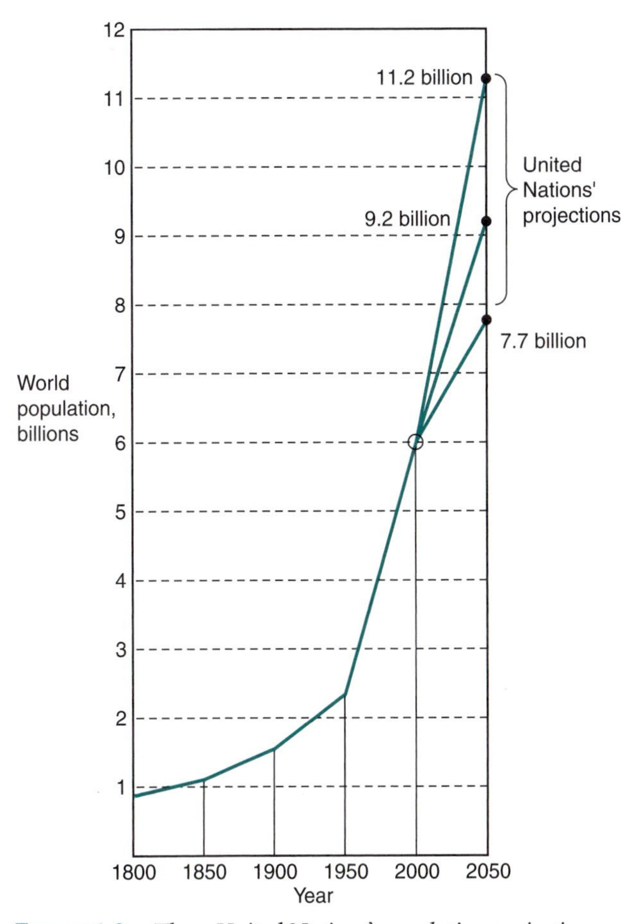

FIGURE 1.2 *Three United Nations' population projections for the twenty-first century. Each is based on a different fertility rate.*

Toward the end of the twentieth century, some poor countries also began to experience economic development—in particular, the populous countries of South and Southeast Asia. This growth was fostered by integration, or globalization, of the world economic system leading to increased investment and trading among developed and developing countries. Wages improved, poverty declined, infant mortality declined, and life expectancy increased in these countries. However, air quality declined, deforestation accelerated, urban sprawl advanced, populations burgeoned, and social conflict increased. Yet in some other poor countries, economic conditions worsened as a result of social and political unrest. Civil upheaval and tribal warfare, for example, have left African countries such as Sudan, Somalia, Chad, Liberia, Rwanda, and others in chaos.

Despite the economic advances in developing countries, for the world as a whole, the gap between rich and poor countries continues to widen. In 1960 the gap between the richest 20 percent of the world and the poorest 20 percent was 30 to 1. In 1992, it was 61 to 1. Today this gap is even greater. The poorest 20 percent of the people in the world today make only 1.1 percent of the global income. Clearly not all countries have benefited from economic globalization and new development. At least one-fifth of the world's people are worse off today in terms of income than they were a generation ago (Table 1.1).

Forecasting economic development trends is risky business, and we would only be guessing if we tried to describe what the world's economy will look like in 2050. However, we can cite some inevitable trends or

TABLE 1.1

*QUALITY OF LIFE DIFFERENCES BETWEEN RICH AND POOR COUNTRIES**

	Poor Countries	Rich Countries
Gross national product per person	less than $200	about $23,000
Life expectancy	49 years	77 years
Birth rate per 1,000 people	45	13
Infant mortality per 1,000 live births	122 deaths	6.4 deaths
Grams protein per day	50	95
Percent safe drinking water	36%	100%
Female literacy	20%	close to 100%

*Based on differences between 10 richest and 10 poorest countries in the world.

Source: World Resources Institute, 1994–95.

conditions that we know will influence economic development. First, world population will increase by 2 to 5 billion people, with most of the increase occurring in the poorest countries most in need of economic development. Pressure on basic resources such as open land and water supplies will increase, and per capita rates of food production may actually fall in these countries. In addition, without substantial income growth in poor countries, most of the additional people will have little or no buying power and will not create new markets for the global economy. Second, selected resources (e.g., fossil fuels) that are necessary for economic development are likely to decline, limiting industrial development and efforts to develop Western-style consumerism. Third, the demand for consumer products will rise, but the capacity to increase consumption will not rise correspondingly in many poor nations. As the new century progresses, the gap between the world's rich and poor countries is likely to grow even wider, with an increasing number of people counted among the poor and deprived.

Environment

The twentieth century saw more deterioration of the environment than any century in human history. Let us begin with a brief review of the impacts on land. As population and farming expanded, soil erosion increased dramatically in the second half of the twentieth century. The annual loss of sediment to the oceans now more than doubles that of preagriculture times (before about 10,000 years ago). With the expansion of farming in the next century, total erosion will rise. Soil loss in some regions is likely to double current rates. Highest erosion rates will occur in poor countries with large, growing populations as people push into lands marginally suited for farming. At the same time, deserts will continue to expand because land-use pressure from grazing, deforestation, and cultivation, coupled with drought, will cause degradation of grasslands and woodlands on their fringes. Again, this trend will be most pronounced in poor countries of South America, Africa, and Asia (Figure 1.3a).

Deforestation in the tropics increased rapidly in the second half of the twentieth century. In this period, the world lost more than 20 percent of its tropical forests. In some West African countries (e.g., Sierra Leone, Ivory Coast, and Ghana), significantly less than half the tropical forests remain. In the Amazon, the world's largest area of tropical forest, most of the forest is intact, although deforestation is advancing at a rate of about 1 percent of the remaining forest area per year. In short, everywhere tropical forests grow they are declining, and some forecasts estimate that as little as 10 percent of these forests will remain in the world by the second half of the twenty-first century. With the loss of tropical forest, Earth's biodiversity will decline precipitously, because approximately 50 percent of our species of plants, animals, and microorganisms live in these forests. On the other hand, in other parts of the world, particularly the midlatitudes, forest cover will likely increase as population stabilizes, land area under cultivation declines, and reforestation and preservation efforts expand. Species lost in these regions during the period of deforestation and settlement will not, of course, recover (Figure 1.3b).

FIGURE 1.3 *Photographs showing (a) desert sand dunes encroaching on a settlement in the Sahara desert of North Africa; (b) tropical forest eradication in South America; and (c) a water-stressed landscape in India.*

Air pollution increased significantly in the second half of the twentieth century with the growth of global population, industry, and automobiles. Although pollution rates for many developed countries declined near the end of the twentieth century, global air pollution is expected to increase in the twenty-first century with population growth, development, and increased consumption in countries such as China, India, Mexico, and Brazil. One outcome of global air pollution is atmospheric warming related to the addition of carbon dioxide, a heat-absorbing gas, primarily from the burning of fossil fuels mainly in developed countries. Atmospheric carbon dioxide increased by 30 percent over the past two centuries and may reach twice the level of the preindustrial era (before 1750) by 2050. Although the whole issue of global warming and its causes are hotly debated, evidence indicates that the warming process has already begun. What the environmental effects will be are far from certain, but it is likely that earth will witness rising sea levels as glacial ice melts, increased storminess in some regions, increased droughts in others, and shifts in bioclimatic zones in the middle latitudes.

In the twentieth century, global water consumption increased six times, more than twice the rate of population growth. Water use rates will continue to rise, and while water supplies are abundant over much of the world, large areas in developing countries lack adequate supplies for expanding agriculture that requires irrigation. Currently, one-third of the world's people have less than adequate water supplies, and that could rise to two-thirds or more by 2035 (Figure 1.3c).

FIGURE 1.4 *The sprawling mass of Djakarta, Indonesia, one of scores of rapidly growing coastal cities.*

Land use

With the growth in world population and economic development, land use changed enormously in the past century. The trend toward urbanization, which began with the Industrial Revolution, continued throughout the twentieth century at an accelerated rate. Today urbanization is advancing fastest in the developing world as poor people, deprived of land and the capacity to farm, are displaced to the urban fringe. Urban population worldwide is growing four times faster than world population, and this trend is expected to continue well into this century. By century's end, Earth may see sprawling urban regions containing up to 40 or 50 million people. Most of these megacities will be in coastal areas of Asia, South America, and Africa (Figure 1.4).

In rural areas, population and farming will increase in frontier lands such as the desert fringe, tropical forests, and mountain lands. Since most of these lands are marginal for farming, the potential for land-use failure is high, and land abandonment and regional migration will be common in the twenty-first century. But many governments, in order to encourage development, will promote settlement in the marginal lands by building roads, dams, and even new towns.

Despite the population forecasts, existing economic stress, and existing and impending environmental difficulties throughout much of the world, virtually every nation on earth, rich and poor alike, is promoting developmen as a means of improving the quality of life. To some observers, this flies in the face of reason because they see development as the root of some of our worst environmental and social problems. However, development is also one way to improve personal and family security and, in turn, reduce the need for large families. Managed properly, development can lower population growth and bring countries into better balance with food production and land and water resources.

1.3 THE CONCEPT OF DEVELOPMENT

We hear and read about development all the time. Sometimes the word *development* is applied to taking up farmland around cities, sometimes to destroying tropical forests in South America, sometimes to investing in the Asian economy. Just what is development? For our purposes, **development** can be defined as economic growth involving a change in land use to a more economically productive activity as, for example, from woodland (forestry) to cropland (farming). **Economic productivity** can be defined by the amount of revenue a land use generates per acre or square kilometer of land. In our economic system, different land uses compete for land, and those able to generate the highest financial return win out, unless, of course, a less productive use is protected by special laws. Public parks are an example of a protected land use.

Other examples of development include replacing cropland with housing, displacing a residential neighborhood to build a factory, and removing a factory to build a shopping center. In almost every instance, development results in a higher land use density—that is, a greater intensity of activities and more built facilities (e.g., buildings, roads and utility systems). **Density** can be measured in various ways, for example, by the amount of surface coverage by facilities, by the number of workers or occupants, by the amount of energy used, or by the quantity of waste produced. Greater-density land use usually produces higher levels of environmental degradation, including more air pollution, more water pollution, more stormwater, and lower biodiversity than lower density ones.

Broadly speaking, development is driven by expanding economic systems and a growing world population. For wealthier countries, however, development is advancing faster than population because per capita consumption rates are rising. In the United States and Canada, each person consumes more goods, services, and resources each decade. If this trend continues, the economy of these countries will continue to grow and development will continue to advance even after these countries have achieved zero net population growth. In many poorer countries, *both* population *and* consumption rates are increasing and will probably continue to do so for many decades (Figure 1.5). There is a strong tendency in poor countries to promote Western style (i.e., North American and European) consumerism and development, despite the environmental, social, and other problems associated with consumer-oriented development.

This means increasing commercial and industrial development. In many instances new development competes with existing agricultural land uses that are essential to maintaining sustainable economies in poor countries. Not only does the new development displace farming, but it fosters urban sprawl, draws people from rural lands, and causes farming to decline over broad areas. The point here is that development for many poor countries does not immediately or inevitably lead to greater personal security and higher standards of living for the population as a whole. In addition, it puts many people at risk, because when the economy slumps and jobs are lost, as inevitably happens, those who moved to cities have no means of income or growing food.

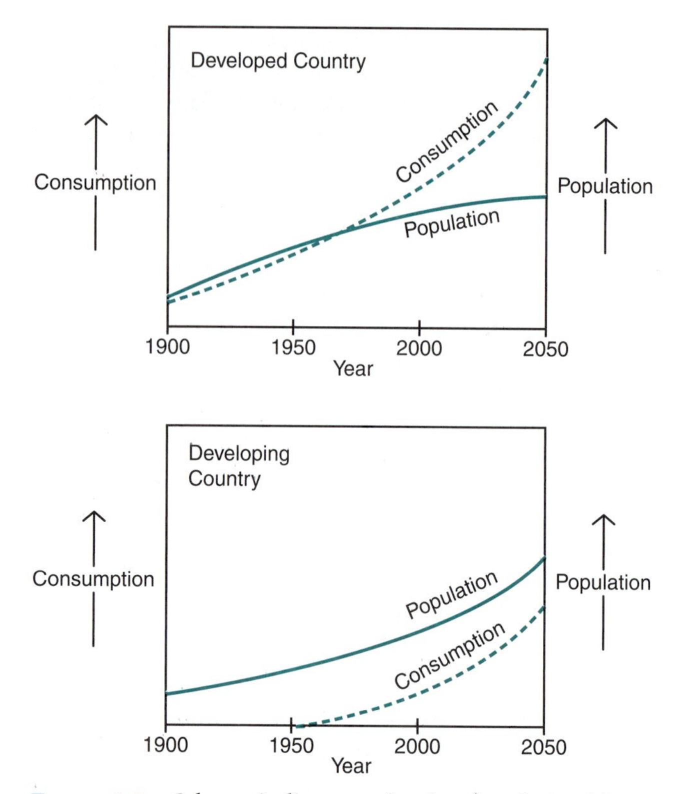

FIGURE 1.5 *Schematic diagrams showing the relationship between population and consumption in a developed and developing country.*

When we apply terms like **developed countries** and **developing countries** to the nations of the world, we are usually referring to the general status of a nation's economy. At this level, these are rather imprecise terms that reflect the general characteristics of national production, consumption, income, and demographic conditions. Rather than falling into discreet categories with well-established definitions, the economies of the world's nations lie more on a broad continuum. For example, some nations with rapidly growing economies have recently attained many of the *economic* characteristics but few of the *social* characteristics of a developed country. Others that are classified as developing countries have stagnant or declining economies, and they might never attain the features of a developed economy.

What are the characteristics of a developed country? Countries with developed economies include relatively high productivity and income, a labor force engaged mostly in service or industrial occupations, and demographic characteristics that include even-aged, nearly stable or slowly growing population that are mostly urban. Nearly all of the western European countries, Japan, Australia, Canada, and the United States have long-standing status as developed economies, while countries like Israel, South Korea, and some eastern European countries such as

Poland and Hungary have more recently been included in tabulations of economically developed countries. These most affluent nations have less than one-fifth of the world's population, yet they generate nearly 80 percent of the monetary income (Figure 1.6).

In sharp contrast to the relatively affluent citizens living in urban areas of the economically developed countries, the majority of the world's people are relatively poor and live in countries where rural landscapes and villages are dominant. In many of these countries, a large agricultural labor force is necessary to produce adequate food supplies. As a result, the industrial and commercial sectors usually play a relatively small but often growing role in the economy of most developing countries.

As you might expect, within this large group of developing countries there are many disparities. Some countries, including the world's two most populous—China and India—support growing economies that are becoming more diversified and productive. In these countries a demographic and economic transition features slowing population growth and urban expansion as their burgeoning service and industrial sectors participate in the rapidly growing global marketplace.

Indeed, in several developing economies the pace of economic growth may have been too frenetic. For several Asian countries, including Thailand, South Korea, Malaysia, and Indonesia, the rapid economic expansion of the 1980s through the mid-1990s ended suddenly in 1997. After suffering severe economic setbacks in the late 1990s, the economies of these countries are now growing again. The breakup of the Soviet Union was another setback for several countries. After the USSR broke apart in 1991, many of the newly created independent nations, including Uzbekistan, Tajikistan, and Kazakhstan, became struggling developing economies. Many have experienced prolonged economic decline as they have struggled to build their own economic policies and infrastructure.

The world's very poorest countries—**deprived countries**—are losing ground every year. These are mostly African nations where political instability sometimes approaching anarchy has led to civil and intertribal warfare. Their economies are in shambles; their production and transportation infrastructure have been damaged or destroyed by ongoing conflict. Until political conditions stabilize there will be little opportunity for economic development. For these nations the term *developing economies* is misleading and inappropriate.

Furthermore, within *any* nation, be it economically affluent or poor, there is a wide range of economic conditions. Millions of people living in the world's wealthiest nations have standards of living well below that of the nation's average citizen. Likewise, in the poorest countries a very small fraction of the people command massive wealth and live in sumptuous luxury (Figure 1.7).

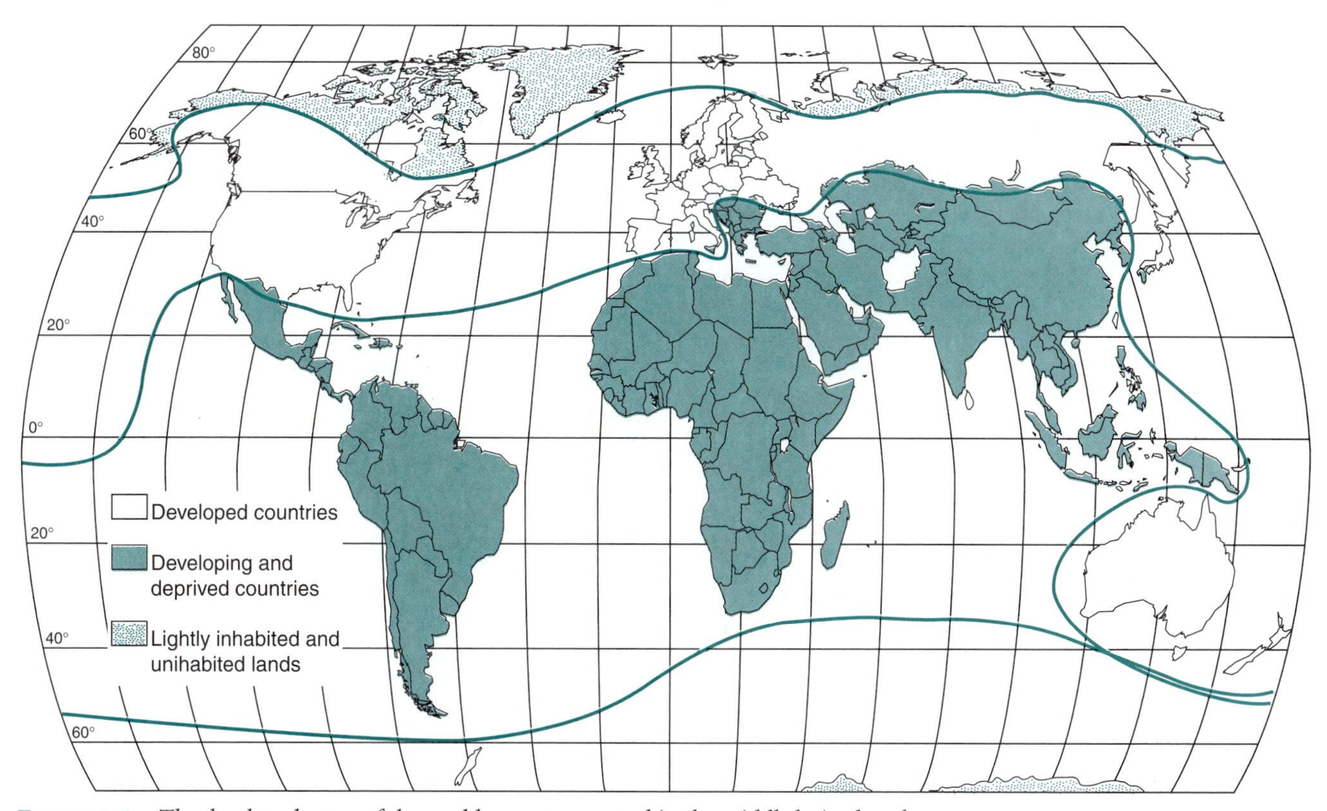

FIGURE 1.6 *The developed areas of the world are concentrated in the middle latitudes whereas the developing and deprived countries are concentrated more in the low latitudes.*

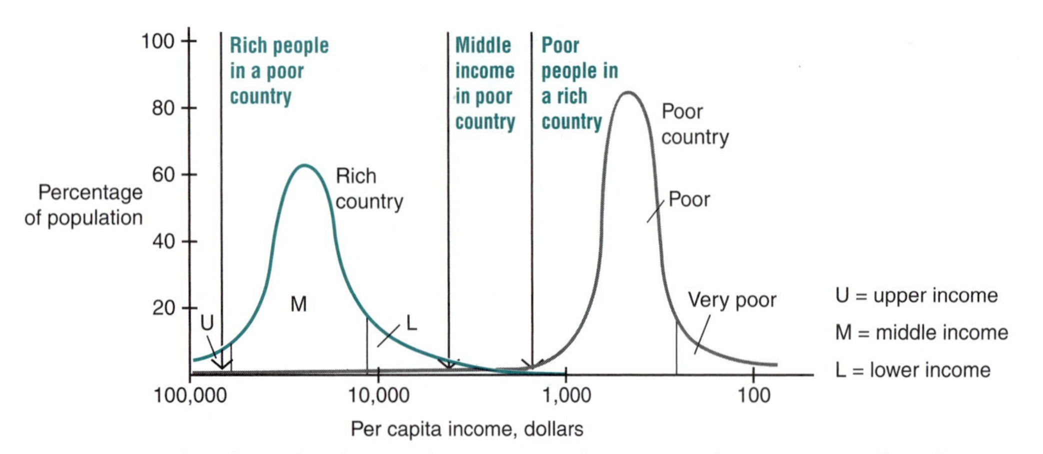

FIGURE 1.7 *The relative distribution of income in a rich country and poor country. The rich country has a large middle income population and small percentage of very poor. The poor country is dominated by poor and very poor people, with a small fraction of rich.*

And while economic disparities have always existed to some degree within all countries, in many countries, income and lifestyle disparities have increased. Thus, the concept of a developed nation, a developing nation, and a deprived nation is based on the degree to which a country has attained certain economic and life-quality characteristics for its population in general. To some extent, there are elements of developed, developing, as well as deprived classes in every country.

1.4 EARTH'S CAPACITY TO SUPPORT HUMANS

Since the 1960s, the world's population has doubled, from about 3 billion people in 1965 to more than 6 billion today. All the major centers of population have grown, and many have spread into new lands. Remarkably, agricultural production has kept pace with population growth. However, since 1965, per capita agricultural production has increased by only 0.5 percent for the world as a whole, and since 1990, per capita agricultural production has by some estimates declined by about 1 percent. Since 1965 more than 200 million people have died of hunger or hunger-related diseases, and the quality of life, according to some observers, actually declined for 1 to 2 billion people. Today, as many as 1 billion people (18 percent of humanity) are incapable of vigorous work and good health because of malnourishment.

Earth must support a huge and rapidly growing number of humans. Each day more land and resources are needed to sustain humanity, and some of the resources being used are nonrenewable (e.g., fossil fuels), while others that are considered renewable are being seriously degraded. The result is a decline in the life-support capacity of the global environment. Deserts are spreading into grasslands, energy reserves such as petroleum and coal are being depleted, tropical forests are being replaced by farms, roads, and settlements, and domes of air pollution are expanding as cities grow and spread. The principal driver is population growth. Although it may sound radical, could the human population machine literally be consuming its own habitat? In the extreme, are we like the farmer whose solution to his hunger problem is to eat his own seed?

It took more than 100,000 years for the human population to reach 2 billion people (in 1930) but only 46 years to raise that number to 4 billion. The current world population is increasing by about 80 million a year. Optimists argue that the planet can support 10 to 12 billion people, twice the current number. Idealists, such as some religious leaders, think the planet is capable of supporting 40 billion. Pessimists or realists—depending on how you define them—argue that world population has already exceeded a reasonable limit, defined as an environmentally sustainable population.

The more pessimistic of this group argue that we are beyond the point of recovery because the momentum of the population explosion is too great and the planet's resources have already been irreparably damaged. Those less pessimistic argue that it might be possible to slow growth and recover the remaining resources, but to achieve a sustainable balance between humans and resources would require a colossal effort with much sacrifice and decades of time.

The Carrying Capacity Concept

The number of people the Earth can support is called its **carrying capacity.** It can be defined in purely biological terms based on the maximum limits of global food production with no consideration to quality of life, other organisms, and environmental disturbance, or it can be defined in cultural terms, taking into account human values for the quality of life, environment, and future generations. **Cultural carrying capacity** is defined by the size of the population that can live in a long-term, sustained balance with the environment at a reasonable quality of life with land use systems that do not degrade over time. It includes not only sustainable food production that conserves soil and water resources, but maintenance of other life forms and environmental features that are not of direct and immediate economic value. It also means living with a capacity to sustain disturbances from nature (e.g., hurricanes and earthquakes) without permanent destruction to life systems.

Is human/Earth sustainability possible in our times? Without population control it is very doubtful, even with optimistic improvements in food production and environmental management of land, water, air, and biological resources. It is especially doubtful when we consider that most societies are also striving to increase consumerism. Thus, not only are there more people, but also each person is consuming more goods and services, which, in turn, demands more resources, creates more waste, and degrades more environment.

Earth is not an infinite bounty of resources. Through experiments with expanding animal populations in finite ecosystems, biologists have shown that population rapidly outgrows resources. This leads to malnutrition, pollution, and social stress. Both environment and organisms decline, resulting ultimately in extermination through massive die-offs. But in animal populations this trend is associated with declining consumption rather than the increasing consumption associated with humans.

In addition, modern consumerism depends on market competition, and competition depends on innovations. Innovation demands technological advancements: new tools, new materials, new techniques—all of which bring to the environment things that it has not seen before. These things—technical processes, machinery, chemicals, and new land uses—are often incongruent with the environment. That is, the environment cannot process them efficiently without damage to ecosystems, water systems, or air systems.

1.5 LIMITS TO GROWTH

As we consider the tenuous state of the environment and humanity, we must highlight four critical factors. First and foremost is *population*, for without massive human populations and the living space, water, energy, and food they demand, problems of the environment would be relatively small or nonexistent. The second is *technology*, represented by all those things of cultural origins—including food production and pesticides, industry and hazardous wastes, expressways and automobiles—that enable us to reach further into the environment in more profound and exotic ways. The third is *consumerism*, the ability of each person to utilize greater and greater shares of resources, including wood, metal, plant fiber, medicines, water, and air itself. The fourth is *land use*, the types of activities carried out by humans on Earth, especially our ability to build land-use systems that displace natural systems, disturb the environment, and distribute pollution to land, water, air, and biota.

These factors are set into a world that is geographically varied with broad differences from region to region in environmental conditions, resources, and life-support capacity. These geographic realities may or may not match with the distribution of people and with their needs, aspirations, political goals, social values, and economic practices. As a result, serious mismatches occur throughout the world between human demands and expectations and what the environment is able to provide and endure. These mismatches are becoming more common and larger with each decade while population and land use grow and expand.

To alleviate mismatches, humans strive to rearrange the balance between the geographic environment and human needs and desires by redistributing water, food, fuel, technology, population, and other things. Most geographic redistribution is cooperative, based on agreements between nations, corporations, and other organizations. Unfortunately, we also resort to war, famine, forced migration, and genocide to satisfy geographic imbalances, political imperatives, and religious ideals. Although humans and the environment have achieved balances in some places, broadly speaking, the imbalances are growing throughout the world.

As we open the twenty-first century, some important questions loom before us. Have we reached our geographic limits on Earth? Are there any places left for people to inhabit where they can build self-sustaining land-use systems? Is the cost of redistributing resources to destitute and expanding populations becoming prohibitive? What happens when resources such as fossil fuels are exhausted, or when our ability to expand a resource system, such as food production, reaches its planetary limit? Or will we continue to raise the carrying capacity of the planet by introducing new technologies that will improve our ability to produce and redistribute food and other essential resources? And if so, will such advances be at the expense of the environment and ultimately the quality of planetary life? Or will Earth's carrying capacity, both biological and cultural, be reduced at some point because of environmental change driven by factors such as climate change and groundwater depletion (Figure 1.8)?

Earth has a limit to the growth and expansion of the population of any organism, no matter how inventive and discerning the organism may be. In the twenty-first century, Earth will probably see a human population of 10 billion or more. Ideas about survival and quality of life will be hotly debated. Some people will propose we move our overflow population to another

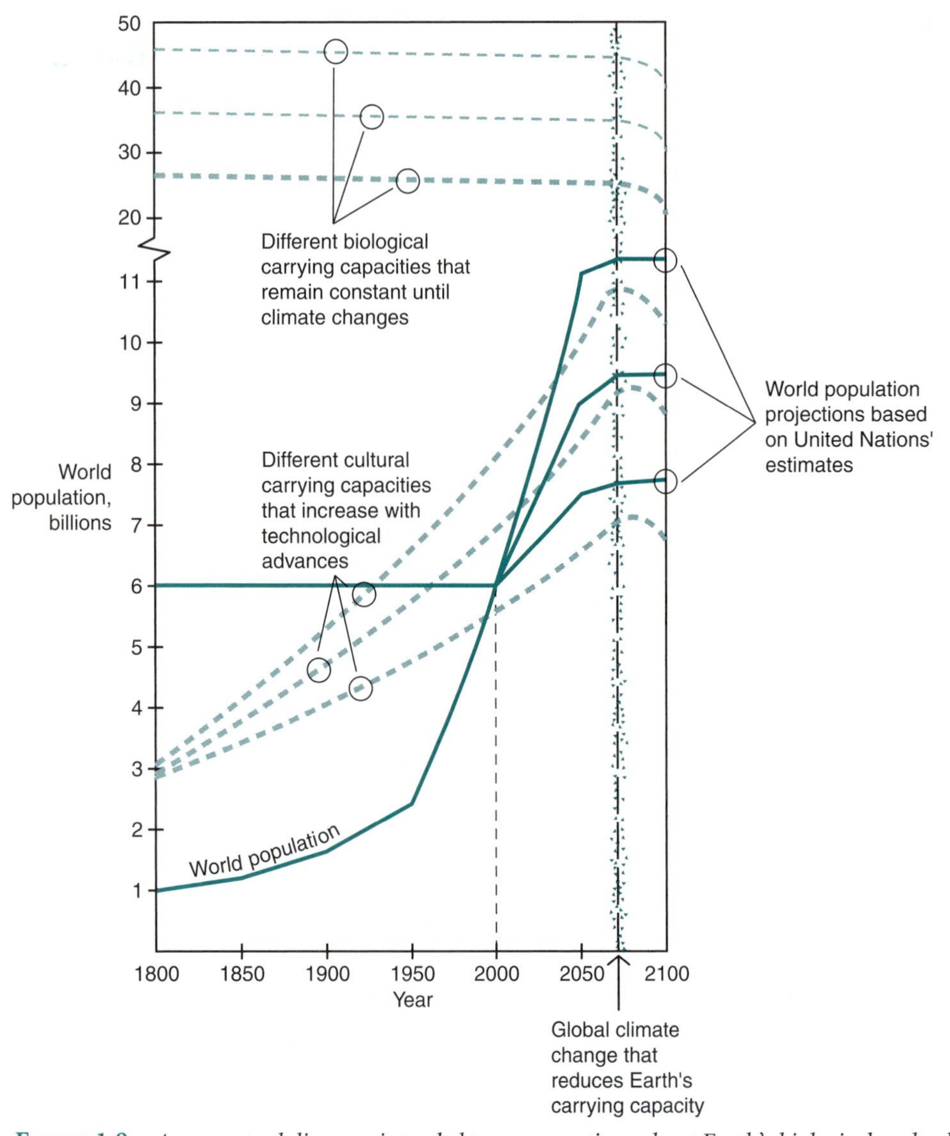

FIGURE 1.8 *A conceptual diagram intended to pose questions about Earth's biological and cultural carrying capacities given different population projections for the twenty-first century. Climate change, a realistic expectation of the twenty-first century, is added as a damping factor on carrying capacity.*

planet, and it appears that we will shortly have the technical capability for such travel. Why cannot we colonize Mars or the Moon when we reach our growth limits on Earth? The answer is quite simple. Earth is the habitat of humans, and the notion that large numbers of humans can spill onto other planets is naive and impractical. The only solution to our environmental dilemma and survival crisis will be found at home, on Earth (Figure 1.9).

Given that, let us also agree that Earth is not an open material system in which all resources are resupplied by nature as they are used. Earth's supplies of water, mineral resources, and fuel are essentially fixed, and although some resources can be cycled and recycled, our dependency on the planet's natural resources must ultimately be balanced with available supplies, their cycle rates, and our ability to reuse them. Thus, Earth's carrying capacity is not infinite. Although it can be expanded through more efficient means of food production and resource management in general, there is an upper limit to the number of humans the planet can support at a reasonable quality of life.

Human population growth is the single greatest root cause of the environmental crisis. It is the single greatest challenge facing humanity in its struggle to find a sustainable balance with the planet. But Earth's carrying capacity and the long-term quality of human life are also closely tied to consumerism. The more we consume, the faster the resource base is expended and the more waste per person is added to the environment. Technology en-

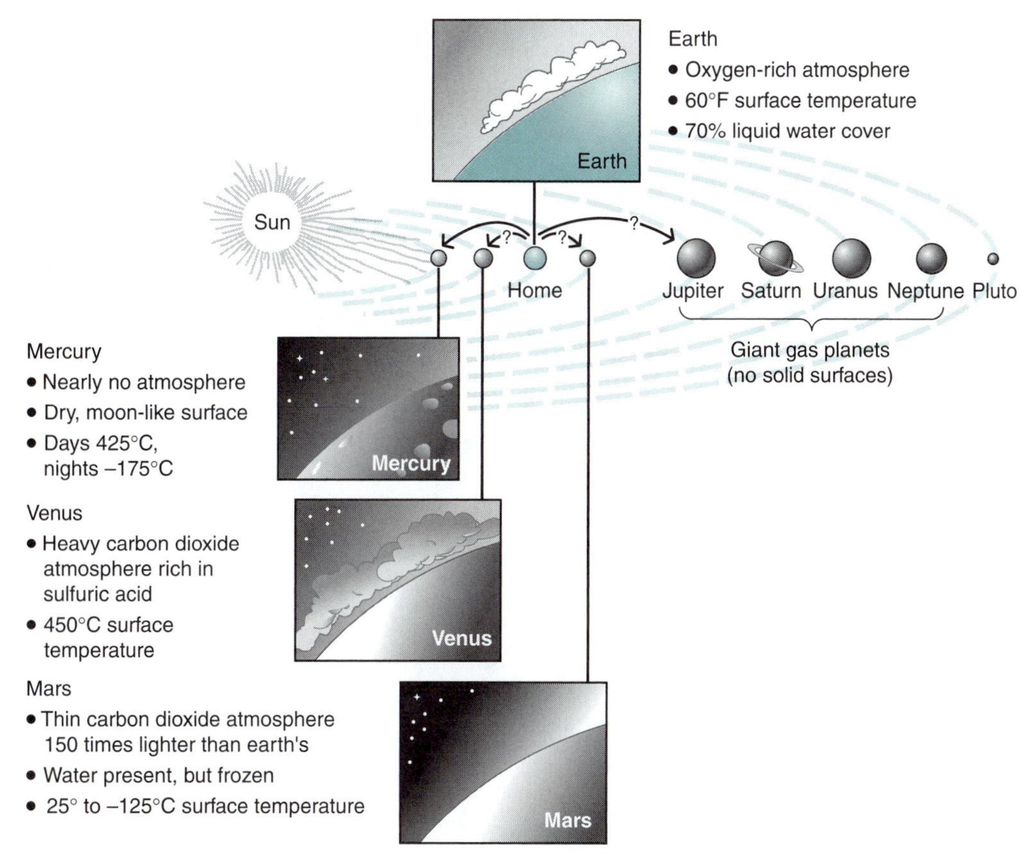

FIGURE 1.9 *A summary of the environmental conditions of the terrestrial planets in the solar system. Earth stands out among its sister planets as the only truly inhabitable planet for humans.*

hances our capacity to utilize resources. When both population and per capita consumption increase, the rate of resource use increases rapidly. In addition, human values are a critical part of environmental dilemma—for both negative and positive reasons. Our values drive wars, population growth, and much of consumerism. On the other hand, a system of environmental values may be emerging in the global community that will lead to changes in international environmental policy, consumer habits, and population trends that will favor a long-term sustainable balance between land use and environment.

Finally, achieving a sustainable system for Earth, its humans, and other organisms will require not only recognizing limits to growth as an inherent condition of the planet, but also engaging in long-term planning programs. Human land use traditionally suffers from the lack of planning and programmatic management. History reveals that when we are faced with a dwindling resource supply, humans tend not to reduce and manage their demands on it, but instead often do just the opposite. In an effort not to lose to the competition, they tend to overuse, leading to environmental damage and reduction in the resource base—in other words, less for all in the long run.

1.6 YOUR ROLE IN THE GLOBAL ENVIRONMENTAL COMMUNITY

We must agree on the need for change in the way we live, consume resources, expand our population, produce waste, and degrade the environment. Even though each of us represents a minute fraction of humanity, we are all investors in the environment and Earth's future. Just as corporate investors have responsibilities for the direction companies take and how they solve major problems in conducting business, we have responsibilities for the direction the global community takes and how Earth's problems are addressed. As an American or Canadian in particular, you are endowed with considerable power through the things you buy, how you vote, what you do, and what you say. These are opportunities, and what you do with them can make a difference in Earth's condition, no matter what your profession and station in life (see Figure 1.10).

As a **consumer**, you will, in your lifetime, spend more than a million dollars on food, fuel, transportation, housing, and services. Much of this is based on a global system of resources, and the choices you make will help shape national and global economic practices. As an educated person sensitive to environmental and survival issues, you will

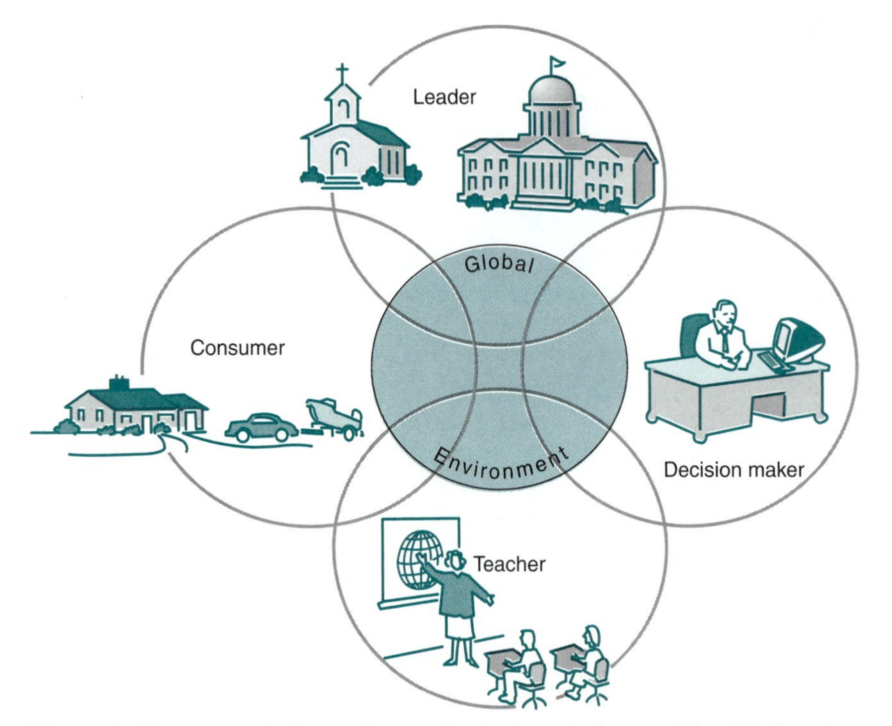

FIGURE 1.10 *A model of our roles as individuals in the future of the global environment.*

make considered decisions on the kinds of food you eat, the amount and types of fuels you consume, and even the sort of house you live in. These decisions will influence the success and failure of businesses, and it is businesses and consumers that make up economic systems, which, in turn, dictate the use of natural resources, the production of waste, and the disturbance of ecosystems.

Second, you are a **leader**. As a college student you are preparing for a leadership role in society, whether you choose a career in business, government, or education. In addition, we are all members and leaders in communities, neighborhoods, and families. The directions you take and the values you express will be followed by others. This includes consumption and lifestyle habits and moral values, including how we treat other organisms and spend our wealth.

Third, you are a **teacher**. As an educated person, you possess knowledge and communication skills that enable you to impart information to others. But in order to teach you need to be informed not only about the critical issues of our times, but about the way Earth's environment functions, what it is made up of and who lives in it, and how things are distributed geographically. Armed with this information, you can teach others how to make rational choices about the ways they live, consume, and vote in matters related to the environment.

Finally, you are a **decision maker** with surprising power to influence public policy, corporate investment, the direction of institutional development, and lifestyle choices in ways that can shape the course of society and its interrelations with the environment. Consider that at this time, more than any other moment in history, Earth's citizens are in the driver's seat, and it is our responsibility to make informed choices if life on the planet is to be sustained at a reasonable quality.

1.7 SUMMARY

The world is more crowded, more polluted, more urban, more biologically stressed, and warmer than ever before in history. It is a world divided into three great camps of nations—the *haves*, the *have-nots*, and the *deprived*. All three camps contribute to the degradation of the global environment. The new century, your century, will see an increase in world population, reaching 8 billion to 11–12 billion people by 2050. Economic development will continue to grow as it did in the twentieth century, but the gap between the world's rich and poor countries will increase, and development will place the environment under greater stress as resource extraction and pollution expand with population and consumption. Deserts are likely to expand, tropical forests will be reduced, soil erosion rates will rise, earth's biodiversity will decline, and the atmosphere will grow warmer. Land use will expand, especially in developing countries, and urban populations will grow substantially in both developed and developing nations.

Development is at the heart of many of our environmental problems. Economic development results in higher land use densities and leads to greater impacts on the environment. The terms *developed* and *developing* countries are based on the general economic status of a country related to national production, consumption, income, and demographic conditions. At the very lowest end of the development scale are deprived countries such as Somalia, Rwanda, and Sierra Leone, where civil strife and intertribal warfare have produced economic anarchy and chaos in land use systems.

Earth's capacity to support human beings is not infinite, but we are uncertain about the planet's true carrying capacity. Cultural carrying capacity takes into consideration quality of life, environmental stability, and sustainable land use, especially agriculture. Whether a sustainable balance between Earth and humans is possible is one of the great questions of the twenty-first century, and the answer will depend on population trends, technological advances and applications, rates of consumerism, and land use development and management. In the end much depends on what we do as individuals in the capacities of consumers, leaders, teachers, and decision makers.

1.8 KEY TERMS AND CONCEPTS

Prospects for the twenty-first century
- population
- economic development
- environment
- land use

Concept of development
- development
- economic productivity
- density
- developed country
- developing country
- deprived country

Earth's carrying capacity
- cultural carrying capacity

Limits to growth

Personal opportunities/responsibilities
- consumer
- leader
- teacher
- decision maker

1.9 QUESTIONS FOR REVIEW

1. List a number of major changes that took place in the global environment during the twentieth century.
2. What are developed countries, developing countries, and deprived countries? Name several examples of each. Can you make a case for the argument that all three camps of countries play a part in the global environmental community?
3. Summarize the global prospects for the twenty-first century in terms of world population, economic development, environment, and land use. What group of countries will likely experience the greatest population increases, and what change is expected in the economic gap between the richest and poorest countries?
4. What is meant by the term *economic development?* What drives economic development, and what are some environmental changes that usually accompany it? Discuss the observation that terms such as *developed countries* and *developing countries* are highly relative and that developed countries, for example, contain poor people whose standards of living are more typical of developing countries.
5. What is meant by Earth's carrying capacity? By cultural carrying capacity? What does carrying capacity have to do with the use of renewable and nonrenewable resources, and with population growth and food production?
6. Discuss the observation that because we live on a geographically varied planet, serious mismatches occur between human demands and expectations and what the environment is able to provide and endure. Can we make up for these differences by redistributing resources and by inventing new technology to help us use more resources? If we reach our limits to growth on Earth, what happens to additional population?
7. No matter how you look at the problems facing humanity and the global environment, we must agree on the need to improve on the ways we live, interact with the environment, and consume resources. Discuss your role in the global community as a consumer, leader, teacher, and decision maker.

2

PREPARING TO UNDERSTAND ENVIRONMENTAL GEOGRAPHY

2.1 INTRODUCTION

Environmental science is rooted in geography. In order to understand the interplay of humans and environment, the effects of humans on the environment, and the nature of land use, we must first know something about the geographic character of our planet. This includes not only the distribution of environmental features, conditions, and resources, but where people live and how they use the environment. In short, we must know where things are and what goes on in the natural and human environment in different places.

We begin this chapter with a brief discussion on the scope of environmental geography, followed by some basic facts about Earth, including the distribution of land and water, the elevation of the land, and the global system for describing geographic location. Next, a case is made for looking at Earth as a geographically integrated planet. The final two sections outline the human use of the planet and the role of human values and environmental policy in governing our treatment of the environment.

2.2 THE SCOPE OF ENVIRONMENTAL GEOGRAPHY

Environmental geography covers the issues of environmental degradation, quality of the global environment, and the condition of life. It is actually made up of many fields of geography—physical geography, human geography, regional geography, economic geography—that deal with different parts of the environment, human population, land use, and their interrelations. The scope of environmental geography is shaped by our concern over the condition of the environment and the quality of life for humans and other organisms. At the center of concern are land-use activities that degrade the environment and reduce its potential to support life. **Land use** is defined as the human activity that takes place on the land—for example, agriculture, residential, transportation, and industry. **Land-use systems** are interconnected, interdependent sets of land uses such as mining, transportation, and industry.

Land uses that deplete resources and degrade the environment without provisions for renewal and remediation are not sustainable. For example, those powered by fossil fuels are not sustainable because the oil, gas, and coal extracted and burned are not renewable (at least in human time frames). In addition, the pollutants released with the combustion of fossil fuels are degrading the environment. As a result, many of our agricultural practices are not sustainable because we use excessive amounts of energy and employ cultivation practices, fertilizers, and pesticides that damage soil and water resources. Of greatest concern is that soil is being lost to erosion much faster than it is being replaced by nature and constructive farming practices. Commercial fertilizer is added to fields to make up for declining soil fertility and to increase crop yields. The damaging effects of such unsustainable land uses tend to fall into two categories of environmental impacts: pollution and disturbance.

Environmental Pollution

Pollution is degradation of the environment as a result of some type of contamination. The contamination may take the form of foreign substances (e.g., pesticide residues in food chains or bacteria in water systems) or it may take the form of greatly increased levels of naturally occurring substances (e.g., sediment in streams or plant nutrients in soil). Pollution may be *acute*, as in the case of highly poisonous substances such as chlorine or arsenic being released suddenly in amounts that seriously damage or kill organisms; or *chronic*, as in the case of less toxic or nontoxic substances like fertilizer residues being released into groundwater at slower rates with less immediately harmful effects. Chronic pollution may, nonetheless, have serious effects, because contaminants or their effects might build up over time or act in combination with other substances to produce harmful impacts.

Environmental Disturbance

Disturbance is physical disruption of the environment, such as forest clearing, strip mining, and soil plowing. As we expand land uses in response to increasing population and growing consumption rates, environmental disturbance is inevitable. Cities spread, forests are removed, hillsides are scraped away to mine coal, and dams are built to serve cities and irrigate cropland. Not only has the economic incentive to clear the land increased enormously, but our technological capacity to do so is now alarmingly efficient. Today's massive machinery can quickly remove forests and prepare land for agriculture and settlements. With our modern machines, we are now the most effective agents of landscape change on the planet.

Most environmental problems involve both disturbance and pollution. In many instances, they are geographically connected—as one often accompanies or follows the other at a particular place. For example, when clearing land for agriculture, disturbance takes place when trees are cut, plants are exterminated, and animals displaced. This is followed by pollution as pesticides and chemical fertilizers contaminate the ground, air, and water as part of the farming operation.

Land Use and Technology

The forces driving both pollution and disturbance include rapid population growth in the developing countries and massive consumption throughout the economically developed world. **Technology** has been used as both a constructive tool and a destructive force in assimilating increased population and consumption. With modern technology we have developed an astounding capacity to clear forests, plow land, build cities, and pollute air and water systems. Technology also accounts for the energy systems, manufacturing processes, and synthetic substances (like plastics and chlorinated materials) that enter the environment through human consumption, work, and neglect.

• Human technology adds a complex twist to the life–environment relationship of land-use systems. Consider for a moment that each year we invent thousands of new chemical compounds, many of which are used in pesticides, cleaning compounds, fuels, lubricants, and medicines, as well as an indescribable variety of manufacturing materials and processes. On top of that, vast cultural differences among societies influence the availability and use of new technology, as do differences in attitudes about the natural environment and how we use it, our place as humans in it, and our impacts on the environment. For example, some American Amish colonies limit themselves to farming, transportation, and household technologies that date before 1850 or so: no tractors, cars, electricity, or mechanical sewer systems (Figure 2.1). Also, consider the difference between the American Indian's traditional view of land ownership and the Euro-American view: To the Indian, land, trees, water features, and game could not be owned. To the European, they *could* be owned, and, by virtue of the right of ownership, could be used, altered, and destroyed if necessary.

Environmental geography has to do with all life, not just human life. It is the science of survival in which we try to understand the interplay of life and environment in a rapidly changing world. Although the fundamentals of the relationships between life and the environment are basically no different than they were, say, 5,000 years ago, there are factors in the modern world—massive economic

(a)

(b)

FIGURE 2.1 *(a) An Amish farmstead where society makes a conscious choice to minimize the commitment to modern technology such as electricity, tractors and cars; (b) A modern commercial farm in the United States where every opportunity is seized to use more and newer technology.*

systems, huge cities, dwindling fuel resources, worn-out farmland, exhausted water supplies, and chemical pollutants in air and water systems—that make today's environment a more difficult place for most life on the planet.

2.3 THE GLOBAL CONTEXT AND SETTING

To look at the issues and problems of the environment knowledgeably, we need to know some basic facts about our planet. Earth has a surface area of about 510 million square kilometers (197 million square miles), of which close to 70 percent is water and the remainder is land (Figure 2.2). The vast majority of the water is saltwater. Freshwater is limited to the land masses, and if glaciers and snowfields are excluded, freshwater covers only about 1 percent of the Earth's surface. Moreover, in the last several centuries, the area of freshwater has shrunk as wetlands have been drained and shallow lakes filled to expand agriculture, settlements, and transportation lines. Over a much longer time frame, even the shapes and sizes of the continents and oceans are changing.

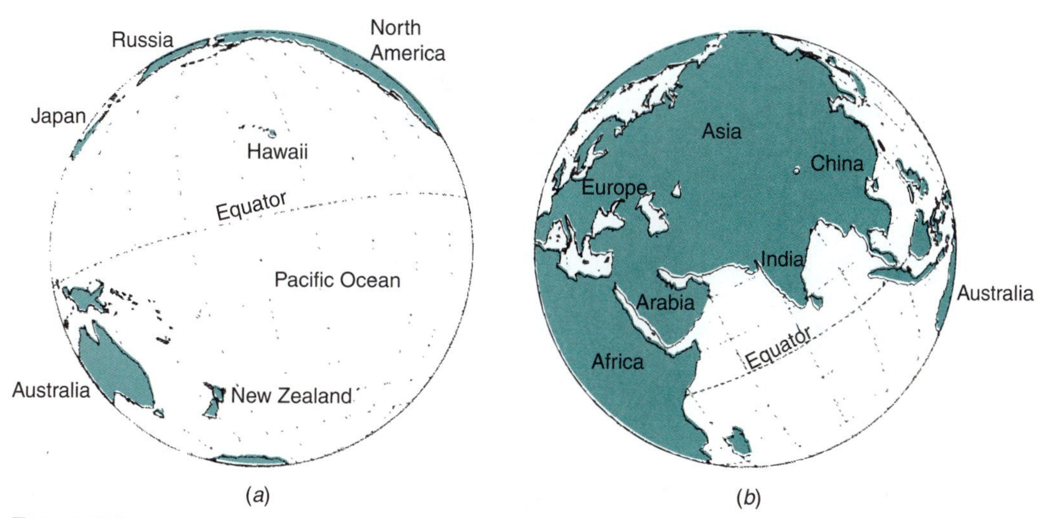

FIGURE 2.2 *Earth is nearly 70 percent covered with water. Land is unevenly distributed geographically, with most located in the Northern Hemisphere.*

Earth's Surface Features

Most of Earth's organisms live on land and in the shallow water fringing the continents, called the **continental shelves.** The continental shelves are the lowest parts of the continents. They form the "shoulders" of the land masses and extend seaward to a water depth of around 200 meters (650 feet). Inland from the continental shelves, the land rises to form orogenic belts, and shields (Figure 2.3a). **Orogenic belts** are the broad bands of mountainous highlands that usually lie along one or two sides of the continents just landward of the continental shelves. In North America, the Rocky Mountains and the Appalachian Mountains are the orogenic belts. **Shields** form the interiors of the continents. Lower and less rugged than the orogenic belts, they are the geological cores of the continents. Shields (so named because they resemble the shape of a war shield, handle side down) form two types of terrain, exposed granitic rock such as in the Canadian Shield and covered granitic rock such as the Interior Plains of North America, where sedimentary rocks (mainly sandstone, limestone, and shale) blanket the granitic rock (Figure 2.3b).

The average elevation of the continents is 840 meters (2,800 feet) above sea level, but the bulk of humanity dwells at much lower elevations, within 300 meters (1,000 feet) of sea level. Although we celebrate our mountains, very little land lies at high elevations. Less than 5 percent of all the continents lies above 3,700 meters (12,000 feet) elevation. The climate at higher elevations is colder and harsher, and the potential to support life declines sharply above 3,000 meters (10,000 feet) elevation throughout the world. Humans can and do survive at very high elevations, but we cannot sustain life much above 5,000 meters (17,000 feet). Not only is food production impossible, but human reproduction appears limited by extreme elevation (Figure 2.4).

All earth environments, land uses, and habitats rest on the planet's **crust**, a solid layer of rock 16–40 kilometers (10–25 miles) thick. The crust and the lithosphere (the thicker layer that the crust caps) are partitioned geographically into seven huge sections, or **tectonic plates.** Because of the gradual flow of hot rock deep within the Earth, the plates slowly shift about on the planet's surface, moving the continents and changing the size and shape of the ocean basins. As the plates move, they push, pull, and scrape against each other with tremendous force. Most of Earth's volcanic and earthquake activity is concentrated along the contacts between the plates (Figure 2.5).

Not surprisingly, tectonic plate borders are some of the most hazardous environments on the planet. No plate border illustrates this better than that of the Pacific plate. Known as the *Ring of Fire*, the Pacific border is a line of earthquake and volcanic activity, continuous in some areas, intermittent in others. Millions of people and some of the world's largest cities—including Los Angeles, San Francisco, Tokyo, and Manila—lie on or near the Pacific border, and each year the number of people living on the coast of the Pacific Basin and other coastlines of the world grows significantly.

The legacy of plate tectonics has given us a world in which land and water are unevenly distributed geographically. Because of differences in their size, location, and geographic situation, the land masses have different climates, soils, vegetation, and economic potential. The Northern Hemisphere, which represents the half of the Earth north of the equator, contains roughly twice the land area of the Southern Hemisphere (Figure 2.2). Not only that, but the land is distributed differently in the two hemispheres. In the Northern Hemisphere, most land (Eurasia and North America) is located in the midlatitudes, whereas most land in the Southern Hemi-

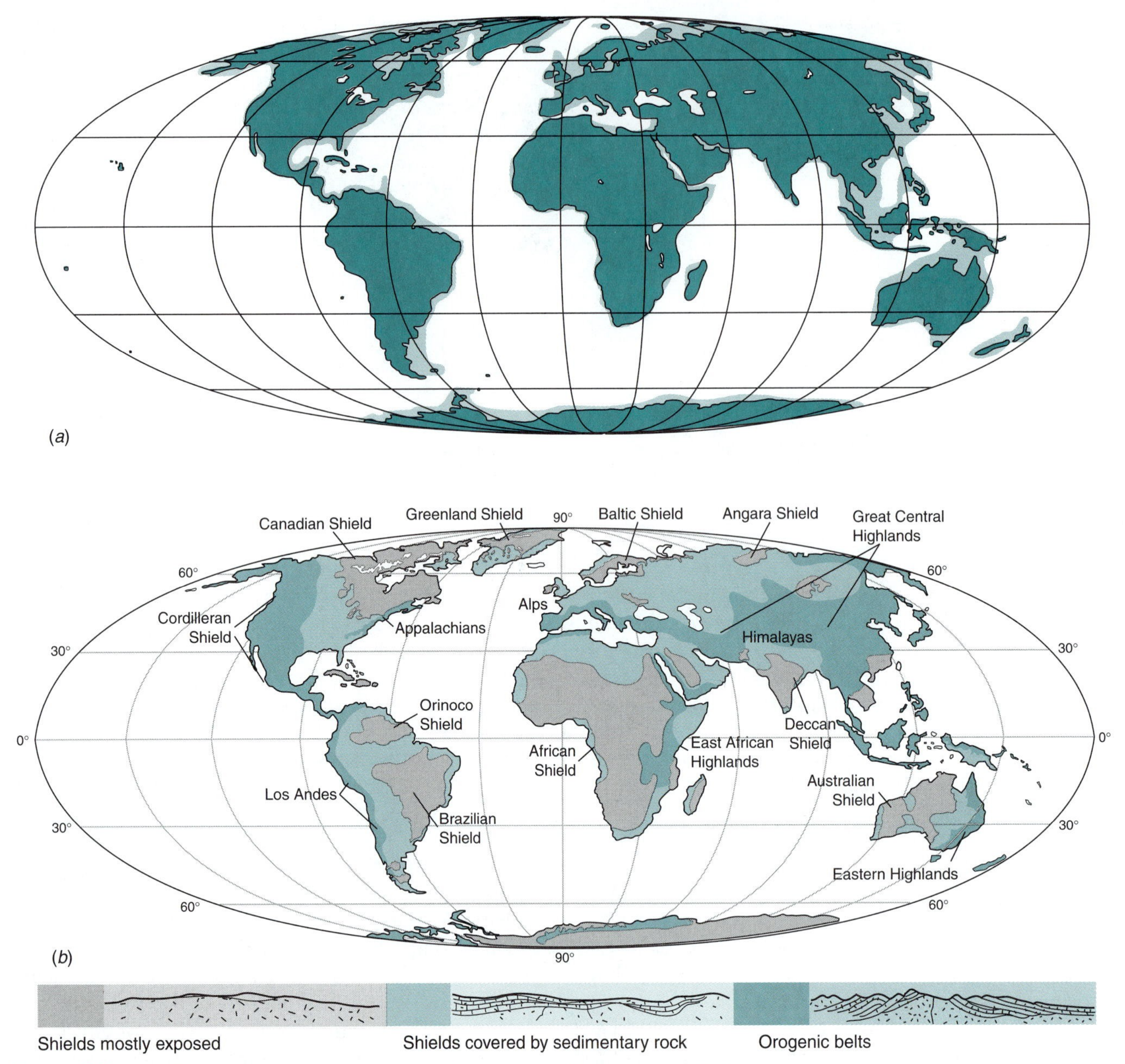

FIGURE 2.3 *(a) The continental shelves of the world range in width from several kilometers to several hundred kilometers. (b) The global distribution of orogenic belts and shields.*

sphere (South America, Australia, and part of Africa) is found in the low latitudes (which helps explain the large areas of tropical wet climate and rainforests in the Southern Hemisphere). The exception is Antarctica, which is centered on the South Pole, and is dry, cold, and ecologically desolate.

Global Coordinate System

The designation of low, middle, and high latitudes is based on reference lines drawn across the globe by map makers. There are two sets of lines: **meridians** run north and south; **parallels** run east and west. These lines form a grid network known as the **global coordinate system.** Both parallels and meridians are numbered in degrees. The equator is the zero-degree parallel. Northward and southward from the equator, numbers increase to a maximum of 90 degrees at the north and south poles (Figure 2.6a). Locations measured according to the system of parallels are designated as degrees latitude north or south of the equator. For example, 45 degrees north latitude is represented by the parallel halfway between the equator and the North Pole in the North Hemisphere (Figure 2.6b).

The system of meridians also begins with a zero-degree line. This line is called **the prime meridian** or Greenwich

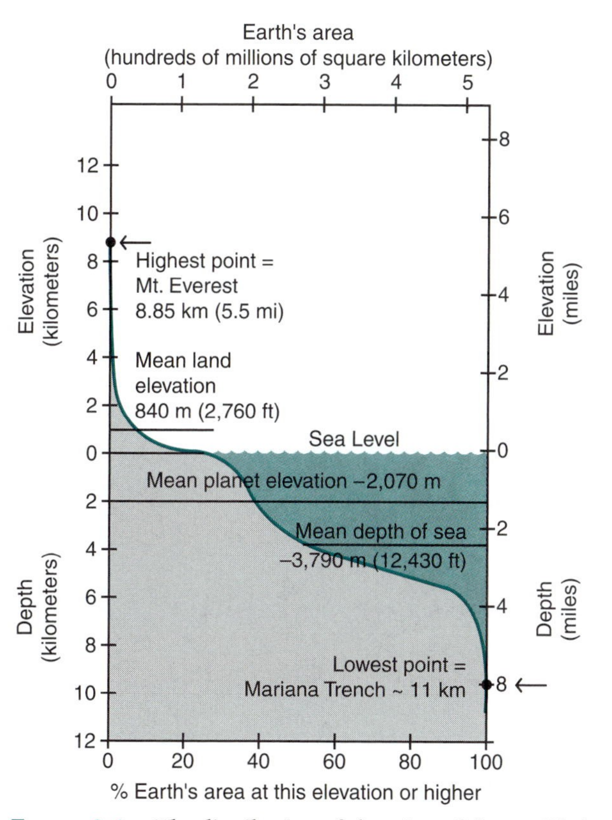

FIGURE 2.4 *The distribution of elevation of the world's land masses and ocean basins.*

Meridian (because it runs through the town of Greenwich, England) (Figure 2.7a). All locations east of this line are designated east longitude, and those west of this line are designated west longitude. The system ends on the opposite side of the Earth from the prime meridian at 180 degrees longitude (Figure 2.7b). With this system, a second set of hemispheres can be defined. The Eastern Hemisphere includes Europe, Africa, and Asia; the Western Hemisphere includes North America, South America, and most of the Atlantic and Pacific Oceans.

Zones of Latitude

Earth's bioclimatic environment is arranged roughly into several great belts of latitude. Much of our discussion in the following chapters uses this framework. Three broad zones of latitude can be defined in both hemispheres: the **high latitudes**, which cover the upper 23.5 degrees latitude (66.5 to 90 degrees); the **middle latitudes** between 66.5 degrees and 23.5 degrees latitude, and the **low latitudes**, which lie between 23.5 degrees and the equator. Latitude 23.5 degrees is significant because it marks the highest latitude that receives direct solar radiation—that is, where the sun's rays hit Earth's surface perpendicularly (at a 90-degree angle). In the Northern Hemisphere this occurs on June 20–22 each year, and this parallel at 23.5 degrees is called the **Tropic of Cancer**. In the Southern

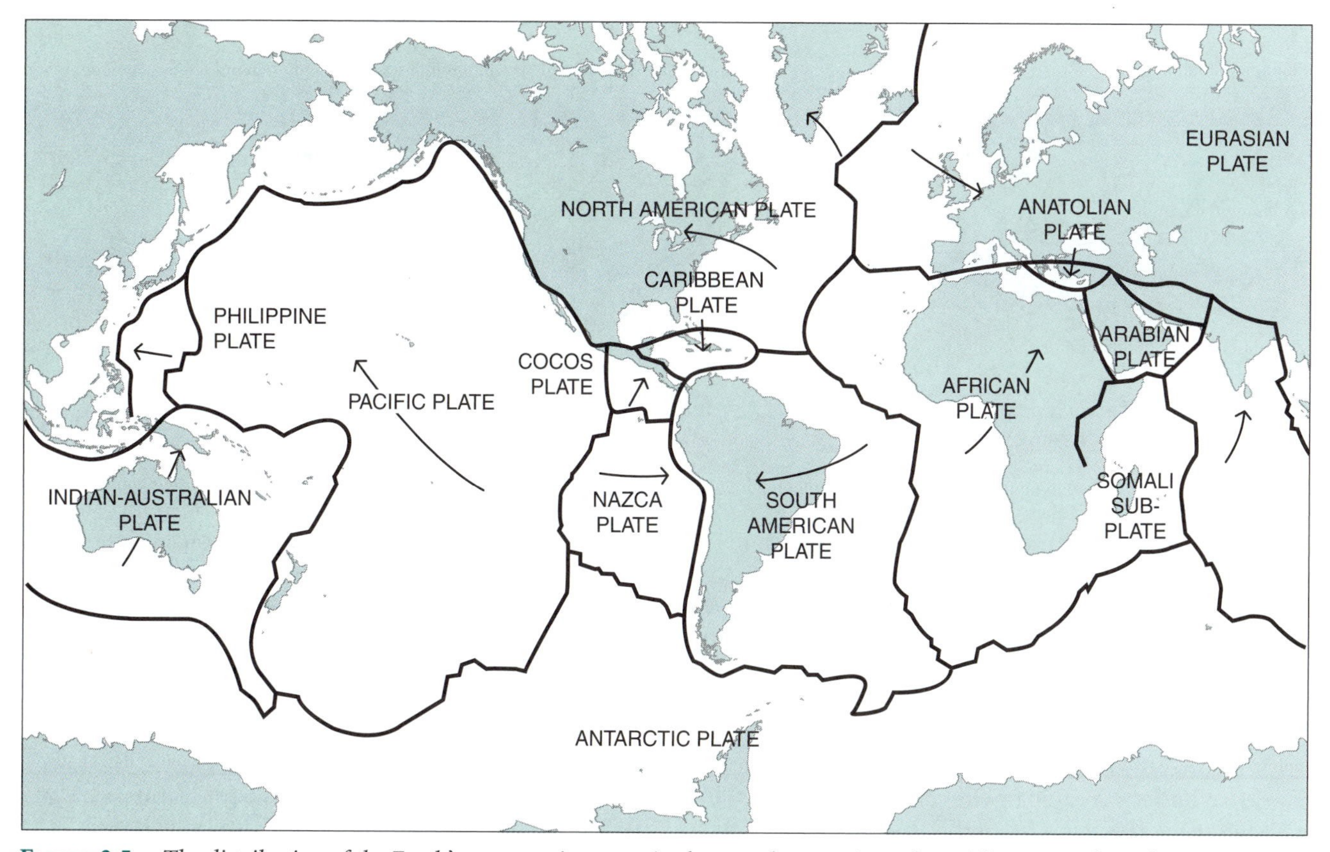

FIGURE 2.5 *The distribution of the Earth's seven major tectonic plates and most minor plates. The arrows show their general direction of movement.*

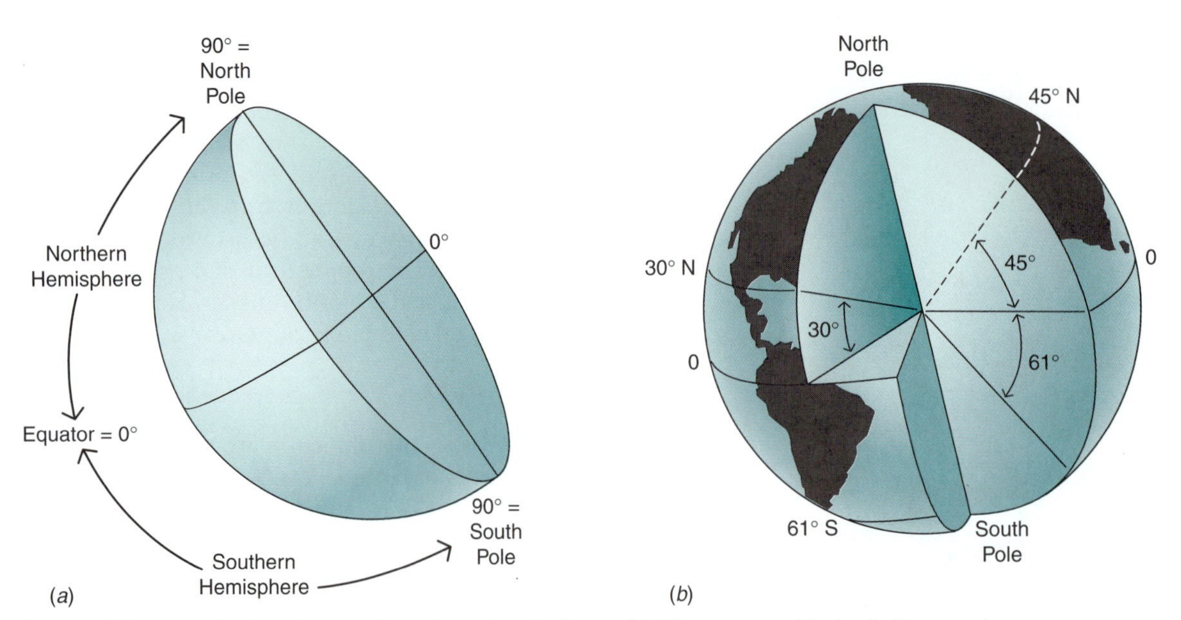

FIGURE 2.6 *(a) The Northern and Southern Hemispheres; (b) The concept of latitude illustrated by angles north and south of the equator.*

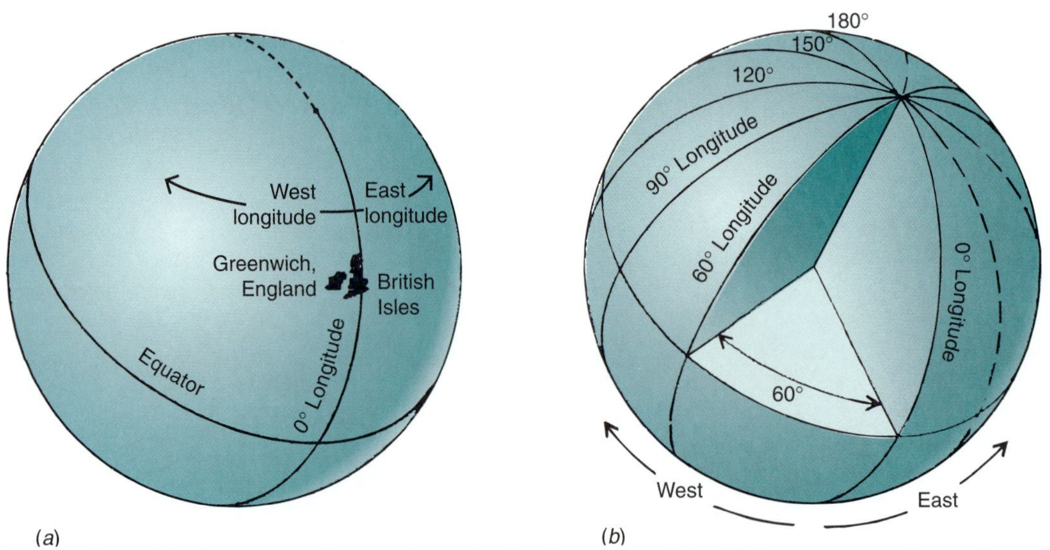

FIGURE 2.7 *(a) The concept of east and west longitude represented by (b) degrees east and west of the prime meridian at 0 degrees longitude.*

Hemisphere it is called the **Tropic of Capricorn**, and it marks the highest latitude in the Southern Hemisphere to receive the direct rays of the sun (on December 20–22) (Figure 2.8). Latitude 66.5 degrees marks the Arctic and Antarctic circles, above which are the only locations on earth to experience day-long (24 hours) light and day-long dark each year.

Used correctly, the term *tropics* refers to the Tropic of Capricorn and the Tropic of Cancer. It follows that the zone between 23.5 degrees south latitude and 23.5 degrees north latitude should be the *intertropical zone*, and indeed, many scientists do follow this convention. However, the term *intertropical* has declined in usage, and today *tropics* or *tropical zone* seems to be the preferred term for this zone. Most of the land area in the tropics is found in South America, Africa, and southern Asia. The equatorial zone is the narrow middle belt of the tropics, extending about 10 degrees latitude north and south of the equator (Figure 2.8).

Within the broad belt of the middle latitudes, three additional zones are often designated, although their locations are somewhat arbitrary. Just above the tropics is the

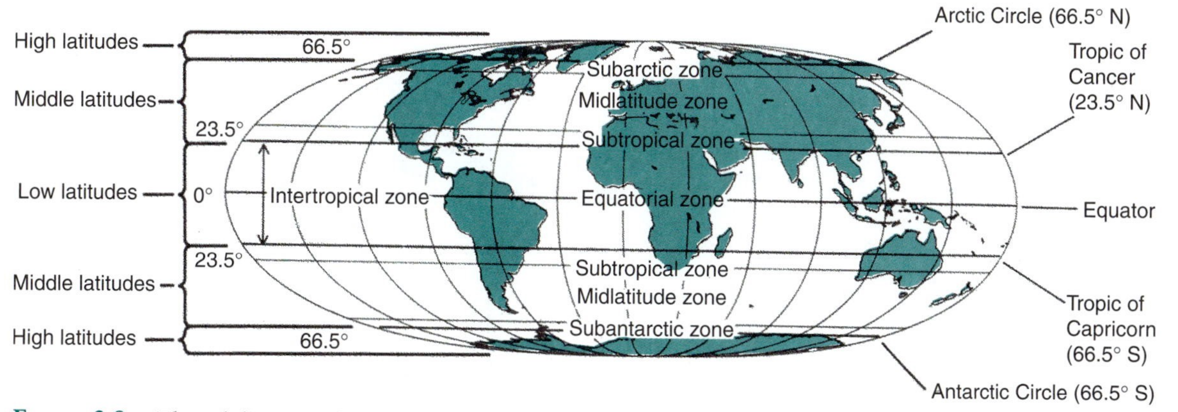

FIGURE 2.8 *The subdivision of Earth into zones of latitude. Notice the relatively large amount of land area in the Northern Hemisphere midlatitude zone.*

subtropical zone, which extends from 23.5 to 35 degrees latitude. Much of the American South, the Mediterranean region, Australia, and southern China lie in this zone.

The midlatitude zone is the center belt. Its limits are not fixed, but 35 to 55 degrees is usually given for it. Very little land in the Southern Hemisphere lies in this zone, whereas large areas of the Northern Hemisphere in North America, Europe, and Asia lie in the midlatitude belt. Beijing, Tokyo, London, Paris, Berlin, Moscow, New York, Philadelphia, Chicago, Montreal, Buenos Aires, and Sydney all lie in this zone.

The subarctic zone lies between 55 degrees and the Arctic and Antarctic circles. The bulk of the land in this zone is occupied by Russia, Canada, and the United States (Alaska), and settlements are sparse throughout. Beyond 66.5 degrees north and south are the Arctic and Antarctic zones, the uppermost parts of which (generally given as above 75 degrees latitude) are the polar zones where populations are exceedingly light or absent over vast tracts (Figure 2.8).

Where People Live in the World

About 90 percent of Earth's 6.1 billion people inhabit roughly 30 percent of its land area. The remaining 70 percent of the land is either lightly inhabited or uninhabited. The lightly inhabited areas are occupied mainly by traditional societies engaged in fishing, hunting, gathering, herding, and some crop farming. These societies are found in three principal environments: deserts and grasslands of Asia and Africa; tropical forests of South America, Africa, and Southeast Asia; and the cold lands of North America and Asia. Although these societies have occupied these remote environments in relative harmony with nature for thousands of years, some are now giving way to modern land uses and other lifestyles as their homelands are being probed and exploited by various forms of development. With the exception of small scientific outposts, Antarctica is uninhabited, and so far as we know, has never been occupied by humans.

The bulk of humanity lives in the midlatitude and subtropical zones. In the Northern Hemisphere, the principal zone of human occupation lies between the twentieth and fiftieth parallels. This area includes most of the United States, Europe, and Russia, as well as Japan, China, and India, and contains nearly 4 billion people. More than half of the world's people live within 100 kilometers (60 miles) of the sea. Nine of the ten largest cities in the world are located on seacoasts, and the world's most populous countries in terms of population density are coastal nations (for example, the Netherlands, Bangladesh, and Japan). If we add inland waters (lakes, rivers, etc.), the relationship to water is all the more impressive. The Mississippi River, the Great Lakes, the Chang River, the Seine River, the Volga River, and the Nile, for example, are the sites of many large cities, including St. Louis, Chicago, Shanghai, Paris, Moscow, Cairo, and the homes of more than a billion people.

2.4 OUR GEOGRAPHICALLY INTEGRATED PLANET

Earth is a geographically integrated planet. It tends to function as one great working whole in which widely separated places are connected by, among other things, common air and water systems. If we lived on a planet less geographically integrated, the patterns and problems associated with pollution and environmental disturbance would be significantly different. Since pollutants would not be dispersed as widely, they would tend to be more concentrated around their sources. Geographically, pollution patterns would tend to be nodal and the related environmental problems more compartmentalized in scope. Societies that created the pollution would be the ones that had to live with most of it. The geopolitics of pollution and environmental management would be less complicated because transfer by nature across political borders would be less of a problem. But nature is not that simple or judicious on Earth.

Earth's systems of air, water, energy, chemicals, sediment, and organisms span such enormous ranges that changes introduced in one place often affect the environment great distances away. This is especially so with the atmosphere, the most mobile of Earth's environmental systems (Figure 2.9). Pollutants such as carbon dioxide and ozone-destroying chlorofluorocarbons (CFCs) from industrial countries, for example, are dispersed across the globe. Note that the major effect of CFC pollution, namely, the seasonal reduction of stratospheric ozone, is found over Antarctica, the cleanest of all continents, and in the United States, the pesticide DDT, which has been banned for more than 30 years, is still falling on land and water because it is being blown in from Central America and other distant regions.

Distance and Scale

The concept of a geographically integrated planet is extremely important to our understanding of the environment, our impacts on it, and our need to manage it as a global concern. However, not all problems and their treatments are global in scale. Most have their roots in much smaller areas, ranging from large regions such as a part of a continent to local areas such as an urban center. Effective understanding and management of environmental problems require recognition of the scale factor.

For example, the most damaging environmental impacts of pollution are at and immediately around the pollution source. At this proximal scale contaminants are usually most concentrated and therefore pose the greatest environmental danger. Beyond the pollution source, most contaminants transported in air and water are dispersed rapidly, resulting in declining concentrations as distance increases from the source. Thus, concentrations at the distal scale, that is, over distance space, are usually much less than at proximal scales (Figure 2.10).

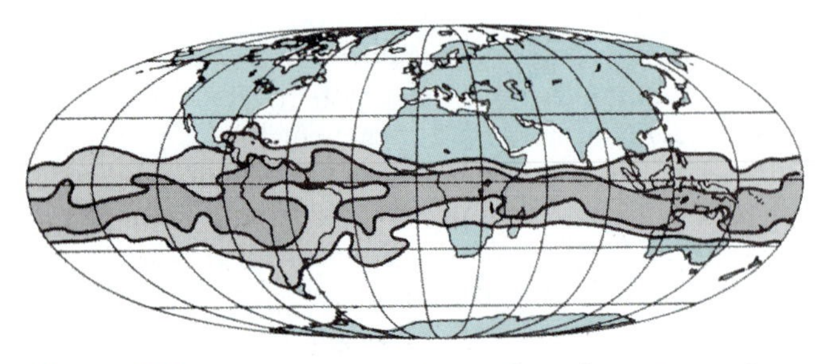

FIGURE 2.9 *Pattern of air pollution from the 1991 eruption of Mt. Pinatubo in the Philippines extended around the world in a matter of only weeks.*

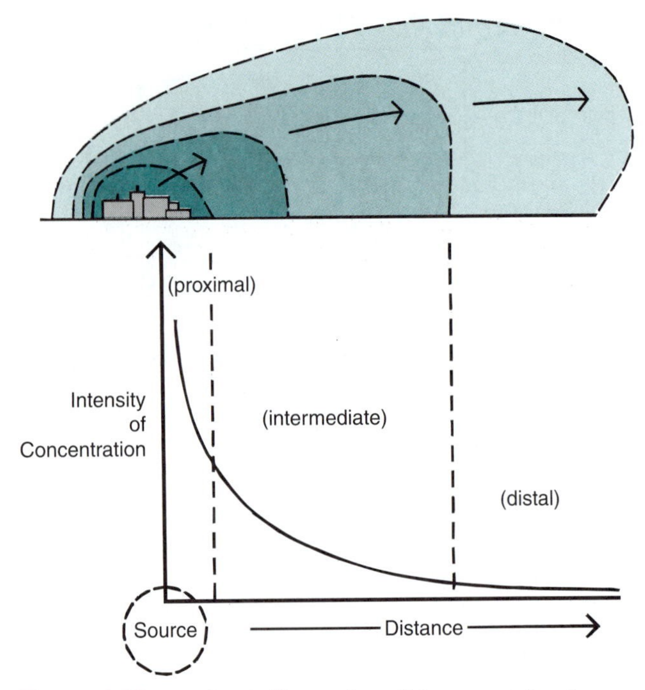

FIGURE 2.10 *A simple illustration of three spatial scales relative to a pollution source. The graph line shows the rapid decline of concentration with distance.*

There is one important exception to the scale/distance principle, and that is found in ecosystems. In ecosystems, the concentrations of many contaminants actually increase with distance or steps along food chains. Contaminants such as pesticides become lodged in the tissue of plants and animals and are passed along to other organisms in the food chain. Because the total amount of tissue gets smaller with each transfer, the concentration of contaminants can get larger. Thus, organisms at the ends of food chains such as eagles, hawks, and gulls often end up with heavier loads of contaminant residues in their flesh than the organisms that originally ingested it.

Uncertainty and Risk

The same processes and systems that link distant parts of the Earth together and redistribute water, air, and energy also pose threats to life because of irregularities in their behavior. In the atmosphere, for example, extreme events such as hurricanes and tornadoes interrupt the day-to-day flow of weather, damage the environment, and disrupt life. Throughout nature there is a distinct pattern to the size and rate of occurrence of atmospheric, hydrologic, and geologic events: The larger the event, the less frequently it occurs. This follows a basic principle called the **magnitude and frequency principle**, in which small, nondestructive events occur frequently and large, destructive events occur infrequently. Large events are not only the most destructive or catastrophic, but also the most uncertain or unpredictable.

All life systems are subject to the uncertainties of the magnitude and frequency principle. We do understand, however, that certain geographic locations are riskier than others. Stream valleys, for instance, are prone to flooding, and some more than others. Certain sea coasts (e.g., the U.S. South) have a long history of destructive hurricane strikes and are more prone to storm damage than others. And as we noted earlier, tectonic plate borders such as the Pacific Rim are dangerous places because earthquakes and volcanoes are concentrated there (Figure 2.11).

We also know that humans can increase risk by altering natural systems such as river channels. If river channels are narrowed through engineering, as is the case along parts of the Mississippi River, then the magnitude of flooding may be greater because water is forced to higher levels, where flow is constricted. This was a contributing factor in the 1993 catastrophic flood in the Mississippi Valley, where levees (banks constructed to contain flood waters) in some areas forced water to rise higher than it naturally would have. In urban areas, engineered systems for removing stormwater from streets, yards, and parking lots have increased the magnitude and frequency of flood flows on local streams, resulting in greater risk to homes and other land uses downstream. On the other hand, engineering can also reduce risk; for example, irrigation systems in dry regions can minimize the risk of crop failures caused by drought.

For Earth as a whole, the effects of natural forces on land use and human population increased throughout the twentieth century, and scientists forecast that losses from earthquakes, floods, hurricanes, and other natural hazards will continue to rise in this century. The reasons for this increase are related first to the great increase in world population and second to the spread of people and land use into hazard-prone environments such as river valleys and coastal lowlands. In addition, some scientists speculate that the incidence of hazardous events such as hurricanes and drought may be increasing as a result of global environmental change related, for example, to atmospheric warming. Both issues are very serious matters facing humanity in the twenty-first century.

2.5 Human Use of the Earth

During its 4.5-billion-year life, Earth has experienced massive volcanic eruptions, huge explosions from asteroid collisions, the die-off of the great dinosaurs, the

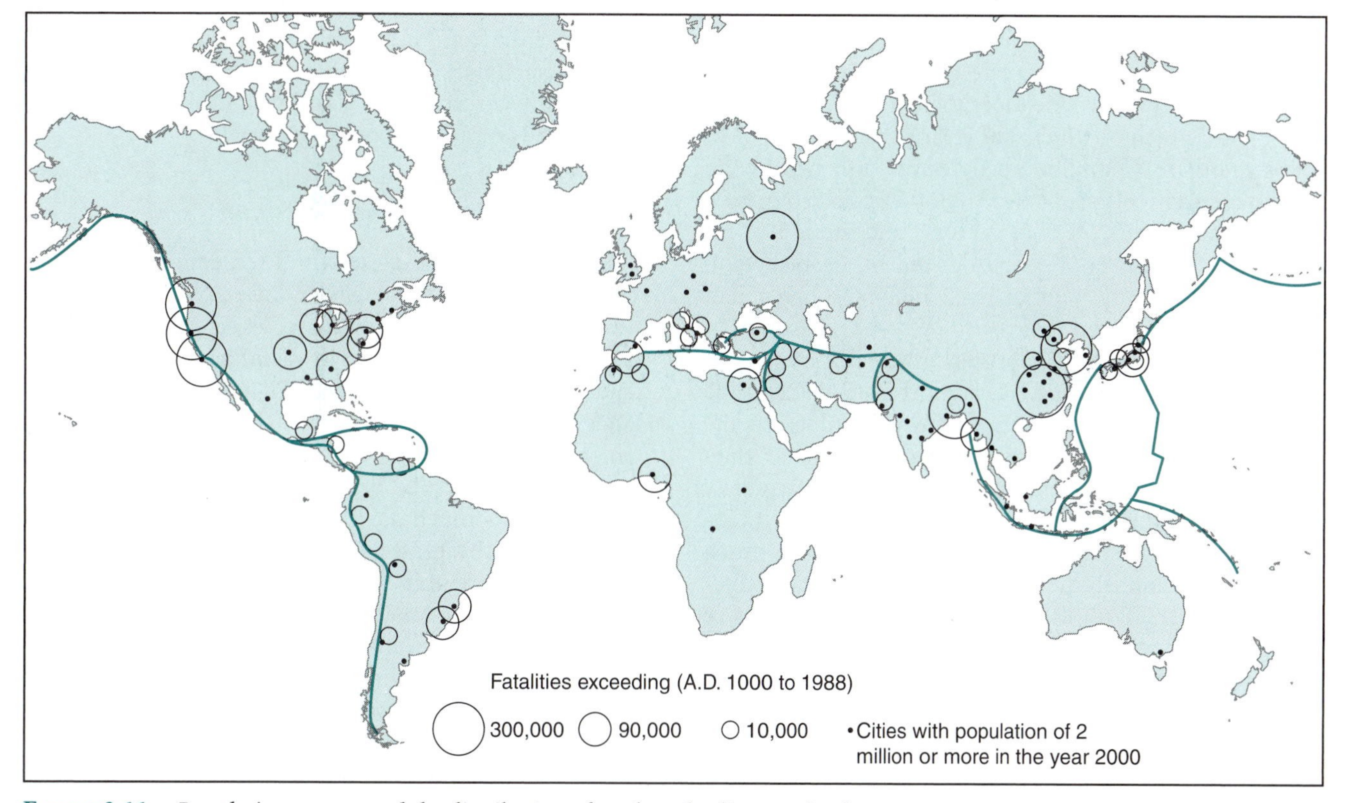

Figure 2.11 *Population centers and the distribution of earthquake disasters for the past 1,000 years. The line represents critical tectonic plate borders.*

emergence and spread of flowering plants, and the rise and fall of continental glaciers. The evolution and spread of humans ranks among the most spectacular of these events. Since the rise of modern humans about 100,000 years ago, our species has spread to every continent on the planet and, with the exception of Antarctica, has established life-sustaining land-use systems on them. In the past 12,000 years, since the invention of agriculture, human population has grown rapidly, reaching such an explosive rate that we now represent a biological event of global magnitude.

Before agriculture, the Earth was sparsely populated. People lived by hunting and gathering. Based on remnant populations that have survived, it appears that population densities were only a few people per square mile in favorable environments and much less in unfavorable ones. With agriculture, population grew rapidly, and in favorable environments, human population density increased by a hundredfold or more.

The Rural World

Today, close to half the people on Earth make a living basically as they did 5,000 years ago, working as subsistence farmers on small plots of ground. In subsistence farming, farmers produce food mainly for their own consumption with little or no surplus for sale or trade. Virtually all subsistence farming is found in poor countries (both developing and deprived) in Asia, Africa, South America, and Central America. The greatest number of subsistence farmers is in China, where more than 500 million people work land, much of which was cleared centuries ago. In India there are 400 million people who also depend on subsistence agriculture. When the other poor countries, such as Bangladesh, Mexico, Pakistan, and Indonesia, are added in, the number of people dependent on small-plot, subsistence farming approaches 3 billion, roughly half the world's population.

The parallel between subsistence farming of ancient and modern times, however, falls short when we examine the conditions of twenty-first century farming. In the past, agricultural populations were much smaller than today, and the stress on the land, though substantial in some locales, appears to have been generally less intensive. Today, subsistence farming must support a much larger population, so there is a push for greater productivity. This usually leads to the use of more intensive farming methods and marginal or poorer quality lands.

Intensive subsistence farming is aimed at achieving greater yields. It commonly includes the use of some chemical pesticides and/or chemical fertilizers. The use of both marginal land (e.g., mountain slopes) and chemical additives (e.g., DDT) enables the subsistence farmer to drive the land to higher short-term levels of performance. But there are drawbacks in the form of environmental degradation and economic uncertainty, because marginal lands are more prone to failures from drought, flooding, and other hazards, and chemical additives increase farming costs and human health risks. Therefore, it is questionable that subsistence farming on these marginal lands is sustainable.

In economically developed countries, agricultural populations are much, much smaller, but the land area devoted to agriculture is relatively large and the farms are commercial rather than subsistence. Cultivated land in the United States exceeds 185 million hectares (450 million acres), significantly more than in China where there are about 96 million hectares (230 million acres) under cultivation. In the United States the labor force devoted to all agricultural land is less than 20 million people, whereas in China the agricultural labor force, as we noted earlier, is more than 500 million people.

This striking difference points up one of the enormous contrasts between the developed countries and the poor countries. In developed countries, comparatively few people produce huge amounts of food because technology greatly enhances the capacity of individuals to successfully work large expanses of land and achieve high yields from it. But there is also a negative side to this massive system. Environmental impacts are considerable, especially water pollution in the form of sediment from eroded soil, nutrients lost from fertilizers, and chemical residues lost from pesticides. In addition, the loss of habitats such as wetlands and the depletion of groundwater are also recognized as serious problems of technologically advanced farming.

Agriculture, Forest, and Grassland

Today, cropland and agricultural settlements occupy 11 to 12 percent of the world's land area. Most of this area was originally forested or grass-covered. Forest now covers about 30 percent of Earth's land area and the grassland covers about 25 percent (Figure 2.12). The grasslands are counted among the Earth's dry environments, which also include the deserts. All together, deserts, grasslands, and related dry lands such as tropical savannas occupy as much as 40 percent of the planet's land area. Where the grasslands are not taken up by crop farming, they are utilized (along with the desert margins) as grazing land or rangeland. Grazing is also an ancient agricultural practice, and although it claims a relatively large area of land, it supports a very small population.

Prior to the spread of agriculture, three great belts of forest covered 40 percent or more of Earth's land area: the tropical forests of the low latitudes, the temperate forests of the midlatitudes, and the boreal forests of the subarctic zone (Figure 2.12). Little of the original temperate forests

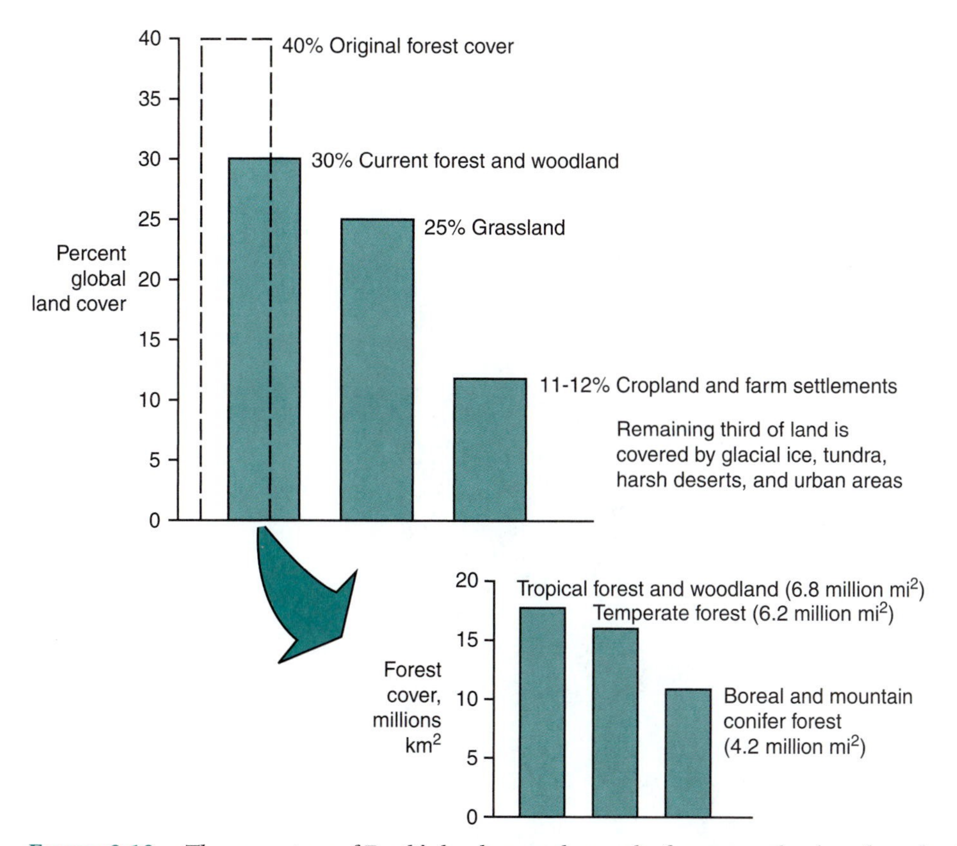

FIGURE 2.12 *The percentage of Earth's land area taken up by forest, grassland, and cropland. Below, the areas of the three major global forests.*

remain, largely because of displacement by agriculture and settlements in Europe, Asia, and North America. Large tracts of temperate forests, however, have become reestablished and today are mixed among farmlands and settlements in the midlatitudes. The boreal forests represent the largest continuous belt of forest in the world. Found only in Eurasia and North America, these forests stretch nearly across both continents, covering 11 million square kilometers (4.2 million square miles) of land. Although they are utilized for lumber and pulpwood, most of the boreal forests have not been displaced by agriculture and settlements owing to their cold climate and poor soils.

Tropical forests are found mainly in Africa, South and Central America, and Asia. They represent about 40 percent of the world's current forest cover, and are decidedly the largest forests on Earth in terms of total mass of living matter, biological diversity, and growth rates. These forests contain 50 percent or more of Earth's species. They are also the principal habitat of most of the world's remaining hunting–gathering societies. The tropical forests and the indigenous people who inhabit them are being sharply reduced in the face of clearing for agriculture, roads, and related land uses. Both the tropical forests and the boreal forests, however, continue to be lightly populated despite pressure from a rapidly growing world population to harvest them and replace them with land use and settlements.

The Urban World

The vast majority (70–80 percent) of the people in developed countries live in urban areas and have little or no direct (livelihood-based) association with the land. In addition, a rapidly growing urban population makes up 30–40 percent of the population of most poor countries (Figure 2.13). Together the world's urban and suburban population totals about 2.5 billion people, about 45 percent of the global population. The 1 billion people who live in cities in developed countries consume nearly two-thirds of the world annual output of natural resources.

The system of resource acquisition and economic support for the urban world is both profound and far-reaching. Cities depend not only on a massive inflow of resources but on a system that extends over much of the world. The United States, for example, imports oil from, among other places, the Middle East, Venezuela, and Europe's North Sea; metals such as copper and gold from Brazil, Malaysia, and Africa; and selected types of lumber from Canada, Brazil, and Indonesia. Japan's system is

FIGURE 2.13 *Rio de Janeiro, Brazil. This city, like many others in the developing world, is overflowing with people and sprawling into the surrounding landscape. Inset shows graph of population change in cities relative to world population, 1800–2000.*

even more extensive in many respects. It not only imports energy, manufacturing, and building materials from a global network, but it also depends on a vast international network for agricultural resources. The same is true of most of the European nations.

Coupled with the vast consumption of resources is the production of massive amounts of waste, much of which reenters the environment as pollution. Urban areas are the world's primary sources of air pollution and hazardous waste production. (Agricultural areas are the primary sources of water pollution.) In the case of air pollution, the contamination and its effects are not limited to cities and their immediate geographical areas. Many pollutants extend over international regions, and even over the entire globe. Global carbon dioxide levels have increased by at least 30 percent in the past several centuries because of increased urbanization, industrial development, and motor vehicle travel. Currently, the world's urban complexes, which occupy less than 5 percent of the Earth's land area, produce as much as 80 percent of the carbon dioxide pollution added to the atmosphere from human sources.

2.6 The Role of Human Values and Environmental Regulations

Attitudinal differences among cultures demand that geographers consider more than the scientific character of the environment and venture into the realm of human values as well. The job of science is to document, analyze, and explain the processes, changes, and effects of various environmental phenomena, such as changes in the chemistry of air or the impacts of floods on land use. Strictly speaking, we cannot rely on science to tell us what constitutes an environmental issue or moral dilemma or where and how much attention we should give to this or that change in the environment. In other words, science has no universal answer as to what level of property damage or loss of life from flooding, for example, constitutes a moral issue or problem to society. Scientists do indeed address the moral side of environmental problems, but only in the capacity of members of society with sensitivities and convictions like anyone else.

Unfortunately, nature provides no guidelines for the care and treatment of the environment. It does not tell us how much use is abuse or how much discarded waste constitutes pollution. For these determinations we must rely on science and our own values. Science helps define thresholds of change—that is, those levels of pollution or disturbance at which significant environmental change begins or increases. Ultimately, however, we must set the standards, whether in the form of our personal attitudes and principles or of formal law legislated by governmental action, by which we judge the health and well-being of the environment. These standards are based on what we as individuals, communities, and nations decide to be appropriate, practical, just, desirable, moral, responsible, and necessary.

If humans possessed a single body of values, the problem of setting environmental standards would be difficult enough. But humanity is made up of diverse, complex, and often contradictory values systems (see Figure 2.1). Witness, for example, that we promote human life and the right to create it, yet realize that our own numbers may eventually exceed the capacity of the planet to support us. We tolerate mismanagement and even malicious treatment of the environment because we support the right of self-determination, including the right of others to consume their resources as they like. Yet we invade other countries and decimate their environments in war because of differences in political and religious ideologies.

We are well into the era when the world must speak to the problems of the environment as a community. No longer are nations able to deal effectively with the problems of territory, resources, population, food, and pollution in geographic isolation. Besides the practical issues, monumental moral questions face humanity as we enter the twenty-first century. For instance, what responsibility should humans take in controlling their own numbers in order to improve the quality of existing and future human life? And what is the rightful role of humans in determining the destiny of other species? Do some societies have the right to use natural resources and pollute the environment at the expense of others? What responsibility should humans assume in the role of stewards of Earth's environments, resources, and organisms? These critical issues and the ways we treat them depend on our values, both informal (nonstatutory) and formal (statutory). Our statutory values are expressed in the environmental policies and regulations we enact at various levels of government from villages and towns through the United Nations.

Environmental Policy and Regulation

Government has become increasingly involved in regulating the activities of people in order to protect the environment and human health and safety. Regulation occurs at all levels of government and applies to all levels of human activity, from the individual person to transnational corporations. Although the vast majority of environmental regulations have been enacted in the past 30 years, the roots of modern environmental policy actually go back to the 1800s with regulations addressing sewage disposal, water supplies, and the work environment.

By the middle of the twentieth century most developed countries had enacted health and safety laws for residential facilities, while industry remained largely free of environmental regulations governing pollution inside and outside factories. In North America, laws setting pollution emission limits were generally weak and ineffective, even though the industrial economy was growing rapidly after World War II. Most U.S. environmental laws applying to industry came with or after the **National Environmental Policy Act** of 1969 (NEPA) and in general were strongly resisted by the business community. NEPA was the first comprehensive body of environmental law, addressing air quality, water quality, hazardous waste, species protection and many other areas. However, business was reluctant to accept laws calling for changes in manufacturing systems—that is, changes aimed at improving resource extraction and handling, material processing, and product development, the heart of industrial operations.

Bowing to pressure from the business community to keep government out of the manufacturing process, U.S. action under NEPA focused on 'clean-up' and 'stop-gap' measures. These were mostly 'end-of-the-pipe' regulations such as sewage treatment plants, exhaust filters on factories and cars, and hazardous waste clean-up programs. This, in turn, created a whole new line of business, namely, environmental technologies and services designed specifically to manage waste, including not only pollution-treatment facilities, but hauling and storage services, remediation services, legal services, and administrative services. Under the U.S. Superfund Program, for example, enacted to clean up hazardous waste sites, more than $30 billion had been spent to remediate only 2,000 sites by the year 2000.

The global market for environmental technologies and services is growing rapidly and is currently estimated to be $300 billion to $600 billion a year. In the United States, the **environment industry** has an annual revenue totaling close to $100 billion, nearly 2 percent of the country's gross national product. Many of the biggest industrial polluters of the 1960s and 1970s are now manufacturers of pollution-control equipment, and critics charge that these companies now profit from cleaning up the messes they helped create. On the other hand, many industries are improving the efficiency of their manufacturing systems in order to reduce the need for expensive pollution treatment technologies and services. The objective of these manufacturers is pollution prevention, with a long-term goal of developing environmentally sustainable manufacturing technologies.

Despite the strides that have been made in legislating environmental policy, a significant number of environmental regulations are unsuccessful for the simple reason that they cannot be adequately enforced. Enforcement requires persistent monitoring, and most governmental agencies charged with oversight responsibilities lack the resources to efficiently monitor large operations such as trucking, railways, and farming, to say nothing of the difficulties entailed in monitoring large and complex public sector systems such as cities. As a result, some people argue that we have approached the regulation problem backward. They suggest that policing-type programs that penalize for infractions should be replaced, at least in part, by incentive programs that reward reductions in pollution. Table 2.1 lists some common types of environmental regulations in use today. None currently offer tax breaks or other incentives.

Enforcement is also affected by the geographic character of a land-use system. Decentralized or geographically diffused land uses such as agriculture and residential development are more difficult to monitor and regulate

TABLE 2.1

SOME COMMON TYPES OF ENVIRONMENTAL REGULATIONS

Regulation Type	Definition	Application
Ambient standards	Pollution limits set according to the background quality of air or water over an area or region	Often used to protect human health or ecosystems such as a river or estuary
Discharge standards	Pollution limits placed on individual municipalities or industrial firms	Used with both wastewater (effluent standards) from pipes and exhaust stacks
Performance standards	Polluters can select the most cost-effective way of meeting requirements	Allows use of prevention techniques as well as recycling and treatment techniques
Best management practices	Polluters are required to use certain management techniques for a particular land use such as farming	Used in water pollution where pollutants are not contained but spread over land, as in pesticides on cropland
Product ban	The requirement of a license to produce and sell certain products	Commonly used to limit availability of products such as pesticides, solvents, and other chemicals

than centralized land uses because the critical points of contact with the environment involve millions of individual operators. How, for example, should we regulate a problem such as homeowners' application of pesticides to their yards? Many property owners saturate yards on a regular basis in an effort to exterminate harmful and annoying insects, but they also kill bees, earthworms, spiders, soil microorganisms, and small songbirds. Lawn chemicals pollute runoff and groundwater and affect others' health. There are more than 50 million residential lawns in the United States, and the average rate of pesticide application (per acre) is 10 times greater than it is on farmland. It would take an army of environmental agents to monitor and regulate this problem. The same holds for other land uses including farming, trucking, and even certain forms of recreation such as motor boating.

On the other hand, centralized land uses such as mining, most manufacturing operations, and air transport (airports) are better suited for monitoring and regulation because activities are clustered in relatively few places. In the case of diffused land uses, we must turn to other regulatory devices. With homeowners, for example, we cannot regulate their individual behavior in lawn and yard care, but we can regulate the types of pesticides manufactured for sale. We can also regulate the information manufacturers distribute to consumers about environmental and health risks associated with these products.

Whether consumers heed regulatory information depends on education, values, and economic well being, among other things. The automobile industry illustrates this point. Car manufacturers have responded to regulatory pressure by producing vehicles that are about 50 times less polluting than cars of the 1970s. New cars can be factory tested to ensure that regulations are being met. However, this is no guarantee that car operators will properly maintain the pollution control devices on their cars, especially as cars age. Indeed, some car owners actually remove these devices to reduce vehicle maintenance costs and improve fuel economy. Others argue that keeping old cars running is a form of recycling that offsets their heavy pollution output.

International Regulation

Finally, there is the international factor in environmental regulation. Nations have the right to legislate and enforce environmental laws as they see fit. Attitudes on environmental matters vary widely among nations, and not surprisingly, so do environmental laws, regulations, and enforcement practices. Moreover, the environment is not neatly partitioned geographically to coincide with political borders. In fact, many environmental systems (e.g., prevailing winds, ocean currents, migratory birds, and large watersheds) are immune to political borders, so pollution created by one country is shared by others, sometimes much of the world. The United States and Mexico, for example, not only have different types and levels of policy and regulation, but vastly different environmental monitoring and enforcement systems. The pesticide DDT, which was banned in the United States more than 30 years ago, is still used in farmlands in Central America. Wind blows some of the DDT residues back into the United States, reaching the Great Lakes and beyond into Canada. Ironically, much of the DDT used in Central and South American countries is manufactured by U.S. chemical companies. Clearly this problem calls for an international agreement between these countries and the United States on the sale and use of DDT. But such policies are typically difficult to negotiate and implement because of differences in cultural values, economic practices, political priorities, cost factors, and many other considerations. A higher authority is called for to establish international environmental policy.

The United Nations (UN) is the world's chief international system of government, and it is very concerned with transnational environmental problems. The UN has enacted laws in a number of critical areas, including nuclear waste dumping in the ocean, export of hazardous waste, and international fishing limits. However, the UN has difficulty enforcing these laws because of the geographic scope of monitoring required and because some nations refuse to recognize certain environmental laws. It has, however, successfully promoted international dialogue on the environment, population planning, economic development, and many related issues. Several large conferences have brought world attention to the condition of the global environment, its future, and the need for international cooperation and joint action. The first truly global-wide dialogue on the environment occurred in 1992 with the Earth Summit conference in Rio de Janeiro. It was attended by representatives of more than 150 countries and more than 1400 nongovernmental organizations.

The highlight of the Earth Summit conference was the signing of two international treaties on climate change and biodiversity and a global action plan called Agenda 21 to guide environmental protection and sustainable development into the twenty-first century. This, of course, was just a first step. During the years since the Earth Summit, many developed countries have continued to expand policies and programs that deal with environmental protection. In the developing world, many countries have begun to establish policies to address environmental issues and to improve environmental management.

However, the effectiveness of policy implementation and compliance monitoring varies greatly among countries and regions. For example, among the developed regions such as Western Europe and North America, legislative and administrative measures to combat environmental deterioration have become well established. And while the institutional structure is in place to protect the environment, much more must be accomplished before sustainable levels are realized. In most developing regions where environmental protection policies are being initiated, implementation and compliance monitoring is constrained by lack of financial resources and availability of skilled manpower.

2.7 SUMMARY

Environmental geography is concerned with the quality of the global environment, the processes and issues of environmental degradation, and the condition of life, both human and other organisms. At the center of our concern is land use, particularly those land-use activities that degrade the environment and deplete resources without provisions for renewal and remediation. The damaging effects of such land-use activities are classified as either pollution, which involves environmental contamination, or disturbance, which involves environmental disruption. Pollution and disturbance are driven by many factors including population growth, increased consumption, expanding land use, and new technology. The role of each of these factors varies from society to society according to human values related to birth control, war, technical innovation, and other factors.

Earth is covered primarily by vast oceans of saltwater. The continents cover about 30 percent of the Earth's surface. Most organisms live on the continents and in the shallow ocean waters surrounding them, the continent shelves. The continental shelves are one of the three major subdivisions of the continent; the others are the shields and the orogenic belts. Both the continents and the ocean basins are part of the Earth's crust, which is partitioned geographically into huge sections, called tectonic plates. Tectonic plates are subject to slow lateral movement that produces earthquakes and volcanoes along their borders.

The global coordinate system is made up of parallels and meridians and is used to reference locations on the earth's surface. It is conventional to divide the Northern and Southern Hemispheres in three zones of latitude, high, middle, and low, and each in turn can be further subdivided into zones such as the tropics and subtropics. For the Earth as a whole, 90 percent of humanity inhabits about 30 percent of the total land area, and most people live within 100 kilometers of a coastline.

Earth is a geographically integrated planet tied together by great systems of winds, water, currents, and organisms. As a result, environmental problems often affect vast regions—indeed, sometimes the whole planet. However, most pollution problems are tied to relatively small source areas, and beyond these areas, contaminant concentration declines rapidly with distance. Only in certain food chains of ecosystems do contaminant concentrations increase with distance (or number of consumers) from the source.

The Earth is also a risky place that can threaten life. The atmospheric, oceanic, and geologic systems are capable of producing huge though infrequent events that can devastate land use and ecosystems. The geographic distribution of such events is very uneven, and humans try to reduce nature's destructiveness by, for example, building dams and embankments to control flooding. Nevertheless, loss of property and human life is rising as people crowd into hazardous areas such as floodplains and seashores.

Close to half of Earth's 6.1 billion people live as subsistence farmers. Virtually all of these people live in poor countries, that is, developing and deprived countries. In developed countries, relatively few people live as farmers. Worldwide, the total area devoted to cropland is about 10–12 percent of Earth's land area. Most of this land is found in forest and grassland regions.

The vast majority (70–80 percent) of people in developed countries live in urban areas. In poor countries 30 to 40 percent of the people live in cities, but the numbers are increasing rapidly. All told, about 45 percent of the world's people live in urban areas and these areas draw on a vast network of global resources to support these people. They also produce most of the pollutants entering air, water, land, and biological systems.

Because we place a high value on the quality of our environment, we have established a vast body of regulation to reduce the impacts of land uses such as industry, transportation, and agriculture. Environmental regulations occur at all levels of government from local community ordinances to international programs of the United Nations. Most industrial regulations are aimed at "end-of-pipe" measures. For other land uses such as agriculture and residential, enforcement is often difficult because these activities are geographically diffused. The increasingly regional and global nature of environmental problems is calling attention to the need for international regulation, an area that calls for UN action. Agreements are being formulated, but they are proving difficult to enforce. On the other hand, international dialogue such as the 1992 Earth Summit has increased sharply in recent years, which is cause for hope that global regulations may be enacted successfully in this century.

2.8 KEY TERMS AND CONCEPTS

- Land-use systems
- Environmental impacts
  - pollution
  - disturbance
  - land use
  - technology
- Earth's Surface Features
  - continental shelves
  - orogenic belts
  - shield's
  - crust
  - tectonic plates
- Global coordinate system
  - meridians and parallels
  - prime meridian
  - zones of latitude
  - Tropic of Cancer
  - Tropic of Capricorn
- Where People Live
  - latitudes
  - water relations
- Geographical Integration
  - distance and scale
  - magnitude and frequency principle
  - global land use
- Environmental regulation
  - NEPA
  - environment industry
  - enforcement
  - international regulation

2.9 QUESTIONS FOR REVIEW

1. Discuss the scope of environmental geography. What role does land use play in this field? Distinguish environmental pollution from environmental disturbance. How does technology influence human land use and the potential for environmental impact?
2. How much of the earth is covered with water, and how much of this is fresh water? The continents are made up of three major surface features. Name and briefly describe each. What is the average elevation of the continents, and how does that relate to the distribution of humans?
3. What are tectonic plates? How do they relate to the distribution of volcanoes and earthquakes? What is the Ring of Fire? Which hemisphere contains the most land area, and in what zone of latitude is most of it concentrated?
4. Describe the global coordinate system and distinguish between meridians and parallels. What are the three main zones of latitude in each hemisphere, and in which are most of the world's people concentrated? Name some of the key countries.
5. What is meant by the concept of Earth as a geographically integrated planet? Why is geographic scale such an important factor in the distribution of pollution? How does the risk of hazards from natural factors relate to the magnitude and frequency principle?
6. What percentage of the Earth's people live in rural lands, and what is their principal livelihood? Are the countries where most of these people live rich or poor? Name some leading rural countries in the world.
7. What percentage of the Earth's land area is taken up by cropland and farm settlements? What kinds of natural vegetation has been displaced by farming over the Earth, and in what belt(s) of latitude is farming most concentrated today?
8. What percentage of people live in urban areas in developed and poor countries? Explain why cities occupy a small percentage of the Earth's land area but extend over the entire planet, more or less.
9. "Science sets the standards that we should use to regulate land use and its impact on the environment." Discuss this statement. What is the National Environmental Policy Act, and how did the U.S. business community respond to new environmental laws after 1970?
10. List and briefly describe several of the problems related to the enforcement of environmental regultions. What was the Earth Summit conference in Rio de Janeiro, who sponsored it, and what did it address?

3

ENVIRONMENT, LAND USE, AND SUSTAINABLE DEVELOPMENT

3.1 INTRODUCTION

World leaders and scholars increasingly agree that the future of humanity and the quality of life in general rests with our ability to achieve a lasting balance with the global environment. In other words, we must achieve a system of global land use that is sustainable generation after generation. Sustainable use of the Earth requires human participation in the natural cycles of air, water, organisms, and other systems without degrading and depleting them. It requires thinking less in terms of consuming the planet's wealth and more in terms of sharing the planet's harvest. It also requires rethinking our notions of progress, growth, development, wealth, and well-being at all scales of human existence from the personal to the global.

The United Nations Conference on Environment and Development (UNCED), the so-called Earth Summit held in 1992, was a milestone for the global community. For the first time, the world's leaders convened to discuss policies and programs to halt environmental degradation and to plan for sustainable economic development and growth. Most importantly, it was a first step toward managing the environment from the global perspective. Since the Earth Summit, several additional global conferences have addressed specific environmental and development issues, including the UN Conference on Population and Development held in 1994 and the UN Framework Convention on Climate Change Conferences, which have been held annually since 1995.

This chapter summarizes current ideas about managing the global environment to achieve a sustainable balance among environment, land use, and development. Most societies throughout the world now recognize that most current economic growth and development patterns cannot be maintained indefinitely. As a result, programs endorsed by most nations are underway to develop new approaches to sustainable land use. The first part of the chapter reviews the widespread scope of environmental changes brought about by human activity and then discusses the initial international and global efforts to reduce environmental damage and to develop sustainable economic programs. The latter part of the chapter examines the constraints, geographic realities, and a grass-roots effort of sustainable land use.

3.2 THE GROWING ENVIRONMENTAL IMPACT OF HUMAN ACTIVITY

Since the beginning of the nineteenth century the explosive growth and expansion of the world's population accompanied by new technological advances have dramatically modified Earth's landscapes. Humans are reaching further into the environment, altering the very systems that cycle energy and material and support life itself. In the Americas and Australia, most of the vast forests and grasslands of the midlatitudes and subtropics were converted to crop and grazing land in little more than a century. Today, vast areas of tropical forest are being cleared, resulting in the eradication of countless species, many with potential value as sources of food and medicine (Figure 3.1).

Worldwide, 9 million square kilometers (3.5 million square miles) of new cropland have been opened up in the last 150 years. Such massive deforestation and grassland conversion has inevitably been accompanied by environmental degradation. Soil erosion has reduced range and cropland productivity and increased the loads of sediment

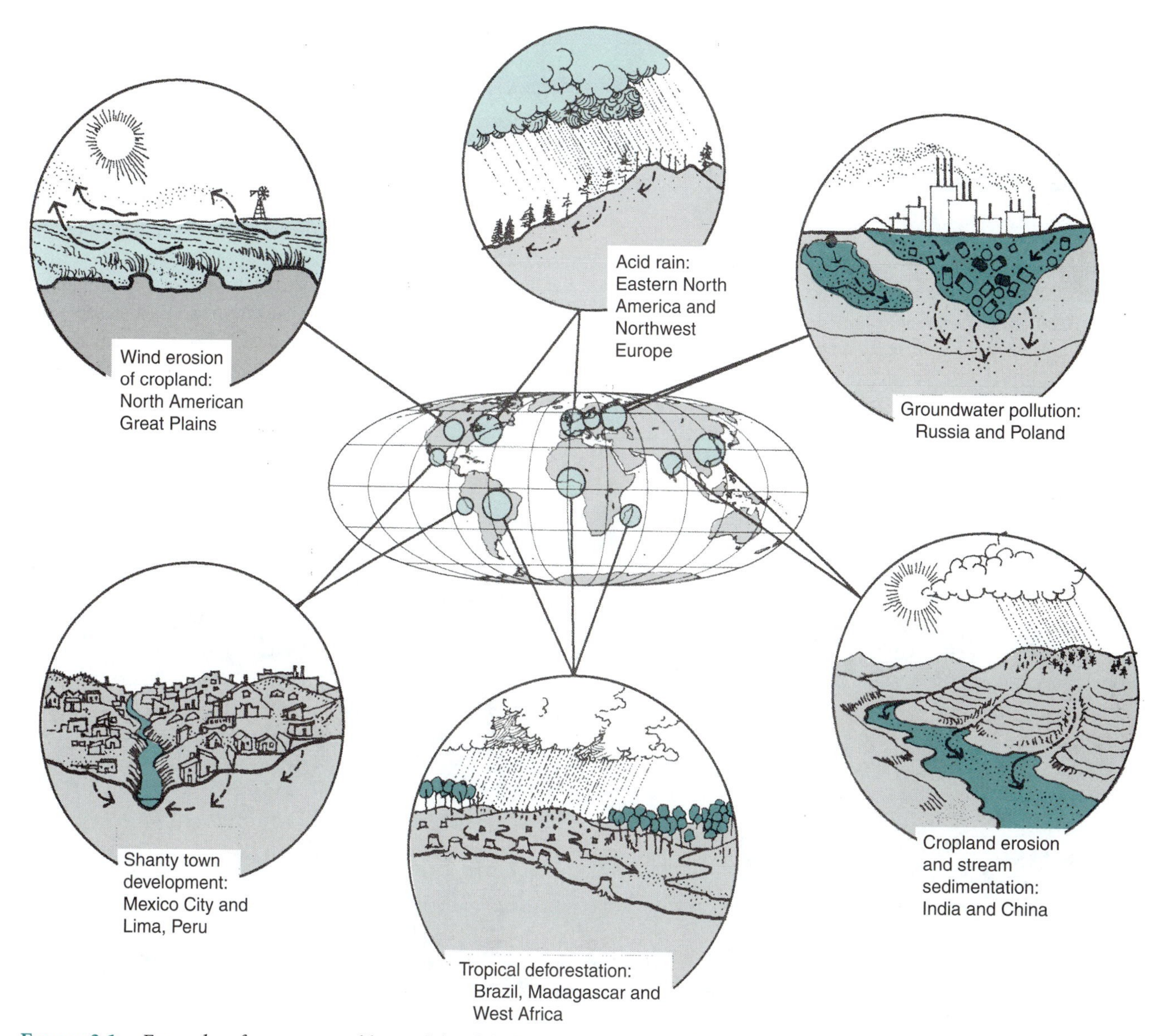

FIGURE 3.1 *Examples of common problems of the global environment at the beginning of the twenty-first century. Most are tied to rapid population growth and economic expansion.*

and other pollutants in streams, lakes, and wetlands. Overgrazing in the arid and semiarid lands has resulted in vegetative cover loss, increased erosion, and even desertification. The tenfold increase in irrigation in this century has modified surface water flows with diversions and impoundments, along with significant drawdown and depletion of groundwater. Furthermore, much irrigated land has been degraded by increasingly saline or waterlogged soils. Finally, agricultural expansion has contributed to acute scarcity or deficits of fuelwood in many parts of the developing world (Figure 3.2).

FIGURE 3.2 *Excessive grazing, browsing, land clearing, and farming have virtually eliminated a wood supply for fuel in many poor countries such as shown here in Afghanistan.*

On top of this, rapid population growth in the countryside has outpaced agriculture's labor needs. Thus, many landless peasants migrate to cities to seek a livelihood. Unfortunately, most do not find permanent jobs and are forced to live in ramshackle, impromptu slums that ring many cities in developing and undeveloped countries. With no sanitation systems, safe water supply, or permanent housing, these congested, squalid slums are unhealthy, criminal-ridden places that perpetrate poverty and chronic poor health. Because these slums are growing so rapidly, they are becoming major sources of water and air pollution, thereby adding further to environmental degradation in the poor countries (see Figure 2.13).

Trends in Poor and Wealthy Countries

At the root of resource degradation in the world's poor countries, both developing and deprived, are the interrelated problems of rapid population growth and poverty. These countries, with more than 75 percent of the world's population, utilize less than 25 percent of the resources. Perhaps one-fifth of those living in poor countries do not even have enough to eat. Chronic poor health and high infant and child mortality rates characterize many poor nations and become basic disincentives to reducing family size. Until poverty is alleviated, large families provide the only social and economic security for jobless, ill, or elderly family members (Figure 3.3). Furthermore, the impoverished will continue to deplete basic resources such as soil, forests, and water as they do what they must in order to

FIGURE 3.3 *Indian farm children, a key part of the peasant work force and family social security system. Population control in the poor countries is more than a simple matter of birth control.*

survive from year to year. Only when people have the means to produce sufficient food or earn a decent wage will they feel comfortable having smaller families, which will, in turn, lead to reduced population growth.

In contrast, rapid **unsustainable economic growth** in the developed world has for a long time, wrought environmental disruption, and that has now reached the global scale. For two centuries the technology-driven consumer economies of Europe and North America have gene-rated wealth and prosperity for about 20 percent of humankind. By consuming 70 to 80 percent of the mineral and fuel resources, this wealthy minority produces large emissions of carbon, sulfur, nitrogen, chlorofluorocarbons (CFCs), and other pollutants that are producing global atmospheric changes, chronic regional air pollution, and degradation of ecosystems.

The environmental impacts of some of the tens of thousands of manufactured chemicals such as DDT and various toxic metal compounds have been documented. Their production, however, continues, as does their use in some countries. Chlorofluorocarbons (CFCs), long used in aerosol propellants and refrigerants, contribute to dramatic seasonal decreases of atmospheric ozone over the polar regions. Ozone depletion subjects the Earth's surface to more intense, tissue-damaging ultraviolet radiation. Even though CFCs are no longer produced, they will remain in the atmosphere for decades. On the biological front, two relatively recent developments have resulted in the loss of floral genetic diversity. The rapid clearing of tropical forests over the last 25 years has eliminated thousands of wild plant species, and many traditional varieties of important agricultural plants have been replaced by fewer high-yielding hybrid varieties.

With world population over 6 billion and economic development initiatives exploiting even more resources, human-induced environmental disruptions are now massive enough to modify geochemical cycles, influence portions of the hydrologic cycle, and significantly alter atmospheric chemistry on global scales. It is not known what lasting effects current environmental changes will have, but with population expected to double and economic activity expected to increase by 5 to 10 times during this century, increasing environmental disruption and further degradation of the world resource base seems inevitable.

3.3 Sustainable Land Use: The Key to Environmental Management

The previous paragraphs paint a bleak picture of deepening environmental degradation in poor and rich countries alike. In many poor countries, basic environmental resources are being exhausted by populations trapped in a cycle of poverty, poor health, and disease that results in short and difficult lives for most. In the developed world, industrialization and unsustainable economic growth practices have built wealth and prosperity for some, while increasing environmental degradation. Consumption has become equated with progress, with little concern or even consciousness that this way of doing business is unsustainable. Moreover, these economic policies and consumer practices have also become a goal of many developing countries, suggesting that the habit of unsustainable development is expanding despite its limitations (Figure 3.4).

Nonetheless, some environmental scientists and policy makers argue that environmental decline is not inevitable if current unsustainable practices are replaced by sustainable ones. Sustainable economies maintain the basic resources upon which humankind depends. In a sense they are designed to thrive on the interest yielded by the environmental bank, such as renewed freshwater, solar energy, and annual plant productivity, without consuming the principal. Such economic restructuring will require that society recognize that changes are not only in its best interest but are also required if living conditions are to be maintained or improved.

Sustainable development has become a key term for economic development strategies that are designed to reduce environmental impacts and maintain the resource base while allowing a certain amount and type of development to take place. This principle has evolved from the recognition that irreversible environmental damage has accompanied many current forms of economic development, degrading the resource base on which we ultimately all depend. In 1987, the UN World Commission on Environment and Development issued a report that called for new development strategies that would meet current needs without compromising the ability of future generations to meet their needs as well.

Although the concept of sustainable development has generally won popular support, it will be difficult to implement. For many developing countries, sustainable economic development efforts will require a major change in political thought and government policy, followed by substantial technological and economic assistance from the wealthy countries. For the developed countries, sustainable economic practices will require many changes in the types and amounts of waste that are released into the environment. This means changes in basic societal values; it means changes in the way business is conducted; it means changes in the way decisions are made and policies are carried out; and finally, it means unprecedented cooperation among nations—developed, developing, and deprived, political allies and foes alike. These fundamental changes are summarized as follows.

Humanity as Part of Nature

There must be a change in the widespread perception that the environment may be modified, manipulated, or even controlled as if it were the exclusive property of

FIGURE 3.4 *Western-style consumerism being promoted in Calcutta, India. An example of pressure toward unsustainable development and consumerism habits.*

humankind. The technological advances that have brought about industrialization, urbanization, and mechanized agriculture have indeed transformed landscapes and thereby fostered the belief of human dominion over nature. These beliefs have led to increasingly unsustainable development throughout the world. New values must be developed that place the human species within the environment as a working and interdependent part of nature.

We must recognize that all land uses, whether urban or rural, are working parts of ecosystems, and like all organisms in ecosystems, we humans are dependent on the environmental resource base. Sustaining that resource base is required if humanity is to thrive. No land use should get more attention in this regard than agriculture. To begin with, agriculture should be assigned to lands that are appropriate for farming (as opposed to other land uses or no uses) by virtue of topography, soil quality, water supply, and climate. Performance should not be based just on short-term economic gains but on long-term maintenance of soil, groundwater supplies, and the quality of local streams and ecosystems.

Accounting for Pollution Costs and Environmental Damage

Environmental accounting methods must be developed to assess the environmental costs associated with producing, transporting, and consuming goods and services. Until recently, common resources such as the atmosphere, the oceans, groundwater, and even public land have been used for private profit with few restrictions and little or no costs. The real and intangible costs associated with air and water pollution and land and ecosystem degradation and destruction have been or will have to be borne by society. Legislation in some countries has begun to consider the environmental costs that the free market has historically ignored. These costs are calculated and integrated into the price structure of the producer or service provider. With such costs allocated directly to the producer (polluter pay laws), there is a built-in incentive to reduce pollution and environmental damage in order to lower costs. In some parts of Europe and North America, air and water quality have measurably improved as a result of environmental legislation enacted since the 1960s.

Natural Resource Accounting

The value of renewable natural resources must be considered in national accounting systems. Measurements of annual economic growth and output have ignored declining productivity of soils, forests, and fisheries, as well as the degradation of parks, historic sites, wild landscapes, and natural ecosystems. Factoring resource accounts in with other economic accounts would provide more realistic and accurate information with which to gauge economic performance. Furthermore, resource accounting would indicate the effect that current economic policy is having on stocks of ecological capital. National resource accounting systems are beginning to emerge in some developed countries.

Reduction of Poverty

Economic development strategies in the developing countries must follow a different course than the environmentally disruptive methods used over the past two centuries in Europe, North America, and other developed regions. In many parts of the developing world, the impoverished, rapidly growing populations are being forced to degrade and deplete their basic stock of capital resources, such as fuel, grasslands, and water reserves, in order to stay alive from year to year. In fact, the rate of depletion often accelerates as the supply dwindles, and poverty increases as people push to get their share out of a declining stock. However, before economic reorientation and sustainable development can begin in these poor countries, poverty must first be reduced. Reduced poverty can open the way to reduction in family size and population growth rates, both essential prerequisites to rational economic development.

For many poor countries where farming remains the dominant economic activity, breaking the poverty cycle depends on reestablishing a productive system of sustainable agriculture. Four basic requirements are necessary (Figure 3.5).

1. *Farmers must grow enough food or earn enough income to meet essential needs.* National market policies must ensure that farm prices provide an adequate incentive for farmers to produce enough food.
2. *Farmers must have secure rights to the land.* Land reform programs that provide ownership or long-term rights to farm the land will be necessary in some countries to encourage conservative land-use practices.
3. *Farmers must be taught how to manage their land properly.* People without managerial experience often lack the insight to build sustainable farms; therefore, educational programs are often needed.
4. *International marketing strategies and some food aid programs must be modified.* Although the influx of cheap surplus grain from developed countries may increase current food supplies, it lowers the price paid for domestically produced grain, thereby discouraging sustainable food production in poor countries.

3.4 Necessary Changes in Resource Policies for Sustainability

Deforestation, desertification, degradation of soil, destruction of habitat, and the decline of water and air quality are some of the byproducts of unsustainable economic policies. Such policies must be replaced with those emphasizing sustainable practices and conservative yet efficient use of resources. These policy changes must occur in all countries.

Reforming Agriculture

Over much of the Earth, agriculture is the dominant economic activity. In fact, for many people it is the way of life. Many traditional agricultural systems have sustained

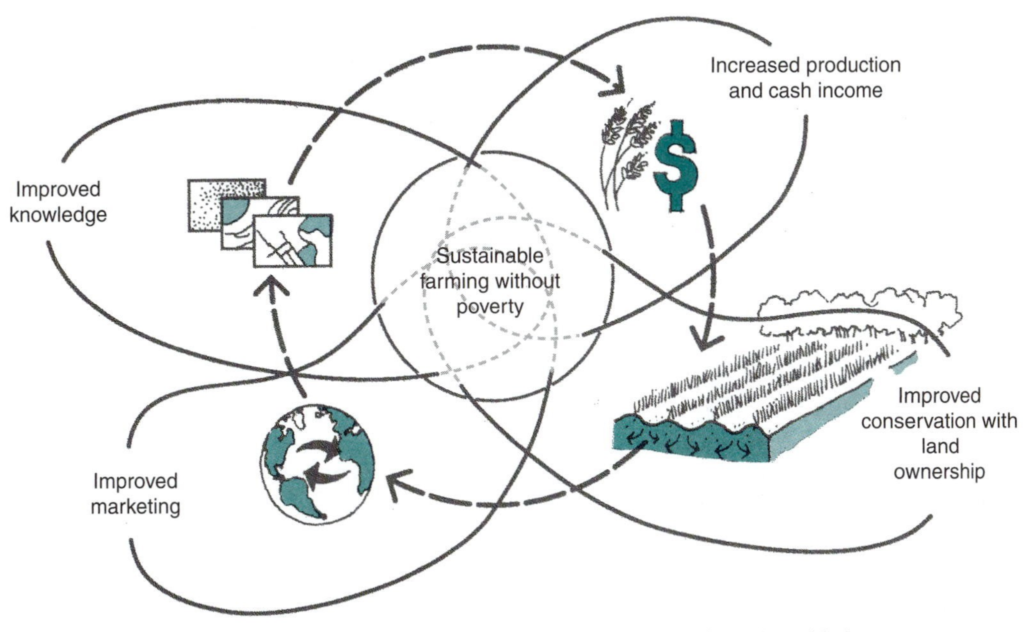

FIGURE 3.5 *The four basic requirements for breaking the poverty cycle and establishing sustainable farming in poor countries.*

themselves for centuries or even millennia, while those that depleted the soil and degraded the land eventually failed, such as occurred over much of ancient Mesopotamia centuries before Christ. With the development of commercial agriculture, competitive market forces, which do not account for the costs of land and water degradation, promote agricultural practices that maximize short-term outputs and profits. As a result, farming methods that conserve long-term productivity of the land itself, but which may have lower yields or higher costs, are less competitive. Until costs associated with soil erosion, fertilizer loss, and groundwater contamination from fertilizer and pesticides are factored into overall production costs, market mechanisms do not favor long-term sustainable agricultural practices.

Reforming agriculture by reintroducing sustainable agriculture will require adopting policies with long-term objectives aimed at developing and disseminating suitable agricultural practices and assisting farmers in implementing them. Initially, required production changes would be subsidized and regulated, but eventually, as all external costs are factored in, market forces would again dominate. In addition, for sustainable agriculture to be reestablished it must be done on a global scale. In other words, farming practices in the developed, developing, and deprived countries alike must be made sustainable, and costs and prices must be based on worldwide standards.

How would such changes be initiated, and how would they be funded? In some developed countries, programs to conserve agricultural land have been ongoing for several decades. The U.S. Department of Agriculture's Natural Resource Conservation Service (NRCS), formerly the Soil Conservation Service, helps property owners implement soil conservation practices such as reduced tillage farming and contour plowing. The Consolidated Farm Service Agency (CFSA), formerly known as the Agricultural Stabilization and Conservation Service, administers the federal soil conservation programs. Perhaps the most comprehensive of these programs has been the Cropland Reserve Program (begun in 1985) that has resulted in the conversion of millions of acres of cropland to grass or forest cover, thereby reducing erosion and soil depletion and expanding wildlife habitat while reducing the huge grain surplus.

By establishing similar cropland conservation programs in other countries and financing them with the subsidies currently paid to farmers to grow grain and other crops, an initial step could be made toward sustainable agriculture. At the same time it would reduce the glut of agricultural commodities being sold at less than market prices or even given away to poor countries as part of politically motivated food aid programs. Currently, annual subsidies for agricultural production in Europe, North America, and Japan cost hundreds of billions of dollars. Rather than continuing to subsidize current unsustainable agriculture, these monies could be allocated to programs to develop and promote sustainable agriculture. Instead of providing nonemergency food aid to poor countries in the form of subsidized agricultural surpluses, the developed countries could provide financial and technical assistance to promote sustainable agriculture based on food crops for domestic consumption.

Promoting Efficient Raw Material and Energy Use

Industrialization, urbanization, and the development of commercial economic systems over the past two centuries have been built on ever-increasing uses of nonrenewable energy and mineral resources. Current world energy production from oil, natural gas, coal, and nuclear fuels has increased by more than 50 percent since the global energy shocks of the 1970s and more than fifteenfold since 1900. Although mineral production and consumption has fallen in many developed countries during the last two decades because of improved efficiency and recycling, world demand for minerals remains voracious as industrialization expands in some developing countries (Figure 3.6). More efficient use of fuel and material has obvious environmental benefits as land and water degradation associated with mining is reduced. With less material to process, transport, and consume, less energy will be expended, and less air and water pollution will be created.

When energy prices rose rapidly during the 1970s and early 1980s, many governments and corporations developed programs to use energy more efficiently. As a result of pricing policies and new energy-efficient methods and technologies, such as improved automobile mileage, energy demands have risen more slowly. Continued improvements in energy efficiency would not only reduce greenhouse gas emissions and acid precipitation but would reduce overall costs to industries, commercial businesses, and consumers. During the 1970s and 1980s, some countries such as Japan, West Germany, and Sweden reduced energy use per unit of production by as much as 40 percent, and rapidly industrializing nations such as South Korea and Taiwan adopted state-of-the-art energy-efficient technologies.

In short, **reforming energy use**—improving energy and material efficiency—saves money and reduces environmental degradation. Furthermore, it provides more time for alternative energy sources to be developed from sustainable (renewable or perpetual) supplies such as hydrogen and other solar-based technologies. In developing countries where increasing demand for fuelwood has resulted in acute shortages and deforestation, it is essential that low-cost, efficient alternatives such as kerosene heaters and stoves be made available.

Implementing steady gains in energy efficiency and an orderly transition to a sustainable energy economy will

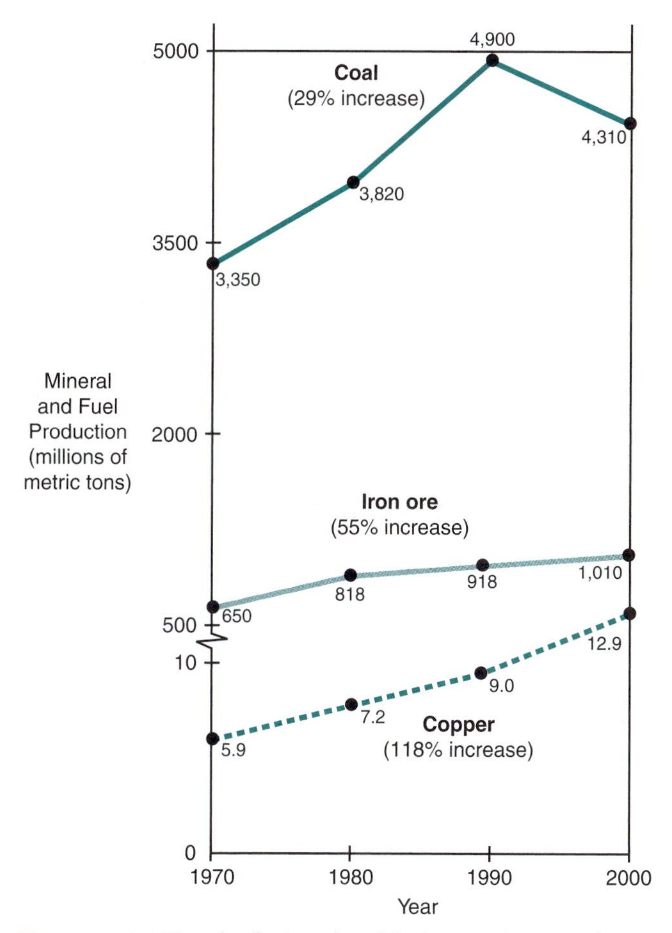

FIGURE 3.6 *Trends of mineral and fuel extraction over the past 40 years, reflecting increasing consumption rates.*

require government-directed policies that may prove politically difficult. In North America and Russia, where per capita energy use is highest in the world, energy taxes are relatively low. In contrast, in developed countries that have imposed high energy taxes for many years, per capita energy use is only about half that of the United States. However, trying to impose conservation pricing through increased energy taxes when market prices are low is politically unpopular in democratic countries such as the United States and Canada. A second government strategy would require more energy-efficient designs in buildings, appliances, heating and cooling systems, machinery, and vehicles. Financial incentives could also be made available to utilities and manufacturers who develop programs or products that exceed efficiency standards.

Finally, the monopolistic control that most energy utilities have over their markets must be modified so that suppliers of alternative energy sources can compete for those energy markets. For example, in California, public utilities are now required to purchase electricity generated from privately owned wind turbines and solar electric facilities. A number of other states (more than 20 by 2000) have passed laws to restructure energy utilities by gradually increasing competition among energy providers while reducing the states' regulatory powers. Will these restructured, more competitive industries favor new or alternative technologies that will be both cheaper and cleaner? It is certainly possible, but it will be some time before the answer is clear.

3.5 CONSTRAINTS TO ACHIEVING SUSTAINABLE LAND USE

To many observers of the global scene, the prospects for achieving a sustainable balance between humans and the environment in the next 25 to 50 years are not good. The root cause of this gloomy forecast continues to be rapid population growth, especially in the world's poorest countries. Currently, world population is increasing by more than 82 million a year, with 98 percent of the babies born in the developing countries, which are least capable of supporting them.

On top of the population problem, mass consumerism, once associated with developed economies, is increasing rapidly in many developing countries as well. This means that per capita use of most resources is rising. As a result, gross consumption for much of the world is actually rising faster than world population. Although the developing world often criticizes the wealthy developed nations for excessive resource consumption, the governments of many, if not most, developing countries strive to develop consumer economies as well.

With rising population and consumerism comes rising output of waste and more widespread disturbance of remaining open lands: forests, grasslands, wetlands, coastal lands, and other unsettled or lightly settled landscapes. In general, those societies that consume the most pollute the most. Therefore, the developed countries, though much smaller than the developing countries in population, return a much larger amount of waste to the environment per capita. But the developing countries are catching up. The strong trend toward urbanization in these countries is resulting in higher rates of per capita pollution. This trend is especially noteworthy because the developing countries lag far behind the developed countries in pollution mitigation technology and controls.

With modern pollution control programs, there is a cause for hope. Many methods and techniques have demonstrated a capacity to reduce air and water pollution from some sources. In the past two decades, many industrial nations reduced air pollution from power plants, industrial operations, and automobiles. These reductions were the result of multifaceted programs usually involving new and stricter pollution laws, cleaner fuels, and new pollution abatement technologies. The resulting improvements, unfortunately, were often offset by economic growth and slow adoption and implementation of new technology.

For example, although pollution emissions from automobiles and light trucks were reduced for individual vehicles, the number of miles traveled in the United States rose by 129 percent between 1970 and 1996. This was a consequence of both more travel per capita and more automobiles and light trucks on the road, and it was a trend followed in all types of economies throughout the world. Furthermore, whereas significant technological advances were made in sewage treatment, for example, few cities outside the developed world have been able to take advantage of them, largely because of costs and governmental priorities. Instead, cities in most developing countries and all deprived countries dump raw sewage into rivers and seas at an increasing rate. For instance, in Alexandria, Egypt, considered one of the world's most polluted cities, the Egyptian government passed the first pollution law in 1994; however, enforcement is unlikely because many of the heaviest polluters are government-operated industries.

Indeed, the governments of most nations recognize the need to develop **sustainable economic systems** with lower rates of resource consumption and waste production, but few seem willing to attempt it. We have only to look at the platforms of major political parties: economic growth is invariably a major goal. Elected officials from the local to national levels of government typically win or lose based on economic trends, principally jobs, wages, and development. Few are able to win on a platform aimed at minimizing growth and building sustainable economic systems. Instead, a platform of *sustained growth*, as opposed to *sustainable development or sustainable land use systems*, has generally proven to be a prerequisite for advancement in the world of politics.

This is not to say that all leaders lack the courage and public backing to move toward sustainable land use systems. A few face the problem and the difficulties that come with it realistically. More, however, embrace a more politically palatable concept called "sustainable development," which calls for reduction in the rates of consumption without sacrificing growth. This would necessitate changing the development process to one that is more compatible with the environment: In a nutshell, conserve resources and environmental quality without curtailing growth. In the words of Jim MacNeill, former secretary-general of the United Nations Commission on Environment and Development (UNCED), this would require "fundamental changes in the ways nations manage the world economy . . . a world with new realities, realities that have not yet been reflected in human behavior, economics, politics, or institutions of government." Is sustainable development possible? Is it possible to have any kind of economic growth at a major scale without sacrificing irreplaceable resources and environmental quality? It certainly will be difficult in the world we see today, especially for developing nations saddled with huge international debts (e.g., Brazil and Mexico), rapid population growth, and rapidly developing Western-style consumer economies. For many of the very poorest countries, burdened with the immediate problems of burgeoning populations, social upheaval, and tribal warfare, there seems to be little reason to even address the sustainable development issue.

In addition, there is the practical problem of how to build sustainable systems at the grassroots level. While operating models of sustainable systems do exist, most are simple and small in scale—for example, certain farms and rural communities. How these models can be integrated into a larger, more complex working whole and balanced with the needs of modern societies for health care, safety, education, and other things is a paramount challenge to the leaders, planners, decision makers, and voters of the twenty-first century.

3.6 Some Geographic Requisites for Sustainability

Economists tell us that in order to raise their standards of living, developing countries need to increase their gross domestic product, and to do this, they need more development. But the results of development efforts in many of these countries do not entirely support this view, because much, if not most, conventional development is not sustainable. For example, the International Monetary Fund (IMF), which assists troubled national economies with large loans, has encouraged Asian peasant farmers in some areas to shift from food crops to cash crops, such as tea, to generate export dollars and improve their purchasing power. But this has made farmers vulnerable to the economic uncertainties of the global market, and when the export market declines or fails, they are left with neither a source of income nor a local food production system. Fields are abandoned, and the land declines. The traditional agricultural system that took many generations to develop is lost, and future generations suffer loss of economic opportunity. For new development to be sustainable in today's world, it must meet several requirements that are not widely understood and supported by modern developers.

First, sustainable development must be built on a sustainable resource base, and this demands recognition of the simple fact that the world's resource systems vary geographically, often radically so. Any attempt at development must recognize a country's geographic uniqueness, both opportunities and constraints. It would be unwise, for example, to promote commercial agricultural development in countries with small and uncertain water supplies (Figure 3.7). On the other hand, commercial lumbering may be appropriate for countries with large forest reserves, but for forestry to be sustainable, rates of extraction must be balanced with rates of reforestation, as well as with competing land uses such as farming.

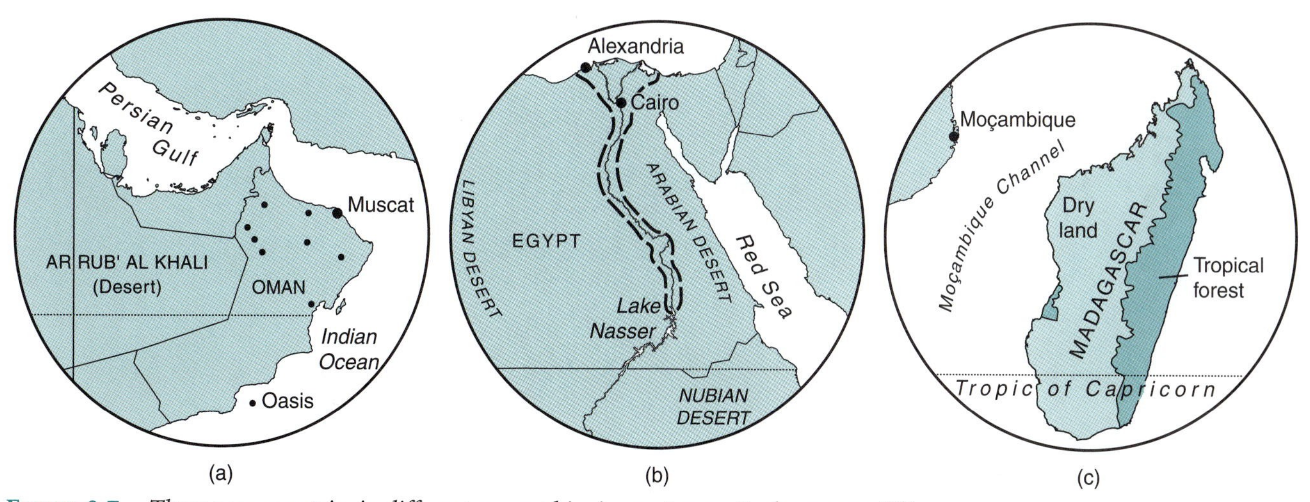

FIGURE 3.7 *Three poor countries in different geographic circumstances. Each presents different opportunities and constraints that must be taken into account in development programs. (a) Oman is very dry throughout with no significant river for water supply; (b) Egypt is also very dry but the Nile provides a major source of water; and (c) Madagascar is wet on one side (east) and dry on the other (west).*

Second, development must be focused on those levels of society and those geographic locations in a country where it will do the most good for the greatest number of people in the long run. For example, development of mining operations—a favorite enterprise in several developing countries—usually benefits relatively few people in only a few places and is typically not sustainable because it draws on a nonrenewable resource base. There is also a risk of countries becoming functionally dependent on the revenue from such operations. Using this revenue to import major food supplies, for example, is questionable because the flow of money is not tied to a sustainable resource system. On the other hand, using mining revenues for environmental restoration, land-use planning, and the development of infrastructure in support of domestic food production and health maintenance systems might be highly advisable.

Third, sustainable development must include a strong educational component focused on the production-level people in a society, people with hands-on responsibility for the use and management of natural and human resources. The information provided should not be narrowly focused and dependent on special technologies, that is, so-called "quick fixes," but rather broadly applicable in the context of local economic practices and traditional technologies. And fourth, in order to be sustainable, development cannot conflict with local culture but must be compatible with folk customs and existing social order including land use practices, local governance, aspects of religion, trading practices, and so on.

How do these requirements for sustainable development match up with modern development led by massive international corporations tied by telecommunications to central management operations in New York, Tokyo, or London? In other words, is sustainable development possible in big, top-down organizations, or must it be a grassroots sort of enterprise focused on communities in local settings?

3.7 A GRASS-ROOTS EFFORT AT SUSTAINABLE LAND USE

The developing world is rife with examples of deteriorating landscapes, declining land use, and rising poverty, but there are places where environment, land use, and the quality of life are actually improving. The Rishi Valley in India is an example of a long-term **grassroots effort** that has achieved remarkable success in restoring a worn-out landscape and building a sustainable village farming economy. Located in south-central India, the Rishi Valley has been farmed for many centuries. By the opening of the twentieth century, the landscape was severely degraded, denuded of most forest, topsoil, and water resources, and with it, the lives of thousands of peasant farmers had become impoverished.

In the early 1930s a boarding school was established in the Valley, and under the direction of an insightful leader, Gidda Krishnamurti, a relationship was forged with villagers to begin the long process of restoring the environment, building a sustainable land-use system, and improving the quality of life. Students worked closely with peasant farmers to arrest soil erosion, rebuild topsoil, conserve water, and revegetate the landscape. At the same time, an educational program was launched for both children and adults addressing not only basic liter-

FIGURE 3.8 *A healthy rural landscape in India. The Rishi Valley is a result of 70 years of landscape restoration and land-use reform.*

acy but community health, conservation, and finance, among other things.

The effort has been slow and arduous, but the outcome very encouraging. Worn out, nonproductive landscapes have been restored over vast areas. Bird species rarely seen in early years have reinhabited the landscape. Farm production, literacy, and life quality have improved significantly for most villagers. And yet the whole program has been accomplished by grass-roots action involving only peasant farmers, their families, and high school students and their teachers (Figure 3.8).

Virtually no outside funding or modern technology has been used in Rishi Valley. The solutions employed in land restoration—such as capturing runoff and returning it to the soil, terracing farmfields, and reforestation—have required little more than traditional knowledge, cooperation among villagers and student teams, and of course, a great deal of hard work. Among the many important lessons from Rishi Valley are the following two:

1. Life in societies dependent on subsistence farming is tied to the land. To improve one, you must also improve the other.
2. Sustainable land-use systems must be built and managed by the people who depend on them. Human and natural productivity can be returned to even the most degraded landscapes using traditional (i.e., nontechnical) methods planned and executed without the imposition of external governmental or corporate programs.

3.8 SUMMARY

A major goal of the twenty-first century is to achieve more effective management of Earth's population, land use, and environment. In most nations, efforts are underway to develop new approaches to land use that will make it more sustainable. But the problem is a daunting one. Indeed, to many observers it is at the very heart of Earth's environmental crisis.

Humans have had made enormous changes in Earth environments in the past several centuries. In just the past 150 years, for example, 9 million square kilometers of new cropland have been opened. To do this, massive tracts of forest and grassland were cleared, which, in turn, has led to increased soil erosion, water pollution, groundwater depletion, and many other environmental problems. As the environment declines, agricultural yields fall and food production costs rise. Costs and environmental impacts may rise still higher with increased fertilizer and irrigation applications. Clearly our agricultural system over much of the world is not sustainable in the long term. And this argument can be made for other land-use systems as well.

All land uses are working parts of ecosystems, and like all organisms in ecosystems, we are dependent on an environmental resource base. For a land use to become sustainable, it must be in balance with that resource base. This will require meeting certain performance standards, including accounting for pollution costs and damages, accounting for the use of natural resources as a part of economic growth and output, and contributing to the elimination of poverty. This, in turn, will require major policy changes, including reforming agriculture by rewarding low-impact, sustainable farming methods, revamping agricultural subsidy programs, and promoting efficient use of energy and raw materials.

Implementing sustainable land-use programs will be difficult, and to many observers, prospects for achieving a sustainable balance between humans and environment are not good. The demand for short-term gains to provide food for a rapidly rising population in poor countries diminishes opportunities for instituting lower output land-use programs aimed at long-term sustainability. Rising consumerism in both developed and developing/deprived countries also stands in the way of changing to more sustainable economic practices. But there are reasons for hope. For example, since the 1980s air pollution has been reduced over much of the developed world, and among the rural poor, the land restoration program of the Rishi Valley India, has improved agricultural production, the quality of life, and the environment.

Among the geographic requisites for sustainable development is the need to adjust planning goals and programs to the resource base of individual countries, followed by programs that focus on those levels of society and geographic locations where development will do the most good in the long run. In addition, sustainable development must include a strong educational component focused on the people directly responsible for the care and management of land use and resources, and the changes employed must be compatible with local culture and customs.

3.9 KEY TERMS AND CONCEPTS

Unsustainable economic growth
Sustainable development
Environmental management
- pollution accounting
- natural resource accounting

Necessary changes
- Reforming agriculture
- Reforming energy use

Sustainable land use
Sustainable economic systems
Geographic requisites for sustainability
Grass-roots effort

3.10 QUESTIONS FOR REVIEW

1. List a number of the environmental and land-use changes that have accompanied the growth and geographic expansion of world population since the beginning of the nineteenth century. Describe some of the resultant trends in the condition of humans and the environment in poor and wealthy countries.
2. It is argued that economic development over much of the world is unsustainable. What does this mean? How would you define sustainable development? In order to be sustainable, land use must be part of nature, that is, part of ecosystems. Name some areas of change that are necessary to improve the performance of land-use systems and make them more sustainable in the long term.
3. Briefly describe some of the changes in resource policy that are needed to achieve sustainability. Discuss several constraints that stand in the way of achieving sustainable land use. Distinguish between *sustained growth* and *sustainable development.* Would you argue that sustainable development leads to sustainable land use, in your opinion?
4. Discuss several geographic requirements for sustainable land use. Argue pro or con that if development is to lead to sustainable land-use systems in poor countries, it must be built on local society and local rather than external resources.
5. Briefly describe the Rishi Valley problem and the program advanced there to build a sustainable system of land uses. What was the general role of external technologies and resources in the Rishi Valley?

4

THE SPREAD AND DEVELOPMENT OF HUMAN POPULATION

4.1 INTRODUCTION

Based simply on the size and growth rate of our population, humans are an enormously successful organism. In the past 10,000 years the number of humans on Earth has grown from perhaps 10 or 20 million to 6,100 million. Humans have spread from one continent, Africa, to all the continents and most islands of the world. What accounts for the astounding expansion of the human species?

Most of the success of the human organism in its settlement of the earth can be attributed to culture. **Culture** is represented by all those things that have been invented by humans and subsequently passed on from one generation to the next: tools, weapons, clothing, houses, language, and agriculture, for example. Culture is all that we possess that is not part of our basic biological makeup. The technology that is so much a part of our lives today, such as automobiles, computers, and factory systems, is the most blatant manifestation of modern Western culture.

Culture has enabled humans to spread well beyond their natural biogeographic limits and to inhabit environments to which they are not purely suited as biological organisms. Accordingly, if humans were stripped of culture, we would have to retreat from much of the area we now inhabit. Take a moment to consider how long we would survive without the physical rudiments of culture such as shelter, clothes, fire, and agriculture. In the middle latitudes, which include the United States, Canada, Japan, China, Russia, and Europe, most of us would perish in the first winter.

The goal of this chapter is to outline the spread of humans over the Earth and the development of areas of concentrated population. Special emphasis is placed on the enviroments where humans have been least successful. These are the frontiers of human settlement. Five frontier environments will be discussed: the wet tropics, the dry lands, the cold lands, the high mountains, and the continental shelves. Currently, these environments support relatively small numbers of people, yet they are increasingly the object of settlement efforts and modern economic development. Many of these efforts have led to debacles that have damaged both the environments and the traditional societies that inhabit them. The resultant degradation not only threatens the very character and even the existence of these environments, but in some instances shapes their role in the larger balance of the global environment as well.

4.2 HUMAN DISPERSAL AND ADAPTATION

Biological *Homo sapiens* is ecologically suited only to the warm climates, the tropical wet and tropical wet/dry climates where temperatures rarely fall below 20°C (70°F) or so. These bioclimatic zones are limited to about 30 percent of Earth's land area. This is the environment where the first populations of our biological ancestors began several million years ago. From Africa early humans spread beyond the tropics into other climatic zones (Figure 4.1).

The process of early **human diffusion** into new lands was, by today's standards, a very slow process involving interplay and experimentation with each new environment. This resulted in the development of new ideas about survival as well as new social practices and new technology for resource utilization. In other words, each new environment required **adaptation**, but the vehicle was mainly culture rather than biological evolution. It was characterized by the invention of new knowledge about environment and resources; new weapons; new forms of social organization for hunting, gathering, and protection; new modes of shelter; and new types of clothing, among other things.

Curiously, this adaptation did not necessarily favor the easiest environments. In the midlatitude lands much of the diffusion took place during the last stage of glaciation, 10,000 to 50,000 years ago, when the climate there was harshest. With the aid of fire, hardy clothing, and new hunting techniques, humans learned to inhabit the cold environments near the edges of the great ice sheets. In North America the date of migration from Asia across the Bering land bridge into North America is uncertain, but there is little question that by 10,000 years ago, when glacial ice still occupied the northern Great Lakes and most of central and southern Canada, both North and South America were inhabited by the ancestors of the American Indians (Figure 4.1).

With the exception of Antarctica, humans had learned to inhabit every large land mass of Earth. Remarkably, in a period of only 50,000 years or so—a mere wink in the scale of Earth time—people moved from Africa to the geographical extremes of the planet. Their populations, however, were extremely light even in the most favorable environments. Most people lived in small, mobile bands supported only by simple hunting and gathering practices. As far as we know, their impact on the environment was quite limited, but not absent. It is argued, for example, that they were responsible for the eradication of some animal species; but as a whole, the total environmental impact was small by today's standards.

4.3 THE RISE OF HUMAN POPULATIONS

About 12,000 years ago this picture began to change with the advent of agriculture. In parts of Asia, and later in Central America and South America, human populations suddenly increased and areas of relatively concentrated population emerged. The main reason for

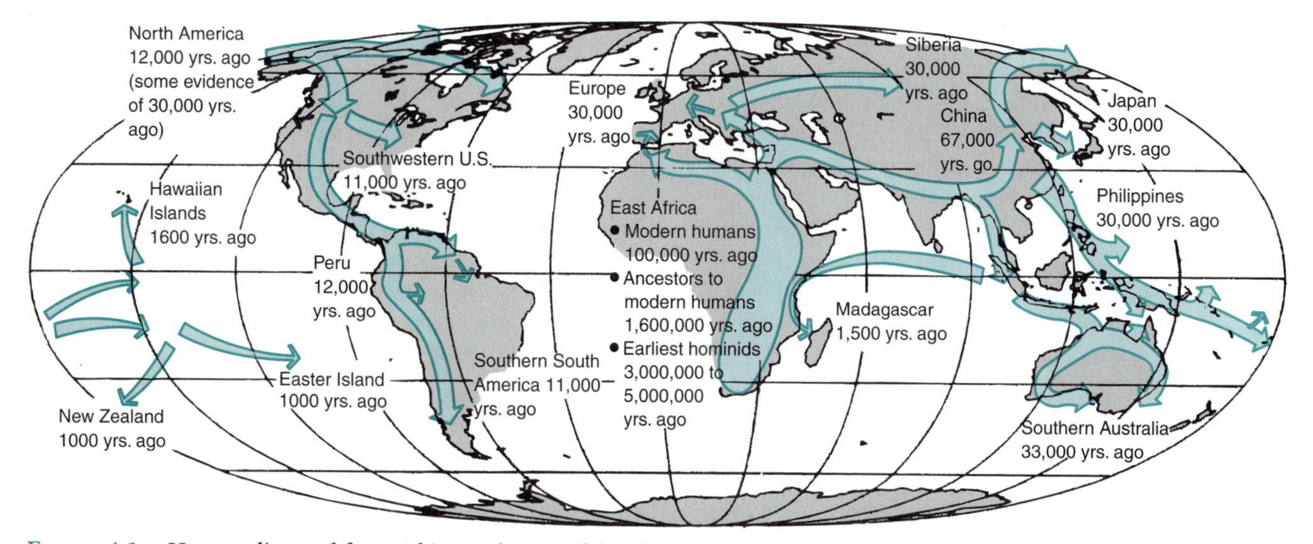

FIGURE 4.1 *Human dispersal from Africa to the rest of the planet with approximate dates.*

the population change was that agriculture enabled a much higher rate of food production than was possible with hunting and gathering. Food could be produced in excess of individual family or band needs, thereby supporting more people per area of land. Agriculture also enabled humans to become sedentary (staying in one place) and establish settlements. In short, farming of such staple crops as wheat, corn, and rice opened the way for a whole new relationship between humans and the environment, but it also started the long and complex drama leading to our current environmental dilemma on the planet.

Agriculture, even in its most elementary form, exacted a toll on the environment. Land had to be cleared of natural vegetation, and with it large areas of plant and animal habitat were altered or destroyed. Water features such as streams and wetlands were also altered, first in minor ways but later with major stream diversions and the development of canals for irrigation. Soil erosion rates increased dramatically because the ground was laid bare to runoff and wind. From the new landscape of farm fields and settlements, streams received their first significant discharges of pollutants: sediments, animal wastes, and human wastes.

The Spread of Agriculture

By 6,000 or 7,000 years ago, several large agricultural centers or regions had emerged (Figure 4.2a). Most were located in tropical and subtropical Asia in what today is China, Iraq, Iran, India, and Southeast Asia. These became the great hearths of ancient civilization in Asia. In North Africa, the Egyptian civilization was well established by this time, and agricultural centers were emerging in Central America. Although political and military power have waxed and waned among these centers for thousands of years, each has remained a major center of agriculture and population.

Commensurate with the establishment and growth of the earlier agricultural centers, farming spread to the midlatitudes. More than 2,000 years ago, most of Europe was occupied by farmers (Figure 4.2b). Major population centers comparable to those in India or Egypt were somewhat slow to emerge in the midlatitudes. By Christ's time, however, large agricultural populations were well established in Europe and midlatitude Asia, and by 1500, Europeans had begun to spread into the Americas, where the native population had reached an estimated 54 million. World population had now grown to about 350 million, of which hunter–gatherers were less than 5 percent. Only 12,000 years earlier hunter–gatherers had represented 100 percent of the world population. Figure 4.2c shows that agriculture had spread to portions of all inhabited continents except Australia by 1500.

By the time of the Industrial Revolution (1750–1850), Australia had been colonized and agriculture was firmly entrenched there. In North America, Euro-American settlement had displaced most Amerindians east of the Mississippi, and more than half the Amerindian population in the Caribbean and Central and South America had been lost to diseases and war. Forests were cleared and plow agriculture was established over much of the vast region from the Great Lakes to the Gulf of Mexico. The same trend had also begun on the West Coast in the valleys of California, Oregon, and Washington. At the same time, tropical forests became reestablished over large areas of Central and South America where Amerindian agriculture had been abandoned as their populations declined sharply after European colonization.

Emergence of Global Power Centers

Between 1500 and 1800 the power centers of the world shifted from the ancient hearths of Asia, Africa, and the Mediterranean region to the midlatitudes. Several European nations, led by England, Spain, and France, emerged as global power centers. These centers, or empires, were built on a global system of resource extraction that utilized a network of colonies and trade relations in Asia, Africa, and the Americas. This system gave rise to the modern industrial economies.

At the heart of this system, which by 1900 included both Europe and North America, was the development of widespread consumerism. In North America the system relied first on exploitation of internal (domestic) resources and later on both internal and external (foreign) resources. One measure of the system's appetite for resources is the consumption of forests. In the United States more than 90 percent of the original forest in the vast area east of the Mississippi River had been cut by 1950 (Figure 4.3). Nearly 1 million square miles of forest was removed over a period of less than 300 years. Never before in history had humans carried out such extensive landscape change in such a short period of time. By mid-twentieth century the United States, with less than 5 percent of the world population, had become the world's leading economic power. It was also the leading consumer of fossil fuels, lumber, minerals, and other resources as well—and, it follows, the leader in waste production, air pollution, and many other forms of environmental degradation.

Global Zones of Human Occupation

Today the world is arranged into roughly three types of geographic areas of human occupation: developed areas, developing areas, and frontier areas or environments. **Developed areas** include all or parts of the economically developed countries: United States, Japan, Canada, Australia, New Zealand, and most European countries. These areas contain about 30 percent of the world's people, consume the majority of the world's resources, and are responsible for the largest share of global environmental degradation. It is noteworthy that several of the developed countries, in

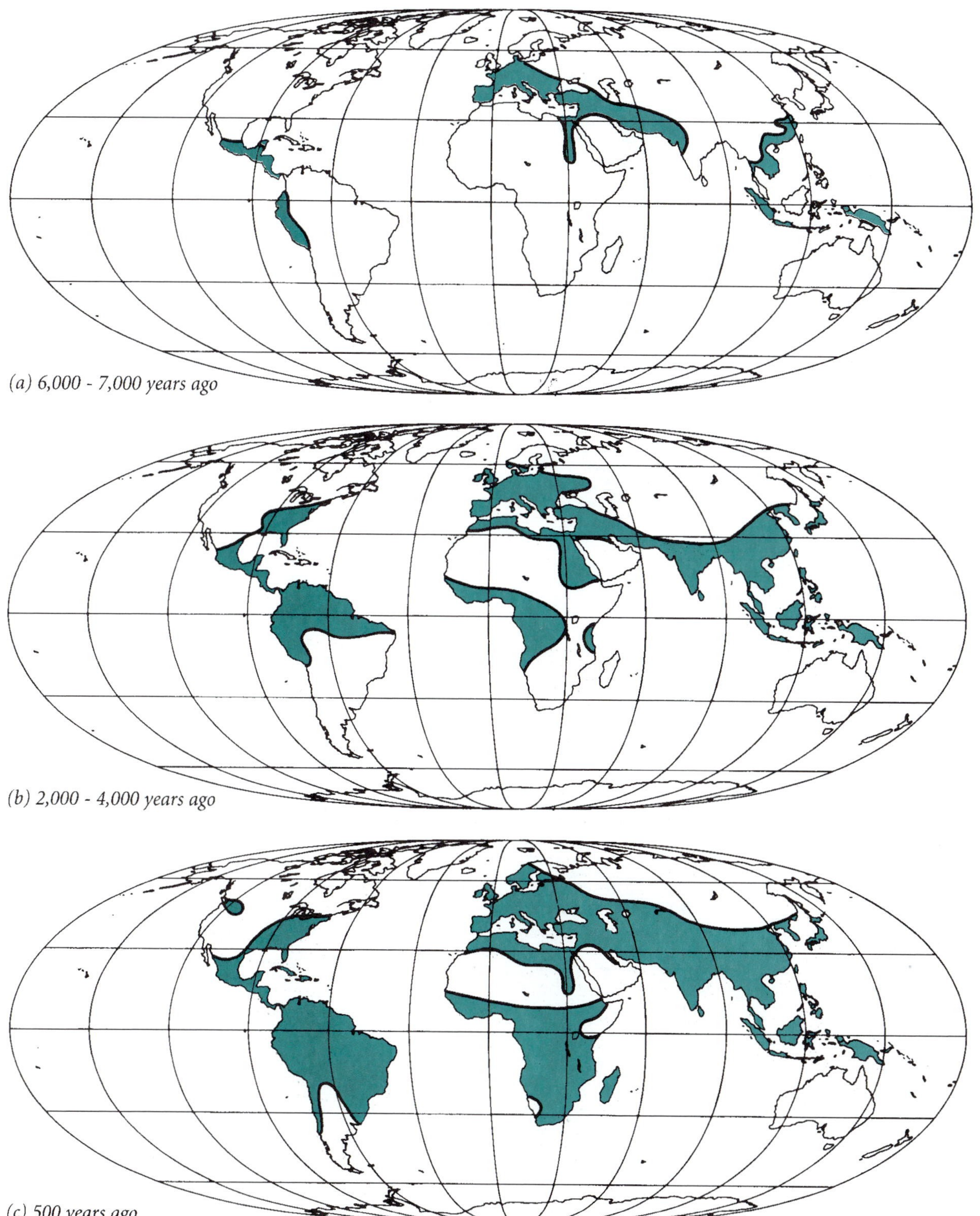

FIGURE 4.2 *The spread of agriculture (mainly crop farming) over the world by 1500 A.D. Also includes areas where both farming and hunting–gathering were widely practiced.*

particular Canada, Australia, New Zealand, and Russia, contain vast, lightly populated areas without well-developed consumer economies. These areas belong to the third type of area, the frontier environments, described later.

The second type of area of human occupation is represented by heavily populated zones with **developing economies.** These represent most of the old agricultural hearths and include all or parts of many developing and

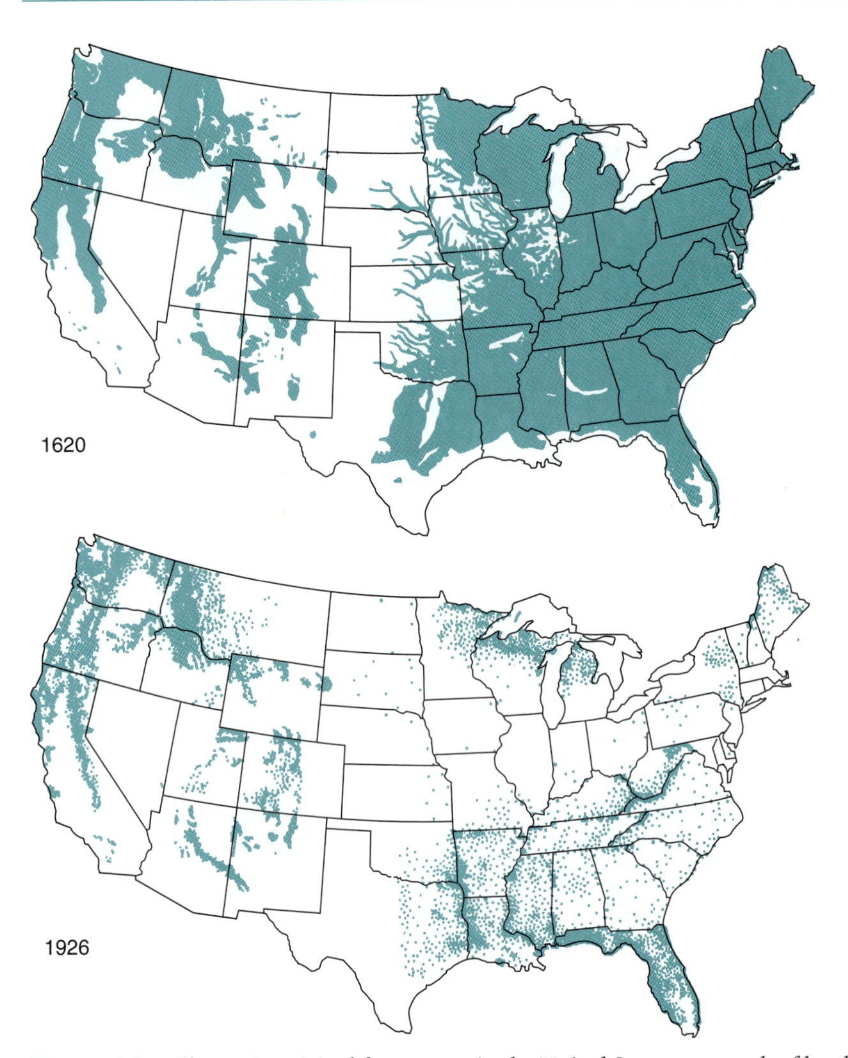

FIGURE 4.3 *Change in original forest cover in the United States as a result of lumbering and land clearing between 1820 and 1926.*

deprived countries including India, China, Nigeria, Bangladesh, Iraq, Iran, Mexico, and the countries of Southeast Asia. These areas contain two-thirds of the world's people and are overwhelmingly dependent on subsistence agriculture. Whereas the developed areas are faced with colossal resource consumption and environmental pollution problems, the developing areas are faced with colossal population, food supply, and disease problems. Increasingly, they are also faced with rapid urban growth and the accompanying problems of pollution, sanitation, and social unrest.

The third type of area in the world today we will call the **frontier environments**. These are vast regions where humanity, especially Western culture, has had only limited success in establishing more than very light populations. The collective frontier environments account for less than 10 percent of the world's population. They are inhabited by mainly indigenous people who practice hunting, gathering, fishing, and some form of subsistence agriculture. Only Antarctica and the interior of Greenland remain truly uninhabited.

4.4 FRONTIER ENVIRONMENTS AND THE MODERN THREAT

Frontier environments are bioclimatic zones where human societies have faced seemingly insurmountable obstacles to occupation and long-term dominion. They are zones where some societies have lived in small numbers for thousands of years practicing traditional land uses, such as herding and hunting–gathering, with little impact on the environment. Our current concept of a frontier environment has its roots in Western culture and is based on technologically advanced, midlatitude societies with limited experience in polar, tropical, and arid lands. This point of view comes to us from the European tradition rooted in the period of world exploration and colonization in the sixteenth, seventeenth, and eighteenth centuries.

To the British, for example, the polar lands were a forbidding frontier. To the Eskimo (Inuit), who had explored and settled this environment several thousands of years ago, polar lands were homelands. To modern societies based on agricultural and industrial economies, the polar lands remain a distinct frontier or wilderness.

Experiments in modern settlement of polar lands and the other frontier environments have sometimes met with limited success but often with abject failure. Herein lies one of the principal sources of many of today's environmental problems: Technologically advanced societies attempt to impose inappropriate methods of resource extraction, agriculture, manufacturing, settlement, and institutional organization on wilderness environments and their indigenous peoples.

The outcome is often a serious mismatch between land-use practices and environment that results in severe environmental disturbance, pollution, failed land uses, and degradation of indigenous societies. In Brazil, where a vast interior region of tropical rainforest is the frontier, it is estimated that the native Indian population has been reduced by more than 90 percent since outsiders entered the region several centuries ago. In the West Indies on the large islands such as Hispanola, Jamaica, and Cuba the original Indian populations, which numbered in the hundreds of thousands at the time of Christopher Columbus, have been totally decimated.

For 500 years Western culture and economic practices have probed and tested the world's frontier environments. Until this century, the rate of economic development and environmental degradation have, for the most part, been relatively modest. In the last several decades, however, both have increased sharply, and it is now clear that unless population growth and economic expansion are curbed and managed, some of these environments will disappear or decline dramatically in the next century. How that process unfolds will be an important barometer of humankind's ability to manage itself and the global environment.

The remainder of this chapter is devoted to an overview of Earth's frontier environments. These environments are surprisingly large, covering about two-thirds of Earth's land area. The continental shelves, the edges of the continents covered by shallow ocean water, cover about 5 percent of Earth's surface and can also be considered a frontier environment.

4.5 THE WET TROPICS

Nowhere is the mismatch between land use and environment more apparent than in the wet tropics, the lands of the tropical forests. These lands are concentrated along the equator with smaller areas extending poleward to 15 degrees or so latitude mainly along selected tropical coastlines. They are consistently warm with mean monthly temperatures in the 20 to 30°C (70–90°F) range year around. Near the equator where the forests are densest, they are wet in all seasons with rainfall typically averaging more than 200 centimeters (80 inches) a year. These conditions foster massive, fast-growing vegetative covers broadly referred to as tropical forests.

Tropical forests are actually composed of several different types of forests, but the largest and ecologically richest is the **tropical rainforest**. The tropical rainforests are found only in Asia, Africa, the Americas, and islands in the Pacific. The largest single area of tropical rainforest is the Amazon River Basin of north central South America. As the map in Figure 4.4 shows, the core areas of tropical rainforest on each continent currently have relatively light human populations.

For centuries tropical forests have been attractive, challenging, and often forbidding to most Western land uses. Agriculture is difficult to maintain because of low soil fertility and erosion problems. Human health is a persistent problem because of pervasive tropical diseases such as malaria, yellow fever, and a host of infectious parasitic worms. As a result, population density has traditionally remained very low, especially in tropical rainforest regions, the heart of the tropical forests. To prospective settlers, however, these drawbacks were often masked by the warm climate, ample water, and lush vegetation, leading to the brand "counterfeit paradise" for these lands.

Tropical Forest Loss

In the past several decades the tropical rainforests have been open to an influx of people and new economic activity. This had led to wholesale destruction of the great forests. The combined rate of tropical forest (rainforests and other tropical forest types) destruction is estimated at 170,000 square kilometers (65,000 square miles) a year, which is equivalent to 460 square kilometers (180 square miles) a day. In the twentieth century alone the tropical forests have been reduced by 2.6 million square kilometers (1 million square miles) or more. Today, about 17.6 million square kilometers (6.8 million square miles) remain. The rate of destruction varies from country to country, but some experts forecast that if current trends continue, humans will have destroyed nearly all the world's tropical forests by the year 2050.

Why is the destruction taking place, and who is responsible? Most of the deforestation is undertaken to provide new farmland and commercial lumber. Forests are also taken for local needs such as firewood, urban sprawl, and reservoir construction, but these are secondary to farming and lumbering (Figure 4.5). The most direct force behind the destruction of tropical forests is population growth and modern economics. Most of the countries with large forest reserves, such as Brazil, are developing countries with rapidly growing populations. Lumbering and the clearing of crop and grazing lands are ostensibly quick ways to increase income and jobs.

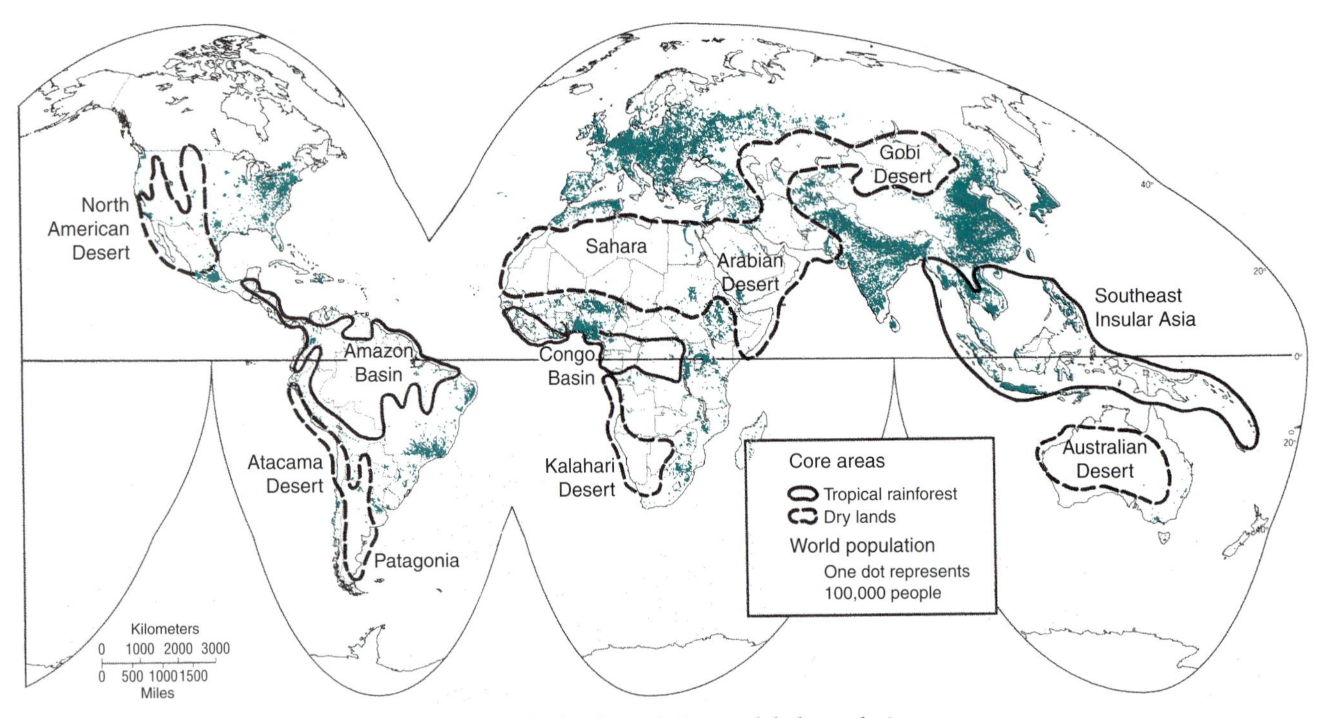

FIGURE 4.4 *Core areas of tropical rainforest and dry land in relation to global population distribution.*

FIGURE 4.5 *Destruction of tropical forest in Papua, New Guinea.*

Economic Pressure

The economic pressure is great to replace traditional land uses such as foraging and shifting cultivation with commercial farms that produce exports and generate revenue (Table 4.1). Most of the new farms, however, are small and of marginal profitability. The wet tropics generally do not carry the agricultural potential that the lush forest covers seem to imply. Soils tend to be heavily leached, and when the forest is removed, the principal source of soil nutrients is destroyed and the soil rapidly deteriorates.

Although much of the responsibility for rainforest destruction rests with peasant farmers and opportunist

TABLE 4.1

SOME CHARACTERISTICS AND CONSEQUENCES OF DIFFERENT LAND USES IN AREAS OF TROPICAL FOREST

	Foraging (Hunting) and Gathering	Shifting Cultivation	Commercial Agriculture
Population density	less than 10 per km^2	25–50 per km^2	up to 500 per km^2
Settlement types	temporary	temporary	sedentary (permanent)
Food production	less than 60*	60–120*	240–480*
Energy requirements	negligible (no imports)	negligible (no imports)	massive (with large imports)
Nutrient cycles	no disturbance	periodic disturbance	massive disturbance
Environmental impact	little to none	localized and temporary	massive and permanent
Species diversity	unaffected	generally maintained	greatly reduced

*thousands of calories per square metre of ground per year
(*Source*: Adapted from Deshmukh, 1986)

settlers (squatters) leaving urban slums, a variety of organizations are also responsible for tropical deforestation. Some are internal to the countries involved, such as government-supported agricultural programs aimed at expanding grazing lands to increase beef exports. These programs, however, are usually motivated by corporate interests from developed countries such as the United States and Japan. Billions of dollars in international debts owed by developing countries such as Brazil and Mexico also promote exploitation of tropical forests. By expanding lumber, mineral, and agricultural production from the forest frontier, these governments can generate reserves for debt repayment.

Environmental Consequences

What are the environmental consequences of tropical forest eradication? One of the most serious is **species extinction**. Tropical forests are the richest reservoir of plant and animal species on Earth, containing, by recent estimates, more than 50 percent of Earth's species. The majority of these species, principally insects and microorganisms, are not yet known to science. The moral dilemma of extinction is incalculable, of course, but there is also a practical side to the problem. Potential sources of food, medicine, and other resources are forever lost with the eradication of species. Today many tropical plants are used for modern medicines, and many more hold promise in treating various forms of cancer. (See Chapter 16, "Biological Diversity and Land Use," for a more complete discussion of these and other issues related to tropical rainforest eradication.)

Tropical forests are also important in maintaining the Earth's climates and the balance of gases. They absorb carbon dioxide and help maintain the heat balance against increasing carbon dioxide production from air pollution worldwide. Clearing and burning these forests *releases* additional carbon dioxide to the atmosphere—as much as 10 percent of the world's production of carbon dioxide from pollution sources—which adds to the global **greenhouse effect**.

Tropical forests are also important in the **hydrologic cycle** in maintaining the supply of moisture to the lower atmosphere. Studies show that widespread destruction of tropical forests can lead to reduced rainfall, increased evaporation of surface moisture, and desiccation of ground level climate. This is often followed by increases in stormwater runoff and soil erosion. Additionally, when the forest cover is eliminated the cycle of nutrients between soil and vegetation is interrupted and soil fertility declines sharply.

In combination, these changes in climate, moisture, and soil might so drastically alter the environment that it might never again be capable of supporting tropical forest. That is to say, even if a large area of land were made available, we could not put a tropical forest back together even if we had the species to stock it with. On the other hand, it is unrealistic to imagine saving the species of the tropical forests if their habitats are destroyed. The evolutionary processes that create new species are irreversible; therefore, once species and their habitats are destroyed we can never recreate the rich mix of organisms that these forests historically supported.

Destruction of tropical forests is widely recognized as a critical global problem. The responsibility for the problem extends well beyond the countries directly involved and is shared by the world economic and political community as a whole. Efforts to curb the loss of tropical forest, limited though they may be, are increasingly being shared by both directly affected nations and world organizations such as the United Nations and the affiliated World Bank, which provides loans and technical assistance for economic development projects in developing countries. After many years of granting loans that resulted in tropical forest destruction, the World Bank is now supporting projects that create national parks and forest preserves to encourage forest preservation.

In another effort, world environmental organizations such as the World Wildlife Fund and The Nature Conservancy are buying out a small fraction of some countries' international debts in exchange for preservation of tracts of rainforest. The "debt for nature" program has recently worked on a limited basis in Bolivia, Ecuador, Costa Rica, and the Philippines. Such efforts provide some hope that at least a fraction of the world's tropical forest may survive the twenty-first century.

4.6 The Dry Lands

In sharp contrast to the wet tropics are the dry lands. Though often as warm as the wet tropics, dry environments receive much less rainfall, typically less than 50 centimeters (20 inches) a year. Virtually all this water is, in turn, lost to the atmosphere through vaporization, with little left in the ground or in stream channels. Dry environments cover a large share of the Earth's surface—as much as 40 percent of the total land area by some estimates. Although all continents, even Antarctica, contain dry environments, the bulk of the Earth's dry lands are found in Asia (including the Middle East), Africa, and Australia (Figure 4.4).

Types of Dry Environments.

About half of the dry land is **true desert**, such as the Sahara of North Africa and the Sonora of North America, where drought conditions persist throughout the year. Precipitation is very light, generally less than 30 centimeters (12 inches) a year, and when it does occur, most rapidly evaporates. Understandably, sustained use of deserts for agriculture and settlements is very limited. Where it is possible, water supplies such as groundwater or exotic streams (those such as the Nile and Colorado Rivers, which flow into the desert from wetter climatic zones) must be utilized.

Bordering the deserts are the **semiarid** lands, or **steppe**. Here annual precipitation is greater (30 to 60 centimeters) and drought is usually relatively mild or absent for part of the year. The semiarid zones are usually grasslands, such as the American Great Plains and the steppes of Russia and Ukraine. Today sustained agriculture and settlement are possible there, but often only with the assistance of irrigation (Figure 4.6). Prior to the twentieth century semiarid lands were the world's great grazing lands; in modern times they are also used for cropland, particularly wheat farming.

FIGURE 4.6 *Aerial view of a modern irrigation system feeding water to cropland (in circles measuring half mile in diameter) in an otherwise brown, semiarid landscape in the southern reaches of the American Great Plains. The wavy pattern is inactive sand dunes.*

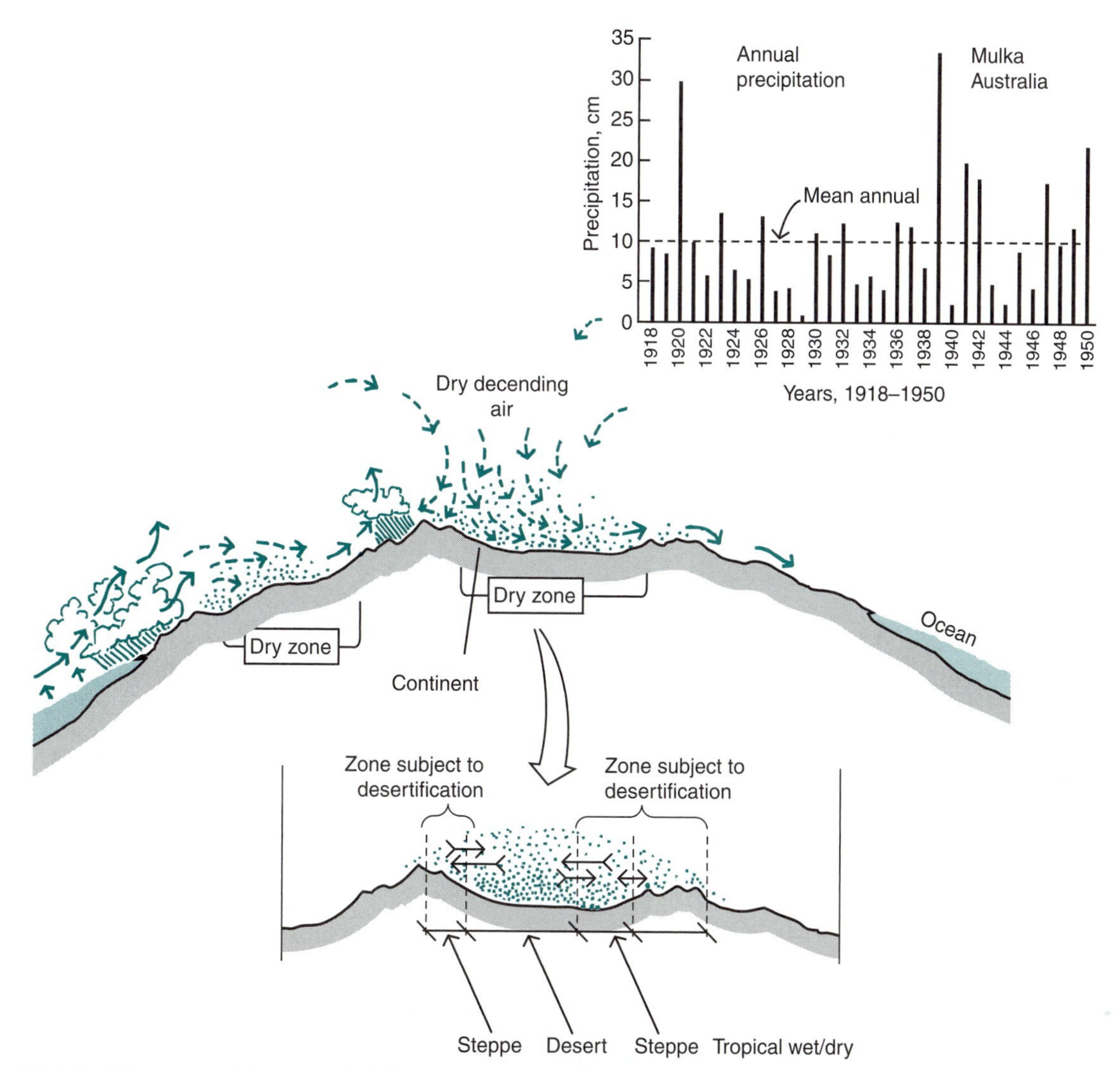

FIGURE 4.7 *A graph (upper right) illustrating the variable nature of precipitation in deserts; below, a schematic diagram showing the flow of air associated with the major deserts and the zone on the desert fringe that is most subject to desertification.*

In the tropics the **tropical wet/dry** lands are characterized by a short wet season in summer, with dry conditions the rest of the year. For our purposes the dry lands can also include much of the tropical savanna landscapes of Africa and South America and the monsoon lands of South Asia, especially India.

Moisture Variability

An important factor in understanding the dry lands as frontier environments is **moisture variability** over time. Variability is a measure of precipitation dependability—that is, how steady the annual moisture supply is from year to year. Climatologists have found that precipitation variability follows a simple rule: The lower the mean annual precipitation value, the higher the year to year variability, and vice versa. Not only do the dry lands receive small moisture supplies on the average, but the rate of delivery is highly irregular from one year to the next. For example, notice the difference between 1939 and 1940 in the graph in Figure 4.7. In short, the mean annual precipitation value is not very meaningful in deserts, and in order for life to survive there, it must be adapted to radical swings in the moisture supply.

Viewed in a larger time frame, we find the moisture supply also varies greatly with climatic fluctuations over periods ranging from several decades to thousands of years. These climatic fluctuations produce expansions and contractions of the dry zones (Figure 4.7). During a shift toward drought, the desert actually expands into the bordering semiarid and tropical wet/dry zones. If the drought is not prolonged, say, more than several years in duration, the semiarid landscape (that is, vegetation and soil) usually survives. In subsequent years, as moister climatic conditions return, the landscape recovers.

Desertification

If, however, people occupy a semiarid zone that is undergoing drought, there is a strong tendency to overwork grazing lands and croplands to offset declining yields. This increases the stress on the landscape, leading to severe damage to soils and plants. After the drought ends, the weakened landscape is unable to recover and the affected area becomes a desert. This illustrates a process that is known as **desertification**—landscape degradation from the interaction of land use and climatic processes. Desertification is marked by three main characteristics:

1. Reduced plant cover with lower production (growth) rates and the loss of certain, often valued, species
2. Reduced soil cover, especially topsoil
3. Reduced soil moisture reserves.

Under desertification agricultural capacity declines sharply, often leaving people stranded on damaged land.

With rising world population and increasing pressure on grasslands for agriculture, it is not surprising that desertification is widespread in the world today. Although reliable data are lacking, the current global rate of desertification may be as great as 200,000 square kilometers (75,000 square miles) annually. The largest and most devastating recent example is found in the Sahel, a region on the southern fringe of the Sahara of Africa, where drought began in 1968 and extended through most of the 1970s. Grazing lands declined dramatically, millions of herding animals died, and more than 100,000 people died of starvation as the landscape deteriorated. Although the Sahel had suffered severe droughts in the past, notably in the late 1600s and in the mid-1700s, the effects of modern drought are especially severe because of the relatively large populations involved.

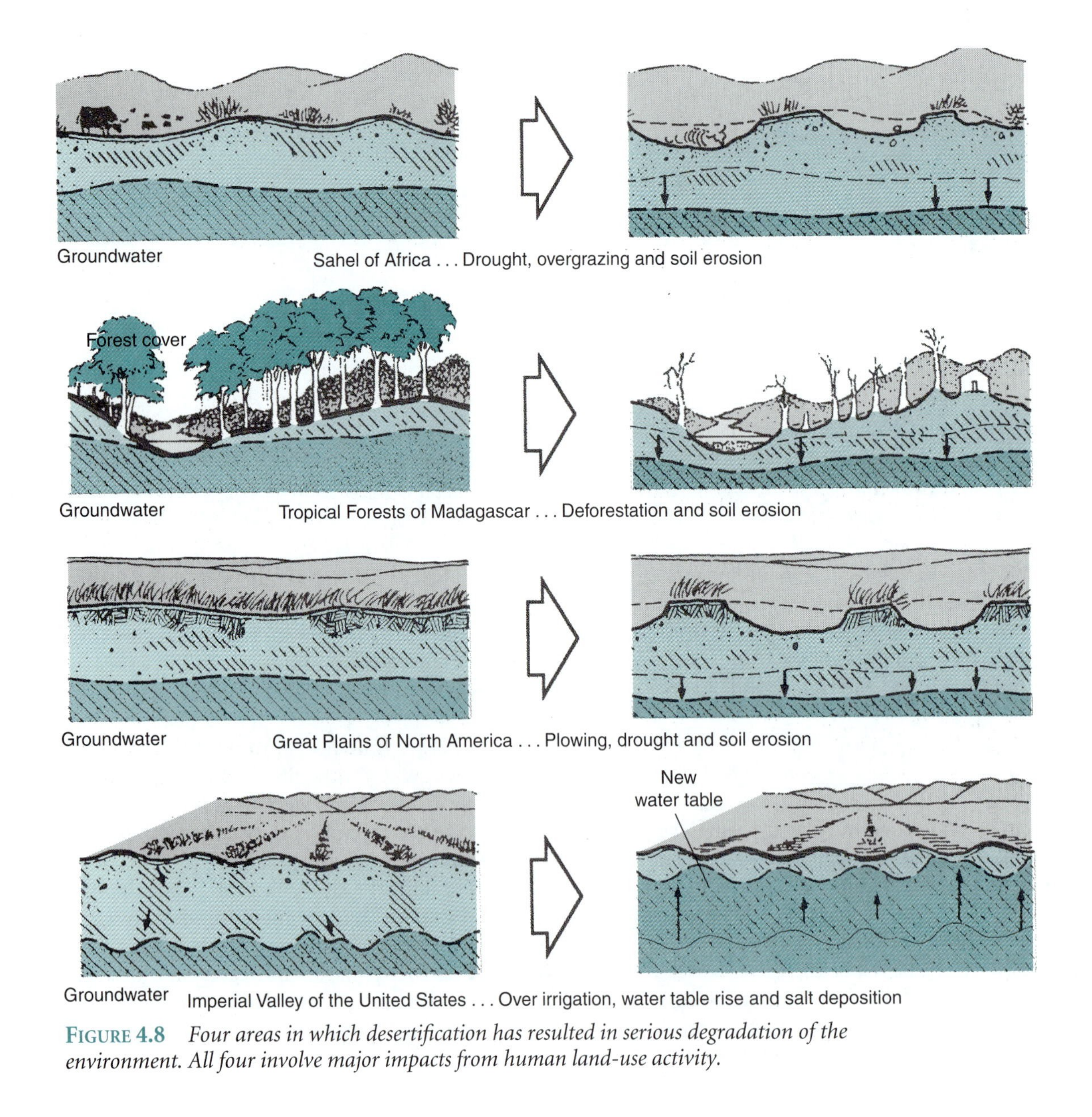

FIGURE 4.8 *Four areas in which desertification has resulted in serious degradation of the environment. All four involve major impacts from human land-use activity.*

Other examples of desertification are found on Madagascar, where 80 percent of the forests have been removed, leading to severe desiccation and erosion of soils; in the North American Great Plains, where in the 1930s large areas of drought-stricken agricultural lands without the protective cover of natural grasses lost their topsoil to wind erosion; and in selected areas of irrigated cropland, such as the Imperial Valley of Southern California, where excessive irrigation has so saturated soils with salt that they can no longer support plants (Figure 4.8).

4.7 The Cold Lands

Cold lands are characterized by long, harsh winters and short, cool summers (less than three months). The climate over most of the earth's cold regions is incapable of sustaining crop agriculture; vast areas of land are underlain by permafrost. **Permafrost** is ground in a permanently frozen state. It is the product of a cold climate in which summer heat is insufficient to melt the deep ground frost of previous winters. Approximately 26 percent of Earth's land area is occupied by permafrost, and these lands represent one of the most difficult environments for modern settlement and land use. There are three principal regions of permafrost in the world: Asia (mainly Siberia), North America (mainly Alaska and part of Canada), and Antarctica.

Types of Cold Lands

Broadly speaking, cold lands fall into three classes: polar, tundra, and boreal forest (Figure 4.9). **Polar lands** are the harshest of the cold environments. Snow and ice are present year around, plants and animals are very scarce, and permafrost extends from the ground surface to depths of hundreds or thousands of feet. Antarctica is the largest single area of polar land on Earth, and until the twentieth century, it was unexplored and unsettled. Even today Antarctica has only a few outposts inhabited mainly by scientific personnel. Only the polar lands of the Northern Hemisphere have traditionally supported human populations, but these are small, widely scattered groups of nomads and hunters.

Tundra lands are a less harsh environment than the polar lands. Winters in the tundra are long and fiercely cold, but there is a short, cool summer that allows the establishment of extensive covers of herbaceous plants, scattered shrubs, and small trees (see the temperature graph in Figure 4.9). The snow cover melts in summer and the ground thaws to a depth of 10 feet or so. Permafrost begins below the thawed layer and extends to depths of 10 to 100 meters (33–333 feet) in most areas. The thawed layer, which is called the **active layer,** freezes in winter and melts in summer, when it often becomes saturated or covered with water because the underlying permafrost retards drainage through the soil. The vast majority of the world's tundra is found in North America and Asia.

The **boreal forests** represent the southern margin of the cold lands. Occupying a subarctic climate and only partially underlain by permafrost, the boreal forests are the largest belt of continuous forest in the world. They stretch across 11,000 kilometers (7,000 miles) of Eurasia and nearly 8,000 kilometers (5,000 miles) of North America from Newfoundland to Alaska. Winters in the subarctic are nearly as fierce as those on the tundra, but summers are a little longer and warmer. This difference is significant, for it permits the establishment of the most northerly forests on Earth. Understandably, the boreal forests are limited to relatively few hardy tree species, principally conifers, and they are slow-growing and often small in stature. Permafrost in the boreal forest zone is discontinuous or patchy and is much thinner than farther north.

Land Use and Environment

When water freezes, it expands by nearly 10 percent of its liquid volume. Therefore, as water in the soil freezes with the formation of permafrost, it causes upward expansion of the ground, called *frost heaving.* When land-use facilities (buildings or roads, for example) are placed over the permafrost, they typically change the annual rhythm of heat flow in and out of the soil. This, in turn, changes the volume of ice in the permafrost, causing the soil to expand or contract. The most common change is partial melting of the permafrost, leading to uneven sagging of the ground. This can result in failure of building foundations, fracturing of highways and bridges, and rupturing of utility lines. These constraints, along with the heavy energy requirements for heating, severely limit settlement and modern development over much of the cold lands, to say nothing of the severely limited potential for agriculture. The coincidence of the southern margin of the cold lands and the northern limits of development in Canada illustrate this (Figure 4.10).

For countries with large expanses of permafrost—namely, Russia, Canada, the United States, and China—cold lands have traditionally been viewed as hostile environments to either be avoided or probed and used where possible for economic gain. Although the attitudes of these countries toward permafrost lands are changing somewhat, the prevailing attitude today is mainly one of exploitation for economic and military purposes. Oil, natural gas, uranium, precious metals, iron ore, and forest products are the principal economic targets. Disruption of the environment, including the indigenous people, has been extreme in many places but nowhere more so than in the vast Siberian region of the former Soviet Union. Under the Communists, the environment and the indigenous peoples in Siberia were often seriously mistreated. In the United States and Canada, **economic development** proposals have led to increasingly hard-fought contests among environmental interests, industrial interests, and governmental operations and decision makers.

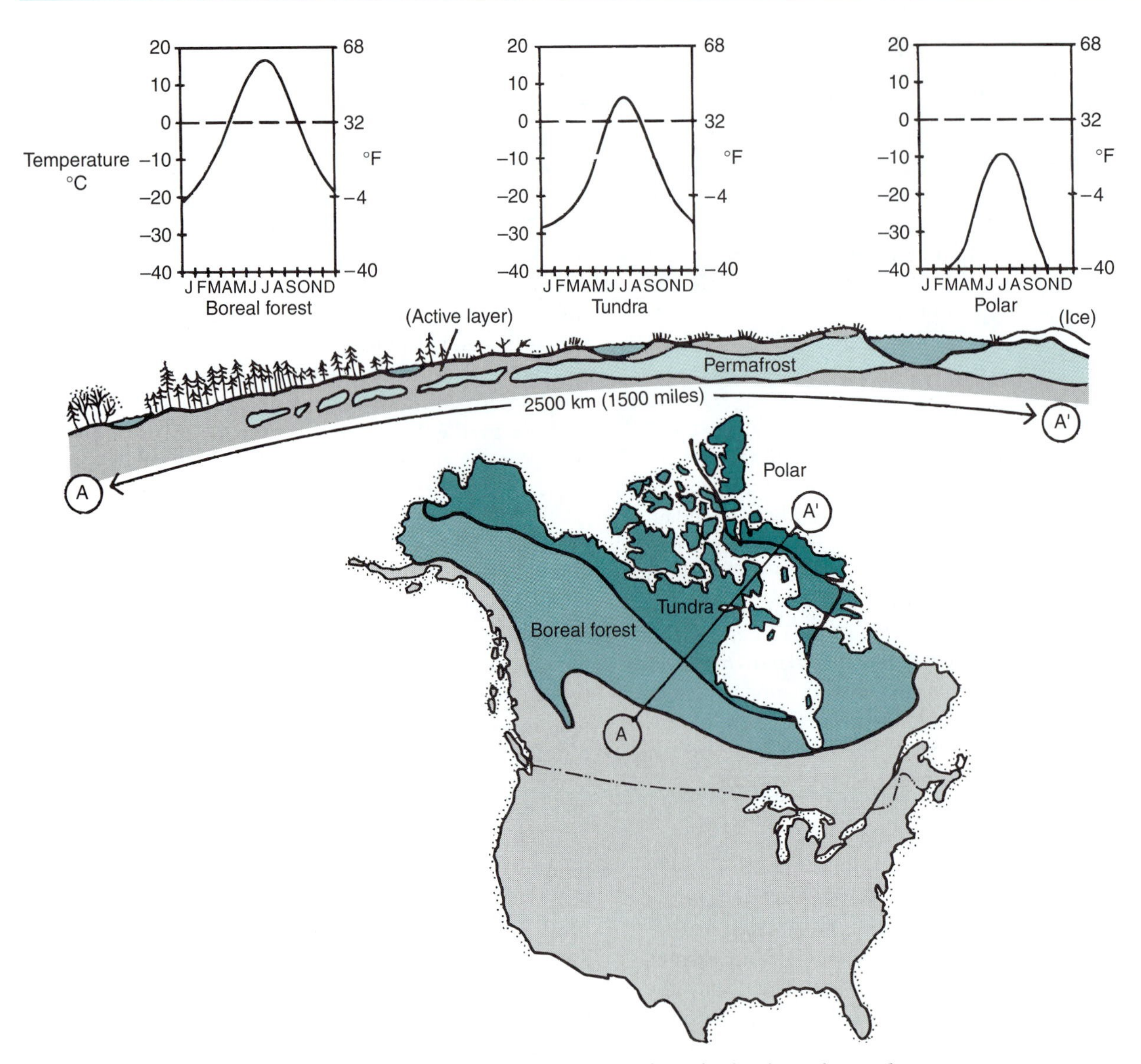

FIGURE 4.9 *A schematic diagram showing the permafrost layer under polar land, tundra, and boreal forest and the atmospheric temperatures associated with each.*

Piping crude oil across the tundra is a great concern because of the potentially dangerous consequences of an oil spill. This concern is all the more serious because the tundra supports a rich complex of relatively undisturbed ecosystems, and as such, represents one of the last great natural reserves remaining on the planet. The development of the 1,100-kilometer- (700-mile) long Trans-Alaska pipeline from the Alaska North Slope to Valdez, Alaska, on the Pacific Ocean, was possible only after extensive environmental planning and special engineering that attempted to reduce the chances of damage to the tundra environment from construction activity and accidental oil spills (Figure 4.11). In Russia, on the other hand, the Communists left a system of poorly planned and maintained pipelines that are leaking millions of barrels of oil a year.

Although tundra and boreal forest together are roughly comparable to the wet tropics in total area, the scope of the environmental problems is different in these two frontier environments. Many of the problems of tropical forest regions are rooted in international economics and politics, whereas those of cold lands, whether polar, tundra, or boreal forest, tend to be more national in scope or shared by two neighboring countries such as Canada and the United States. Only Antarctica, which has not yet been pressed with economic development, is international in scope. Antarctica is the only continent without an indigenous population. It has been an object of scientific inquiry throughout the twentieth century, and this has led to, among other things, the discovery of considerable mineral reserves there, notably petroleum, coal, and metallic minerals. In 1991 an international agreement was struck to restrict economic exploitation of Antarctica. Nonetheless, economic and political pressure to extend mining and drilling operations to this continent still exist in many

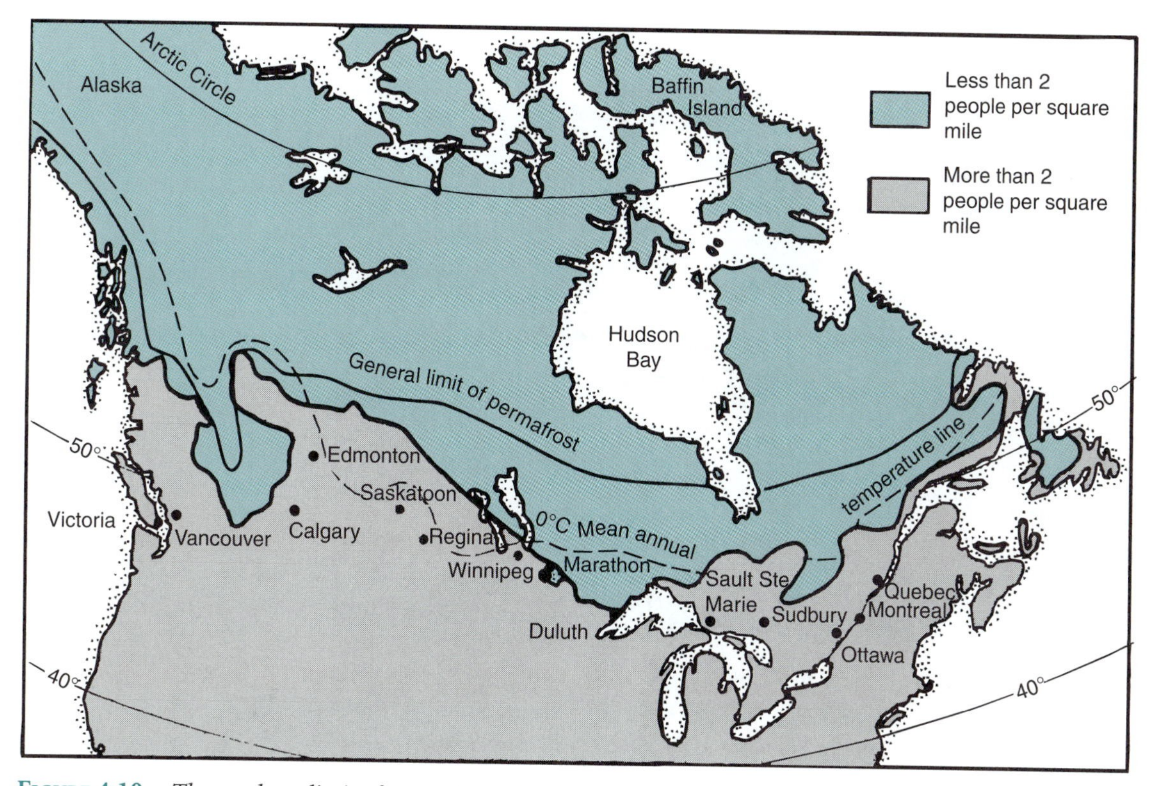

FIGURE 4.10 *The southern limit of permafrost, the 0°C mean annual temperature line, and the northern fringe of development in Canada.*

FIGURE 4.11 *The Trans-Alaska Pipeline crosses the tundra of Alaska connecting the Alaska North Slope with the Pacific port of Valdez.*

developed countries. The opposition holds hope that Antarctica may become a world preserve or park.

4.8 THE MOUNTAIN LANDS

Each of the continents contains at least one major belt of high mountains. The total area of mountainous land lying above 2,000 meters (6,500 feet) elevation is 20 million square kilometers (7.7 million square miles), 13 percent of the world's land area. Although various peoples have shown remarkable success in mountainous lands—for example, the Swiss and Incas—most mountain environments pose serious limitations to survival and land use (Figure 4.12). Above 5,500 meters (18,000 feet), humans cannot sustain life on a permanent basis. Among other things, it is not possible for humans to reproduce at such elevations because of difficulties with full-term pregnancies.

Environmental Barriers

With respect to land use, the first problem with high-mountain environments is **climate**. Under average conditions, air temperature declines significantly with increasing altitude. At an elevation of 3,000 meters (10,000 feet), a modest height for a mountain, air temperature is around 20°C (35°F) lower than at sea level in the same location. Winters are longer, the growing season is much shorter, storms are fiercer, and snowfall is greater at higher elevations. As a result, agriculture is

FIGURE 4.12 *Successful human habitation of the Altiplano of the Andes of Peru in South America, 3,000–4,000 meters (10,000 to 13,000 ft) elevation. Above an elevation of 5,500 meters, the environment is too severe for sustainable land use, even in mountains close to the equator.*

limited mostly to grazing activity and to hardy crops such as potatoes, oats, and barley. Of course, crop productivity rates are low.

Mountainous lands are also limiting to land use because of their rugged terrain and large **slopes**. Above the valley floors, slopes are steep, soils are thin, and only small patches of arable land are normally available. Attempts to push farming onto mountain slopes have in many instances met with disastrous results. In Nepal, which lies in the rugged Himalayas north of India, population pressure in some areas has led to deforestation and cropping on slopes traditionally viewed as too steep for agriculture. Without the stabilizing forest cover, these slopes are prone to failure and erosion that damages not only fields on the slopes but farmland nearer the valley floor as well. It is also argued that flooding has increased in the valleys and further downstream in India because of higher runoff rates brought on by the deforestation. See "Topography and Land Use" in Chapter 15 for more information on mountainous topography and land use.

4.9 THE CONTINENTAL SHELVES

All continents are partially fringed by lowlands called coastal plains. Coastal plains slope gently toward the sea, and although we map the sea coast as their outer border, in reality they continue beneath the sea to a water depth of about 200 meters (650 feet). The underwater portion of the coastal plain is the **continental shelf**. Based on its geology, the continental shelf belongs to the continents rather than the ocean basins. It is made up of a thick mass of sedimentary rock that has built up over millions of years with the accumulation of river and marine sediments. Ecologically, however, the continental shelves belong to the sea by virtue of their abundant and rich marine ecosystem.

Natural Resources

The continental shelves are surprisingly large, covering about 30 million square kilometers (11 million square miles) worldwide. This represents about 5 percent of the Earth's surface, an area somewhat larger than that of Earth's mountainous land. The continental shelves average 75 kilometers (45 miles) in width, but this dimension varies greatly from one coastline to another (Figure 4.13). On the North American east coast, the continental shelf is more than 250 kilometers (150 miles) wide, whereas in some places on the West Coast it is only a few kilometers wide. For the United States the continental shelves are equivalent to nearly one-fifth of the nation's land area.

The continental shelves are the focus of much economic attention and land-use development. About 50 percent of the world's human population lives within 100 kilometers (60 miles) of the sea coast. Nine of the world's ten largest

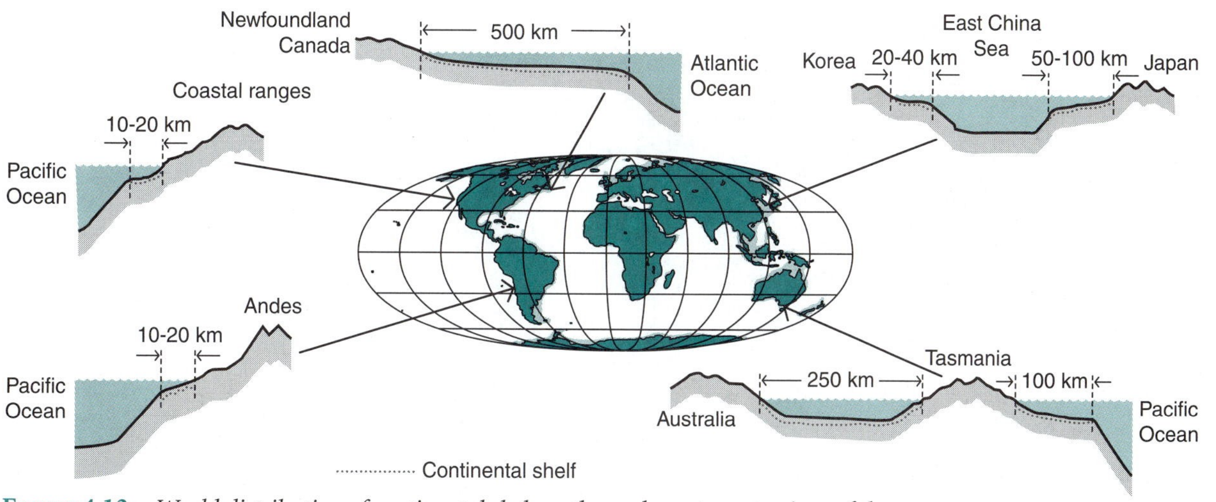

FIGURE 4.13 *World distribution of continental shelves, the underwater extensions of the continents.*

urban centers are found on the sea coast, and each decade more people crowd into the narrow fringe of land along the sea. This brings them into contact with some of the most biologically productive habitats on Earth, such as coastal lagoons, estuaries, and wetlands. Although the total productivity and the species diversity generally declines with distance from shore, the outer parts of the continental shelf, such as the Grand Banks of the North Atlantic, support the ocean's largest fish populations. Nearly all of the world's marine fish catch, which totaled 81 million tons in 1997, was taken from the continental shelves. In recent years, however, except for the recently exploited fishery in the Indian Ocean, all fishing grounds including the Grand Banks have shown serious declines in catches. Since 1991 the world catch has declined by 30 percent, suggesting to some observers that the productive continental shelves are being overfished.

In the past several decades the continental shelves have been tapped for another major resource, fossil fuels. In some areas the sedimentary rocks underlying the shelves contain large reserves of oil and natural gas. These reserves have given rise to offshore drilling operations, which have brought the petroleum industry into conflict with environmental and fishing interests over the issue of accidental oil spills. In the United States, 12 percent of the domestic oil production and 25 percent of the domestic natural gas production come from the continental shelves, mainly along the Gulf of Mexico and the Southern California coast.

Environmental Threats:

Oil and natural gas are extracted with the aid of platforms constructed on stilts over the water. Although platforms are prone to storm damage, explosions (blowouts), and mechanical failures, spills from oil shipping are a greater threat to the continental shelves (Figure 4.14). In 1983 there were 5,490 recorded spills in U.S. ocean waters involving a total of 8.3 million gallons of oil, virtually all from tankers, barges, and related facilities. In 1989 one of the worst oil spills occurred in Prince William Sound at the Pacific terminus of the Alaska Pipeline. A tanker spilled 11 million gallons of crude oil into productive fishing waters. In a matter of only a few weeks, the oil spread along 1,200 miles of shoreline. The environmental damage included the death of an estimated 500,000 birds and serious damage to Alaskan commercial and sport fisheries. In 1994 Alaskan fishermen were awarded $250 million in compensation.

Also serious but less visible are the effects on coastal environments from pollution generated by terrestrial land uses. About 40 percent of the precipitation falling on the land is discharged by streams to the oceans. With it comes pollution from agriculture, cities, and other land uses. More petroleum waste is discharged onto the continental shelves by stormwater runoff from highways and cities than from oil spills. Coupled with other pollutants such as air pollution, sewage, garbage, and landfill materials, it appears that the ecosystems of large parts of the world's continental shelves are being seriously threatened.

Change in the balance of freshwater and saltwater is another concern in coastal areas. Near some urbanized areas, coastal ecosystems are being deprived of freshwater inflows by reservoirs built on rivers for city and agricultural water supply. Beyond urban areas, the destruction of rainforest near the coasts in some tropical areas is leading to increased soil erosion and sediment loading of the continental shelf. Where the shelf supports coral reefs, many reef organisms, including the corals themselves, are being destroyed by sedimentation and the reef ecosystems are declining.

Pressure to use the resources of the sea floor is increasing rapidly in the world today. In the United States it is widely believed by government and petroleum industry specialists that the largest remaining domestic reserves of

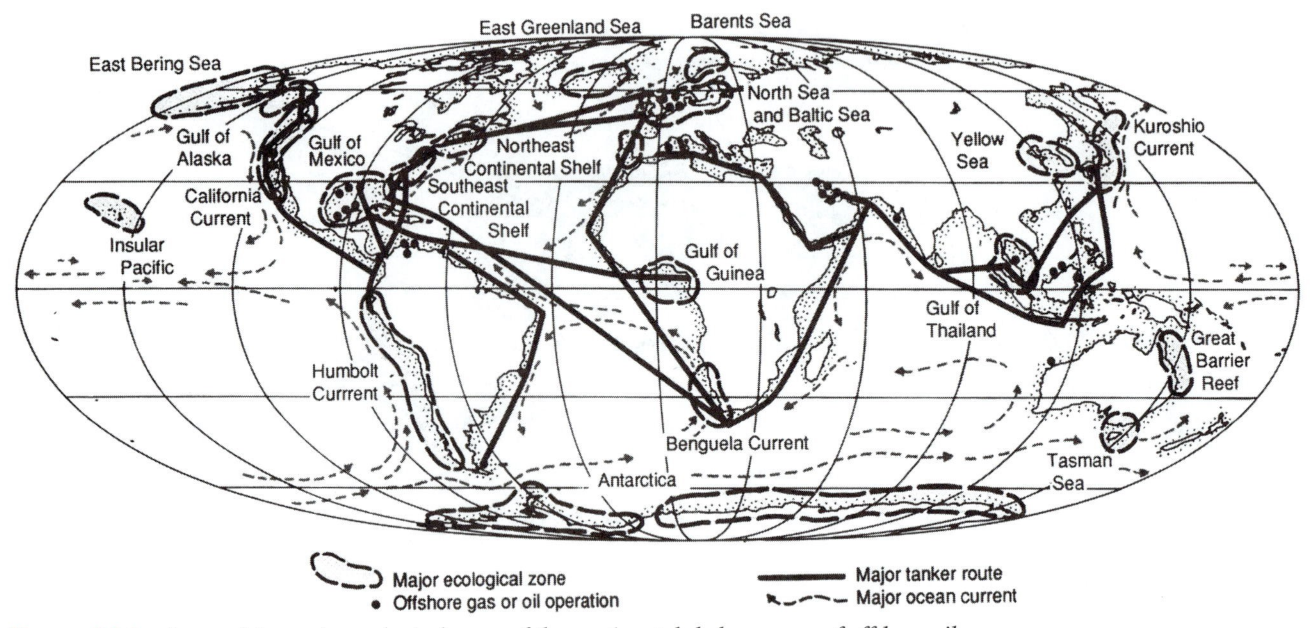

FIGURE 4.14 *Some of the major ecological zones of the continental shelves, areas of offshore oil and natural gas extraction, and major tanker routes.*

fossil fuels lie under the continental shelves. In 1983 the federal government extended United States national jurisdiction beyond the traditional territorial limits (about 5.5 kilometers) to a distance of 370 kilometers (230 miles) offshore. This expanded zone, called the *Exclusive Economic Zone*, gives the United States sovereign rights and control over nearly 8.1 million square kilometers (3.1 million square miles) of ocean floor bordering the United States, its territories, and its possessions.

4.10 SUMMARY

Over the past 100,000 years humans have settled over most of the world. Their success has been due principally to the evolution of culture, which has enabled human populations to adapt and invent means to survive and flourish in environments much different than their biological hearth in the tropics. Agriculture revolutionized the relationship with environment, and human population soared. By the twentieth century, humans dominated much of the terrestrial world outside the cold regions, harsh deserts, rugged mountains, and tropical forests. These open lands are the frontier environments, the earth's remaining wilderness and its margins, and today they represent a large but rapidly shrinking proportion of the Earth's land area.

Although success at establishing and sustaining large populations in the frontier environments has eluded us for most of human history, that picture is beginning to change. Modern development is pushing into these regions and is producing some serious mismatches between land use and environment, resulting in severe environmental disturbance, pollution, failed land uses, and the demise of indigenous peoples. The problem was at first somewhat localized but now has become broadly regional with serious implications for the global environment and the long-term capacity of the planet to sustain life. The environmental concerns include alteration of the global climate, pollution of the oceans, depletion of soils and grasslands, extensive species eradication, and loss of indigenous cultures. The long-term implications are reduced carrying capacity of Earth and lowered quality of life expectations because of, among other things, losses in food production capabilities, losses of medicine sources, and higher costs in managing the environment as a whole.

The world's tropical forests are being eradicated at alarming rates for pasture land and farmland, commercial lumber, and various other land uses. The environmental consequences are not only regionally significant, but of global significance because these forests contain Earth's richest reservoir of species and appear to be important in maintaining the balance of atmospheric gases. The issue is deeply rooted in both the problems of individual tropical countries and in international economics and politics, especially because of the massive debts owed by many of these poor countries to developed nations.

Dry lands are extensive and include the true deserts, the semiarid lands, and the tropical wet/dry lands. Because of the extreme variability of precipitation in dry lands, prolonged droughts are common. When drought occurs, harsh desert conditions expand into the semiarid and wet/dry zones, placing stress on the landscapes and dependent land uses. Desertification results from the combined effects of land use pressure and the natural stress of drought. It is manifested by a degraded landscape with a reduced life-support capacity. According to some current estimates, the world's dry land may be expanding by as much as 3 percent every decade.

The cold lands occupy about 30 percent of the world's land area. Besides harsh climatic conditions and the obvious limitation on agriculture, these lands are underlain by permafrost, which hampers modern development involving structures and utilities. Of special concern is the development of oil fields and the transportation of oil across the tundra, one of the few large wilderness reserves remaining on the planet.

Mountain lands have varied environments, but in general they share relatively harsh climatic conditions and very rugged terrain. These conditions severely limit the land-use carrying capacity of mountainous lands. Attempts to push land use beyond the limits more or less set by traditional societies have resulted in serious environmental degradation characterized by deforestation, slope failure, soil erosion, and increased flooding.

The continental shelves are large underwater extensions of the continents that support major marine ecosystems. They are among the richest ecological environments on the planet and have long served as the world's primary fishery. In recent decades the continental shelves have been subject to increased pollution from terrestrial runoff, atmospheric fallout, and industrial and shipping activities. This is mainly a result of rising coastal population and land-use development throughout much of the world. In addition, petroleum production and shipping have given rise to increased accidental oil spills and widespread conflict with both environmental and fishing interests. Finally, increased commercial fishing has resulted in a precipitous decline in the once-abundant fish stocks of these fertile waters.

4.11 KEY TERMS AND CONCEPTS

Culture
Human diffusion
Adaptation
Agriculture
- origin of agriculture
- early environmental impact
- spread of agriculture

Global land use zones
- developed areas
- less developed areas
- frontier environments

Wet tropics
- tropical rainforest
- species extinction
- greenhouse effect
- hydrologic cycle

Dry lands
- true desert
- semiarid, steppe

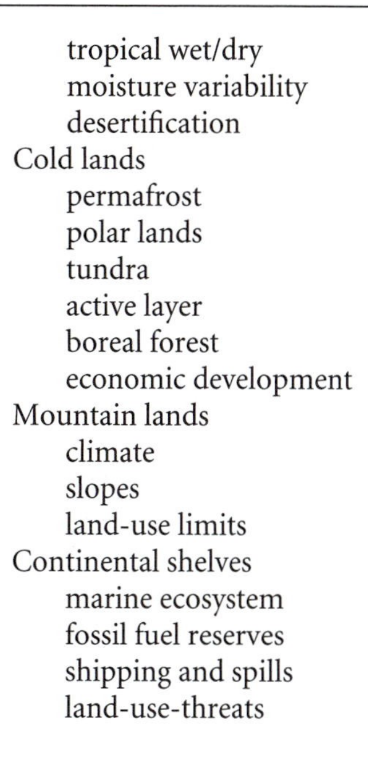

- tropical wet/dry
- moisture variability
- desertification

Cold lands
- permafrost
- polar lands
- tundra
- active layer
- boreal forest
- economic development

Mountain lands
- climate
- slopes
- land-use limits

Continental shelves
- marine ecosystem
- fossil fuel reserves
- shipping and spills
- land-use-threats

4.12 QUESTIONS FOR REVIEW

1. As biological organisms, what are the climatic limitations of the human habitat on Earth? Define this climatic zone on the map of world climates in an atlas.
2. What is culture, and how did it enable early humans to spread from Africa to the rest of the world? What enabled these people to spread into harsh environments such as the cold, glacial environments of the European and North American midlatitudes?
3. Where were the great centers of early agriculture in the millennia before Christ? What sorts of environ-

ments were these? Name the changes in human settlement, population, and environmental impact that accompanied early agriculture.

4. When did Europeans begin to spread to the Americas? What was the estimated Indian population there at the time of European contact? What changes took place in Indian population and land use? What landscape changes were caused as a result of European settlement in the Americas?
5. Name the three types of zones or areas of human occupation in the world today. What are the distinguishing characteristics of each? Identify on a world map several major areas of these zones of human occupation.
6. How would you define the term *frontier environment*? Identify several environmental, land use, and cultural characteristics of frontier environments. What is meant by the term *mismatch* in reference to Western settlement of frontier environments?
7. Describe the characteristics of the environment of the tropical forests, and explain why this environment has been referred to as a counterfeit paradise. What land uses are the chief causes of destruction of tropical forest today, and why are we so concerned about this loss?
8. How do you define dry lands? What are the different types of dry environments and how much of the Earth's land area do they cover? What is meant by precipitation variability and how does it work in combination with land use to produce desertification? Finally, where are some areas of modern desertification and what are its effects?
9. What are the overriding limitations of cold environments to modern land uses? Where are most cold environments found in the world, and how much of Earth's land area do they cover? What is permafrost, and why does it pose a threat to pipelines, roads, and settlements? What is the current geopolitical status of vast reaches of cold lands in Antarctica, Eurasia, and North America?
10. In mountainous lands, what are the principal environmental limitations to life and land use? How much of the Earth's land area lies above 6,500 feet elevation, and what are some of the consequences of pushing agriculture into marginal mountain lands?
11. What are continental shelves, and why have they traditionally been important environments for humans? Why are the continental shelves the focus of increasing economic interest, and why is there such conflict between development interests and environmental interests? What are Exclusive Economic Zones?

5

GLOBAL CYCLES AND SYSTEMS: THE ENVIRONMENTAL FRAMEWORK

5.1 INTRODUCTION

As we contemplate the next wave of human exploration and colonization, we look not to other continents as the Europeans did five centuries ago, but to other planets in our solar system. In particular, we look to the other terrestrial planets—Mars, Mercury, and Venus—and certain satellites such as our moon. The emerging picture of these planets indicates that our neighbors are probably desolate and inhospitable places. Among the benefits of our scientific insights into our neighboring planets is a clearer perspective of Earth as a planet and as a place to live. Little by little we are beginning to see that Earth may indeed be the Eden planet of our solar system.

What sets Earth apart from the other **terrestrial planets** and makes it a good place to nurture and sustain life? There is no single factor that explains Earth's uniqueness—the explanation lies in a combination of factors. Three factors in particular set the foundation for Earth's surface environment: (1) intermediate distance from the sun, (2) a dense, oxygen-rich atmosphere, and (3) abundant water. The result is a planet with these surface conditions:

- Moderate radiation intensities with little toxic radiation.
- A moderate heat balance with an equilibrium surface temperature of 15°C (60°F).
- Widespread moisture in all three physical states: vapor, liquid, and solid.

These conditions, in turn, set the framework for the global life-support system.

This chapter examines the basic infrastructure of Earth's life-support system. We open with a brief description of its major environmental divisions and then turn to a discussion of energy and matter. We are especially interested in energy as the driving force of ecosystems. This is followed by an examination of matter as it is cycled between ecosystems and the physical environment.

5.2 THE GENERAL ORGANIZATION OF THE EARTH'S ENVIRONMENTS

It is traditional to begin the description of the Earth environment by defining four broad divisions or spheres. The **lithosphere** is the zone of rock; the **hydrosphere** is the zone of water; the **atmosphere** is the zone of air; and the **biosphere** is the realm of living matter. These spheres are organized vertically according to density into a gravitationally stable configuration. Air, at the lightest density of 1.3 kilograms per cubic meter, is on top; rock, at the heaviest densities, greater than 2,000 kilograms per cubic meter, is on the bottom. Water and living matter are of intermediate densities and so occupy an intermediate position among the spheres (Figure 5.1).

Lithosphere, Hydrosphere, and Atmosphere

The lithosphere is by far the Earth's most massive sphere It is made up of two main parts, the **mantle** and **core**, each of which is about 3,000 kilometers (2,000 miles) thick (Figure 5.1). Only the upper 100 kilometers (60 miles) or so of the mantle interacts directly with the biosphere, hydrosphere, and atmosphere. As we noted earlier, this relatively thin shell of rock (which contains the Earth's crust) is subdivided geographically into huge sheets called tectonic plates, which shift slowly across Earth's surface (Figure 2.5).

This process, known as **plate tectonics**, is responsible for shaping the ocean basins and the continents over hundreds of millions of years, as well as influencing the com-

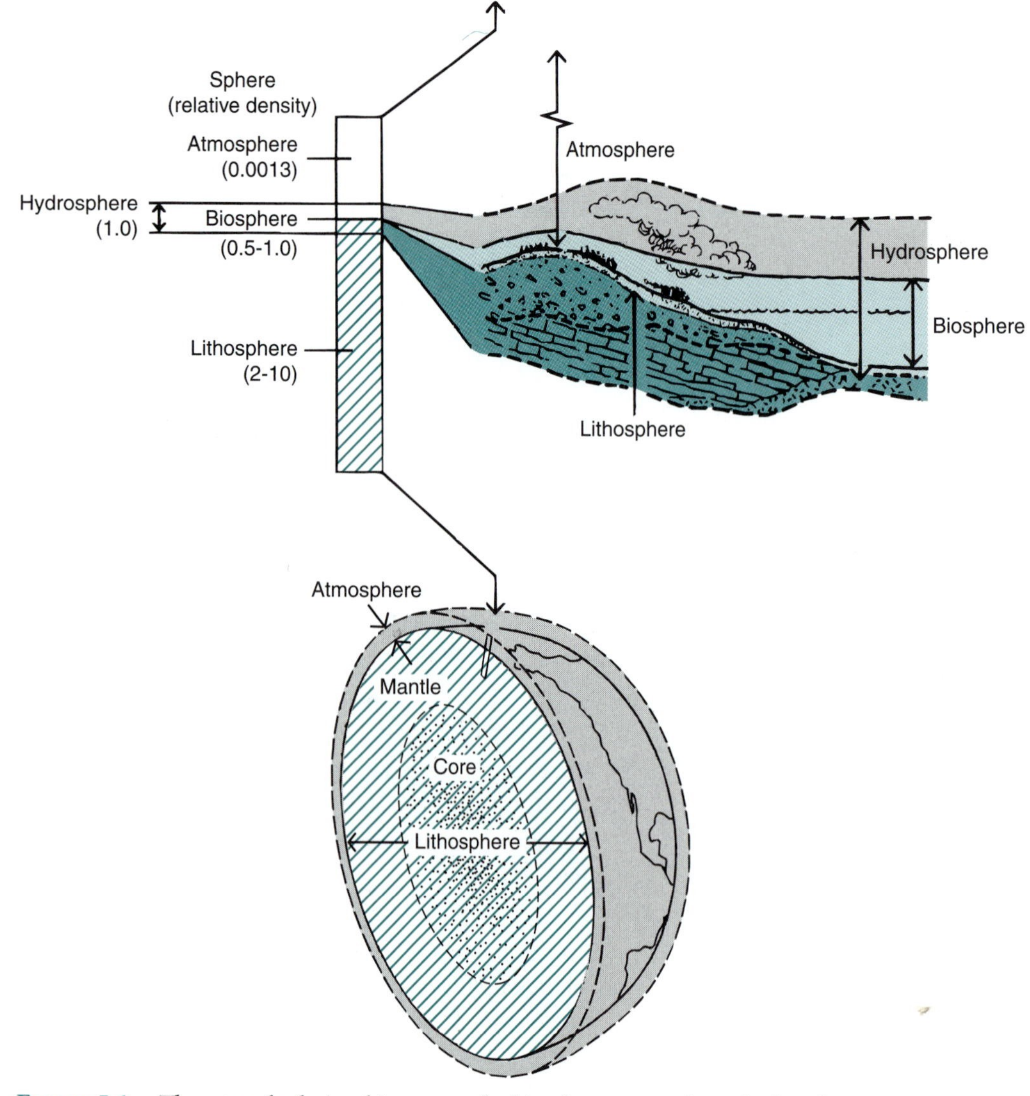

FIGURE 5.1 *The general relationship among the biosphere, atmosphere, hydrosphere, and lithosphere.*

FIGURE 5.2 *Outgassing of the lithosphere through volcanic activity. Volcanoes are constantly active on Earth, especially on the Pacific Rim.*

position of the planet's surface environment. Through a process called **outgassing**, the volcanoes around and within the plates release carbon dioxide, water vapor, and other gases into the atmosphere (Figure 5.2). On the ground, weathered fragments of rock contribute to soil formation, and through related chemical processes minerals are released to plants, ecosystems, and the global water system. The load of dissolved minerals in seawater (35 parts mineral per thousand parts water) is testimony to the effectiveness of these processes.

With more than 97 percent of the Earth's total water supply, the oceans represent the heart of the hydrosphere. All other water on Earth—glaciers, groundwater, lakes, streams, and vapor—are derived from water supplied by the oceans. The oceans feed the atmosphere with a steady flow of water vapor that is quickly redistributed and precipitated back to the Earth as freshwater. About 20 percent of this precipitation falls on the land. Part of it runs off to the ocean, completing a system called the **hydrologic cycle,** which is discussed in detail in Chapter 12.

The atmosphere is a mixture of gases dominated by nitrogen (78 percent) and oxygen (21 percent). Although atmospheric gases can be traced to altitudes of more than 100 kilometers (60 miles), more than 95 percent of the atmosphere's mass is compressed into a layer only 10–12 kilometers (6 miles) deep over the Earth's surface, called the **troposphere**. This layer holds virtually all the atmosphere's heat and moisture and is responsible for the Earth's weather and climatic processes including wind systems and hurricanes. Above the troposphere is the **stratosphere,** where the air is very thin and ultraviolet radiation is absorbed by ozone, a minor atmospheric gas that protects the biosphere from this harmful form of radiation.

The Biosphere

At the surface of the land and the oceans, the spheres blend into each other and the simple four-part designation breaks down. As the landscape portrayed in Figure 5.3 suggests, the biosphere is the most tenuous of the spheres. The total mass of living matter worldwide is only 8 trillion metric tons as compared to 5,140 trillion metric tons of air and 1,500,000 trillion metric tons of water. Viewed in global perspective, the biosphere is little more than a thin web stretched over the land and water surfaces. Although microorganisms can be detected scores of meters underground and pollen, spores, and other tiny biological agents can be detected thousands of meters into the atmosphere, the bulk of the life web is concentrated on the surface itself. At its strongest points on land, represented by the great forested landscapes of the tropics and humid coastal lands, the central mass of the life web is barely 80 meters (250 feet) deep. At its weakest points, in the desert and polar landscapes of the Earth's harshest climatic zones, the web is less than 3 meters (10 feet) deep and frequently as thin as only several centimeters.

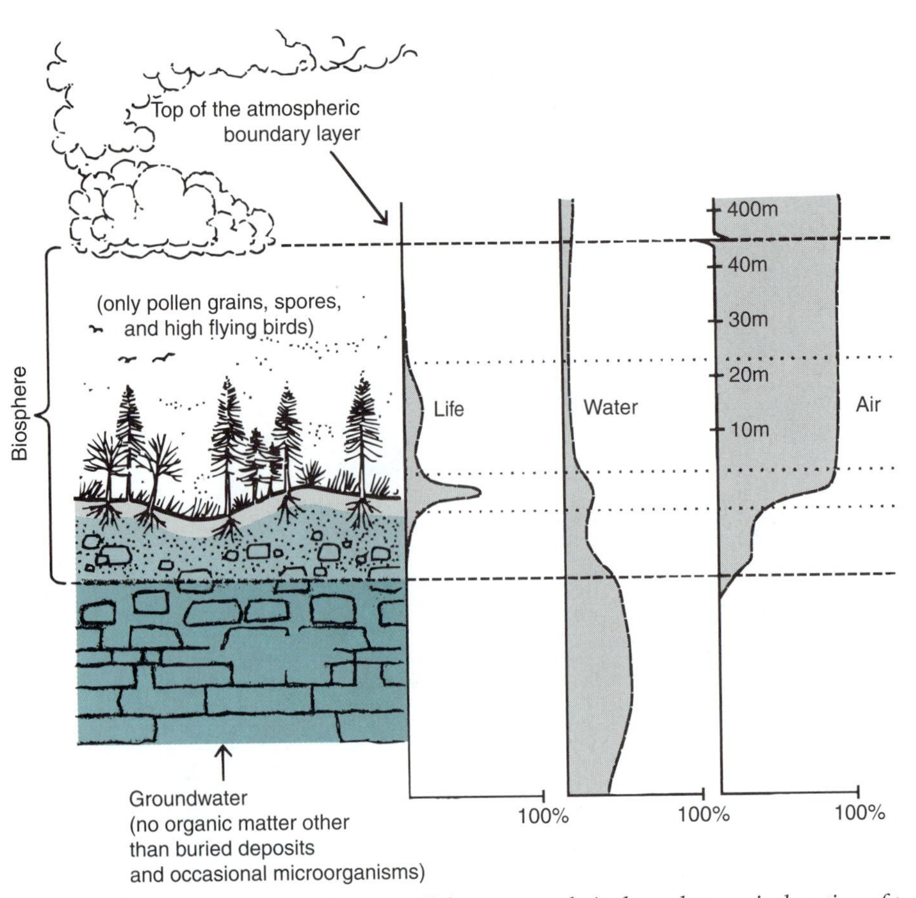

FIGURE 5.3 *The relative distribution of life, water, and air through a vertical section of terrestrial biosphere.*

In the oceans the biosphere web is also concentrated in a thin layer at the surface. The bulk of marine life lies near the surface, in a zone about 100 meters (330 feet) deep, called the **photic zone**, or zone of light. At greater depths, the water is too dark for much plant life, particularly the tiny phytoplankton that represent the bulk of the sea's vegetation, and the biosphere quickly thins out. Life is present to depths of 1,000 meters (3,300 feet) and deeper, but it is lightly distributed over a vast volume of water. This deeper zone of life on the faint fringe of light rises and descends with day/night changes in light penetration. At the bottom of the deep ocean, where organisms feed on the debris raining down from the photic zone above, the abundance of life increases somewhat, though the increase is not great (Figure 5.4).

The vast majority of the ocean basins consists of **abyssal plains,** or deep ocean floors, where typical water depths range from 4,000 to 5,500 meters (13,000–18,000 feet). Life is extremely sparse on these vast, dark plains, comparable to that of the harshest climatic zones on land. By contrast, the shallow waters on the margins of the oceans are abundant with life. The continental shelves occupy about 10 percent of the oceans with water depths generally less than 200 meters (650 feet). The continental shelves average about 75 kilometers (45 miles) in width and become gradually shallower toward the shore (Figure 5.4). At the shore itself the surface and bottom zones of marine life merge to form one of the strongest segments in the planet's life web. This zone, which is characterized by wave action, tidal fluxes, and inflowing streams, is called the **littoral zone.**

In broad geographic terms, the web of life occupies just two classes of environment: terrestrial and aquatic. Terrestrial environments take up about 29 percent of the planet's surface and aquatic about 71 percent. Most of the Earth's 10 million or more species live in the terrestrial environment and the adjacent waters of the continental shelves. Within this environment, however, there is tremendous variation in the abundance of organisms. Very little life resides in the cold lands and the harsh deserts, and if they are subtracted from the terrestrial environment we are left with most living organisms occupying only about 30 percent of the Earth's surface. This area supports all but a few percent of the Earth's 6.1 billion humans. Herein lies one of our most fundamental environmental problems: competition for basic resources (space, soil, water, and air) among humans and other organisms.

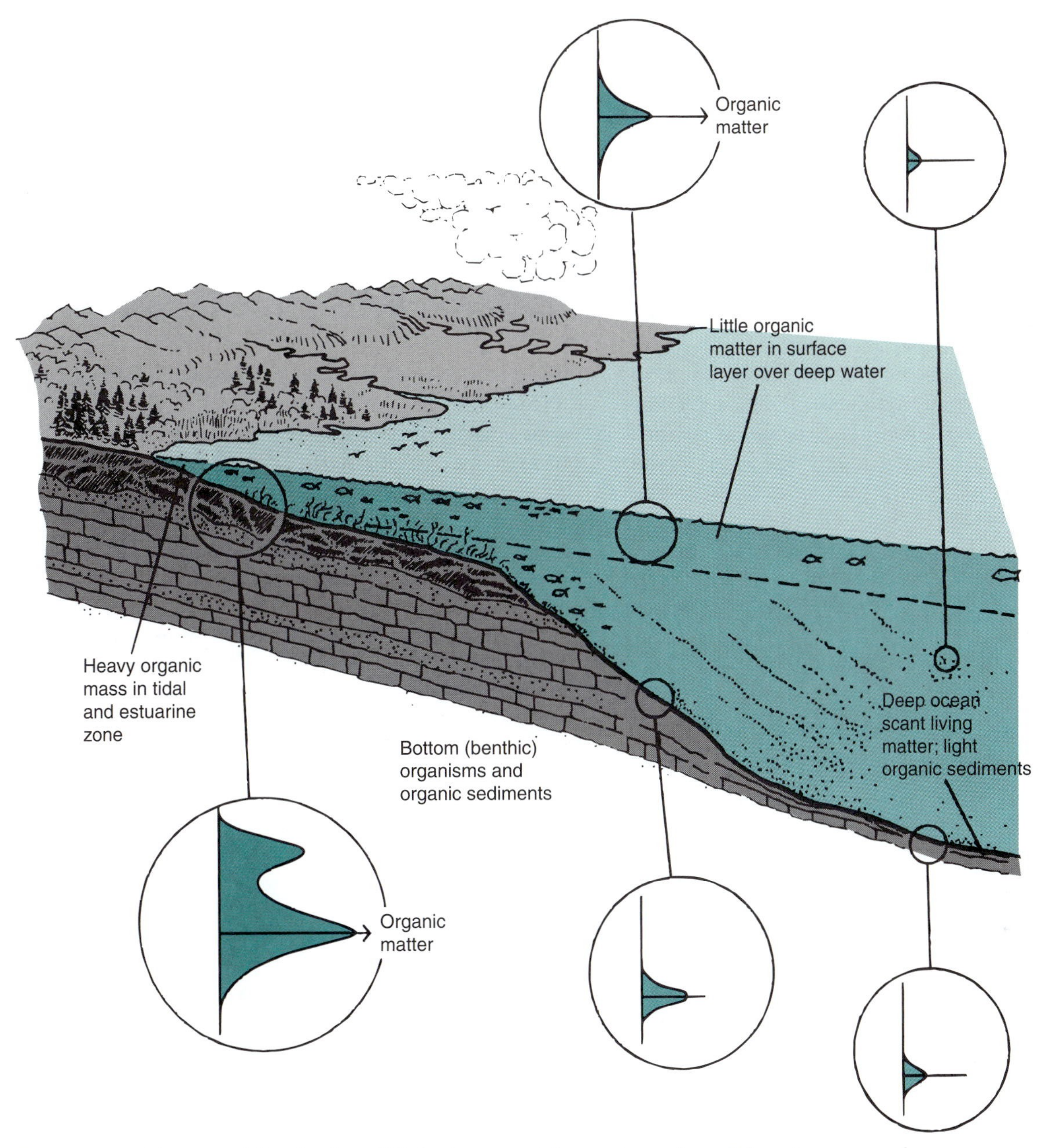

FIGURE 5.4 *The general distribution of life in the ocean. The graphs are intended to give a sense of the relative distribution of organic matter at different depths.*

The web of life is not only spread unevenly, but is also constantly changing over time. It fluctuates with the change of seasons and with oscillations in climate and ocean circulation over periods ranging from decades to many thousands of years. It also changes with disturbances, especially radical events caused by volcanic eruptions or meteorite impacts or caused by human disruptions such as major oil spills, war, and urban development.

These disturbances are able to tear gaping holes in the biosphere. The holes heal over in time; however, if they occur with great magnitude and frequency, there is reason for concern that the entire web or large sections of it may not achieve full recovery. Environmental scientists look for indicators of such decline in the biosphere and generally agree that accelerated extinction related to tropical forest eradication, the buildup of toxic contaminants in food chains, and the impoverishment of dry landscapes as a result of desertification are high on the indicator list.

The Ecological Context

No organism on the planet exists independent of other life forms. All life belongs to networks or systems of other organisms. These systems are called **ecosystems**, and they are the basic organizational units of the vast complexes of life found on land and water over virtually all of the Earth.

An ecosystem is an energy system made up of different groups or levels of organisms. Each level of organisms depends on the previous level for its energy (food) supply, such as herbivores on plants and carnivores on herbivores. Figures 6.1 and 6.2 in Chapter 6 will help explain this idea, and we will explore it in greater detail later in the chapter. All ecosystems occupy space on Earth's surface, and each ecosystem is tied functionally (or operationally) to that space. This space constitutes the habitat of the ecosystem from which it draws its essential resources: energy (sunlight and heat), water, and nutrients (Figure 5.5). Science recognizes three basic classes of ecosystems based on the type of habitat or environment they occupy: terrestrial, saltwater, and freshwater. Within these classes there are innumerable types of ecosystems overlapping and interconnecting in multitudes of ways.

The distribution of ecosystems corresponds not only to the distribution of land and water but to the geographic patterns and availability of the essential resources. Where dependable supplies of sunlight, heat, water, and nutrients are abundant and the habitat is not subject to frequent disturbances from outside forces such as hurricanes or land clearing, ecosystems are rich in organisms and are highly productive. Productive means that they are able to draw large quantities of resources from the environment and convert them into organic matter. The most important of these resources is solar energy.

5.3 THE SOURCE OF ENERGY FOR THE EARTH'S ENVIRONMENTS

Surface environments, both organic (biotic) and inorganic (abiotic), run almost entirely on **solar energy**. Energy from the Earth's interior, called **geothermal energy** is negligible by comparison, representing only a tiny fraction (about 1/6,000th) of the solar contribution to Earth. Except for geological processes such as volcanism and earthquakes, all the essential processes and systems on the earth surface are driven by solar energy. These include (1) *basic life processes*, such as plant growth, reproduction, and the ecosystems themselves; (2) *life-support processes*, such as precipitation and soil formation; and (3) *life-threatening processes*, such as hurricanes and tornadoes.

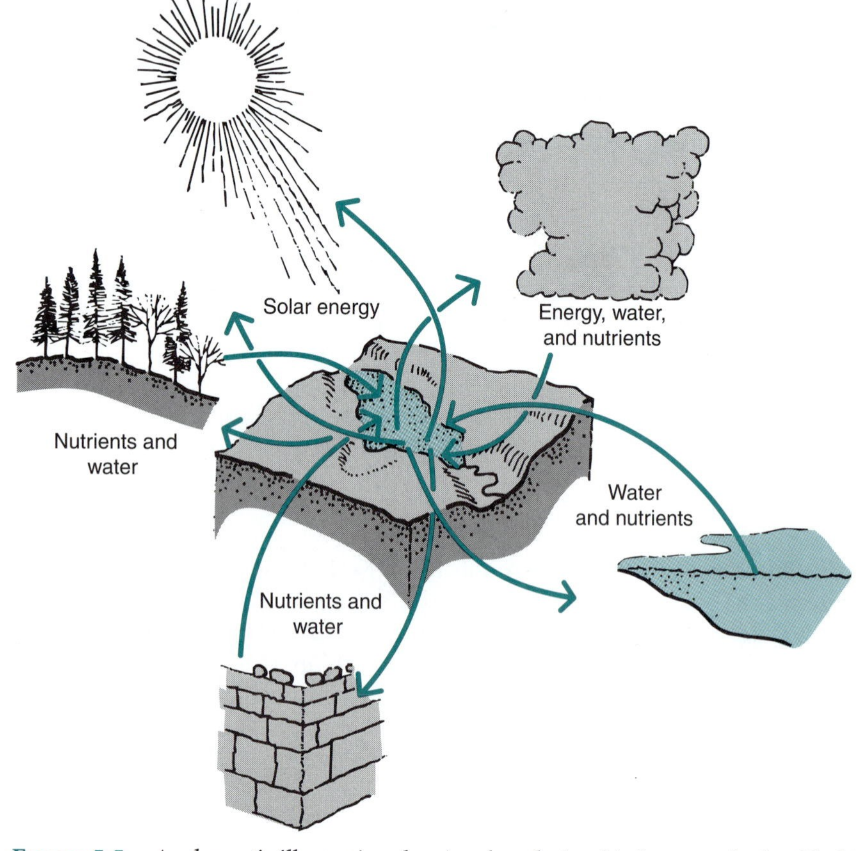

FIGURE 5.5 *A schematic illustration showing the relationship between the local habitat of an ecosystem and its principal resources.*

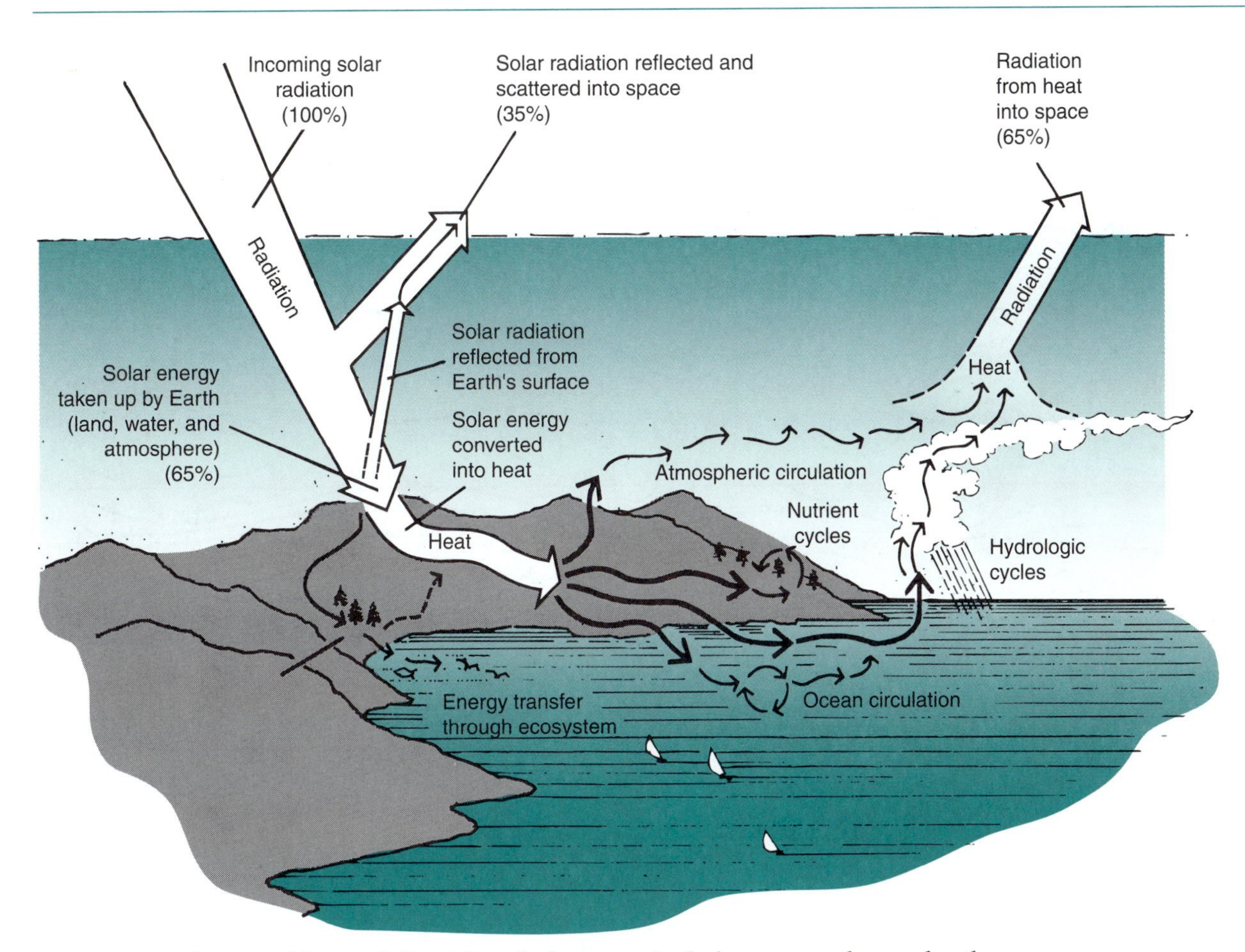

FIGURE 5.6 *The general flow and disposition of solar energy in the lower atmosphere and at the Earth's surface. Less than 1 percent of this energy goes to the direct support of Earth's life system.*

Available Solar Energy

Not all the solar energy received by Earth, however, is utilized, or taken up, by its environments. As solar radiation passes through the atmosphere, about 31 percent of it (based on average figures worldwide) is thrown back to space by the reflection of rays off clouds and scattering of rays off gas molecules and particles in the air. At the Earth's surface an additional 3 or 4 percent is reflected off water, ice, snow, and land and this, too, is lost to space. Thus, the Earth's environments, including the atmosphere, are powered by about 65 percent of the total solar energy contributed to Earth (Figure 5.6). This is a huge amount of energy, and it flows to Earth in a continuous stream.

Global Distribution of Solar Energy

The solar energy delivered to the Earth's surface is not evenly spread over the planet, however. Because of the Earth's curvature (and other factors), the polar ends of the planet receive much less energy than the equatorial zone. Measured in millions of kilocalories of solar energy per year,[1] the high latitudes (above the Arctic and Antarctic circles) receive less than 60 million kilocalories of solar energy. The midlatitudes in both the Northern and Southern Hemispheres predictably receive an intermediate amount of solar energy, 80 to 120 million kilocalories a year.

Equatorward of the midlatitudes, however, an unexpected reversal occurs in the solar energy values. Instead of the highest values occurring along the equator, the highest values, at 180 to 220 million kilocalories a year, are found in the subtropics, around 20 to 30 degrees latitude. The values along the equator, which range from 120 to 160 million kilocalories a year, are lower than those of subtropics because heavy cloud cover

[1]This flow, or *flux*, of solar energy is based on a measurement of kilocalories of solar radiation received over a sample area of one square centimeter of Earth's surface over a period of one average year. One calorie represents the amount of heat energy required to raise the temperature of 1 gram of water 1°C when the water is at a temperature of 14.5°C. One kilocalorie is equal to 1,000 calories.

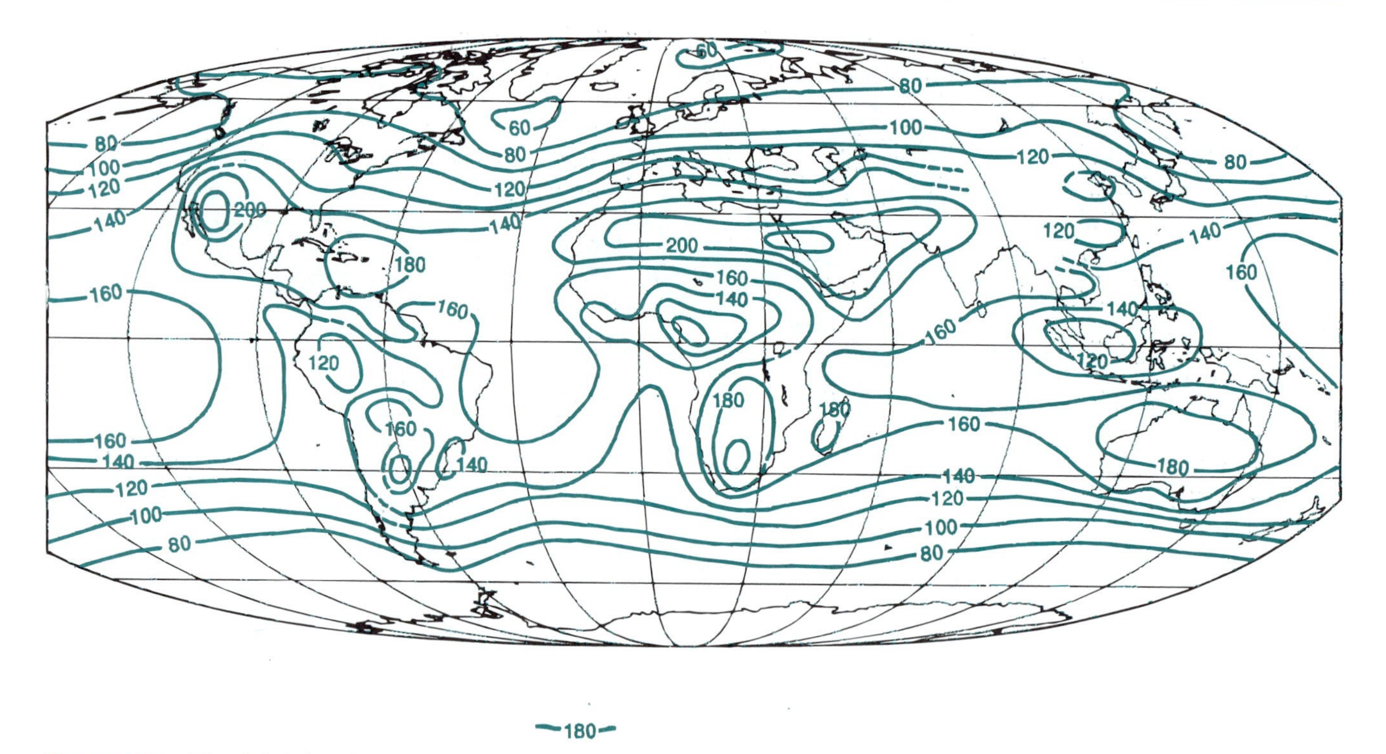

FIGURE 5.7 *The global distribution of solar radiation at the Earth's surface. Note that the radiation values peak in the subtropics and not at the equator.*

near the equator reflects 30 to 40 percent of the incoming solar radiation. The subtropics are the zones of the great deserts where there are few clouds to block the flow of solar radiation to the Earth's surface. This accounts for the intense surface heating in the Sahara and Arabian deserts and other large deserts of this zone (Figure 5.7).

Radiation and Moisture: Implications for Life Support

To gain a general concept of Earth's capacity to support life, we need to consider the distribution of moisture as well as solar radiation. The distribution of moisture is also broadly zonal. Based on mean annual precipitation (that is, all moisture falling to the surface in the average year), there are two great zones of high precipitation and two great zones of low precipitation.

The tropics, located in a belt centering on the equator, is decidedly the wettest zone on Earth (Figure 5.8). Annual precipitation ranges from 100 centimeters (40 inches) to more than 200 centimeters (80 inches). Combined with abundant solar radiation, the tropics are warm, wet, and well-lighted—ideal conditions for life. It follows that the wet tropics support the greatest mass of living matter and the greatest number of plant and animal species. On their margins, the wet tropics give way to a wet/dry zone where life is sharply limited for several dry months a year.

The subtropics receive heavy solar radiation but relatively little precipitation. The surface is well-lighted and warm but exceedingly dry. This zone is noted for the world's great deserts, where precipitation is not only light—generally less than 30 centimeters (12 inches) annually—but also highly variable and subject to rapid evaporation. Life is sparse and very slow-growing in the subtropical deserts (Figure 5.8).

The midlatitudes are damp in both the Northern and Southern Hemispheres. However, because summers are warm, evaporation rates can be high, limiting moisture in some areas for several months a year. In addition, the midlatitudes are heat-limited in the winter, decidedly the most severe season. As a result the capacity for life support is essentially semiannual in this zone. Poleward of the midlatitude zone is the polar zone, where heat and light are severely limited most of the year. Precipitation is surprisingly light in this zone, not much more than in the deserts, but heat is the critical limiting factor with fewer than thirty frost-free days a year over vast areas. The polar zone, like the harsh deserts, supports relatively sparse life.

5.4 THE MAJOR ENERGY SYSTEMS OF EARTH

When we say that solar energy is utilized, or "taken up" by Earth, we mean that it is available to do work or that it is stored in the environment. In either

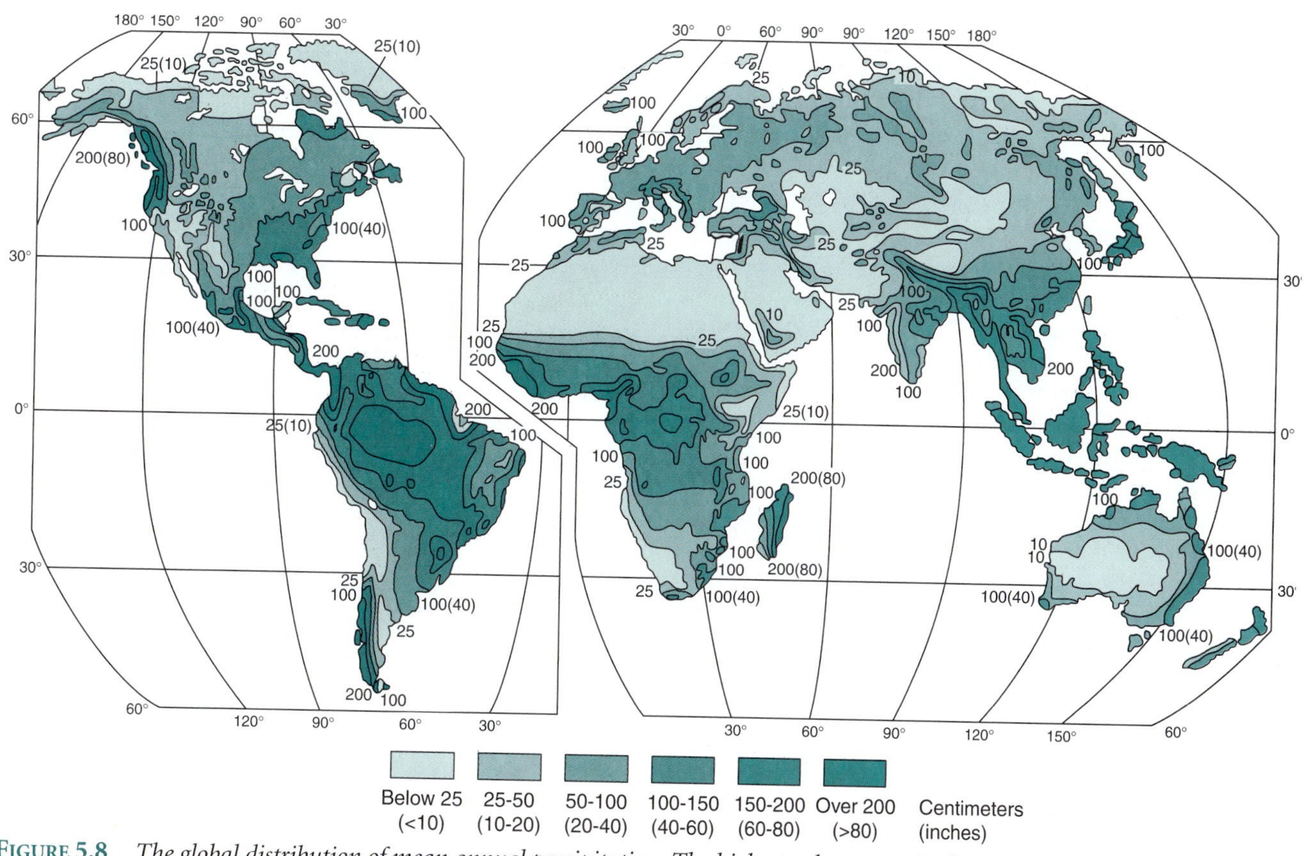

FIGURE 5.8 *The global distribution of mean annual precipitation. The highest values occur in the tropics and midlatitudes.*

case, the solar energy taken up by the Earth is always converted from radiation to other energy forms, specifically **heat** and **organic** (*chemical*) **compounds.** This energy then enters the Earth's environmental systems and we can trace it through the environment along two separate paths, heat and organic, as was shown in Figure 5.6.

Within these systems, heat energy is converted into other forms of energy: for example, **kinetic energy**, the energy of motion represented by winds and ocean currents; and **potential energy**, represented by a mass of snow or rain precipitated on a mountain slope high above sea level (Figure 5.6).

The Heat Path and Geophysical Systems

More than 99 percent of solar energy takes the heat path. This path starts when solar energy is absorbed and converted to heat by the atmosphere and by the surface layers of the oceans and continents. Here the heat is stored for a short time and thus represents a temporary reservoir of energy in the environment. This heat energy drives the physical processes and systems of the Earth's surface, including:

1. the **hydrologic system**, the evaporation and cycling of water from the Earth's surface through the atmosphere;
2. the **geochemical** or **nutrient system**, the cycling of chemicals such as phosphorus and nitrogen through water, soil, and air;
3. the **atmospheric circulation system**, the formation of pressure cells and wind systems in the atmosphere.

The Organic Path and Life Systems

The organic energy path begins with the conversion of solar radiation into plant materials by photosynthesis. **Photosynthesis** is the process by which green plants utilize solar radiation to manufacture organic compounds that go into the building of vegetation. Although this path utilizes less than 1 percent of the available solar energy, it is absolutely essential to the entire system of life on the planet. The organic compounds manufactured by plants are passed on to other organisms via **food chains**, defined as a series or sequence of organisms that rely on each other as sources of food. Thus, all life rests on a foundation of solar energy. We will say more about the nature of ecosystems in the next chapter. Here we simply want to establish the idea of energy transfer by organisms.

The systems of plants and animals that are linked together by food chains are ecosystems. As the energy in an ecosystem is passed from one level to the next, only a small share of it (roughly 10 percent) is retained in the bodies of the living organisms in the food chains. Most energy is converted into heat, which, in turn, is given up to the atmosphere. Because of the high rate of heat loss from organisms, the supply of energy in ecosystems attenuates very rapidly with transfer through each group of eaters.

In the end the biosphere gives up nearly all the chemical energy in ecosystems to the atmosphere as heat. But the total contribution is very small. Compared to the amount of heat in the atmosphere generated directly from the absorption of solar radiation, the amount of heat energy contributed to the atmosphere by organisms is so minuscule that it has no measurable effect on the atmosphere's temperature. But not absolutely all of the energy of ecosystems is given up as heat, for a tiny fraction is retained by Earth as stored chemical energy.

Chemical energy is stored in the form of organic deposits in soil and rock formations. Most of this is represented by topsoil and fossil fuels (coal, oil, and natural gas). Worldwide fossil fuel reserves are estimated to represent a mass of energy equivalent to 3,600 years of photosynthesis. This represents all the energy the Earth has been able to place in storage over the past 2 or 3 billion years. Set in this time scale, it is apparent that our energy reserves of fossil fuels are very small indeed. When this organic material is used as a source of energy in industry, automobiles, or homes, it is converted into heat, and, along with chemicals produced in the burning process, is then released into the atmosphere.

5.5 The Cycle of Matter in Ecosystems

All organic material is made up of molecules of **matter**. These molecules are made up of atoms of different elements that are drawn from the environment by organisms. Three elements account for almost all (99.47 percent) living matter: hydrogen, carbon, and oxygen. Most of the remaining half-percent or so is represented by twelve elements (Table 5.1).

TABLE 5.1

The 15 Most Abundant Elements of Living Matter

Element (Nutrient)	Percentage
Hydrogen (M)	49.740
Carbon (M)	24.900
Oxygen (M)	24.830
Nitrogen (M)	0.272
Calcium (M)	0.072
Potassium (M)	0.044
Silicon	0.033
Magnesium (M)	0.031
Sulfur (M)	0.017
Aluminum	0.016
Phosphorus (M)	0.013
Chlorine	0.011
Sodium	0.006
Iron	0.005
Manganese	0.003

(M) = Macronutrient

Nutrients

The three major elements—plus nitrogen, calcium, potassium, sulfur, and phosphorus—are called **macronutrients** because all of them are needed in relatively large quantities for life. The remaining elements in Table 5.1, as well as several others including boron and copper, are called **micronutrients**. They are generally required in much smaller and more variable quantities among different organisms. Both sets of nutrients are drawn from the spheres (litho-, hydro-, and atmo-) of the environment, cycled through ecosystems, and ultimately returned to the environment.

Within organisms the atoms of the nutrient elements are used to build molecules. The molecules can take on many different forms as a result of different chemical and biochemical processes. The atoms themselves, however, are not altered or destroyed by these processes, but remain constant within the different molecular arrangements. Thus, within an ecosystem, matter is conserved as atoms are used and reused, or *cycled*, within organisms, and among organisms and the environment. As with energy, matter can be neither created nor destroyed, only converted into different forms. Unlike energy, however, which flows from space through the Earth environment and then back to space, matter is not lost to space but is recycled within the Earth's environmental systems.

Critical to our thinking about nutrient cycling is the study of changes brought on by human actions. Through air and water pollution, disturbance of soil, deforestation, desertification, and other actions we are altering the flow of both macro- and micronutrients within the global environment. Some ecosystems are becoming overloaded with certain nutrients, while others are being deprived of certain nutrients. The results are imbalances that accelerate or retard ecological processes and change the function and composition of ecosystems. For example, many freshwater ecosystems, such as those of inland lakes, are experiencing biological overgrowth because of nutrient loading from agricultural runoff and fallout from air pollution. At the same time, other freshwater ecosystems are experiencing decline because of chemical changes from acidic precipitation generated by air pollution in urban industrial regions.

Nutrient Cycles

The cycling of atoms and molecules through ecosystems is termed a **nutrient cycle** or *biogeochemical cycle.* Two types of nutrient cycles are recognized: sedimentary and gaseous. In the **sedimentary cycle** nutrients held in rock are released by weathering. Weathering involves chemical decomposition of rocks that frees nutrients into the hydrosphere. The nutrients, dissolved in water, are extracted by plants and animals and transferred through the ecosystems, eventually ending up as a sedimentary residue when these organisms complete their life cycles. This residue may be deposited as sediment in the sea or on the margins of the land masses, where it may be reincorporated into the lithosphere as rock. When the rock is uplifted by geological forces and once again subjected to weathering, the cycle is completed. The time involved in this cycle is very long, normally tens or hundreds of millions of years, and the rock segment is decidedly the slowest and largest part of the cycle. Therefore, the lithosphere functions as a *sink,* or storage bin in the system.

In the **gaseous cycle** nutrients are exchanged between the biosphere and the atmosphere without going into the lithosphere. This cycle is much faster, usually hundreds of years or less, and is the chief means of circulation for carbon, hydrogen, oxygen, and nitrogen. Phosphorus, calcium, potassium, and magnesium, on the other hand, are circulated by the sedimentary cycle.

The Carbon Cycle

Carbon is moved in both the gaseous and sedimentary cycles (Figure 5.9). In the gaseous cycle, carbon occurs in the form of carbon dioxide (CO_2), a free gas in the atmosphere and a dissolved gas in freshwater and saltwater. Although CO_2 represents only 0.033 percent of the atmosphere, the atmosphere is the principal reservoir, or pool, of Earth's carbon dioxide. Between the atmosphere and the Earth's surface a great exchange of CO_2 is continuously in motion, involving billions of tons annually.

The atmosphere gives up its CO_2 by only two processes: (1) photosynthesis by land and ocean plants; and (2) absorption by seawater. The oceans and the continents together take in an estimated 203 billion tons of CO_2 from the atmosphere annually. The atmosphere gains CO_2 from four sources. The chief input comes from the respiration of plants and animals, both terrestrial and aquatic. Volcanic activity also supplies CO_2 to the atmosphere through the outgassing process and the oceans release CO_2 directly from the water with wave spray, evaporation, and other surface processes. These natural sources, respiration and outgassing, release an estimated 200 billion tons of CO_2 to the atmosphere annually. The third source is **air pollution** from the burning of fossil fuels and wood, which contributes an estimated 7 billion tons of CO_2 to the atmosphere annually, giving us a current total balance of 207 billion tons in and 203 billion tons out each year. The extraction and burning of fossil fuels (coal, oil, natural gas) is

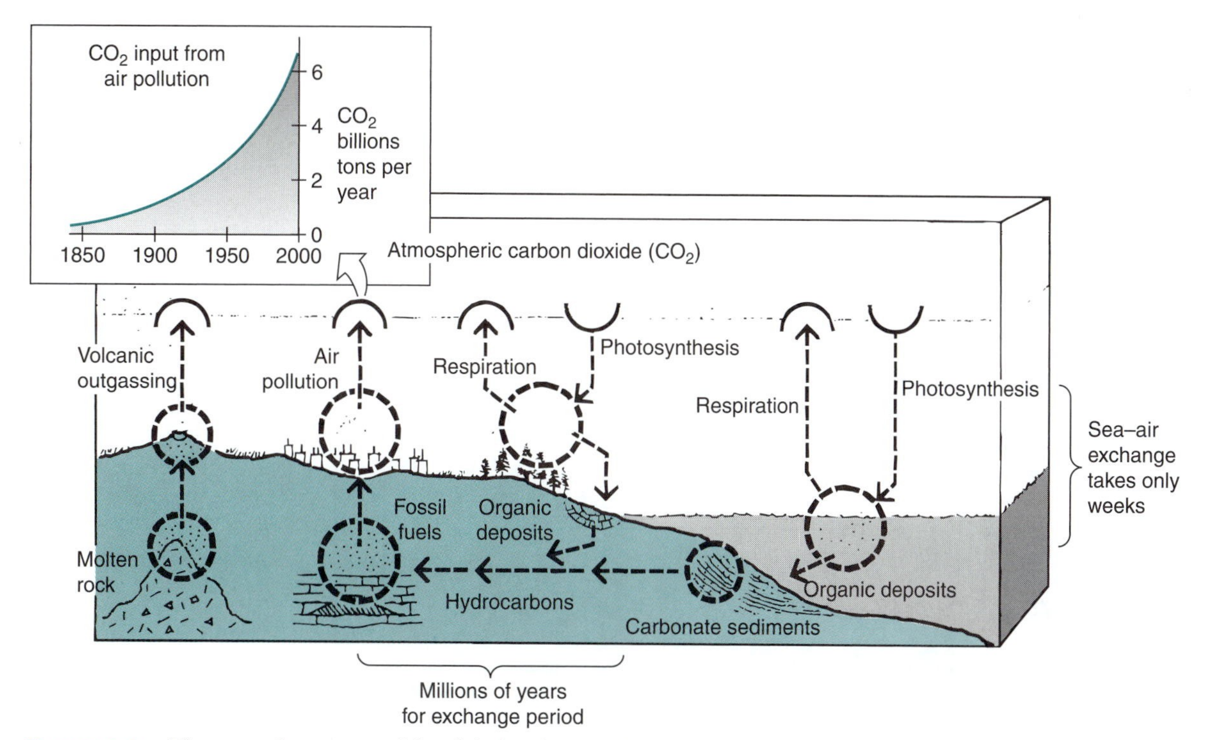

FIGURE 5.9 *The general structure of the global carbon cycle in modern times.*

the principal link between the sedimentary and gaseous cycles of carbon (Figure 5.9).

The CO_2 Balance

The relative balance of the carbon cycle is one of the serious questions being examined by environmental scientists today. We have known for a long time that the sedimentary cycle of carbon is far out of balance because we are consuming fossil fuels at an accelerated rate compared to the meager rate of natural replacement with the formation of new fossil fuel reserves. The carbon released in the combustion of fossil fuels goes directly into the atmosphere as CO_2. In addition, deforestation in the tropics, which involves massive burning activity, is also releasing CO_2 into the atmosphere. Of the 7 billion tons added annually from combustion sources, the oceans are taking up a large part of this surplus, possibly 3 billion tons. The remaining 4 billion tons remain in the air, building up year after year. On balance, the atmosphere's CO_2 content is increasing significantly: Since the eighteenth century it has increased by at least 30 percent. Scientists generally agree that if air pollution trends continue, atmospheric CO_2 will have doubled by the year 2050.

The rise in atmospheric CO_2 is a major source of environmental concern, because carbon dioxide is an important heat-absorbing gas. As CO_2 increases, the potential for atmospheric warming also increases. With this comes the very real prospect of global climatic change leading to the likelihood of increased drought in some regions, higher sea levels, and coastal flooding from increased rates of ice-cap melting, and many other changes that will affect Earth's life-support capacity. More discussion of the carbon dioxide problem follows in Chapters 10 and 11.

The Oxygen Cycle

As with carbon, the gaseous cycle for oxygen (O_2) involves massive exchanges between the Earth's surface and the atmosphere. The atmosphere is made up of nearly 21 percent oxygen, and this pool is fed by two sources: a recycle system and a primary supply. Photosynthesis (both on land and in the oceans) is the recycle source. Volcanic outgassing, which releases oxygen as a part of CO_2 and H_2O molecules, is the primary source of new oxygen (Figure 5.10). The principal flow of oxygen to the atmosphere is provided by photosynthesis, and it is noteworthy that over the past 3 billion years the oxygen content of the atmosphere has grown significantly as plant life has evolved. Under current biological conditions, it takes about 7,600 years for the atmospheric oxygen to be completely recycled through the Earth's plant cover.

The Oxygen Balance

Oxygen is removed from the atmosphere by the respiration of animals and microorganisms and by oxidation of minerals on or near the Earth's surface. In addition, oxygen is removed by fires from the burning of fossil fuels and wood, including forest fires. In the oceans some oxygen is removed in calcium carbonate sediments (such as shells), which are the principal ingredient in limestone (Figure 5.10).

Although the oxygen content of the atmosphere has increased over geologic time, the last century has seen a slight decline in atmospheric oxygen. The source of the decline is all the burning activity by humans (industrial, domestic, forest fires, and others), the same activity that is causing the increase in atmospheric CO_2. Oxygen, however, is 700 times more abundant than CO_2 in the atmosphere; therefore, the relative change in the two with the exchange of gases in the combustion process is pro-

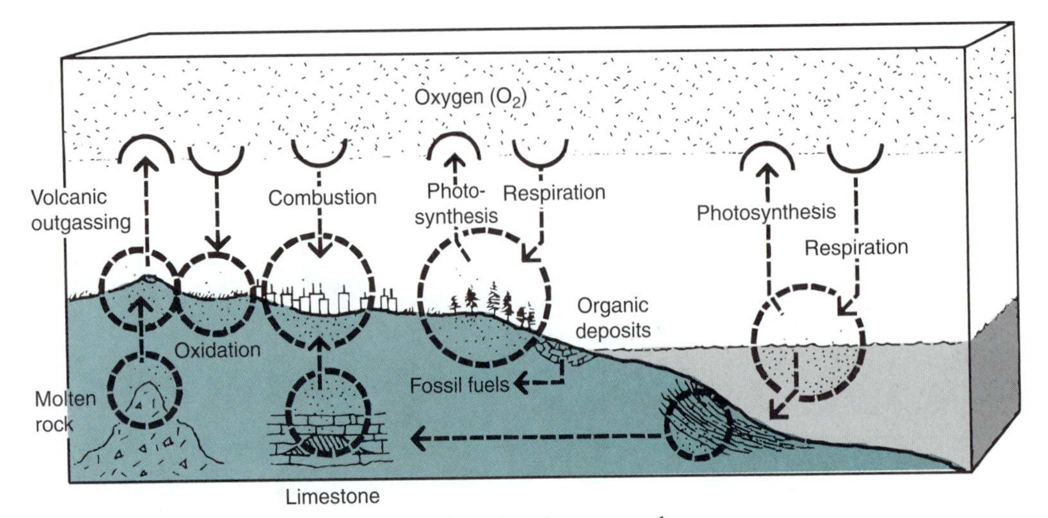

FIGURE 5.10 *The major components of the global oxygen cycle.*

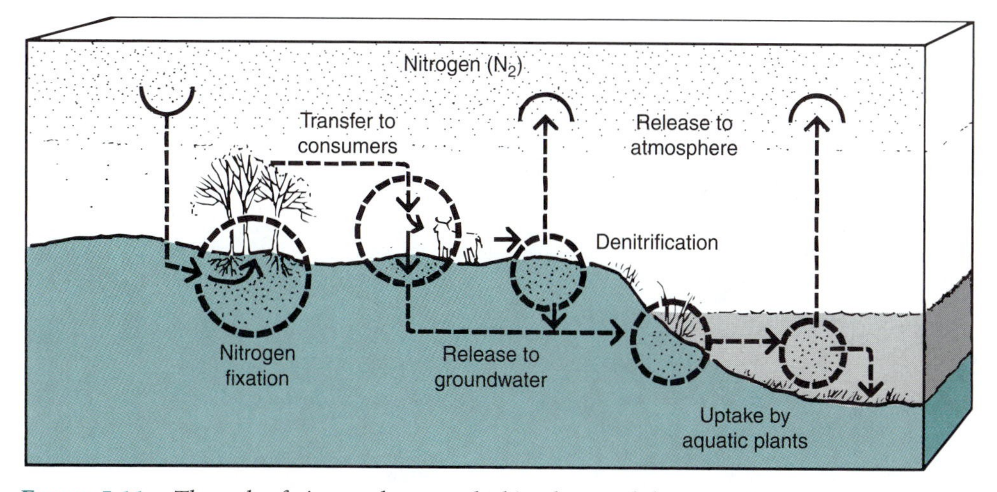

FIGURE 5.11 *The cycle of nitrogen between the biosphere and the atmosphere as a function of nitrogen fixation and denitrification.*

portionally much greater for CO_2. The decline in oxygen is not a matter of great concern at this date.

The Nitrogen Cycle

The atmosphere is made up of 78 percent nitrogen and is the principal pool of this gas. The exchange of nitrogen with the Earth's surface is much slower than that of oxygen. Coupled with the fact that it is also more than three times more abundant in the atmosphere than oxygen, it takes a very long time, millions of years, to recycle all the nitrogen in the atmosphere. Nitrogen is extracted from the atmosphere chiefly by plants through a process called **nitrogen fixation** (Figure 5.11). Nitrogen fixation is a two-part process involving certain bacteria living on the roots of selected species of plants, including certain trees, shrubs, and most members of the pea family. The bacteria extract gaseous nitrogen from the atmosphere and convert it into forms usable by the plants, most importantly ammonia, which goes into the building of amino acids and protein.

Through this process, nitrogen enters the various food chains, such as those of the grazing animals and their predators and the microflora, insects, and other organisms in the soil. Ultimately, the vast majority of the organic nitrogen of plants and animals in terrestrial environments ends up in the topsoil. With the decay of organic matter in the soil, the nitrogen undergoes "unfixing." *Denitrifying bacteria* convert it back into the gaseous form and release it into the atmosphere.

The global nitrogen cycle would be balanced between nitrogen fixation and denitrification, were it not for interference by modern human activity. Large amounts of nitrogen are converted by industrial processes into solid and liquid forms in the production of, among other things, commercial fertilizers. Total yearly production by industrial fixation amounts to about 50 percent of the annual production by natural fixation. Thus, nitrogen is building up somewhere in the global cycle, probably in the water system (for example, in groundwater) and in organic matter added to lakes, streams, and the oceans.

The Phosphorus Cycle

Phosphorus is also essential to life, but in much smaller amounts than nitrogen. In addition, phosphorus has no gaseous loop in its cycle and is therefore limited to a sedimentary cycle. This cycle is exceedingly slow, which explains why the supply of phosphorus to ecosystems is usually very limited. When phosphorus is taken up by plants, it moves through the food chains and is eventually returned to the soil. In the soil phosphorus, unlike nitrogen, is slow to enter the water system and return to the ocean.

Humans are also altering the phosphorus cycle through the application of phosphorus-rich *fertilizers* to agricultural lands and the use of detergents containing phosphorus. Phosphorus is manufactured from phosphate rock, which is mined in various parts of the world. As with natural phosphorus, manufactured phosphorus is slow to move from the land to the ocean; therefore, it is currently building up in the soil and local water features such as lakes and wetlands (Figure 5.12). Phosphorus is a highly effective fertilizer for both agricultural and nonagricultural plants; therefore, when added to lakes, ponds, and wetlands, it often causes accelerated growth of aquatic plants that may, in turn, eventually choke out the water body. Since 1945, phosphorus applications as a part of commercial fertilizers have increased fourfold in the United States. Most is for agriculture, but part is due to increased residential lawn applications. There are now more than 50 million lawns in the United States.

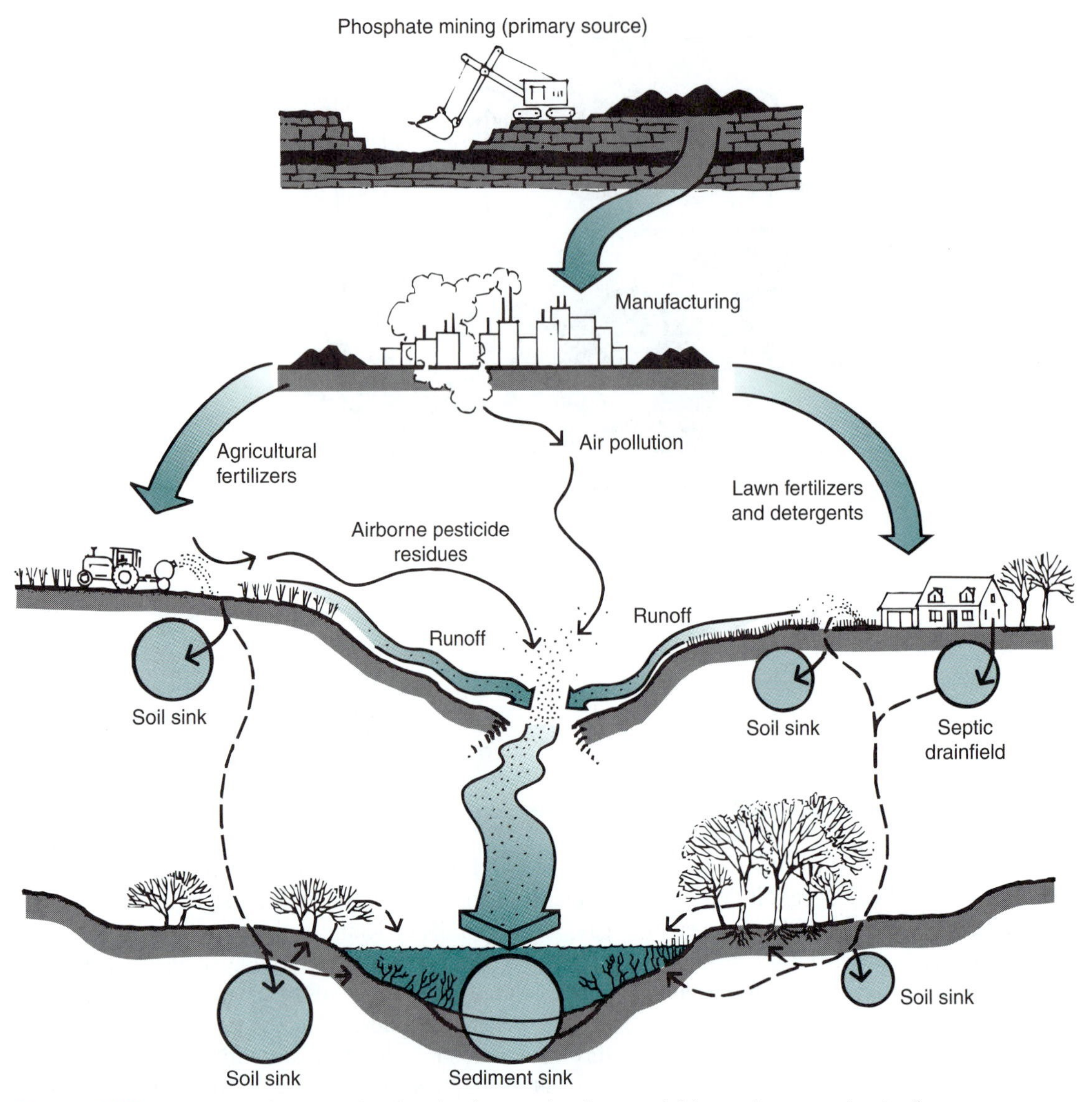

FIGURE 5.12 *The phosphorus cycle related to human land use activities and storage sites in the environment. Since 1945, phosphorus sales for fertilizers have increased fourfold.*

5.6 INTERRELATIONS IN ENVIRONMENTAL SYSTEMS

The Earth is run by a complex set of interdependent and interlocking systems. In their most basic forms, these systems are merely energy and material cycles. All of these systems are maintained by inflows and outflows, as in the input and output of energy in the atmosphere, or the input and output of organic compounds in ecosystems. The balance of environmental systems, which is essential to the balance of life, is made possible through two types of relationships in systems: (1) **primary**, or direct cause and effect; and (2) **secondary**, or *feedback.*

Primary Relationships

In their simplest form, direct cause-and-effect relations work like this: An organism, such as a species of freshwater algae, is dependent on water temperature and sunlight or growth and reproduction (given that all other conditions such as nutrients and water conditions are suitable). When water temperature and sunlight intensity rise above some critical level, the organism grows, reproduces, and increases its population. Beyond this, imagine another organism, such as a snail, which is dependent on the algae as its principal food source. As the algae increases, the population of the snail also increases—given, of course, that all other factors necessary for its growth

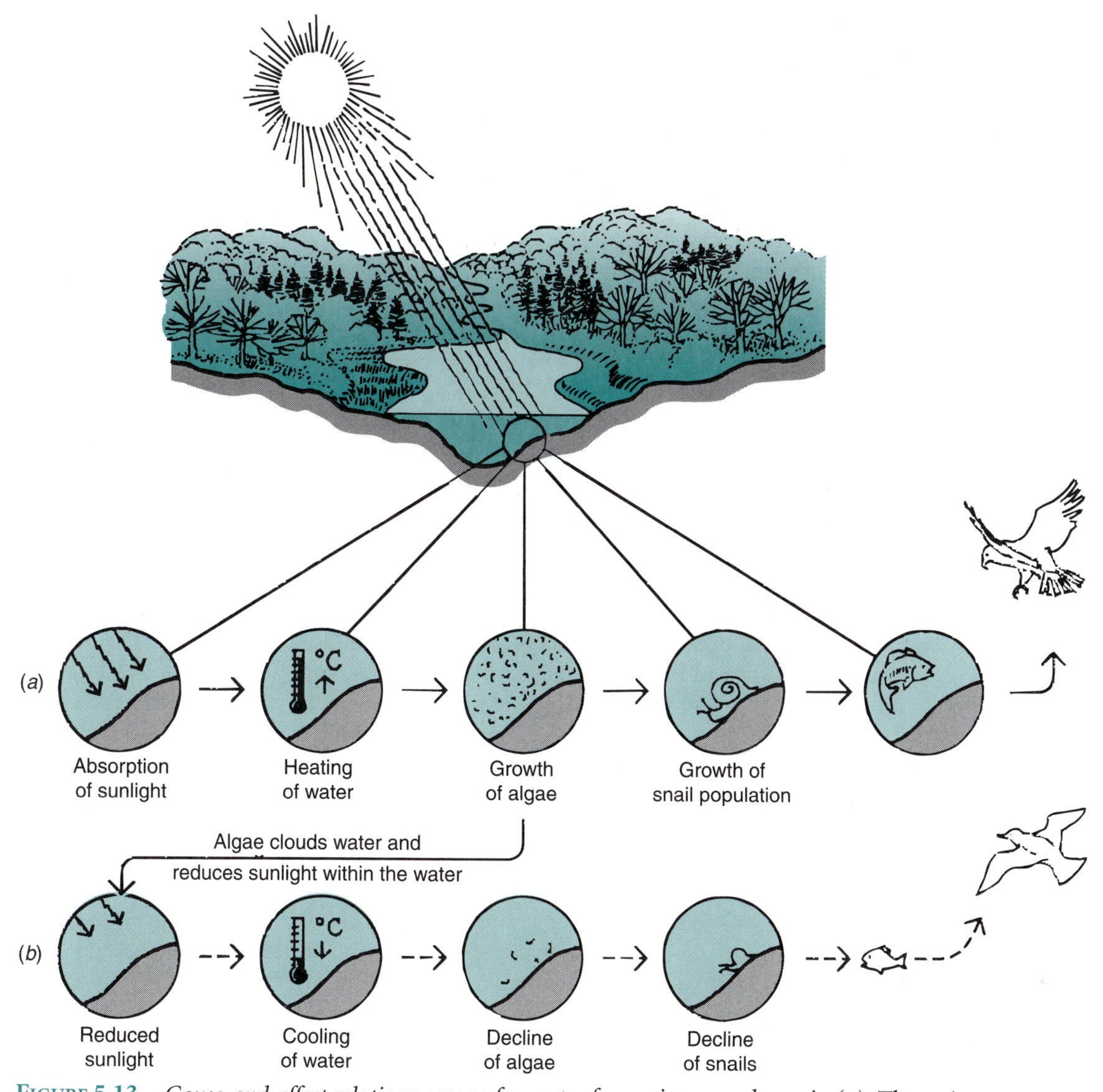

FIGURE 5.13 *Cause-and-effect relations among four sets of organisms are shown in (a). The entire system relies on algae, whose growth and production are dependent, among other things, on water temperature, and sunlight. In (b), the concept of feedback is illustrated in which the growth and emergence of algae block the penetration of sunlight into the water column.*

and reproduction are also met. The immediate and direct cause of the snail's change is the algae's change; therefore, the algae's change is the *cause* and the snail's change is the *effect.* But the root cause of change for both organisms was the change in water temperature and sunlight intensity; therefore, from a broader perspective you can argue that the population dynamics of both organisms are effects of the thermal and solar causes (Figure 5.13a).

If you add several other organisms to this sequence, such as fish feeding on snails and a predator such as an osprey feeding on the fish, it is easy to see a chain of cause-and-effect relationships emerging. In the environment such chains may involve dozens of steps, or phases, among organisms, among organisms and the physical environment, and among different parts of the physical environment. Where they involve organisms, such relationships are parts of ecosystems.

If we are to gain scientific understanding of the workings of the environment, we must be able to define and analyze the various chains of cause-and-effect relations, particularly the root causes of change sequences. In order to account for a change in the osprey population, for example, it may be necessary to understand, among other things, the conditions controlling the growth and abundance of algae.

In addition, it should be apparent that an organism such as an osprey, which is situated at the end of its food chain, is dependent on a long sequence of cause-and-effect relations. This is one of the reasons why such animals are ecologically vulnerable under even natural conditions and why, for example, the osprey, unlike the snail, has evolved the capacity to utilize several food sources.

Secondary Relationships

Feedback is another type of cause-and-effect relationship. For example, the algae, having undergone a population increase because of a rise in water temperature and sunlight intensity, can alter the water temperature and sunlight by physically blocking the penetration of sunlight into the water column. As it blooms and clouds the water, the algae reduces sunlight and water temperature, causing a decline in further algae production (Figure 5.13b). Thus, feedback is characterized by a reverse trend in the cause-and-effect sequence.

Feedback may be positive or negative, depending on whether the alteration caused even higher water temperatures and more sunlight (positive) or lower water temperatures and less sunlight (negative). Also, various organisms working individually or in concert could exercise a level of feedback that brings water temperature and sunlight to an optimum level for maintenance of the whole system. Under these circumstances, the feedback process functions as a mediator, or thermostat of sorts, never allowing light and water temperature to fall too low or rise too high to impair the survival of the system. In a single organism such balance is called **homeostasis**, or the tendency toward constancy.

On the other hand, the feedback can have disruptive rather than mediating effects. Take the example of humans attempting to improve and use dry lands through irrigation. Initially the application of water improves productivity and plants flourish. But in time excess water may raise the water table and salt content of the soils, rendering them toxic to most plants and less productive than they were before irrigation. Hence, positive feedback (more soil salt) has a negative result.

5.7 The Gaia Concept

The concept of mediation brings us to an idea called the **Gaia Concept**, a controversial concept named for the Greek goddess of Earth. The basic premise of this concept is that life, through the feedback process within environmental systems, is able to regulate the Earth environment at a life-sustaining level. The idea holds that Earth functions as an integrated whole and that life itself is capable of rendering broad adjustments in the global environment, particularly the lower atmosphere, to serve its own maintenance and survival. The atmosphere, it is argued, is an integral and necessary part of life, and that life acts as a thermostat on the temperature and composition of the atmosphere. Proponents of the Gaia concept cite the contribution of plants to the development of an oxygen-rich atmosphere and to the reduction in atmospheric carbon dioxide as evidence for life's ability to change and regulate the global environment to its own advantage. If Earth was lifeless, such as Mars, the atmosphere would have very little oxygen and instead would be dominated, as is the Martian atmosphere, by carbon dioxide.

The Gaia perspective is one of life as a coshaper of the global environment in terms of the atmospheric heat balance, the gaseous composition of the atmosphere and oceans, and the cycling of matter. This view contrasts with the more conventional view that life is dominated by the geophysical environment and is little more than planetary cargo. The life-as-cargo perspective implies that life is largely a product of the physical environment, and as such is more or less subject to whatever turns the environment happens to make. This includes fluctuations that originate from natural disturbances, such as volcanic activity, or human accidents, such as environmental pollution, that upset the balance of the environment and reduce its capability to support life.

The Gaia concept, by contrast, proposes that life is somehow able to maintain the environment for its own good. Life is capable of making the necessary adjustments to bring the global system into a life-sustaining balance. The chief spokesman and founder of the Gaia concept, James Lovelock, argues that we should look on Earth as an organism, not as a ball of rock moistened by water and sprinkled with organic matter. Lovelock calls this physiological perspective of Earth *geophysiology*, and though the mechanisms by which "organism Earth" regulates itself are not yet well understood, the system is supposed to act like the thermoregulatory system of the human body. There appears to be a grand purpose to the whole scheme, namely, the maintenance of life; therefore, little is left to chance.

Herein lies one of several major disagreements scientists have with the Gaia concept. Evidence is abundant of unexpected outside forces, nonbiological ones, shaping the environment and the development of life. For example, there is strong evidence that the massive extinctions that marked the end of the age of dinosaurs were caused by climatic cooling from an asteroid collision and explosion on Earth. Similarly, the global cooling leading to the last ice age was probably related to several geophysical factors, including variations in the Earth's orbit and volcanic activity. Conversely, there is no evidence that the return to warmer conditions about 10,000 years ago was brought on by biological factors. On balance, the Gaia concept seems to be a wonderful metaphor for Mother Earth that helps to bring into focus the important interplay between the biological and nonbiological parts of the environment. But the Earth system is huge and extraordinarily complex, and the mechanisms and the interrelationships between them that drive it toward equilibrium are not well enough understood by scientists to test the Gaia concept rigorously.

5.8 SUMMARY

Among the terrestrial planets, Earth appears to be unique for its life-nurturing environment. Earth receives moderate levels of solar radiation and maintains a hospitable thermal condition over much of its surface. Water is abundant as a liquid, and where it is available with ample supplies of solar radiation and heat, the potential for life is good.

The zone of life, the biosphere, forms a thin web over the land and water surfaces. Based on relative density, the biosphere is intermediate among the atmosphere, hydrosphere, and lithosphere, and this explains why the biosphere is concentrated on the Earth's surface at the interface between air, water, and rock. The biosphere is maintained by resources drawn from the adjoining spheres: heat and light from the atmosphere; water and nutrients from the hydrosphere; and nutrients from the lithosphere.

These resources vary greatly over the planet. Therefore, the life web also shows great geographic variation over both land and water. In the harsh environments such as the deserts and cold lands, as well as the vast interior of the oceans, the biosphere is thin and tenuous. It is thickest and strongest in the great forested landscapes, on the continental shelves, and in the littoral zone at the border between land and water. The distribution and composition of the life web is altered by human activity; for example, it is torn by war and urbanization and weakened by deforestation and desertification.

Earth's surface processes, including those of the biosphere, are powered almost entirely by solar energy. The flow of solar energy to the Earth's surface is first filtered and reduced appreciably by the atmosphere. That which is finally available is channeled along two paths: heat and organic. Most solar energy reaching the surface is converted into heat, which in turn drives the essential physical processes of the planet such as evaporation of water and the movement of winds and currents.

A small amount of the solar energy is also taken up in photosynthesis and converted into organic energy. This energy runs the Earth's life system as it is passed along the food chains through the various ecosystems. Ultimately most of the organic energy is converted to heat, and along with heat from the absorption of solar radiation, this energy is released into the atmosphere and then back into space.

In addition to energy, matter is necessary to life. Organic material is made up of molecules that are built from the atoms of certain elements. These elements are called nutrients, and they are grouped into two major classes: macronutrients and micronutrients. The nutrients are cycled through the global environment as gases or as sediments, and plants extract them from air, water, and soil and pass them to the food chains. Human activity is seriously altering several nutrient cycles. The most alarming change is taking place in the carbon cycle, where carbon dioxide, a heat-absorbing gas in the atmosphere, is increasing significantly with the burning of fossils fuels and wood worldwide. This poses a threat of global climatic change with potentially grave environmental impacts. Nitrogen and phosphorus, two important plant nutrients, are also on the rise in terrestrial and freshwater environments as a result of human activity.

Earth's environment is run by a complex of interlocking energy and material systems. The systems are made up of chains of cause-and-effect relations and the balance of environmental systems often depends on feedback within these chains. Feedback may be negative or positive, depending on whether it causes a decrease or increase in the originating factor. The Gaia concept proposes that through feedback processes life is able to regulate the global environment to serve its own maintenance and survival. This concept opposes the viewpoint that life is merely cargo on spaceship earth and argues that Earth functions as a great organism capable of adjusting its environment to ensure its own survival. The Gaia concept is an exciting metaphor for life and the Earth environment, but as a scientific concept it has several shortcomings, including the inability of life forces to regulate outside forces such as extraterrestrial ones.

5.9 KEY TERMS AND CONCEPTS

Terrestrial planets
Earth's surface environment
- radiation
- heat
- water

Lithosphere
- mantle
- core
- plate tectonics
- outgassing

Hydrosphere
- hydrologic cycle

Atmosphere
- troposphere
- stratosphere

Biosphere
- photic zone
- abyssal plains
- littoral zone
- ecosystems

Solar energy
- geothermal energy
- distribution of solar energy
- life-support potential

Earth energy systems
- heat
- organic
- hydrologic system
- geochemical (or nutrient) system
- atmospheric circulation system
- kinetic energy
- potential energy
- photosynthesis
- food chains

Matter
- macronutrients
- micronutrients
- nutrient cycles
- sedimentary cycle
- gaseous cycle
- carbon cycle
- air pollution
- oxygen cycle
- nitrogen cycle
- nitrogen fixation
- phosphorus cycle

Systems interrelationships
- primary
- secondary
- feedback
- homeostasis

Gaia concept
- geophysiology

5.10 QUESTIONS FOR REVIEW

1. As we explore the other planets in the solar system, Earth's uniqueness has come into clearer focus. In general, what makes the Earth's surface environmentally distinctive and conducive to the support of life?
2. What are the four great divisions or spheres of the Earth's environment, and what is meant when we say they are arranged in a gravitationally stable configuration?
3. Describe the general (global) distribution of the biosphere over land and water. Where is it strongest (densest) and weakest (thinnest)? What are the environmental conditions associated with dense and thin zones, and is the distribution of the biosphere affected much by human action?
4. What are the basic types of ecosystems? What are the essential resources that ecosystems draw from the environment?
5. What is the primary source of energy that powers the Earth's surface processes? Of the solar energy that enters the atmosphere, what percentage is available to Earth environments? What happens to the rest of it? How is solar energy distributed by latitude?
6. Considering the global distribution of both solar energy and water, which geographic zones are most and least limited in terms of life support? What zones are seasonally limited?
7. What is meant by the concept of energy paths? What are the two main energy paths fed by solar radiation, and what systems are driven by this energy?
8. What are the major elements making up living matter? What are macronutrients, micronutrients, and nutrient cycles?
9. Describe the carbon dioxide cycle, the essential processes involved, and how human activity is altering it.
10. What processes are responsible for maintaining the atmosphere's oxygen balance? Why is it that the oxygen cycle is not the source of concern that the carbon dioxide cycle is?
11. Briefly describe the nitrogen and phosphorus cycles. Where is the major pool of each, and how are human actions affecting these cycles?
12. The balance of Earth systems is maintained by two types of relationships: primary and secondary. Describe these relationships and provide a simple example of how each works.
13. What is the Gaia concept and the concept of geophysiology? What are some of the arguments for and against Gaia?

6

ECOSYSTEMS AND THE BIOCLIMATIC ENVIRONMENT

6.1 INTRODUCTION

Although the Earth is not the center of the solar system as the ancients thought, it is clearly unique among the planets in the solar system. It appears, however, that it did not start out as a unique planet. Earth and the neighboring terrestrial (solid) planets appear to have been formed by the same processes and of the same materials at the beginning of the solar system 5 to 4.5 billion years ago. They all developed hard surfaces that were subject to volcanism, folding, faulting, and impacts by meteorites. Their atmospheres were heavy in carbon dioxide and most probably contained water. But with the evolution of life, Earth developed along a different line than Mars, Mercury, and Venus.

Life began on Earth within the first billion years of the planet's formation. For 3 billion years or so it remained concentrated in the sea and then, about 500 million years ago, life emerged onto the land. Simple plants were the first terrestrial organisms, and in the next 100 million years, they evolved into thousands of species with large statures and circumglobal distributions. Terrestrial animals also evolved, but measured in terms of total mass, plants remain the dominant group of organisms on the planet. Measured in terms of numbers of species, however, insects are clearly the dominant organisms on Earth, numbering in the millions of species. The rest of us—the reptiles, the birds, and the mammals—represent the smallest group in terms of total species, total numbers, and total mass.

Although we are recognized as different classes of organisms with different evolutionary histories, we are all tied together in ecosystems. If humans are to understand our role in the great scheme of life on Earth and the role other organisms play in our lives, we must understand the makeup, operation, and effects of ecosystems. In this chapter we are interested in learning how ecosystems are organized and how they function in the environment. We also want to learn about the biogeographical aspects of ecosystems, that is, how they are distributed over land and water and what controls their distributions and operations. Finally, we need to examine some of the alterations of ecosystems at the hands of humans.

6.2 BASIC FORM AND FUNCTION OF AN ECOSYSTEM

The basic concept of ecosystems as energy systems is easy enough to understand. We see examples of them around us every day as in our basic food/life system. All ecosystems are made up of **food chains** that begin with energy, principally sunlight, extracted from the physical environment and converted into organic matter by plants. Animals that eat plants synthesize a portion of the plant material in their own bodies. In turn, the flesh of the plant-eating animals provides nutrition and energy when eaten by other animals. Thus, energy is passed along, and the route it takes from one organism to another defines a food chain.

Organization of Food Chains

Each food chain is organized into different segments or levels. Each level is defined by the point of energy transfer from the environment to an organism and then from that organism to another (Figure 6.1). All food chains are organized in basically the same fashion, with three or four levels of energy transfer called **trophic levels** that represent the basic framework for all ecosystems. Each trophic level represents a particular function in the ecosystem.

1. Plants are the **producers** of organic energy. This is followed by a series of **consumers**.
2. Herbivores, or *primary consumers*, make up the second level. These include cattle, buffalo, squirrels, and other plant-eating animals.
3. Carnivores, or *secondary consumers*, such as wolves, lions, and weasels, are the third level. These are meat eaters.
4. Specialized carnivores, or *tertiary consumers*, are the fourth level. Hawks, eagles, and sharks are among this group that eats other carnivores.

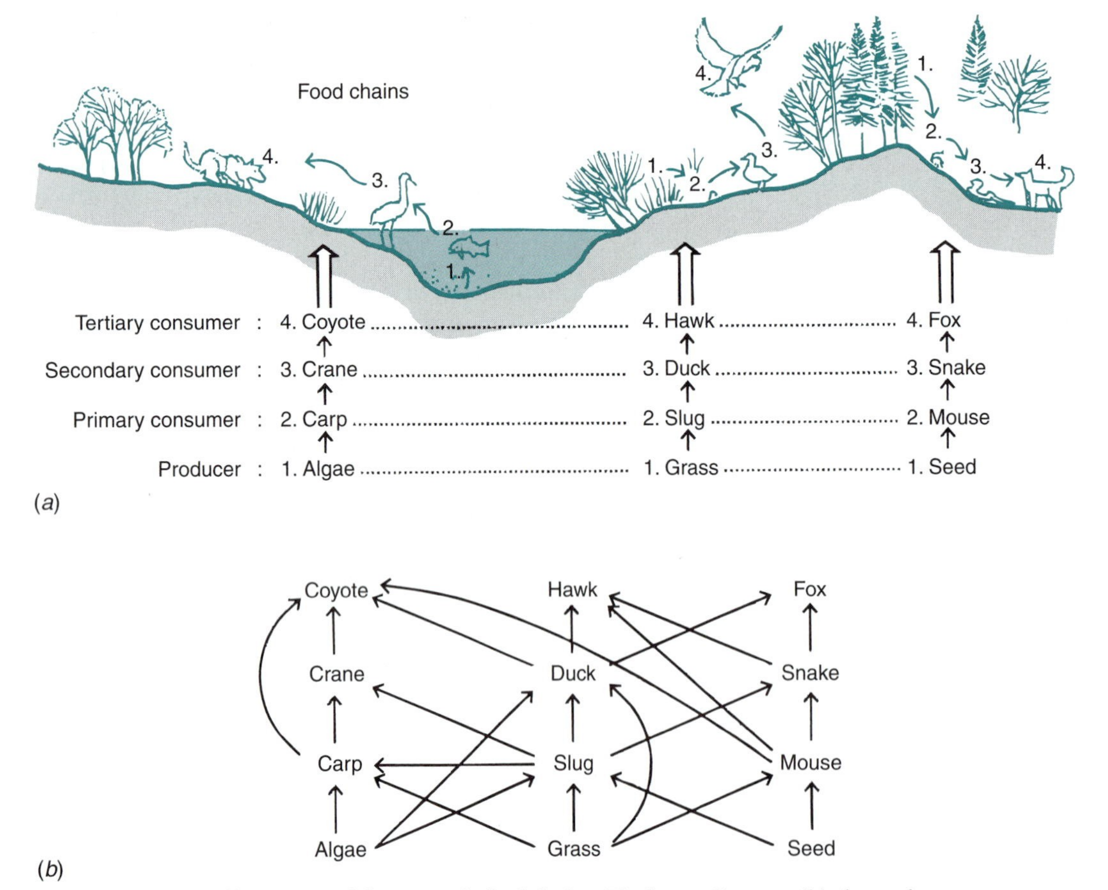

FIGURE 6.1 *(a) Illustration of three simple food chains. The lower diagram (b) shows the interconnections among the food chains that form food webs.*

Omnivores such as humans, raccoons, and black bears function as both carnivores and herbivores. Figure 6.1 shows several examples of food chains, as well as a set of integrated food chains, called a **food web**.

Energy Flow

In addition to trophic levels, all ecosystems share two other distinctive traits: (1) the presence of **detritivores**, which are organisms that live on the refuse of ecosystems. Detritivores include *scavengers* such as vultures and crabs that eat the remains of dead animals and *decomposers* such as fungi and bacteria that break down plant debris, animal droppings and other dead organic matter; and (2) **energy attenuation** among the trophic levels; that is, the massive loss or release of energy because of organism respiration. **Respiration** represents the maintenance processes of plants and animals. In most ecosystems it accounts for a discharge of 90 percent or more of the energy at each trophic level. The energy discharged is mainly *heat*, which goes into the atmosphere, as we noted in the last chapter. Taking the attenuation factor into account, it is apparent why the energy flow in ecosystems is often characterized as an **energy pyramid** in which the tiers, represented by the trophic levels, shrink rapidly toward the apex of the pyramid (Figure 6.2).

Let us review our thinking to this point. Ecosystems are energy systems comprised of many food chains that link together a vast number of organisms with the physical environment. Therefore, all ecosystems incorporate both living organisms, or a **biotic** component, and a nonliving, or **abiotic** component such as sunlight, soil, heat, air, and water. The biotic component is made up of three groups of organisms: producers (plants), consumers, and detritivores. Ecosystems are organized into trophic levels and energy is passed from one trophic level to the next. Most of the energy at each level is given up as heat in respiration; therefore, the energy budget of an ecosystem dwindles rapidly with each energy transfer. Of the energy that is retained as tissue, which amounts to about 10 percent of that received at each level, some is passed onto the next trophic level and some falls to the ground as organic debris to feed the detritivores. Because of this rapid attenuation of energy, the available energy (that is, energy usable to organisms) at higher trophic levels is very small.

Where do humans fit into this scheme? Humans are omnivores, functioning as primary, secondary, and tertiary consumers. As meat eaters, humans sacrifice a great deal of food energy. For example, most of the calories in grains that could be eaten directly by people are fed to primary consumers such as cattle. Not only is most of the energy taken in by a cow lost in maintaining the animal's metabolism, but a large proportion of the animal itself is not edible and is discarded. But with people as plant eaters, human ecosystems are more efficient, even though plants as a whole provide fewer calories and less protein per ounce than meat. In a manner of speaking, when we eat plants the middle man (the cow) is eliminated and we save the energy it would consume.

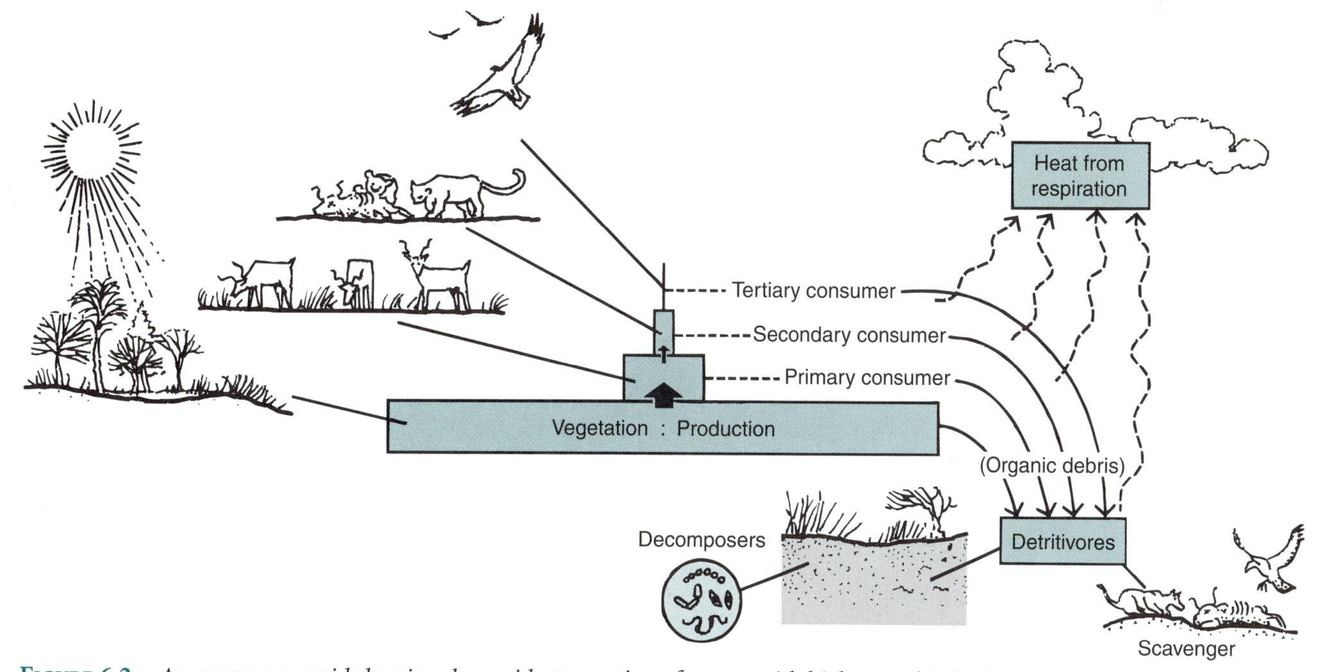

FIGURE 6.2 *An energy pyramid showing the rapid attenuation of energy with higher trophic levels, the release of heat energy, and the place of detritivores in the system.*

Can this concept be extended to the capacity of the Earth to support human population? The United Nations estimated that with food growing and processing technologies of the 1990s Earth can satisfactorily nourish about 5.5 billion healthy people if they all lived on a vegetarian diet. However, if 15 percent of the world diet relied on animal products, Earth's sustainable carrying capacity for a healthy human population would be about 3.7 billion. If people derived 25 percent of the calories from animals, Earth could support only 2.8 billion people at a healthy level. Although these figures may be debatable, they starkly illustrate the principle of energy attenuation in ecosystems and underscore the safety of living at the lowest and largest possible trophic level in the food system.

6.3 PLANT PRODUCTIVITY: FEEDING ECOSYSTEMS

The most fundamental process of ecosystems involves plants capturing solar energy and converting it into forms usable to themselves and to other organisms. Living plants are the only organisms on the planet capable of performing this process. As such, they are the gateway to the sun's energy for all creatures, from the smallest detritivore to the largest herbivore or carnivore. **Photosynthesis** is the specific process by which plants take in solar radiation, combine it with carbon dioxide, water, and heat, and convert it into chemical energy. This chemical energy takes the form of organic compounds made up of sugars and carbohydrates that go to the cells in leaves, flowers, wood, bark, seeds, and other parts.

As with any organism, plants need energy to maintain themselves. Therefore, part of the organic energy manufactured in photosynthesis is expended within the plant in respiration. Ultimately, the energy used in respiration is converted to heat and slowly released to the atmosphere. That which is left over, called **net photosynthesis**, goes into the building of new plant cells. These new cells represent energy added to the ecosystem. We measure it in grams of organic matter added to the landscape, a lake bed, or the ocean floor per year and call it *net primary production*, or *plant productivity*.

The percentage of solar energy that plants are able to convert into organic energy through photosynthesis is termed **efficiency**. Despite the fact that plants physically absorb a great deal of solar radiation, relatively little sunlight, less than 1 percent on the average, is actually taken in for photosynthesis. Most solar radiation absorbed by

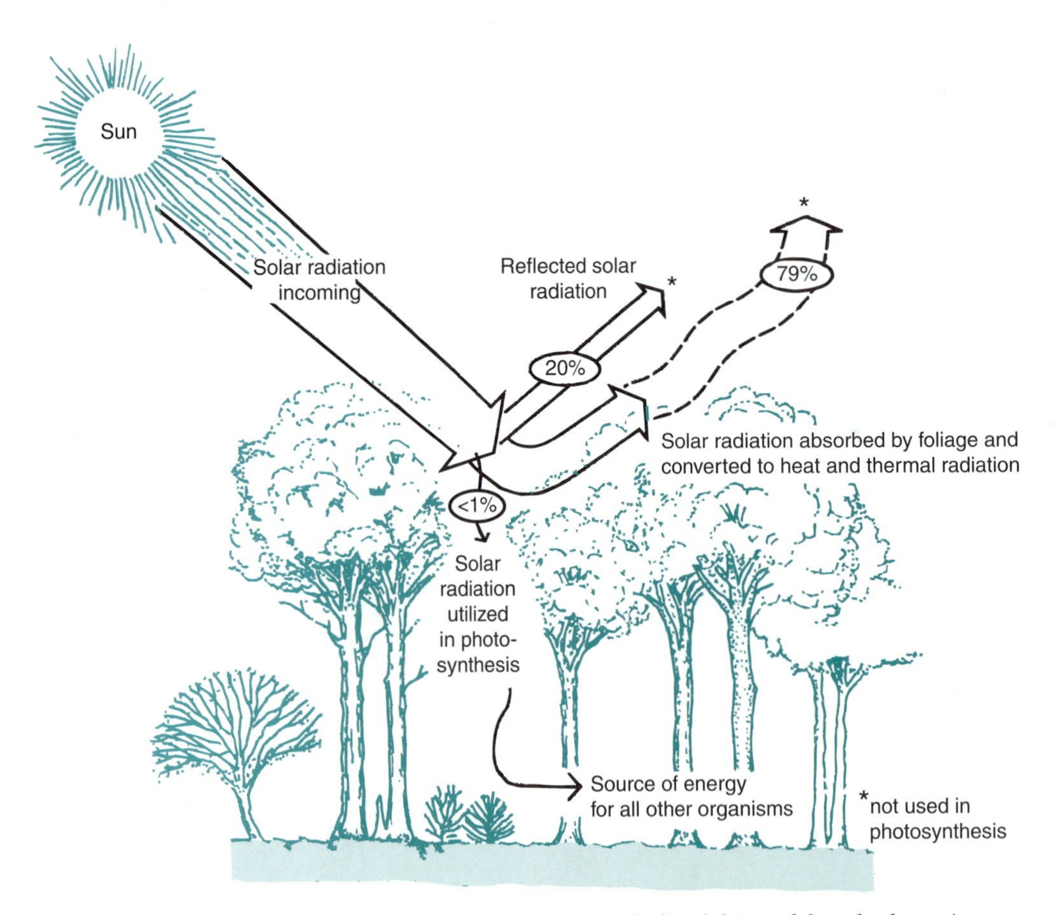

FIGURE 6.3 *The receipt of solar energy by vegetation with the breakdown of the solar beam into heat, radiation, and organic compounds.*

foliage is converted into heat on the surface of leaves and immediately returned to the atmosphere, bypassing the photosynthesis energy path (Figure 6.3). Thus, the pathway from solar radiation to the production of organic energy by the plant operates through a very tiny window. As we shall see, it is influenced by many other factors besides solar radiation.

Limiting Factors

Productivity is an important measure of an ecosystem's vitality. Productivity depends on two sets of factors: (1) the **limitations** on the availability of basic resources for photosynthesis, and (2) **disturbances** imposed by the external environment. To maximize photosynthesis, five basic ingredients (light, heat, water, carbon dioxide, and nutrients) must all be available to the plant at certain levels. If any one ingredient is in short supply, the level of photosynthesis is limited by that one quantity. This defines a basic biological rule called the **principle of limiting factors**.

For a local example of the limiting factors principle, let us look at a freshwater lake or pond. Aquatic plant growth in summer, when sunlight and heat are plentiful, is often limited by one nutrient, phosphorus. Where phosphorus is in short supply not much growth takes place; however, when it is introduced from pollution sources, growth accelerates and a water body may become overgrown and choked with weeds in a matter of years. Recognizing this principle, efforts to manage lakes and ponds are often aimed at one or two critical ingredients rather than many, and phosphorus is usually the prime target. The main sources of phosphorus are agricultural fertilizers, urban stormwater, sewage, and atmospheric pollution.

An excessive quantity of a basic ingredient, such as extreme heat or pollution, can also be a limiting factor. Within the range between the lower and upper (excessive) limits lie the **optimum conditions** for an organism. The actual levels at which the organism's life processes are arrested are called **tolerance thresholds**. Tolerance thresholds may be *acute*, such as extreme low temperatures brought on by a sudden change in weather; or *chronic*, such as the gradual buildup of toxic substances in the living cells of plants and animals in polluted environments.

External factors that influence productivity are those that the plant does not depend on for photosynthesis. These disturbances often take the form of events such as fires, floods, diseases, droughts, storms, land use, and pollution. These processes either disrupt the flow of basic resources to the plant, as in the loss of nutrients when soil is eroded, or physically damage the plant and reduce its capacity for photosynthesis, as when foliage is stripped by a

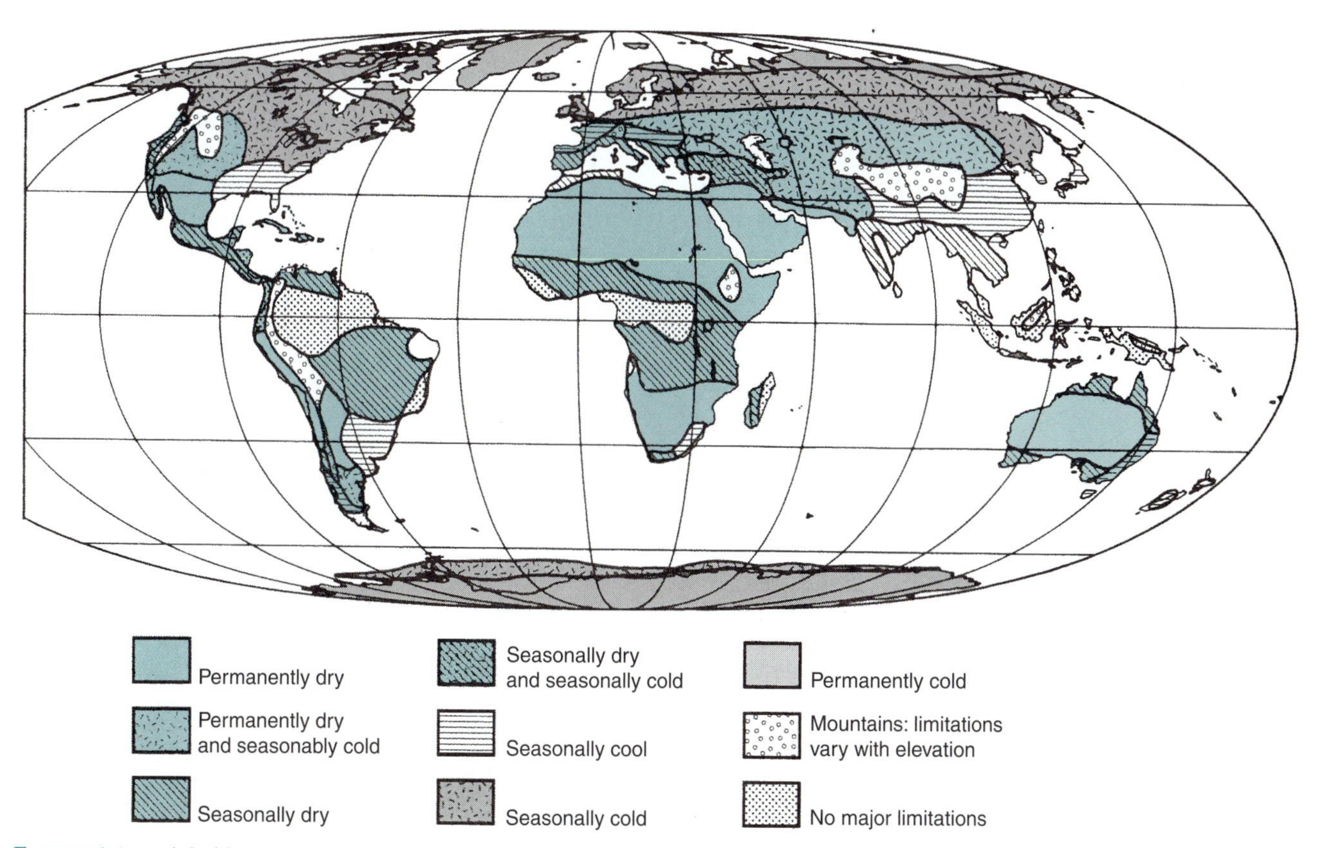

FIGURE 6.4 *Global limitations on plant productivity related to climate.*

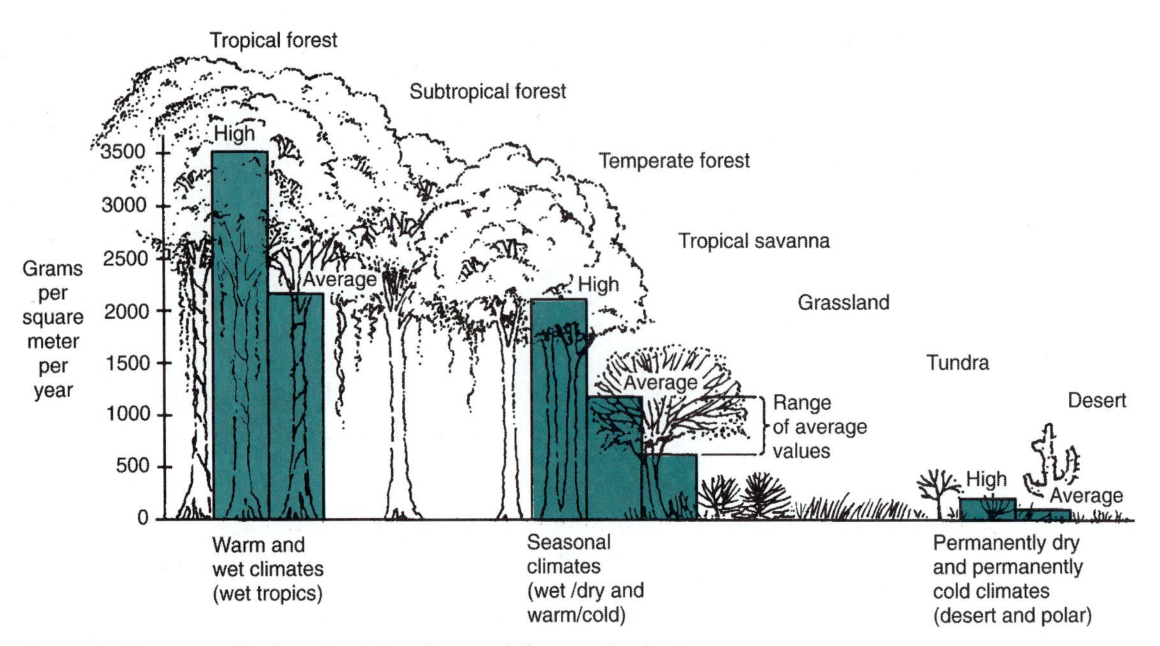

FIGURE 6.5 *Generalized productivity of terrestrial vegetation in grams per square meter per year.*

powerful storm. When we examine both the limitations from external disturbances and the limitations from the availability of resources, we can see why certain environments are easier places for plants and thus favor much higher levels of productivity than others. As we shall see later, human activity has become a major source of disturbance to ecosystems over broad regions of the planet.

Productivity and Climatic Limitations

The worldwide patterns of **climatic limitations** on productivity are shown in Figure 6.4 They fall into three broad categories: (1) environments with no limitations, (2) environments with seasonal limitations, and (3) environments with permanent (year around) limitations. Stable environments with a dependable flow of resources are most favorable to high productivity rates. On the land, none is more favorable than the wet tropics, home of the tropical rainforests. Rainfall and heat are abundant in all months, sunlight varies little with the seasons, and until humans entered the picture, external disturbances such as geologic upheavals and drastic climatic changes, broadly speaking, seem to have been less severe here than in most other climatic zones. The productivity rates of the tropical forest bear this out. At an annual average of 2,200 grams of organic output per square meter of surface (4 pounds per square yard), the tropical forests have the highest productivity of any large terrestial ecosystem on Earth (Figure 6.5).

Beyond the wet tropics, all the remaining terrestial environments on Earth pose limitations of various types and magnitudes related principally to climate. Many climates are seasonally limited with insufficient heat, water, or light over several months of the year. The midlatitude climates in the center of North America and Eurasia, for example, are limited by both heat and light in the winter, especially heat. The tropical wet/dry (savanna) climate, which borders the tropical forest, and the Mediterranean climate of southern Europe and southern California, are limited by a rainless season. Annual productivity rates in the seasonally limited zones are markedly less than those of the tropical forests, between 600 and 1,200 grams per square meter (1.1–2.2 pounds per square yard) (Figure 6.5).

Three climatic zones are consistently limited year around: arid, arctic/polar, and high mountains. These are Earth's harshest environments. In the latter two, the warmest month of the year has a mean temperature below freezing, and of course, only small numbers of selected small plants can grow there. Deserts are water-limited in all seasons, and the plants there are small, scarce, and slow-growing. The productivity rates for these three climatic zones are very low, less than 5 percent that of the tropical forests, ranging from 5 grams to 90 grams per square meter (0.01–0.2 pounds per square yard) a year (Figure 6.5).

Productivity also varies widely in aquatic environments. Over the broad expanses of the deep oceans average productivity is low, less than 250 grams per square meter per year, which is about one-tenth the tropical forest rate. Within this vast area, which covers nearly 90 percent of the ocean basins, there are broad regional variations with notably higher rates in tropical latitudes and lower ones in polar latitudes. In the shallow water of the continental shelves, productivity rises dramatically, reaching the highest rates in lagoons and wetlands along the shore. Locally annual productivity in nutrient-rich coastal waters may be as great as 6,000 grams per square meter (11 pounds per square yard). We will say more about the oceans later in this chapter, including the influence of ocean currents on ecosystems.

Global Productivity and Human Needs and Impacts

Each year the world's vegetation is capable of producing 225 billion metric tons of organic matter as net primary production. Of this quantity, land plants produce 135 billion tons and ocean plants about 90 billion metric tons. How much of these quantities are consumed by humans? Direct consumption for food, firewood, lumber, and livestock feed is relatively modest: about 4 percent (6.1 billion tons) of the terrestrial productivity and 2 percent (2.2 billion tons) of the aquatic productivity. But this is not the whole picture.

Humans also account for large indirect losses of productivity because of damage and reduction of ecosystems related to land clearing, overgrazing, soil erosion, wetland eradication, fires, and pollution. This quantity is difficult to estimate, but some scientists place the loss in productivity as high as 30 to 40 percent for the Earth's terrestrial ecosystems. The loss for saltwater ecosystems, related mainly to pollution and sedimentation of coastal waters, is significantly less, probably under 10 percent.

As the human population grows and approaches twice its current number sometime in this century, direct consumption of global productivity will surely double. On land it is likely to reach 8 to 10 percent. More important, however, is the question of indirect loss in productivity. Will losses to terrestrial ecosystems reach 60 to 80 percent? And if so, will such heavy losses in productivity cut into the capacity of ecosystems to provide the necessary levels of food, animal feed, fuel, and lumber to support a reasonable quality of life? If we take just one indirect loss, soil erosion, as an indicator, the answer is yes. Soil erosion of cropland and grazing land is increasing worldwide and there is no reasonable prospect for a reversal in this trend for several decades (see Chapter 15 for details on soil erosion).

6.4 TYPES OF ECOSYSTEMS AND THEIR CHARACTERISTICS

As we noted in Chapter 5, there are three basic classes of ecosystems: Terrestrial, saltwater, and freshwater. Ranked in terms of productivity, terrestrial ecosystems are the most active, accounting for about 60 percent of the Earth's total photosynthesis. Saltwater ecosystems, which occupy about 70 percent of the Earth's surface area, rank second, accounting for about 39 percent of the Earth's total annual photosynthesis. Freshwater ecosystems account for the remaining 1 percent or so.

Terrestrial Ecosystems

Each class of ecosystems is made up of many different types of individual ecosystems or groups of ecosystems. The terrestrial class is subdivided into five great groups based on the general structure and composition of the vegetation: forest, savanna, grassland, desert, and tundra. Each of these is differentiated geographically according to moisture and temperature into biomes. A **biome** is a large biogeographical unit characterized by a particular combination of vegetation and animals whose distribution is associated with a general climatic type. You can also think of a biome as a complex of ecosystems with broadly similar biogeographical characteristics over the region occupied (see Figure 6.8).

The Earth's principal biomes in terms of geographic coverage, productivity, and species diversity are the forest biomes. These biomes are made up of many different ecosystems, and today forests of all types cover about 30 percent of the Earth's land area and account for about 67 percent of all terrestrial productivity. Among the forest biomes, tropical rainforests head the list; they have the greatest mass of living matter (biomass), the largest numbers of species, and the highest productivity. The coverage of tropical rainforest is estimated at about 5 percent of the Earth's land area and tropical rainforest productivity totals about one-third that of all the terrestrial vegetation on Earth. (Tropical forests, of which rainforests are a part, make up about 40 percent of the Earth's forest cover.)

Other forest biomes include temperate or midlatitude forests, which originally covered broad belts of land in the midlatitudes of Europe, Asia, and North America but have been largely destroyed by lumbering and agricultural and urban development. Among these are relatively limited areas of marine west coast forest, which lie in the moist coastal climates of the midlatitudes and are the only forest ecosystem to rival the tropical rainforests in physical stature and productivity. Virtually all of today's temperate forests are reestablished (second-growth) forests. With the exception of the marine west coast forests, temperate forests are intermediate in height, biomass, and productivity among the forest ecosystems. On their northern border the temperate forests give way to the boreal forests, which stretch across North America and Eurasia in a great subarctic belt. Limited by the long, harsh subarctic winters, the boreal forests are the smallest and least productive of the large forest biomes. However, in terms of geographic coverage, they are the largest continuous forest biome in the world.

The most severely limited ecosystems are found in the desert and tundra biomes, where moisture and heat are the major controls and productivity, species numbers, and biomass are very low. The grassland ecosystems are somewhat intermediate; moisture is the key limiting factor, but it is available in the soil in quantities sufficient to nurture grasses for several months a year. The savanna biomes are also intermediate, generally characterized by a mix of tropical grasslands and forests. These ecosystems, which lie on the margins of the tropical forests, are the products of a strongly seasonal rainfall regime that nurtures vegetation with abundant summer rain and severely limits it with winter drought coupled with warm temperatures (Figure 6.6).

Compositional Traits

Although productivity and climatic setting are important features of ecosystems, several other traits also need our attention. Three important ones are biological organization, species diversity, and biomass. *Biological organization* refers to the order or arrangement of organisms in an ecosystem. **Species diversity** refers to the total number of species occupying a given area, and **biomass** refers to the total weight of living matter in that area. At the simplest level of biological organization is the individual **organism,** which may be any form of living matter.

Organisms of a single **species** (those individuals that can freely interbreed and reproduce) form a population. The population of every species occupies a particular geographic area, called its **range** (see Figure 16.2). The North American bison, for example, had a range originally extending over most of the grasslands of the prairies, plains, and forest margins before it was hunted to near extinction in the 1800s. A number of different species with overlapping ranges that live together in an interdependent fashion is called a **community.**

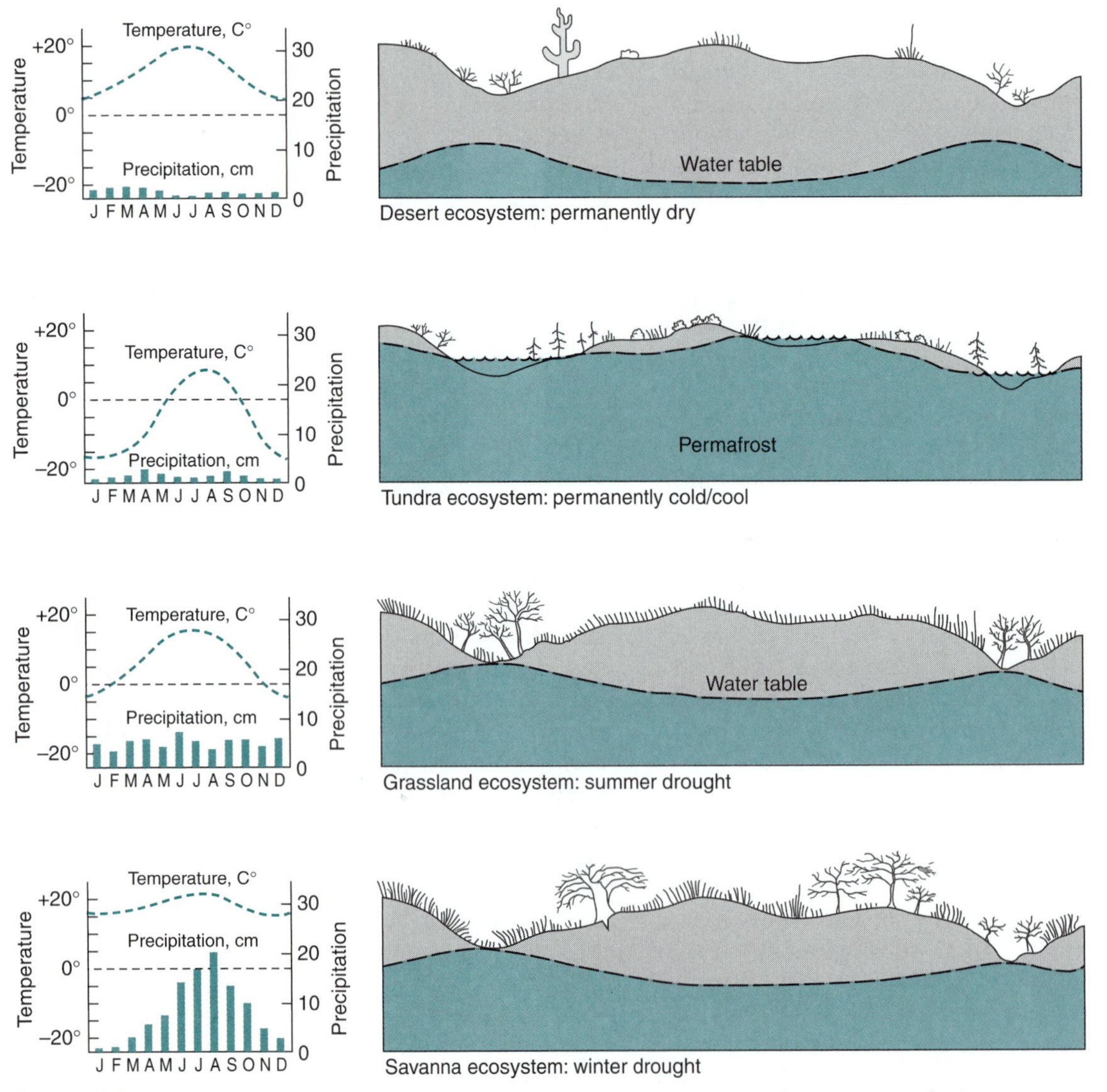

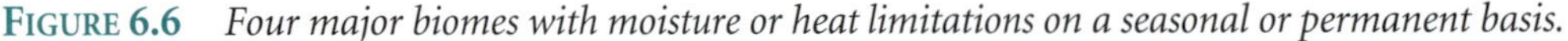
FIGURE 6.6 *Four major biomes with moisture or heat limitations on a seasonal or permanent basis.*

A simple example of a community is illustrated by the bison, the prairie grasses, and the buffalo-hunting Indian tribes such as the Comanches. The U.S. government recognized this relationship, and when the military struggled to bring Comanches under control in the 1860s and 1870s, it encouraged eradication of the bison by buffalo hunters (see Figure 16.4). By the 1880s the Comanches were living on reservations. Most ecological communities are far more complex, involving hundreds of species ranging from invisible microorganisms to the great macroorganisms such as grass, trees, fish, and mammals.

Some Interrelations within Communities

There are various forms of cooperation, or symbiosis, among species in a community. **Symbiosis** is characterized by an intimate association between dissimilar species. It may result in benefits for both species, called **mutualism**, as in the case of fungi and algae, which together form lichens. The algae produce carbohydrates used by both fungi and algae and the fungi provide moisture for the algae. Other forms of symbiosis include **commensalism**, in which one species benefits and the other is neither benefited nor harmed, and **parasitism**, in which one is benefited and the other harmed, a common trait of many tropical diseases in humans. **Predation**, though not cooperative, is another form of species interaction that ties species such as lions and wildebeests together in communities. Many predatory and nonpredatory species also practice various forms of intraspecies cooperation involving organized social structure within groups. Wolves, ants, and bees are outstanding examples.

Biogeographical Trends

Some communities, such as an aquatic community in a freshwater pond, form relatively discrete geographic entities and can be identified in the landscape as individual ecosystems (Figure 6.7a). Most communities are not physically discrete, however, and overlap and interlock with other communities to form large, continuous ecosystems occupying vast areas. Temperate forests are a good example with beech/maple communities, oak/hickory communities, and conifer communities merging with each other to form relatively continuous and extensive ecosystems stretching over vast areas (Figure 6.7b).

The species composition and diversity of communities and ecosystems differ enormously over the Earth. Species diversity follows a basic principle of **biogeography**: The harsher the environment, the fewer the species and the larger the populations of individual species. Accordingly, diversity is greatest in the tropics, especially in the tropical rainforest, and least in the harsh climatic zones such as the arctic and the deserts (Table 6.1). In the tropical rainforest, species number in the tens of thousands per square kilometer. Most are small organisms, principally insects, invertebrates, and microorganisms.

TABLE 6.1

Species Diversity and Latitude

	LATITUDE			
Organisms	27°N Florida	42°N Massachusetts	54°N Labrador	70°N Baffin Is.
Beetles	4,000	2,000	164	90
Land Snails	250	100	25	0
Reptiles	107	21	5	0
Amphibians	50	21	17	0
Flowering Plants	2,500	1,650	390	218

(*Source:* Clark, G. L. 1954. *Elements of Ecology.* New York: John Wiley.)

Beetles are the most abundant insects. They function as detritivores, primary consumers, and secondary consumers. Microorganisms, which include algae, bacteria, and fungi, function mainly as **decomposers**, breaking down organic litter on the forest floor. The best estimates of science place 50 percent or more of the Earth's species in the tropical rainforests. The vast majority of these, as we discuss in Chapter 16, have not been discovered and recorded.

Biomass follows the same geographic trend: The harsher the environment, the smaller the mass of living matter. The biomasses of tropical rainforests with massive trees, vines, and a plethora of ferns, orchids, and other nonwoody plants average more than 250 metric tons per hectare[1] (220,000 pounds per acre). Temperate forests have smaller biomasses and boreal forests have the lowest biomass among the Earth's major forests. The very lowest biomasses are, as expected, found in the harshest environments: the desert, arctic, and high mountains.

In global perspective, the trend is clear: Poleward from the tropical rainforest, species diversity and biomass decline. The decline in diversity, however, is offset by larger populations of individual species. Among the forest ecosystems, for example, the boreal forest occupies the harshest environment in the subarctic climatic zone. The diversity of tree species here is very low, usually not more than ten species or so per square kilometer. By contrast, tree species diversity in the tropical rainforest is several hundred or more per square kilometer. In the boreal forest, however, the populations of the individual trees are massive, with one or two species, such as spruce and tamarack, often forming continuous covers over tens of square kilometers of land (Figure 6.7c).

[1]Biomass is a measure of living matter in an area based on its dry weight. For trees, living (wet) weight would be 20 to 30 percent heavier than dry weight.

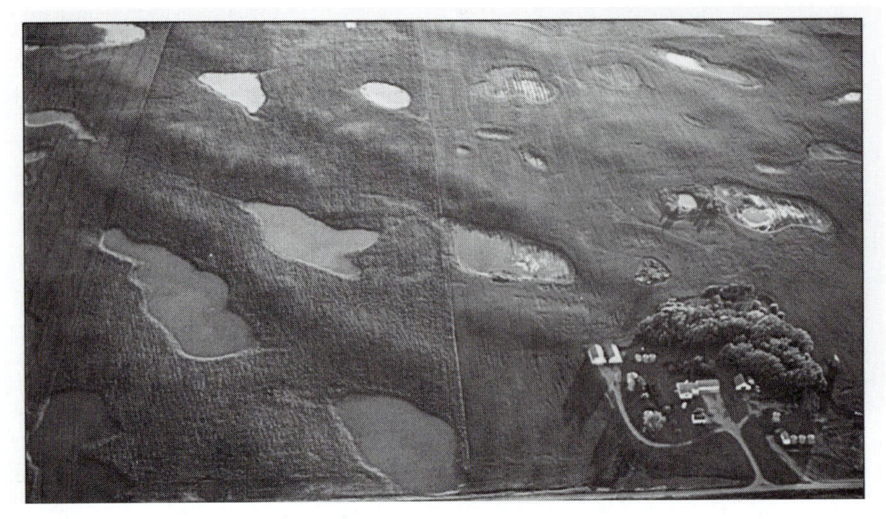

(*a*)

(*b*)

(*c*)

FIGURE 6.7 *Examples of geographical isolation and scale in plant communities: (a) prairie potholes, (b) mixed temperate forest, and (c) boreal forest.*

6.5 Ecosystem Patterns and Distributions

There are many controls on the distributions of ecosystems over land and water. Prior to the advent of agriculture, humans appear to have had little influence on terrestrial ecosystems and their distributions. The controlling factors were climate, drainage, soil conditions, and large-scale destructive forces such as continental glaciers, fluctuating sea levels, storms, and fires. Today there is little doubt that *both* nature and humans shape the patterns of ecosystems.

The magnitude of human influences, however, is highly variable from place to place. In this century it has been greatest in the midlatitude and subtropical land areas where humans are found in greatest numbers and exercise the most intensive land uses. In the oceans humans have had less influence on the distributions of the major ecosystems; however, through pollution, fishing, and hunting humans have certainly influenced the composition and productivity of saltwater ecosystems.

Global Terrestrial Patterns

At the broadest scale, climate is the chief control on both terrestrial and aquatic ecosystems. The terrestrial biomes are distributed in great belts that generally correspond to global climatic zones (Figure 6.8). Most of these belts trend east–west, but along windward coasts and adjacent mountain ranges some have a distinct north–south orientation. Where major mountain chains such as the Andes and the Rockies extend north–south across climatic zones, they tend to depress biomes equatorward at higher altitudes.

In some regions, the fit between climate and biomes is not a good one, because in the past several thousand years humans have significantly altered the size and patterns of certain biomes. The terrestrial biomes most seriously altered are the temperate forests in China, Europe, and much of North America, where large sections have been replaced by cropland and urban development; the tropical forests currently being cleared for cropland, grazing land, and lumber; and the grasslands where overgrazing and erosion are advancing desertification. Least altered are the tundra and the boreal forest. Beyond the terrestrial environments, freshwater ecosystems also exhibit extensive alteration, especially wetlands, which have been reduced dramatically by agriculture throughout the world.

Global Marine Patterns

In the oceans the picture is somewhat different. The distribution of ecological activity here is only broadly zonal, with a wide belt of high productivity in tropical latitudes and lower productivity toward the poles (Figure 6.9). But

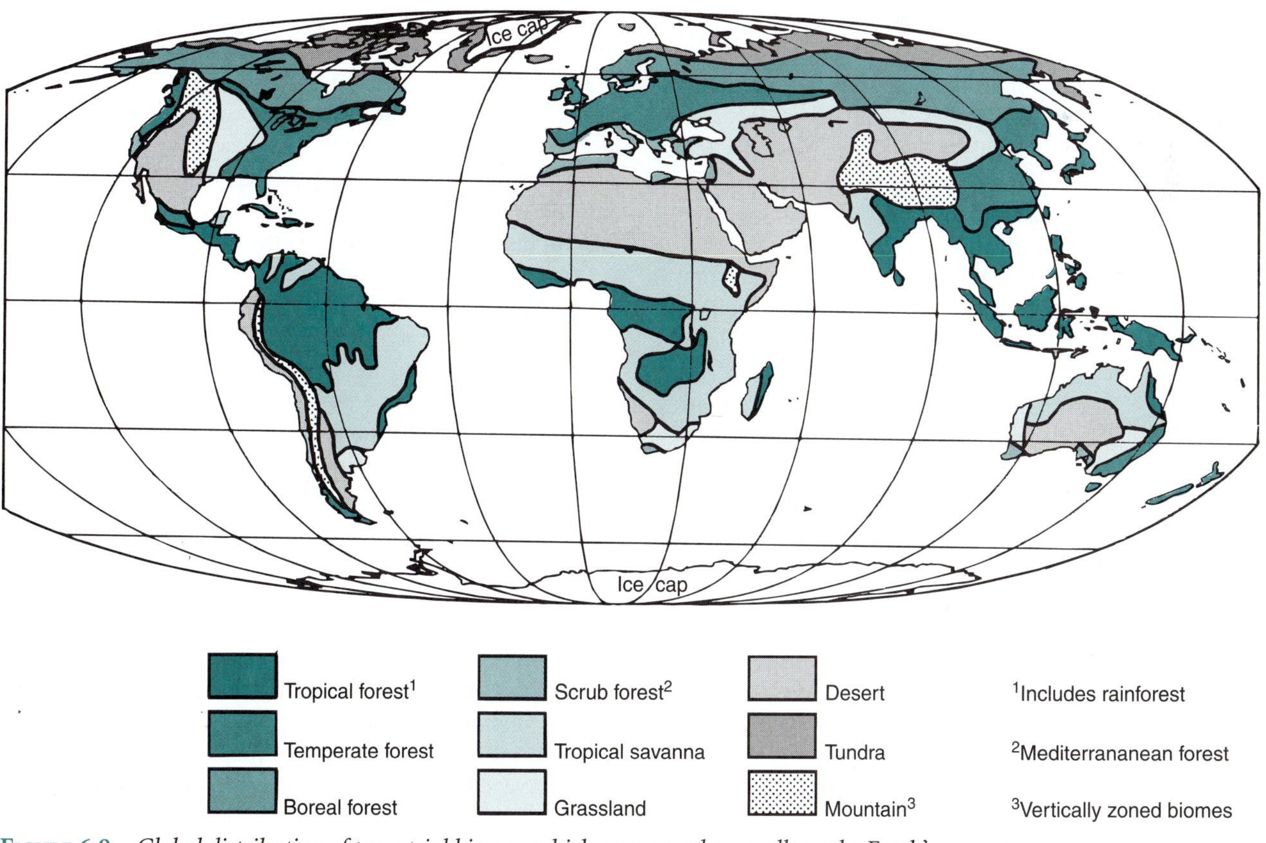

Figure 6.8 *Global distribution of terrestrial biomes, which correspond generally to the Earth's climatic zones.*

FIGURE 6.9 *Global patterns of ecological activity as revealed by satellite imagery. The light areas represent low levels of activity.*

within this framework there is significant variation related to water depth and ocean circulation. Productivity, biomass, and species diversity are greatest in the shallow waters over the continental shelves. Nearer shore, in the relatively quiet waters of bays, lagoons, and wetlands (the littoral zone), ecological activity and diversity, as we noted earlier, can be exceptionally high (Figure 6.10). The total area of ecosystems in the littoral zone is very small, less than 1 percent of the ocean's surface area, but it is very important ecologically.

Narrowing the focus to the shore itself, we find that productivity varies with the shoreline's physical diversity. Where shorelines are relatively uniform in shape and composition, the diversity and productivity of life tend to be relatively low. For example, the broad, straight, sandy shorelines that are so popular for recreation are not so popular with lots of other organisms. By contrast, irregular coastlines such as most river deltas are extremely diverse and productive biologically because they provide, among other things, an extraordinary mix of habitats. Their geographic configurations are complex, which gives rise to long and complex border arrangements among habitats that also enhance diversity (see Figure 6.12b). These complex habitat configurations include both saltwater and freshwater bodies and intermediate areas of brackish water. In many areas these habitats are subject to serious alterations by land-use activity. Filling of wetlands and dredging of harbors and river mouths are two common and highly destructive activities. Oil spills and pollution in terrestrial runoff are also significant.

Farther out to sea, over the deep basins of the oceans, most life is concentrated in a thin layer of surface water. Species diversity, biomass, and productivity tend to be relatively low compared to the near-shore waters. Productivity here, as we noted earlier, is also low by global standards, about equivalent to that of a semiarid grassland. It is maintained largely by tiny floating plants called **phytoplankton,** which live mainly in the upper 100 meters (330 feet) of the ocean, the photic zone (see Figure 5.4). Phytoplankton are the foundation of oceanic ecosystems. They are found in all oceans, but their abundance varies appreciably with latitude and other factors.

Within the broad zonal patterns of ecological activity, the next most significant level of variation in the oceanic province is associated with boundaries between different types of ocean water. These habitats occur, for example, where cold currents and warm currents mingle, as in the North Atlantic at the contact between the Gulf Stream and the Labrador Current (Figure 6.11).

Where ocean currents bring cool waters along the continental shelves, ecological activity also flourishes. Some of the Earth's richest saltwater ecosystems are found where cold currents run along midlatitude and subtropical coasts. This is explained largely by the abundant oxygen and carbon dioxide in the cool waters and the availability of sunlight in these latitudes. The map in Figure 4.14 shows the locations of a number of such cur-

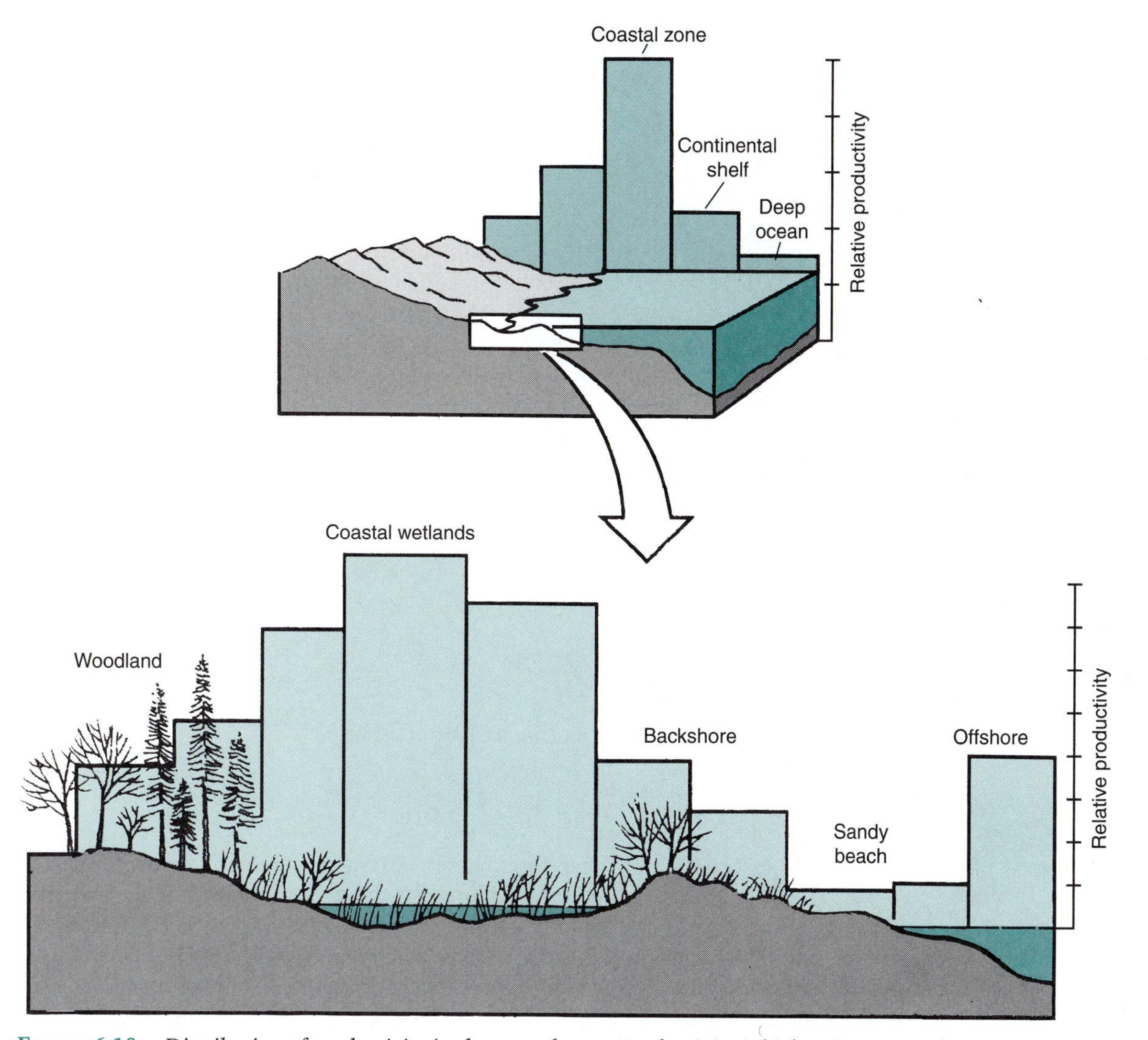

FIGURE 6.10 *Distribution of productivity in the coastal zone. Productivity is highest in protected waters such as lagoons and wetlands.*

rents and related ecosystems. Note the California Current, which flows north to south, and the Humboldt Current in South America, which flows south to north. Each is fed by relatively cold water originating in the subarctic zones of the Pacific Ocean and each extends to the margins of the tropical zone only about 20 degrees from the equator.

Patterns within Climatic Zones

Within individual climatic zones on land, we can illustrate several additional controls on the distributions of terrestrial ecosystems. One is **elevation**, because as the land rises, climate grows colder and harsher. For example, temperatures change about 35°F between sea level and the summit of a 10,000-foot mountain peak. In addition, mountains protrude into the more active levels of the atmosphere where winds and storms are stronger. Together these factors produce dramatic changes in bioclimatic conditions with increased elevation.

The pattern of the bioclimatic environment on mountain slopes is characterized by **altitudinal zonation**, or belts with different ecosystems at successive levels (Figure 6.12a). On very high mountains in the tropics, such as the Andes, the vertical zonation of life approximates that of the Earth's biomes between the equator and poles. The base of the mountain is occupied by tropical rainforest, the middle zone by temperate forests, and the upper zone by arctic-like ecosystems resembling tundra and arctic fell. Above 15,000 to 17,000 feet (4,600 to 5,200 meters) elevation there is little life indeed, even in tropical latitudes. Although mountaineers climb above 20,000 feet and have even reached the peak of Mt. Everest—the highest point on Earth at 8,848 meters (29,028 feet) elevation—even with the aid of shelter and special clothing humans themselves can-

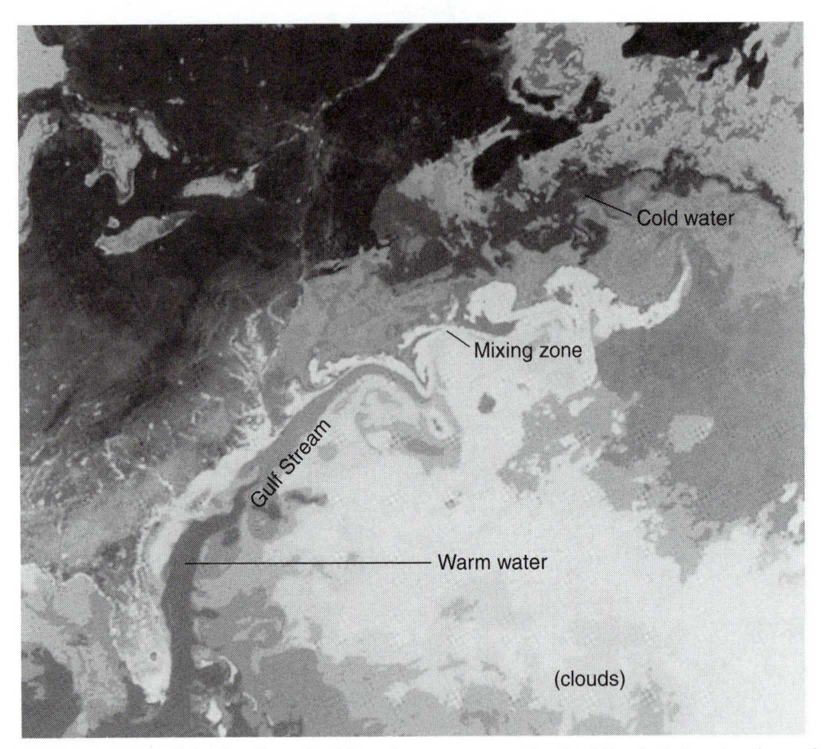

FIGURE 6.11 *The border of cold and warm water in the North Atlantic along the path of the Gulf Stream, a marine environment noted for ecological activity.*

not inhabit lands above 5,200 meters (17,000 feet) on a permanent basis.

Streams flowing through the landscape also control the distribution of ecosystems. All land areas (with the exception of ice-covered ones) are drained by streams that are linked together in systems called **drainage networks.** Each network drains an area of land called a **watershed** (Figure 6.12b). As the streams flow toward the central and lower portion of the watershed, they grow larger as more and more water is added. The largest stream in the network, the trunk stream, discharges directly into the ocean. The Mississippi, the Nile, and the Amazon are large trunk streams that drain huge land areas of North America, Africa, and South America, respectively.

Within most watersheds, the distribution of ecosystems is strongly influenced by the availability of a dependable water supply. As a result, freshwater ecosystems thrive along the trunk streams and major tributaries where adequate water supplies are generally available. Wetland ecosystems inhabit the neighboring lowland environments such as floodplains. Thus, both the aquatic and wetland ecosystems are distributed along stream corridors. In map view they have the pattern of great branching networks extending across the land (Figure 6.12b).

In the upper parts of the watershed, water supplies are generally less abundant. These areas lie on higher ground between the stream corridors. Ecosystems here must depend directly on atmospheric precipitation for water. Therefore, these ecosystems tend not to be aligned in linear corridors like those of the lowlands, but are distributed in broad, irregularly shaped zones among the winding lowland corridors. In addition, productivity and species diversity tend to be lower in these upland ecosystems. Not only is there less moisture, but water supplies are more variable than in the stream corridors.

Where large watersheds extend across two or more climatic zones, ecosystems may assume more complex distributional patterns because they are influenced by both climate and stream corridor networks. Especially interesting are *exotic rivers* such as the Colorado and Nile, which rise in a relatively moist climatic zone and then cross a desert before reaching the sea. The stream so drastically alters the otherwise arid surface environment that it enables aquatic and wetland ecosystems to thrive under a desert climate (Figure 6.12c).

The streams that cross the North American deserts and plains reveal similar patterns. The Missouri River, for example, once (before agriculture and the construction of reservoirs) supported a corridor of forest in its valley where it crossed the grasslands of the semiarid Great Plains. As a result, the grassland ecosystem of the Great Plains, which forms a great north–south belt stretching from Texas to Saskatchewan, was in fact broken by the east–west corridors of forest along the valleys of the Missouri and other large streams flowing eastward from the Rocky Mountains.

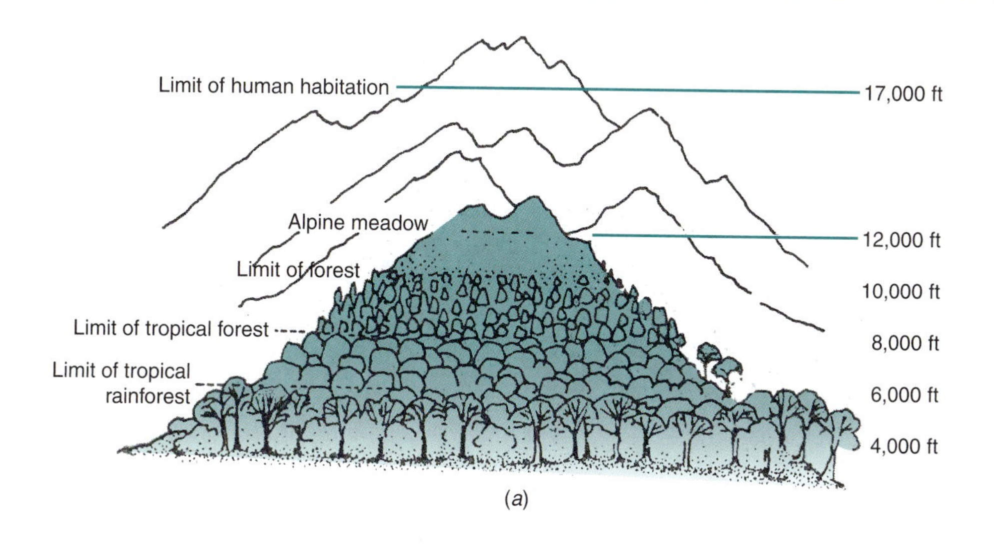

(*a*)

(*b*)

(*c*)

FIGURE 6.12 *Biogeographical patterns in three different geographic settings: (a) mountains with altitudinal zonation; (b) stream network, watershed and river delta; and (c) exotic river (the Colorado River) in an arid landscape.*

Finally, where the trunk streams meet the sea, special ecological environments are formed. These are **deltas** where freshwater and saltwater mingle and ecosystems are fed with nutrients brought from the land by streams. The delta environment is physically diverse, with a wide array of wetland, aquatic, and terrestrial habitats, and each usually exhibits greater productivity, biomass, and species diversity than other coastal environments (Figure 6.12b).

6.6 MODELS OF SPATIAL DYNAMICS IN ECOSYSTEMS

Maps tend to stabilize or fix the distribution of things in our minds, but this tendency is misleading with ecosystems. In reality the spatial expressions, or patterns and distributions, of ecosystems are almost constantly changing in response to environmental changes. Scientists have devised several models to help explain these dynamics. We briefly examine two here: succession and disturbance.

Succession Theory

Succession is based on the observation that communities of organisms have the capacity to alter the surface environment (especially soil, moisture, and ground-level climate) making it suitable for other communities. The succession model usually begins with a denuded (barren) surface, such as a surface freshly exposed by a retreating glacier, abandoned farm field, or an eroded grassland, into which a community of hardy organisms, called **pioneers** becomes established. These are usually small plants such as algae, mosses, lichens, and grasses. They help stabilize the environment and in turn give way, or are succeeded by, another community, usually larger and more complex than the first. Each community or **successional stage**, renders change to the surface environment favorable to another community, until eventually a relatively steady condition, called a **climax community**, is established that is in balance with the larger bioclimatic condition of the biome.

Does succession actually work in nature? It does in the sense that ecosystems *fill in* spaces left open in the environment. Abandoned farm fields, burned forests, or degraded rangelands, if left alone, are usually reoccupied by plants and animals, which in time assume a balanced state within the larger biome. But the notion of an orderly succession of communities, changing sequentially from small and simple to large and complex communities, is not supported by the scientific evidence. Instead, it appears that any one of many communities may start the process. The change from one group of organisms to the next may follow various sequences, depending on the forces operating in the environment. In the Hill Country of central Texas, for example, juniper trees invaded rangeland following degradation of grasses and soil by cattle overgrazing and rancher control of range fires in the late 1800s. In this case the controlling force—severely degraded, dry soils—discouraged grasses but not junipers, which are hardier and more drought-resistant (Figure 6.13). The junipers remain today as the apparent climax community. Consideration of controlling forces brings us to the second concept, disturbance theory.

Disturbance Theory

Disturbance theory holds that the external environment is forceful and variable and causes constant change in ecosystems. A host of forces, including fires, storms, drought, diseases, land-use activity, and pollution, operate at different magnitudes and frequencies around and within ecosystems. The organisms that make up ecosystems survive or perish depending on whether the impact of these forces, acting individually or collectively, exceeds their tolerance limits. When these forces are strong, such as with a hurricane, volcanic explosion, or construction activity, ecosystems in the affected area may be damaged, reduced, or destroyed.

Disturbance theory argues that the Earth's surface is characterized by a more-or-less continuous string of such events operating at different intervals, and that ecosystems are in a state of continuous adjustment to these forces. In other words, the patterns of ecosystems we see on the Earth's surface are in a constant state of flux, and that progressive change ending in a climax state, as the succession concept argues, is too simple an explanation for this process. Thus, the disturbance concept views the environment's behavior as more chaotic.

Much of what we see as evidence of succession or disturbance depends on our perspective of time. If we look at the segments of time in the intervals between powerful events, succession appears to be a reasonable model. These intervals are relatively quiescent periods when regrowth and infilling can take place. However, if we take a longer view of time, then ecosystem changes look more like fluctuations in response to environmental perturbations and a particular spatial trend may or may not be apparent.

The Human Factor

When we consider the actions of humans, the time scale is especially critical. Set into geologic time (millions of years), the human epic on Earth is a single event or disturbance of global proportions, the end of which we have yet to see. It is one of the major events of the Earth's recent geologic history. Set into historical time (several thousand years) however, the human epic can be broken down into thousands of events in different places with different effects. Examples include the land degradation in the

(a) (b)

FIGURE 6.13 *The original vegetative cover of the Texas Hill Country was a parkland with abundant grasses (a) until overgrazing and fire control (b) allowed the invasion of junipers.*

Mediterranean Basin in ancient Greek times, industrialization and urban development in Europe in the eighteenth and nineteenth centuries, plowing of the American prairies in the nineteenth and twentieth centuries, and the destruction of large areas of tropical rainforest in the twentieth and twenty-first centuries.

In many instances, there is some ecological recovery or succession. After major events, however, it is rarely complete. The Mediterranean lands can no longer support the forests and grass covers that existed before Christ. The lands in the Great Plains ravaged by erosion during the North American Dust Bowl of the 1930s once again support grasses, but the cover is weaker with fewer species (see Figure 8.12a). Overall, the trend in historical times has been toward increased magnitude and frequency of disturbance by humans as our numbers, consumption of resources, and occupancy of the planet's surface have expanded. The periods or windows of time available for recovery, in turn, are shorter, leading to a lower order of quality in ecosystems with fewer species, lower productivity, and often less resilience to future disturbances.

Some observers argue that the global environment has become more chaotic in the past several decades. They cite human-induced environmental change such as increased runoff and flooding and global warming as the cause. Scientists are uncertain about this, but we are certain that today more people occupy disturbance-prone environments than in the past. Marginal environments that were traditionally beyond the bounds of significant agriculture and settlement, such as steep mountain slopes, stormy coastlines, and flood-prone river valleys, are clearly subject to more erratic changes and disturbances than nonmarginal environments. As population growth and economic development drive land uses farther into marginal environments, our susceptibility to disturbance rises. Natural disasters seem more common and more destructive. Not surprisingly, as these experiences mount, nature appears to many societies as less knowable, less predictable, and more chaotic (Figure 6.14).

6.7 LAND USE AS ECOSYSTEMS

Although modern urban society tends to have an abstract and seemingly distant relationship with nature, all humans are functionally bound to ecosystems. In fact, the land-use systems we build are themselves ecosystems or component parts of ecosystems. As ecosystems, we are learning that many modern land uses are not very durable or sustainable, for several reasons. First, most require inordinate amounts of outside energy and maintenance. For example, with mechanized, commercial farming operations, it often takes more calories of energy (in the form of technical assistance such as tractor fuel and manufactured fertilizer, for example) to produce a pound of food than there are calories in the food itself. Second, the resource base, mainly in terms of soil and water supplies, is progressively depleted by land use with each season of activity. In the parlance of banking, this practice spends part of the principal as well as the

(*a*)

(*b*)

FIGURE 6.14 *With more people crowded into marginal environments, disruptive events such as volcanic eruptions (a) often seem more frequent and devastating (b).*

annual interest earned by the ecosystem account. In the long run, such land uses consume themselves. Third, as land use is pushed into more marginal environments by population growth and economic expansion, the risk of serious disturbance from natural forces increases. As we noted in the previous section, we see this trend in coastal environments prone to hurricanes, in river valleys prone to flooding, on mountain slopes prone to heavy erosion, and in mountains prone to volcanic eruptions.

The sense of a dependable, productive environment, which is so deeply imbedded in the thought of most developed countries, is a tenuous concept for much of the developing world. Increasingly, the marginal environments of many developing and deprived countries are being viewed as undependable and chaotic, where disturbance is more common than order and failure more common than success. Imagine the world envisaged by a Bangladeshi farmer trying to eke out a living on a small plot of delta land. He has already seen two or more devastating typhoons and floods in his lifetime in which thousands of people were killed and his village and farm were damaged or wiped out (see Figure 10.12b). To the south lies the Indian Ocean; to the north lies only more overcrowded lowland and an impenetrable political border with India. The government of Bangladesh can neither manage the flood waters coming from India, the storms coming from the Indian Ocean, nor the growth of its own population. As an ecosystem this poor agricultural country is trending toward serious imbalance, perhaps chaos, in this century.

6.8 HUMAN IMPACT AND DISTURBANCE OF ECOSYSTEMS

Humans have had a profound effect on ecosystems over much of the Earth. Although human impacts can be traced back thousands of years, the most serious ones, as we noted earlier, have come in the past several centuries. They are concentrated in the areas of heaviest land-use activity, but to some extent they can be detected over much of the globe. We can group the human impacts on ecosystems into six major classes:

- Reduction
- Fragmentation
- Substitution
- Simplification
- Contamination
- Overgrowth

Reduction refers to the loss in area or coverage of an ecosystem as a result of burning, agricultural development, urbanization, and lumbering. In the European midlatitudes, for example, the great forests of preagricultural times have been reduced to mere remnant patches; by 1600 or so Great Britain had less than 1 percent of its forest cover remaining. In Central Africa the core area of tropical rainforest is a fraction of its size of several thousand years ago. The apparent cause of the reduction in Africa was clearing for agriculture, especially from burning on the forest perimeter for grazing land. As grazing societies advanced over the years, the rainforest ecosystem steadily declined in area. The reduction process continues today at a massive rate over most of the world's tropical forests. As we noted in Chapter 4, the chief causes are agricultural expansion and lumbering. If the destruction continues at its current rate, the world's tropical forests will be largely gone by the middle of this century. Destruction of the tropical forests is this in greater detail in Chapter 16.

Fragmentation occurs when ecosystems are broken down from large, continuous areas into smaller parcels. This sort of disturbance is clearly illustrated in ecosystems distributed in corridors along stream valleys. Agriculture, cities, highways, dams, and reservoirs are the primary causes of stream valley fragmentation. Nearly 80 percent of the major streams in North America and Eurasia are interrupted by dams and related facilities. Like cities built in stream valleys, dams either destroy segments of the corridor environment or create such constrictions in it that many of the ecosystem's functions are hampered or eliminated (Figure 6.15).

This is particularly apparent with the ranges and migration patterns of certain animals that follow the corridors. Among the animals most seriously hampered are migratory fish such as the salmon, whose spawning routes have been severely restricted or denied by the construction of dams. The dams on the Columbia River of the American Northwest, considered one of the principal salmon spawning streams on the continent, so severely impair fish migration that the stream may soon be eliminated as a spawning environment. Many California streams have already been eliminated as spawning environments because of damming, ditching, diverting, and rerouting for agricultural irrigation. In the United States there are more than 5,000 dams that are 50 meters (165 feet) or more high. In China there are more than 18,000!

Substitution involves replacing one set of organisms in an ecosystem with another. When agriculture moved into grasslands such as the American prairies, for example, native grasses and wildflowers were replaced by wheat, corn, and other crop plants. In addition, weeds moved in with the farmers and displaced many of the native flowering plants. The extent of the replacement process in the North American grasslands has been enormous; of an area covering more than 1.3 million square kilometers (500,000 square miles) of the Great Plains and Midwest

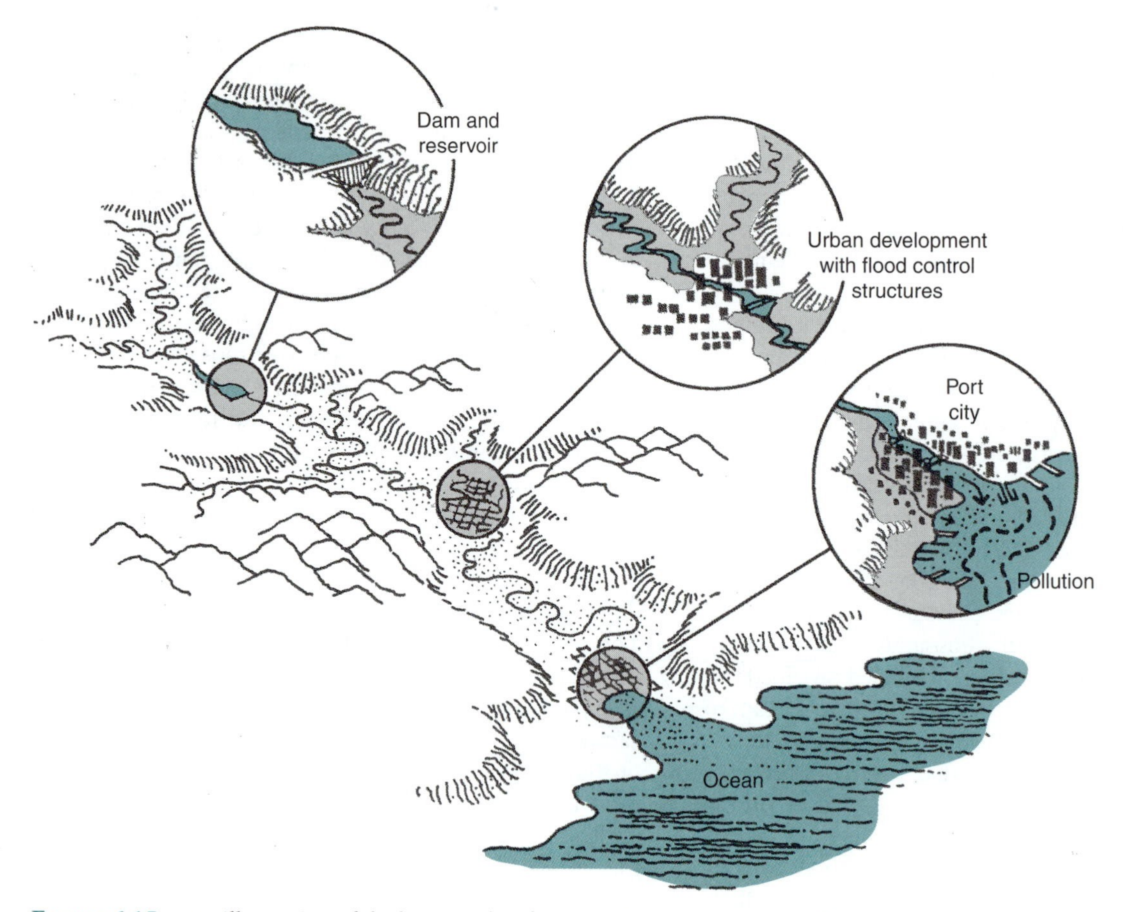

FIGURE 6.15 *An illustration of the barriers that fragment habitat corridors such as stream valley.*

Prairies in 1850, less than 1 percent of natural grassland are thought to remain today. The original grasses have survived in many areas, of course, but they are usually mixed with introduced plants. Just to the east in the Great Lakes, various animals invaded waterways in the twentieth century, displacing native species. One of the most recent invaders is the zebra mussel, a small mollusk brought to the region by ocean ships via the St. Lawrence Seaway. It has spread into streams, wetlands, and inland lakes in the Great Lakes watershed.

Substitution is usually accompanied by **simplification** in which the makeup of the replacement set of organisms is significantly less diverse than the original set of organisms. In other words, certain species are not only changed but the numbers are thinned out. Simplification is common with both agriculture and modern forestry. Natural stands of timber comprised of many tree species are cut and often replaced with planted stands comprised of one or two tree species. The ecosystem remains forest, but its species composition is greatly reduced. Because of its simple composition, the term *monoculture* is applied to such ecosystems. With a reduction in tree species many other dependent organisms—animals, insects, microorganisms—are also reduced in numbers or lost altogether. Much of the controversy about cutting old-growth (prehistoric) forests in the American Northwest has to do with the effects on other species, including higher carnivores such as owls. Besides the loss of dependent species, monocultures tend to be more vulnerable to destruction from disease and other threats because they lack the strength inherent in species diversity.

Extinction is also a source of simplification in ecosystems. The primary cause of extinction is habitat loss from human actions. Area reduction, replacement, fragmentation, and pollution all contribute to species eradication. The current rate of extinction related to human activity is estimated to be more than a thousand times greater than the natural rate. Most extinction today is taking place in the wet tropics with the clearing of rainforest. Both natural and human-related extinction will be discussed in Chapter 16.

Contamination involves the incorporation of pollutants into the ecosystem. Pollutants can enter ecosystems via many pathways: air, water, soil, or organisms. Once in the system, they pass through the food chains in the tissue of plants and animals. Certain pollutants such as heavy metals and the synthetic organic molecules from pesticides tend to be retained in the food chain, with larger concentrations in animals, including some humans, that draw on multiple sources of contaminated tissue.

Many of the contaminants are **persistent chemicals**, meaning that they are not quickly broken down in the environment but linger for months or years. Once in organisms, persistent chemicals like PCBs and DDT are stored in fatty tissue and various organs. At low levels the animal survives with certain limiting (mostly unknown) effects, but some animals are weakened and killed. The striped bass that migrate into San Francisco Bay on their annual spawning runs accumulate such heavy loads of toxic matter from urban and agricultural runoff that an annual die-off event has developed. When fish or their remains are eaten by other organisms, such as gulls and eagles, the contaminants are passed up the food chains and the concentrations grow with each trophic level. This process is known as **bioaccumulation** or **biomagnification**, as illustrated in Figure 6.16, based on DDD (a derivative of DDT) applications to a lake near San Francisco.

In the Great Lakes some fish species contain concentrations of contaminants such as DDT, PCBs, and dioxin 10,000 to 100,000 times greater than the concentrations in surrounding lake water. In secondary and tertiary consumers, the result is often genetic alteration of certain species leading to weakened populations. The effects on some birds, which include thin egg shells and physical malformities, have been widely publicized. Some of these effects are no doubt due to a phenomenon called **synergism**, in which the effects of two things such as two persis-

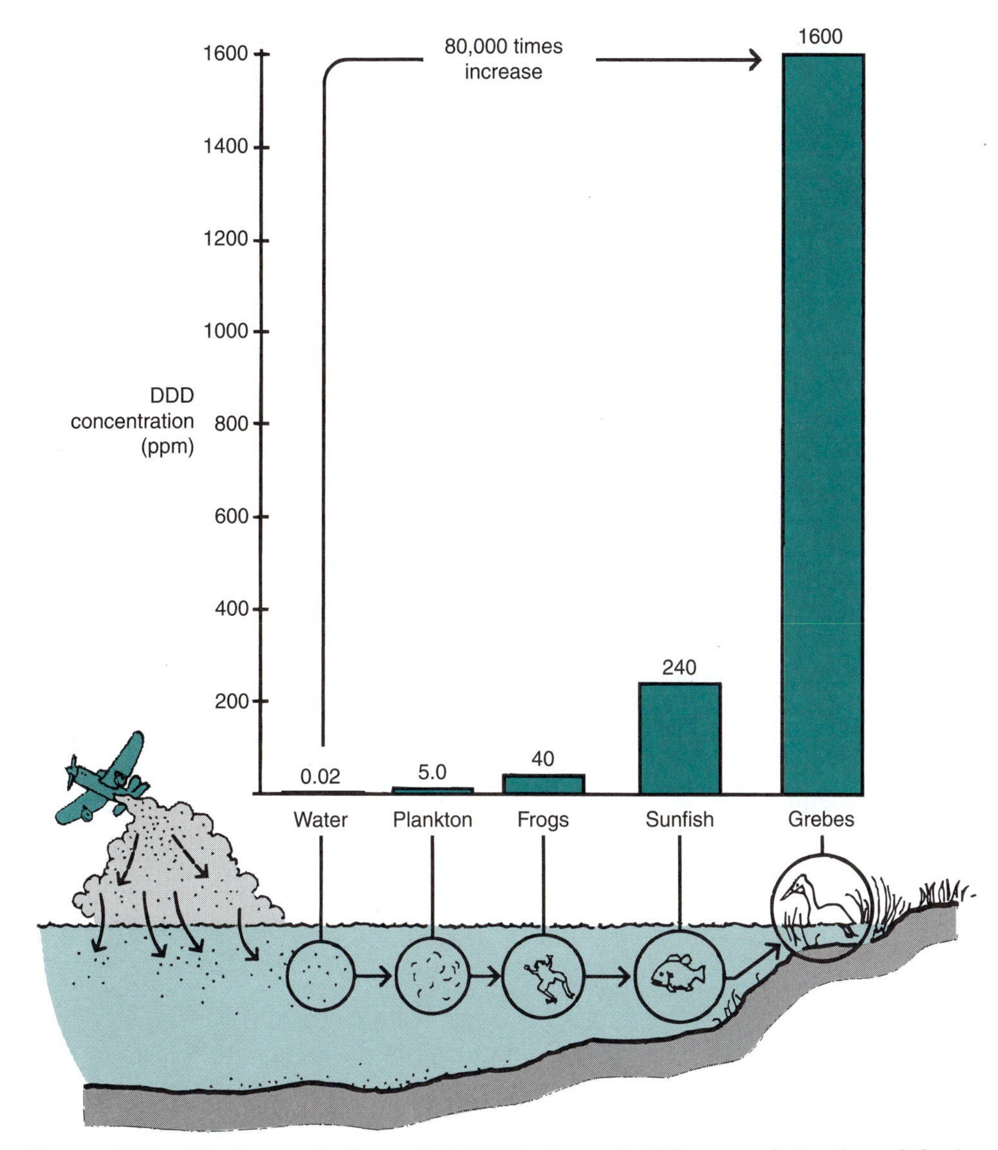

FIGURE 6.16 *The increase in the pesticide DDD as a result of bioaccumulation through food chains. This pesticide was applied to Clear Lake, near San Francisco, to control gnats in the 1940s and 1950s.*

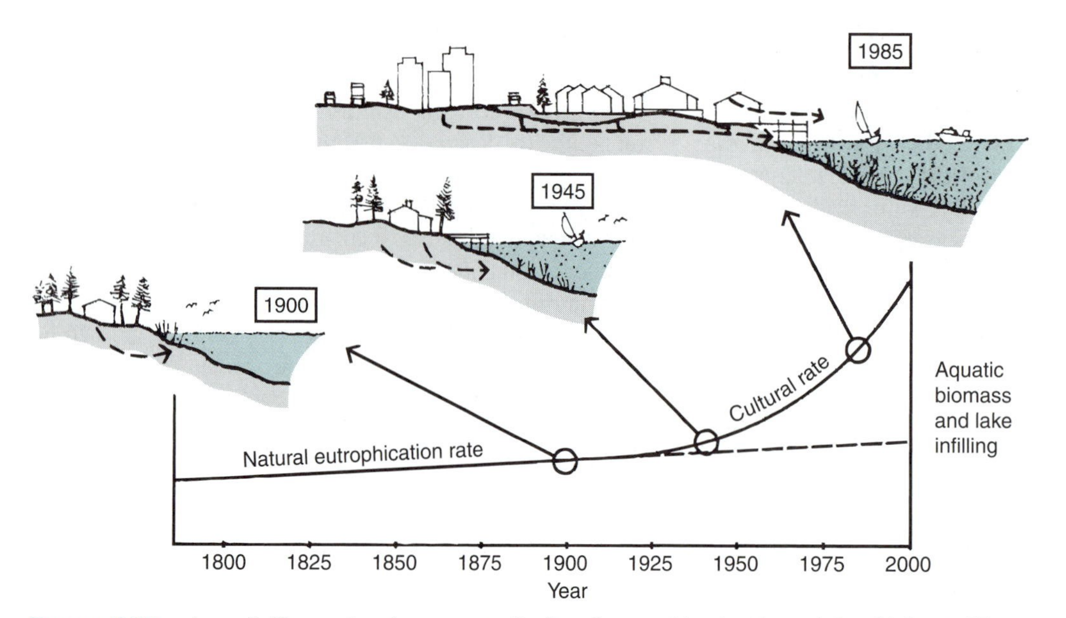

FIGURE 6.17 *A graph illustrating the concept of cultural eutrophication in an inland lake. Infilling accelerates with nutrient loading from development in the watershed around the lake.*

tent chemicals acting together is greater than the sum of the two if they acted separately. Since little is known about the chronic impacts of individual contaminants, the impacts related to synergistic effects involving several contaminants are very confusing indeed, but they do exist.

Contamination can also lead to **overgrowth** of ecosystems. In this case the contaminants are usually plant nutrients such as nitrogen and phosphorus instead of toxic substances such as pesticide residues or heavy metals. The nutrients enter the water system and cause accelerated plant growth leading to exceptionally high productivity. One of the most striking examples of overgrowth takes place in freshwater ecosystems. Nutrients released to lakes, ponds, and streams with runoff from fertilized farm fields and dirty city streets and from discharges by city sewage treatment plants and industry induce rapid growth in algae and other aquatic plants, resulting in large accumulations of organic debris in waterbodies.

As the organic matter decomposes, the consuming bacteria draw large amounts of oxygen from the water. The reduced oxygen limits fish species such as trout, and they, in turn, are replaced by more resistant and rougher fish such as carp. On balance, the water body declines as it fills with organic matter, loses oxygen, and changes fish species as well as many other organisms.

In nature, this process, known as *eutrophication*, usually takes place gradually. However, eutrophication can take place at an accelerated rate with human intervention. The accelerated version, called **cultural eutrophication**, is common to water bodies worldwide, but the rate is usually greater in warm climates and in areas of heavy urban and agricultural land use (Figure 6.17). In the Great Lakes, for example, eutrophication is greater in Lake Erie than in Lake Superior because of, among other things, the heavier nutrient loading in Lake Erie's watershed from three sources: (1) a large population more than 10 times that of Lake Superior's; (2) extensive agriculture; and (3) extensive urban and industrial development. In addition, Lake Erie is much smaller than Lake Superior, so the pollutants are more concentrated in Lake Erie.

6.9 SUMMARY

Life has been a part of Earth's marine environment for 3 billion years or more. Terrestrial life emerged about 500 million years ago and has since become the dominant life on the planet. Terrestrial vegetation accounts for nearly 60 percent of the Earth's photosynthesis, whereas saltwater plants account for 39 percent and freshwater plants for about 1 percent.

All life is organized into three main classes of ecosystems: terrestrial, saltwater, and freshwater. Each ecosystem is made up of food chains and functions as an energy system. Energy is supplied by solar radiation. Through photosynthesis less than 1 percent of this energy is converted into organic compounds and transferred through the ecosystem via the food chains. At each level of transfer, or trophic level, the vast majority of the organic energy is given up as heat to the atmosphere. Trophic levels are defined according to their function in the ecosystem: producers, primary consumers, secondary consumers, and tertiary consumers. Because of the rapid attenuation of energy, available energy is very small at higher trophic levels.

Humans are inextricably bound to ecosystems and function only as consumers. Human societies currently utilize only about 4 percent of the global productivity for food, fuel, and other purposes. In societies where vegetarian diets are the rule, as in most poor countries, much less energy is needed to provide adequate nutrition because they utilize producers (plants). In contrast, the animal-rich diets of most rich countries require much higher energy expenditures because they utilize primary consumers (animals). Simply put, in ecosystems vegetarian diets are more efficient than those that depend on plant-fed livestock.

Ecosystems can be characterized in several different ways. Productivity is the primary measure of an ecosystem's vigor as an energy system. The wet tropics and warm coastal waters exhibit the highest productivity rates worldwide because there are relatively few limitations in the flow of vital resources and relatively few disturbances from outside forces in these environments. Species diversity is another important characteristic of ecosystems, and the tropical forests exhibit the greatest biodiversity of any major ecosystem. On the other hand, biodiversity declines sharply in the harsher environment.

The global distribution of ecosystems is controlled chiefly by climate, but the fit between climatic regions and ecosystem patterns is not a good one everywhere. One of the main reasons for this in terrestrial environments is that humans have drastically altered the distributions of many ecosystems, particularly the forests in the temperate, subtropical, and tropical zones. Variations from the regional climatic framework also appear where mountain belts and river corridors cut across climatic zones. Where stream corridors and climatic zones intersect, the pattern of ecosystems is often complex. In the oceans, ecosystem patterns correspond broadly to climate at a global scale, with a belt of high productivity in the tropics and low productivity in the cold latitudes. Within this framework, however, there is considerable variation, with zones of high productivity associated with the continental shelves and certain current systems. Human influences on the distribution of ecosystems have not been as great in the oceans as on land.

Scientists use two principal models to help explain the spatial changes that occur in ecosystems: succession and disturbance. In the succession model, organisms change the environment, making it more suitable for other communities of organisms. As a result, one community succeeds another until a balanced state, or climax community, is attained. In the disturbance model, the external environment represented by fires, floods, and land use, for example, is constantly altering the distribution of communities. Therefore, communities are in a constant state of flux, depending on the magnitude and frequency of external events. The environment is seen as more chaotic from the disturbance model perspective. From the standpoint of land use, marginal environments are more chao-tic than nonmarginal ones. This has serious implications for humanity as more people spread into marginal lands with population growth in the developing and deprived countries.

Humans have had profound influences on ecosystems. These influences are increasing rapidly and can be grouped into six major classes: reduction, fragmentation, substitution, simplification, contamination, and overgrowth. Reduction is the loss in ecosystem area as illustrated by the shrinking coverage of tropical forests in Africa, Asia, and South America. Areal reduction in forests and increased disturbance of ecosystems in general from grazing, soil erosion, wood cutting, wetland eradication, and other factors decreases global productivity. By some estimates, these actions may have already reduced global terrestrial productivity in natural ecosystems by 30 to 40 percent.

Fragmentation is the breaking down of ecosystem corridors, as occurs when dams are built on streams blocking the spawning routes of migratory fish. Replacement involves substituting one set of organisms for another, as in the case of farm plants and weeds for native grasses and trees. Replacement almost always involves simplification, or the reduction in biodiversity. Monocultures such as plantation forests are simplified versions of natural forests. Extinction is also a source of simplification, and it is increasing rapidly worldwide.

Contamination involves the incorporation of pollutants into the ecosystem. Certain contaminants, such as pesticide residues and heavy metals, are retained in the food chains and passed along to the higher consumers, where they are known to weaken populations. Contamination by nutrient pollution leads to plant overgrowth, especially in aquatic ecosystems. Overgrowth is characterized by excessive productivity, which often leads to cultural eutrophication with species changes and filling of water bodies.

6.10 KEY TERMS AND CONCEPTS

- Ecosystems
  - food chains
  - trophic levels
  - producers
  - consumers
  - food web
  - energy flow
  - detritivores
  - energy attenuation
  - respiration
  - energy pyramid
  - biotic
  - abiotic
- Photosynthesis
- Net photosynthesis
- Efficiency
- Productivity
  - limiting factors
  - optimum conditions
  - tolerance thresholds
  - climatic limitations
- Global productivity
- Terrestrial ecosystems
  - biome types
  - composition
  - biological organization
  - species diversity
  - biomass
  - organism
  - species
- Interrelations
  - symbiosis
  - mutualism
  - commensalism
  - parasitism
  - predation
- Biogeographical trends
  - biogeography
  - decomposers
- Ecosystem patterns
  - global terrestrial
  - marine
  - elevation
  - altitudinal zonation
  - drainage networks
  - watershed
  - deltas
- Spatial dynamics in ecosystems
  - succession theory
  - pioneers
  - successional stage
  - climax community
  - disturbance theory
  - human factors
- Land uses as ecosystems
- Human impacts
  - reduction
  - fragmentation
  - substitution
  - simplification
  - extinction
  - contamination
  - persistent chemicals
  - bioaccumulation
  - synergism
  - overgrowth
- Cultural eutrophication
  - nutrient pollution

6.11 QUESTIONS FOR REVIEW

1. Review the concept of an ecosystem. What are food chains and how are ecosystems organized into trophic levels? Define the three classes of consumers in ecosystems and give some examples of organisms representing each class.
2. Is it appropriate to think of ecosystems as energy systems? If so, how are they organized, where does energy come from, and where does it go? Why is a pyramid used as a model for ecosystems? Where do humans fit into the energy pyramid? Are there differences in the earth's human carrying capacity related to the percentage of vegetation and meat in our diets?
3. Define the differences between photosynthesis (or total photosynthesis) and net photosynthesis in terms of plant productivity. What is the principle of limiting factors and how does it relate to plant productivity?
4. Climate poses limitations on plant productivity. Describe the differences in limitations among the major climatic zones. What are the productivity rates associated with different climatic zones?
5. How do humans fit into the scheme of global productivity? What are the forms of direct consumption, and how do indirect losses in productivity take place? What is likely to happen to global productivity with human population growth in the next century?
6. What are biomes? Can you name and define the principal forest biomes? Which are the largest geographi-

cally and the most productive? What limits forests from growing over the tundra and savanna biomes? Where have humans had the greatest impact on forest cover?

7. What is a community? What are some of the means by which species cooperate in communities? Do communities form discrete spatial or geographic entities? Explain.
8. Describe the biogeographical trends in species diversity, population, and biomass associated with climate and latitude. What is the pattern of species diversity and productivity in the oceans?
9. In terrestrial ecosystems, geographic distribution varies with different controls. Name several of these controls and the sorts of geographic patterns that result in ecosystems.
10. Compare and contrast the succession and disturbance models of biogeographical change, and indicate how human influences on the landscape fit into these schemes. How do you explain the observation that to many societies nature seems to be growing more chaotic?
11. Name and briefly describe the six major types of human impacts on ecosystems.
12. What are persistent chemicals? How does synergism operate in an organism? Why is it that animals higher in the food chain tend to have the greatest concentrations of contaminants in their flesh?
13. Illustrate how a land-use system functions as an ecosystem. Why are many modern land uses in both developed and developing countries not sustainable as ecosystems?

7

THE HUMAN POPULATION: TRENDS AND PATTERNS

7.1 Introduction

By most measures, *Homo sapiens* has become extremely successful competing with other forms of life. During the twentieth century alone, world population nearly quadrupled—from 1.6 billion to over 6 billion—and we continue to add more than 80 million people each year. Over the past 500 years, human population has increased more than tenfold. Although we are not the most abundant organisms on Earth, we are the dominant species. We can manipulate the environment. We can develop and use a wide range of biotic and mineral resources while purposefully or inadvertently displacing or eliminating other species. And because of our widespread distribution and sheer numbers, our presence has produced significant environmental impacts.

Our unique capacity as human beings to create resources and build civilizations has enabled us to influence, alter, and direct the flow of energy and matter in environmental systems. As our numbers increase and our technologies become more powerful, we produce greater and more lasting changes in the environment. Although our efforts are generally focused to improve the human condition, we have been woefully unaware and neglectful of the overall effect on the interdependent natural systems (air, water, soil, and biotic) that together form the environment. As a result, we are beginning to suffer from the feedback of our actions and massive populations. Furthermore, conditions are sure to worsen if our population increases as projected to more than 10 billion during this century.

In this chapter we will see how human population growth is linked directly to our ability to develop and expand our resource base through technological revolutions. We will examine the demographic processes that affect population size, distribution, and trends, as well as how they are influenced by economic growth and development. Finally, we will focus on the economic and environmental problems caused by rapid population growth in poor countries and the growth of resource consumption in affluent countries.

7.2 Historical Perspective of Population Trends

We became the dominant species on Earth by expanding our resource base through a variety of technological revolutions. Such technological milestones began with the development of agriculture about 12,000 years ago, when the entire human population probably numbered no more than a few million.

Resources, Technology, and Population Growth

With the domestication of plants and animals and the development of herding and tilling practices, farmers were able to provide a reliable food supply to relatively large, permanent settlements where some members of the population could now be involved in culture-building activities such as government, applied science, trade, art, and religion. As agricultural civilizations developed and spread, population grew slowly but steadily, reaching approximately 250 million by the time of Christ. During the next 17 centuries, world population doubled, so that at the dawn of the **Industrial Revolution** around 1750, the world had about half a billion people (Figure 7.1).

The stage was set by exploration, settlement, and expanding trade in the sixteenth and seventeenth centuries, but in the eighteenth century world population truly exploded. A series of dramatic technological changes rapidly expanded the resource base and provided a foundation for accelerated population growth that continued for more than two centuries. The term Industrial Revolution is not sufficiently comprehensive to describe the simultaneous revolutions in agriculture, transportation, and medicine that were as significant to population growth as the changes in industrial and energy technologies.

The development and refinements in the steam engine that harnessed the energy of combustion literally powered the technological revolutions of the nineteenth century in Europe and North America. For the first time, humankind had a more powerful energy source to supplement or replace human and animal energy and the mechanical energy of water and wind. Steam engines were applied to new mining, manufacturing, transportation, and agricultural machinery, dramatically increasing human productivity. Cities—linked by more efficient transportation systems to regional, national, and even international markets—became centers of production and commerce. At the same time, as agricultural production became more mechanized and productive, the proportion of the labor force that farmed began to decline and rural to urban migrations began throughout Europe and North America.

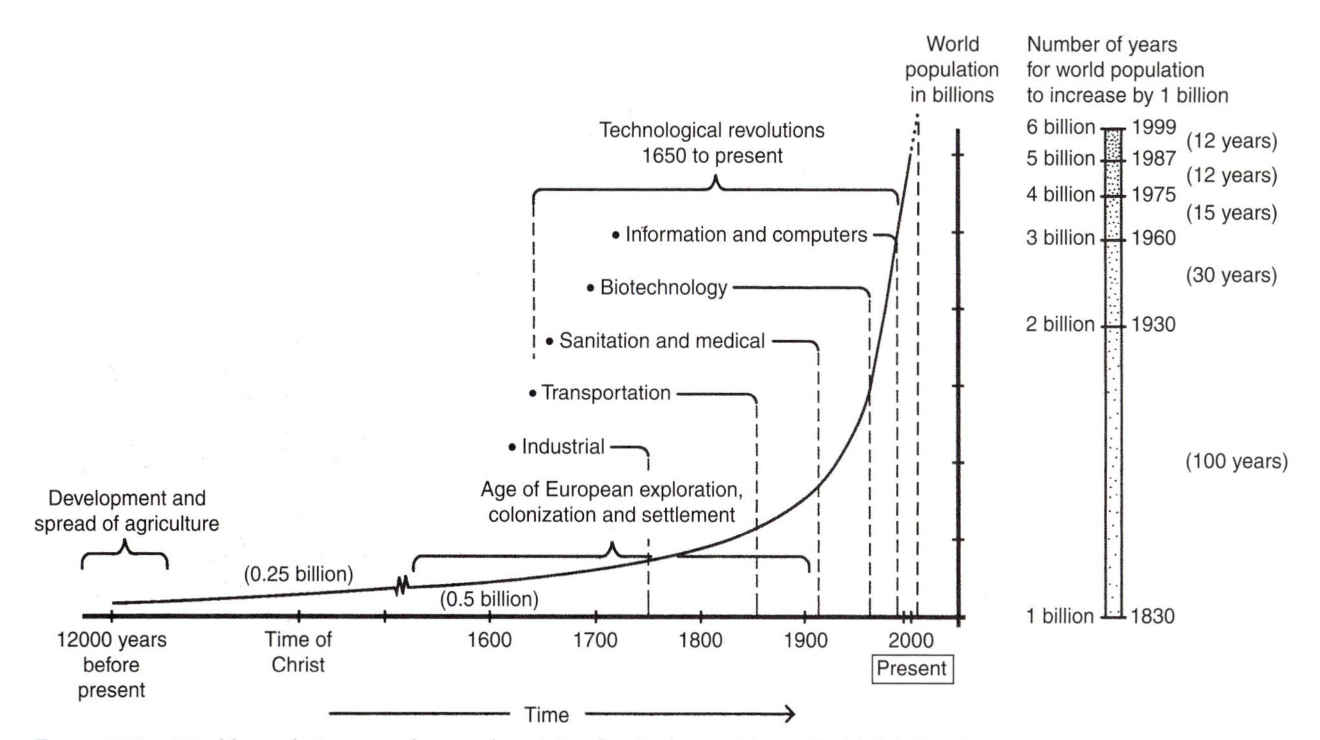

FIGURE 7.1 *World population growth since the origin of agriculture with a scale (right) showing years and dates for the world's population to grow by 1 billion.*

Although mechanization in economically developed countries has increased agricultural productivity and reduced farm labor, additional technological revolutions have also raised food production in developing countries providing at least temporary support for global population growth. During the last century, agricultural yields have increased because of advances in plant and livestock breeding. New breakthroughs continue, as current research unlocks some of the secrets of plant and animal genetics (see Chapter 8). Highly responsive hybrid crops such as corn and rice have been developed, and chemical fertilizers, pesticides, and herbicides have been improved.

Medical advances have also changed global population dynamics quickly and dramatically. Inoculations against epidemic and other communicable diseases, suppression or elimination of many disease vectors (carriers and spreaders), and improvements in sanitation have all contributed to the rapid decline in death rates in virtually all parts of the world. The result of these changes in the human condition can be seen in Figure 7.1. World population at first grew very slowly as agriculture spread, civilization developed, and commercial and regional trading patterns evolved. But spurred by industrialization and simultaneous revolutions in medicine, sanitation, transportation, and agriculture, world population surged to 1 billion by 1850 and 2.5 billion by 1950.

When the Industrial Revolution began, world population was growing at about 0.12 percent a year, but then the rate began to accelerate dramatically. By 1930 population was increasing by 1 percent each year and by the early 1960s when the world population reached 3 billion, the annual growth rate had reached 2.1 percent. Since then, the growth rate has slowly declined to the current annual level of about 1.4 percent.

Demographers tell us that the world's population reached 6 billion during 1999. With the current annual growth rate, in just four years the cumulative increase will produce a new population equal to that of the entire world at the time of Columbus's voyages 500 years ago! Or to put it in a current context, the 400 million added in about 5 years is nearly equal to the combined population of the United States and Japan.

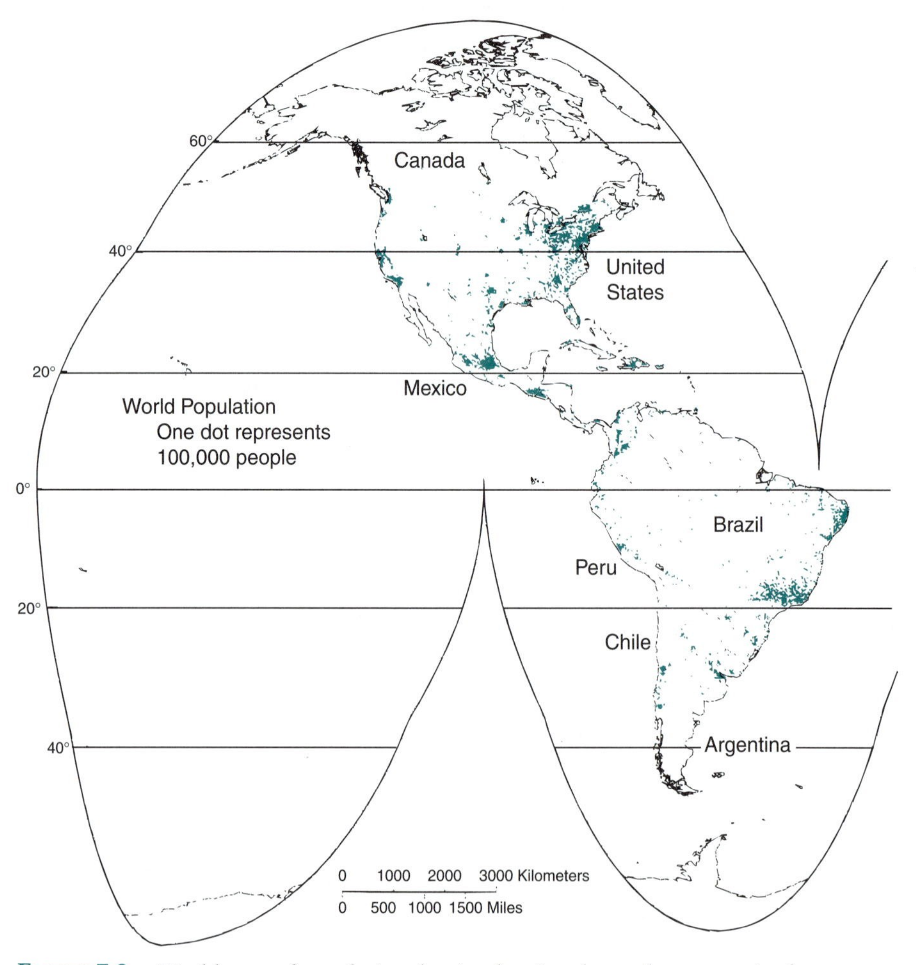

FIGURE 7.2 *World map of population density showing the stark extremes in the distribution of humans. Notice the Nile Valley of Egypt and the massive populations of the great river valleys of India and China.*

Future Population Changes

It is likely that this trend of slowing growth rates will continue. Population growth in the developed countries has slowed to 0.1 percent a year, and in many developing countries, the rate of growth is also slowing. According to some estimates, world population will reach 6.8 billion by 2010 and will be about 8 billion by 2025. These projections, based on recent trends in census data, fertility rates, and mortality rates of countries throughout the world, indicate a population increase of about 2 billion in the next 25 years, with over 98 percent of that growth in the developing countries.

Accordingly, the developed countries, which currently have 20 percent of the world population, will have only 15 percent by 2025. The worldwide growth rate is projected to decline to about 0.8 percent by 2025. However, with a projected total population of 8 billion, there will be 64 million people added each year. If fertility rates continue to decline toward replacement levels (2.1 to 2.2 children per woman) throughout the middle decades of this century, world population will stabilize between 9 and 11 billion sometime around 2100.

7.3 POPULATION PATTERNS AND TRENDS

Figure 7.2 shows that the human population is spread unevenly across the continents. Why do a few places support such large concentrations of humanity while vast areas support few people or none at all? This pattern represents the dynamic forces that affect natural population changes within an area, differences in the quality of the environment, and factors that have induced migration between areas. The current population distribution is markedly different from that of 6,000 years ago or even 500 years ago, just as future population maps will reflect the rapid growth still occurring in many of the developing countries as well as the stabilization of population growth in most of the developed countries.

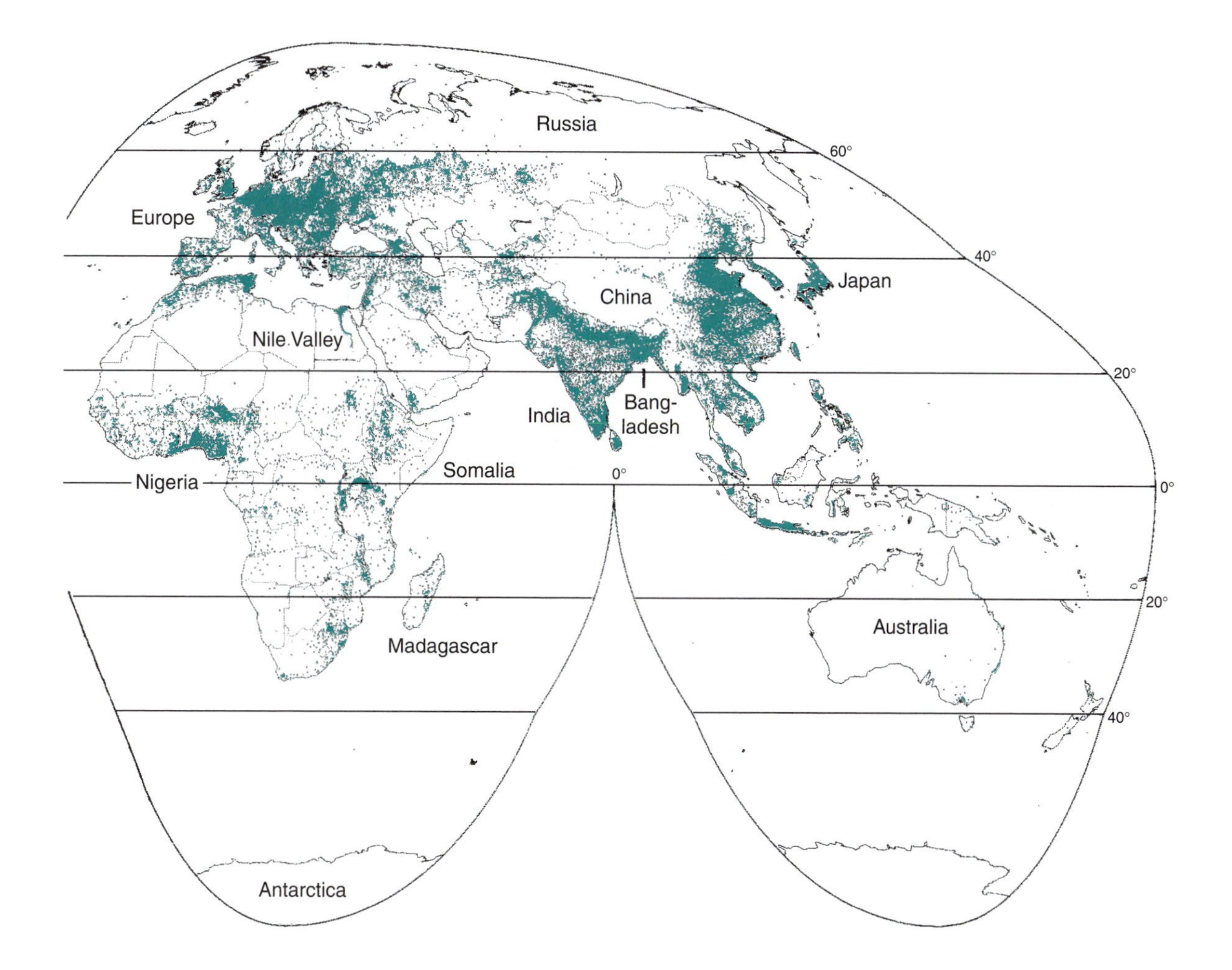

Temporal and Spatial Variations in Population

What are some conditions and circumstances that help explain the spatial distributions depicted on the population map? The physical environment has always influenced the spatial distribution and density of population, both in terms of supporting our basic biological needs and providing sufficient means or opportunities to support our livelihood. In general, the most favorable environments—that is, those with humid climates, fertile soils, and long growing seasons—have sustained the majority, while lands with harsh or difficult environments supported few, as was illustrated in Chapter 4.

But human modification of the physical environment has greatly altered the pattern of favorable and unfavorable environments at the local and even regional scale. The earliest examples come from the harsh desert lands of the Middle East and Egypt, where thousands of years ago irrigation transformed forbidding, barren desert into productive agricultural landscapes and the Nile Valley and the Tigris–Euphrates Lowland became early population centers. The geographic pattern of population associated with the Nile is remarkable even at a global scale of observation, as the map in Figure 7.2 reveals.

The cultural component of humanity also exerts an important role in population patterns and movements. Traditions and behaviors associated with common ancestry, religion, and language influence population concentrations and dispersals, just as political control and policies of national governments have also fostered population growth, decline, or migration. People may move from an area as a result of one or more *push factors* such as difficult economic conditions; religious, ethnic, or political intolerance; or even civil or international wars. On the other hand, *pull factors,* such as improved economic opportunities, may attract people to a place. A combination of such push and pull factors are evident in the settlement patterns that evolved in North America. For instance, some emigrants seeking political or religious freedom established colonial or pioneer settlements. Others came because of difficult economic conditions in their homelands. The more than 1 million Irish that came to America after the 1846 potato famine provides a most striking example.

Once immigrants obtained inexpensive farmland or jobs in factories or mines, news of plentiful economic opportunities traveled back to Europe and Asia through friends and relatives, and the stream of immigrants became a flood. Specific ethnic groups settled in certain areas. For example, early German immigrants concentrated in southeastern Pennsylvania, where they came to be known as the *Pennsylvania Dutch.* Later, Scandinavians and Finns homesteaded around the upper Great Lakes and Northern Plains. As industrialization expanded rapidly at the beginning of the twentieth century, vast numbers from southern and eastern Europe sought plentiful manufacturing and commercial jobs. Ethnic enclaves of Poles, Italians, and Greeks characterized North American industrial cities during the first half of the twentieth century.

Today, perhaps more than ever before, economic hardship, political unrest, and war result in significant population movement. Events during the past decade, for example, have created tens of millions of refugees. Among the most notable events are the Persian Gulf War; civil wars in Zaire (Congo), Ethiopia, Sudan, and Chad; ethnic reprisals and revolution in Rwanda and Sri Lanka; military coups in Haiti; the dissolution of the Soviet Union and the creation of 15 independent nations; and the fragmentation of Yugoslavia and Czechoslovakia into several republics based on ethnic differences.

Current population patterns are dynamic, but they reflect both recent demographic trends as well as those that have evolved over long periods of time. For example, productive agriculture in the valleys and deltas of the great rivers of China, India, and Southeast Asia have long supported large populations, whereas the dense urban populations of western Europe and the northeastern United States emerged as the result of the technological revolutions, economic development, and large-scale migrations during the nineteenth and twentieth centuries. Today the most rapid population growth is occurring in parts of Africa and Latin America, where death rates have fallen sharply while birth rates remain relatively high (Figure 7.2).

International Comparisons of Population Size and Density

A practical way to study population trends and patterns is to consider **population size** of individual countries. A country is the political and geographical unit in which decisions relating to population, environment, and resources are made. Figure 7.3 charts the world's 10 most populous countries. Together these countries make up nearly 60 percent of the world's population. The political, economic, and environmental decisions made in these few large nations directly affect the majority of humankind. Note that 6 of the 10 are Asian countries, and that 1 in every 5 persons in the world lives in China and 1 in 6 in India.

Population density is another measure to analyze population and environmental issues by relating the number of people in a country to its land area. The most straightforward is the crude or arithmetic population density, which is the number of people divided by the total land area. Although this statistic ignores differences in population distribution within a country, it is a better way to compare population characteristics of countries with large and small areas. For example, the United

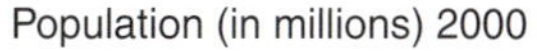

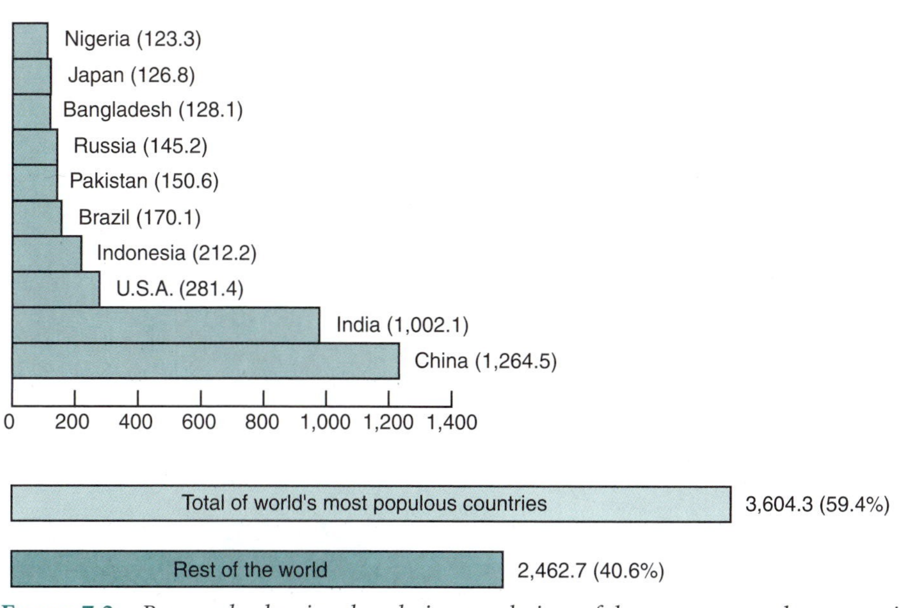

FIGURE 7.3 *Bar graphs showing the relative populations of the ten most populous countries in the world in 2000. (Source: Population Reference Bureau.)*

States is the third most populous nation but it also has the third largest area, so its population density is relatively low—about 28 per square kilometer (73 people per square mile). In contrast, no individual European country is among the 10 most populous (Germany, with about 82 million people, ranks twelfth). Yet Europe, excluding Russia, has 40 independent countries with a combined population of 582 million, or more than twice that of the United States, living in an area only about half the size of the United States. So Europe has a population density of 104 per square kilometer (273 people per square mile), nearly four times the number of people per unit area as the United States. In fact, four European countries are among the most densely populated in the world, and all but four European countries (Iceland, Finland, Sweden, and Norway) have higher population densities than the United States.

Physiologic density relates population to cropland area. In developing countries where subsistence agriculture remains the most important economic activity, this measure of population density reflects the intensity of agriculture. For example, in nearly all of the populous developing countries in Asia—including India, Indonesia, Pakistan, and Bangladesh—there is less than one acre of cropland per person. In China, the world's most populous country, each acre of cropland must support 5 people (12 people per hectare) (Figure 7.4). In most of these agricultural countries, virtually all the land that is suitable for crop production is being utilized. So as populations grow, more and more people must be supported from the existing cropland.

Because agricultural productivity varies from place to place, physiologic density provides only a rough measure of population pressure. For example, the relatively infertile tropical soils and inefficient land tenure practices contribute to low agricultural productivity over much of Africa and Latin America. In contrast, the developed countries with very high physiologic densities, including the European Union and Japan, support relatively small but highly productive commercial agricultural sectors. Furthermore, in most of these countries nonagricultural economic endeavors provide the means to purchase food produced elsewhere.

Variations in Population Density Patterns

As noted, population densities for a continent or even a country can be misleading because people are not usually evenly distributed. Generally there are a few favored areas where large numbers concentrate and extensive areas where few live. In the developed countries, the greatest population concentrations are clustered in urban areas, such as the megalopolis region in the U.S. Northeast, where much business and commerce occurs. Although cities are growing rapidly in many developing countries as well, huge populations remain concentrated in productive agricultural areas where livelihoods depend on intensive subsistence agriculture. In contrast, the rural areas of developed countries that support productive agriculture have relatively low and declining population densities because of the low labor requirements and large farms

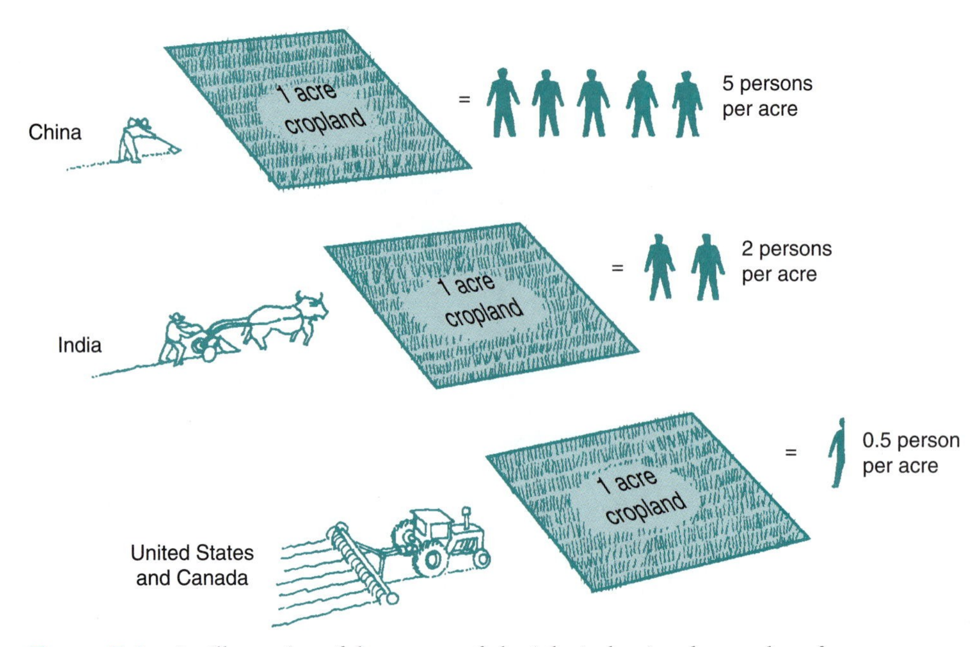

FIGURE 7.4 *An illustration of the concept of physiologic density, the number of person supported by an acre of cropland. This reflects one of the great differences between subsistence and commercial agricultural nations.*

that characterize commercial mechanized agricultural production.

Because current population patterns in most of the world still reflect the traditional ties to areas where food can be produced, areas that are unsuited for agriculture support relatively few people. These lightly populated or uninhabited nonarable areas, which are identified as the frontier environments in Chapter 4, occupy more than 60 percent of the Earth's land. These include the following:

- **Dry lands**, where lack of precipitation is the limiting factor and where irrigation has not been feasible.
- **Cold lands at the high latitudes**, where frigid temperatures preclude agriculture.
- **Major mountain ranges**, and other mountainous areas where climate is limiting and terrain is too rugged to be cultivated.
- **Wet tropics,** where heavy precipitation and high temperatures combine to produce relatively infertile soils that do not support intensive permanent cultivation, as well as a high incidence of debilitating diseases such as malaria.

Industrialization and modern technologies have played an important role in modifying settlement and population patterns over the past two centuries. Nearly three-quarters of the population in the developed countries now live in urban environments, with many more living near major metropolitan areas. Populations of northern and western Europe are among the most urbanized; more than 80 percent of their 275 million residents live in urban communities. About 75 percent of North Americans are city dwellers. Relatively large numbers also live in city-states such as Hong Kong and Singapore, which have virtually no domestic rural or agricultural hinterlands yet have come into prominence as international commercial centers.

Not only have industrialization and commercialization processes caused a population shift from the countryside to the city, but technologies have created the artificial environments of modern cities and their sprawling metropolitan areas. Rising from the deserts of southern California and Arizona, Los Angeles, San Diego, Phoenix, and Tucson are sprawling, rapidly growing metropolitan areas that are sustained only by importing water via complex systems of canals and aqueducts. In fact, every modern city is a human-made environment that can insulate itself from harsh or inhospitable aspects of the surrounding natural environment.

Over the last century or so, permanent settlements have been established in remote, difficult environments where few, if any, people had lived prior to the development of some local mineral or forest resource now required by technology-based economies. For example, in the 1920s and 1930s major industrial and mining cities were carved out of the vast, virtually uninhabited Siberian forests in the Soviet Union. In arctic North America, permanent settlements such as Prudhoe Bay sprang up as the large oil fields on Alaska's North Slope began production during the 1970s.

Changing Perceptions of Population Trends

Despite the epidemic of HIV/AIDS that is expected to kill more than 15 million people by 2005, Africa's annual population growth (2.5 percent) was the highest of major world regions. In Nigeria, Africa's most populous country with about 123 million people, the annual rate of population growth was 2.9 percent—one of the highest in the world. At this rate, Nigeria's population will double in less than 25 years. At the other extreme, several of the largest European countries, including Germany—the most populous in Western Europe—have experienced small but steady population declines over the last decade or so.

The deteriorating environmental and social conditions in many republics of the former Soviet Union have spurred a devastating demographic decline. Serious environmental contamination and degradation that has persisted for decades in many Soviet industrial and mining centers, coupled with the prolonged post–Soviet period of economic instability, has resulted in a rapid rise in death rates and a continued decline in birth rates. Life expectancy has dropped sharply for both men and women and because the health of women of reproductive age is declining, the infant mortality rate has increased as has the death rate during childbirth. In the two largest republics, Russia and Ukraine, natural population change is currently minus 0.6 percent per year. In 2000 infant mortality rates averaged 16 per thousand—nearly twice as high as the European average.

Although these represent the extreme cases of current demographic trends, population change has always been taken seriously in any organized society, so it is an important issue both in countries where populations are growing and in the negative-growth countries. Perhaps surprisingly, most concern in the past has been with population decline, which was perceived as a weakness of the society or nation as a whole. Population decline indicated that resources that had supported a population at a given level had become insufficient to maintain that population. Unless the population decline could be reversed, it followed, the basic structure of the society itself might become unstable. In addition, periods of population growth usually signaled societal prosperity and progress as the resource base grew. A growing population would, in turn, support specialized and more efficient organization and production strategies, resulting in overall societal advancement.

In more recent times, population growth has been viewed ambivalently. If the resource base is sufficient or appears underutilized, a nation may promote population growth through incentives for natural increases such as "baby bonuses" or substantial tax exemptions for large families or through policies that accept immigrants or promote internal migration. U.S. policies in the nineteenth century that allowed millions of Europeans and Asians to immigrate and the Homestead Acts that subsidized settlement and development of vast portions of the Midwest were examples of programs that promoted population growth.

On the other hand, if land and other critical resources are scarce, population growth may be considered a distinct problem and governments will enact policies to control or contain population expansion. In fact, birth control programs sponsored by governments of many developing countries reflect the desire and need to reduce birth rates in order to slow the rate of natural population growth. China has gained world attention for its national birth control program, which has a goal of one child per family. In 1994, at the United Nations International Conference on Population and Development (ICPD), most nations endorsed a plan to stabilize world population over the next two decades. The World Programme of Action (WPOA) would achieve this goal by expanding the woman's role in family planning through literacy education and through reproductive health and child health care programs that would be available to all.

The mechanisms responsible for temporal or spatial population changes are **birth rates, death rates,** and **migration.** The relationship between the number of births and deaths during a year determines the annual rate of natural increase or decrease. Migration between countries and continents, which played such an important role in demographic changes during the nineteenth and early twentieth centuries, is less important today; however, migration within nations continue to produce significant population shifts.

Natural Population Changes

It is obvious that if births exceed deaths within a given year there will be a net population increase, and it deaths exceed births, population will decline. If the relationship between deaths and births changes drastically, then populations can explode or crash over relatively short periods. On the one hand, epidemic disease, or prolonged famine may result in a rapid increase in death rates within a country or region. On the other hand, widespread inoculations against chronic or communicable diseases, safe supplies of drinking water, and urban sanitation systems can dramatically lower death rates within a generation.

In order to compare demographic data between large and small countries, the number of births and deaths are commonly expressed in terms of yearly births and deaths per thousand people. Those statistics are termed **crude birth rates** or **crude death rates.** The difference between the crude birth and death rate is the population's annual

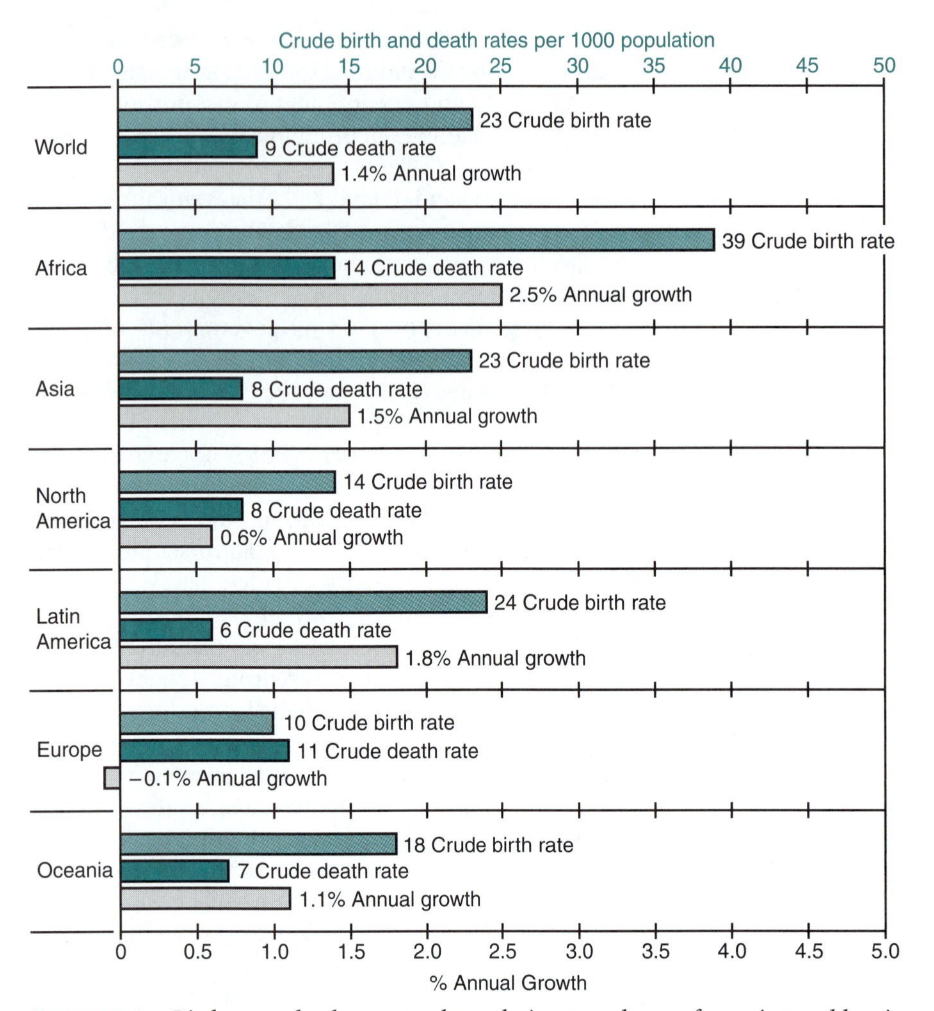

FIGURE 7.5 *Birth rates, death rates, and population growth rates for major world regions. (Source: World Population Data Sheet, 1999, Population Reference Bureau.)*

natural increase or decrease. By multiplying this number by 10, the annual percentage of population change can be calculated. Figure 7.5 charts the crude birth and death rates and the natural population increases for major world regions.

Note that among the major world regions, Africa has the highest natural population growth rate at 2.5 percent per year, down from 2.9 percent in 1990. Poorer countries' rates are the highest—Nigeria alone adds more than 3 million babies each year. In contrast, the relatively few African countries that are more economically developed (e.g., South Africa and Morocco) have annual growth rates less than 1.7 percent. Unfortunately, part of this decline is the result of increasing death rates due to the AIDS epidemic in sub-Saharan Africa.

Across the Mediterranean Sea, Europe's demographic profile is much different. As this century began, Europe was experiencing more deaths than births, so that the annual natural population change was –0.1 percent. This is in part due to the significant increase in death rates as Europe's predominately affluent, urban population has aged.

Although annual population change rates seem small, they can be deceptive for two reasons. First, when a small annual rate is applied to a very large population, it will yield a large absolute change. With the current world population of nearly 6.2 billion, growing at 1.4 percent, about 82 million people are added to the world population in one year. If these additional 82 million represented a new country, it would have nearly the same population as Germany and would be the thirteenth most populous country in the world. Second, the changes are cumulative, for even if the growth rate continues to decline slowly over a period of years, the base population continues to grow each year. By the year 2010, world population will likely be about 6.8 billion.

Another way that demographers compare population growth rates is by calculating the time it takes for a population to double using the current annual growth rate. Figure 7.6 shows **doubling times** for selected coun-

Annual population growth, percent		Population doubling time, years
3.5		
3.4		
3.3	Chad	21
3.2	Congo, Nicaragua	22
3.2	Mali, Liberia	23
3.0	Nigeria, Niger, Saudi Arabia	
2.9	Guatemala, Yemen, Ghana	24
2.8	Iraq, Syria, Pakistan	25
2.7	Cameroon, Paraguay	26
2.6	Laos, SUB-SAHARAN AFRICA	27
2.5	AFRICA, Afghanistan, Ethiopia	28
2.4	Algeria, Cambodia	29
2.3	Philippines, El Salvador	31
2.2	Peru, Sudan, Mexico	32
2.1	Malaysia, Kenya, Ecuador	33
2.0	Egypt, Columbia, Myanmar	35
1.9	India, Tajikistan	36
1.8	LATIN AMERICA, Bangladesh, Iran	38
1.7	Morocco, Malawi, Jamaica	41
1.6	Indonesia, South Africa, Israel	43
1.5	ASIA, Turkey, Vietnam, Brazil	46
1.4	Chile	49
1.3	Mongolia	50
1.2	WORLD, Sri Lanka, Argentina, Zimbabwe	58
1.1	OCEANIA, Thailand, Azerbaijan	63
1.0	China, South Korea, Singapore	70
0.9	Puerto Rico, Iceland	78
0.8	New Zealand, Uruguay	83
0.7	Australia, Taiwan, Cuba	104
0.6	NORTH AMERICA, United States	116
0.5	Kazakhstan, Armenia	151
0.4	Canada, Netherlands	162
0.3	France, Norway, Switzerland	210
0.2	Japan, United Kingdom, Finland	318
0.1	Poland, Belgium, Portugal	900
0.0	Spain, Austria	
−0.1	EUROPE, Italy, Germany	
−0.2	Czech Republic	
−0.3	Estonia	
−0.4	Belarus, Hungary	
−0.5	Russia	
−0.6	Ukraine, Bulgaria, Latvia	

FIGURE 7.6 *Doubling times for selected countries and major world regions. (Source: World Population Data Sheet 1999, Population Reference Bureau.)*

tries and regions at current rates of natural increase. Seventy-one countries with current growth rates between 2 percent and 2.9 percent will double their populations in 24 to 35 years and 14 countries with growth rates ranging from 3 percent to 4.4 percent will double in 16 to 23 years. Together these 90 countries have a combined population of nearly 1.3 billion, which means that about one-fourth of the world's population live in countries whose population may double in one or two generations. India provides a sobering example, for if its current 1.9 percent natural increase continues, its population of over 1 billion would double in about 36 years.

Note that Figure 7.6 also shows that there are a number of countries—most economically developed, urban, and all in Europe—where natural population change is near zero or declining. As a region, Europe is currently experiencing a small natural population decline. If the current demographic trend continues, Europe's population of almost 730 million, which represents about 12 percent of the world's population, will fall to less than 720 million by 2010.

As already mentioned, most eastern European countries that were part of the former Soviet Union have experienced difficult political and economic conditions since the dissolution of the Soviet Union. Such hard times have resulted in population decline. Birth rates in these countries are now the lowest in the world, and death rates are among the highest in the developed countries. Specifically, Eastern Europe, which includes Russia, currently has a population of about 306 million and has average birth rates of 9 per thousand population and death rates of 13 per thousand, that the annual rate of population change is −0.4 percent. As a result, the population of this region is projected to decline by about 5 million over the next decade.

Elsewhere in Europe, outside the former sphere of Soviet influence annual population change fluctuates from just above to just below zero population growth (ZPG). Of the five most populous countries in Western Europe, Germany and Italy have had declining populations for over a decade while Spain is currently at ZPG. France and the United Kingdom continue to add population slowly with annual growth rates of 0.3 and 0.2 percent.

HIV/AIDS Epidemic Slowing Population Growth

For the past decade an **HIV/AIDS epidemic** has been sweeping through parts of sub-Saharan Africa. By 2000 nearly 4 million new HIV cases were occurring annually, with more than 500,000 babies infected at birth or through breastfeeding. In this region more than 5,500 people were dying from AIDS-related causes each day, and since then the death rate has increased. By 2005 the daily toll is expected to reach 13,000. The outbreak has been most severe in Africa's southernmost countries. Over one-third of the adult population in Botswana suffers from HIV. In South Africa, Zimbabwe, and Zambia, more than one-fifth of the adult population has the HIV virus and in several additional countries, infection rates are over 15 percent. As a result, death rates have increased dramatically while life expectancy has plummeted. For example, in Zimbabwe, where HIV infects

TABLE 7.1

CHANGES IN DEATH RATES AND LIFE EXPECTANCY (1990–1999) IN SOUTHERN AFRICAN NATIONS AS A RESULT OF HIV/AIDS EPIDEMIC

Country	% of Adult Population with HIV (1999)	Death Rate per 1000 Population		Average Life Expectancy	
		1999	1990	1999	1990
Botswana	36	33	11	40	59
Malawi	16	24	18	36	49
Namibia	20	36	12	42	56
South Africa	20	27	8	58	63
Swaziland	25	42	15	39	50
Zambia	20	23	14	37	53
Zimbabwe	25	20	10	40	58

25 percent of adults, death rates have doubled in just 10 years, from 10 deaths per thousand people to 20 per thousand. Average life expectancy has fallen from 58 to 40 years. Table 7.1 compares death rates and average life expectancy between 1990 and 1999 among the African countries most affected by the epidemic.

The devastating impacts of the epidemic affect virtually every aspect of these societies. The region's modest economic progress has reversed. Health care systems are being overwhelmed as hospitals fill with AIDS patients. Already scarce capital and medical personnel are being diverted to treat the epidemic. Education has been severely affected, for it is often the better educated population who suffer the highest infection rates. At the start of the century, 30 percent of teachers in Malawi and Zambia were HIV positive. The epidemic has led to labor shortages, and agricultural production has dropped significantly in parts of sub-Saharan Africa. It is expected that over the next 20 years the AIDS epidemic will reduce the economy of sub-Saharan Africa by nearly 25 percent. But most tragic is the specter of the 23 million Africans who will likely die from AIDS over the next 10 years.

Although sub-Saharan Africa has been the focus of the epidemic, by 2000 HIV/AIDS had also become a serious health issue in Asia, where about 25 percent of the world's new cases were occurring. Significant numbers of cases were also being reported from Russia and Ukraine. The virus has established a strong presence in southern India, as well. Infection rates there are expected to double every 14 months. By 2005, AIDS deaths in Asia are expected to reach levels currently being experienced in Africa.

In summary, the deadly HIV/AIDS epidemic in Africa and rising mortality from AIDS in Asia and parts of the Soviet Union is measurably slowing population growth. Recent UN population projections made for 2050 have been reduced by more than 170 million as a result of rising mortality from AIDS.

The Demographic Transition

Current demographic trends reveal that the annual average population increase among the developing countries is more than 20 times that in the economically developed world. A check shows that although the average crude death rates in both groups are low, the average birth rates in developing countries are nearly three times those of the developed countries. Why is there such a huge disparity in the birth rates while death rates are relatively comparable? Demographers recognize close historical ties between the processes of economic development and those of population growth. As a rural agrarian society evolves into a technology-based urban society, there are changes in demographic trends as well. The **demographic transition** model correlates changes in population dynamics with industrialization and urbanization processes associated with economic development. This model, shown in Figure 7.7, illustrates the trends in birth and death rates and the corresponding population change through the period of economic development.

The first phase of the model represents the demographic trends before the processes of economic development began. It portrays the demography of Europe prior to the Industrial Revolution or that of Japan in the mid-nineteenth century, or perhaps even that of a few tribal bands who today still remain isolated in the tropical forests. The common demographic characteristic in each society during this phase is that population is relatively small and stable over time. Note that both birth rates and death rates are very high but that death rates decline during periods of prosperity and rise during times of famine, disease, or war (Figure 7.7).

The second phase begins at the onset of the technological revolutions that characterize the early stages of economic development. In eighteenth- and nineteenth-century Europe and North America, it was the Industrial Revolution that begot transportation, agricultural, and

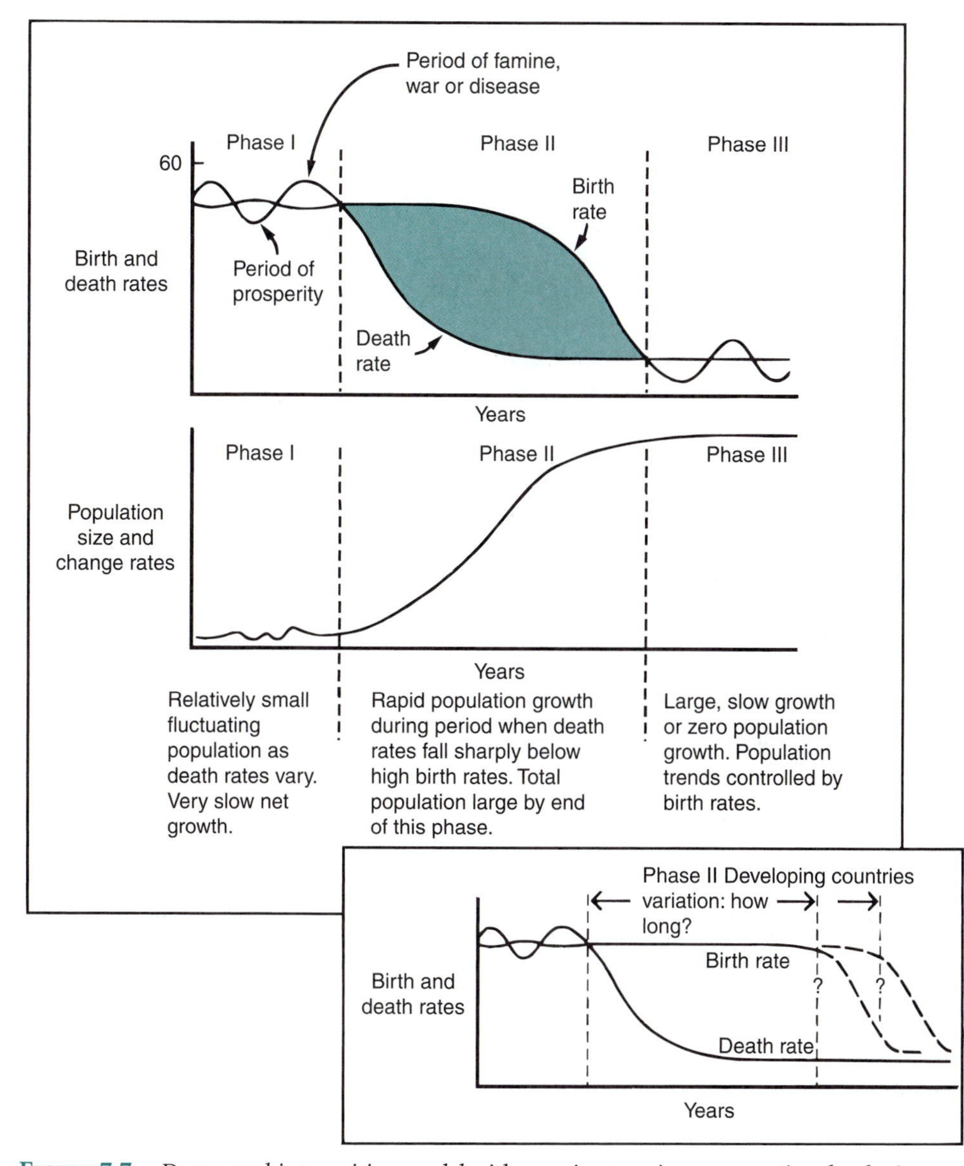

FIGURE 7.7 *Demographic transition model with a variant version representing developing countries in the inset.*

medical revolutions. Together, these gave rise to increased economic productivity that resulted in better diets and higher living standards, particularly in cities, as sanitation and public health systems developed. In demographic terms, such progress led to steady declines in death rates. Birth rates, however, remained high during these initial periods of prosperity and only later began to decline as living conditions continued to improve.

As the gap between high birth rates and declining death rates widens, population grows rapidly (Figure 7.7). When this occurred in Europe during the nineteenth century, literally millions were able to emigrate to the Western and Southern Hemispheres, where immigrants were needed to settle new lands and fill jobs in mines and factories. Unfortunately, the many Asian, African, and Latin American countries that are currently in this phase of rapid growth do not have the option of large-scale migration abroad, so these nations must develop different strategies to deal with their burgeoning populations.

As more people followed jobs to towns and cities, the birth rate began to decline. Urban life simply did not require the labor provided by children on the farm. As laws were written against the abuses of child labor, an additional child became an economic liability rather than an asset. At the same time, urban living conditions and medical care reduced infant and child mortality, eliminating this traditional incentive for high fertility. As decisions were made to limit family size and birth control and family planning measures were adopted, birth rates eventually slipped to levels comparable to the already low death rates. During the later stages of phase two, as the gap between birth and death rates narrows, the rate of growth slows, but now total population is high (Figure 7.7).

In the final phase of the demographic transition model, the momentum for rapid population growth has ended and the population grows slowly, if at all, as demonstrated by recent demographic trends in Europe. Today the number of deaths in Europe's aging populations is actually greater than births. On the other hand, birth rates surged between 1946 and 1964 in the developed countries, producing the baby boom generation. Therefore, it appears from twentieth-century demographic trends in Europe and North America that birth rates may fluctuate—turning lower during periods of economic difficulty such as the Great Depression of the 1930s and higher during periods of prosperity such as the 1950s and 1960s and mid- to late-1980s. During the 1990s birth rates in Eastern Europe and Russian plunged in response to the economic and political restructuring that has been ongoing since the demise of communism in the Soviet-dominated countries and the breakup of the Soviet Union.

Demographic Transition in the Developing World

The demographic transition model is based on the economic and demographic histories of Europe and North America. It may or may not be an accurate predictor of recent demographic trends in the developing world. Here population has grown rapidly during the past several decades due to measures that improved health and increased longevity. In recent years however, social and economic advances and family planning programs have contributed to declining growth rates in many developing countries. Between 1990 and 2000, the birth rates in the developing world dropped from 31 per thousand to 26 per thousand, while death rates declined from 10 per thousand to 9 per thousand. This translates to a net decline in annual growth from 2.1 to 1.7 percent. The most significant reduction in population growth has occurred mainly in some Asian and Latin American countries, which have more or less followed the demographic transition model in which birth rates have declined in response to urbanization and the acceptance of family planning. For example, in Asia, where nearly 60 percent of the world's population lives, population growth rates declined from 1.9 to 1.5 percent during the past decade. Over the same period Latin America's rate of growth dropped from 2.1 to 1.8 percent.

However, most of Africa and some of the poorest Asian and Latin American countries remain, as they have for several decades, in the high-growth phase of the demographic transition because cultural traditions of large families and high fertility have remained strong (Figure 7.6 inset). These societies have not yet experienced the urban-based economic growth that led to the improved standards of living and smaller families in the developed countries and more recently in parts of the developing world. Many families in these poor countries depend primarily on subsistence agriculture, so a large family remains an economic asset and affords a measure of security and care in lieu of social security, medical insurance, pensions, or other financial reserves.

Without the significant economic incentives and social changes that occurred in the developed countries during their demographic transition from rapid growth to slow or zero growth, many poor developing countries seem stuck in phase two, where the demographic gap between birth rates and death rates remains wide and where population growth rates are rapid. Demographic trends in Africa over the last 20 years exemplify how much of the developing world remains in the high-growth stages of the demographic transition model. At the turn of the century, birth rates for Africa averaged 39 per thousand, only 7 less than in 1980, while average death rates had dropped to 14 per thousand, down 5 from 20 years ago. In the same period, the African population has increased by 70 percent, growing at almost 3 percent a year. Currently, Africa continues to grow at about 2.5 percent, in part due to the AIDs epidemic. Although it is unlikely that such high growth rates will continue indefinitely, demographers project that Africa's population will grow faster than that of any other continent and will increase by 30 percent to 980 million by 2010.

Given these trends, there is no assurance that the world's poorest countries will experience the economic and societal changes that led to reduced birth rates in the increasingly urbanized economically developed countries. So for at least a significant part of the developing world, the sharp decline in births that occur at the end of the second phase in the demographic transition model is still speculative, as is the entire third phase where birth rates fluctuate slightly above or below the nearly steady death rate in response to economic or other social factors.

Even though the demographic transition model may not provide a precise scenario of future population trends, particularly in the poorest countries, several features remain valid. First, virtually all nations have experienced a decline in death rates sometime before birth rates began to fall. This *demographic gap* between birth and death rates accounts for the trend of accelerating population growth that the world has experienced for most of the last two centuries. Second, until recently, population changes in the developing countries mainly reflected changes in death rates. However, now that the average death rates for the developing world stand at about 9 per thousand and more than 90 developing countries with youthful populations have death rates that are currently *below* the average death rates experienced in the mature populations of the developed countries, birth rate trends in the developing countries will be a principal determinant of population size, just as they have been for decades in the developed world.

7.4 FERTILITY, AGE STRUCTURE, AND POPULATION MOMENTUM

In addition to crude birth and death rates, two variables play an important role in predicting demographic trends. The first, **total fertility rate (TFR)**, is the average number of children born to a woman. In countries where the crude birth and infant mortality rates are low, the TFR is also generally low. Today, the TFR average for the developed world is 1.5; such a low fertility level is below that needed to simply maintain population levels by replacing both parents. If the TFR remains below 2.1, which is considered to be the **replacement fertility rate** for developed countries that have low infant and child mortality rates, population will eventually begin to decline. As might be expected, the TFR in most developing countries is considerably above the replacement levels. With current rates at about 3.8 children per couple, there remains a strong impetus for continued population growth in the developing world. The potential for rapid population growth in Africa remains high because TFR for the continent is currently 5.4. Because of relatively high mortality rates among infants and children in many developing countries, their replacement fertility rate is about 2.5 children per woman.

In most of the developed countries, current TFRs are below the replacement level. Yet population continues to grow. For example, in the United States, the TFR has been at or below replacement rate for nearly three decades, yet the natural increase in population is about 0.6 percent a year. This growth continues because the number of children now being born to the large cohort of women from the baby boom generation remains well above the number dying each year.

A second useful illustration of current and future demographic trends is to structure population by age groups and sex. The U.S. population and **age structure** from 1970 to 2030 (Figure 7.8) illustrates how the large number of children born during the high TFR years of

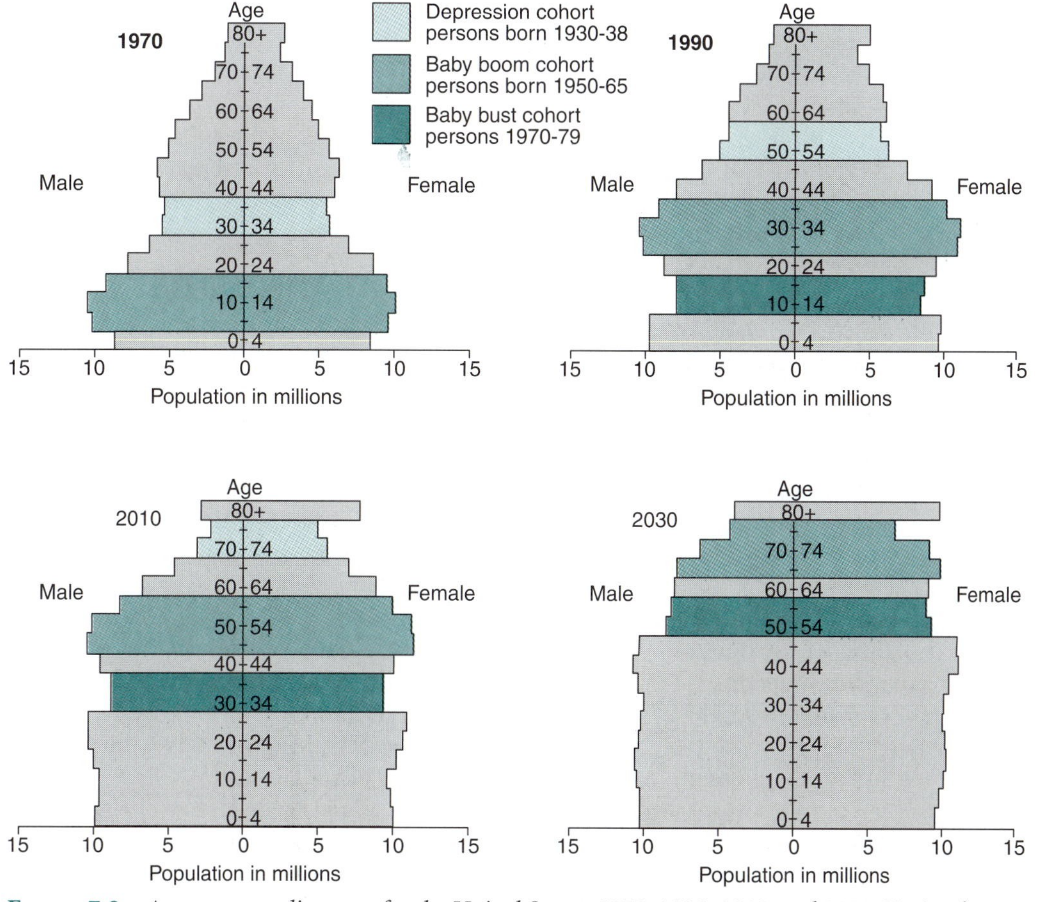

FIGURE 7.8 *Age structure diagrams for the United States: 1970, 1990, 2010, and 2030. Notice the baby boom generation in each diagram.*

the baby boom now represent a large cohort in the child-bearing, family-raising ages. So even as TFRs have dropped below the replacement rate, the large number of women now having children explains the natural population increase. However, when these demographic trends are projected to 2010 and 2030 (see lower diagrams in Figure 7.8) the resulting age structure reveals a large population of elderly baby boomers and relatively small cohorts in the child-bearing years.

An aging population will have a crude death rate that slowly increases as the population ages. Unless the TFR increases, a crude birth rate below that of the crude death rate will provide a natural decline. Nearly all of the European countries currently have aging populations and TFRs below the replacement level, which means that most have, or will soon experience zero population growth followed by slow natural population declines. For example, none of the five largest Western European countries—Germany, United Kingdom, France, Italy, and Spain—have TFRs above 1.7. Population change in these countries ranges from 0.3 percent annual increase in France to −0.1 percent in Germany.

In contrast with the aging population in the developed world, many developing countries have extremely youthful populations due to high fertility and declining child and infant mortality. At the turn of the century, Nigeria, the most populous country in Africa, had a TFR of 6.2. Its population structure illustrates two important demographic characteristics (Figure 7.9). First, the broad three-tiered base indicates the large proportion of the population that is less than 15 years old. Nearly half of the Nigerians are children who, for the most part, depend on their families for support. This means that the number of people in the economically productive age groups (15 to 65 years) is about the same as the total dependent population. Such a large dependent population puts Nigeria and other fast-growing poor countries at a severe economic disadvantage, even compared with other developing countries where the average dependent population is about 34 percent. In contrast, the developed countries have, on average, two-thirds of their population in the economically productive age groups, with 19 percent under 15 years and 14 percent 65 years or older. The second demographic characteristic implicit in such youthful populations is their substantial growth potential for several decades even if their TFRs should decline quickly as a result of effective family planning programs or other cultural change that would promote smaller families.

These large youthful populations in the developing world guarantee continued momentum for significant population growth as they pass through their reproductive years. Demographers estimate that this phenomenon, called **population momentum**, will account for about half of the world's projected population growth during the first 25 years of this century, as some 3 billion people will enter their reproductive years while only about 1.8 billion will leave that phase of life. Even if each couple had only two children—a fertility rate well below the current average in the developing world of about 3.2 children—world population would still expand by about 1.7 billion.

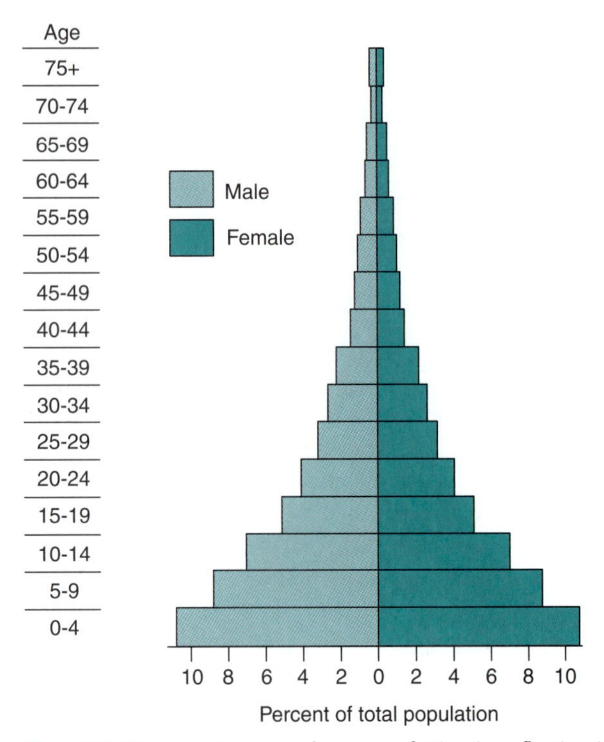

FIGURE 7.9 *Age structure diagram of Nigeria reflecting its very high birth rate. Nearly half the population is less than fifteen years old.*

7.5 FAMILY-PLANNING PROGRAMS IN THE DEVELOPING WORLD

The poorest countries do not have the sustained economic growth and subsequent social changes that induced the birth rate decline among developed countries during the latter stages of their demographic transition. Family-planning programs initiated by the governments of many developing countries have helped to reduce birth rates, but overall, birth and fertility rates remain quite high. The most successful family-planning programs have been in countries with growing urban economies, where the economic incentives for smaller families have been recognized by most.

South Korea provides an excellent example of a country where rapid economic development and urbanization along with effective family planning has resulted in dramatic declines in both fertility and birth rates. Since the mid-1980s South Korea's birth rates have dropped from

23 per thousand to just 15 per thousand, and the TFR has fallen from 2.6 to 1.6. Natural annual population growth has declined from 1.6 to 1.0 percent. As indicators of economic and urban development, per capita income rose more than five times, and its proportion of urban population increased from about 50 percent to nearly 80 percent during the same period. By most demographic and economic indicators, South Korea has made the transition from a developing to an economically developed country.

Elsewhere in Asia, Indonesia and China are predominantly rural countries where strong family-planning programs have been successful in lowering the fertility rate. Even in some of the poorest countries, viable family-planning programs and social change have resulted in declining fertility rates. For example, in the mid-1980s the east African county of Kenya had the world's highest TFR (8.0) and crude birth rate (54 per 1,000). Based on its annual population growth rates of over 4 percent, Kenya's population was forecast to double in less than a generation. However, as a result of effective family-planning education campaigns and clinics that dispense inexpensive contraceptives, birth rates have declined steeply. By 2000 population growth had been cut in half (about 2 percent) as birth rates had dropped to 34 per thousand and the TFR was down to 4.7.

By contrast, many developing countries have ineffective or no family-planning policies. In Latin America, few countries have effective family-planning programs, and fertility rates average nearly 3 children per woman. In many Islamic countries such as Bangladesh, Pakistan, and some Arab countries, birth rates remain high and family planning programs are weak or nonexistent. For many African countries, where fertility rates exceed five children per woman and infant mortality rates of more than 100 per thousand births are common, the large family tradition remains so entrenched and economic conditions are so poor that family-planning programs have had little success.

7.6 A Worldwide Program to Stabilize World Population—A New Approach

In 1994, the International Conference on Population and Development (ICPD), sponsored by the United Nation, was held in Cairo, Egypt. Unlike the earlier UN population conferences held in 1974 and 1984, there was no debate on the need to limit world population growth. In fact, the primary objective of the conference was to develop a plan to stabilize the world's population over the next two decades. This plan, called the **World Programme of Action (WPOA)**, would achieve population stabilization by giving priority to the education of women, providing access to reproductive health care, and by improving child health care and survival rates. Such a program based on education and health care services represents a departure from past plans that focused directly on family planning and set specific demographic targets. By improving the status of women through education and economic opportunities, women will be able to make informed and safe decisions about reproduction and effective contraception, and about their own health as well as their children's health.

There is evidence from some developing countries that such a strategy will work. Thailand provides a striking example. Forty years ago Thailand's population growth rate was over 3 percent, and on average, each woman gave birth to six children. Over the last four decades Thai government policies and programs have emphasized women's education and health. Today, female literacy is 90 percent and women make up 45 percent of the labor force. Women's social and economic empowerment has resulted in dramatic demographic changes as well. More than 70 percent of Thai married women use contraceptives, so the TFR has dropped to 2.0 and annual population growth is now at 1.1 percent and falling.

The WPOA was developed and negotiated by participants from more than 150 countries, including official governmental delegations and unofficial participants from nongovernmental organizations (NGOs) and international agencies. Implementing programs that improve women's literacy and provide health care services will be lengthy and expensive processes. While nothing in the WPOA is binding on governments, this program is endorsed by the world community and is beginning to shape national and international population policies.

7.7 Migration: Where Do People Go?

Migration has played an important role in shaping the social, political, economic history, and geography of many nations. In the past, large-scale migrations between countries and continents resulted in dramatic changes in population patterns over relatively short time periods. In recent decades **international migration** has again been increasing. For most, voluntary migration offers improved economic or other opportunities, but a significant number involuntarily move to other countries as refugees from civil war, political unrest, or (less commonly) because of environmental degradation. At the beginning of the century, the UN estimated that about 120 million people worldwide were living outside of their native countries, including about 15 million refugees.

Internal migration is an even more widespread demographic process. It involves hundreds of millions of people leaving the countryside for cities, or from overcrowded or depressed areas to other regions offering better opportunities.

(a) (b)

FIGURE 7.10 *Refugees of war: (a) Rwandans living in refugee centers after intertribal warfare forced them from their country. (b) A Bosnian refugee camp in Albania after mass exodus from Serbia in 1998 and 1999.*

Some Reasons for Migration

As we noted earlier in this chapter, migrations occur for many different reasons. Sometimes people are forced to leave their homelands, but more often migration is voluntary and sometimes illegal. Today, most voluntary migration is for economic reasons where people temporarily leave their homelands hoping to find work abroad. Such is the case in the Philippines, where several million are legally working in more than 150 other countries. In contrast, thousands of Mexicans illegally cross the U.S. border each year hoping to earn a better living. Some migration, as in the case of Israel, is to establish or return to an ethnic and religious homeland. In a number of developing countries, many are leaving their homelands because of political and social unrest and civil war. For example, in 1994 a million people fled the central African country of Rwanda to escape anarchical conditions that had led to the slaughter of as many as 5 million. During the late 1990s the intertribal strife spread to neighboring Burundi, and a civil war in nearby Zaire (now the Democratic Republic of the Congo) resulted in hundreds of thousands of refugees relocating in Tanzania. Although thousands of these refugees have returned to their homelands, many still reside in "temporary" refugee centers (Figure 7.10a).

Elsewhere, political upheavals in southern and Eastern Europe and in parts of the former Soviet Union have produced scores of thousands of refugees. Longstanding disputes among ethnic groups have erupted as political restructuring occurred after the demise of communist governments. When multiethnic Yugoslavia split into several republics based more or less along the ethnic lines, brutal ethnic cleansing campaigns were initiated. The result was mass murder and refugees streaming from embattled Bosnia-Herzegovina and Kosovo province in Serbia. After the United Nations intervened in both of these regions, some refugees returned, but many thousands remain displaced in refugee camps (Figure 7.10b).

Probably more than 50 million Europeans, Africans, and Asians have come to the Americas over the past 500 years. Nearly all of the 10 million or so Africans who came to the New World before the mid-1800s came as captive slaves, whereas many early Europeans voluntarily sought refuge in America to escape religious or ethnic intolerance and to establish colonies based on their political or religious ideologies. Yet the vast majority of Europeans came to the Americas during the 80 year period of unrestricted immigration that preceded WWI. Most left their poor, overpopulated rural homelands with hopes of finding a better life. Initially they came from the British Isles and Scandinavia and later from Germany, Poland, Italy, and other European countries. During this same period Chinese immigrants provided much of the labor to build cities, roads, and railroads along North America's Pacific shores. Although most emigrated primarily for economic reasons, kinship, religion, and ethnic ties drew groups who shared these cultural characteristics to specific regions, cities, and even neighborhoods.

North American Internal Migration

As North America's settlement frontier spread westward during the nineteenth century, two related but very different internal migrations were taking place. First was the migration of pioneers who settled the vast interior as land was made available for homesteading by the federal government and the transcontinental railroads. At the same time, many Native American people were forced from their traditional tribal lands and made to resettle in remote reservations. In the United States both of these internal population movements were instigated by government policies and played an integral role in the development and settlement patterns of nineteenth-century America.

At about the same time, the frontier was closing, another internal migration trend was emerging: Factories and other businesses were growing in cities. With these new jobs came people from rural areas, and before the twentieth century was a decade old, more than half of North America's population lived in cities. This rural-to-urban migration continued through the first half of the twentieth century, with many people streaming from the rural South to northern industrial cities such as Detroit and Chicago.

As these industrial cities aged and the U.S. economy expanded and matured, service-oriented businesses that were not tied to traditional industrial centers sprang up in southern and western cities. This encouraged migration from the industrial Frost Belt to new growth centers and job markets in the Sunbelt, in cities such as Atlanta, Dallas, and Orlando. Since the 1960s the influx of retired people has added another, different, demographic component to many Sunbelt communities in Texas, Florida, and Arizona.

Just as significant migration followed industry to the U.S. South and West in the 1960s and 1970s, a similar internal migration is occurring in northern Mexico today. Since the North American Free Trade Agreement (NAFTA) went into effect, U.S. companies that built manufacturing facilities in Mexico during the late 1990s have hired more than 600,000 new workers. Because most of these factories are in the lightly populated, arid regions of northern Mexico, many thousands have migrated to cities such as Torreon, Chihuahua, and Reynosa.

In less than a decade the population of Torreon has doubled to 1 million as more than a score of U.S.-owned factories have opened there. Such rapid, mostly unplanned urban and industrial growth has been accompanied by urban sprawl, air and water pollution, and depletion of surface and groundwater resources (Figure 7.11).

Some Demographic Effects of Migration

Because the demographic characteristics of an area are modified by immigration and emigration, significant economic and social impacts often occur. Those most likely to participate in voluntary migration are the young, the enterprising, and often the well-educated job seekers. As a result, areas that have experienced significant out-migration are left with an older, less productive, poorer population that must bear the burden of declining tax bases and services. Although areas receiving an influx of new people experience economic growth, they also face the expensive task of upgrading, expanding, or even building new transportation, water, and sewage-treatment systems, as well as providing education and health-care systems required by a young, growing population. The combination of massive populations and high rates of consumption has also placed great stress on the environment in many expanding urban centers such as Los Angeles, the San Francisco Bay area, and Miami. In retirement communities, such as those in Florida and Arizona, expanded medical services and care facilities for mature and elderly residents are required.

FIGURE 7.11 *Urban sprawl in northern Mexico in response to more than a half million new industrial jobs associated with the North American Free Trade Agreement.*

Migration and Environmental Impact

The term *Dust Bowl* emerged to describe parts of the western Great Plains in the drought and depression years of the early 1930s. People who had homesteaded these lands and built towns and service centers just decades earlier were forced to abandon the land and move on. Much of the land, ill-suited for annual cropping, suffered severe erosion and was no longer productive. This area—once a burgeoning region of economic development—became a crippled, unproductive land, deserted by those who had inadvertently ruined it. Like many other settlement schemes before and after, it resulted in huge economic losses as well as untold human suffering, and it inflicted long-term damage to the environment.

Today in many developing countries, similar scenarios are unfolding. Forests are cleared, hillsides are planted, and grasslands are plowed as these places become home for impoverished, landless settlers forced from their overcrowded homelands. Most of this new development does not appear to be sustainable and is having a severe environmental impact.

Similarly, the migration of the poor to the cities results in the expansion of the unhealthy, teeming slums that overload the urban infrastructure. The result is increased water and air pollution and overall environmental decline in the midst of abject poverty and human suffering. Such migration is taking place on a massive scale in South and Central America, West Africa, and Southeast Asia, and the resultant environmental and health problems are overwhelming (Figure 7.11).

Even large cities designed to support a million or more people, are being overwhelmed by migrants. For example Lima, Peru, now has a population exceeding 6 million, an increase of 5 million since 1960. Even faster growing and more crowded is Lagos, the former capital of Nigeria, which has experienced a fifteen-fold population increase in the past 40 years. With population now estimated at over 11 million, it is the world's most crowded city, with a population density of about 100,000 per square kilometer (260,000 per square mile).

7.8 Problems Faced by Developing Countries with Rapidly Growing Populations

Recent experiences of poor countries in different parts of the world provide examples of a wide range of social, economic, political, and environmental problems that are associated with rapid population increases. Earlier in this chapter we saw that high fertility rates and large families are a natural human response to basic survival instincts in poor countries where child mortality is high and life expectancy is shortened due to malnutrition, disease, and inadequate health care, and where the labor provided by an additional child will provide another increment of food or income that might help sustain the family or provide support for elderly family members no longer able to work.

This scenario, which occurs in the world's poorest countries makes real economic growth and improved living standards difficult or impossible to achieve even when absolute production increases, because each year there are more people to share the increments of food, goods, or money. As a result, per capita measures of production might have no increase and, in some years, might even decline. So it seems that rapid population growth, poverty, hunger, poor health, and high child mortality rates are interrelated components of a self-perpetuating cycle. This **poverty cycle** is the fundamental reason for many of the specific consequences of rapid population growth. The poverty cycle is summarized in four interrelated phases as follows and is illustrated in Figure 7.12.

- **Reduction in the quantity and quality of food consumption.** Although enough food is produced each year to adequately feed humankind, allocation and distribution inequities occur and hundreds of millions are poorly nourished—some face starvation. It is the poor who are not adequately fed, as they have neither adequate resources to produce their food nor money to purchase it. Furthermore, most live where food marketing and distribution systems are poorly developed or nonexistent (Figure 7.12a).

- **Expansion onto less productive land use.** In fast-growing, poor countries where most arable land is already under cultivation and much of the labor force is engaged in agriculture, there is increased pressure to expand agricultural land and to use land resources more intensively. Often this means that crop production expands onto erosion-prone hillsides, onto soils ill-suited for continuous cropping, into semiarid rangelands where precipitation is often unreliable or inadequate to support agricultural production, or into deserts where irrigation projects must divert water to permit production. Expansion onto such marginal lands often results in only short-term production gains that are not sustainable, while at the same time damaging or destroying other resources (Figure 7.12b).

- **Increasing unemployment and underemployment.** Without rapid economic growth in poor countries there are far too few new jobs available for the increasing number of young people entering the labor market. The result is widespread unemployment among those seeking jobs in the relatively underdeveloped manufacturing and service sectors and significant underemployment among those seeking agricultural labor. Such massive unemployment and underemployment further swells the already large ranks of the economically nonproductive portion of the population and further retards real economic growth. Coupled with land degradation and

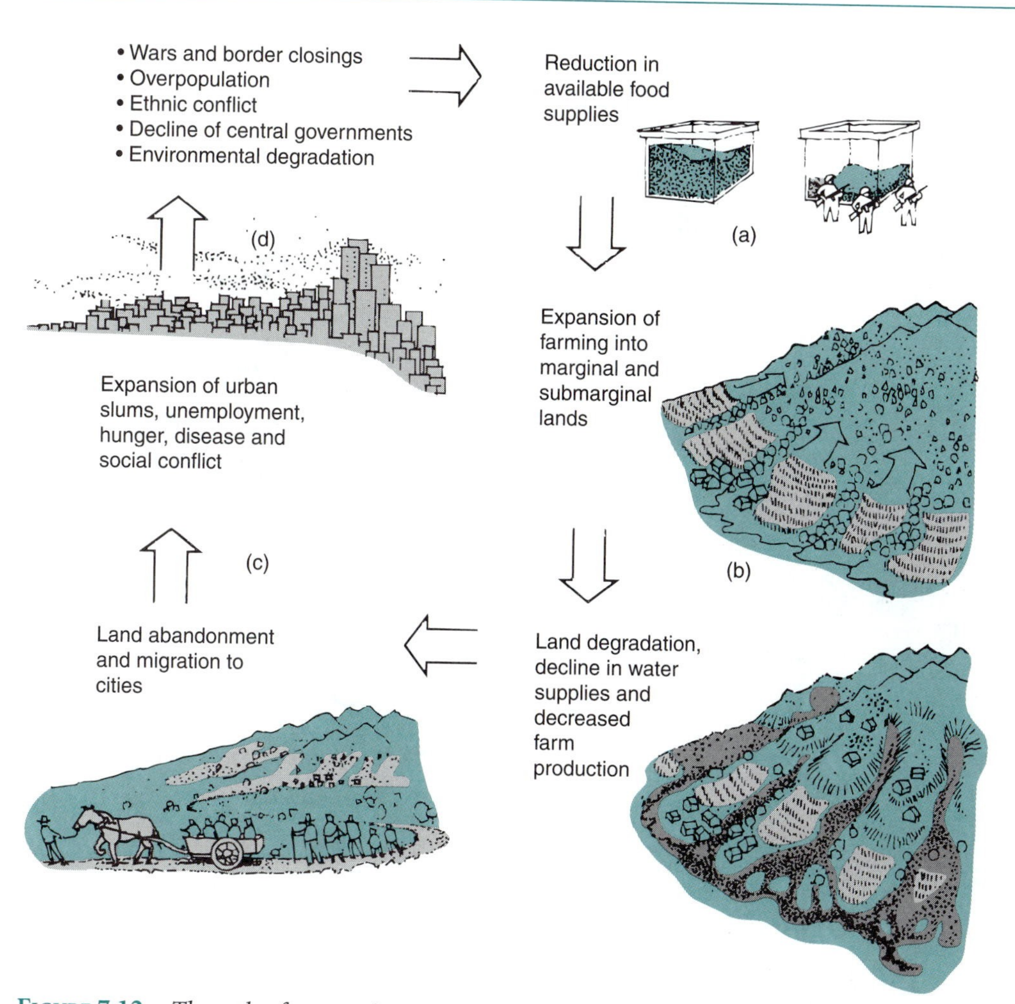

FIGURE 7.12 *The cycle of poverty in poor countries beginning on the upper right with declining food supply.*

the absence of real opportunities to farm additional land, the only option is to move (Figure 7.12c).

- **Expansion of urban poverty.** In many rapidly growing countries, underemployment and unemployment among landless agricultural workers has resulted in large-scale migration to cities to seek employment. Most who come to the cities have little education and few skills but have no other choice but to compete for the limited number of low-paying service and unskilled manufacturing jobs available in large cities. Most will not find regular jobs and will settle in the overcrowded shantytown slums or squatter settlements that have grown up on the outskirts of many cities in the developing world. Makeshift housing is fashioned from scraps of metal, wood, plastic, and even cardboard. Generally there is no dependable water supply or sanitation system, so outbreaks of disease and chronic poor health are characteristic of these areas. These teeming slums become the breeding grounds for additional sets of economic and social problems and may also contribute to social unrest, political instability, and civil disorder (Figure 7. 12d).

Economic and Environmental Problems of a Fast-Growing Developing Country—A Case Study of Brazil

With more than 170 million people and a rapidly growing youthful population that may double in size in the next 50 years, Brazil provides an excellent case study of the problems that face many developing countries worldwide (Figure 7.13). In Brazil's impoverished northeast, deforestation and desertification are driving malnourished refugees from the worn-out land to seek jobs in cities or to government-sponsored settlement programs in the Amazon basin. While in the productive agricultural regions of the south, the landowners' decision to grow export crops such as soybeans and citrus on large mechanized farms has displaced thousands of tenant farmers, who have also migrated to the cities or the Amazon frontier in search of a future. For many, the future looks as bleak as the *favelas* or shantytown slums that ring Brazil's two great cities of Sao Paulo (28 million) and Rio de Janeiro (11 million) and the other Brazilian cities having a million or more residents (inset Figure 7.13).

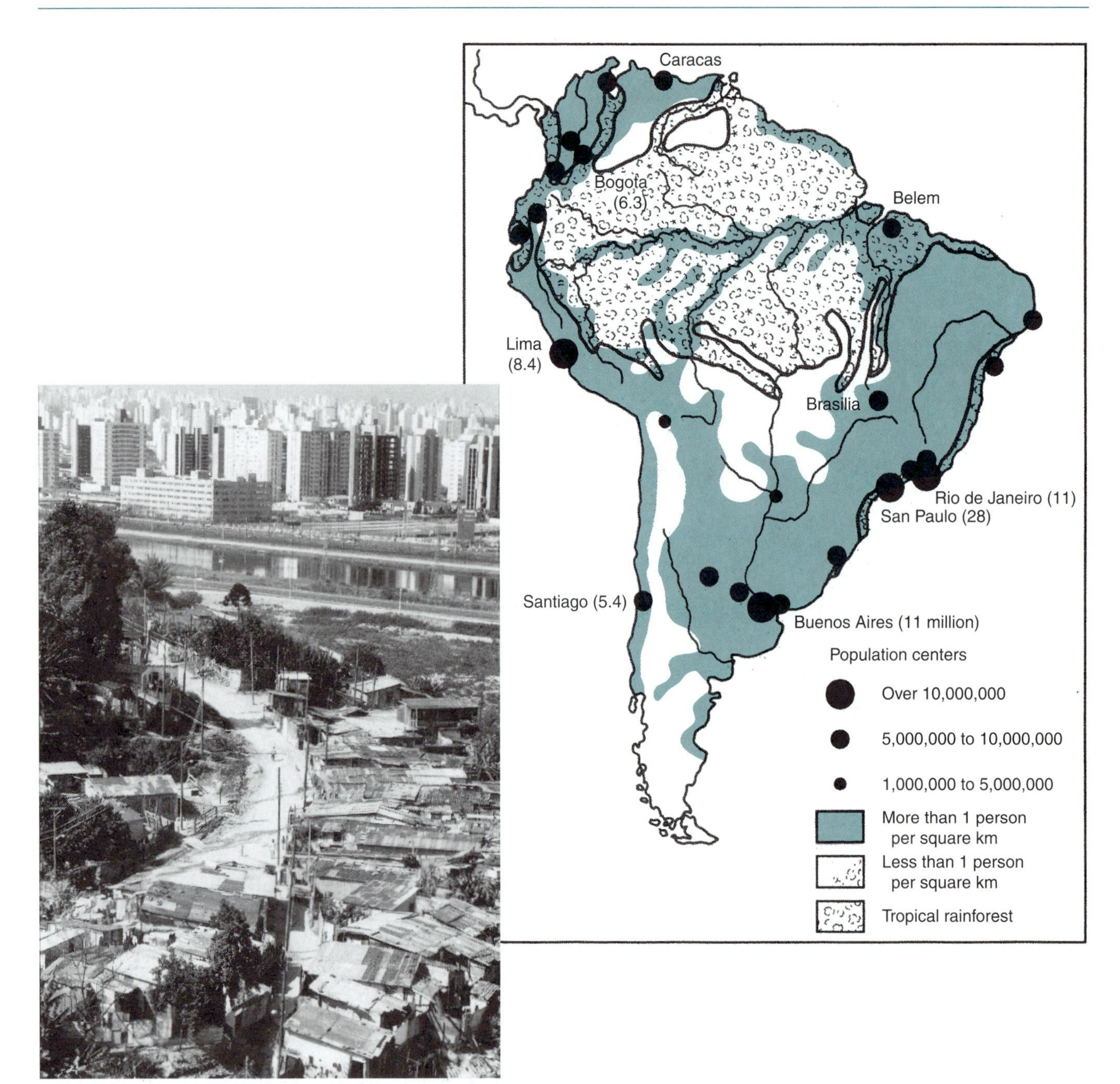

FIGURE 7.13 *The distribution of population, major urban centers, and tropical rainforest in South America. The inset photo shows a shantytown (foreground) in Sao Paulo, Brazil.*

Life seems nearly as desperate in the Amazon basin, where settlers try to scratch out a living growing subsistence crops on the infertile tropical soils. The government-sponsored settlement projects were conceived as a way to bolster economic growth and provide an outlet for Brazil's growing population. As a result, settlers have cleared as much as 25 percent of the rainforest during the past 20 years. However, little sustainable agriculture has been realized because the soils in most areas cannot support continuous crop production. Much of the cleared land has been converted to rough pastures where cattle are raised for low-grade beef to export to North America and Europe.

Because of its large area and vast resource base, Brazil's economy is still growing and its per capita income puts it in the upper tier of developing countries. However, its future economic prospects remain uncertain with its population growing by nearly 1.5 percent per year and about one-third of its people under the age of 15. Where will the 30 to 40 million Brazilians who will be entering the labor force over the next 15 years find employment, and where will the 2.5 million added to Brazil's population each year live? The rainforest frontier is already beginning to close, and the cities where 78 percent of the population lives are already overcrowded and ringed with slums.

A Neo-Malthusian Specter

About the time the Industrial Revolution began to affect changes in the economic and demographic structure of Western Europe, an economist named Thomas Malthus observed that while human population could, in theory, grow at ever-increasing rates (near-geometric), food production could not. With population beginning to grow very rapidly as industrialization transformed the economy, Malthus was concerned that a surging population would eventually outpace food production.

Although technological advances unforeseen by Malthus have kept overall food production slightly ahead of population growth for most of the last two centuries, Malthus's concerns may still be valid. Some very disturbing trends are beginning to emerge in some of the world's poorest countries. National governments have been unable to cope with surging population growth, deepening poverty, declining health, and frequent food shortages. Without effective government policies, economic and environmental conditions worsen. Displaced from overcrowded and worn-out agricultural areas, rural refugees crowd into shantytowns and squalid slums. Many suffer from debilitating diseases such as malaria. As already noted, in sub-Saharan Africa, AIDS is spreading rapidly though the entire population. Such chaotic conditions result in crime and social unrest. Deep-seated tribal, ethnic or clan rivalries and unresolved disputes can erupt in violence and, in some cases, civil war. Such conditions of anarchy have already beset several African countries including Somalia, Sierra Leone, and Rwanda, while several other countries, particularly in West Africa, teeter on the brink of anarchy.

Does the current decline and possible demise of these increasingly poor and dysfunctional countries portend a grim future for other poor countries with rapidly growing populations? Indeed, any nation may face a similar plight if demographic, social, economic, and environmental conditions fall to the same depressed levels that have led to chaos and anarchy in the aforementioned African countries. As rural populations grow rapidly, only a minority can be absorbed into the agricultural work force. Most who are landless and impoverished will become rural refugees who will crowd into ramshackle shantytowns that spring up around major cities. Most will not find productive work. Such squalor and poverty foment dissatisfaction, crime, and anarchy, especially among the young who do not share past values and traditions of their culture and do not see hope for the future.

7.9 Perceived Problems in Developed Countries with Stable or Declining Populations

By the beginning of the twenty-first century Europe's population had essentially stopped growing. In fact, virtually all of the Eastern European states that were part of the former Soviet Union have declining populations. Even in Western Europe two of the largest countries, Germany and Italy, currently have negative population trends. Elsewhere in the developed world, natural population growth continues to slow. As the rest of the developed world moves toward stable populations and the population ages, what types of problems might they face? A common concern relates to the increasing population of dependent elderly who will have to be supported by an aging and shrinking labor force. Such demographic trends could result in declining rates of economic growth.

However, a case may be made that depicts some positive effects of a stable population. With lower birth rates there will be fewer children to support and educate, and a stable or slow-growth economy might result in less pressure on the domestic and global resource base, which, in turn, might reduce environmental deterioration and pollution. Although some economic policy makers and politicians fear that stable populations and slowing economic growth may lead to declining natural prosperity and security, creative national strategies that focus on the well being of a stable population may prove beneficial for society and environment alike.

7.10 Two Alternative Views of Future Population Trends

In the preceding pages we have presented current population trends and projections from the perspective of most demographers. There are, however, other views about future population trends that differ from the convention of the demographic transition model.

The **fertility opportunity** hypothesis maintains that rising living standards in the developing world will generate expectations of a better life, which, in turn, will lead to an increase in fertility and population growing faster than anticipated by the demographic transition model. Proponents cite the examples of rapid population growth in nineteenth-century England that occurred along with industrialization and the post–WWII baby boom in North America that coincided with a generation of economic prosperity. Critics, including most demographers, concede that perceived affluence likely boosts birth rates but point out that many other factors also influence fertility, such as the availability of contraceptives and the growing percentage of women in the workplace. Indeed, a comparison of birth rates and total fertility rates of the economically developed countries between 1986 and 2000 (by most measures one of the most affluent periods in North America, Western Europe, and Japan) showed significant declines for both measures over the period. In 1986 birth rates in the economically developed world were 15 per thousand but had dropped to 11 per thousand by 2000. During the same period total fertility rates fell from 1.9 to 1.5.

The contrasting **birth dearth** view contends that the decline in fertility rates in most countries over the last 25 years represents a trend that will likely continue over the next several generations and result in a *declining* world population by the second half of the century. Since the mid-1970s, the total fertility rate in the developed world has remained well below the replacement rate of 2.1 children per woman. Furthermore, during this generation-long period, the TFR has continued to fall in almost all countries in the developed world. As already noted, the average TFR in the developed world at the beginning of this century was 1.5 children per woman.

In the developing world similar trends are occurring. During the past quarter of a century the average TFR among the developing countries has dropped from 4.7 to 3.1 and in several populous countries including China, Thailand, South Korea, and Taiwan, the TFR has fallen below the 2.1 replacement rate. Today nearly 45 percent of the world's people live in countries where the TFR is below the replacement rate. It appears that as economic development expands in many developing countries, TFRs will continue to decline. In fact, if the average TFR in the developing world continues to decline at about the same rate as the past 25 years (4.7 to 3.1), the world average TFR will be well below the replacement rate and world population will begin to decline over the next few generations.

In closing, we must point out that while neither of these views of future population trends represents the current conventional or majority view of demographers, population projections are at best educated guesses. Barring nuclear wars or global epidemics, world population will continue to grow, with more than 98 percent of the growth occurring in the developing world. As living standards improve and consumption rates rise in many if not most countries, even small increments of population growth will put additional pressure on the environment. To maintain a viable environment during this period of population growth it is essential that the global community focus on sustainable economic and resource development strategies.

7.11 SUMMARY

Some of the poorest countries are experiencing environmental degradation as domestic food production intensifies or expands onto marginal land in response to increased demands of a rapidly growing population. This is the typical example of overpopulation—too many people living in an area for the resource base to support without being degraded or depleted. But perhaps a more pervasive and far-reaching environmental impact, having even global implications, comes from the very affluent populations of the developed countries. For even though the population of these countries represents just one-fifth of the world's population, these same few people use 70 to 80 percent of the world's fuel and mineral resources.

The question then comes to mind: Are these affluent countries overpopulated too? If overpopulation implies degradation and depletion of the resource base, then they must be seriously overpopulated. The lifestyle of those living in most developed countries produces environmental impacts not only where they live, but also wherever their resources are being extracted and processed.

The clearing of tropical forests in Central and South America may have been undertaken primarily to provide cropland for the exploding population of landless peasants, but as most agriculture there is not sustainable, much of the land is now being used as rough pasture to produce livestock for hamburger or even pet food for the developed world. Other products from the tropical forest, such as exotic hardwood lumber and wood pulp from less desirable species, also find their way to markets in developed countries. On the one hand, overpopulation in poor countries establishes a cycle of poverty to which most people are inextricably bound and results in significant environmental degradation on a local and regional scale. On the other hand, slow population growth in the wealthy countries, coupled with increasing consumption of more commodities, products, goods, and services, is threatening ecosystems worldwide.

Populations and resources have been closely linked throughout human history. The development of agriculture marked a significant expansion of the human resource base, so that relatively large populations could be supported where a reliable and plentiful food supply could be produced. Through the millennia as agriculture spread and improved, the world population grew slowly, but it was the series of technological revolutions that began about 250 years ago that provided the resources to support the population explosion that continues today. Quite simply, these revolutions in transportation, production, communication, and medicine have lowered the death rate and provided the resource base to support huge populations.

Each year the huge and growing human population taps vast quantities of renewable and nonrenewable resources at rates that cannot be sustained indefinitely. As we noted these resources are not equally shared among the world's nations; the vast majority go to the affluent countries. By this measure, the environmental impact produced directly or indirectly by the relatively small proportion of people living in wealthy countries is far greater than that produced by the much larger population of the developing world. Nevertheless, the problem is shared by all humankind and will grow as world population continues to grow over the next several decades. Projected family size data reveal that the total fertility rate will likely remain high in Africa, Latin America, and Asia well into the next century (Figure 7.14).

As a result, it is likely that over 98 percent of people added to the earth in the next thirty years will be born in developing countries. Coupled with increased consumerism in many of these countries, the impact on resources and environment will be extreme. As a result, nonrenewable fuel and mineral resources will be depleted at alarming rates and resources such as soil, groundwater, and rangelands, which are normally renewable, will be damaged or even depleted under ever-intensified use. Even vital resources thought to be unlimited, such as the atmosphere and the oceans, will be profoundly affected by the vast human presence.

Of course, the bleak future portended here is based only current demographic projections. Future demographic trends may be more positive if the population stabilization program endorsed in 1994 by the world community at the International Conference on Population and Development (ICPD) can be implemented. Emphasizing women's education and universal health care for women and children, such as women's empowerment programs, have been responsible for dramatically reducing population growth in some emerging countries such as Thailand. Whether these comprehensive and expensive social and economic reforms can or will be implemented throughout the world is unknown. However, this new paradigm for population policy may provide a practical solution to out most pressing environmental problem—overpopulation.

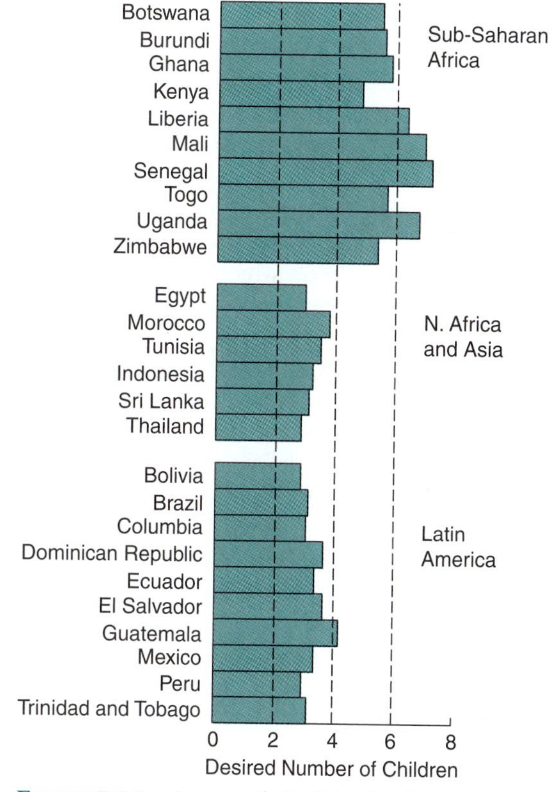

FIGURE 7.14 *Average desired family size among married women in selected developing countries.*

7.12 KEY TERMS AND CONCEPTS

Technological revolutions
- Industrial Revolution
- medical advances

Population patterns
- temporal and spatial changes
- population size
- population density
- physiologic density
- patterns in developed and developing countries

Natural population changes
- birth rates
- death rates
- migration
- crude birth and death rates
- doubling times
- HIV/AIDS epidemic

Demographic transition

Fertility and age structure
- total fertility rate (TFR)
- replacement fertility rate
- age structure
- population momentum

Family-planning programs in developing countries

World Programme of Action (WPOA)

Migration patterns
- international migration
- internal migration

Poverty cycle

Neo-Malthusian specter

Perceived problems associated with stable or declining populations

Alternative views of future population
- fertility opportunity
- birth dearth

7.13 QUESTIONS FOR REVIEW

1. After millennia of relatively slow population growth, the human population began to increase rapidly about 300 years ago. Discuss the reasons for this population explosion.
2. Describe the role that the physical environment (climate, soils, terrain) has played in shaping current population patterns. Explain why the correlation between population density and productive agricultural lands has changed.
3. Figure 7.2 depicts areas with dense population. Describe the differences between the densely populated areas in the developing world and those found in the developed countries.
4. In some parts of the world, most people now live in cities. Where are the most urbanized populations, and why have cities grown rapidly in recent centuries? Explain why even in predominantly rural countries cities are growing rapidly.
5. The demographic transition model correlates population change with the industrialization and urbanization processes associated with economic development. Describe the basic stages of the demographic transition model. Explain why this model may not be a good predictor of demographic trends in some of the poorest countries.
6. Describe the differences in the total fertility rates (TFR) and age structures between the developing and developed countries. Why are these two variables useful in predicting demographic trends?
7. With population growing by 2 to 3 percent each year, it is difficult for many developing countries to realize real economic growth and improved living standards. As a result, most people live in poverty. Describe some of the specific consequences of rapid population growth that perpetuate this poverty cycle.
8. During the past 50 years, world population has doubled and population patterns have changed. Which regions have experienced the most rapid growth? Which have had moderate to slow growth? Are these trends likely to continue? How will world population patterns likely change during the next 50 years?

8

AGRICULTURE, FOOD PRODUCTION, AND HUNGER

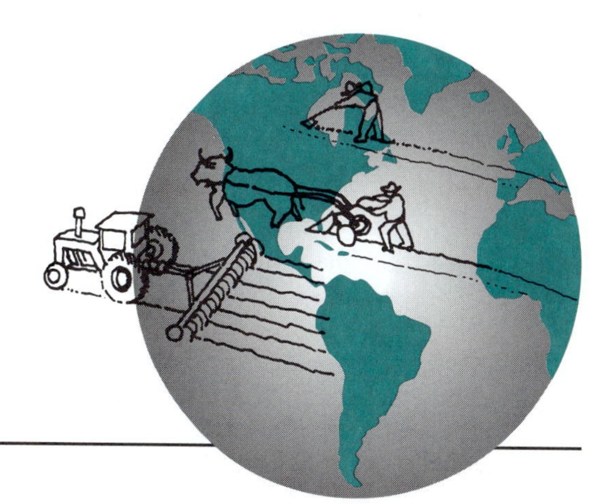

8.1 INTRODUCTION

Without question, the most widespread and environmentally significant human impact on the environment results from food production. As a result, most of our landscapes bear the imprint of crop or livestock production. Indeed, in areas where soils and climate are most favorable, the managed food production system that we call agriculture defines the landscape. Today nearly 37 percent of the world's land area is given over to agricultural production as cropland, pasture, or rangeland.

Before agriculture, people gathered their food from the environment, utilizing wild plants and animals. Since relatively few people could be supported through these naturally limited food resources, humans had little impact on the environment. Not until agriculture began to evolve about 12,000 years ago did human beings begin to make significant environmental alterations. Not only did agriculture alter the environment directly, but it revolutionized the way humans lived. Agriculture provided a bountiful and dependable food supply that led the economic and cultural advancement. Once established, agriculture set into motion processes that led to rapid population growth, urbanization, and eventually the specialized production and transportation systems that have shaped the economies and landscapes of the world. For many of us, our ways of life seem only vaguely related to agriculture, but without today's global systems of specialized agriculture our lives would be very different. Through agricultural food production, a large share of the population is freed from the land for other economic pursuits.

In this chapter we will see how the evolution and spread of agriculture supported population growth and the rise of civilization. We will trace the development and dispersal of the major agricultural systems that feed the world. Next, we will examine the flaws in our agricultural systems that result in widespread hunger and environmental degradation. Finally, we will discuss recent efforts to expand food production, reduce environmental impact, and make our agricultural systems more sustainable.

8.2 Cultural Evolution and the Development of Global Agriculture

Before agriculture humans had little ability to manage, manipulate, and modify the energy and material cycles that govern plant and animal growth. But once agriculture was established and humans learned to manage and husband food sources, the seeds of human ability to modify the environment were sown. Now humans would replace the natural vegetation cover with their cultivated crops and would alter soil characteristics as their crops drew out nutrients while their irrigation water concentrated minerals and saturated soils. Soils left bare before and after planting were exposed to the erosive forces of wind and water, and crops planted in rows channeled runoff and hastened erosion.

With food resources stabilized, permanent human settlements grew up around these food-growing locales. As food supplies became more plentiful and dependable, more people could be supported and societies became more complex. Urban centers began to evolve whose residents, freed from food-production labors, developed specialized occupations required in complex societies—merchants, craftsmen, laborers, and religious and government officials (Figure 8.1).

Great civilizations were built on the foundation of successful agricultural systems. Agricultural bounties reaped from the irrigated valleys of the Tigris and Euphrates rivers gave rise to the great Sumerian and Babylonian cultures between 5000 and 2500 B.C. Somewhat later, the Egyptian civilization reached its zenith by using the annual flood of the Nile to nurture a productive agricultural system. Advanced Oriental civilizations also evolved in the fertile irrigated valleys and deltas of the great Asian rivers from the Indian subcontinent to China. Just as these cultures grew and flourished with an efficient productive agricultural infrastructure, most declined or stagnated as their agricultural systems grew unsustainable before Christ's time (Figure 8.1).

Over the next two millennia, successful food production systems were developed in most parts of the world. In areas where soil and climate would support crops year after year or where irrigation could be readily undertaken, **sedentary agricultural systems** with permanent fields and villages emerged. Sedentary farming was usually based on only a few main plants and animals and a dozen or so secondary ones. In the arid lands of Asia and Africa, the nomadic form of agriculture became established. **Nomadic herding** is a form of animal husbandry in which herders move their livestock with the seasonal changes in pasture forage. With relative freedom of movement over established grazing lands, an equilibrium developed between humans and the environment that ensured the sustainability of this system.

In the forested lands of the wet tropics the heavily leached, nutrient-poor soils could only sustain short-term crop production cycles. In response to this limitation, a form of **shifting cultivation** emerged based on temporary fields that produced several seasons of crops before the soils were exhausted. The fields were then left fallow for several decades as the forest cover and soils reestablished themselves and the area again became available for crop production. Although this is a sustainable agricultural system well-suited to the tropical forests, it can support only relatively small populations. Nomadic herding and shifting cultivation will be further, discussed in the section on subsistence agriculture, later in this chapter.

Prior to the Industrial Revolution, increases in agricultural productivity were achieved in Europe by adopting a crop rotation system that made it unnecessary to leave up to one-third of the agricultural land idle each year. In *crop rotation*, fields that have been exhausted are planted peri-

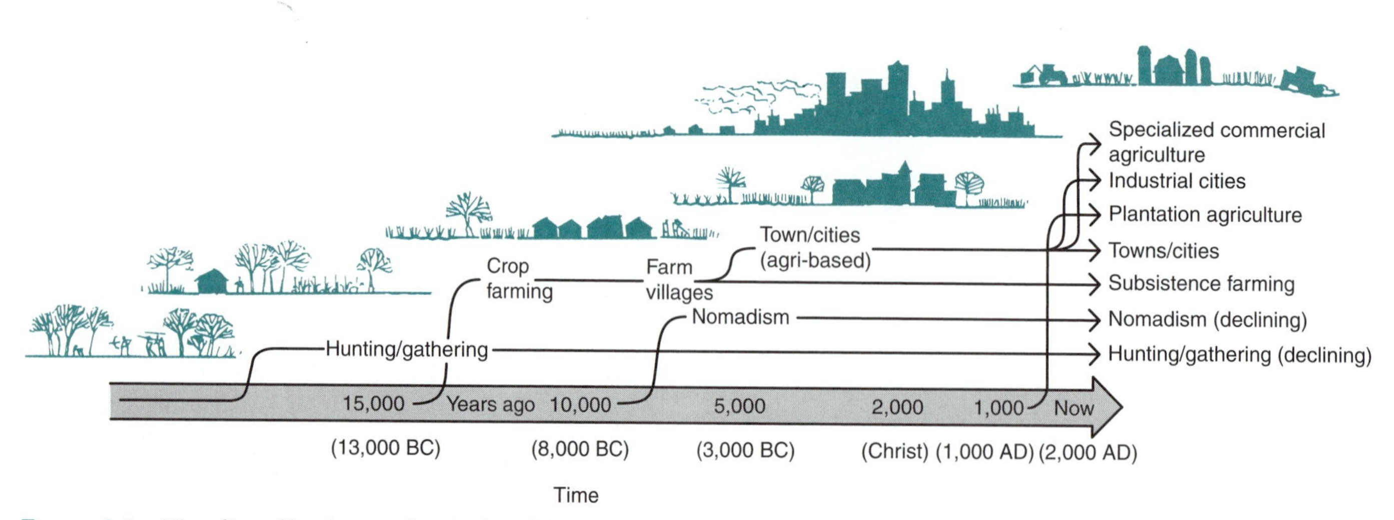

Figure 8.1 *Time line of land use and agricultural development over the past 15,000 years.*

odically with a fertility-restoring crop such as clover. This important advance resulted in immediate productivity increases and played a central role in the Renaissance of Western Europe beginning in the fourteenth century. Indeed, this agricultural revolution provided an important base for the technological revolutions that began in the eighteenth century and continue today.

At about the same time, the European powers began to establish colonies in the Americas as well as parts of Africa and Asia. Their explorations, conquests, and subsequent settlement led to the first system of interregional and transoceanic trade in agricultural commodities. These colonial outposts would specialize in crops such as cotton, sugar cane, and rice that could not be produced in Europe. As demand for these exotic crops grew, the **plantation system** emerged, which marked the beginning of large-scale, specialized commercial agriculture (Figure 8.1).

Exploration and settlement resulted in many other transformations in agriculture. Worldwide *diffusion and exchange* of important agricultural species occurred as explorers and traders returned home with plants and animals integral to indigenous agricultural systems, while immigrants and settlers took with them the staples of their agricultural traditions. For example, potatoes, which were native to the Andean highlands, flourished in the cool, damp environment of northern Europe and quickly became an agricultural staple. Likewise, the American grain staple, corn (maize), spread across the world to become the third most widely grown grain after rice and wheat.

8.3 INDUSTRIALIZATION AND THE CHANGING NATURE OF AGRICULTURE

The technological revolutions that spread through Western Europe and across the Atlantic to North America in the eighteenth and nineteenth centuries resulted in revolutionary changes in agricultural practices. More efficient and more specialized agricultural implements changed the character, scale, and geography of agricultural production (Figure 8.1). Innovative agricultural machines like McCormick's reaper and Deere's plow, for example, were made in large factories and sold to farmers throughout Canada and the United States as the agricultural economy became dependent on factory-built machinery. By the end of the nineteenth century, steam engines provided a power source for threshing machines. Later, internal combustion engines were used in tractors and other machinery.

In North America, mechanization enabled farmers to expand and to specialize in the production of commodities that could be sold for the most profit. Geographic regions emerged where specialized **commercial agricultural systems** became the dominant component of the agricultural landscape (Figure 8.2). In the Midwest, cen-

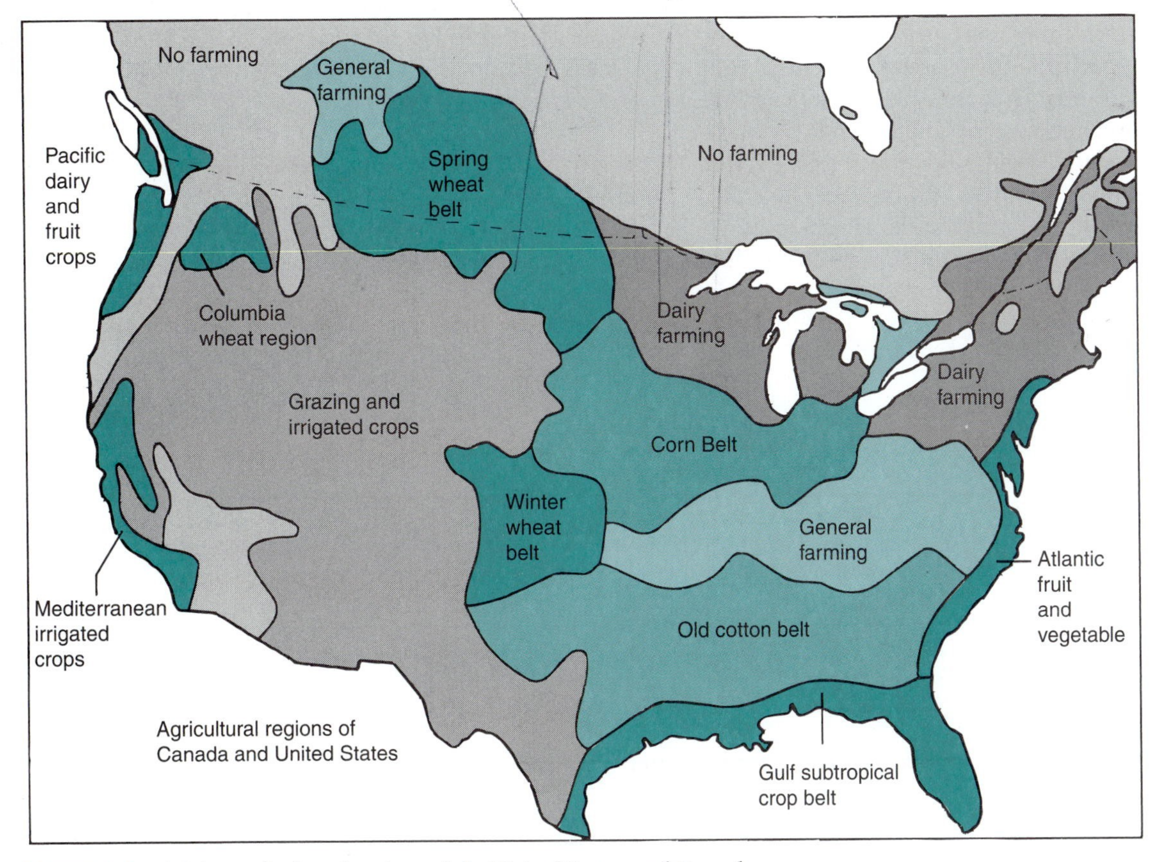

FIGURE 8.2 *Major agricultural regions of the United States and Canada.*

tering in Illinois and Iowa, a system of commercial agriculture emerged that raised hogs and fattened beef cattle on locally raised corn and soybeans and then shipped livestock and meat products to national markets (Figure 8.2). This heartland of American agriculture, known as the Corn Belt, exemplifies agriculture's adaptation to *mechanization* and *specialization* that characterizes the technology-based economies of the economically developed countries. Citrus and fresh vegetable production for winter markets became crops of choice in portions of Florida, southern Arizona, and California, as well as the lower Rio Grande Valley of Texas, where year-round growing seasons are the norm. As national and international grain markets expanded and became accessible, large tracts of fertile grasslands were opened for cultivation. In particular, the semiarid western Great Plains from Canada in the north through Kansas, Oklahoma, and the Texas panhandle became the granaries for wheat and other drought-tolerant grains (Figure 8.2).

Today, agriculture occupies 55 percent of the land in the 48 contiguous United States. More than 30 percent, mostly in the West, consists of nearly 600 million acres of open lands used for pasture and range, while about 460 million acres, or 25 percent of the land, is used for crop production. Corn, wheat, and field crops such as soybeans are dominant in the Midwestern states (corn and wheat belts) and fruits and vegetables are specialties in the southeast and irrigated lands in the West. Livestock ranching, of course, occupies much of the dry and mountainous land in the western states, while dairy and beef cattle, hog, and poultry production is becoming more concentrated in large confinement facilities in the Southeast, Midwest, and western United States.

In other parts of the world, similar technological and economic revolutions occurred. In most of Western Europe, where traditional small-scale agricultural systems had evolved slowly over centuries, mechanization and production strategies were downscaled and otherwise modified for efficiency on the smaller farms and fields. Adoption of hybrid seeds, chemical fertilizers, and pesticides have resulted in dramatic yield increases in Europe since the middle of the twentieth century.

8.4 FOOD CHOICES: THE PLANTS AND ANIMALS THAT FEED THE WORLD

Food energy for the majority of the world's population is obtained almost entirely from plants. Of the immense variety of plants, less than two dozen species are major food sources. These species have three characteristics in common:

- **High yield**: High production per unit area of land. This characteristic is essential to the millions of subsistence farmers in the developing countries. They often have to support an entire family on production from a farm no larger than a hectare (2.47 acres), about the size of two football fields.
- **High food value**: In terms of total calories and the essential nutrients of carbohydrates, proteins, fats, and vitamins, staples have high food value. Most subsistence farmers produce a staple grain or tuber crop that furnishes most of their caloric intake, along with several secondary crops such as vegetables and fruit that provide additional nutrients.
- **Storage ability**: Most crops can usually only be harvested during a limited time in the year. Since the food must last until the next harvest, the "fruits," or seeds, must be sufficiently hardy so they can be stored without significant deterioration.

The world's food supply is dominated by only five crops—three cereal grains and two tubers (Table 8.1). While they all share the three characteristics just discussed, each has evolved in a particular and often different environment. In combination, they provide basic foodstuffs for nearly all the inhabited world. The differ-

TABLE 8.1

PRODUCTION CHARACTERISTICS OF THE FIVE DOMINANT FOOD CROPS (1999)

Crop	Area Harvested (millions of hectares)	Annual Production (millions of metric tons)	World Average Yield (tons/hectare)	Average Yield in Developing Countries (tons/hectare)	Average Yield in Developed Countries (tons/hectare)	Type
Wheat	215.27	583.6	2.71	2.70	2.72	grain
Corn (Maize)	139.21	600.4	4.31	2.93	7.08	grain
Rice	155.13	596.5	3.85	3.78	6.19	grain
Potatoes	17.99	294.3	16.36	16.09	16.57	tuber
Cassava	16.58	168.1	10.1	10.1	n.a.	tuber
All Cereal Grains	679.88	2,064.2	3.04	2.76	3.55	

Source: FAO Statistics. United Nations Food and Agricultural Organization Statistics Division, 2000.

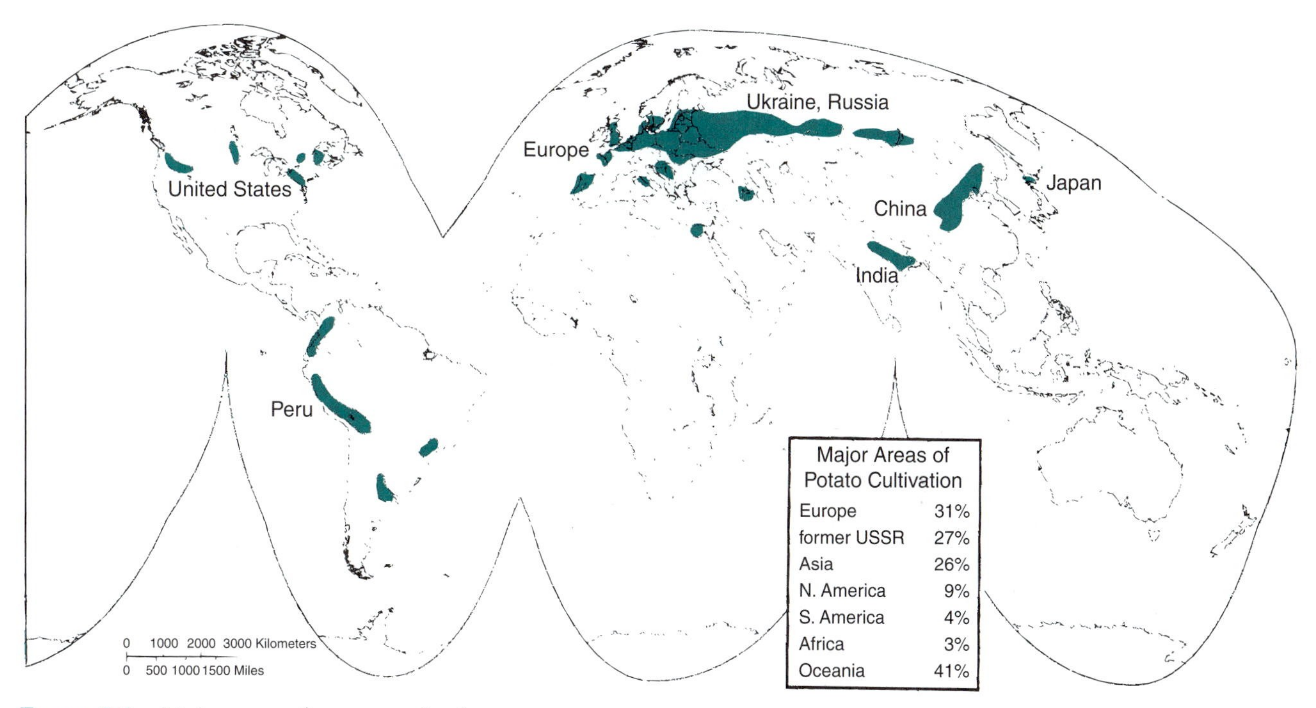

FIGURE 8.3 *Major areas of potato production.*

ences in yields between the developed and the developing countries (see columns 3 and 4 in Table 8.1) reflect the relative access to agricultural technology. Most developing countries do not have the range of pesticides, fertilizers, hybrids, and equipment routinely available in the developed countries.

For the most part, these five crops do not compete directly with each other because of their different environmental requirements. *Potatoes*, for example, originated as an Andean highland crop in South America. They grow best in a mild, humid environment. Today they are grown throughout the humid midlatitudes, with the former Soviet Union and Eastern European countries producing more than 50 percent of the world's crop (Figure 8.3).

The second tuber of the group, *cassava* or *manioc*, also originated in South America, but is strictly a tropical crop. Compared to the other four food crops, it is deficient in proteins and minerals, and most varieties contain a cyanide-based poison that must be removed. However, there are several compensating advantages. First, cassava will grow under a variety of tropical conditions where other crops will not, and it is relatively immune to most food crop pests. Second, the ripe tubers can be left in the ground for long periods without deteriorating—an extremely advantageous attribute in a tropical area. For these reasons, it is a staple crop in much of Southeast Asia (e.g., Malaysia, the Philippines, and Thailand), as well as in Central Africa and tropical South America.

Wheat is the most widely grown of all the cereal grains. With fair amounts of protein as well as carbohydrates, it is one of the most nutritious grains. Although wheat is hardy, it does not grow well under conditions of high heat and humidity, so it is grown principally in the midlatitudes. Although yields are highest in the humid midlatitudes, the major wheat belts are in drier semiarid climates. The areas of greatest production are the Great Plains of the United States and Canada, Ukraine and steppe regions of the former Soviet Union, and the North China Plain. Large-scale commercial production also occurs in Australia and on the Pampas of South America (Figure 8.4).

Most varieties of *rice* are grown in paddies flooded with 10 to 25 centimeters (4 to 10 inches) of water. Its nutritional value is good, especially when the outer hull layer containing important vitamins is not removed in processing. Unlike other cereal crops, rice can tolerate high temperatures combined with high humidity. Therefore, it is well suited to the humid tropics and subtropics. Ninety percent of the world's rice is grown in East and South Asia, and it is the principal food crop for half the world's population (Figure 8.5).

Corn (maize) is another New World crop that has spread over the world from its origins in Central America. It is a fairly high-yielding crop that grows best where summers are warm and humid. Corn is less nutritious than wheat and rice (not as much protein). It is an important subsistence food in Central America, South America, Africa, and to a lesser degree in India and

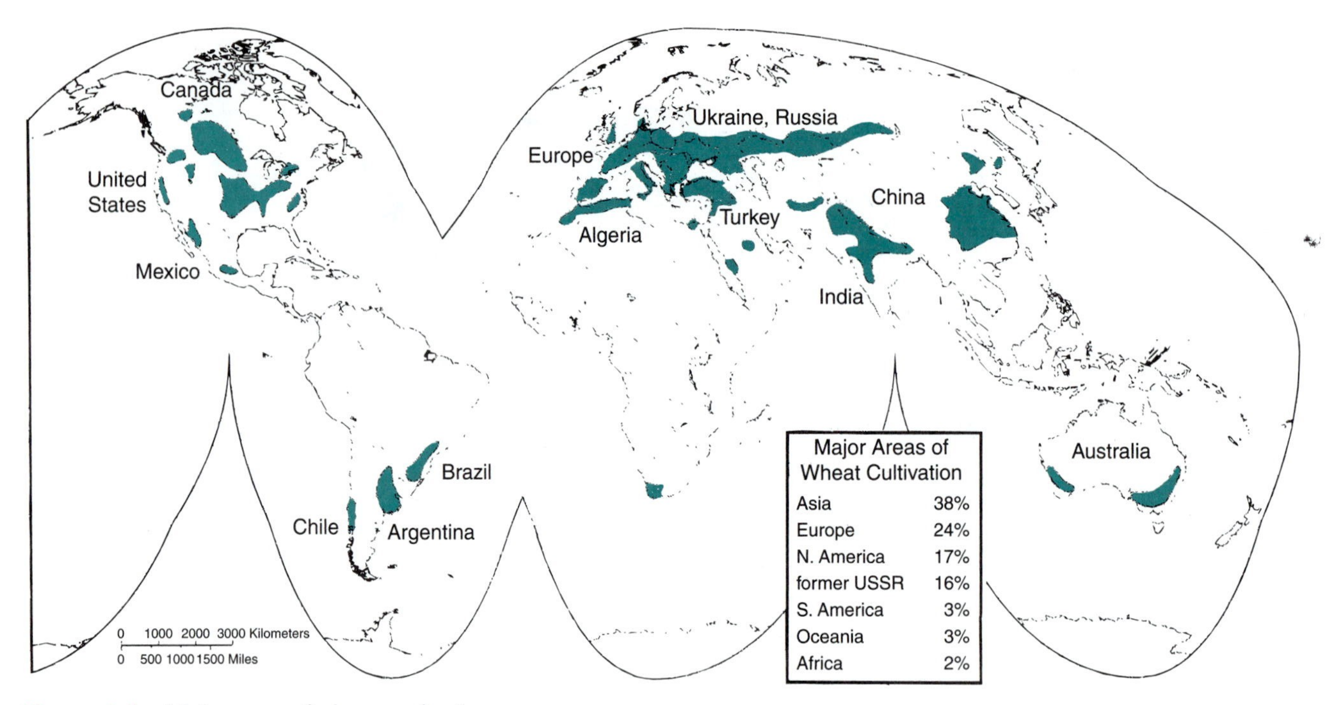

FIGURE 8.4 *Major areas of wheat production.*

China. About half of the world's corn is grown in the United States, but 80 percent of this is used for animal feed and corn oil, not for direct human consumption (Figure 8.6).

Although almost 90 percent of human food comes directly from plants, livestock products (e.g., meat, milk, and eggs) play an important role in the diet of the affluent countries and a minor but nutritionally important role in the developing countries. In the crop-based subsistence agriculture of the developing countries, the hogs, poultry, and fish that are raised convert materials inedible to humans (e.g., crop and manure residues) into nutritious, protein-rich food, while their wastes, in turn, are used to fertilize the fields. From an ecological standpoint, it is noteworthy that developing countries are more efficient than developed countries because they rely overwhelmingly on the lowest, and therefore the most abundant, trophic level in the energy pyramid (see Figure 6.2).

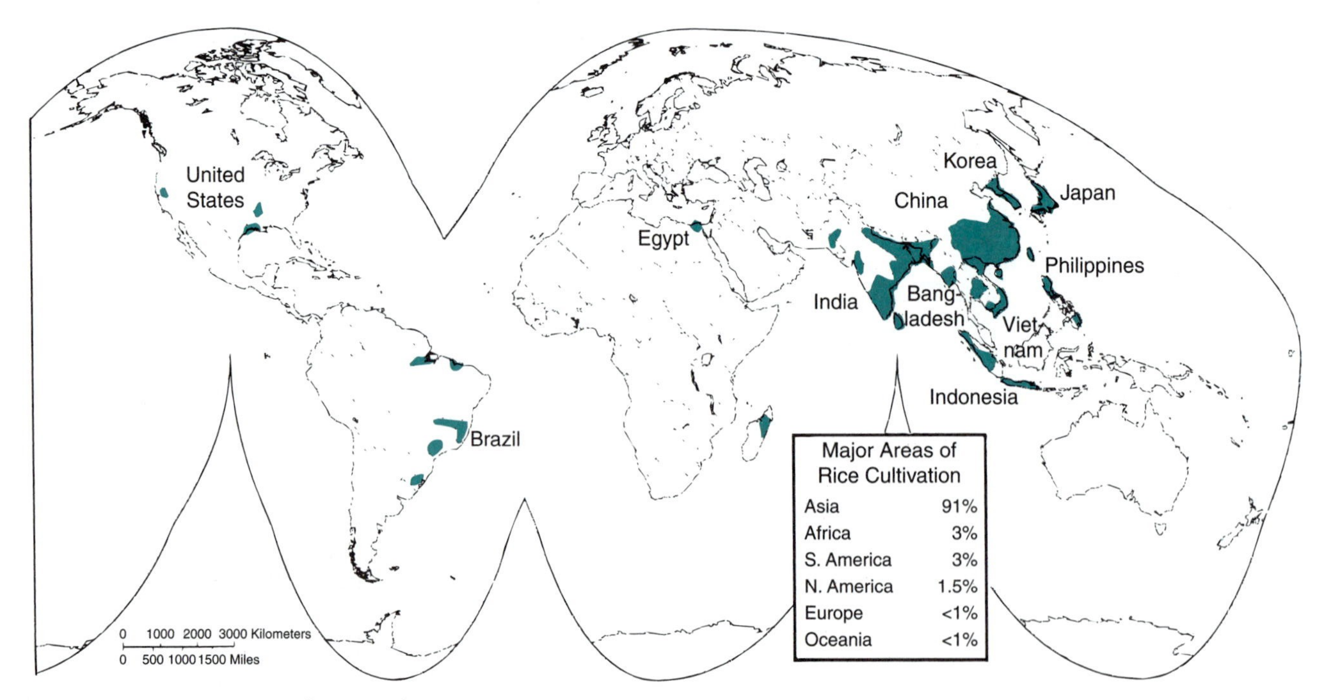

FIGURE 8.5 *Major areas of rice production.*

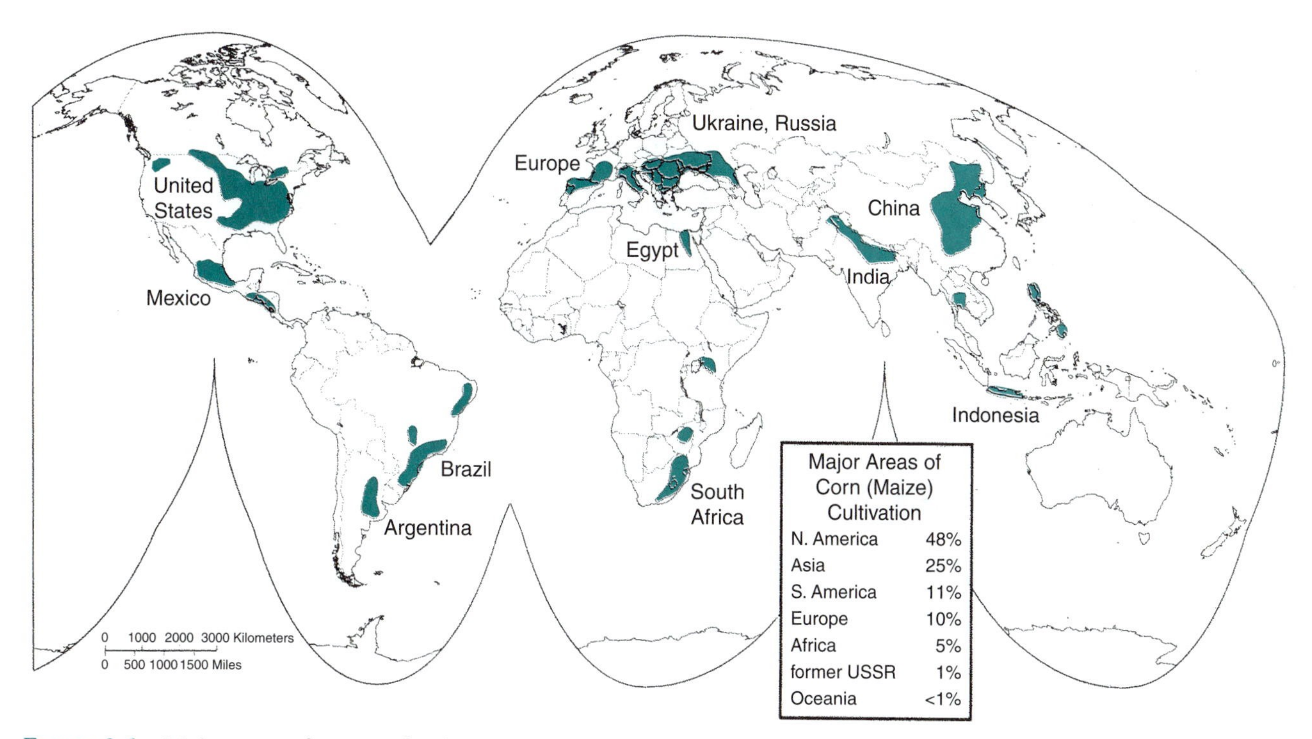

FIGURE 8.6 *Major areas of corn production.*

In most developed countries, the major emphasis of agriculture is the production of livestock and the crops that feed them. As a result, huge amounts of food energy are expended in producing animal protein. For example, in the United States two leading crops, corn and soybeans, are primarily used for animal feed. Significant amounts of other grain (including oats, wheat, millet, and sorghum) are also used for animal feed. Additional crops such as alfalfa and other hay crops and grasses are planted for animal feed or for forage on pastures and rangeland. As we noted in Chapter 7, using the developed countries' agricultural system, the world would be able to support far fewer people at a sustainable level. Worldwide, less than a dozen species of domestic livestock produce the bulk of meat, dairy products, and eggs that account for about 10 percent of the global food supply. Cattle and hogs provide approximately 75 percent of the world's meat supply, with poultry providing 20 percent.

8.5 SYSTEMS OF AGRICULTURAL PRODUCTION

Subsistence Agriculture

Subsistence agricultural systems are the most widespread forms of agricultural production today. For almost half of the world's population—some 2.9 billion people—these systems remain the way of life (Figure 8.7). So important is food production that in most developing countries, a majority of people in the work force are subsistence farmers. Typically, the production units (farms, fields, or livestock herds) are small and relatively self-sufficient, so that in good years, basic needs are met with a small surplus to trade or store. The three traditional subsistence systems are *intensive subsistence agriculture, shifting cultivation,* and *nomadic herding.* Their relative efficiencies are compared in Table 8.2.

More than 2.4 billion people are supported by **intensive subsistence agriculture** (Figure 8.8a). This system, which is the economic base of the densely settled portions of China, India, and the rest of monsoon Asia, produces relatively high yields per unit of agricultural land as a result of heavy inputs of labor. Wet or paddy rice is the principal crop in regions with long, warm, and rainy growing seasons. Wheat, upland rice, and other small grains are the staple crops in areas with cooler, drier climates. *Multiple cropping,* which produces two or even three rice or other grain crops each year on the same field, is common in a few areas where temperature, moisture, and soil conditions are most favorable. Such intensive food production is also illustrated by vegetables and fruits, *intercropped* or grown along paddy dikes, and by fish raised in the flooded rice fields. Poultry and hogs are also raised because they can scavenge and feed on material inedible to humans. Over the last two decades, significant productivity increases have occurred in areas of intensive subsistence agriculture where *hybrid varieties* of rice and wheat were adopted. As a result, this traditionally subsistence form of agriculture has developed some characteristics of commercial agriculture, because the hybrid seeds, chemical fertilizers, and pesticides must be purchased and more of the crops must be sold.

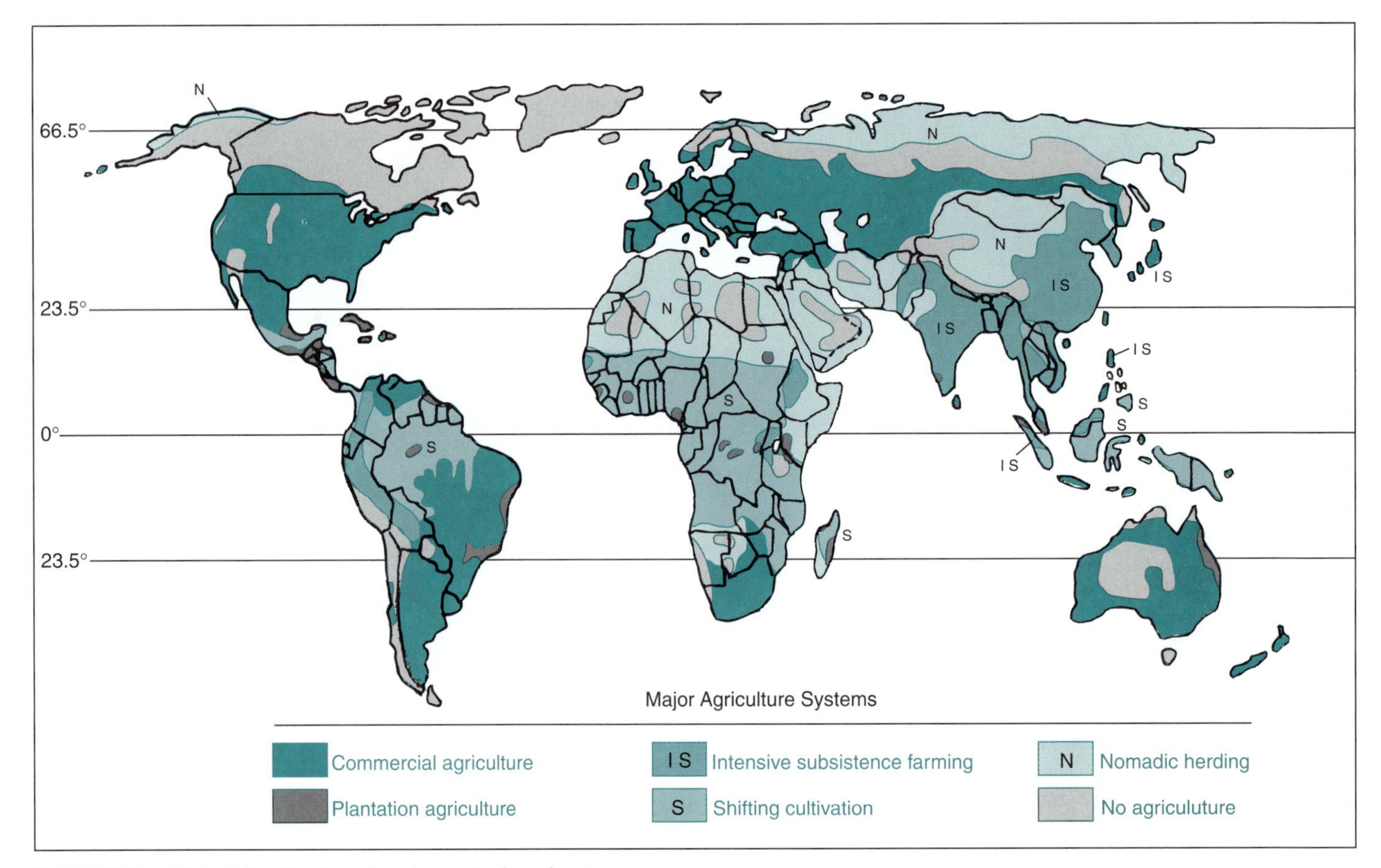

FIGURE 8.7 *Global distribution of major agricultural systems.*

TABLE 8.2

RELATIVE RESOURCE REQUIREMENTS FOR SUBSISTENCE AND COMMERCIAL AGRICULTURAL SYSTEMS

	Inputs				Outputs (Food Produced)			
	Land	Labor	Purchased Materials*	Mechanization	Per Hectare	Per Person	Total Food Produced	Use
INTENSIVE SUBSISTENCE	Low (small fields, farms, but intensive use—intercropping, multiple cropping)	High	Low to moderate (Increasing)	Low (Small-scale)	High	Low to moderate	High	Most used by farm family but more being sold
SHIFTING CULTIVATION	Large total (but only small "patches" farmed at one time)	Low	None	None	Low	Low	Low	Used by producers, little or no trade
SUBSISTENCE HERDING	Large areas required for pasture (usually arid to semiarid)	Low	None	None	Very low	Low	Low	Nearly all used by herders, some barter, trade with farmers
COMMERCIAL	Small to large	Low to moderate	High	High	Low to high	Very high	High	Nearly all sold

*seed, fertilizer, pesticides, etc.

(a) (b) (c) (d)

FIGURE 8.8 *(a) Intensive subsistence agriculture in Asia; (b) Shifting agriculture in the Brazilian rainforest; (c) Traditional nomadism in the Middle East; and (d) Modern commercial farming in North America.*

In contrast with the high population densities supported by intensive subsistence agriculture, **shifting cultivation** supports relatively small populations (Figure 8.8b). Only about 200 to 300 million people live in large areas of tropical Africa, Southeast Asia, and Latin America, covering 30 to 35 million square kilometers (12–13 million square miles), where shifting cultivation is practiced (Figure 8.7). Vast areas are required for shifting cultivation because worn-out fields are abandoned after a few years and revert to forest until soil fertility is replenished. As with other subsistence cropping systems, a variety of plants are grown simultaneously or intercropped.

Today **nomadic herding** supports the smallest population, only 15 to 20 million people (Figure 8.8c). Most herders tend livestock (goats, sheep, camels, and cattle) over vast areas of the arid and semiarid lands stretching nearly 13,000 kilometers (8,000 miles) from the Sahel and Saharan Africa, across the Arabian Peninsula and Southwest Asia, into central Asia, then north and east into China and Mongolia (Figure 8.7). Nomadic life is based on regular migration patterns that are dictated by the seasonal condition of grass and shrub rangelands grazed by the livestock. This is leading to a decline in the nomadic lifestyle because international border restrictions and settlement programs are forcing nomads to abandon traditional migration routes and grazing lands.

Commercial Agriculture

In contrast with subsistence agricultural systems, commercial agricultural systems emphasize specialized production of crops and livestock to sell (see Figure 8.7). Most commercial farms are relatively large and utilize specialized machinery, seeds, fertilizers, and other products developed to increase production efficiency. Through commercial farming a single farmer can produce enough food to feed many other people, which accounts for the fact that less than 10 percent of the population of developed countries are directly engaged in farming. For instance, each U.S. farmer produces enough to feed more than 60 additional people.

This production efficiency is realized in two ways. Improved inputs such as seeds, fertilizers, and pesticides promote higher yield. Specialized machinery speeds up production and reduces the human labor required for cultivation, harvesting, irrigation, and other farm tasks. In the United States, the leading commercial farming country in the world, agricultural output has doubled over the past half-century, while its agriculture work force has declined by more than three times. At the same time the number of farms has dropped from 6.5 million to just over 2 million, with less than 500,000 full-time farms today.

Fewer farms and farmers producing more food reflects the trend toward larger farms, fields, and livestock herds. At such relatively large scales, using even more specialized or larger equipment or by purchasing or shipping large quantities can create more savings in labor and production costs. The full-time commercial farm in the developed countries has become much more a business enterprise than a tradition or way of life (Figure 8.8d). Agricultural operations and management strategies must consider production costs and market prices that are driven by the interplay of economic, political, and institutional forces at work in the national and global economies.

FIGURE 8.9 *Famine victims in Somalia, Africa. In the 1990s, hundreds of thousands of people starved in Africa.*

8.6 A GEOGRAPHICAL PERSPECTIVE ON WORLD HUNGER AND MALNUTRITION

Despite the significant increases of agricultural production and data that show there is more than enough food produced most years to adequately feed the world, current estimates indicate more than 1 billion people—about one in every six—suffer from chronic hunger and nutrient deficiencies. For these people, hunger means that their daily diet does not provide the quantity and type of food needed to maintain health, normal growth and productive work. Four indicators are used to compile global estimates of hunger; **starvation**, **undernutrition** and the hidden hunger of **micronutrient deficiencies** and **nutrient-depleting diseases and parasites**.

Famine

Widespread starvation most often occurs as a consequence of **famines**—the acute shortage or absence of food within a region due to crop failure or destruction or by withholding or blocking food shipments into a country or region. Although famines are associated with widespread crop failure, most are the result of social or political processes that disrupt traditional agricultural production strategies or that interfere with food relief efforts.

Famines resulting in widespread suffering and death in the Sahelian region of sub-Saharan Africa in the 1970s and 1980s were associated with recurrent droughts and subsequent failure of crops and lack of forage for livestock. However, changes in the traditional agricultural practices, prompted by government policies to increase production of nontraditional crops for export, resulted in the shortfall of subsistence food crops to carryover during the string of dry years. Further exacerbating the food shortage situation in some areas were ongoing military conflicts and civil unrest. Elsewhere in Africa during the 1990s, brutal and sometimes lengthy civil wars prevented farmers from planting crops and hindered relief organizations from delivering food. As a result, hundreds of thousands starved in Sudan, Ethiopia, Somalia, Angola, and Rwanda. Throughout the 1990s it was estimated that between 150,000 and 200,000 people starved each year. Today, about 15–35 million are at risk of starvation in any given year (Figure 8.9).

Although recent famines have gained widespread attention and have been truly devastating, many larger famines have occurred during the past century. Millions starved in each of the great famines in the Soviet Union (1932–1934), Bengal, India (1943) and China (1958–1960), which were caused by flawed or deliberate government policies.

Malnutrition

Although Africa has suffered the most from recent famines, the greatest hunger problem throughout the developing world is **undernutrition**. Today as many as 800 million to 1 billion people—about one in five in the developing world—do not have access to adequate food for a healthy life. Although the *proportion* of undernourished has continued to decline over the past two decades, the *absolute* numbers of hungry have risen in Africa, Latin America, and South Asia. South Asia and China have the majority of the world's undernourished, but their share of the population lacking sufficient food has declined in the last decade to around 20 percent (Figure 8.10). Even within the most affluent countries there are significant numbers of hungry people. For instance, in the United States as many as 30 million people may experience difficulty in obtaining adequate food sometime within a given year.

Dietary deficiencies in essential micronutrients such as iodine, iron, and vitamin A also play a significant role in the geography of hunger. **Micronutrient deficiencies** often occur where poverty precludes a varied and nutritious diet. Soil conditions and other geographic factors often influence the amount of these micronutrients found in food. Iodine, which is available from plants grown on soils of former seabeds or from seafood or seaweed, is often lacking in food produced in mountainous regions. Vitamin A is available from many vegetables, fruits, and animal foods, but these foods may be only seasonally available in some regions and unavailable in others. Iron deficiency or anemia is the most common micronutrient disorder, due in part to the fact that most grains, vegetables, and legumes are low in iron. As a result, nearly 50 percent of women in developing countries are anemic. A 1997 World Health Organization report indicated that 88 percent of pregnant women in India suffer from anemia.

Nutrient-depleting illnesses having environmental causes are another source of malnutrition for millions worldwide. Parasitic worms, which absorb nutrients in the human digestive tract, cause debilitating illness in approximately 450 million people in developing countries. The most serious worm infection is *schistosomiasis*, which causes chronic ill health in over 200 million people in Asia, Africa, and South America. Diseases such as diarrhea (where ingested food is not absorbed) and measles and malaria (where energy is dissipated by fever) are responsible for nutrient-depleting malnutrition among hundreds of millions of people, mostly in the developing world.

Why Does Hunger Persist in the Twenty-First Century?

The goal of the 1996 World Food Summit sponsored by the UN Food and Agricultural Organization was to dramatically reduce world hunger and malnutrition. The target was to reduce the numbers of malnourished by 50 percent within 20 years. However, as we begin the twenty-first century, the prospects for achieving this goal look bleak, given current trends that worldwide hunger is declining by only 10 percent a decade. The FAO estimates there will still be 680 million malnourished people by 2010, with over one-third (280 million) in sub-Saharan Africa.

Why are so many millions hungry and malnourished when there is more than enough food produced in the world each year to adequately feed everyone? Several important social, political, economic and environmental conditions help to explain why hunger abounds in the developing countries. They include famines caused by war, the ownership of land and the structure of agriculture, commercialization, poverty, and the geography of surplus food production and food aid.

Although famines caused by drought or other natural disasters are less likely to occur today because of early warning systems and emergency relief measures, widespread hunger and even starvation occurs as a result of armed conflicts that disrupt food production and cause displacement of people who are fleeing the war. Even today, withholding food or blocking relief efforts is a common tactic of war.

War refugees are often deprived of adequate food as they move *en masse* into neighboring countries or regions. Without the means of producing or procuring food, refugees are at substantial risk of hunger or even starvation. Since there seems little likelihood of stopping regional conflicts in Africa and elsewhere, politically induced famines and food shortages will likely continue and millions more will suffer and starve as a result.

Perhaps the most important reason for hunger is related to the structure of agriculture and land ownership in many developing countries. In many Latin American and Asian countries a small minority of owners control most of the land. The majority are farm laborers or tenant farmers who must sharecrop or rent small plots. These forms of **land tenure** offer little security to those who actually produce the food. For example, tenants may lose their plots and laborers their jobs if the landowner opts to mechanize and expand production of commercial crops. These displaced, underemployed, and landless farmers may migrate to cities to join the growing class of urban poor who are even less able to procure adequate food (see Figure 7.12). Furthermore, the land itself becomes less productive under tenancy and hired farm labor. With no long-term commitment to the land, there is little incentive to maximize production and even less to protect and conserve the land itself. Furthermore, agricultural production is usually inefficient where few owners control the land. Large landowners need not be efficient producers; they need not even use all of their land for farming. In Latin America, where large estates occupy most of the land, much land available for crop production is idle or used for pasture.

The hungry throughout the world have a common trait: they are poor. Poverty is inexorably linked with

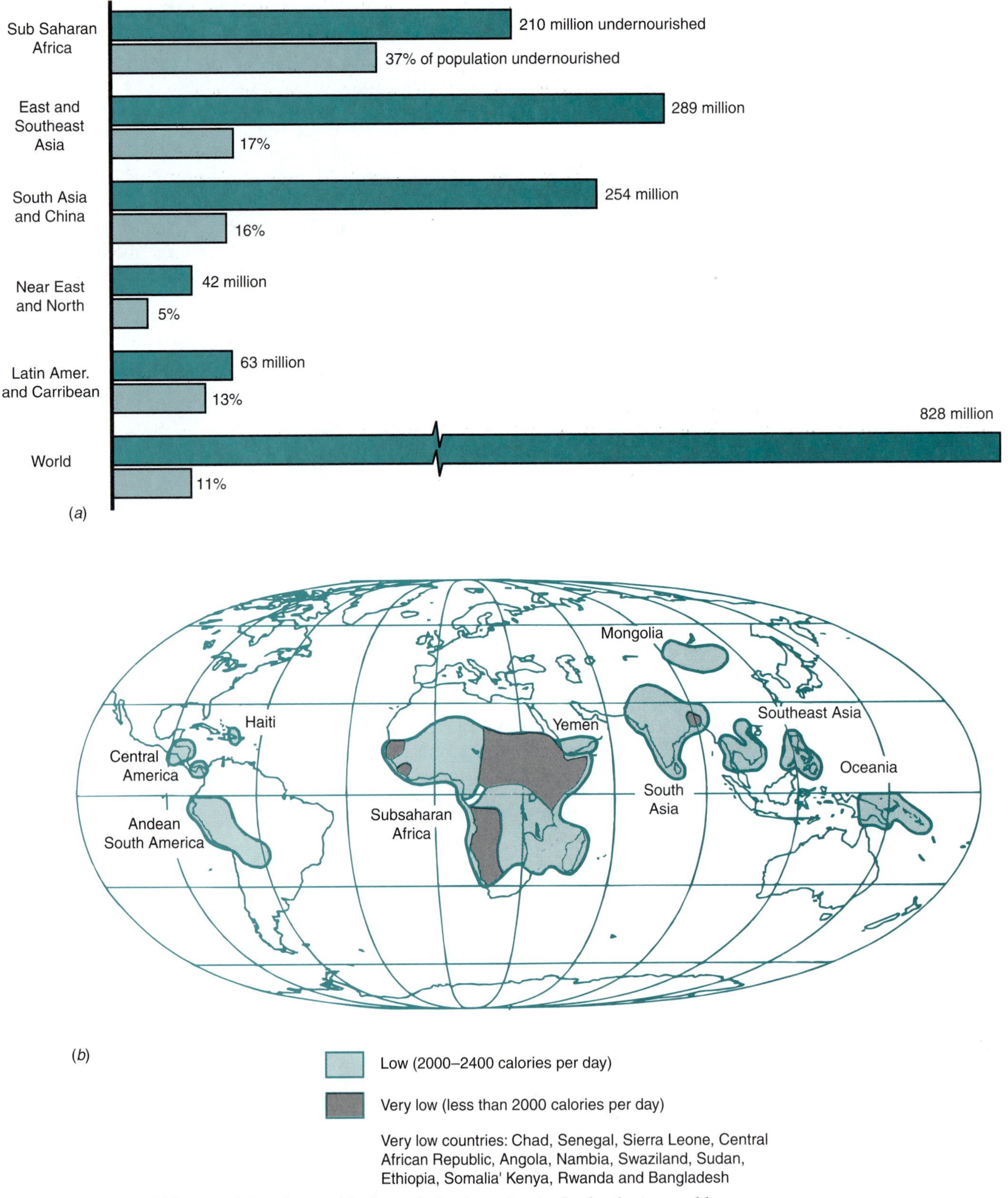

FIGURE 8.10 *World hunger: (a) undernourished population by region in the developing world; and (b) the major regions of undernourished population.*

hunger, because the landless, underemployed, or unemployed do not have the means or money to acquire food. With commercialization of agriculture in many developing countries, more food is being sold to distant markets rather than providing subsistence food for the local population. It also means that nonfood or luxury crops, which have greater market value, may be grown instead of traditional food crops. For example, Brazil is among the world's leading agricultural exporters, shipping soybeans, coffee, citrus products, and meat to developed countries, yet many Brazilians are undernourished. And food exports from the famine-plagued Sahelian countries Ethiopia, and the Sudan, were greater than food imports even during the famine years of the early 1970s and mid 1980s. The point is that the poor in developing economies cannot compete for food in the global marketplace. The result is that hundreds of millions are chronically hungry, and some even starve.

The world's food supply is unevenly distributed. Currently only a few regions produce large grain surpluses—North America, Western Europe, and Australia. These economically developed regions seek to sell their grain at world market prices. Most of the grain sales are to other relatively affluent countries, often for livestock feed. The poor developing economies can afford little or none of the grain offered on the commercial markets. Only a small fraction of the grain entering international trade is given as food aid. Generally, when the world grain supply is low, demand and prices are high and food aid is scarce. *Food aid*, as with other foreign assistance programs, is often provided mainly to benefit foreign policy, rather than given where and when it is needed. For instance, during the mid-1980s when famine revisited the Sub-Saharan countries in Africa, the United States shipped nearly four times more food aid per capita to politically allied Central American countries than to the entire famine-ravaged region in Africa.

8.7 NEW HORIZONS IN WORLD AGRICULTURE: THE GREEN REVOLUTION

When hybrid varieties of wheat and rice were introduced to subsistence agricultural economies nearly 50 years ago, the world entered a new era of agriculture called the **Green Revolution.** These new varieties, developed at international agricultural research centers, produce superior yields with adequate inputs of fertilizer and water. Because of this, they were quickly adopted in parts of many developing countries. During the 1970s and 1980s the Green Revolution varieties of rice were widely planted in the paddies of Southeast Asia and China, as were new varieties of wheat in the drier regions of Asia and Latin America. Today more than half of Asia's rice and wheat land are planted with these hybrids, and 95 percent of China's rice fields grow these varieties. In Latin America more than 80 percent of the wheat land now grows Green Revolution varieties.

Because hybrid wheat and rice varieties were adopted by farmers throughout the developed countries and by many farmers in developing countries touched by the Green Revolution, grain yields increased by over 2 percent annually between 1950 and 1990, resulting in nearly a threefold increase in world grain production during that 40-year period. However, since 1990 world grain yields have risen only about 0.5 percent annually and grain stockpiles, often used as a measure of *food security*, have declined. One reason for the slowing pace of world grain production relates to the steep production declines in the grain belts of the former Soviet Union. Total production in the newly independent states dropped from 180 million metric tons in 1990, the year before the Soviet breakup, 116 million metric tons in 1997—a decline of 36 percent (Figure 8.11).

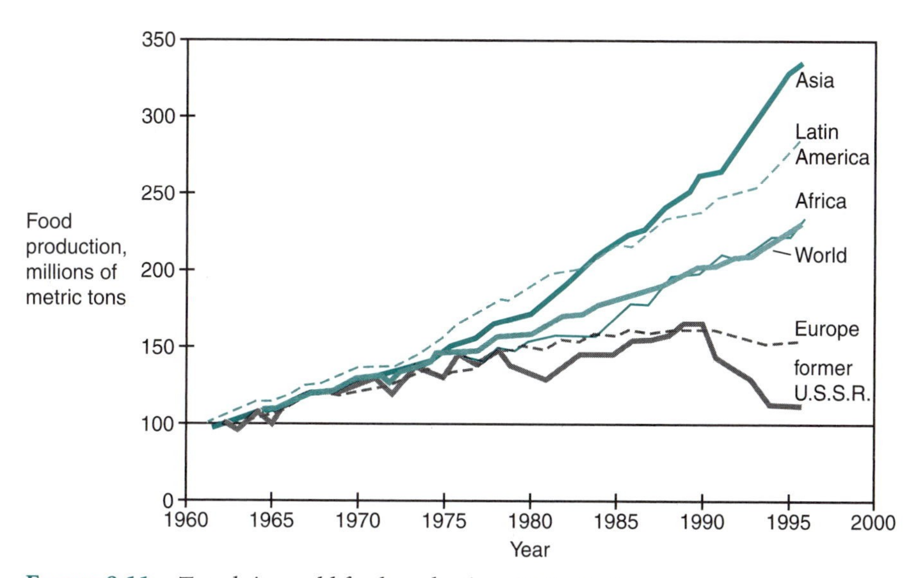

FIGURE 8.11 *Trends in world food production, 1961–1995.*

The impressive long-term production gains outside the former Soviet Union have barely kept ahead of population increases in most of the developing world. A significant exception is China, where national economic reforms and family-planning programs operating simultaneously with the Green Revolution have resulted in per capita food production increases that in most years provide adequate food supplies for its 1.3 billion people. The Green Revolution achieved the objective of expanding food resources by focusing on major rice- and wheat-growing regions where suitable land and climate had supported productive traditional agriculture. It emphasized planting the new, hybrid rice and wheat varieties that would produce substantially higher yields than traditional varieties in response to high inputs of chemical fertilizers, irrigation water, and pesticides. As a result, the Green Revolution strategies were most effectively carried out by focusing on the most productive areas and most productive farms.

While the Green Revolution has resulted in dramatic production increases in the developing world where rice and wheat are grown, hunger, malnutrition, and poverty remain chronic problems throughout many developing regions. Furthermore, the capital- and land-intensive methods required to produce the Green Revolution varieties have increased environmental impact and reduced genetic diversity. Some of the specific shortcomings are summarized here:

- **Limited participation by small, subsistence farmers.** In many developing countries the average farmer lacks credit or capital to obtain seed, fertilizer, pesticides, and other inputs necessary to produce the hybrid Green Revolution varieties.
- **Increased mechanization and farm size.** The consolidation of small farms and fields required for mechanized farming displaces many tenant farmers and reduces farm labor needs. The resulting unemployment and underemployment intensifies poverty in the countryside.
- **Increased commercialization.** The harvest from the larger, mechanized farms that now produce Green Revolution varieties are sold at market prices that many local poor cannot afford. Furthermore, as commercial production expands, more profitable luxury or nonfood crops replace staple food crops.
- **Loss of genetic diversity.** A handful of hybrid rice and wheat varieties are replacing thousands of traditional varieties that had been selected and developed for local conditions over long periods.
- **Reduction in soil fertility and increased erosion potential.** Traditional crop rotations, intercropping, and the use of organic fertilizers that maintain soil fertility and protect against erosion are being replaced by mechanized production of grain monocultures grown with inorganic fertilizers that provide less soil protection.
- **Soil damage and water resource depletion from increased irrigation.** Expanding irrigation to meet increased water needs of Green Revolution varieties has produced increased soil waterlogging and salinization. In areas where irrigation wells have been drilled, water tables are declining and some areas face groundwater depletion.
- **Many regions initially bypassed by the Green Revolution.** Many of the poorest regions such as Africa, where wheat and rice are not staple crops, have not yet experienced significant food production increases. Although progress has been slow in developing improved varieties of other important food crops, by the turn of the century, hybrid varieties of other grain staples were being introduced in some areas.

A New Chapter in the Green Revolution

During the 1990s, progress had been made to introduce improved hybrid varieties of important food grains such as sorghum and millet to African farmers. In politically stable areas of Africa, a private philanthropic organization working with supportive governments has established several hundred thousand demonstration plots to test the new varieties. Preliminary results are promising—they indicate that the hybrid varieties produce at least double the previous average yields. To date, the current efforts to bring this new wave of Green Revolution technologies have been limited, because of the chronic political and social unrest in many parts of Africa. In Ethiopia, however, the post-revolution government has supported agriculture by providing hybrid seeds and fertilizer to millions of subsistence farms. In response, Ethiopian grain production, which emphasizes sorghum, maize (corn) and a traditional grain, teff, nearly doubled between the 1995 and 1997 harvests.

Food production will undoubtedly increase as this phase of the Green Revolution spreads into more areas of Africa. However, much of this vast continent lacks the productive soils and benign climates that foster bountiful agriculture. Furthermore, only if and when social and political stability becomes the norm in Africa will governments begin to invest more in national food production in itiatives. Unfortunately, as the new century unfolds across Africa, turmoil and tumult is more common than tranquility.

Most likely the greatest potential for expanding food production will be in areas that already are the granaries and breadbaskets of the world. Because the productivity of most varieties of important grain crops is already very high, future increases will likely come in small increments. However, plant breeders are currently beginning to exploit the long overlooked genetic resources of *seed banks*—repositories that contain more than 6 million varieties of the seeds of some 100 crop species and their wild ancestors. As the hidden genes from these vast archives are discovered, new, promising hybrids are being developed. One such experiment in China has produced rice varieties that may yield 20 to 40 percent more than current hybrids by using genes from uncultivated rice varieties.

There are additional strategies being explored to expand agricultural production and perhaps open a new chapter in the Green Revolution, as well:

- Developing varieties that tolerate acidic or metallic soils. Field experiments using aluminum-tolerant varieties of corn, soybeans, rice, wheat and pasture grasses are being conducted on heavily leached tropical soils in Brazil.
- Developing varieties that are more resistant to diseases and pests. One notable past success was the development of resistant varieties of corn in the United States using the trait from a wild variety of maize that was resistant to the Southern corn blight fungus that threatened the U.S. crop in 1970.
- Maximizing potential crop yields for given agricultural conditions by matching appropriate inputs (seed, fertilizer, and water) with local site conditions.
- Discovering more useful genes by mapping the genomes of wheat, rice, and corn.

Outside the Green Revolution: New Strategies for Traditional Agriculture

Although farmers in many areas have achieved significant productivity increases in recent years, perhaps as many as 1.5 billion people in Asia, Africa, and Latin America depend on crops produced by traditional methods little affected by the Green Revolution. For the most part, these people live and work on small-scale subsistence farms in remote or marginal agricultural lands that lack access to markets or to the fertilizer, fuel, or even irrigation water that are required to successfully grow hybrid rice or wheat varieties. Furthermore, wheat or rice are not the staple food crops in many areas where traditional, subsistence agriculture remains. How can these farmers increase their food supply, given their limited resource base?

New strategies focus on the most efficient use of their limited resources, coupled with the traditional intercropping methods that have sustained land productivity for centuries. Intercropping offers both environmental and economic advantages over monoculture. When certain crops are grown together, they share soil nutrients, moisture, and light and contribute to pest control in complementary ways, so that they produce higher yields while enhancing soil fertility and controlling erosion. Also, intercropping several crops with different growth requirements reduces the risk of total crop failure while spreading labor inputs more uniformly over the growing seasons. This strategy forms a dependable, sustainable local agricultural system that emulates natural ecological systems as some nutrients are returned to the soil via plant and animal wastes, which protects the soil and maintains plant genetic diversity (Figure 8.12a).

Shifting agriculture is a traditional way of life for several hundred million people in portions of the humid and subhumid tropics. While specific crop mixtures and rotations and planting and fallow periods vary by region and local custom, the common thread of this subsistence agricultural system is based on a sustainable cycle where a patch of forest is cleared then planted for short period and then abandoned to allow the forest to restore soil fertility.

Agroforestry and nutrient recycling employ variations of traditional intercropping to increase the productivity in areas of shifting cultivation:

- In the subhumid Sahel region of west Africa the traditional shifting agricultural system that required 15 to 20 years of brush fallow is being modified to an **agroforestry** system in which nitrogen-fixing acacia trees are intercropped with traditional millet and sorghum crops. The trees improve the productivity of the soil in several ways. Soil fertility is enhanced through nitrogen fixation as well as the upward transfer of nutrients that the acacia roots draw from deep soil layers. Leaf-drop returns organic matter and nutrients to upper soil layers, making them available to crops while also improving soil texture, moisture-holding capacity, and the efficiency of fertilizer. The results are higher crop yields and shorter or no fallow periods (Figure 8.12b).
- In Peruvian Amazonia a low-input **nutrient-recycling** cropping system for the region's relatively infertile acid soils has been developed in which high-yielding, acid-tolerant rice and nitrogen-fixing cowpea varieties are rotated without fertilizer, lime, or tillage. Crop residues are returned to the fields and human labor and commercial herbicides are used to control weeds. After several satisfactory crops of rice and cowpea grain, a cover of tropical kudzu is planted to choke out invading weeds. After one year, the kudzu cover is buried, restoring soil nutrients and leaving the fields weed-free and ready for another cycle of agricultural production.

Beyond the Green Revolution; Biotechnology

Biotechnology, or the use of scientific techniques to develop more productive crops and livestock has been used for more than a century. For example, crossbreeding different varieties of the same crop species has resulted in high-yielding hybrids such as the Green Revolution varieties of wheat and rice. In recent years a new

(a)

(b)

FIGURE 8.12 *New strategies for traditional agriculture; (a) intercropping; and (b) agroforestry, aimed at increasing production, food variety, and improving soil.*

biotechnology, called **genetic engineering**, has systematically altered the genetic structure of plants and animals. Genetic engineering uses a technique called **gene splicing** in which a desired gene is isolated, extracted, and then inserted into specific cells of the targeted crop or livestock species. These genetically transformed cells are induced to grow into individual plants or animals that carry the desired trait.

Biotechnology can be superior to traditional plant- and animal-breeding techniques in three ways.

- Desirable characteristics can be achieved quickly. Once genetic material is successfully introduced into the plant or animal, that trait will appear in successive generations.
- Biotechnology concentrates on individual genes. As a result, only a single economically important characteristic is changed.
- Genetic materials can be transferred among organisms that cannot be crossed sexually. This enables specific traits to be transferred among unrelated plants or even between plants and animals.

Using genetic engineering, growth hormones, veterinary drugs, and vaccines have been developed that have enhanced livestock production. In recent experiments, human genes for medically useful proteins have been successfully spliced into sheep, cows, and goats. The milk from these genetically altered animals produces a steady supply of these valuable proteins used for drugs.

Since the mid-1980s genetic engineering has been applied to more than a hundred crop species to make them more resistant to herbicides, pests, diseases, drought, and frost. By the mid-1990s the first bioengineered food products had been approved by the U.S. Food and Drug Administration. Cotton, corn, and potato varieties with bacterium genes, which produce a natural insecticide (*Bacillus thuringiensis* or *Bt*) that effectively controlled larvae pests, were ready for widespread use. Transgenic soybeans, corn, sugar beets, cotton and oil seeds engineered to tolerate specific types of herbicides have also been widely adopted by North American farmers. By 2000, about 50 percent of U.S. soybeans and cotton acreage and 25 percent of the corn was planted with engineered varieties. In addition, varieties of fruits and vegetables have been genetically engineered to taste better and to resist spoilage, and efforts are underway to fashion crops with superior nutritional properties. Even a bioengineered caffeine-free coffee plant has been developed and is undergoing field trials in Hawaii and Mexico.

With huge profits already being realized, much genetic engineering research and development continues to be done by private corporations in the developed countries. Small firms that specialize in biotech products—as well as a few large American and European firms that manufacture agricultural chemicals—have invested heavily in biotechnology. Public-sector research in genetic engineering is being conducted at university, state, and national government agricultural experiment laboratories in the developed countries. A few developing countries, including Thailand, Indonesia, and the Philippines, have also established national biotechnology programs. The Rockefeller Foundation, which supported development of the Green Revolution rice hybrids a generation ago, has invested millions of dollars to engineer more nutritious, disease-resistant strains of rice.

Although proponents and purveyors of these biotechnologies are optimistic, some people express concern about these new genetic technologies and their rapid, unrestricted growth. Two concerns are focused on the development of herbicide-resistant crops. First, these new crops would encourage increasing herbicide use, making farmers even more dependent on costly agricultural chemicals. Second, these crops might transfer their herbicide tolerance to closely related weed species through pollination. Critics also warn that transgenic crops with built-in pesticides will promote rapid evolution of pests resistant to the pesticide.

Furthermore, a number of consumer groups, scientists, and some farmers have expressed concern over the safety of genetically altered foods. In the United States, where the development and use of genetically modified crops has been most widespread, a federal government task force is reviewing a proposal to require labeling of genetically altered foods as consumer concern has grown. In Europe the fight against "Frankenstein Foods," as genetically altered foods are sometimes called, has essentially driven these crops and food products from the vast European Union Market. For example, U.S. corn and soybeans containing genetically modified varieties have virtually been eliminated from European markets—formerly the largest export market for these important feed grains.

In addition, growing concern among U.S. consumers about the widespread use of genetically modified crops in food has resulted in several major food manufacturers and restaurant chains to stop using food made from genetically modified crops. In response, U.S. farmers have sharply reduced plantings of genetically engineered seeds by as much as 25 percent in 2001.

Beyond these concerns, the growth and spread of genetically engineered seeds and the chemicals to use with these seeds can add to production costs. For example, engineered seeds cost several dollars more per bag because of a technology fee imposed by the company that developed the seed. Furthermore, some companies require that farmers purchase the patented seed each year. To ensure that farmers will comply, a technology has been developed that renders seeds from the current crop sterile, preventing farmers from planting seeds saved from the preceding harvest. Not only does this "no replant" policy add significant cost to small and middle-sized commercial farms in developed countries, but if applied worldwide it would affect the majority of the world's farmers who rely on saved seeds. Furthermore, some organizations dealing with global food security issues worry that only a handful of North American and European corporations will have control of most of the world's certified seed supply.

8.8 Agriculture and the Environment

We have seen that increasing demand for food and other agricultural commodities has led to both to an expansion of cropland and rangeland and to techniques that produce higher yields. These changes have been significant in both commercial and subsistence agriculture worldwide. On the positive side, they have resulted in large increases in agricultural output. On the negative side, they have also caused worldwide environmental degradation. Table 8.3 provides estimates of potential productivity declines for each continent. The most important types of land degradation caused by agriculture include topsoil loss through wind and water erosion, as well as waterlogging and soil salinity (salinization) due to inadequate drainage on irrigated land.

In addition, insufficient application of fertilizer or shortening of crop rotation or fallow periods can lead to a loss of essential soil nutrients, whereas too much fertilizer can cause soil acidification. In areas of commercial agriculture, compaction by heavy machinery causes

Table 8.3

Soil Degradation Related to Agriculture and Deforestation

	Total Degradation as a Percentage of Vegetated Land Area	Chief cause of Degradation as a percent		
		Crop Farming	Grazing	Deforestation*
Africa	22.1	24	49	27
Asia	19.8	27	26	46
Central America and Mexico	15.0	45	45	40
Europe	23.1	29	23	38
North America	5.3	66	30	4
Oceania and Australia	13.1	8	80	12
South America	14.0	26	28	46
World	17.0	28	35	37

*Includes wood cutting for fuel

Source: World Resources Institute; from International Soil Reference and Information Centre 1990.

(a) Dust Bowl of 1935

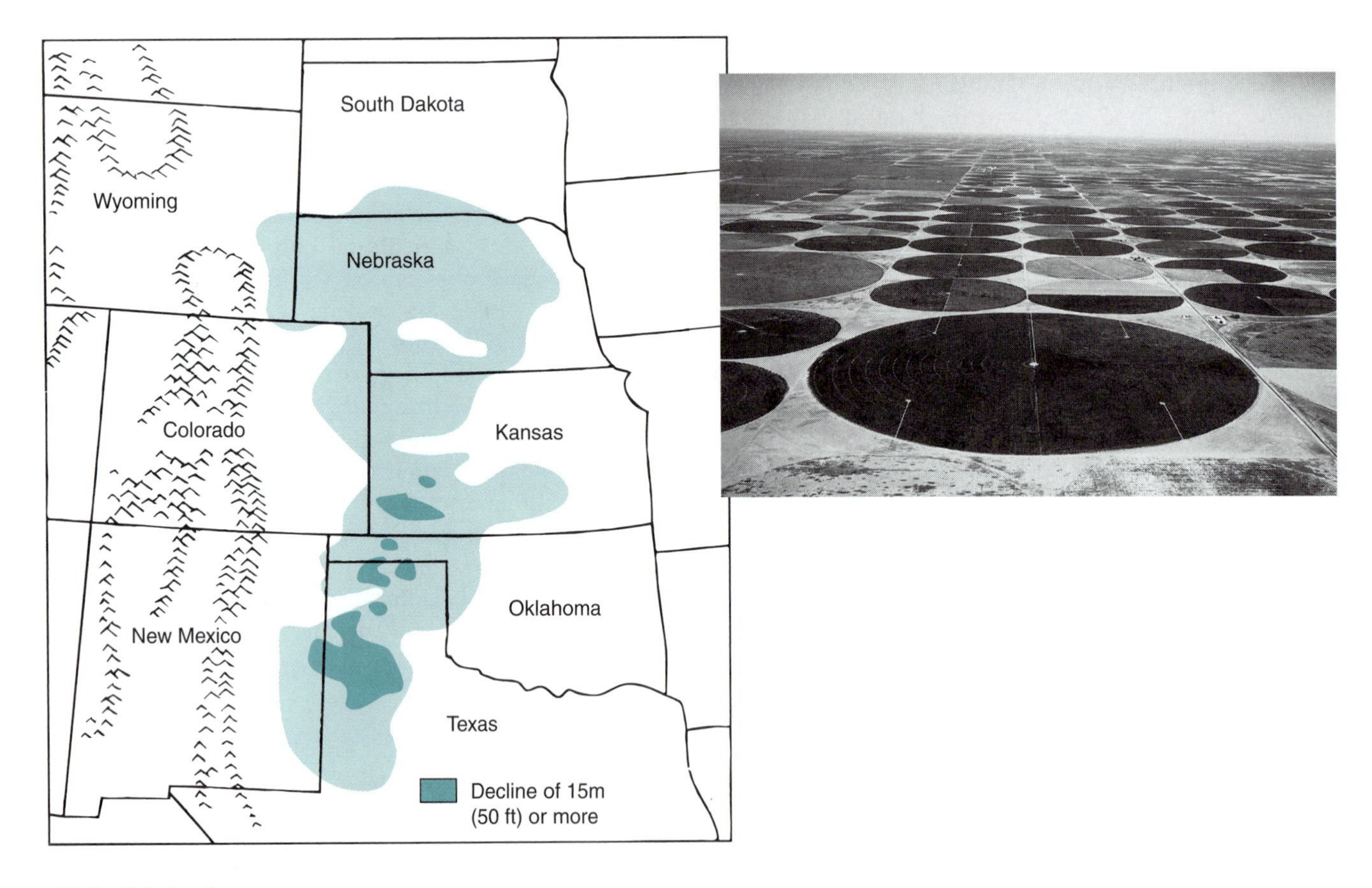

(b) Ogallala Aquifer

FIGURE 8.13 *Two impacts of agriculture on the Great Plains: (a) the Dust Bowl, an area of massive wind erosion in the 1930s; and (b) the Ogalalla aquifer, which is being drawn down in some areas by irrigation pumping. Inset shows fields using center pivot irrigation in northern Texas.*

physical damage to soils and reduces its capacity to retain water. Worldwide the most pervasive soil damage is caused by livestock overgrazing, which compacts soil, reduces vegetation, and exposes the soil to wind and water erosion. Table 8.3 indicates that agricultural activities (including grazing) are responsible for more than 60 percent of soil degradation worldwide while in North America these activities account for over 95 percent of soil damage.

Furthermore, modern agriculture is associated with the depletion of groundwater and the contamination of both underground and surface water from runoff and infiltration of pollutants from agricultural land. These and other environmental problems related to agriculture are discussed in Chapters 12, 13, and 14. Here we will briefly summarize some of them.

Soil Erosion

Today significant soil erosion occurs in both commercial and subsistence agricultural systems. In large-scale mechanized commercial agriculture, most fields are bare or nearly so for significant parts of the year. During the planting season, when plants are small, and throughout the dormant season if plowing occurs after harvest, soils are exposed and vulnerable to wind and water erosion. In the southern Great Plains of the United States, such agricultural practices, coupled with overgrazing, led to serious soil erosion problems when a string of dry years during the 1930s resulted in successive crop failures, dust storms, and millions of acres of ruined agricultural land that came to be known as the Dust Bowl(Figure 8.13a).

Likewise, the former Soviet Union plowed and planted huge tracts of semiarid grasslands with wheat and other small grains during the 1950s. The fertile soils of the *Virgin Lands*, as the experiment was known, at first produced good yields, but drought conditions and erosion of the exposed soils resulted in successive crop failures and abandonment of much of the land within 10 to 15 years. Parenthetically, the failures also contributed to major changes in Soviet political leadership.

Land degradation due to soil erosion occurs in subsistence agricultural systems as more land is brought into production to feed rapidly growing populations. In parts of monsoon Asia, newly cultivated land is carved from marginal and submarginal land such as steep, forested slopes. This results in increasing runoff, soil erosion, flooding and sedimentation of reservoirs, stream channels, and irrigation projects. Deforestation in the Himalayan foothills has resulted in more severe flooding downstream, in India and Bangladesh. In 1988 and 1991 much of the rice crop in Bangladesh, which lies at the mouth of the Ganges and Brahmaputra rivers, was destroyed by two of the worst floods on record.

Salinization and Waterlogging

Excessive irrigation of cropland can result in salty, saturated soils and crop failure. **Waterlogging** occurs when the water table rises into the root zone inhibiting crop growth and contributes to **salinization** as water from saturated soil close to the surface evaporates and leaves behind increasing concentrations of salt. As salinity builds, crop yields decline, and at excessive levels, crop production stops unless the salt can be flushed downward through the soil. Diminished crop yields and damaged soils eventually occur virtually everywhere irrigation agriculture is practiced. Although worldwide estimates of the amount of land damaged or destroyed by salinization are not known, estimates for the five countries with the most irrigated land are summarized in Table 8.4. In addition, literally millions of acres once irrigated have become salty wasteland.

Groundwater Depletion and Contamination

In many parts of the world today, water tables are declining because groundwater is being pumped from aquifers faster than it is being replenished or recharged. During

TABLE 8.4

IRRIGATED LANDS AND AREAS DAMAGED BY SALINIZATION IN COUNTRIES WITH MOST IRRIGATION

Country	Gross Irrigated Areas (million hectares)	Share of Irrigated Cropland (percent)	Estimated Area Damaged (million hectares)	Share of Irrigated Land Damaged (percent)
India	55.0	33	20.0	36
China	46.6	48	7.0	15
Former Soviet Union	21.0	9	2.5	12
United States	19.0	10	5.2	27
Pakistan	16.0	77	3.2	20
TOTAL	157.6		37.9	24
WORLD TOTALS	250.2	17	60.2	24

Source: State of the World 1990, Table 3.1, p. 40, and Table 3.2, p. 45.

the 1990s studies indicated that groundwater levels were falling rapidly on the North China Plain (1.5 meters/year) and in India (1 to 3 meters/year), and in a coastal area of India, overdraft has resulted in contaminating saltwater being drawn into the aquifer. Excessive groundwater pumping occurs on nearly 1.6 million hectares (4 million acres), or about one-fifth of all irrigated land in the United States.

In the U.S. Great Plains, groundwater from the vast Ogallala aquifer supports irrigated agriculture from South Dakota to Texas and New Mexico (Figure 8.13b). Throughout this semiarid area, the rate of groundwater pumping from the Ogallala has averaged eight times that of recharge, and in some areas withdrawal has exceeded natural recharge by 20 times. Since 1978 irrigated cropland has fallen from 13 million acres to less than 10 million acres in the Ogallala region.

In response to the rapid drawdown of the Ogallala aquifer, groundwater conservation practices have been introduced to reduce depletion, but whether long-term irrigation can be sustained remains a question. In addition, agricultural activities can contaminate groundwater as fertilizers and pesticides leak through the soil. The most serious groundwater pollution is caused by excess nutrients (mainly nitrogen) from fertilizer and animal waste. In a recent U.S. study, 12 percent of domestic wells in agricultural areas exceeded the maximum limits set for nutrients. In addition, low concentrations of several different pesticides were found in nearly three-fifths of the shallow wells tested in agricultural areas. Although little is currently known about the health risks of this "pesticide soup," this chemical witches' brew may very well become a toxic pollutant of domestic wells.

Surface Water Impacts

Surface water and wetland resources are also altered or diminished as a result of irrigation projects. When reservoirs are built and streams are diverted for irrigation, lakes, wetlands, and ecosystems downstream may decline or disappear. Such an environmental catastrophe has been developing since the 1960s in the Aral Sea shared by the former Soviet republics of Kazakhstan and Uzbekistan. Once the world's fourth largest inland lake, the Sea's surface area has decreased from 67,000 km^2 to less than 30,000 km^2 and its volume has diminished by more than two-thirds due to irrigation diversions from the only two rivers flowing into it (Figure 8.14). Less freshwater inflow and high evaporation rates have nearly tripled the salinity of the water, so that most of the native fish species have disappeared. A once-productive fishery has been lost, and as the shoreline retreats, thousands of acres of dry seabed are exposed to wind erosion. An estimated 40 to 50 million tons of windblown seabed salts are being deposited each year on neighboring cropland, damaging soils and crops. Plans initiated in the late 1980s to increase streamflow have failed; by 1997, the flow of one of the tributaries had been diverted into a system of inland lakes in order to establish fish farming.

Elsewhere, increased diversions of irrigation water for agriculture and other uses have resulted in major rivers running dry. Today as a result of withdrawals from the Colorado River in the southwestern United States and Mexico, the river often runs dry before it reaches the Gulf of California. Since 1985, China's Yellow River has run dry for part of every year. In 1997 it failed to reach the sea for more than six months, as many new projects diverted its waters for irrigation, industrial, urban, and hydroelectric power production.

Agriculture is the leading cause of water pollution throughout much of the world. Soil sediments, fertilizer, pesticide residues, and other agricultural pollutants are common to streams, rivers, and lakes in both developed and developing countries. In the United States, many of these contaminants flow into the Mississippi River and are transported to the Gulf of Mexico where a huge area of oxygen-depleted water develops each summer. This *dead zone*, as it is now called, is caused by high concentrations of phosphates and nitrates coming mostly from agricultural sources.

Simplification, Substitution, and Environmental Change

Large-scale, permanent agriculture with its few domestic crop plants can disrupt or destroy entire **ecosystems** or *biomes*. A good example occurred when the tall grass prairies of the eastern Great Plains were settled during the latter decades of the nineteenth century. The varied associations and numerous species of herbaceous prairie plants were plowed under and replaced by a few domestic agricultural plants. Today no virgin prairie remains; instead, fields of soybeans and corn, the most profitable crops, dominate.

This biotic **substitution** and **simplification** brought about by agriculture has produced both economic advantages and disadvantages. For instance, when a single crop is grown in a large field, it can be planted, fertilized, cultivated, and harvested using specialized techniques. This economically efficient system known as *monoculture* has become the standard production method of commercial agriculture. A disadvantage of monoculture is that diseases or pests can quickly devastate entire crops. A historic monoculture disaster was the 1840s Irish potato famine, when a virus destroyed the staple crop of the Irish peasants. Since there were no food alternatives, thousands starved and many more emigrated. In 1970, corn blight devastated portions of the United States' most widely grown crop. Fortunately, blight-resistant corn varieties were available for replanting during the next growing season and the disease affected only a small portion of the crop.

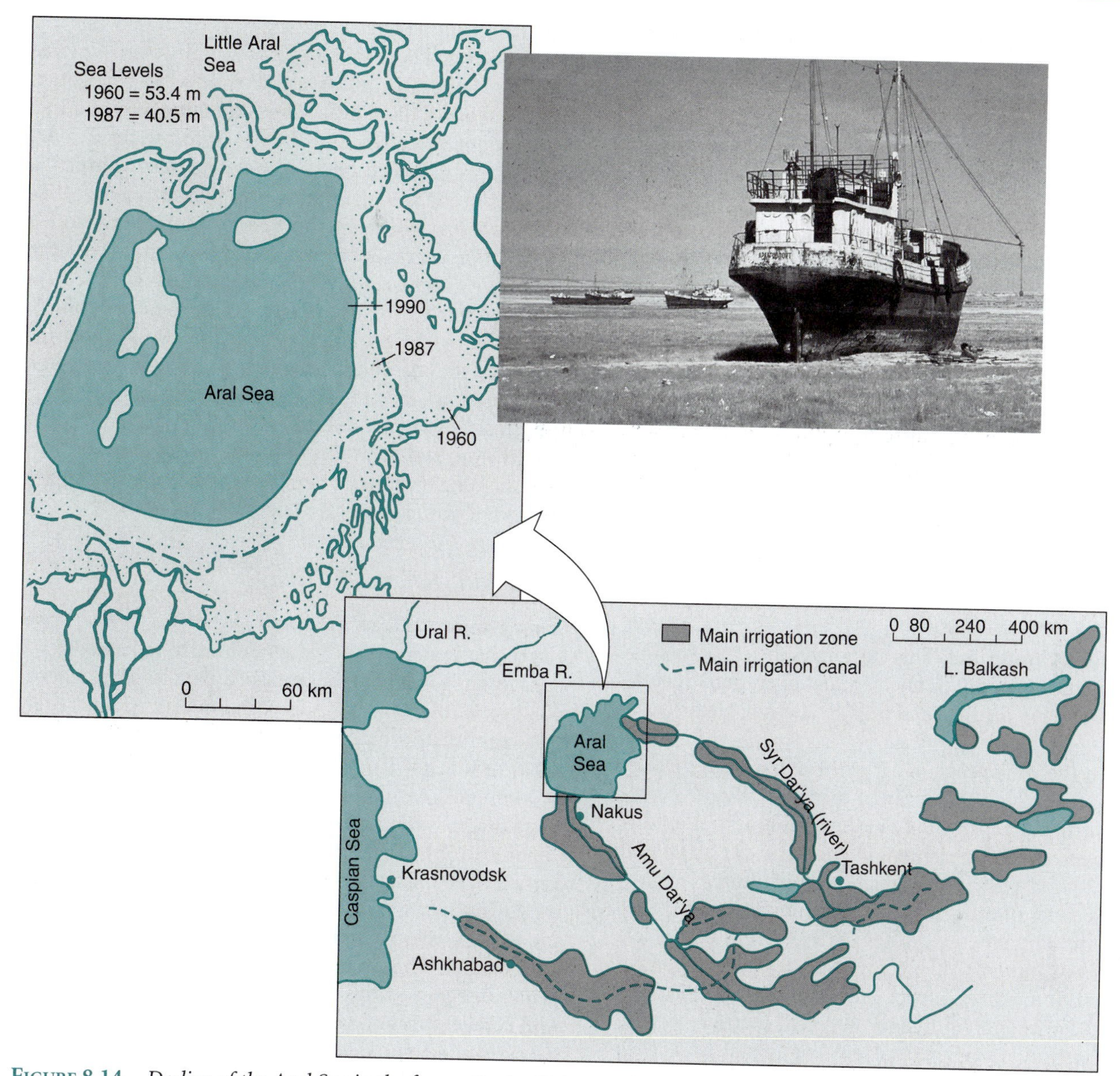

FIGURE 8.14 *Decline of the Aral Sea in the former Soviet Union as a result of diversion of inflowing streams for irrigation.*

In addition to cropland conversion of grasslands and wetlands, additional practices associated with contemporary commercial agriculture such as increasing field sizes and eliminating the brushy edges and fence rows between fields have further reduced native plant and animal populations. By 2000, more than 50 grassland wildlife species were listed as threatened or endangered and even animals and birds that are well-adapted to agriculture such as cottontail rabbits, pheasants, and quail have declined as field edges, pastures, and hayfields are converted to field crops.

On a more positive note, the **Conservation Reserve Program (CRP)** established in 1985, which pays farmers to take selected cropland out of production for 10 years, has retired more than 36 million acres of U.S. cropland. As this land has grown back to wild vegetation, significant improvements in the amount and suitability of habitat have occurred. Subsequent studies have noted dramatic increases in wildlife in areas where CRP enrollments are high. However, not all land will remain in the CRP. As enrollment contracts expire, much of the land will return to crop production, with the apparent loss of wildlife habitat.

Environmental Regulation and Agriculture

Commercial agriculture in the developed countries has not had to meet the same environmental standards required of other commercial businesses and municipalities. For example, in the United States pollutants from nonpoint sources such as stormwater runoff from farm fields are not regulated by the Clean Water Act, and most soil conservation programs are voluntary. There are, however, additional federal environmental laws that regulate certain U.S. agricultural practices:

- **Federal Insecticide, Fungicide, and Rodenticide Act (FIFRA).** Because pesticides used in agriculture have caused serious damage to the environment and human health, FIFRA requires that every agricultural pesticide used on U.S. farms meets minimum risk criteria that are established by the Environmental Protection Agency (EPA).
- **Federal Water Pollution Control Act.** Farmers are required to obtain federal permits to drain wetlands. This is controversial and opposed by agricultural organizations, but only about 3 percent of the applications to alter wetlands are denied.
- **Endangered Species Act.** This act provides federal authority to restrict use of agricultural lands as part of a species recovery program.
- **The Clean Water Act.** This authorizes the states to control pollution from confined animal operations such as beef and dairy feedlots and poultry and hog factories.

Given the wide range of agricultural practices that reflect the diversity of landscapes, climate, soil and watershed characteristics throughout the nation, it is often difficult to effectively implement the one-size-fits-all federal environmental policies that relate to agriculture.

These types of programs might be more effectively legislated and implemented at the state or even local levels where environmental problems related to farming have direct and immediate impact and satisfactory political solutions have a sense of urgency. A recent example of state responsiveness to agricultural–environmental issues has been the regulation of industrial hog farms by states that had actively recruited these operations a few years earlier.

At the international level, recent agricultural trade accords call for decreased domestic agricultural subsidies except for programs that emphasize environmental protection and conservation. In the future, other measures that directly regulate agriculture or others that promote improved environmental protection through sustainable agricultural practices will likely become more common throughout the developed countries.

8.9 MOVING TOWARD SUSTAINABLE AGRICULTURAL PRODUCTION

We have noted that both strategies to increase food production—expanding the cropland base and intensifying agricultural production—have caused such significant environmental problems that some agricultural production cannot be sustained. Such examples include newly cultivated land that is so susceptible to erosion that it cannot remain productive; cropland where irrigation uses more water than the aquifer's recharge rate; and irrigated soil that becomes too salty or wet to sustain production. The Worldwatch Institute estimates that as much as one-sixth of the U.S. grain harvest is unsustainable due to such practices. Although short-term food production is boosted, these are doomed strategies as the long-term productivity of the land resource is diminished.

Given these scenarios, what might be done to promote more sustainable agricultural practices while retaining high productivity? First, the renewable land, soil, and water resources that are the bases of agricultural production must be protected. Farming techniques that reduce soil erosion and land degradation must have priority over farming practices that produce only short-term gains while damaging or destroying the land itself.

Conserving Soil Productivity

Conservation tillage practices have been developed for mechanized, commercial agriculture to reduce soil erosion. In one form of **strip cropping**, parallel strips of erosion prone row crops such as corn are planted alternately with crops that form a more continuous plant cover, such as hay or wheat (Figure 8.15a). In hilly areas this technique is combined with **contour plowing**, where crops are planted across the slope parallel to the elevation contour (Figure 8.15b). Together these techniques retard runoff erosion by reducing the volume and velocity of water moving downslope. Strip cropping produces lower runoff rates and checks velocity, while the cross-slope furrows trap water rather than channeling water flow downslope. In semiarid regions, strips of drought-tolerant crops alternate with parallel strips of bare soil. This strip cropping *summer fallow* method, in which only half of the land is planted each year, enables the bare soil to absorb most of the annual precipitation. When planted the following year, the soil moisture accumulated over two years is available to the crop. By alternating the strips of crops and fallow perpendicular to prevailing winds, erosion of the bare soil is minimized, as any soil blown from the fallow strips is normally deposited in the adjacent planted strips downwind.

Windbreaks or **shelter belts** of trees or shrubs planted in rows act in a similar manner to retard windblown soil movement and reduce wind velocity immediately downwind from them (Figure 8.15c). **Stubble-mulching**, a technique developed in the midlatitudes to control erosion during the long dormant period after harvest, leaves crop residue or stubble on the field to make a rougher, more stable surface that provides more resistance to erosive forces of wind and water than does bare soil. Farmers are increasingly using additional conservation tillage techniques. By reducing or eliminating some plowing or culti-

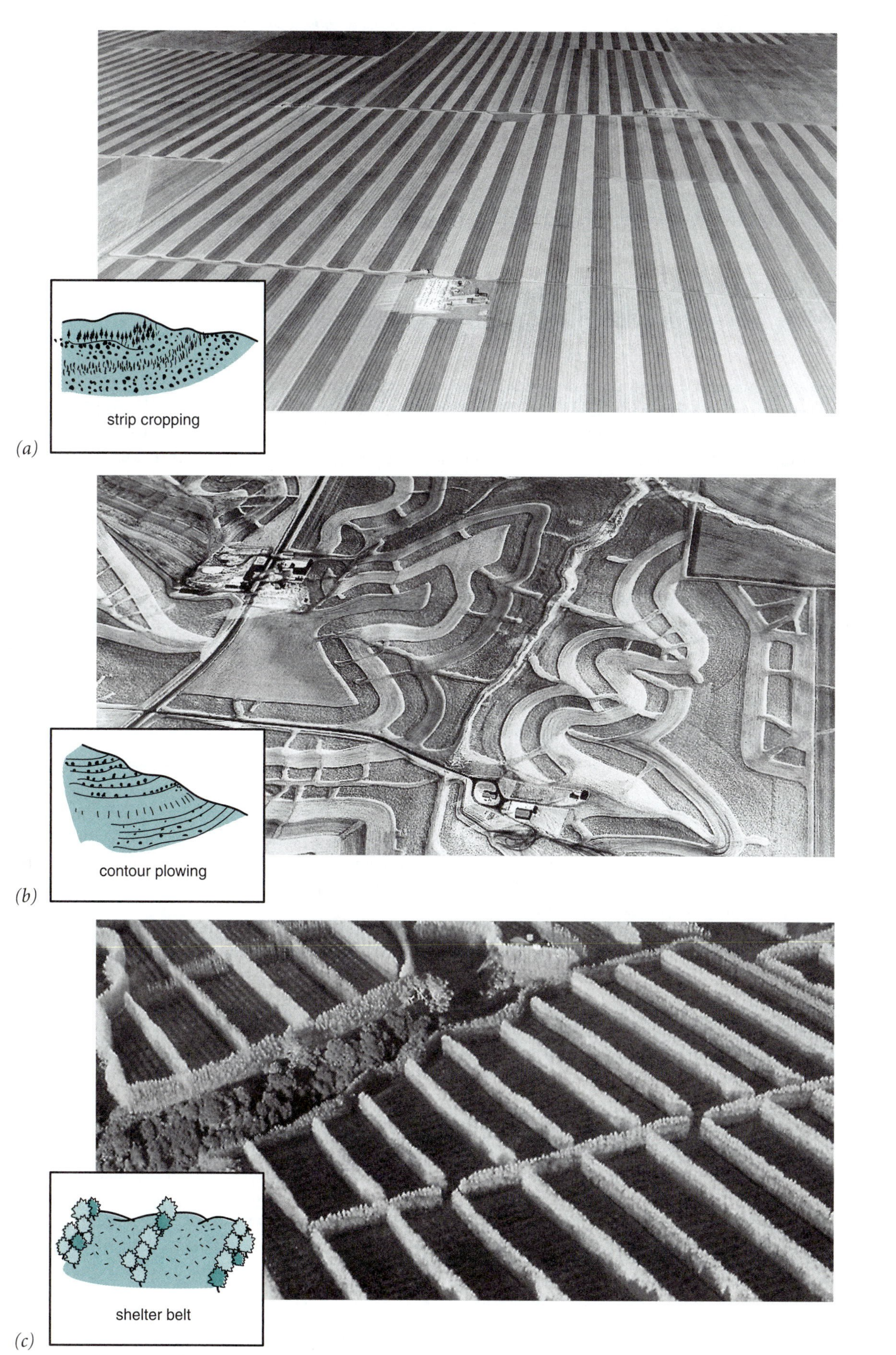

(a)

(b)

(c)

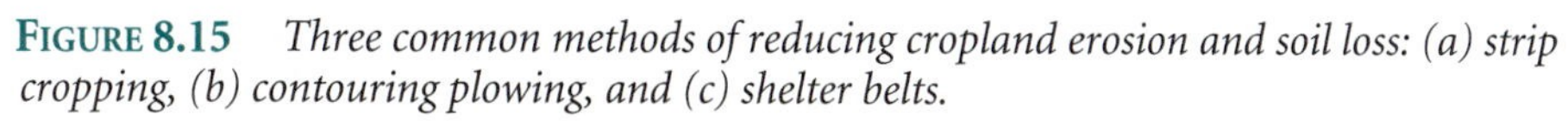

FIGURE 8.15 *Three common methods of reducing cropland erosion and soil loss: (a) strip cropping, (b) contouring plowing, and (c) shelter belts.*

vating, the soil surface is less prone to erosion and significant amounts of energy and labor are saved. However, these techniques do depend on the use of chemical herbicides to control weed growth, and planting and cultivating equipment must often be modified.

Crop rotation is another technique that contributes to sustainable agriculture. Soil fertility can be enhanced and erosion can be controlled by systematically planting a variety of crops on the same field over a period of years. Cover crops such as wheat may be rotated with row crops. Soil-building crops such as alfalfa or soybeans, which add nitrogen to the soil, are rotated with nitrogen-demanding crops such as corn. Cover crops may be planted and then plowed under as a *green manure* to add organic matter to the soil. Another option is to leave fields fallow for a year to restore moisture or nutrients. The variety of crops used in rotation systems provides some biological diversity, as opposed to monocultures of the same crop season after season in the same field. Because a variety of crops are grown in the rotation, the most profitable crop cannot always be grown. Unfortunately, many systems of commercial agriculture consider the short-term profits associated with monocultures over the long-term benefits of sustainable agriculture provided by crop rotation.

Conserving Energy Resources

Sustainable agriculture requires that we reduce or make more efficient use of nonrenewable resources such as fossil fuels. Conservation tillage techniques have become increasingly popular because they reduce energy costs. Other strategies that use renewable biological resources, such as **organic fertilizers** and **biological pest controls**, move toward production practices that are sustainable while reducing costly inputs of chemical fertilizers and pesticides that require high nonrenewable energy inputs in their production.

Conserving Water Resources

New technologies can now utilize irrigation water much more efficiently by delivering water directly to the plant root zone below the soil's surface through **drip and trickle irrigation** systems. Such techniques reduce the large evaporation and infiltration losses that occur with conventional irrigation systems. Because plants use virtually all of the water delivered by the drip and trickle systems, there is little likelihood that salinization or waterlogging will occur.

Use of Perennials

All of the world's staple grain crops are **annuals**—their entire life cycle is completed during a single growing season. Growing annuals means that each year seeds must be planted and young plants nurtured. Such requirements have significant economic and environmental costs. Research is underway to evaluate the use of **perennial** crops that would eliminate annual planting costs and reduce the high erosion risks during the planting period when soils are bare.

Adoption of Natural Resource Accounting

Conventional agricultural accounting systems do not include the costs of environmental degradation produced by farming. They therefore ignore the natural resources on which agriculture is based. Furthermore, by not being accountable, unsustainable agricultural practices will continue to accrue and defer costs that eventually will be borne by future generations. **Natural resource accounting** not only includes the normal agricultural production costs related to material inputs planting, cultivating, and harvesting, but considers both off- and on-site costs incurred by environmental damage such as erosion and runoff. Such accounting systems must be used in order to fairly compare the costs and benefits of resource-conserving farming practices with conventional production systems.

Developing Agricultural Policies that Support Sustainable Agriculture

In both developed and developing countries, policies that support sustainable agricultural systems must be established. These policies must promote productive and profitable farming practices that also protect the food-producing resources. Examples of such policies are summarized in this section.

Setting Specific Measurable Environmental Objectives

Agricultural policies such as the Conservation Reserve Program (CRP) lack sustainable environmental objectives because they have also been tied to reducing production and supporting farm prices. In contrast, a recent program in Oregon to reduce agricultural pollution in rivers establishes *total maximum daily load* (TMDL) standards for selected rivers. To meet these standards, the state Department of Agriculture works with farmers to devise land management practices that will help keep agricultural effluents within the TMDL standards.

Offering Tangible Participation Incentives

To ensure that farmers participate in agricultural environmental programs, it is essential to have adequate incentives. Like the CRP, most past programs have paid farmers to participate. Future programs may include such positive incentives as state or local tax reduction or negative incentives such as pollution taxes from excessive animal waste runoff from a feedlot or hog factory.

Using Market-based Incentives

Just as taxes on excessive pollution from farm operations act as a negative incentive, payments can be made to farms that have kept pollution levels below an established minimum permissible limit. Such a **negative pollution tax** provides a monetary incentive for establishing and maintaining sustainable agricultural production. Another strategy is to establish a market to trade **pollution rights**, as has been done in the electric utility industry. For example, after overall pollution targets are established for a watershed or estuary, pollution rights are allocated to each farm. Farms, which have reduced pollution, can then sell their unused pollution rights to other operations. Such a program provides both an incentive to reduce agricultural pollution and an economically efficient system of distributing pollution costs among producers. The full costs of pollution generated by agriculture are seldom calculated in production costs or market prices. In order to account more fully for these costs, use taxes can be levied on fertilizers or pesticides that are known to pose environmental risks. In Iowa, for example, revenues generated from a fertilizer tax are used to improve water quality within the state.

Using Production Methods that Improve or Sustain Environmental Conditions

Several of these methods are identified as follows:

- **Conservation tillage**. This reduces soil erosion by as much as 90 percent and conserves fuel, labor, and equipment. It is now being used on approximately 40 percent of U.S. cropland.
- **Integrated pest management (IPM)**. These programs reduce or replace the use of pesticides.
- **Plants and animals**. Plants and animals can be established in areas such that they better match the productive potential and physical limitations of the land.
- **Precision farming**. This farming method computes and applies the precise amount of fertilizer and other chemicals necessary for desired yields. This still-evolving technology requires soil and crop conditions to be monitored frequently. Computer programs would calculate the approximate mix of chemicals, and specially designed equipment would dispense the correct chemical application.

In summary, agricultural programs must promote profitable farm practices that also conserve agriculture's basic assets—the land and water resources on which production relies. Unlike current farm policies in many countries, which base payments on the quantity of program crop produced, new programs must mandate agricultural practices that ensure sustainable yet productive agriculture.

8.10 Can Sustainable Agriculture Supply Future Global Demand?

With world population projected to reach nearly 9 billion by 2050 and reach about 11 billion by 2100, global demand for food and fiber during this century will eventually double current production levels. Can the world's agricultural systems supply these huge increases? More importantly, can these production levels be sustained indefinitely? We can attempt to answer these questions by considering the resources available for increased agricultural production. Environmental resources include land, water for irrigation, energy, climate, and genetic materials, while the most important human resources are knowledge-based technologies and social institutions. We will briefly discuss each of these resources here.

Land

In addition to the 1.5 billion hectares (3.7 billion acres) of cropland currently in production, the UN Food and Agricultural Organization (FAO) estimates that there is an additional 1.8 billion hectares (4.4 billion acres) with climate and soils suitable for cultivation. However, most of this potential cropland is in the tropics of Africa and Latin America and is inherently less productive than currently cultivated land. Even with relatively high inputs, it is unlikely that most of this land would produce sustainable crop yields. Furthermore, much of this is tropical forest land, and likely has greater value as reserves for plant and animal genetic diversity. International efforts are now underway to protect more of the vast genetic resources of the tropical forests. Finally, if degradation of existing cropland continues, increased production costs will be needed to maintain production levels.

Water for Irrigation

Nearly one-third of the world's food and fiber is produced on the one-sixth of the cropland that is irrigated. About 60 percent of this land is located in monsoon Asia, and much of this land has supported traditional rice-based agricultural systems for centuries or even millennia. In contrast, many irrigation projects developed over the last century in arid regions are plagued with salty or waterlogged soils and may not be able to sustain production. Even less sustainable are the recently irrigated lands that utilize nonrenewable groundwater pumped from deep aquifers. As future irrigation projects are developed in more difficult locations, both environmental and economic costs will surely be higher. Furthermore, growing urban and other nonagricultural water demands will likely reduce opportunities to expand irrigation for agriculture. Even with improved technologies (such as drip irrigation) that use water much more efficiently and reduce potential for soil waterlogging or salinization, the potential for significantly increasing agricultural production by expanding irrigated cropland seems quite limited.

Energy

With global agricultural production becoming more commercial and specialized, the increased use of chemical fertilizers and pesticides, hybrid seeds, water and machinery result in even higher energy inputs. Given this trend, agriculture's energy demands most likely will continue to increase in the decades ahead. However, global energy supplies are likely to be constrained by price increases and the environmental impacts of fossil fuels. Furthermore, this supply is not indefinitely sustainable, so agriculture, just as the rest of the world's economy, must eventually shift to alternative energy sources. Expanding production in light of increasing energy costs will be a difficult challenge for global agriculture in the decades ahead.

Climate

The consensus among scientists is that global climate patterns will change during the next century. However, there is little consensus about how this will affect global agricultural production. Some studies suggest that hotter, drier conditions in important agricultural areas such as the U.S. Midwest might reduce global food production by 15 percent or more. Yet others indicate that technological adjustments could be made to reduce the negative impacts of climate change. Furthermore, a warmer climate would bolster agricultural production in some areas. On balance the net impact on world agriculture from climate change is expected to be relatively modest.

Genetic Materials

Most of the recent increases in agricultural yields are attributable to the development of more productive crops and livestock. To further expand agricultural production will require even more intensive efforts by plant and animal breeders. Such work will include traditional crossbreeding as well as genetic engineering, which requires access to a broad range of genetic materials. To ensure that the genetic resource is protected, gene banks store and distribute genetic materials from original plant stocks. Given the past successes of crossbreeding, along with the recent advances and future promise of the new biotechnologies, forthcoming productivity gains in agriculture will rely heavily on these genetic resources.

Expansion of Knowledge

If agricultural production is to double or even triple in the decades ahead without depleting its natural resource base, new technologies and institutions must be developed and employed to utilize and husband resources more efficiently. In other words, even more productive Green Revolutions will need to offer new strains of high-yielding yet hardy crops that utilize less fertilizer and water. Traditional sustainable techniques will have to be modified to enhance productivity yet remain protective of vital resources. Once such innovations or technologies are developed, they must be transferred to the farms and fields. Successful technology transfer requires new and more facile institutions that can deliver training to farmers. Transferring new knowledge through agricultural organizations and educational institutions to rural communities throughout the world presents a formidable challenge. For many societies, this will require changing cultural beliefs and practices deeply imbedded in village life, religion, and political governance systems.

On the positive side, such agricultural innovations and institutions have helped triple global agricultural output in the last half-century. It may be unrealistic to expect that expanding knowledge can overcome natural resource limitations and the still-burgeoning population in the developing countries to raise food production levels two to three times above current record levels. Nevertheless, the dire consequences of *not* developing a sustainable global agricultural system that adequately meets human needs provides an extremely strong incentive to succeed.

8.11 SUMMARY

No other human activity has resulted in as much environmental modification as food production. Over a period of only 12,000 years, crop agriculture and grazing have spread over one-third of the Earth's land area and developed a capacity to support more than 6 billion people—but not without great costs to the environment. As sedentary agriculture has evolved and spread across the continents, it has altered and even eliminated natural ecosystems and replaced them with simplistic and—to a significant extent—unsustainable plant and animal food production systems. At the same time, natural water systems have been extensively altered with the development of irrigation facilities, channelization of streams, eradication of wetlands, and the pumping of groundwater.

These disturbances have been accompanied by extensive water pollution—initially from sediment loading of surface waters, but now from contamination of both surface and groundwater resources with fertilizer residues and pesticides.

As human population soars, the demand for additional agricultural production also soars. The pressure for increased production is pushing agriculture in two directions: (1) to use existing farmland more intensively, and (2) to convert open land into more farmland. Because most of the good farmland has already been taken, new farmland must be drawn from areas marginally or poorly suited for agriculture. These are floodplains, wetlands, grasslands, tropical forests, and mountain slopes, which not only pose risks to agriculture from storms, flooding, and drought but also promote environmental degradation from the presence of agriculture itself. The result is a downward spiral in which land quality, farm productivity, and environmental quality decline together.

As economies grow and living standards improve in some developing countries, demand for more food and other agricultural products will grow faster than population. Huge productivity gains have been realized on existing farmland by way of mechanization, expanded irrigation, hybrid seeds, chemical fertilizers, pesticides, and herbicides, but much of this gain may not be sustainable. This is because productive agricultural land is being damaged or lost due to soil erosion and salinization, and production has become increasingly dependent on fossil fuels and other nonrenewable resources that will become increasingly scarce and costly. Arguably, much of today's agricultural land is being pushed beyond the limits of sustainability. Productivity is often based on propped up systems, where increasingly large inputs of chemical fertilizers, pesticides, irrigation water, and machinery are needed to offset a declining resource base of soil and water.

And even while trends in food production continue to increase, perhaps one-fifth of the world's population is undernourished. This is not because there is a shortage of food—plenty of food is produced today, and much is wasted—but because of the inequitable distribution of food-producing resources and food itself due to poverty and government policies. The most glaring examples are related to conflicts within and between countries where ethnic pride, civil war, and political intransigence block commercial trading and food redistribution and thereby promote famine. Somalia are Bosnia and two recent examples.

Can we achieve highly productive, sustainable agriculture for most of the world? It is possible only if some important changes are made. First, improved soil-, water-, and energy-conserving techniques must be incorporated into all types of farming throughout the world. The current trends of soil loss, water use, and energy expenditure are eating into the gains made by improved crop types and farming techniques, which in turn are prohibiting the development of stable agricultural economies.

Second, more renewable resources such as organic fertilizers and biological pest controls are needed. Like so many nature-dependent land uses, agriculture cannot rely solely on technological fixes (e.g., a new chemical for every new bug or disease) to solve every problem. Instead, crop types and farming methods will have to be balanced with environmental conditions, respecting the constraints and building on the opportunities presented by nature.

Third, traditional farming techniques need to be more productive. Among other things, water supplies have to be used more effectively, especially where use rates far exceed recharge rates and where political borders interfere with equitable water allocation. Such progress would require the development of national and international agricultural policies that are supportive of long-term sustainable agricultural production. Such sustainable systems will be very difficult to design and implement because of the increasing pressure for short-term economic gain as commercial agriculture expands throughout the world.

8.12 KEY TERMS AND CONCEPTS

Evolution of agriculture
- agricultural system
- crop rotation
- plantation system
- diffusion and exchange

Industrialization
- commercial agricultural systems
- mechanization and specialization
- specialized production regions (e.g., Corn Belt)

Characteristics of food crops
- high yield
- high food value
- storage ability
- staple crops

Production systems
- small-scale subsistence farming
- intensive subsistence agriculture
- multiple cropping
- intercropping
- hybrid varieties
- shifting cultivation
- nomadic herding

Hunger
- starvation
- undernutrition
- micronutrient deficiencies
- nutrient-depleting diseases and parasites
- famine
- schistosomiasis
- land tenure

Green Revolution

New strategies for traditional agriculture
- agroforestry
- nutrient recycling

Biotechnology
- genetic engineering
- gene splicing

Environmental impacts
- soil erosion
- salinization and waterlogging
- groundwater depletion
- surface water loss
- substitution and simplification
- ecosystem losses
- Conservation Reserve Program (CRP)
- environmental regulation

Sustainable agriculture methods
- strip cropping
- conservation tillage
- contour plowing
- windbreaks and shelter belts
- stubble-mulching
- crop rotation
- organic fertilizers
- biological pest controls
- drip irrigation
- annuals/perennials
- natural resource accounting

Policies to promote sustainable agriculture
- negative pollution tax
- pollution rights

8.13 QUESTIONS FOR REVIEW

1. Briefly describe the characteristics of the traditional agricultural systems that have supported people for hundreds or even thousands of years. Why have these systems been successful for so long?
2. Discuss how industrialization and urbanization during the nineteenth and twentieth centuries in North America and Europe led to the development of specialized commercial agricultural systems.
3. What characteristics do most important agricultural crops share? List the five (5) most important food crops. Which world regions lead in the production of each of these crops?
4. The three traditional subsistence agricultural systems continue to support a majority of people in some parts of the world. Where is each still a way of life for the majority of people? Why are these systems gradually changing in most places?
5. Compare modern commercial agricultural systems with traditional subsistence systems. Where are most commercial farms located? Although agricultural production continues to expand, why is the number of commercial farms decreasing?
6. During the past 50 years, worldwide food production has increased even more rapidly than population. Discuss the factors that account for these significant increases.
7. Although there is enough food produced each year to adequately feed the world, hunger and malnutrition are widespread in several less developed regions. Where are food shortages most widespread? Discuss several reasons for this disparity.
8. Summarize the positive and negative aspects of the Green Revolution.
9. To some, genetic engineering is a new technology that will enable more bountiful agricultural production. To others, it is a potentially dangerous technology that may result in environmental damage and health risks to consumers of transgenetic products. Present both potential contributions and the liabilities of agricultural genetic engineering. Develop a list of genetically engineered food products currently on the market.
10. The rapid expansion, mechanization, and commercialization of agriculture have resulted in widespread environmental degradation. What measures are being taken to reduce environmental damage caused by agriculture? What is being done to initiate sustainable agricultural production?
11. With world population projected to climb to nearly 12 billion, demand for agricultural products will double or even triple during the next century. What is the outlook for achieving sustainable agricultural production that will meet these huge demands?

9

ENERGY GENERATION, USE, AND THE ENVIRONMENT

9.1 INTRODUCTION

If only we were able to tap a small fraction of incoming sunlight, we could easily meet all of our energy needs, because the amount of solar radiation reaching the Earth's surface is about 15,000 times greater than our current energy use. Unfortunately, we now use only tiny amounts of *direct* **solar energy** to heat buildings and water and to generate small amounts of electricity. However, nature converts about one-fifth of incoming solar radiation into energy that drives processes such as winds and currents. For many centuries humans have used these *indirect* forms of solar energy to sail, pump, and mill. Today hydroelectric power generated by water supplies about 7 percent of the world's commercial energy, while wind-generated electricity produces less than 0.2 percent.

For most of human history, plants have provided most of our energy requirements in the form of food and fuel for heating and cooking. Even today *biomass fuels*, principally wood, provide the main energy sources for nearly half the world's people. In fact, *fossil fuels* were ultimately derived from sunlight millions of years ago when plant remains were preserved, concentrated, and transformed in the Earth's crust into the solid, liquid, and gaseous fuels. Thus, we are bound to ecosystems, both modern and ancient, for many of our energy resources.

Even though the sun provides most of our energy, there are a few **nonsolar energy sources**. In recent decades we have discovered the enormous power stored in the atom and have employed controlled nuclear reactions to produce electricity from *nuclear power*.

Today, approximately 6 percent of the world's total commercial energy is produced in nuclear power plants. Also, small amounts of *geothermal power* from the Earth's interior are used to generate electricity or to heat buildings. *Tidal flows*, produced primarily by the Moon's gravitation, are also harnessed in a few locations to generate small quantities of electricity.

Figure 9.1 shows that the current world's commercial energy economy is overwhelmingly dependent on **nonrenewable** fossil fuels and nuclear power with only about one-sixth of our commercial energy derived from potentially **renewable** or **perpetual** sources. Reliance on nonrenewable energy resources has grown steadily and rapidly as the technological revolutions of the last two centuries have been based on large quantities of relatively inexpensive, easy to use, and readily available fuels. Today the developed countries with energy-dependent technologies use nearly four-fifths of this commercial energy, while the developing countries, with nearly 80 percent of world's population, only use about one-fifth. The United States, with less than 5 percent of the population, uses more than 20 percent of the world's commercial energy—of that, more than 90 percent comes from nonrenewable sources (Figure 9.1b). In contrast, the developing countries rely heavily on potentially renewable energy resources because the majority depend on fuelwood for heating and cooking. While the developed countries will likely face dwindling supplies of oil and natural gas over the next several decades, in parts of the developing world many are already experiencing a fuelwood shortage where rapidly growing populations demand more wood than can be grown.

Although only a tiny fraction of Earth's natural energy budget is sold and bought in the world marketplace, the commercial energy sector has become a huge economic enterprise. This sector provides most of the fuels, steam, and electricity that drive our machines and heat and illuminate our homes and businesses. This commercial energy sector will be the focus of this chapter. We will explore how humans developed technologies to harness these inanimate forms of energy, how we develop and use these energy resources, and how these vast energy expenditures drive the world's economies and affect the environment.

We will discuss the significant environmental impacts of energy production and consumption. For example, combustion of fossil fuels is a major cause of local smog

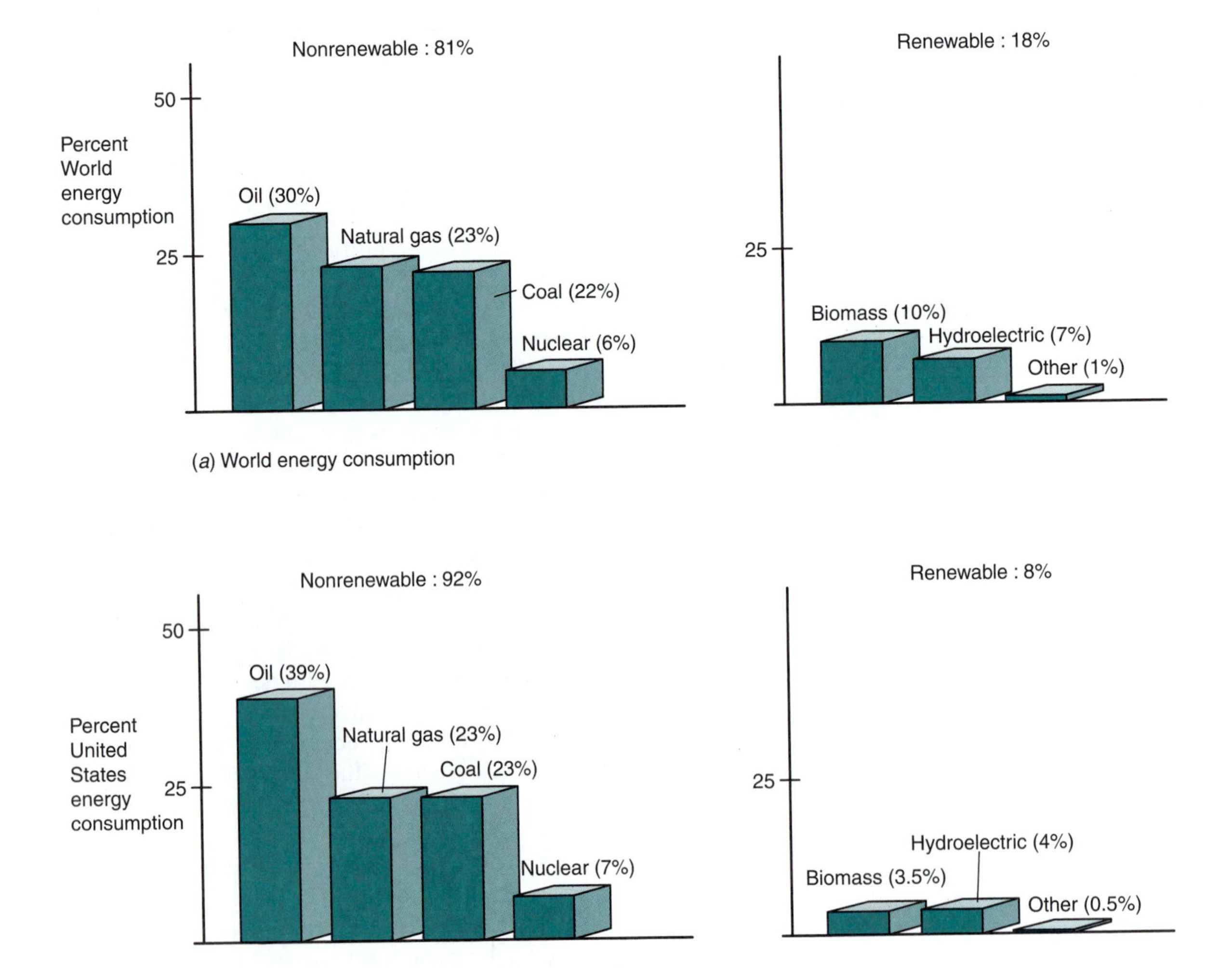

FIGURE 9.1 *Nonrenewable and renewable energy sources for (a) the world and (b) in the United States. (1998)*

episodes, produces acid rain that affects regional ecosystems, and emits greenhouse gases that may play significant roles in modifying global climate. Large hydroelectric dams and reservoirs disrupt riverine and flood plain ecosystems and displace human settlements, while radioactive waste from nuclear power plants creates potential long-term environmental hazards. In addition, mining, processing, and transporting energy resources result in significant environmental impact.

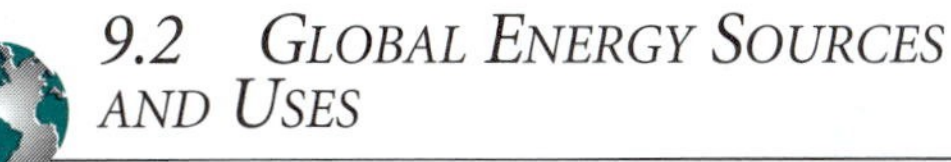

9.2 GLOBAL ENERGY SOURCES AND USES

Throughout human history **fuelwood** has been the major fuel source for heating and cooking, and it remains the dominant fuel in many rural areas in the developing world. Although coal eventually replaced wood as the fuel of choice during the nineteenth century, huge quantities of wood were used to drive steam engines and make charcoal to fuel iron and steel smelters. In fact, charcoal is still used as a smelter fuel in parts of the developing world.

Coal was replaced by oil as the most important fuel around 1950, mainly because of the worldwide expansion of vehicles that use petroleum fuels. Currently oil provides about 30 percent of world commercial energy supplies, down from its peak of 40 percent in the early 1970. Although oil's share of the world's energy supply will likely continue to decline, clean-burning natural gas will probably play a larger role as pipeline networks and liquefied natural gas tankers are built to distribute it to markets.

Figure 9.2 shows that electricity is supplying an increasing share of global energy. Today most of the electricity that supplies industry and the expanding urban-based business and consumer service sectors is generated in large, thermal electric or hydroelectric generating facilities. Thermal electric power plants burning coal and other fossil fuels currently generate 62 percent of the world's commercial electricity, while nuclear thermal electric facilities produce 17 percent. Hydroelectric power plants account for slightly less than 20 percent of the world's electricity; all other renewable energy sources account for only 1.5 percent.

Overall, natural gas and oil together account for 55 percent of the world's energy supply. Unfortunately, we are rapidly depleting this store of fossil fuels. Furthermore, acquiring and using fossil fuels produces significant environmental damage. As a result, it is imperative that current evolving technologies be applied to use fossil fuels more efficiently and develop alternative energy sources based on renewable and perpetual energy resources that will eventually replace fossil fuels.

Even though the world has experienced several energy shocks during the last three decades, technology has not made significant progress in providing widespread energy systems based on renewable or perpetual resources. Will energy policies continue to follow the "business as usual" strategy that depends primarily on fossil fuels? Or will policies emerge that encourage energy efficiency and conservation for the near term while developing sustainable energy resources for the intermediate and long term?

A hundred years ago, most energy resources were still used to heat buildings. Today the majority of energy is used by business, industry, and transportation. The changes in energy consumption have been just as radical as the changes in energy sources. Figure 9.3 illustrates current patterns of commercial energy production and

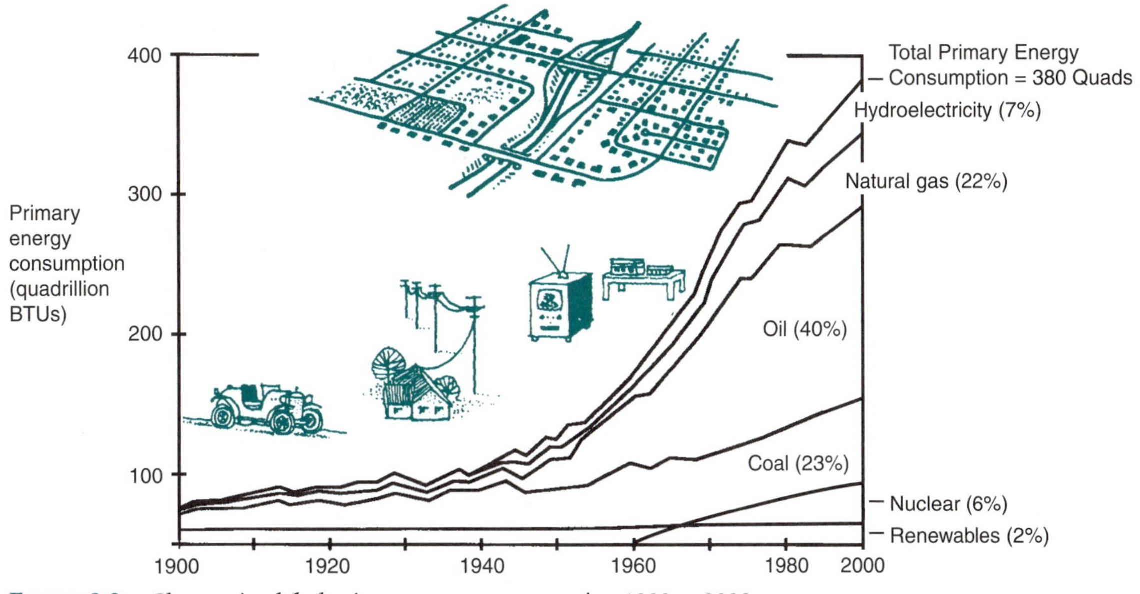

FIGURE 9.2 *Changes in global primary energy consumption 1900 to 2000.*

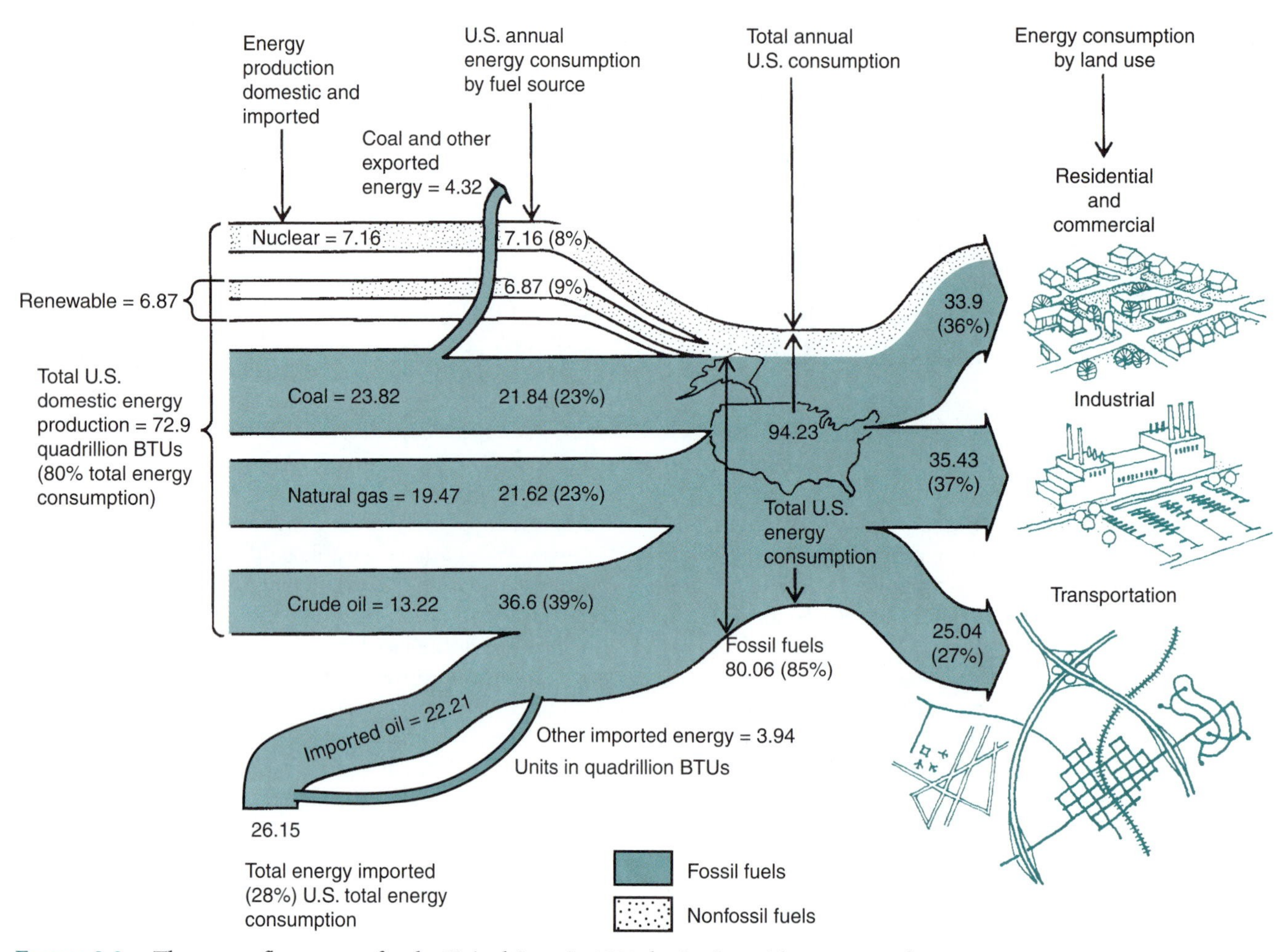

FIGURE 9.3 *The energy flow system for the United Sates in 1998, beginning with sources on the left and ending with consuming land uses on the right. (Units are quadrillion BTUs.)*

consumption as energy flows through the U.S. economy. Let us briefly examine how energy is used in the four sectors of the economy.

Given the long distances to be covered moving people and freight around the United States, it is no surprise that the transportation sector consumes about 27 percent of total energy. About two-thirds of this energy is used to transport people, with automobiles using more than half of the sector's total. Trucks, which can move freight quickly, account for only about 25 percent of the intercity freight, yet consume more energy than the slower but more energy-efficient bulk cargo haulers—trains, ships, barges, and pipelines, which handle the majority of the intercity freight. Increasing energy efficiency in this sector is extremely important because there is so much energy waste in the system and because more than half of the vehicle fuel comes from imported oil.

Another 37 percent of U.S. energy is consumed by the industrial sector, which, along with manufacturing, includes agriculture, mining, and construction. U.S. industries have become more efficient energy users in recent years but still trail Japanese and European industry in efficient energy use.

Heating, cooling, and lighting homes, institutions, and businesses consume almost 36 percent of U.S. energy. Space heating, air conditioning, and lighting account for the majority of energy used both in the commercial sector (offices, schools, hospitals, and stores) and in the residential sector (about 90 million households). With such huge numbers of buildings involved, this heavy energy-using sector provides an excellent opportunity for energy conservation efforts.

9.3 FOSSIL FUELS

As the name implies, fossil fuels were derived from the remains of plants and animals buried millions of years ago in swamps, lakes, and seabeds. Through geologic time, these materials—composed of carbon, oxygen, hydrogen, and other elements—were transformed by heat and pressure within Earth's crust into coal, oil, and natural gas.

There are several reasons why we have come to depend on these nonrenewable resources to fuel the technological revolutions of the last two centuries.

- **Accessibility:** Fossil fuels are accessible in the Earth's crust and under the continents and continental shelves. We have been able to economically find, extract, and transport them in large quantities at thousands of locations.
- **Utility:** Fossil fuels are easy to use. The technologies of controlled combustion have been developed at all scales to provide the energy services we need, ranging from furnaces for home heating and the internal combustion engine for motor vehicles to the steam-driven turbines and generators that produce most of our electricity.
- **High energy content:** Fossil fuels provide relatively large amounts of chemical energy per unit mass. For example, a gallon of gasoline contains about 36,000 kilocalories (a kilocalorie is equal to a calorie of nutrition).
- **Transportability:** Fossil fuels can be transported relatively easily and economically. Coal is transported long distances overland by unit trains or over water by barges and bulk carriers. Specialized tankers now carry multimillion dollar shipments of crude oil, petroleum products, or even liquefied natural gas along transoceanic trade routes and pipelines move petroleum and natural gas from remote production locations to major markets.
- **Conversion to different fuels or feedstocks:** Fossil fuels can be converted from one form to another. For example, liquid or gas fuels can be obtained from coal. And fossil fuels are excellent feedstocks for plastics, fiber, and other chemicals.

Unfortunately, burning these fuels in large quantities has a significant environmental impact. Most notable is air pollution in the form of oxides, particulates, carbon dioxide, and other pollutants that contribute to smog, acid rain, and climate change. Even if the environmental impact caused by fossil-fuel combustion does not constrain or preclude their use, their increasing scarcity and cost might: Within the next century, we may deplete the supply of these finite resources that can be economically exploited. Whatever our energy future will be, it cannot be indefinitely sustained by these fossil fuels. After a decade of relatively low prices, energy prices, led by oil, and natural gas have increased dramatically. Does this rapid increase in petroleum and gas prices signal the end of cheap energy, as some experts claim? As the century unfolds, new energy resources will be needed as traditional fuel supplies shrink and as prices rise.

When we discuss the amount of fossil fuels (or any other nonrenewable resources) that may be available for use, we need to distinguish between the *reserves*, or *economic resources*, and the *resource base*. Figure 9.4 illustrates that the economic resources include identified stocks that can be profitably produced with existing technology at current prices. The total resource base also includes known stocks that are currently too costly to exploit, as well as all undiscovered stocks. Some of these materials in the resource base may become economic resources as they are discovered, as prices increase, or when improved technology lowers production costs.

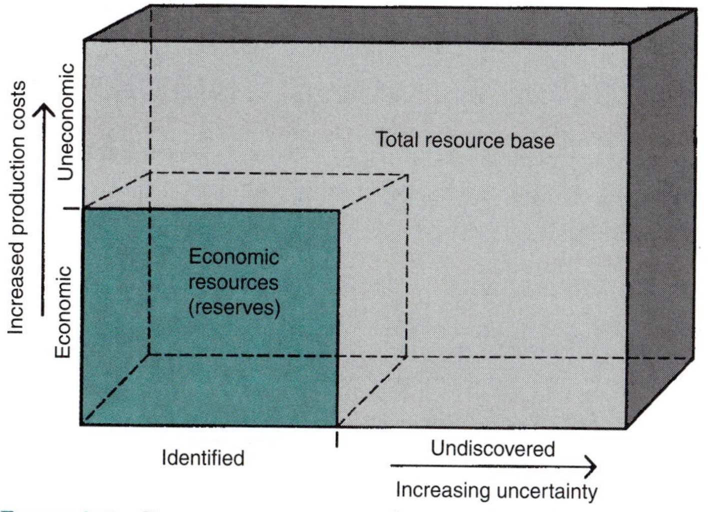

FIGURE 9.4 *Known energy reserves relative to the total resource base.*

Coal

As industrialization spread during the nineteenth century, **coal** became the dominant fuel. At its peak in the 1920s, coal was providing about 70 percent of the U.S. energy supply and even more in industrialized Western Europe. Coal's dominance as a commercial energy source slowly declined, and by 1950, oil was supplying the most energy (Figure 9.2). Even so, as worldwide energy consumption has grown, world coal production reached record production levels around 1990. Since then, production has declined about 5 percent.

Since the oil shocks of the 1970s, coal has experienced a renaissance in the United States. Coal use increased by 40 percent since 1980. Today it supplies about 23 percent of U.S. energy needs. Most coal (85 percent) is used to generate electricity in large thermal electric power plants, with industry utilizing most of the remaining to produce steam, high-temperature heat, and coke. The transportation, commercial, and residential sectors were major consumers of coal in the nineteenth and first half of the twentieth century, but now use less than 1 percent.

Geography of Coal

Coal is the most abundant and widely distributed fossil fuel, with reserves found in many parts of the world (Figure 9.5). During the nineteenth century, the coal fields of Great Britain and northwestern Europe became the world's foremost manufacturing centers as industries located near their chief energy source. Today, the European fields continue to produce but no longer lead world coal production.

With nearly three-quarters of world's coal reserves, China, the United States, and the former Soviet Union currently produce over half of the world's coal. Annual production now exceeds 1.4 billion tons in China, making it the leading producer (the United States is second, with 1.1 billion tons). China, like Europe a century ago, relies heavily on coal. Major industrial centers have been developed close to mining areas.

The geography of coal in the United States has changed significantly since the period of rapid industrialization a century ago when raw materials and energy resources largely determined where new industrial growth would occur. The vast Appalachian coal fields stretching from Pennsylvania to Alabama were developed during the nineteenth century to provide steam power for industry and transportation and coke for iron and steel. The region continues to produce more coal than the rest of the United States, with three states accounting for 37 percent of the country's coal production: West Virginia (16 percent), Kentucky (14 percent) and Pennsylvania (7 percent).

Much of the recent growth in the U.S. coal industry has occurred in the West with the development of large deposits of low-sulfur coals that can be **surface mined**. Large-scale production in Wyoming, Montana, Colorado, and Utah began in the 1970s in response to environmental restrictions that favored low-sulfur western coal over the high-sulfur coals of the East and Midwest. Currently,

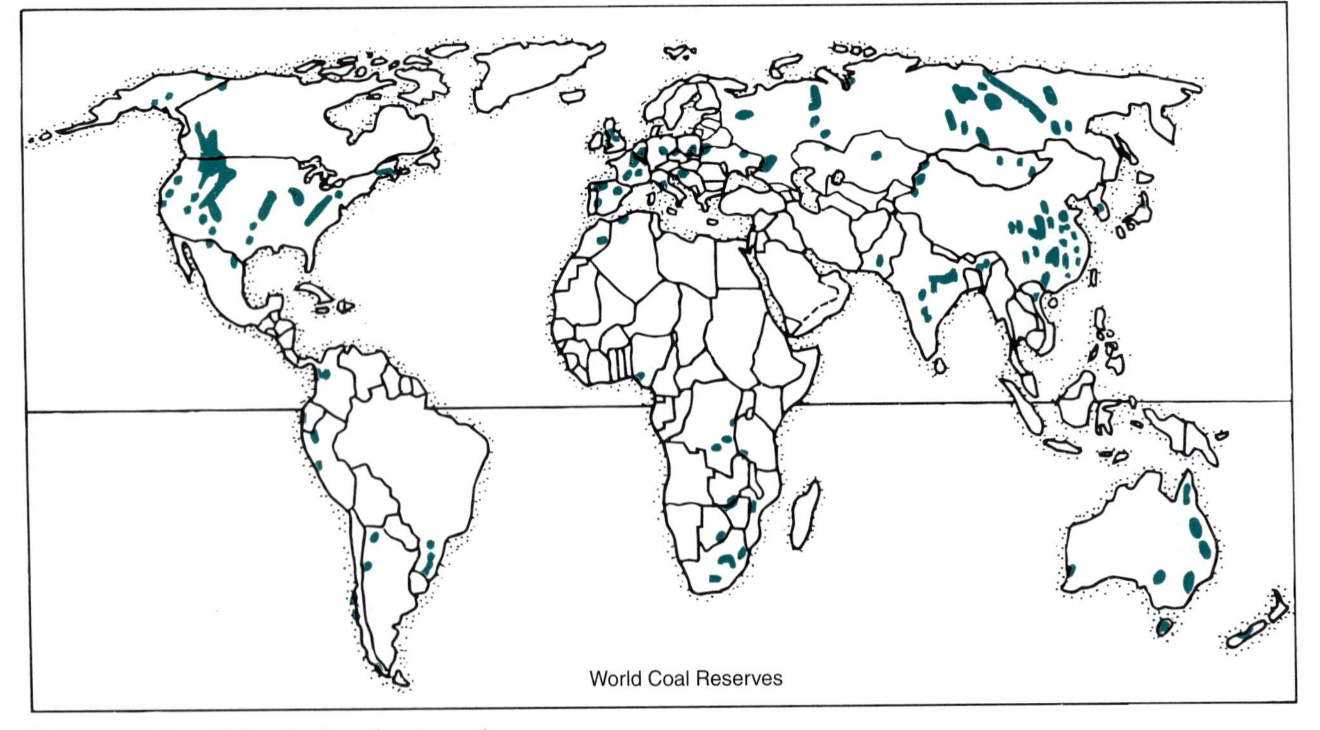

FIGURE 9.5 *World distribution of major coal reserves.*

Wyoming produces more coal than any other state, producing over 25 percent of the U.S. total. Trains more than one hundred cars long, called *unit trains* now haul western coal to power plants in many parts of the country.

Advantages and Disadvantages of Coal

Because of coal's relatively low cost, use will likely continue to increase in areas where it is abundant. Coal has some advantages over other commercial energy sources:

- **Coal is the most abundant fossil fuel.** Identified world coal reserves have nearly three times the energy equivalent of all known oil and natural gas reserves. Current production uses less than 0.5 percent of these reserves annually. The energy equivalent of U.S. coal reserves, the largest of any country, is nearly twice that of the identified oil and gas reserves of the Persian Gulf countries!
- **Coal is cheap.** It is the least expensive fuel for generating electricity and producing high-temperature heat or steam. As a result, coal is the most widely used fuel in thermal electric power plants worldwide. Currently 50 percent of U.S. electricity is generated in coal-fired plants.
- **Coal can be converted into or gaseous fuels called synfuels.** Coal **gasification** and **liquefaction** technologies have been available since the early 1900s. Currently, South Africa uses its relatively large coal reserves to produce synthetic fuels and expects to produce enough synthetic gasoline to eventually meet about half of its petroleum needs.

Although coal is an abundant, accessible, and relatively versatile fuel, its production and use cause significant environmental and health problems. Some of coal's disadvantages are listed here:

- **Coal mining is dangerous and unhealthy work.** Mine explosions and collapses have killed and injured hundreds of thousands while thousands more have been disabled by *black lung disease* caused by breathing coal dust in underground mines.
- **Coal mining can cause land subsidence and acid drainage.** If mining tunnels collapse, the land surface above may subside or sink, disrupting surface land uses. Mine acid drainage produced when water flows through coal or other minerals exposed during mining can pollute groundwater or nearby lakes and streams.
- **Surface or strip mining coal produces vast spoils (waste materials) and damaged land.** Because it is now more economical to strip-mine coal, more than 60 percent of U.S. coal is produced from surface mines (Figure 9.6). **Reclamation** of damaged land is now monitored in the United States by federal law. The Surface Mining Control and Reclamation Act of 1977 requires that the mined land must be restored and revegetated (Figure 9.6c).
- Burning coal produces more atmospheric pollution than other fuels. The nitrogen (NO_x) and sulfur (SO_x) oxides from coal combustion are major contributors to **acid precipitation** and combine with particulates to produce a foul-smelling **sulfurous smog**. The smog aggravates the human respiratory system and can be fatal for people suffering from chronic bronchitis or emphysema. In addition, coal combustion worldwide accounts for about 36 percent of the carbon dioxide (CO_2) emissions added to the atmosphere as a result of burning fossil fuels. China and the United States—the two leading producers and consumers of coal—generate 50 percent of the CO_2 emissions from coal. The buildup of atmospheric CO_2, considered a major contributor to global warming is discussed further in Chapter 10.

Crude Oil or Petroleum

Oil is a liquid hydrocarbon mixture formed when heat and pressure transform decomposed organic material in sedimentary rocks. Both oil and natural gas migrate toward the surface through permeable rocks until trapped by an impermeable formation, where it concentrates or *pools* in cracks and pores of rocks (Figure 9.7).

Although humans have used oil from surface flows and shallow wells for millennia, the age of oil really began in the 1860s, when large-scale production and simple refining processes made *kerosene* the illuminating fuel of choice. By 1910, oil supplied about 10 percent of the U.S. energy as *fuel oil* was being used for heating and steam production and *gasoline* was fueling the internal combustion engine. Its versatility as a fuel and for lubricants, petrochemicals, and asphalt resulted in increased use. By 1950 oil had replaced coal as the world's leading energy source. After world oil production peaked in 1979 at more than 62 million barrels per day, production levels slipped by 10 to 15 percent during the economic downturns of the 1980s. As global economies expanded during the 1990s, oil demand increased again. Currently world production is at its highest level ever and averages more than 67 million barrels of crude oil a day, supplying about 30 percent of the world's energy (see Figure 9.1).

Recent Patterns and Trends of World Oil Production

Throughout the first century of the oil industry, the United States was the world's leading producer. Figure 9.8 illustrates that until 1950 domestic production had kept pace with demand. Although production continued to increase until 1970, consumption rose even faster. By the late 1970s imports accounted for nearly half of the oil consumed in the United States. To bolster domestic production, more inaccessible and costly reserves had to be tapped. An 800-mile, multibillion-dollar pipeline from the shores of the Arctic Ocean to the Pacific Ocean had

(a)

(b)

(c)

FIGURE 9.6 *Conditions associated with surface or strip coal mining; (a) mining; (b) degraded land after mining; and (c) reclaimed land after landscape restoration.*

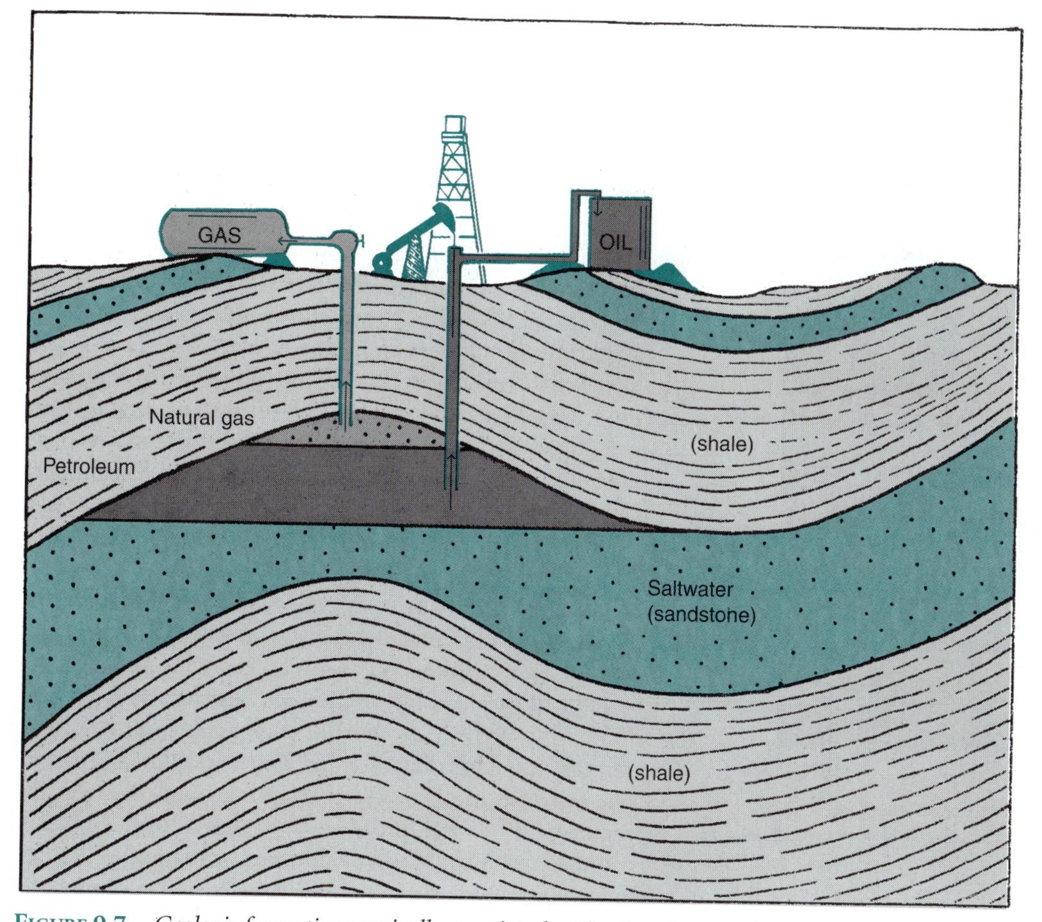

FIGURE 9.7 *Geologic formations typically associated with oil and natural gas reserves.*

to be built before Alaska's rich oilfields could be fully developed (see Figure 4.10). On the continental shelves, offshore production from billion-dollar platforms had to move to deeper waters. Onshore, deeper, costlier wells have been drilled and expensive secondary and tertiary recovery methods that inject materials into wells have been developed to increase oil output from older fields.

Spurred by higher oil prices and these new technologies, U.S. oil production increased during the late 1970s and early 1980s. But over the last 15 years, total production has fallen as both Alaska and offshore production slowed and imports rose to meet increased demands (Figure 9.8). Domestic production has continued to decline so that by 2000, oil production was at its lowest level since the 1950s. U.S. **petroleum reserves**, defined as the known oil deposits that can be recovered at current prices, have been declining annually since the huge Alaska fields were discovered in 1970. This means that each year more oil is being produced than is being discovered. In 2000, total U.S. reserves were estimated at about 21 billion barrels, or about seven times the amount of oil produced domestically in that year.

While U.S. production was declining, Soviet production continued to increase. By 1974, the USSR had become the world's leading petroleum producer. Since political dissolution of the Soviet Union in 1991, oil production has been declining in major oilfields in Russia, Ukraine, and the former Soviet republics of Turkmenistan, Kazakhstan, and Azerbaijan along Russia's southern border. By 2000 the amount of oil produced throughout the former Soviet Union had declined by more than 30 percent from 1990 levels.

Today the former Soviet Union and the United States together produce only a fifth of the world's oil. In contrast, the eleven countries that make up the **Organization of Petroleum Exporting Countries (OPEC)** produce nearly two-fifths (Figure 9.9) and Saudi Arabia has become the world's leading oil producer. As major suppliers of oil to the world markets, the OPEC consortium has played a major role in establishing world oil production levels and prices since the first "oil shock" in 1973.

In recent years, the OPEC consortium began to reassert their influence on world oil production and prices. By cooperating to reduce production, world oil supplies

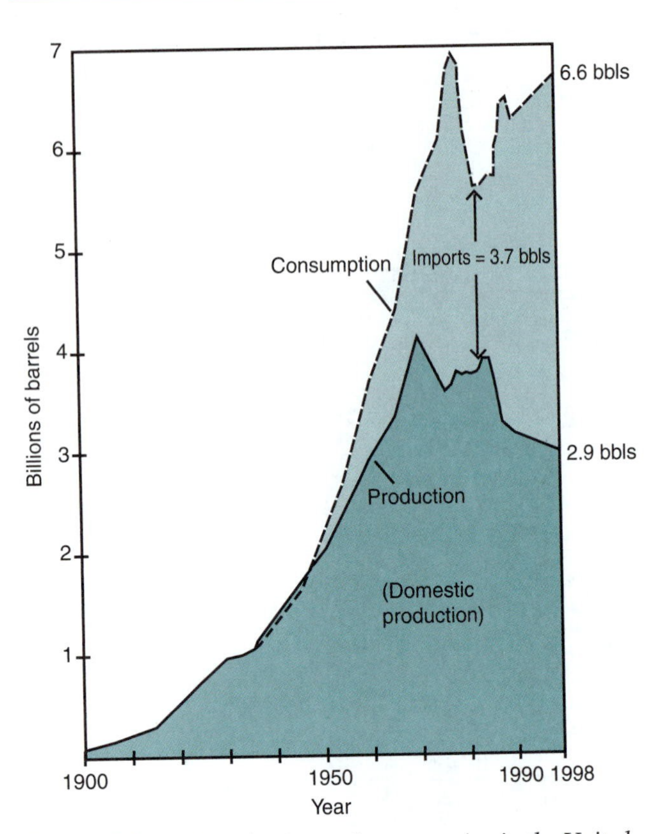

FIGURE 9.8 *Oil production and consumption in the United States, 1900–1998.*

began to tighten while demand continued to increase. During 1999 the OPEC countries—along with several other major oil exporting countries including Mexico and Russia—cut oil production by almost 4 percent. The predictable result was a rapid and significant increase in oil prices. In just a year crude oil prices nearly tripled, rising from less than $12 a barrel in early 1999 to over $32 a barrel in 2000. Furthermore, the OPEC countries control more than three-quarters of the world's proven oil reserves, with almost two-thirds located in the nations surrounding the Persian Gulf (Figure 9.10). The significance of these reserves to the global economy and especially to Japan, the United States, and many other developed countries that import large amounts of oil is evidenced by the huge military expenditures, particularly by the United States, to assure that oil from the Persian Gulf region flows to world markets.

Iraq's 1990 invasion and takeover of Kuwait and impending threats to neighboring Saudi Arabia led to the outbreak of the Persian Gulf war in 1991. This military action, led by United States forces, cost tens of billions of dollars and added huge, if indirect, costs to the price paid for oil from the Persian Gulf region. Beyond the huge military expenditures, literally billions of dollars' worth of oil were destroyed when retreating Iraqis blew up the majority of Kuwait's oil wells. It took more than six months and $2.5 billion to douse the burning wells. The environmental cost of the well fires, oil flows, and spills in the desert and in the Persian Gulf has been immense.

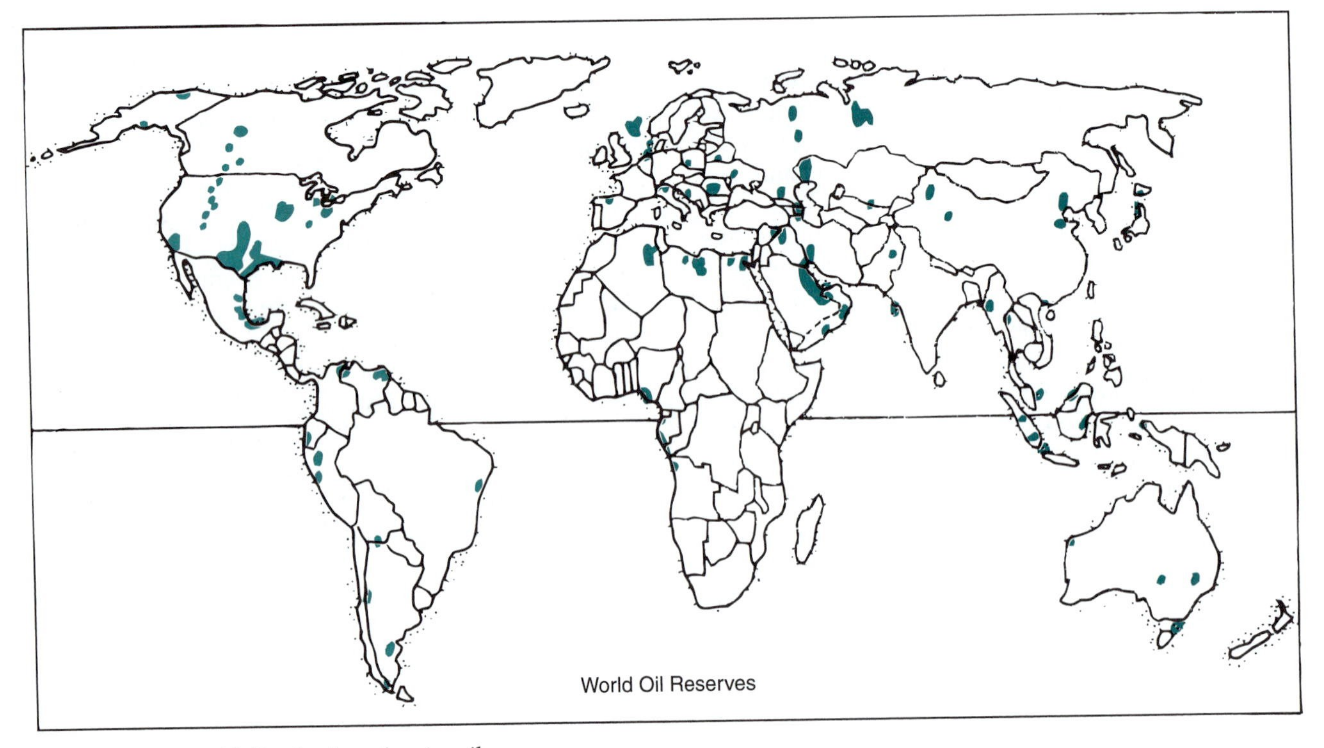

FIGURE 9.9 *World distribution of major oil reserves.*

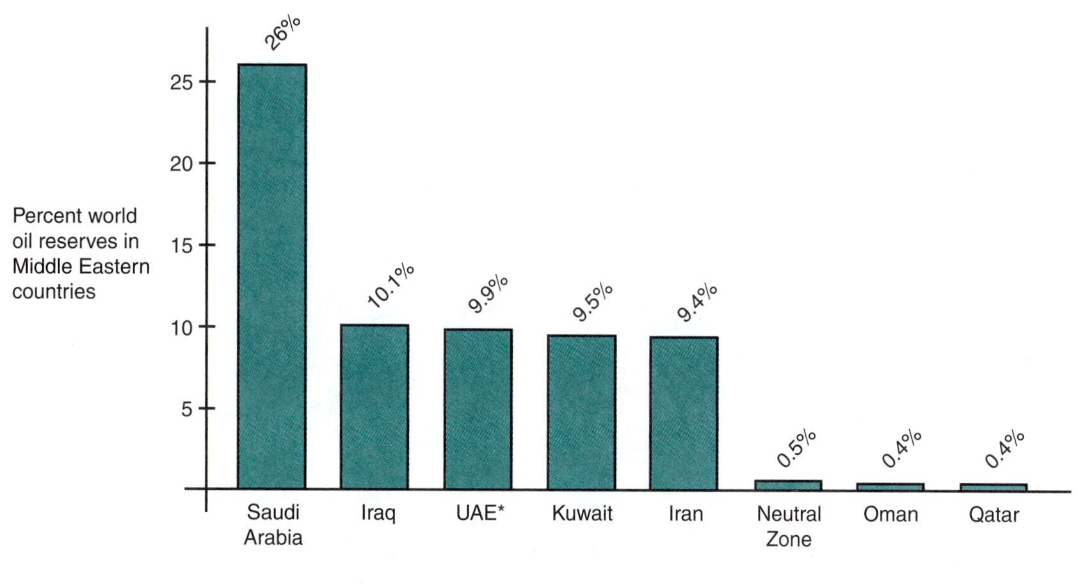

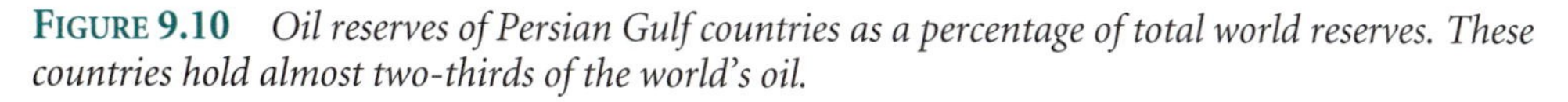

FIGURE 9.10 *Oil reserves of Persian Gulf countries as a percentage of total world reserves. These countries hold almost two-thirds of the world's oil.*

Most of the developed countries are oil deficient and import large amounts from the Persian Gulf countries and other OPEC members. In the 1990s nearly 40 percent of the oil exported to world markets was shipped from Persian Gulf countries, with nearly two-thirds going to Europe, Japan, and the United States. Figure 9.11 shows the patterns of international crude oil trade; note that the United States, the largest national market for oil, imports from most exporting regions. Europe, the largest regional market, relies heavily on oil imported from the Persian Gulf, the former Soviet Union, North Africa, and the North Sea fields controlled mainly by the United Kingdom and Norway. Japan purchases most of its oil from the Persian Gulf, Indonesia, and China.

Some Advantages and Disadvantages of Oil

Oil's versatility, availability, and transportability have made it the world's most popular fuel for the last half-century. Several of the advantages of oil are noted as follows:

- **Its has high energy and stores easily.** Its high energy content and ease of handling and storage make petroleum the chosen fuel for the world's motor vehicle fleet. Transportation fuel is the major use for oil.
- **Oil burns more cleanly.** It burns more cleanly than coal and produces less carbon dioxide, so it is widely used to generate heat, steam, and electricity.
- **It is versatile.** Oil has important nonfuel uses for petrochemicals, lubricants, and asphalt.
- **It is easy to transport.** Oil and petroleum products can be economically transported across oceans by supertanker and across continents by pipeline.

As with all fossil fuels, there are significant environmental impacts and other drawbacks associated with oil production, transportation, and use. When petroleum products are burned, oxides of nitrogen and sulfur contribute to air pollution and acid rain and additional carbon dioxide is released into the atmosphere. In 2000 the consumption of petroleum produced more carbon dioxide (CO_2) emissions than any other fossil fuel. The United States consumes more than 25 percent of the world's petroleum and produces about the same proportion of the CO_2 emissions.

Perhaps the most publicized environmental impacts have occurred while transporting or producing oil offshore. The world's largest marine oil release occurred during the 1991 Persian Gulf War, when 6 to 8 million barrels of oil were jettisoned from an offshore oil terminal in the Persian Gulf. One of the largest and most infamous spills from a supertanker occurred in 1989, when the *Exxon Valdez* struck a reef and sent 250,000 barrels of crude oil into Alaska's Prince William Sound. For nine months in 1979, a well blowout in Mexico's Campeche Bay spewed more than 3 million barrels of crude oil into the Gulf of Mexico before it was capped. But the single largest pollution event caused by petroleum was the detonation of more than 600 Kuwaiti oil wells by Iraq during the Gulf War.

Perhaps this fuel's biggest disadvantage is that oil is being depleted quickly. Because the world consumes

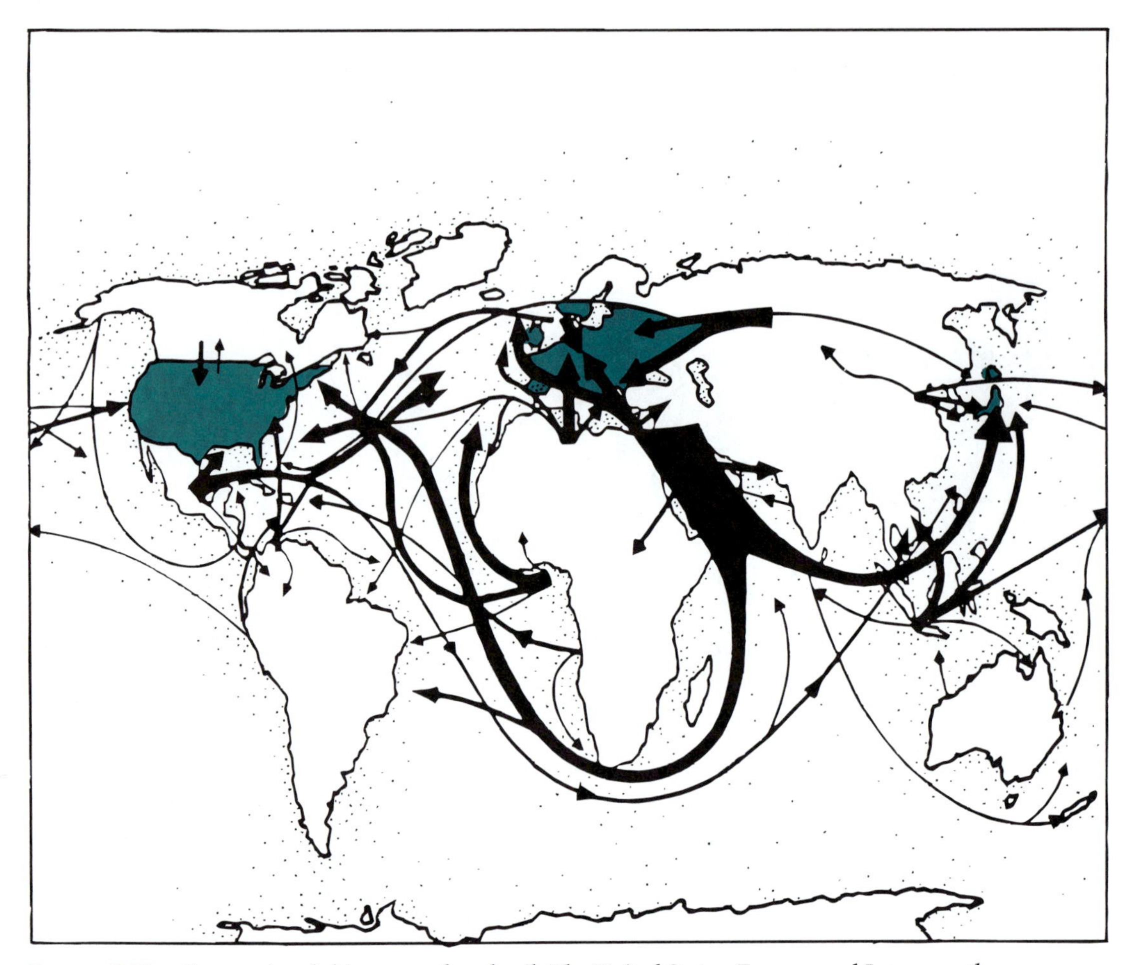

FIGURE 9.11 *International shipments of crude oil. The United States, Europe, and Japan are the destinations for most of the world's oil shipments. The major sources are the Middle East, Africa, Mexico, Venezuela, Southeast Asia, and Russia.*

such huge amounts of oil, it is likely that petroleum as a fuel resource will be available for only about another century. The United States, whose reserves have been declining steadily for the past 20 years and represent only about 2.5 percent of the total world reserves, will likely lose its prominence as a petroleum producer within the next two to three decades. The Persian Gulf's reserves will last about 120 years at current production levels, but it is expected that production rates will rise significantly in the next few decades as oil production and reserves in the United States and Russia decline.

Unconventional Sources of Oil

To date most oil has been produced in liquid form from wells. However, there are other forms of petroleum mixed in rock or sand that can be extracted and processed into *synthetic crude oil.* The two most abundant sources are **tar sands** and **oil shale.**

After mining tar sands, which contain an asphalt-like hydrocarbon called **bitumen**, the material is heated to liquefy and separate the bitumen. The bitumen is then processed to remove chemical impurities and upgraded to synthetic crude oil. The Canadian government sponsors the world's largest commercial production of synthetic crude oil from the Athabasca tar sands in northern Alberta. During the past two decades, this project has supplied about 10 percent of Canada's oil at prices comparable to world oil prices. However, when pollution control and environmental cleanup costs are factored in, the real cost of synthetic crude oil from tar sands is considerably higher. Nonetheless, these immense Canadian deposits are estimated to contain about as much petroleum as the world's total conventional crude oil reserves.

In the United States, projects to develop oil shale technology have located in Colorado around the richest deposits of oil shale, containing a solid hydrocarbon called **kerogen**. Although experiments have been going on for several decades, no commercial production has yet occurred because of high production costs and the severe environmental impacts of oil shale mining. Nevertheless, if developed, synthetic crude oil produced from the large domestic conventional oil shale resources could

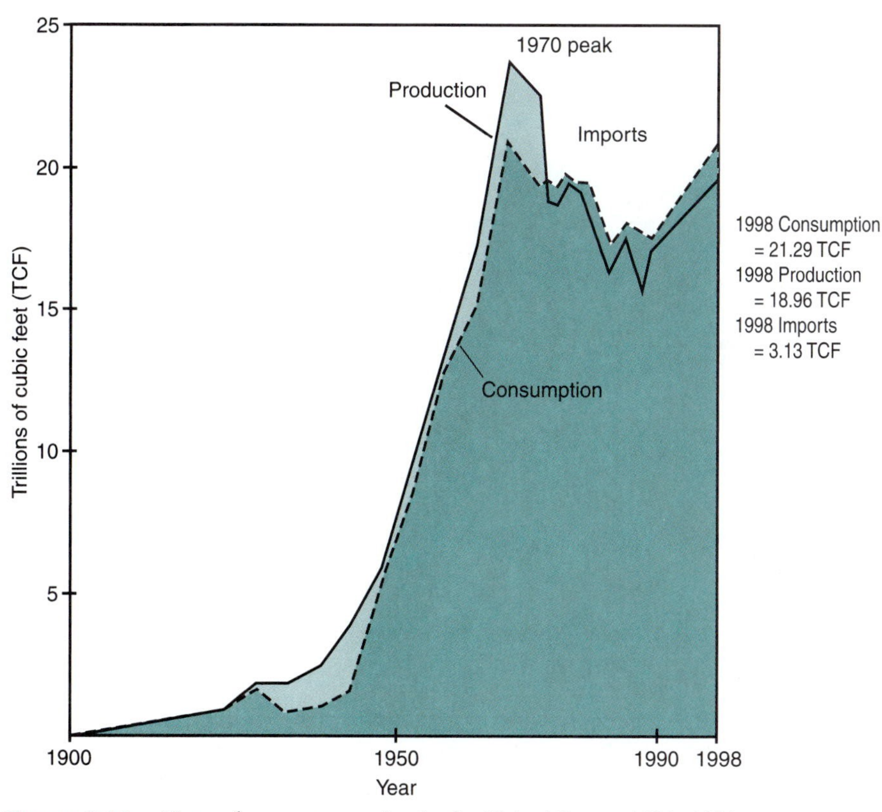

FIGURE 9.12 *Natural gas consumption in the United States, 1900–1998.*

probably meet the United States' petroleum demand for at least several decades.

Natural Gas

Natural gas is composed mainly of the simple hydrocarbon **methane**, with smaller amounts of heavier hydrocarbons that liquefy at normal atmospheric pressure. Two of these natural gas liquids (NGL), **butane** and **propane**, are widely used bottled gas fuels. Although natural gas is often found in association with crude oil (see Figure 9.7), its importance as a fuel has lagged behind. It is costly to store and transport, so much natural gas has been burned off at the well as a wasted byproduct of crude oil production.

Natural Gas Production Patterns and Trends

Once storage facilities and pipeline networks linked production sites in Texas, Oklahoma, and Louisiana with major population centers in the Midwest and East, natural gas became a premium fuel for heating in the United States. By the early 1970s when production and consumption peaked, natural gas provided one-third of U.S. energy. After slipping in the 1970s and 1980s, production slightly increased during the 1990s. Currently natural gas accounts for about 30 percent of U.S. energy production and about 22 percent of the total energy consumed in the U.S. (Figure 9.12). Texas, Louisiana, and Oklahoma still supply the majority of U.S. natural gas, with current production accounting for more than two-thirds of domestic output. Canada currently provides the bulk of U.S. natural gas imports, about 15 percent of U.S. consumption.

World demand for natural gas has grown as links between gas fields and markets have been established. Since 1980 world production has increased by 50 percent. Natural gas currently supplies about 23 percent of the world's commercial energy (Figure 9.13). The Persian Gulf nations and Russia have about 54 percent of the world's reserve. The United States, which currently produces 23 percent of the world's natural gas, has less than 4 percent of proven reserves.

International trade of natural gas accounts for only about one-sixth of total production. Large European markets are served by pipelines that import gas from Russia and the Netherlands, while most U.S. imports are piped in from Canada. Liquefied natural gas moves to European markets from Algeria and Libya. All of Japan's gas is imported by tankers, with most currently coming from Indonesia.

Advantages and Disadvantages of Natural Gas

In may ways natural gas is the premium fossil fuel. It burns cleanly in automatic furnaces, and where available, is the

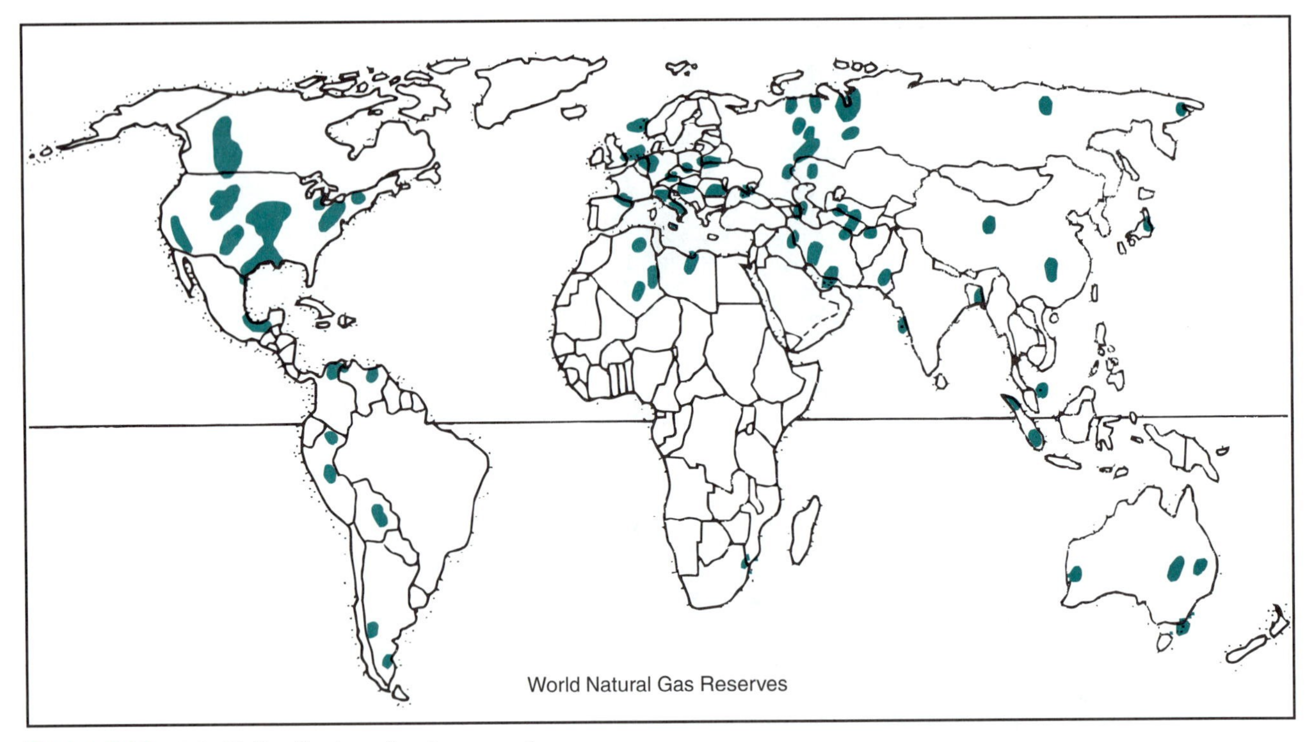

FIGURE 9.13 *World distribution of major natural gas reserves.*

fuel of choice for residential and industrial heating. Natural gas-fired turbines are the most efficient and cleanest method to generate electricity in fossil-fuel thermal electric power plants. When burning in high-efficiency furnaces, more than 90 percent of its energy supplies useful heat, compared with electric hearing systems where converting electricity to heat reduces efficiency to less than 30 percent.

Where pipeline networks and storage facilities have been built, transporting gas is inexpensive and relatively safe. However, ocean transport by liquefied natural gas tanker is expensive compared to crude oil moved in supertankers. Pipelines, ships, and storage facilities are expensive to build, yet are required if natural gas is to be used as a fuel. Gas pipelines supply most major markets in North America and Europe, with more being built and planned, but such facilities are prohibitively expensive for most developing countries. Even in the Persian Gulf region most natural gas is flared at the wellhead, with relatively little used domestically or exported by pipeline or LNG tanker.

Unfortunately, natural gas is the least abundant fossil fuel. So with worldwide demand increasing steadily, natural gas reserves may be exhausted in less than a century unless new sources of natural gas can be developed.

9.4 NUCLEAR POWER

Generating electricity from controlled nuclear reactions began nearly 50 years ago and has been wrapped in controversy from the very beginning. It once was heralded as an unlimited source of electricity that would be "too cheap to meter." Ironically, nuclear power is currently among the most expensive sources of electricity in the United States. Since the nuclear reactor accident at Three Mile Island in 1979, there have been no new orders for nuclear power plants in the United States. Yet throughout the 1980s and 1990s the share of electricity produced by nuclear power increased as reactors under construction finally came on line. In 2000, 20 percent of U.S. electricity was generated in 104 nuclear power plants (Figure 9.14). Today the U.S. produces about 30 percent of the world's electricity generated by nuclear power.

Worldwide there are similar disparities. Switzerland and Germany currently have moratoriums on additional nuclear power development and Sweden is calling for termination of its nuclear power program by 2010, whereas France has built more than 55 nuclear power stations since the mid-1970s to supply more than 75 percent of its electricity. Even after the world's worst nuclear power plant accident in Chernobyl in 1986, the Soviet Union continued to expand its nuclear power program. However, with the political disintegration of the Soviet Union, such costly nuclear power expansion has nearly come to a halt. On the other hand, Russia and the independent republics have continued to operate existing nuclear power plants considered to have unsafe reactor design. In Asia, energy-poor countries such as Japan, Taiwan, and South Korea have come to rely heavily on electricity generated from nuclear power as their economies have expanded rapidly. More

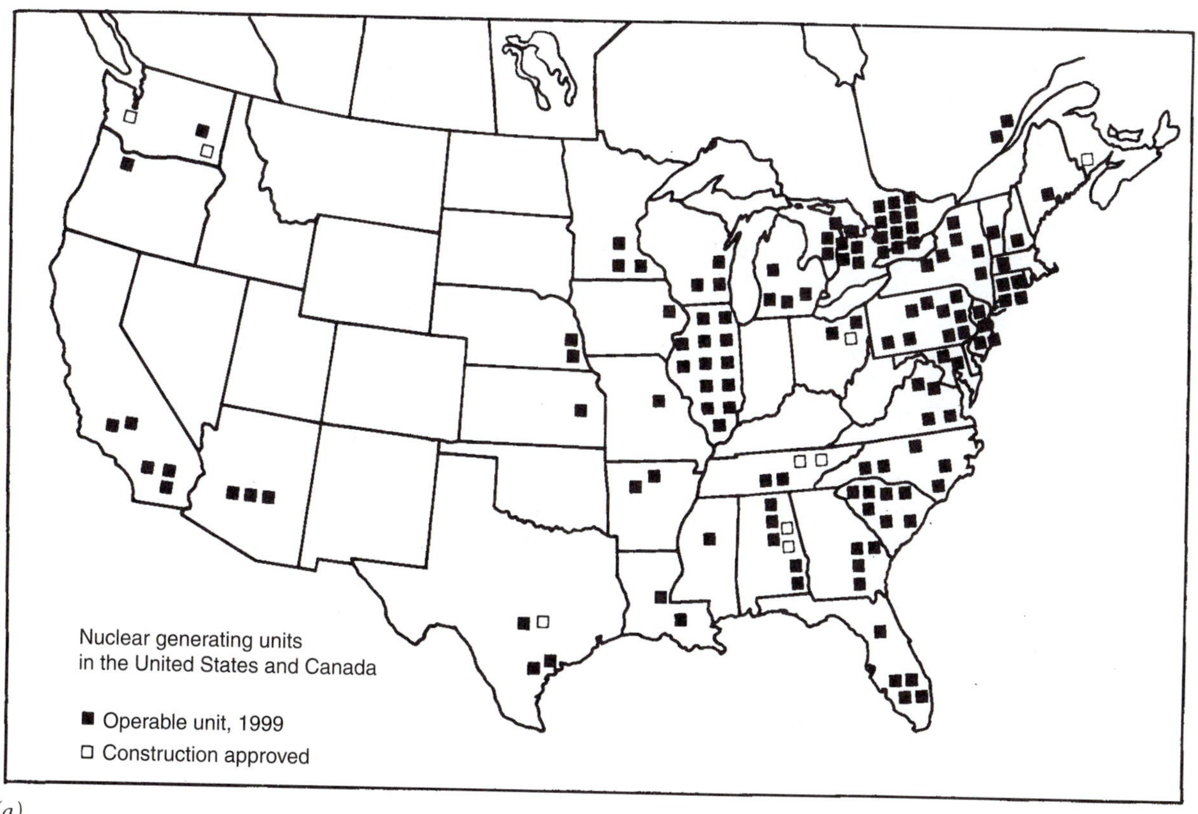

(a)

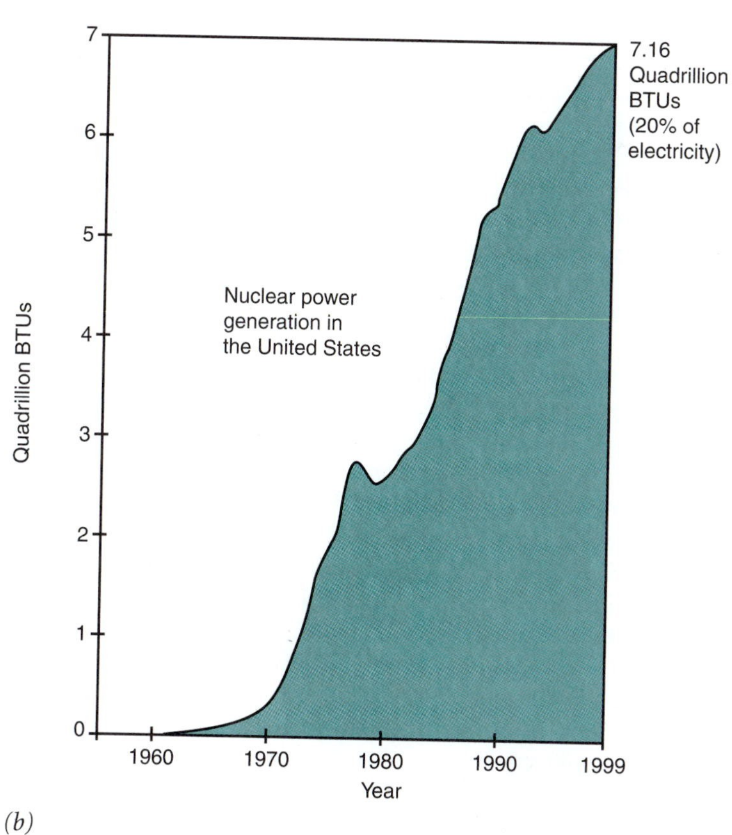

(b)

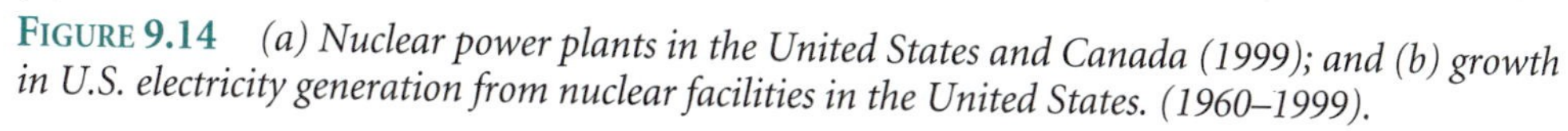
FIGURE 9.14 *(a) Nuclear power plants in the United States and Canada (1999); and (b) growth in U.S. electricity generation from nuclear facilities in the United States. (1960–1999).*

than half of South Korea's electricity is generated by nuclear power plants, while about 40 plants supply nearly one-third of Japan's electricity. Elsewhere in North America, Canada produces about one-fifth of its electricity from more than 20 nuclear reactors (Figure 9.14).

From this summary we can see that nuclear power is viewed in some quarters as a most appropriate technology for generating electricity while elsewhere it is perceived as inappropriate. What is certain as the nuclear industry grows in some areas and contracts in others is that controversy over its use will continue. Here is a brief description of how nuclear power plants work and some advantages and disadvantages of this energy source.

Generating Electricity in a Nuclear Power Plant

The **nuclear fission** reactions that occur in a power reactor involve a controlled-chain reaction in which the nuclei of radioactive isotopes, usually uranium 235, are split or fissioned when bombarded by neutrons. The energy released by splitting the nuclei of U^{235} atoms is used to heat water. The steam from this heated water, as in conventional thermal electric power plants, drives turbines that turn electric generators. It is ironic that this extraordinarily complex and costly technology is used to accomplish the very simple task of heating water!

More than 80 percent of the world's 420-plus nuclear power plants have **light water reactors (LWR)**, which use ordinary water as the *moderator* to determine the critical speed at which neutrons must strike the uranium atom (U^{235}) in the nuclear *fuel elements* in order for fission to occur. Most of the remaining reactors use either **heavy water** (deuterium oxide) or solid **graphite** (a form of carbon) as the moderating medium.

Some Advantages and Disadvantages of Nuclear Power

Perhaps the most important advantage for using nuclear fission to generate electricity is that it emits no carbon dioxide. As evidence mounts that increasing atmospheric greenhouse gases such as carbon dioxide are inducing climate change, some scientists think that nuclear power plants must be built to replace carbon-dioxide-emitting fossil fuel plants. Discussion of additional advantages follows:

- **Air pollution is minimal.** When nuclear power plants are operating as designed, they emit only minuscule amounts of radiation—in fact, less than some coal-fired power plants. Furthermore, they emit no smoke, particulates, or nitrogen and sulfur pollutants, as do fossil-fuel power plants.
- **Less environmental impact from mining and transportation of nuclear fuels.** Compared with coal, the most used fuel for electricity production, much less uranium ore is needed to generate equal amounts of electricity. This means there is considerably less land disturbance, fewer transportation problems, and lower costs with uranium.
- **Fissionable fuels may last indefinitely.** To meet the demands for commercial nuclear power, the known reserves would provide enough reactor fuel for a century or more. In addition, fuel reserves for reactors now in use can be extended by *reprocessing* spent nuclear fuel elements. Worldwide there are reprocessing facilities currently operating in England, France, India, and Japan. Finally, if safer, more reliable **breeder reactors** (which produce more fissionable fuels than they use) can be developed, then the fuel supply for nuclear fission will be unlimited.
- **Nuclear power is an alternative method for generating electricity.** In places where fossil fuels are scarce or where hydroelectric sites are unavailable, nuclear power provides an alternative for generating large quantities of electricity.

Of course, there are many real and perceived problems associated with nuclear power, with the issue of reactor safety the most controversial. Although there is no danger of a nuclear explosion in the case of a reactor accident, the release of radioactive materials into the environment endangers people living in surrounding areas. The 1979 accident at the Three Mile Island nuclear plant near Harrisburg, Pennsylvania, the worst in the United States, released some radioactive gas into the atmosphere and forced the evacuation of thousands of nearby residents. Fortunately there were no deaths, injuries, or illnesses attributed to the accident; however, cleanup costs and litigation expenses have risen to more than $1 billion. Most of all, the added societal concern generated by this near catastrophe literally halted expansion of the U.S. nuclear power industry.

Adding to worldwide concern over reactor safety was the 1986 explosion at the Chernobyl plant in Ukraine, which released huge amounts of radiation that contaminated nearly 10,000 square kilometers (4,000 square miles) in Ukraine and neighboring Belarus and Russia. Additional radioactive fallout was detected in Sweden and Norway more than 2,000 kilometers (1,200 miles) from Chernobyl. The accident killed at least 31 people and more than 200 were hospitalized with radiation sickness. Thousands more were exposed to radiation levels that cause cancers, thyroid conditions, and cataracts. By the late 1990s, more than 9,000 people had died from these aftereffects. Billions of dollars have already been spent by the governments as the cleanup continues, and the ill-fated reactor has been entombed in steel and concrete.

While concerns over reactor safety seem foremost, there are other problems and disadvantages facing the nuclear power industry:

- **Radioactive waste must be disposed.** Many of the fission byproducts from the nuclear reactors will remain radioactive for hundreds or thousands of years. With more than 435 nuclear power plants in operation worldwide and nearly 100 more being built, the problem of long-term disposal of radioactive waste is literally growing every day. To date there is not a single permanent disposal site in the world for high-level waste, but plans are being developed for one at Yucca Mountain in Nevada.
- **Nuclear power plants must eventually be decommissioned.** Even though nuclear power plants have been operating only since the 1950s three dozen or so have already been retired from service and 200 or more built in the 1960s and 1970s will have to be decommissioned during the next decade. Since the worn-out plants are radioactive, they must be disposed of in a safe way. Two methods have been proposed. *Entombment* surrounds the entire structure with reinforced concrete, while *dismantlement* involves taking the radioactive plant apart and burying it. Both procedures are very expensive, so decommissioning a plant may actually cost more than building it.
- **Nuclear weapons are a possible byproduct.** Nuclear weapons can be made from plutonium produced in fuel reprocessing facilities and in breeder reactors, so any country that develops these facilities will have the raw materials to build nuclear weapons.

To address public concern over the safety and high cost of nuclear energy and to resurrect the moribund nuclear industry, new reactors are being designed that are safer, smaller, and less complex and costly to build and operate. These reactors will rely on **passive safety systems** where reliable forces such as gravity and convection would regulate coolant flow inside the reactor rather than the myriad pumps, valves, and other control devices that must be switched on or off. They are also being designed in smaller *modular components* in order to reduce construction costs and to better match electricity demands. One module installed to meet near-term electricity needs could be linked with additional modules as needed. It remains to be seen whether this new generation of nuclear reactors—with names like PIUS (for Process Inherent Ultimate Safety) and PRISM (for Power Reactor Inherently Safe Module)—will someday provide the world with safe, inexpensive electricity.

9.5 RENEWABLE AND PERPETUAL ENERGY RESOURCES

This section provides a brief overview of how we use renewable and perpetual energy resources such as sunlight, wind, flowing water, and biomass to supply nearly one-sixth of the world's energy and, perhaps more importantly, how these renewable energy resources can be used to supply many of our vast and diverse energy needs while providing a sustainable energy future.

Biomass

Through photosynthesis plants convert sunlight into **biomass**, leading some to call biomass fuels "second-hand" solar energy. Because biomass is renewable, widely available, and easy to burn, wood, other plant material, and animal dung have been the most widely used fuels for nearly all of human history. And while these fuels have limited utility today in most of the developed world, they remain the primary energy source for about half of the world's population living in developing countries. In the poor countries of sub-Saharan Africa, nearly 75 percent of all energy used is derived from biomass, and about 10 percent of the energy used worldwide is provided by biomass fuels. In contrast, these fuels currently produce less than 4 percent of U.S. energy.

Wood

Wood, the most widely used biomass fuel, is potentially a renewable resource. However, in many developing countries where wood and charcoal are the primary fuels for heating and cooking, *fuelwood* needs are met only by cutting wood faster than it is growing back, while in other places there is a severe fuelwood shortage even with overcutting. So for nearly half of the people who depend on fuelwood, the supply is not sustainable. In places with critical fuelwood shortages, crop residues and animal dung are burned instead of being returned to the soil. Without these nutrients soil fertility and crop yields decline.

In response to these growing shortages, several strategies have been developed to meet expected fuelwood needs. They include increasing the productivity of natural forest lands by more *selective cutting* and *reforestation*; establishing fuelwood plantations or *agroforestry* projects that combine the production of trees and crops; and *burning wood in more efficient cook stoves and kilns.*

In the developed countries wood was an important industrial and space-heating fuel well into the industrial era. Currently, about 3 percent of energy consumed in the United States is provided by wood. Yet the amount of energy produced by wood today is about the same as in 1890, when wood was supplying about one-third of the United States' energy.

Gas and Liquid Fuels from Biomass

Plants and other organic materials such as animal waste and garbage can be converted into liquid and gas fuels. *Methane* is produced through bacterial digestion of organic materials in the absence of oxygen. Through this process methane can be obtained from sewage and feedlot manure. Small-scale biogas digesters have widespread use in China, where the methane is used locally for heating and cooking. For the last 15 years or so methane produced in large metropolitan landfills in the United States has been collected, purified, and distributed in natural gas pipelines. Although still in early stages of development, **methane recovery systems** might possibly be employed in several thousand large landfills in the United States.

Alcohol produced by fermenting plant sugars is a high-quality liquid fuel that can be mixed with gasoline or burned alone. Methyl alcohol, or *methanol*, can be produced from biomass as well as from natural gas and coal. Ethyl alcohol, or *ethanol*, can be made from grain, sugar beets or cane, milk byproducts, certain plant wastes or even garbage. *Gasohol*, a high-octane unleaded fuel, is a mixture of 90 percent gasoline and 10 percent ethanol. It currently supplies about 10 percent of the gasoline fuels burned in the United States. While virtually all of the gasohol produced in the United States is from surplus grain, Brazil's large biomass fuel program relies on sugar cane grown specifically to produce ethanol fuels. In the early 1990s nearly half of Brazil's automotive fuels were supplied by ethanol, with nearly one-third of its cars designed to run on pure ethanol and the remainder running on gasoline blended with 20 percent ethanol.

To meet national fuel economy standards set by the Environmental Protection Agency, U.S. vehicle manufacturers are currently building hundreds of thousands of vehicles that can burn ethanol fuel blended with 15 percent gasoline. The problem is that there are only about 100 stations in the United States that sell this 85 percent ethanol fuel. So only a small percentage of these ethanol-ready vehicles will likely ever burn ethanol-based fuel.

Because most ethanol is produced from crops such as sugar cane and corn, which are expensive to grow, it costs about three times as much as gasoline. However, new technologies are being developed to allow ethanol production from low-cost wood and agricultural crop byproducts and waste. In 2000 several commercial ventures in North America were gearing up to produce ethanol from biomass waste such as corn stalks, rice straw, sugar cane pulp (*bagasse*) and even sewage sludge and organic waste from garbage. Some researchers predict that ethanol production from these inexpensive waste biomass sources will be cost-competitive with gasoline during this decade.

Methanol-based fuels can also be burned in modified automobile engines. At the beginning of this century more than 20,000 U.S. vehicles burned an 85 percent methanol and 15 percent gasoline fuel, and a few hundred more used pure methanol. Like ethanol, it is a clean-burning fuel and would significantly reduce vehicle emission levels of smog and ozone-forming pollutants. Currently, most methanol is produced from the least expensive sources—gas or coal. However, methanol from fossil fuels will contribute to the build-up of atmospheric carbon dioxide, whereas methanol produced from sustainable biomass would not.

Some Advantages and Disadvantages of Biomass Fuels

The use of biomass fuels in the developed countries may expand significantly as new technologies reduce production costs while fossil fuels become more expensive because of depletion or because of mandated pollution controls. There are four main advantages to this fuel source.

- **Biomass is renewable.** If the amount of biomass used for fuel does not exceed biomass growth, it can provide a sustainable resource that can meet a significant share of global energy needs.
- **Biomass is widely available.** Unlike oil and coal, biomass is locally available over much of the planet.
- **Biomass is a cleaner fuel.** When biomass is burned in efficient stoves or is converted to gas or liquid fuels, it produces less pollutants than fossil fuel. If used on a sustainable basis, the carbon dioxide released when biomass is burned equals the amount of carbon dioxide taken up by photosynthesis, so there is no net addition to atmospheric greenhouse gases.
- **Biomass waste and byproducts can be used for fuel.** Resources are extended when crop residues, scrap wood, or other biomass waste can be used to produce energy. The pulp and paper industry meets most of its energy needs by burning its wood waste and byproducts. Sugar cane pulp, corn stalks, and rice straw are examples of crop residues that have been burned at processing facilities to produce electricity, heat, and steam and are now being processed to produce ethanol.

There are also several disadvantages to biomass fuels:

- **Quantities of biomass for fuel are limited.** Since about half of the world's annual biomass production is used for agricultural crops or wood products, the contribution of biomass fuels to the global energy mix will be limited. Estimates of the total energy from available sustainable biomass in the United States amount to only about 7 percent of current domestic energy consumption.
- **There is an issue of fuel crops versus food crops.** While the use of waste biomass to generate energy is sensible, the cultivation of fuel crops or fuelwood plantations is controversial. In many poor countries where land and food are in short supply, planting crops or trees for fuel may be inappropriate.
- **Removal of biomass may deplete soils.** When biomass is removed for fuel, topsoil may be depleted, soil fertility declines, and runoff and erosion may increase.

- **It yields low net energy**. The relatively low energy-to-weight ratio of biomass and the energy used to collect and process it may reduce the net energy usefulness of biomass fuels.

Hydroelectric Power

As we learned in Chapter 5, solar energy drives the heat engine that sets the hydrologic cycle in motion as moisture is evaporated and carried aloft. As this atmospheric moisture condenses and falls back to the surface, some of this water flowing in rivers can be used to generate **hydroelectric power**, a nonpolluting, renewable, and relatively straightforward energy source. Today hydroelectric power, or simply **hydropower**, produces nearly one-fifth of the world's electricity and about 7 percent of the total energy supply.

Hydropower Potential and Development

Although most hydropower capacity remains in developed countries, most current development is occurring in the developing world where less than 10 percent of the hydropower resources have been utilized. For example, China, Venezuela, and Brazil have more than doubled their hydropower output in the last decade, and China and Brazil are currently building two of the world's largest hydroelectric plants.

In contrast, much of the developed world hydropower potential has already been developed. Today hydropower provides nearly two-thirds of the electricity in Switzerland and Austria and nearly all of Norway's electric power. The leading producers—Canada and the United States—generate more than 25 percent of the world's hydropower. Canada, which produces nearly 70 percent of its electricity from hydropower, has surpassed the United States in output each year since 1985 (Figure 9.15a).

Hydropower production in the United States, which grew rapidly during the first 70 years of the twentieth century, has remained relatively steady since then with only modest year-to-year variations, due to fluctuations of precipitation and runoff. Today hydropower produces about 10 percent of the U.S. electricity, or about 4 percent of all commercial energy consumed. Although only about one-third of the potential hydropower resources are developed in the United States, it is unlikely that there will be any significant growth because many of the potential sites are protected by legislation, such as the Wild and Scenic Rivers Act that prohibits dam construction.

Another type of hydropower installation called *pumped storage*, being developed in some countries, is used during *peak load* periods when demand for electricity exceeds the generating capacities of conventional power plants. In a pumped storage system, a pumping/generating facility located on a body of water pumps water to a reservoir located high above a generating plant. When electricity is needed the water stored in the reservoir flows back down through the turbines to generate electricity.

Some Advantages and Disadvantages of Hydropower

Hydropower provides electricity without the air pollution of fossil fuels or the sophisticated technology, high costs, and potential radiation hazards of nuclear power. Furthermore, efficient hydropower facilities can be built at the scale appropriate to the site and electric power needs Finally, since hydropower is based on flowing water, it is renewable as long as there is adequate water to spin the turbines. However, hydroelectric dams located on sediment-laden rivers may have their useful lifetimes shortened as their reservoirs fill with sediment, reducing the volume of water needed to generate power. Even so, most hydropower facilities have expected lifetimes that are much longer than thermal electric fossil fuel or nuclear plants.

Large-scale hydroelectric facilities, with their costly dams and huge reservoirs, produce significant environmental impact. Reservoirs drown stream valleys, eliminating large areas of habitat, forest, and agricultural land. Downstream the sediment-free discharge from the dam may significantly increase channel erosion, while the dam itself may block fish migrations (see Figure 12.17). While many large and small hydroelectric dams continue to be built especially in developing countries, some dams that have been generating electricity for decades are being removed.

In the United States the federal agency that licenses privately owned hydroelectric dams must now consider not only power generation but fish and wildlife and other environmental issues before renewing licenses. In the Pacific Northwest, California, and Maine, dams that have blocked migratory fish from reaching spawning grounds are being removed. In the last decade more than 80 dams have been removed from American rivers, and others have had to modify operations so as to lessen their impact on fish and other aquatic species. This trend will continue over the next decade as more dams face more stringent relicensing procedures.

Wind Power

Wind power is another source of indirect solar energy that provides a perpetual if temporally variable source of energy that uses the kinetic energy of moving air to spin windmills or wind turbines. **Windmills,** used for centuries to mill grain and pump water, can now be designed as **wind turbines** to generate electricity. In the 1920s and 1930s small wind turbines were used on thousands of farms across the United States to provide enough current for a radio and an electric light or two, but they disappeared as electric lines were extended to most rural areas.

(a)

(b)

FIGURE 9.15 *(a) A large hydropower facility in Canada and (b) a wind farm along the North Sea Coast in the Netherlands. (Note the size of the wind generators relative to the houses.)*

Wind Power Potential and Development

Since the 1980s the wind turbine industry has experienced a renaissance in many parts of the world. In many parts of the developing world thousands of small wind turbines have been installed in remote villages to meet basic electricity needs. But it has been in the developed countries where wind-generated electricity has experienced the most growth. Today Germany is the world's leading wind energy producer. Denmark generates more than 8 percent of its electricity from wind turbines. And it is in Europe where most commercial growth occurred during the 1990s, because many European countries subsidize wind-generated electricity. By 2000, the total installed capacity of wind-generated electricity throughout the world was approaching 10000 megawatts (equivalent to about 10 nuclear power plants, or 15 to 20 conventional coal or gas-fired plants). Four countries—Germany, United States, Denmark, and India—have nearly 75 percent of the world's wind-generated capacity.

In the United States, California continues to produce the majority of the wind-generated electricity. During the early 1980s California and federal tax subsidies promoted heavy private investments in groups of small wind turbines called *wind farms* that could sell electricity directly to California utility companies (Figure 9.15b). The largest of these wind farms has nearly 6,000 wind turbines, and most are located in windy mountain passes near San Francisco and Los Angeles. Currently, the nearly 18,000 wind machines in the United States generate only as much power (1,800 megawatts) as about two conventional large thermal electric plants. Today the greatest concentration of commercial wind farms in the U.S. are being built atop a ridge (glacial moraine) that extends from northern Iowa across southwestern Minnesota and into South Dakota. If these wind farms, which are subsidized by state and federal incentives, prove successful, the windy Midwest may surpass California in wind power capacity within the next few years. Additional large wind farms are being built in western Texas.

During the last decade the cost of wind-generated electricity has steadily declined and is nearly competitive with thermal electric power plants. A new generation of large wind turbines has been developed that generates more power at lower wind speeds and operates near peak efficiency over a range of wind speeds. When installed at sites with moderate wind potential, these machines generate electricity at about half the cost of older, smaller machines. With lower costs, wind power has significant potential for worldwide growth.

Some Advantages and Disadvantages of Wind Power

Although the amount of electricity currently produced by wind turbines represents a minuscule fraction (less than 0.1 percent) of worldwide electricity, this technology will likely become more important in the future as it has several advantages.

- **It is a perpetual energy source.** As such, efficient wind turbines can be installed to produce small to moderate amounts of electricity almost anywhere it is needed at relatively low cost.
- **Wind speed increases when demand is greatest.** Since wind speed is usually fastest during the daytime, the peak generating potential for turbines will normally coincide with peak electricity demand.
- **Turbines are inexpensive and take up little space.** Since wind turbines use only the wind and occupy little space, they can share the land with agriculture. In California and the Midwest, many of the wind farms are located on crop or grazing land. Since most wind turbines are simple, relatively inexpensive machines, they can be installed and ready to produce electricity very quickly.
- **Wind is a clean source of energy.** Wind turbines emit neither radioactive waste nor air pollutants so they are a safe, clean source of electricity.

The major disadvantage of wind power is that it is an intermittent power source that can generate electricity only when there is enough wind to spin the wind turbine. Some critics also perceive large wind farms to be eyesores that visually mar aesthetically pleasing landscapes such as seashores and mountain passes. Finally, wind farms may pose a serious threat to birds flying through.

Solar Energy

Humankind has always depended on sunlight for warmth, but during the last several decades we have developed a variety of simple and complex technologies to more effectively use this inexhaustible and free resource to meet many energy needs. Building materials that absorb sunlight have long been used to collect and store its warmth during the day and slowly radiate it long after the sun has set. Such traditional **solar architecture** is used today along with more recently developed active and passive solar heating systems. *Active systems* use fans or pumps to circulate heat through the building, whereas *passive solar systems* collect, store, and transfer heat without mechanical assistance (Figure 9.16). Today there are more than 250,000 passive solar homes and more than 1 million homes using active solar systems in the United States.

Passive solar buildings are designed to allow winter sunlight to enter through specially glazed windows and skylights and be absorbed by walls and floors that will slowly radiate their warmth within the well-insulated building throughout the day and night. Overhangs, window awnings, shades, and deciduous plantings are used to minimize heat gain during the summer. In most active

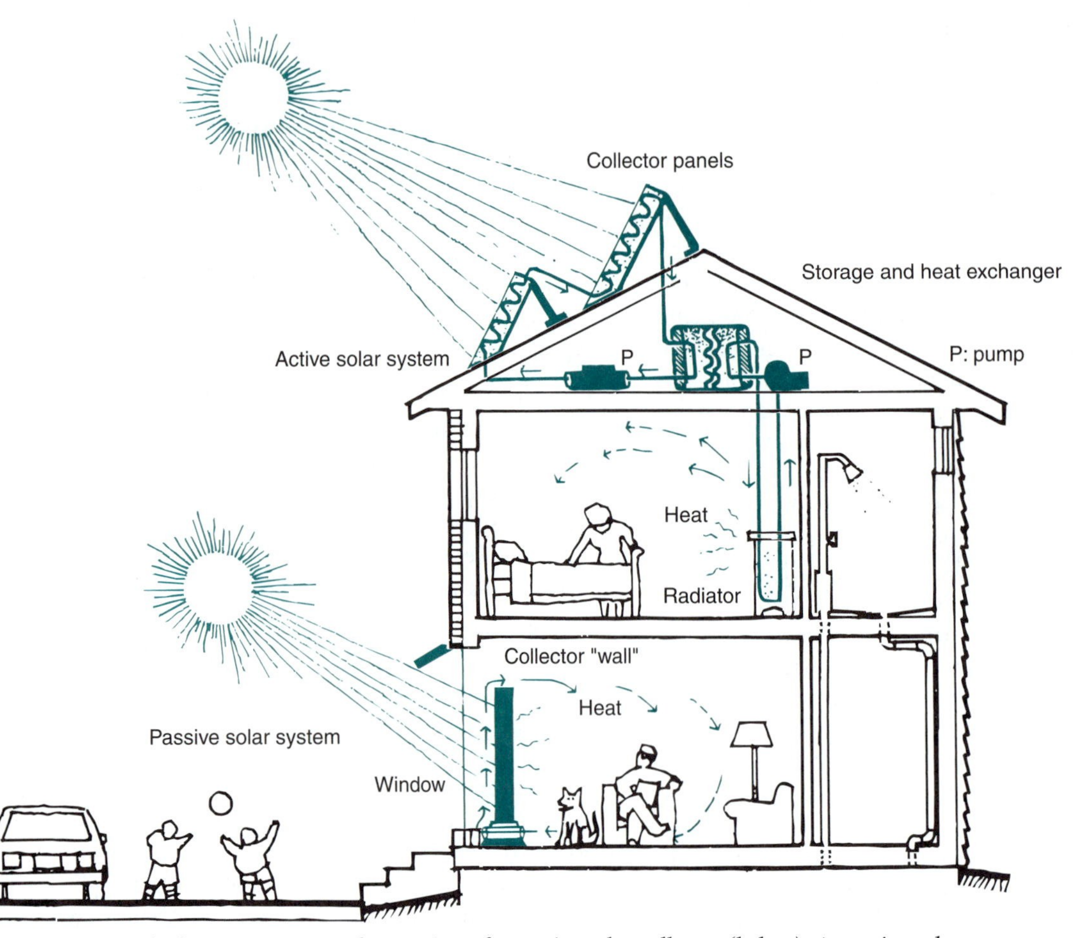

FIGURE 9.16 *The basic structure and operation of a passive solar collector (below). An active solar system (above) uses a fan or pump to circulate air or water through the collector and building.*

solar heating systems, specially designed roof-mounted solar collectors absorb radiant solar energy, which heats liquid or air as it circulates through the collector and on to insulated storage containers. From there, heated air or water is moved by thermostatically controlled fans or pumps through conventional ductwork or hot-water distribution systems that heat the building. In a less costly system that provides heat only when the sun is shining, a fan forces the heated air from the collector directly through heat ducts or registers.

Solar Hot Water Systems

Either passive or active designs are now widely used in sunny climates, with more than a million systems installed in the United States—mostly in the south and southwest. There is even more interest in solar hot water systems where energy costs are higher. For example, more than 10 percent of Japanese and one-third of Australian homes depend on solar-heated water. In Israel, 75 percent of the residences have domestic solar hot-water systems, and in Cyprus the figure is near 95 percent.

Using solar energy to heat buildings and water is an attractive option that reduces our reliance on traditional nonrenewable fuels. Because of the limited winter sunlight in most densely populated midlatitude regions, solar heating would supplement rather than replace existing heating systems. As designs for both passive and active solar heating improve and costs become more competitive with traditional heating systems, the role of supplemental solar heating will continue to expand in these areas.

Solar Energy to Generate Electricity

In sunny parts of the world solar energy can be used to generate electricity at both large and small scales. Promising **solar-thermal electric technologies** are being developed using large arrays of reflective solar collectors called **heliostats** that provide high-temperature heat to drive large steam turbines and electric generators. During the 1970s and early 1980s several solar-thermal facilities were built, such as the *solar furnace* in southern France and the commercial-scale *solar power tower* in southern California's Mohave Desert (Figure 9.17). In 1996 the power

FIGURE 9.17 *Solar electric facility in the Mohave Desert, California.*

tower was modified so that it can also generate electricity even when the sun is not shining. Instead of heating water, the heliostats focus sunlight to heat molten salt to much higher temperatures (over 565°C.). When electricity is needed, the molten salt is used to boil water in a steam-powered electric generator. Generating electricity at the 10 megawatt facility costs about two to three times more than conventional power plants. However, this technology may well be applicable in the world's deserts where other sources of electricity are not available. Figure 9.17 shows a large-scale solar electric facility that uses arrays of photovoltaic cells to create electricity.

Photovoltaic cells which convert solar radiation directly into electricity, have been used to produce small amounts of electricity since the 1950s. Made of thin slices of silicon material, each photovoltaic cell produces a tiny flow of electric current when energized by sunlight. Series or arrays of individual solar cells are wired together as **solar panels** in order to produce useful amounts of electricity. These solar panels have long been used to provide electricity for spacecraft and are now used in isolated areas to power telecommunication equipment, lighting, water pumps, and other electrical equipment. Today, most electronic calculators are powered by a tiny array of photovoltaic cells.

Although still more expensive than conventional electricity, the cost of electricity from photovoltaic cells has been dropping dramatically. Today's technology provides electricity at about two to three times the cost of conventional electricity. In 2000 about 75 megawatts of photovoltaic modules were sold with about two-thirds of the capacity for use in developed countries.

Converting Sunlight to Fuel

While heat and electricity generated from direct or indirect solar energy will eventually meet many energy needs, fuels that can be stored and transported conveniently are needed to power vehicles and provide heat. By converting sunlight into *liquid, gaseous,* or even *solid chemical fuels* that can easily be transported and stored, a sustainable energy supply can be made available wherever and whenever it is needed.

To date, chemical fuel production using solar-generated electricity has focused on *hydrogen,* which can be produced by passing electricity through water to separate hydrogen and oxygen. Hydrogen is a clean fuel-producing water vapor when burned and can be used for transportation and heating. It can be converted back into electricity by combining oxygen and hydrogen in a *fuel cell.* **Hydrogen fuel cells** could power electric vehicles or could stand alone to provide a remote source of electricity and heat. An experimental solar-hydrogen facility in Saudi Arabia uses electricity from photovoltaic cells to produce hydrogen by electrolysis. Although solar-hydrogen fuel is expensive, some experts estimate that it could become cost-competitive with gasoline in the next two decades when solar-hydrogen fuels reach the equivalent price of $3 per gallon of gasoline. By 2000 experimental fuel-cell powered buses were in service in several developed countries and several automakers were developing vehicles powered with hybrid fuel-cell and gasoline engines. Fuel cells are currently being used to provide electric power. A $40 million fuel cell power plant is operating in Santa Clara, California, producing two megawatts of electricity. The world's largest utility, Tokyo Electric Power Company, operates an 11-megawatt fuel cell power plant.

In 2000, scientists reported a promising discovery that may provide an unlimited source of hydrogen. When green algae is deprived of oxygen and a key nutrient, sulfur, this simple plant can convert water and sunlight into hydrogen.

Solar Power Satellites

Another solar technology is being considered to provide inexhaustible and uninterrupted sources of electricity. This electricity would be generated in space from a constellation of low orbiting satellites and beamed via microwaves to antenna arrays on Earth. Although sounding a bit farfetched, the **solar power satellites** seem more plausible in the context of the current and proposed fleets of commercial telecommunication satellites worldwide. These satellites could be pressed into double duty to gather solar energy and relay electric power to Earth. In Japan, a satellite has been designed to demonstrate the concept's feasibility. Although still only a concept, solar power satellites could become a reality if and when telecommunication and electric utility companies explore such a joint venture.

Geothermal and Tidal Energy

The Earth itself contains two sources of energy that can be converted and used to meet human energy needs. **Geothermal energy** is heat energy from the Earth's interior that is brought toward the surface in molten rock (magma). **Tidal energy** is the kinetic energy in the oceans' tides caused primarily by the gravitational attraction of the moon on the oceans. Although both sources of energy have begun to be tapped, there are a number of sites where geothermal energy is already being utilized, and there are dozens of other "hot spots" where this source of heat might someday be used. In contrast, only three tidal electric-generating facilities are in operation worldwide, and there are no definite plans to develop any of the other potential sites.

Since 1904, electricity has been produced in Larderello, Italy, from geothermal steam and today about twenty countries produce electricity from hydrothermal sources. At the Geysers geothermal steam field in northern California, the world's largest **geothermal-electric** facility is currently generating more than 1,000 megawatts of electricity and there are plans to double production over the next decade. In southern California, extensive hot-water geothermal deposits beneath the Imperial Valley are being tapped. In 2000 several small plants continue to generate geothermal electricity. Unfortunately, these shallow **hydrothermal** reservoirs of steam and hot water that provide inexpensive heating and electricity are very limited in their geographic extent. Even if developed fully, they will only provide local energy needs.

The Transition to Renewable Energy Resources

The transition from fossil fuels and nuclear energy to renewable, sustainable energy sources has already begun. Problems as diverse as resource depletion, atmospheric pollution, climate change, and radiation hazards, along with technological advances employing direct and indirect solar energy, are driving these changes that will inevitably produce a more diverse and decentralized set of energy sources.

However, the rate of this energy transition depends in large measure on the energy policies developed and followed by developed and developing countries alike. To date, most government energy research funds have been spent on nuclear and fossil fuels. With policies that shift research funds toward renewable energy technologies and with incentives for installing and building renewable energy systems, national governments could speed this energy transition.

Renewable energy technology such as solar- and wind-based electricity are still not generally cost-competitive with traditional nonrenewable sources of electricity. As a result, when economic incentives usually provided by government subsidies have been made available to develop renewable energy technologies, facilities have been built and capacity expanded. However, when support has been withdrawn, solar and wind energy development has slowed. For example, during the 1980s California became the world's leading producer of solar, wind, and geothermal generated electricity because of generous tax incentives and a guaranteed market for electricity produced from these sources. During the 1990s as energy prices eased, incentives and subsidies were withdrawn and alternative energy development stagnated, facilities closed, and capacity declined.

Over the next two decades renewable energy development is projected to increase substantially. As a result of the 1997 Kyoto Climate Change Protocol, proposals to cut greenhouse gas emission levels have resulted in commitments to expand the development of nonpolluting renewable energy sources such as hydro, wind, and solar. The rapid increase in petroleum and natural gas prices since 2000 has also provided an important economic incentive to develop alternative energy sources.

The mix of renewable energy sources would vary geographically, depending on the availability of sunlight, biomass, wind, geothermal, and hydropower. In humid, cloudy Europe and eastern North America, biomass, wind, and hydropower would supplement decentralized solar systems, while in dry, sunny regions sunlight would be the primary energy source to provide heat and electricity that are needed regionally and solar-chemical fuels that could be transported to other regions.

9.6 USING ENERGY MORE EFFICIENTLY

In the preceding paragraphs we have discussed strategies and technologies that would expand and change our energy resource base. Some have great promise and will be developed to meet future energy needs, but all are expensive and take time to develop and implement. There is, however, an energy strategy available today that provides the same energy services at lower costs. Even more attractive is that using this strategy reduces the economic, environmental, and social costs associated with the production, transportation, and consumption of energy resources. This strategy simply uses our currently available energy supplies more efficiently and reduces energy waste.

Over the last two decades living standards have improved in many parts of the world even as per capita energy use has declined or stabilized. Huge savings have already been made, but many, many opportunities remain to make more efficient use of energy resources. Some economic, social, and environmental benefits associated with improving energy efficiency include lower energy costs to consumers; less environmental impact, especially less smog, less acid precipitation, and less carbon dioxide; lower economic and social costs to phase in new energy technologies; and less reliance on military force to ensure energy security. The potential for military conflict will decline as dependence on Middle East oil declines.

Reducing Energy Consumption in Buildings

Heating, cooling, and lighting commercial and residential buildings consumes over a third of U.S. energy Current technologies are available that could double building energy efficiency Integrated heating, ventilating, and air conditioning (HVAC) systems and fluorescent lighting can be designed into new energy-efficient commercial buildings. Retrofitting HVAC and efficient lighting systems may be more costly but can reduce energy costs enough that payback times of two to five years make economic sense for buildings that will last for decades.

Energy-efficient new homes may cost 5 to 10 percent more for heavily insulated walls and ceilings, fluorescent lighting, low-emissivity windows, and efficient heating and air conditioning systems, but after a two to three-year payback period, these investments begin to save money as well as energy. Older residences can be made much more efficient with relatively small investments to insulate ceilings and to weather strip and caulk around doors and windows. Additional energy savings accrue when worn out appliances and heating and cooling systems are replaced with energy-efficient ones.

Incorporating efficient building technologies varies from country to country. Sweden leads the world in the percentage of energy-efficient buildings and has developed international markets for its energy-efficient building technology. In addition, most Western European countries have *energy performance standards* for commercial or residential buildings. In contrast, the United States has no legislated national standards, although a few states have developed building efficiency norms.

Improving Energy Efficiency in Industry

During the last two decades the industrial sector, which consumes about two-fifths of the energy in the developed world, has realized significant energy savings even as industrial output has continued to expand. For example, efficiency improvements in U.S. industry have led to a 1 percent average annual decline in energy use even as industrial production increased by 2 percent a year. Several strategies have been used to achieve these energy savings:

- Installing equipment and facilities that reduce energy consumption yet are cost-effective.
- Adopting refinements in industrial processes or products that save energy and may reduce labor and material costs as well. For example, the automotive industry has developed a lightweight alloy steel to replace heavier conventional steel for automobile bodies.
- Producing products from recycled materials, which usually requires only about half as much energy as production using new raw materials.
- Developing new technologies that reduce energy and material requirements. The basic oxygen steel-making process that cuts energy inputs by nearly one-half has been implemented by steel manufacturers worldwide.

Saving Energy in Transportation

Since 1980 the world's fleet of motor vehicles has more than doubled to more than 675 million. More than three-fifths of these are registered in North America and Europe. Although the rate of vehicle growth in the developed countries has slowed in recent years, it has accelerated in Asia and Latin America. This vast global fleet consumes well over half of the world's oil, with U.S. vehicles alone using about 14 percent of world oil output.

Not only does this burgeoning vehicle fleet put large demands on oil supplies, but it also produces huge regional and global environmental impacts. In the developed countries pollutants from motor vehicles are the major source of ozone, smog, and acid rain. Furthermore, vehicles are responsible for significant quantities of carbon dioxide and chlorofluorocarbons, which are implicated in global climate change.

As a result, improving motor vehicle fuel economy is a critical economic and environmental issue for both the developed and developing countries. Fuel economy in the developed world improved significantly during the past two decades. In the United States, average fuel economy has doubled. New U.S. passenger cars average more than 28 miles per gallon, while the smaller European and Japanese new vehicles travel more than 30 miles per gallon of gasoline. At the same time, however, the number of automobiles has grown and the miles driven per vehicle have increased (see Figure 11.15b).

During the past decade improvements in new motor vehicle efficiency have been offset in the United States by increasing demand for heavier, less fuel-efficient sport utility vehicles, pickup trucks, and minivans. By the end of the 1990s these vehicles, which averaged less than 20 miles per gallon, made up over 44 percent of the new passenger vehicle market. The significant increase in fuel prices that began in 2000 has reduced demand for these large vehicles.

Current and future efforts to build more fuel-efficient vehicles include using more lightweight materials, such as thinner alloy steel, aluminum, or plastic; using two-stroke engines and continuously variable transmissions for more efficient energy conversions and improving fuel economy; and improving aerodynamic designs by streamlining exteriors to reduce air friction.

Even with improvements in vehicle efficiency, the growing vehicle fleet fueled by petroleum is the largest single source of air pollution. Cleaner automotive fuels that add no net carbon dioxide to the atmosphere include hydrogen and electricity produced from non–fossil-fuel sources and fuels made from biomass such as ethanol, methanol, and methane. Unfortunately, these alternative fuels remain more costly than petroleum. Although the relatively low gasoline prices for most of the 1990s did little to stimulate the growth of alternative fuel vehicles (AFV), United States negotiators have agreed in principle to the Kyoto Protocol, an international agreement that has established goals to reduce greenhouse gas emissions. By increasing the use of AFVs, the United States could reduce these emissions.

By 2000 there were less than half a million AFVs in the United States, with about two-thirds of them using liquid petroleum gas (LPG). Because LPG is a petroleum-based fuel, this type of AFV offers little advantage for reducing emissions or petroleum dependence. The 100,000 or so vehicles powered by natural gas burn fuel cleanly, but compressed or liquid gas fuel storage require elaborate and expensive fuel tanks.

Although ethanol has been used to replace up to 10 percent of gasoline in gasohol fuels, which is burned in conventional gasoline engines, using pure ethanol or methanol requires significant engine modification. In the United States ethanol is generally distilled from grain using fossil-fuel-based energy-intensive methods, so it currently offers little net petroleum or CO_2 reduction. Today the most economical way to produce methanol is from natural gas, while this liquid fuel is portable and burns cleanly, it simply converts a fossil fuel and is expensive to produce and distribute. Currently, only about 50,000 U.S. AFVs are powered primarily by ethanol and methanol, and projected growth remains small.

The much-publicized electric vehicles (EVs) have long been an attractive alternative because they are the only zero-emissions vehicles as they burn no fuel directly. Because the bulky batteries hold much less energy than liquid fuels, EV performance is inferior to conventional vehicles. They have a limited range (around 100 miles) and require lengthy battery recharge periods (several hours). General Motors Corporation, which began leasing EV-1 in 1996, halted production in 2000 after only about 600 of the cars were put into service. Because of these practical limitations, the 7,000 or so EVs in U.S. service are found primarily in government and electric utility fleets.

Interest has now turned to hybrid vehicles running on both a gasoline engine and electric motor. In Japan, vehicle manufacturers have successfully introduced an electric-gasoline hybrid and these cars are now being sold in the United States and Europe. At slower speeds the silent electric motor usually provides all of the power while the gasoline engine kicks in when more power and speed are needed. The electric-gasoline hybrid gets about 55 miles per gallon at freeway speed and travels more than 700 miles between fill-ups. Also, recharging the electric motor does not require an electric outlet.

There is also a need to shift to more energy-efficient freight transportation systems. For example, truck hauling has grown significantly in the past three decades but is relatively inefficient for bulk hauling compared to trains and ships.

9.7 SUMMARY

With rapid population growth and economic expansion worldwide, demand for energy has increased rapidly. Over the last three decades global energy consumption has grown by 90 percent, and the world has become even more reliant on nonrenewable fossil fuels and nuclear-generated electricity. Today 81 percent of the world's commercial energy comes from these sources with oil (30 percent), natural gas (23 percent), and coal (22 percent) accounting for 75 percent and nuclear power plants contributing about 6 percent.

Although oil provides about 30 percent of the world's energy, its relative share has slipped while natural gas share has risen. Since 1980, oil's share of the world energy market has slipped from 45 percent, while overall production has increased slightly in recent years. At the same time, natural gas has increased its share of the world's energy supply. Coal's relative share of the global energy market has declined slightly while production has stabilized in recent years. Electricity produced in nuclear power plants increased nearly threefold as its share of world energy increased to about 6 percent. However, in many countries nuclear power development has slowed or stopped, so production levels will likely slip within a few years as older plants are decommissioned.

Electricity accounts for an increasing fraction of energy demand. More of the world's population have come to rely on lighting, refrigeration, electric motors, mass communication, and information systems and other energy services provided by electricity. While industrial output has expanded worldwide, its share of energy use has declined significantly. However, energy use in commercial and residential buildings has increased, just as it has in the world's growing fleet of motor vehicles.

Currently, renewable energy sources supply only about one-sixth of the world's energy, with most supplied by fuelwood, which is used for cooking and heating by a majority of the population in many developing countries. In the developed countries, where the vast majority of the world's energy is consumed, less than 10 percent of the energy is based on renewable resources; of that 10 percent, hydroelectricity and biomass supply all but a tiny fraction.

As the twenty-first century begins, the world continues to depend overwhelmingly on the dwindling stock of nonrenewable energy resources. As more fossil fuel is burned, the atmosphere becomes more polluted and carbon dioxide levels increase. As more electricity is generated from nuclear power plants, more toxic radioactive waste accumulates in temporary storage.

At the same time, both production and reserves of the most versatile fossil fuel—oil—continue to decline in all but a few countries. Current oil production in the United States continues to decline, and the former Soviet Union has experienced huge production declines as its economy collapsed and its political union shattered. Today nearly two-thirds of the world's proven oil reserves are located in the strife-torn countries surrounding the Persian Gulf, where the constant condition of political tension has been punctuated repeatedly by armed conflict.

Yet even a decade after the Persian Gulf War, the world is even more reliant on the oil reserves of a few countries at ever-increasing new costs. Billions of dollars were spent to wrest Kuwait from Iraq's control, billions more were spent to extinguish and cap Kuwait's wells that were set ablaze, and much more money has been spent to clean up the world's worst case of ecoterrorism. Huge military expenditures will continue to ensure that this oil remains accessible to world markets.

Such trends are unsettling and potentially disastrous. Clearly we cannot stake our future on these dwindling and tenuous supplies of environmentally damaging nonrenewable fuels. We must wean ourselves from profligate energy use by employing technologies based on renewable and perpetual energy sources that will sustain an energy future. The world must commit to long-term energy policies that will guarantee the development of nonpolluting, renewable, energy sources while using energy more efficiently. We must learn to accomplish more while using less!

9.8 KEY TERMS AND CONCEPTS

Energy sources
- direct and indirect solar energy
- nonsolar energy sources
- nonrenewable, renewable, and perpetual energy resources
- fuelwood
- energy and technology revolutions

Fossil fuels
- reserves (economic resources)
- resource base

Coal
- surface mining
- advantages and disadvantages
- gasification/liquefaction
- reclamation
- acid precipitation
- sulfurous smog

Oil
- fuel oil
- gasoline
- production and reserves
- Organization of Petroleum Exporting Countries (OPEC)
- advantages and disadvantages
- tar sands and bitumen
- oil shales and kerogen

Natural gas
- methane, butane, and propane
- advantages and disadvantages

Nuclear power
- nuclear fission
- types of reactors
- advantages and disadvantages
- passive safety systems

Biomass fuels
- liquid and gaseous fuels
- methane recovery systems
- advantages and disadvantages

Hydroelectric power

Wind power

Solar energy
- solar architecture
- solar-thermal electric technologies
- heliostats
- photovoltaic cells
- solar panels
- hydrogen fuel cells
- solar power satellites

Geothermal energy and tidal energy
- geothermal-electric
- hydrothermal heat

Energy conservation

9.9 QUESTIONS FOR REVIEW

1. Describe the role that the energy revolution of the eighteenth and nineteenth century played in the industrialization and urbanization of the developed countries. Discuss the changes in energy resources and energy uses that occurred with the energy revolutions.
2. What fraction of the world's commercial energy does the United States consume? What are our sources of commercial energy? Which energy sources are primarily produced and used in the United States? How much of the energy resources used by the United States comes from foreign sources? Why is energy use less efficient in the United States than in most other developed countries?
3. Although coal is no longer the most widely used fossil fuel, world coal production continues at near record levels. What is the most important use for coal today? Which countries are the leading producers and consumers of coal? Why has coal demand remained high in recent years? What are the economic advantages of coal? Discuss the environmental impacts associated with the production and use of coal.
4. Discuss the advantages of oil over other fossil fuels. Describe the geographic differences between oil production and reserves and consumption. What is the economic and political significance of these differences? Why is the real cost of oil much higher than the actual market price for oil? What are some of the unconventional sources of oil? What are the potential environmental impacts of producing these unconventional oil resources?
5. Although natural gas burns cleanly and is the premium fuel for heating buildings, it has limited markets. Why is most natural gas consumed within the country or region where it is produced? Which countries produce and consume the most natural gas? Which countries have the greatest reserves? Why is much natural gas burned off at the well in some countries?
6. Today there are more than four hundred nuclear power plants generating electricity in more than thirty nations worldwide. Why has nuclear power production caused a great deal of controversy in many countries? What are the economic and environmental advantages of nuclear power over traditional thermal electric technologies? What are the disadvantages?
7. Biomass fuels provide much of the energy for cooking and heating in developing countries. What is the most widely used biomass fuel? Where is it in short supply? What steps are being taken to reduce these shortages?
8. The use of liquid and gaseous fuels from biomass to supplement fossil fuels is increasing in many parts of the world. What are the sources of these fuels, and how are they being used today? Discuss the environmental and economic advantages and disadvantages of increased biomass fuel use.
9. In what world regions is most hydroelectric power generated today, and where is the greatest potential for future hydroelectric development? Discuss some of the environmental impacts that result from hydroelectric power.
10. Although perpetual energy sources such as wind and sunlight are used to meet only a tiny amount of our energy demands, new technologies are reducing the costs of converting these sources into useful energy forms such as electricity. Provide several examples of recent technological developments that use perpetual energy sources. What role will these perpetual energy sources play in the future?
11. What must be done now to ensure an orderly transition from nonrenewable sources, which supply nearly all of our current energy needs, to an energy future based on renewable or perpetual energy sources?
12. Why must energy conservation strategies be an important component in current energy policy? What must be done by government and the private sector to promote more efficient use of energy? In which economic sectors have energy conservation measures been most effective? Where has energy conservation lagged?

10

THE ATMOSPHERIC ENVIRONMENT AND LAND USE

10.1 Introduction

Earth is a glorious planet, unimaginably diverse and constantly changing. Much of its diversity and dynamics is driven by the atmosphere, the great sea of gas that envelops the entire planet. Comparing Earth's atmosphere with that of our sister planet Mars underscores just how different Earth is. The Martian atmosphere is very light, less than one-hundredth the mass of the Earth's atmosphere, and provides a poor buffer for the planet's surface against solar radiation coming in and heat going out. On Earth the atmosphere functions as a highly effective regulator of heat and radiation, the principal sources of energy for life.

In its role as a radiation regulator the atmosphere protects the biosphere from harmful forms of incoming solar radiation. As a heat regulator, the atmosphere slows the loss of heat from the Earth, a process called the greenhouse effect. This process has recently gained worldwide attention in connection with carbon dioxide pollution because carbon dioxide is a good heat absorber. The greenhouse effect is increasing on Earth, and the lower levels of the atmosphere may be heating up as a result.

This chapter is concerned with the atmosphere, its composition, structure, and essential processes. We are interested in understanding the controls on heat and radiation in the atmosphere, including the breakdown of incoming solar radiation, the Earth's heat balance, and the greenhouse effect. This is necessary in order to understand many important environmental problems such as climate change, acid rain, and drought. Urban climate, which is taken up last, is a well-documented example of human-induced climatic change that has a pronounced effect on the health and well being of the people living in cities.

10.2 The General Composition of Air

Air is a mixture of gases. **Nitrogen** and **oxygen** constitute close to 99 percent of pure dry air. In the atmosphere, these two gases are completely mixed together so that they have the character of one gas. Only oxygen, however, is directly critical to life because it is taken in by animals and various microorganisms in respiration. Plants, on the other hand, release oxygen as a part of their respiration. Over the past several billion years, the oxygen content of the atmosphere has grown with the increase in plant life on the planet. The oxygen content of the modern atmosphere is nearly 21 percent; nitrogen is close to 78 percent (Table 10.1).

Carbon Dioxide and Ozone

Most of the remaining 1 percent of pure, dry air is composed of argon, an inactive gas of little importance to natural atmospheric and life processes. Of the small fraction of 1 percent that remains, tiny amounts of many gases are detectable in the atmosphere, but two must be singled out because they are of special significance: carbon dioxide and ozone. **Carbon dioxide** (CO_2) is an important heat-absorbing gas that plays a major role in creating the to Earth's greenhouse effect. Since the beginning of the Industrial Revolution in the eighteenth century, atmospheric carbon dioxide has increased by about 30 percent (from 280 to 365 parts per million) as the rapidly growing global population burned increasing amounts of carbon-based fossil fuels and wood. The greenhouse effect will be discussed in more detail later in this chapter.

Ozone constitutes only 0.000007 percent of the atmosphere, but it plays a very important environmental role. It absorbs ultraviolet radiation, which in large concentrations is lethal to humans and many other lifeforms. Ozone is concentrated high in the atmosphere, 15 to 55 kilometers (10 to 35 miles) altitude, in a zone called the *ozone layer*, which functions as an ultraviolet screen over the Earth. As a result of modern air pollution, the protective ozone screen is actually being damaged. The main culprit is *chlorofluorocarbons* (CFCs), a group of chemical compounds used in spray cans and cooling systems, which destroys the ozone molecule and breaks down the ozone screen. This problem will be discussed in greater detail in Chapter 11.

Water Vapor and Airborne Particles

Of course the atmosphere is not made up entirely of pure, dry air. There are two other important constituents: **water vapor** and **airborne particles**. The particles in the atmosphere are minute liquid-and-solid masses of various compositions. The liquid particles are water droplets in clouds and fog. The solids include dust from soil erosion and volcanic eruptions, pollen from plants, salt crystals from ocean spray, and air pollutants from a host of sources. These particles are important in the precipitation

Table 10.1

Composition of the Atmosphere

Gas	Percentage by Volume in Troposphere	Notes
Nitrogen (N_2)	78.084	
Oxygen (O_2)	20.946	Has developed principally with the evolution of plant life in past 2 to 3 billion years.
Argon (A)	0.934	
Carbon dioxide (CO_2)	0.033	Principal longwave (heat) absorbing gas. Has increased since nineteenth century with population growth, industrialization, and burning of fossil fuels.
Neon (Ne)	0.00182	
Helium (He)	0.000524	
Methane (CH_4)	0.00016	
Krypton (Kr)	0.00014	
Hydrogen (H_2)	0.00005	
Nitrous oxide (N_2)	0.000035	Absorbs longwave (heat) radiation.
Important Variable Gases		
Water vapor (H_2O)	0–4	Principal heat-absorbing gas. Absorbs longwave (heat) radiation
Ozone (O_3)	0–0.000007	Absorbs ultraviolet radiation in upper atmosphere; currently declining in the stratosphere.

process because they function as nuclei around which droplets form. This helps explain the higher precipitation rates over some large urban areas where the atmosphere is heavily polluted with particulates. In addition, airborne particulates also reduce the atmosphere's transparency to solar radiation, decreasing the receipt of solar energy at the Earth's surface.

Water vapor ranges from nearly zero to as much as 4 percent of the atmosphere. Most of this tasteless, odorless gas is found in the lower 16 kilometers (10 miles) of the atmosphere where the bulk of the Earth's clouds (which are tiny liquid particles, not vapor) and precipitation (all particle forms of falling moisture) occur.

Clouds and precipitation are formed by the cooling of air containing water vapor. Water vapor is supplied to the atmosphere mainly through evaporation from the oceans. The capacity of the atmosphere to take on water vapor is dependent on air temperature. The warmer air is, the greater the amount of water vapor it can hold; conversely, the colder air is, the smaller the amount of water vapor it can hold. Considering the range of air temperatures that exist over Earth helps explain why water vapor is so variable in the atmosphere.

The graph in Figure 10.1 shows the relationship between air temperature and the vapor-holding capacity of air. At a temperature of −30°C (−20°F), a typical winter temperature in the Arctic and Antarctic, a cubic meter of air can hold barely one gram of moisture; at 0°C (32°F) it can hold 5 grams of vapor; and at 30°C (86°F), a typical temperature in the tropics, it can hold more than 30 grams per cubic meter. However, we also know that not all tropical air is humid; indeed, some is very dry. So there must be other factors influencing the distribution of water vapor. One of the most important of these factors is the availability of moisture supplies from source regions over the oceans. Most tropical deserts, such as the Sahara and the Great Australian Desert, are dry because of one or both of two factors:

1. They are fed with descending dry air from the upper atmosphere rather than moist air from the surface (see Figure 4.6).
2. They are geographically isolated from moisture source regions over the oceans because of mountain barriers or great distance from the sea.

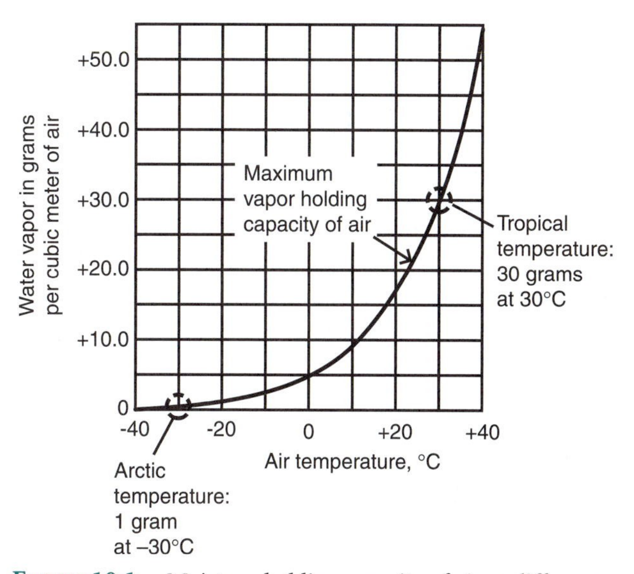

FIGURE 10.1 *Moisture-holding capacity of air at different temperatures. The capacity of cold air is extremely small, which helps explain the low precipitation rates in polar regions.*

10.3 THERMAL STRUCTURE OF THE ATMOSPHERE

The atmosphere's thermal structure refers to the changes in temperature with altitude from the Earth's surface upward. Curiously, temperature values do not decrease progressively with altitude, but are arranged in several contrasting zones, called the troposphere, stratosphere, mesosphere, and thermosphere (Figure 10.2).

The Troposphere

The lowermost zone, the **troposphere**, is warm at the bottom next to the Earth's surface, and cools upward at an average rate of −6.4°C per 1,000 meters (−3.5°F per 1,000 feet). At an altitude around 12 kilometers (7.5 miles), where the temperature is a frigid −60°C to −70°C, thermal conditions stabilize, marking the top of the troposphere. This is the *tropopause* (Figure 10.2).

Unlike the stratosphere above it, which lacks the capacity for vertical motion, the troposphere literally boils with it. This is related, in part, to the fact that the troposphere heats from the bottom up mainly as a result of solar heating of the Earth's surface. As surface air heats, it expands, becomes buoyant, and rises, thereby initiating vertical motion that causes the troposphere to turn over. Of course this action varies greatly over the Earth, but where Earth's surface is warm, as over tropical lands or midlatitude lands in summer, it is very active. Other factors such as converging wind systems or weather fronts also cause tropospheric mixing. On balance the troposphere is very active, exchanging surface air with air aloft almost continuously.

The mixing motion of the troposphere is an essential part of Earth's weather. All the weather we experience at the Earth's surface originates in the troposphere. When weather forecasters speak of air masses, fronts, hurricanes, jet streams, and the like, they are referring to phenomena of the troposphere. The troposphere contains more than 95 percent of the atmosphere's mass and all but a tiny fraction of the atmosphere's water.

All life resides at the base of the troposphere. All the vital exchanges of gases radiation, heat, and water between

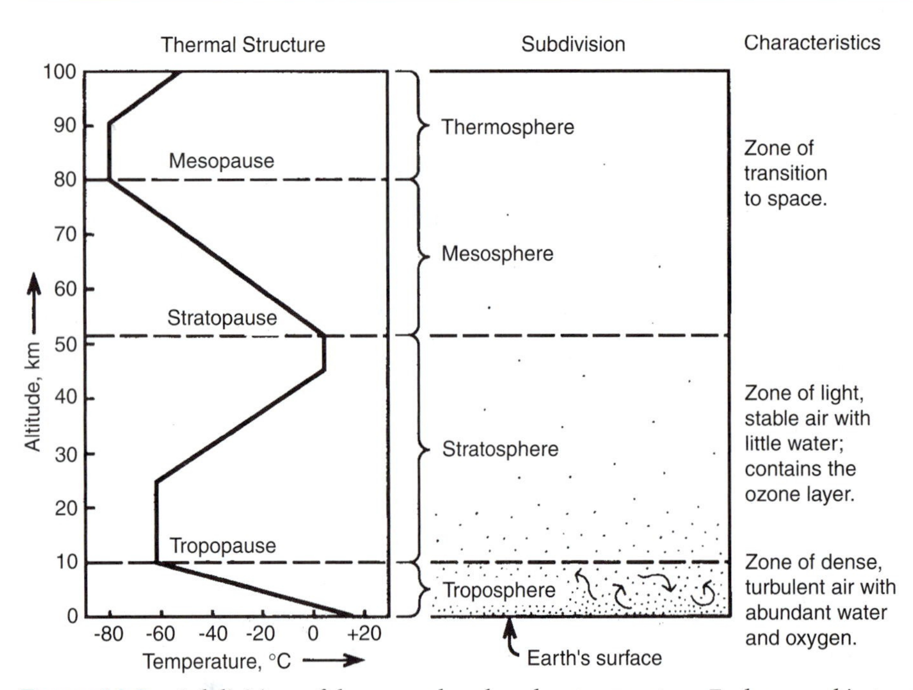

FIGURE 10.2 *Subdivisions of the atmosphere based on temperature. Each reversal in temperature marks a "pause." The troposphere contains the vast majority of the atmosphere's mass.*

Earth and its atmosphere take place in this layer, including air pollution. With the exception of pollution from high-altitude aircraft, virtually all contaminants found in the atmosphere enter through the base of the troposphere. Once in the troposphere, however, they are redistributed by winds and turbulent mixing entering both surface and upper atmospheric wind systems. These wind systems are driven by air pressure differences, which we take up in the next section.

The Stratosphere and Above

Above the troposphere lies the **stratosphere**, where the temperature trend reverses—that is, warms with altitude. The stratosphere is a deep zone made up of very thin air, a little frozen moisture in light, wispy clouds, and the ozone layer. The warming in the stratosphere is caused by ozone's absorption of ultraviolet radiation, which is part of the incoming solar radiation. At an altitude of approximately 50 kilometers (30 miles) the warming trend stops and reverses again as we enter the **mesosphere**. This zone and the one above it, the **thermosphere**, are exceedingly light and have little, if any, known influence on the lower atmosphere and its life forms.

10.4 ATMOSPHERIC PRESSURE AND GENERAL CIRCULATION

The atmosphere has the physical character of a light fluid. As in any fluid arranged in a deep layer such as the atmosphere, the deeper one is in the layer, the greater the pressure from the overlying mass. Measuring downward from the top of the atmosphere, air pressure should therefore be greatest at the bottom (that is, at Earth's surface). At sea level, average atmospheric pressure is 1013.2 millibars (mb). A millibar is a unit of force. Air pressure can also be measured in pounds per square inch. In these units, average atmospheric pressure at sea level is equivalent to 14.7 pounds per square inch of surface area (Figure 10.3). In weather forecasting the designation of *high pressure* and *low pressure* is defined according to whether pressure at the surface is above or below 1013.2 millibars when adjusted to sea level.

Vertical Distribution of Pressure

The atmosphere, unlike water, is a highly compressible fluid. Under the force of Earth's gravity, the bulk of the atmosphere is compressed into a thin envelope immediately over the planet's surface. As a result, about half of the atmosphere is found below an altitude of only 5.5 kilometers (18,000 feet) (Figure 10.4). At that altitude atmospheric pressure is about 500 millibars, that is, about one-half sea level pressure (Figure 10.3). At the cruising altitude of large commercial jetliners, around 12,000 meters (39,000 feet), air pressure is about 200 millibars, far too low for human survival without an artificially pressurized living environment. In mountainous regions, very few humans live above 3,500 meters (11,500 feet) elevation, where air pressure is around 650 millibars, and there is no permanent human habitation above 5,200 meters (17,000 feet).

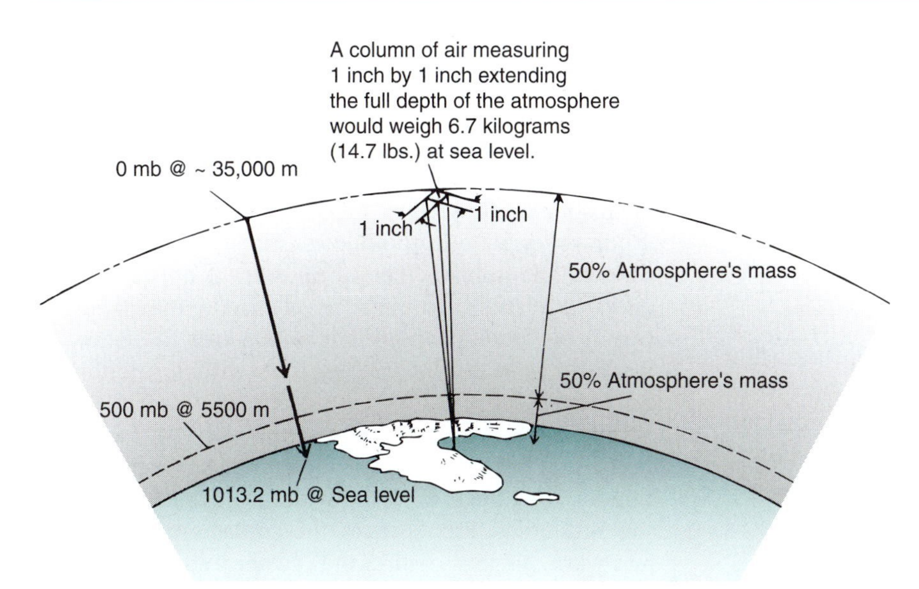

FIGURE 10.3 *Concept of atmospheric pressure in pounds or kilograms per square inch and in millibars.*

Wind and Pressure Systems

The Earth's wind systems are driven by pressure differences from place to place in the atmosphere. High pressure produces an outflow of wind, whereas low pressure produces an inflow of wind. The greater the differences in pressure between adjacent areas of high and low pressure, the faster the resulting winds. In an extreme case of low pressure such as hurricane, the pressure change, or **pressure gradient**, from the perimeter to the center is typically as great as 50 millibars (for example, from 1016 millibars to 964 millibars) and the resulting wind velocities reach 160 kilometers per hour (100 mph) or more.

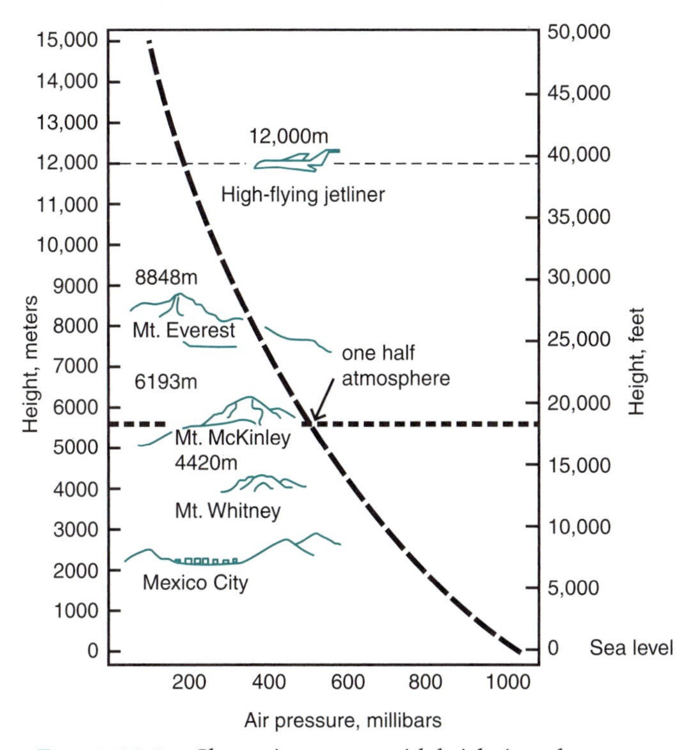

FIGURE 10.4 *Change in pressure with height into the atmosphere. Pressure is one-half sea level pressure at 5,500 meters (18,000 feet).*

Pressure cells such as hurricanes and thunderstorms are examples of **transient pressure systems**, or short-lived, mobile systems. They are an important part of seasonal weather and climate but they do not give us much insight into the global picture of atmospheric pressure and wind systems. For this we must turn to the great pressure belts, called the **semipermanent pressure systems.** These systems form belts of pressure that more or less encircle the globe within certain zones of latitude. They are called *semipermanent* because they are present most of the time and because they wax and wane in strength with the seasons.

There are four great **global pressure belts**, and each is associated with a distinctly different set of climates and wind systems (Figure 10.5). Their origins are attributed to two types of causes: thermal and dynamic. Those of *thermal origins* can be related to heating and cooling of the Earth's surface and lower atmosphere. Warm surface and air conditions produce light, buoyant (or unstable) air and low pressure. Cold surfaces and related cold air produce relatively heavy, stable air and high pressure. Pressure systems of *dynamic origins* are related to the mechanics of the atmosphere's fluid flow, as when two wind systems meet and the air is forced up, forming low pressure, or when flow complications in upper atmospheric winds force air toward the ground, resulting in high pressure.

Global Pressure and Prevailing Wind Systems

The geographic arrangement of global pressure and prevailing winds follows an alternating pattern of high and low pressure and east-flowing and west-flowing wind systems. Beginning at the equator, the first belt of pressure is the **Equatorial Low pressure** system (Figure 10.5). Where it lies over land, this belt is thermal in origin, the result of intensive surface heating along the equator. Over water it is mainly dynamic in origin, the result of converging wind systems. The convergence of wind actually takes place over the whole length of the Equatorial Low, resulting in the term **Intertropical Convergence Zone** (ITC Zone) to describe the circulation associated with this belt of pressure.

The Equatorial Low is associated with abundant precipitation, delivered principally by thunderstorms generated by surface heating and convergent airflow (see Chapter 12 for a description of thunderstorms and convergent precipitation). Near the equator, this pressure system gives rise to the **tropical wet** climate, which is the home of the tropical rainforests. On the poleward sides of this climate lie the **tropical wet/dry** climate, where vegetation, called *savanna*, is mainly grasses and small, scattered trees (see Figure 6.8). This wet/dry climate is found in both the Northern and Southern Hemispheres and its wet season, which occurs in summer in each hemisphere, is brought on when the Equatorial Low shifts into these zones in response to intensive solar heating of the land surfaces.

The next belt of pressure is the **Subtropical High pressure** system, located at 25 to 35 degrees latitude. This belt is dynamic in origin, the result of complications in the upper atmospheric winds, and is characterized by air descending toward the Earth's surface. This air is dry, and as it nears the Earth's surface and heats up it becomes severely arid, giving rise to the world's great deserts. These include the Sahara of North Africa, the Sonora of North America, and the Great Australian Desert. The air that enters the high pressure cells from aloft flows out as surface winds (see Figure 4.6). Those winds that flow toward the Equatorial Low form a wind system called the **Easterly Trade Winds**, whereas those that flow poleward (on the other side of the belt) form the **Prevailing Westerlies** (Figure 10.5). Although both these wind systems begin as dry air, especially over the continents, they rapidly gain moisture as they move toward their respective low pressure cells to the north and south.

The next major belt of pressure is found poleward of the Prevailing Westerlies around 45 to 65 degrees latitude (Figure 10.5). It is a low pressure system called the **Subarctic Low**, and is the result of large-scale mixing of cold and warm air along an atmospheric boundary called the **Polar Front**. The Polar Front marks the edge of a mass of cold air that caps the polar zones of both hemispheres (see Figure 12.4a). Disturbances along this front give rise to storms that are set into swirling motion by the deflecting effect of the Earth's rotation, called the **Coriolis effect**. The Coriolis effect causes a right deflection in the Northern Hemisphere and a left deflection in the Southern Hemisphere; thus, the direction of wind circulation around pressure cells (both lows and highs) is opposite in the two hemispheres. The Coriolis effect not only influences circulation and pressure cells in the mid-latitudes, but in the tropics as well, but only above 10 degrees or so latitude, where it causes the swirling motion in hurricanes.

The storms that form along the Polar Front are low-pressure cells called **midlatitude cyclones**, which migrate rapidly along the Polar Front with the flow of the Prevailing Westerlies. Fronts mark the contacts between cold and warm air from within the cyclone and are responsible for producing large amounts of precipitation, rain in summer and snow in winter. Midlatitude cyclones occur frequently—so often, in fact, that they tend to dominate this zone of latitude, especially in winter. They are especially persistent over the oceans, where they create stormy, cloudy conditions throughout much of the year. The wet, windy climates of Iceland and the Aleutian Islands in the North Atlantic and North Pacific exemplify conditions dominated by cyclonic systems (see Chapter 12 for a description of frontal precipitation).

The fourth belt of pressure is found at the polar caps of the planet where cold conditions produce high pressure, called the **Polar High**. This system is mainly thermal in origin, but unlike the Equatorial Low, it is caused by cold, stable air. The Polar Highs over both the Arctic and Antarctic are also fed by descending air from aloft, and like all major pressure systems, are tied into upper atmospheric circulation that functions as a return air system. Temperatures under the Polar Highs are well below freezing in most months of the year, and because of the coldness of the air, precipitation rates are also low.

Global Circulation and Air Pollution

In addition to revealing some of the basic factors underlying the global pattern of climates, the large-scale system of pressure and winds also suggests some things about global air-pollution patterns. The first is that pollutants released into a wind system such as the Prevailing Westerlies can be carried enormous distances within one belt of latitude. Streams of polluted air, called *plumes*, emanating from volcanoes, large metropolitan areas and massive fires have been traced thousands of miles downwind. The plume from New York City, for example, has been observed by aircraft over Iceland 5,000 kilometers (3,000 miles) away in the North Atlantic. In some instances, pollutants and the plumes from major volcanoes such as Pinatubo in the Philippines have been traced

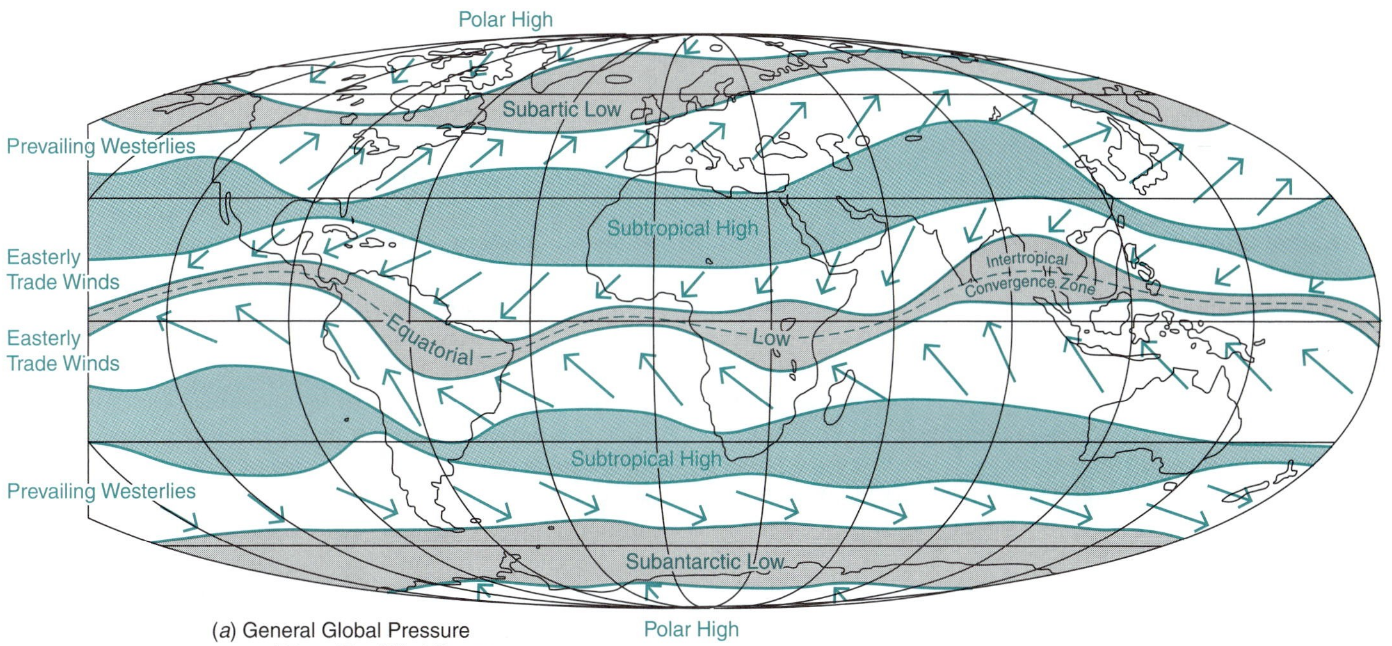

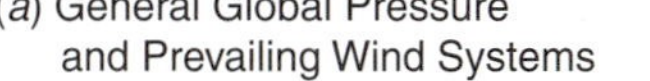
(*a*) General Global Pressure and Prevailing Wind Systems

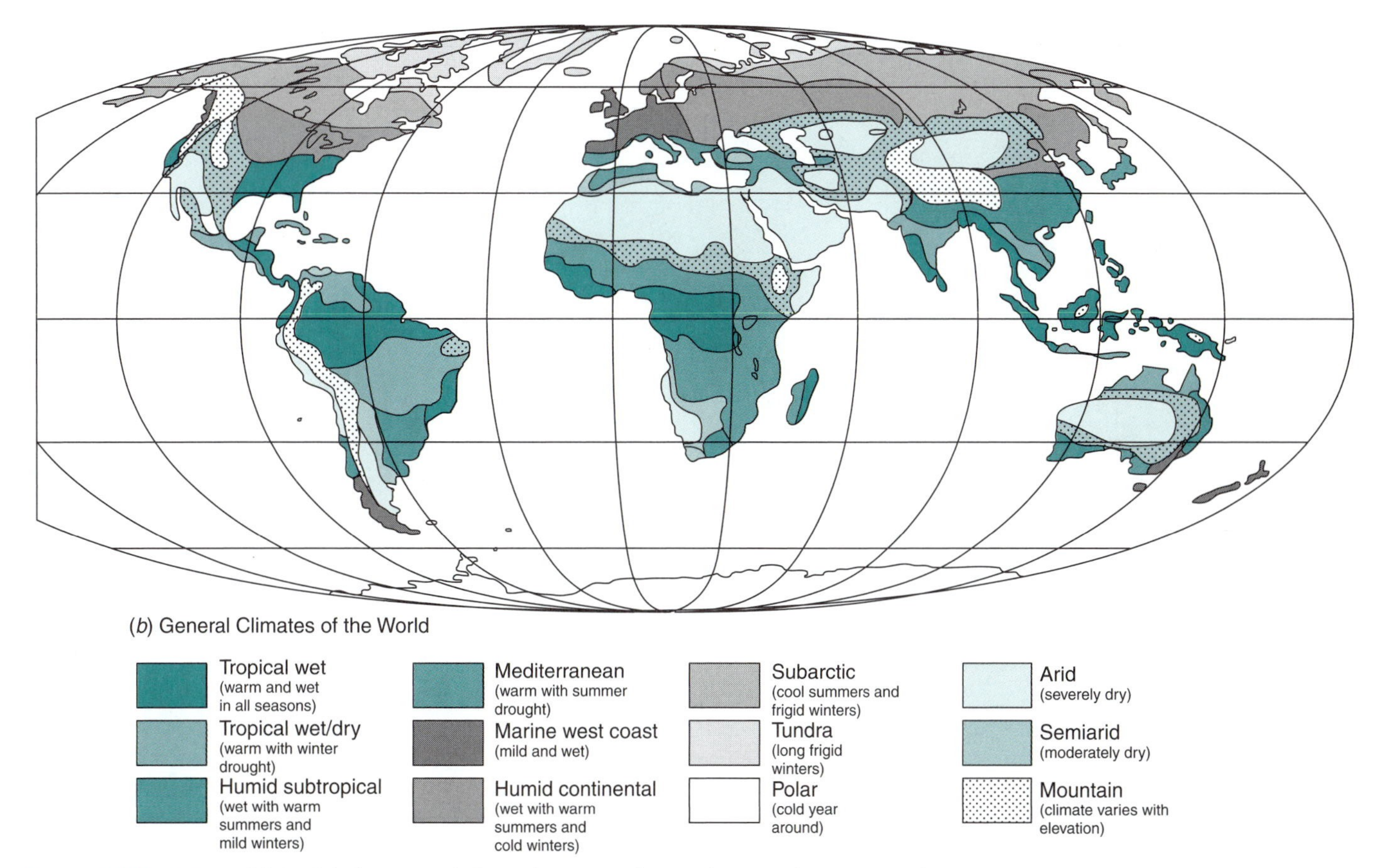

FIGURE 10.5 *Maps of (a) Global pressure belts and associated prevailing wind systems; and (b) general climate types of the world.*

around the world after they become entrained by prevailing winds (see Figure 2.9). The next chapter provides additional information on global dispersal of air pollutants.

The second thing suggested is that pollutants can be lifted high into the atmosphere and introduced to upper-level winds. These winds, in turn, carry the pollutants to other pressure and wind systems. Polluted air raised in low pressure systems can be recycled from above into high pressure cells and brought back to the surface in a different climatic zone. Undoubtedly a large part of the air pollution recorded in the Arctic and Antarctic comes from pollution driven into the upper atmosphere by pressure systems of the mid-latitude/subarctic zone. The startling decline of the ozone layer over Antarctica, and more recently over the Arctic, is related to such large-scale circulation of pollutants emanating from midlatitude sources.

10.5 SOLAR RADIATION IN THE ATMOSPHERE

To help us understand the role of solar radiation in creating the life environment on Earth's surface, let us briefly consider what conditions would be like without the atmosphere. First, the amount of solar radiation received directly at the surface would be about 100 percent greater worldwide and Earth's surface itself would be much hotter during daylight hours. Second, the surface would be bombarded by heavy doses of lethal radiation: not only ultraviolet radiation, but also x-rays and gamma rays that are normally intercepted by the atmosphere. Third, with no gases to regulate the outflow of heat from the planet's surface, day-to-night differences in surface temperature would be extreme, with night temperatures falling hundreds of degrees below daytime temperatures. It would be a desolate planet, not unlike the moon or Mars, where conditions would be dominated by solar radiation and where weather, such as it would be, would be reduced to a stark day/night cycle.

When solar radiation strikes the outer edge of the Earth's atmosphere, its strength (or energy level) is equivalent to about two calories of heat over an area of one square centimeter measured in a time span of one minute (abbreviated: 2 cal per cm^2 per min.). This flow of solar energy changes very little over the seasons and years, and is therefore termed the **solar constant.** As solar radiation passes through the atmosphere, it is reduced in strength and selectively filtered. Reduction and filtering take place by two processes: **absorption** and **reflection.** To understand these processes it is necessary to examine the composition of solar radiation.

The sun emits energy known as **electromagnetic radiation.** This form of energy can be described as a collection of electromagnetic waves, all of which travel through

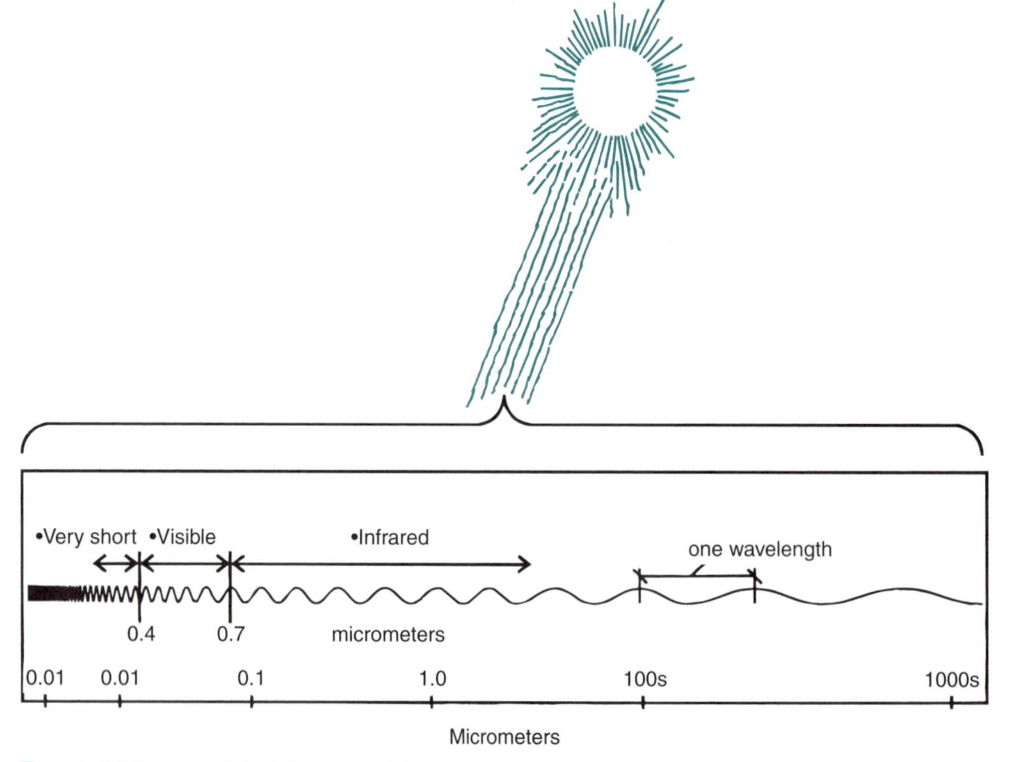

FIGURE 10.6 *Simplified diagram of the electromagnetic spectrum showing three classes of radiation: very short, visible, and infrared.*

space and the atmosphere at a rate of 300,000 kilometers a second, the speed of light, taking only 8 to 9 minutes to reach Earth. Electromagnetic waves are classified according to their wavelengths in a scheme called the **electromagnetic spectrum**. The unit of measure for wavelengths is the micrometer, equivalent to one-millionth of a meter (Figure 10.6).

The behavior of electromagnetic waves in the atmosphere varies with wavelength and the composition of air. For example, atmospheric ozone in the stratosphere absorbs most of the short ultraviolet wavelengths of sunlight while visible wavelengths pass through clear, dry atmosphere. For longer wavelengths, called infrared, the atmosphere is more opaque, absorbing the radiation mostly in the molecules of gases. This also holds for outgoing radiation, the radiation transmitted upward into the atmosphere from the Earth's surface, which is mostly infrared.

10.6 BREAKDOWN OF THE SOLAR BEAM

For our purposes, the electromagnetic spectrum for incoming solar radiation can be divided into three classes: very short, visible, and infrared. The *very short* class represents wavelengths less than 0.4 micrometers and is made up primarily of ultraviolet radiation. This class of radiation constitutes about 9 percent of the solar beam and also includes ionizing radiation in the form of x-rays and gamma rays. These waves are capable of destroying living cells, altering genetic material within cells, and causing cancer. Almost all very short radiation is absorbed by the atmosphere before reaching Earth's surface (Figure 10.7).

Reduction by the Troposphere

As the remaining solar radiation—*visible* (0.4 to 0.7 micrometers wavelength) and *infrared* (longer than 0.7 micrometers wavelength)—penetrates the atmosphere, it encounters an increasingly denser medium with higher concentrations of gas molecules and particles. Absorption of mostly infrared wavelengths takes place in the troposphere as the incoming radiation encounters carbon dioxide, water vapor, and particles. These substances **absorb** about 14 percent of the total incoming solar radiation on the average worldwide. This energy is converted into heat and warms the atmosphere (Figure 10.7).

Another large block of solar radiation is reflected and scattered as various wavelengths are intercepted by gas molecules and particles in the atmosphere. There are two components to these processes: (1) radiation turned away from the Earth and sent back into space, and (2) radiation rerouted to Earth's surface. The first component is a large quantity about 27 percent of the solar constant. The second component is smaller, about 21 percent, and is broadcast to Earth as **diffuse solar radiation**. The diffuse radiation provides the indirect light that illuminates shadowed areas. Most diffuse radiation is visible light and is absorbed when it reaches the surface.

The remaining radiation, which is also in the visible class, reaches Earth's surface as direct (shadow-making) or **beam solar radiation**. Over the entire planet, a small fraction of this radiation, about 3 percent of the original incoming solar radiation, is reflected back toward space by Earth surface materials: water, ice, sand, and vegetation. Earth surface reflection is referred to as **albedo**. For some materials such as ice and snow, albedo may be as great as 50 to 70 percent—that is, 50 to 70 percent of the solar radiation hitting the surface is reflected—but for forests, grasslands, and most water it is less than 20 percent.

The balance after albedo amounts to about 22 percent of the *original* incoming solar beam entering at the top of the atmosphere. This energy is absorbed by the Earth's surface, whereupon it is converted into heat. Coupled with the diffuse radiation, the grand total of solar radiation absorbed by Earth's surface is close to 44 percent worldwide, a little less than half of the solar constant on the average (Figure 10.7). Most absorption takes place in the oceans, chiefly because they cover most of Earth's surface. In addition, absorption is very high in the tropical latitudes because the albedo of water is very low at high sun angles. At sun angles above 70 degrees, albedo is less than 10 percent, meaning that tropical seas consistently absorb 90 percent or more of the solar radiation that strikes them.

Cloud and Pollution Factors

Of all the processes we have mentioned, the most effective natural means of reducing solar radiation worldwide is **cloud reflection**. Cloud reflection reduces the receipt of solar radiation by an average of 22 percent over the Earth as a whole. But cloud reflection varies enormously with different weather systems and different climatic zones over the Earth. In most places cloud reflection may range from zero on clear days to as much as 80 percent on heavily overcast days.

Cloud cover also varies appreciably among the major climatic zones. One of the most significant influences is found in the wet tropics, where without clouds, solar radiation at the surface would be the highest in the world. As it turns out, the wet tropics are actually second to the dry subtropics because the cloudless desert air allows the highest level of solar radiation receipt on Earth (see Figure 5.7). By far the most cloud reflection happens in the coastal marine climate of the subarctic, such as the Aleutian Islands of the North Pacific, where only several sunny days can be expected a year.

Air pollution also increases reflection and scattering. Over large cities, for example, solar radiation may be re-

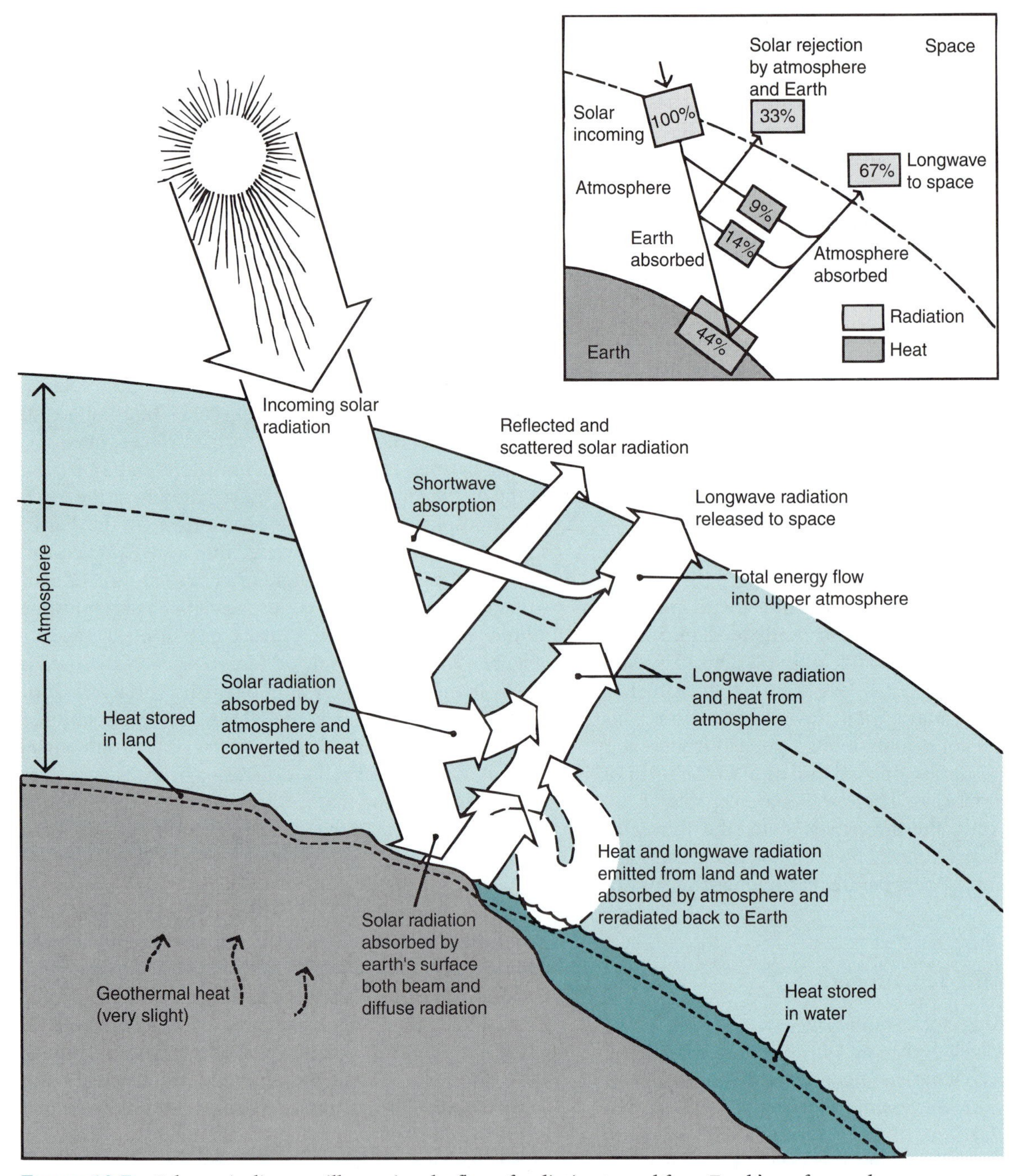

FIGURE 10.7 *Schematic diagram illustrating the flow of radiation to and from Earth's surface and atmosphere. The inset gives the quantities represented by heat and radiation in the flow scheme. Incoming solar radiation absorbed by the atmosphere and Earth's surface is converted to heat and ultimately radiated back into space as longwave (infrared) radiation.*

duced 50 percent or more because of pollution in the lower several thousand meters of the troposphere (Figure 10.8). Air pollution also increases cloud cover by enhancing condensation in the atmosphere. If cloudy conditions (natural and/or human-induced) are combined with a dirty atmosphere, the solar beam can be reduced to as little as 10 percent by the time it gets to Earth's surface (Figure 10.8). Scientists are able to demonstrate that solid and liquid pollution particles and pollution-induced cloud cover are currently reducing incoming solar radiation over much of Earth. The resultant cooling in the lower atmosphere is reducing the full effects of global warming from heat-retaining pollutants such as carbon dioxide. This will be covered in more detail later in this chapter.

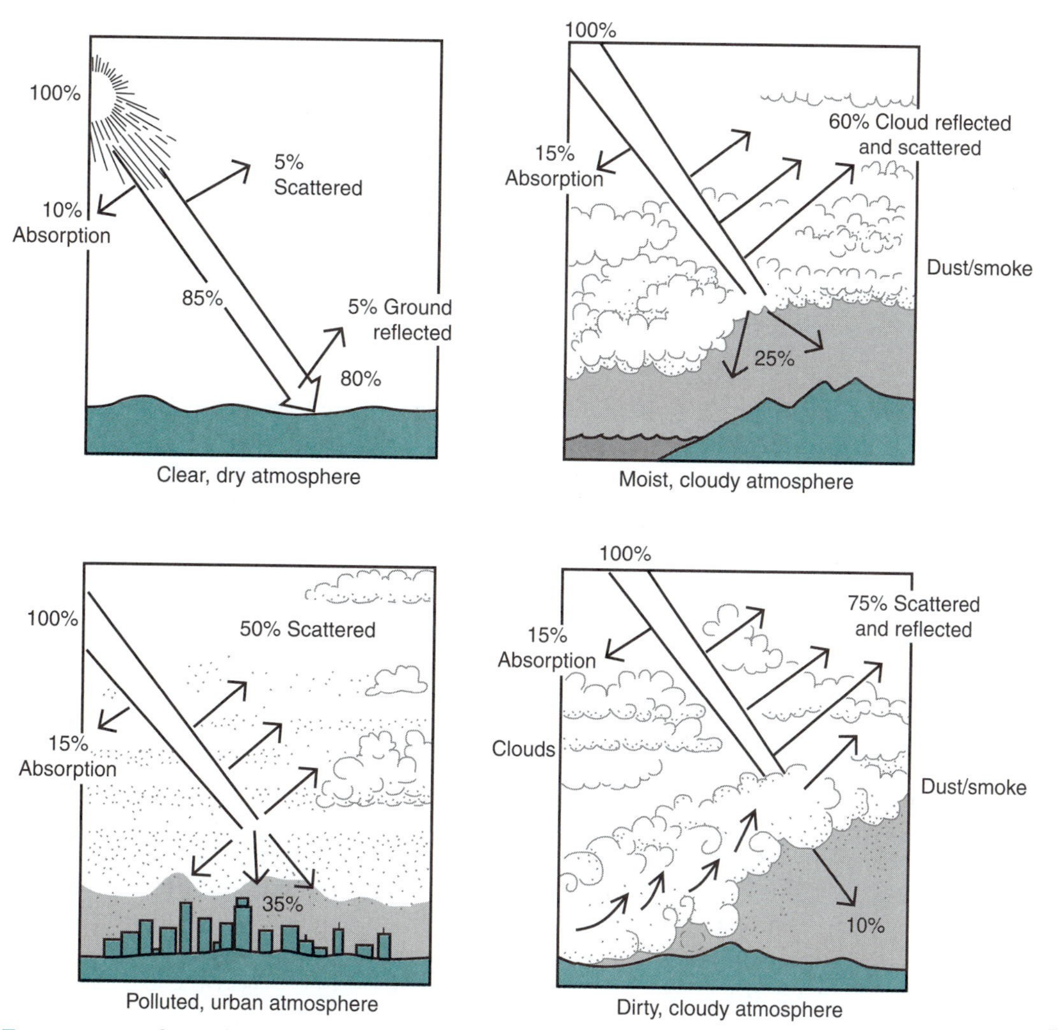

FIGURE 10.8 *Scattering and reflection of solar radiation under four different atmospheric conditions. Note effect of clouds and air pollution.*

10.7 EARTH'S HEAT BALANCE

The solar radiation absorbed by the atmosphere and Earth's surface materials is converted into heat energy. In the atmosphere the heat is held in gases and particles, whereas on Earth's surface it is held in a thin layer of water and soil (Figure 10.7). Ultimately, the heat absorbed in land and water is transferred into the overlying atmosphere by conduction, convection, and infrared radiation, whereupon much of it is absorbed, adding substantially to the atmosphere's heat supply.

Thus, the atmosphere is heated from two sources: incoming solar radiation from above and Earth's surface from below. These two sources account for about 25 percent and 75 percent, respectively, of the atmosphere's heat supply. They are the energy sources that warm the atmosphere and drive the essential atmospheric processes such as winds, evaporation, and precipitation. The only other source of heat, geothermal heat coming from Earth's interior, is so slight (less than 0.01 percent) that it is of no consequence in atmospheric heating.

The Distribution of Solar Heating

The pattern of atmospheric heating through the absorption of solar radiation varies with altitude in the atmosphere and location on Earth. Most of the very short wavelengths (UV and ionizing radiation) are absorbed in the stratosphere more than 16 kilometers (10 miles) above the Earth's surface. It warms the stratosphere but has little influence on the troposphere. As the absorbing capacity of ozone is reduced by air pollution, specifically by CFCs, more ultraviolet radiation is reaching the planet's surface, although it has little effect on surface heating. All the remaining heating of the atmosphere, which is the vast majority indeed, takes place in the troposphere.

The global distribution of tropospheric heating is highly uneven. Generally speaking, the tropical and subtropical latitudes, between 25 degrees north and south latitude, together receive the bulk of Earth's heat. In this great belt, nearly 5,600 kilometers (3,500 miles) wide and 40,000 kilometers (24,000 miles) long, the atmosphere is deeper and holds more heat-absorbing gases than in any other belt of latitude. Average annual receipt of solar radiation by land and water surfaces in the tropics and subtropics ranges between 120 and 220 kilocalories per square centimeter (see Figure 5.7). The heat produced by this energy, coupled with that from the absorption of incoming solar radiation in the atmosphere above, is the primary supply of energy driving Earth's atmospheric engine.

Recycling Heat and Radiation

Once the atmosphere is heated, the air itself becomes a radiating body. In contrast to the sun, which emits mainly visible and near-infrared radiation, the atmosphere emits radiation in much longer infrared wavelengths (Figure 10.6). This radiation is broadcast both skyward and earthward. That broadcast to Earth is absorbed by land and water, and along with the energy absorbed directly from solar radiation, is reemitted into the atmosphere as heat and infrared radiation. This energy may then be reabsorbed by the atmosphere and the cycle repeats itself.

How much energy is recycled between the atmosphere and Earth's surface depends on the makeup of the atmosphere. For the Earth as a whole, the energy is recycled one to two times, but there is considerable variation seasonally, geographically, and with different weather events. Where the atmosphere is clean and dry, very little is recycled. Where the atmosphere is laden with clouds and heat-absorbing gases, large amounts are recycled. This is another way in which pollution comes into the picture. Carbon dioxide and several other gases emitted from pollution sources absorb infrared radiation and, in turn, help retain heat in the atmosphere.

Earth's Equilibrium Temperature

In order for the atmosphere to maintain its heat balance, it must give up (lose) as much energy as it gains over some period of time—say, a year (inset Figure 10.7). Energy loss is represented by the infrared radiation that escapes skyward. We know that this actually takes place because Earth maintains a fairly constant basal temperature worldwide, called the **equilibrium surface temperature**, at 15°C (60°F). The process by which the atmosphere gives up energy involves the transfer of infrared radiation to the upper atmosphere from which it is released into space.

Stepping back from the scene and concentrating on the top of the atmosphere, the Earth's energy system can be characterized by two huge energy flows: solar radiation coming in and infrared radiation going out. Most of the solar radiation is delivered to the tropical latitudes, whereas a relatively large amount of infrared radiation is released from the atmosphere in the higher latitudes (Figure 10.9). Alterations in this system could change the

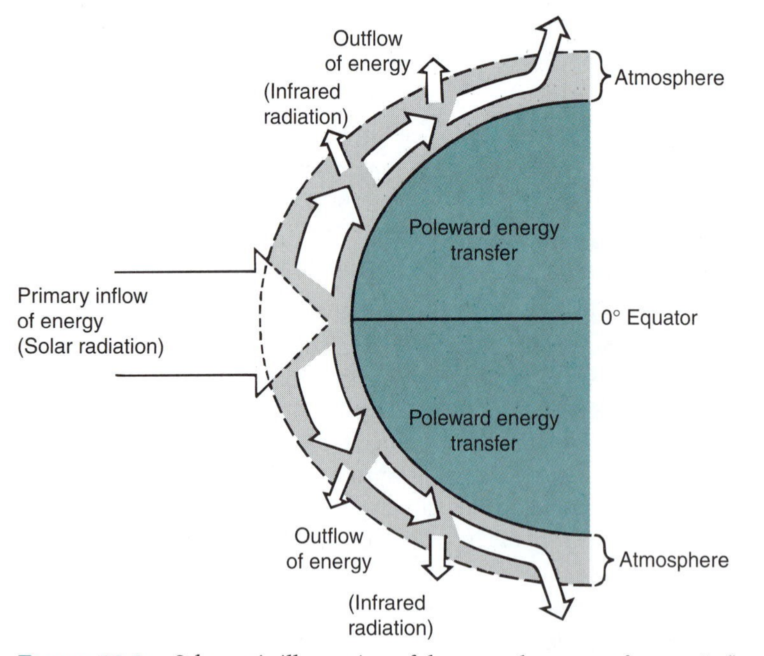

FIGURE 10.9 *Schematic illustration of the general pattern of energy inflow, latitudinal transfer, and outflow over the Earth.*

Earth's energy balance. Should the general conditions of the atmosphere change, such as an increase in heat-absorbing gases or cloud cover, the result would be a shift in the balance of the energy inflows and outflows. The atmosphere's heat storage capacity could, in turn, change, and this would be reflected in a higher or lower equilibrium temperature.

10.8 THE GREENHOUSE EFFECT AND GLOBAL WARMING

Many factors can cause changes in the Earth's heat balance, and we know that changes large enough to cause major episodes of warming and cooling have taken place in the past. One cause of change of great concern to scientists is the amount of heat-absorbing gas in the atmosphere, in particular carbon dioxide (CO_2).

Heat-Absorbing Gases

Carbon dioxide, water vapor, and several secondary gases absorb thermal infrared radiation released from the Earth's surface and the lower atmosphere. When absorption takes place, the gas molecules take on heat and air temperature rises. If the atmosphere is heavy with water vapor and carbon dioxide, then heat tends to be retained in much the same way as the glass of a greenhouse retains long-wave infrared radiation and heat. Hence, the term **greenhouse effect** is used to describe the role of CO_2 and H_2O vapor and various secondary gases in holding heat within the atmosphere. To underscore the importance of the greenhouse effect, scientists estimate that if it were eliminated the Earth's equilibrium surface temperature would fall 33°C to −18°C!

Since the 1700s the carbon dioxide content of the atmosphere has been increasing as a result of air pollution (Figure 10.10). Pollution has also caused a significant increase in **secondary greenhouse gases**, namely *methane* from animal digestion and bacteria, *ozone* and *nitrogen oxides* from urban air pollution, and *chlorofluorocarbons* from spray cans and fugitive refrigerants. The increase in CO_2 and the four secondary greenhouse gases is a global trend. According to the U.S. National Academy of Sciences, the biggest root cause is worldwide population growth.

Carbon dioxide is produced mainly by natural processes such as volcanoes and rock weathering, and is recycled through the atmosphere by the biological processes of respiration and photosynthesis. Over geologic time atmospheric CO_2 has varied appreciably, causing shifts in the Earth's heat balance. Over periods of thousands of years, however, the CO_2 cycle can maintain a fairly stable balance, thereby helping maintain a relatively stable global temperature regime. Carbon dioxide from human sources; however, represents a surcharge on the natural system.

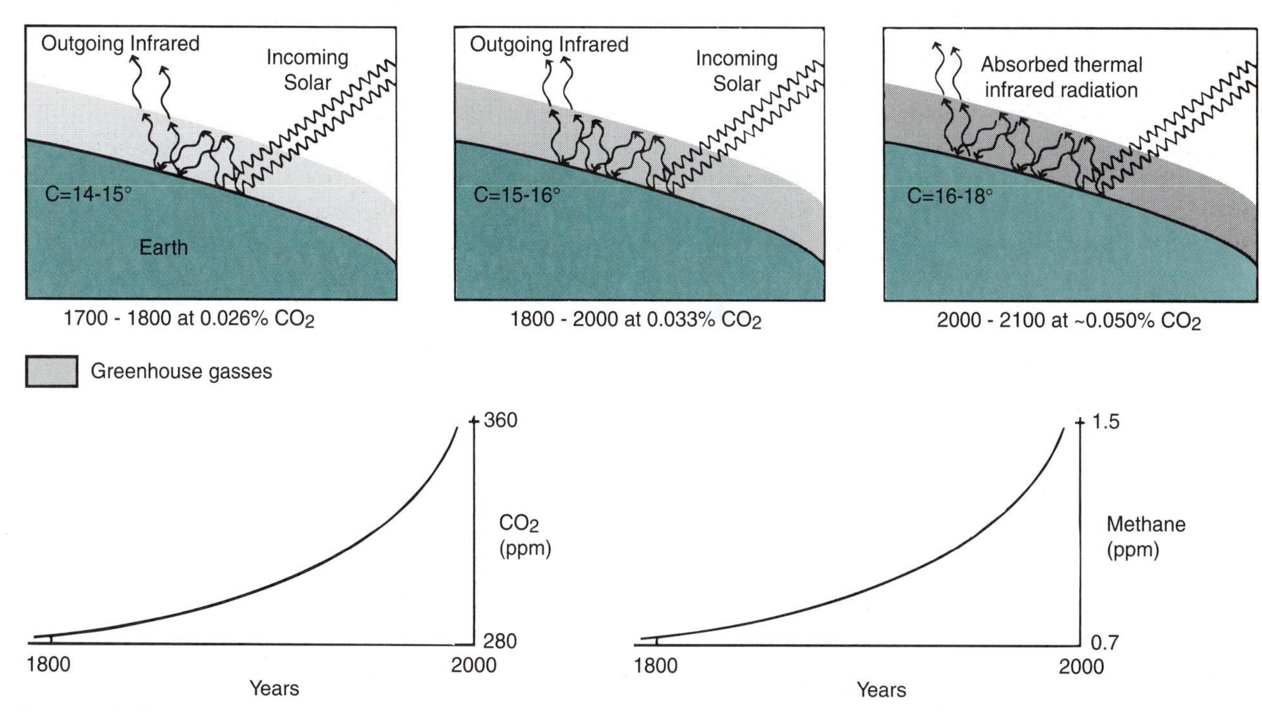

FIGURE 10.10 *Changes in the carbon dioxide and methane content of the atmosphere over the past two hundred years. The upper diagrams illustrate the possible corresponding changes in the Earth's greenhouse effect and equilibrium surface temperature.*

Human sources of carbon dioxide fall into roughly two major classes: fossil fuel combustion and open fires. The fossil fuel sources are industry, power plants, automobiles, and households, and the developed countries are the principal contributors. The open-fire sources are associated with the clearing of tropical forests, forest fires in general, grassland fires, household cooking and heating, and wars. The rate of increase in atmospheric CO_2 approximates that of global economic growth, about 2 percent annually. Current CO_2 levels are about 30 percent above the eighteenth-century (preindustrial) level. Many scientists argue that warming has already begun, and the vast majority *are reasonably certain* that if air pollution trends continue, **global warming** will be significant in the next century (Figure 10.10).

Trends and Forecasts

Actually there are signs that warming is underway, because the global atmospheric temperature has increased at least 0.5°C in the last 100 years (Figure 10.11). This increase corresponds to the measured increase in atmospheric CO_2. Although short-term temperature trends on the order of 50 or 100 years generally have little meaning with respect to global-scale warming, it is curious to note that studies of global temperature records dating from the late 1800s indicate that the 1980s and 1990s were the warmest two decades on record.

One of the reasons for scientists' uncertainty about how much global warming will take place is related to the rate of removal of CO_2 from the atmosphere. Carbon dioxide is removed from the atmosphere by the two main absorbing agents, vegetation and the oceans. Under natural (or at least preindustrial) circumstances, we can assume that the oceans annually extracted 90 billion tons of CO_2 from the atmosphere, an amount equal to the quantity they put into the atmosphere yearly. But with the rise of global air pollution, the oceans have been absorbing part of the 7 billion tons of excess CO_2 put into the atmosphere annually from human sources. The amount may be as great as 3 billion additional tons a year.

The net amount of carbon dioxide left to accumulate in the atmosphere is about 4 billion tons a year. At this rate, coupled with allowances for population growth, atmospheric CO_2 is expected to double by 2050. Whether the oceans and land vegetation will increase or decrease their intake of CO_2 in the future is difficult to say. Recent research reveals that our estimate of 3 billion tons of annual ocean intake may be high. The oceans may actually be taking in closer to 2 billion tons a year and some other agency is responsible for extracting the remaining 1 billion tons of carbon dioxide.

What are the candidate agencies? Perhaps the world's forests have increased their intake. Some studies suggest that air pollutants such as carbon dioxide and sulfur dioxide may actually increase forest growth and CO_2 intake in some regions. However, scientists are uncertain, and many additional questions loom on the horizon. Will tropical rainforest destruction reduce global CO_2 intake? The annual rate of rainforest destruction, as much as 65,000 square miles a year, is a considerable loss, especially when we recall that these forests have very high annual CO_2 intake capacities. On the other hand, can reforestation in the midlatitudes offset rainforest losses? Will the oceans reach a CO_2 saturation limit and take in less excess CO_2 at some point in the next several decades? Will the developed countries reduce their CO_2 output with pollution abatement programs? Will the increase in world population (currently more than 80 million people a year) and increased consumerism in developing countries offset any reductions that the developed countries might achieve?

If global warming proceeds as anticipated, by 2050 Earth's equilibrium surface temperature will be at least 3°C (5.5°F) higher than currently. This means that more thermal energy will be available to drive atmosphere processes such as winds, air mass movement, and evaporation. How all of this will interrelate to shape the broad

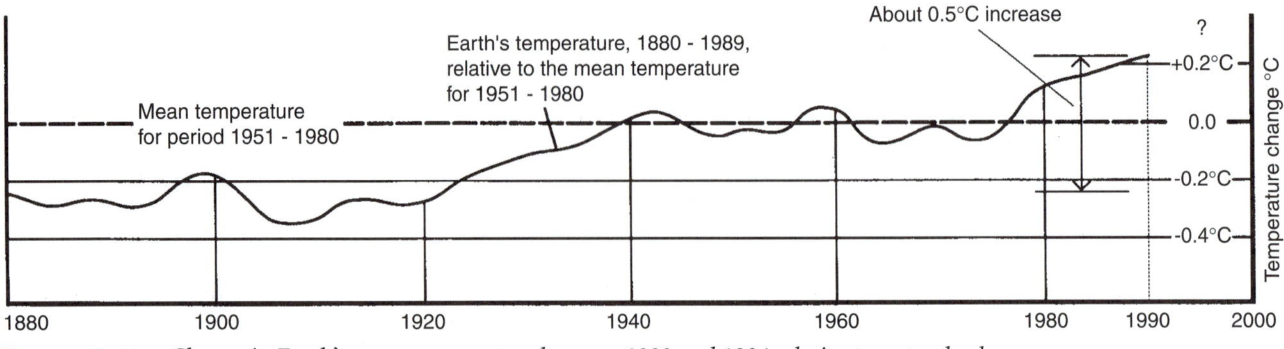

FIGURE 10.11 *Change in Earth's mean temperature between 1880 and 1984 relative to a standard mean temperature based on the period 1951–80. The trend shows an overall rise of 0.5°C.*

picture of global climate is difficult to forecast. Based on computer simulation models illustrating anticipated changes in atmospheric circulation, scientists forecast that warming will not be geographically uniform; in some regions it will be significantly greater than others, and the climatic changes will likely be different in different parts of the world. We will say more about the effects of global warming in the following sections.

A Counterargument

As we noted earlier there is a counterargument, supported by scientific data, that casts doubt on the global warming forecasts. It is based on the blocking effect on solar radiation caused by aerosols. **Aerosols** are particles of solids and liquids suspended in the atmosphere that backscatter incoming solar radiation in the manner depicted in the lower diagrams of Figure 10.8. Aerosols are produced from both natural sources such as volcanoes and pollution sources such as forest fires and urban emissions. They occur in both primary and secondary forms, meaning that some enter as particles whereas others develop into particles from gaseous pollutants. In addition, as we noted earlier, pollutants increase condensation and cloud formation, thereby increasing reflection of solar radiation. Since aerosols and clouds reduce solar radiation receipt in the lower atmosphere and at the Earth's surface, they also reduce solar heating and may cause temperatures to decline as a result.

With the dramatic increase in global air pollution in this century, the atmosphere has grown dirtier and less transparent to solar radiation. Less solar energy reaches the Earth's surface, and if there were no counterbalancing factors, the lower atmosphere would probably be growing cooler. But there *are* counter factors in the form of CO_2 and the other warming gases. Scientists attempt to define the relative balance between the warming and cooling effects of these pollutants. At the current time, the international scientific community leans toward the CO_2 argument and forecasts that warming will outweigh the cooling effects of aerosals, resulting in general global warming.

Some Consequences of Atmospheric Warming

Given a trend toward global warming, what will the environmental effects be? The increased heat content of the troposhere will strengthen certain air masses, increasing their moisture content, leading to greater storminess in some regions. We already know, for example, that storminess and precipitation on the Pacific Coast of the Americas increase substantially with only 1–2°C of atmospheric warming associated with El Niño, the periodic buildup of warm water in the east Pacific near the equator. In other regions, such as under the Subtropical High Pressure cells, the increase in heat may have a stabilizing and drying effect. And these climatic changes, in turn, will lead to biogeographical changes as temperature and moisture sensitive plants and animals shift with the changing habitat.

Circulation Changes

Large-scale circulation may also be altered by global warming. The Equatorial Low Pressure system, which is driven in large part by heat in the lower troposphere, could intensify, forcing more air into the upper troposphere to feed wind systems aloft. The return flow for much of this air is the great Subtropical High Pressure cells on the poleward margins of the tropics. Returning air enters these high pressure cells from aloft, and it is very dry by the time it sinks to the ground, which accounts for the arid conditions at the surface (see Figure 4.7). If the flow of returning air is increased, then the intensity and geographic size of the arid zones may increase, as well as shift poleward.

Forecast based on complex models of global circulation show poleward shifts in the major climatic zones, including expansion of dry zones and perhaps an increase in the magnitude and frequency of drought periods. As some deserts expand, the fringing semiarid landscapes will be transformed into arid ones with drier soils and sparser plant covers. Where lands in these regimes are already under stress from overgrazing, deforestation, and expanding agriculture, desertification may accelerate. By contrast, some regions may experience greater precipitation, longer growing seasons, and increased plant productivity. These regions may also experience increased runoff, soil erosion, and flooding, however. Overall, current scientific forecasts of the location, extent, severity, and effects of regional climatic changes are very uncertain. Other changes, however, are more certain; one is a rise in sea level.

Hydrological Changes

Atmospheric warming is expected to cause a shift in the balance of global ice and ocean water. Several degrees of atmospheric warming will increase rates of melting of the ice caps and glaciers, which hold most of the planet's water supply outside the oceans. Warming will also cause thermal expansion of the ocean water mass. These two changes will produce higher sea levels and flooding of low-lying coastal lands worldwide. Melting of just half the world's current volume of ice would produce a rise in sea level of 40 meters (130 feet).

Given the latest forecasts of atmospheric warming, by 2050 sea level will be 3 to 6 meters (10 to 20 feet) higher. Some population centers would be greatly affected. Witness, for example, that most of New Orleans is at or

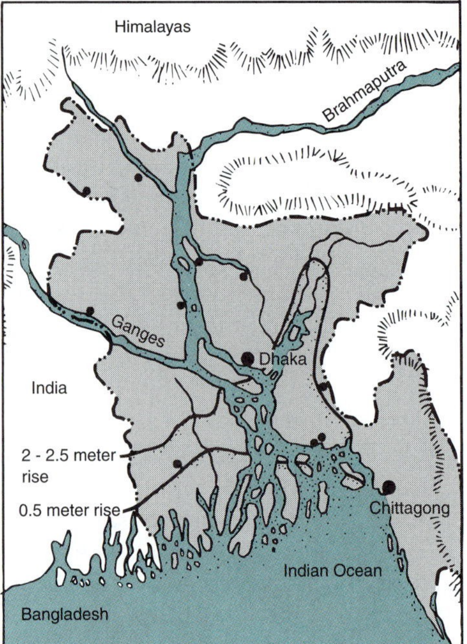

FIGURE 10.12 *Hurricane (typhoon) flooding and devastation in Bangladesh. Hundreds of thousand of people were killed by hurricanes in Bangladesh in the 20th century. Inset shows area flooded by a sea level rise up to 2.5 meters.*

below sea level; most of the 14 million people of the Netherlands live at elevations ranging from a few meters above to several meters below sea level; Alexandria, Egypt, lies only 1 meter above sea level; and most of New York City and Hong Kong lie within 20 meters of sea level.

A modest sea-level rise of 2 to 3 meters would be devastating, affecting several billion people worldwide by 2050. What is more, a rise in sea level would also lead to greater coastal damage from hurricanes and other storms, especially in low areas such as Bangladesh, because storm waves would penetrate the shoreline with greater force (Figure 10.12) Situated on the delta of the Bramaputra and Ganges rivers, about 28 percent of Bangladesh and tens of millions of people lie less than 2 meters (7 feet) above sea level. Millions of people will be displaced, and in this crowded, poor country, there is nowhere for them to go to reestablish their lives.

Biological Changes

The effect of a sea-level rise on coastal ecology could also be devastating, because the forecasted rate of rise will be too fast for many intertidal, wetland, and other shallow-water communities to adjust. These communities live in delicate balance among many factors (e.g., light, water temperature, water motion, and sediment supply) that are controlled directly or indirectly by water depth. And just as with land use in low-lying areas, storm damage, wave erosion, and sediment movement will increase as water deepens, which will also weaken and eliminate many coastal communities of plants and animals.

Humans will also be affected by changes in the geographic distribution of infectious diseases such as malaria and dengue fever (Table 10.2). These diseases are spread by organisms, called **vectors**, such as mosquitoes, flies, and snails, that are sensitive to temperature and moisture. Most are tropical and subtropical in origin, and it is likely that their geographic ranges will expand in some areas with global warming. Malaria, for example, which is spread by mosquitoes, will likely increase as temperatures and rainfall increase in equatorial Africa, South America, and Southeast Asia.

Biodiversity is another concern related to climate change. Climate change will result in habitat changes that will reduce the ranges of many species, weaken their populations, and reduce their competitive advantages with other species. This may drive susceptible organisms, such as threatened and endangered species, to rapid extinction. If the rate of climate change is rapid, some species may not be able to shift their ranges quickly enough to

TABLE 10.2

LIKELIHOOD OF CHANGE IN INFECTIOUS DISEASE DISTRIBUTION WITH GLOBAL CLIMATE WARMING

DISEASE	Vector	People at Risk (millions)	Number Infected per year	Geographic Distribution	Likelihood of Distribution Change with Climate Change
Malaria	Mosquito	2400	300 million to 500 million	Tropics/subtropics	Highly likely
Schistosomiasis	Water snail	600	200 million	Tropics/subtropics	Very likely
Lymphatic filariasis	Mosquito	1094	117 million	Tropics/subtropics	Likely
African trypanosomiasis	Tsetse fly	55	250,000 to 300,000 cases/year	Tropical Africa	Likely
Dracunculiasis	Crustacean (copepod)	100	100,000/year	South Asia/Middle East/Central and West Africa	Unknown
Leishmaniasis	Phlebotomine sandfly	350	12 million infected, 500,000 new cases/year	Asia/South Europe/ Africa/Americas	Likely
Onchocerciasis	Blackfly	123	17.5 million	Africa/Latin America	Very likely
American trypanosomiasis	Triatomine bug	100	18 million to 20 million	Central and South America	Likely
Dengue fever	Mosquito	2500	50 million/year	Tropics/subtropics	Very likely
Yellow fever	Mosquito	450	<5,000 cases/year	Tropical South America and Africa	Very likely

Source: World Health Organization (WHO), *Climate Change and Human Health*, A.J. McMichael, *et al.*, eds. (WHO, Geneva, 1996).

survive. Land-use barriers such as highway corridors, urbanized areas, and belts of cropland stand in the way of plants and animals as they migrate in response to changing climate patterns (see Figure 10.14).

10.9 THE EFFECTS OF URBAN CLIMATE

Climatic change can also be caused by alterations of the landscape as a direct result of human land-use activity. In the natural or rural landscape, climatic conditions near the ground are strongly influenced by vegetation, soils, and moisture. These materials control much of the landscape's ability to absorb solar radiation, conduct heat, and generally exchange energy with the atmosphere. When they are destroyed or covered over by asphalt, concrete, and buildings, the flow of energy is altered at the base of the atmosphere, resulting in climatic change, that is, long-term changes in temperature, precipitation, wind, and related factors.

Thermal Conditions

The magnitude of urban climate change depends on the type of land use and the size of the area covered. With the possible exception of war, urban development is the most consuming land use yet devised by humans. Not only are forests, soils, and wetlands eradicated or greatly reduced by urbanization, but a new topography in the form of tall, closely spaced buildings made up of artificial materials is added to the landscape. The urban topography alters the flow of air over the landscape by lowering wind speed near the ground and, in turn, reducing the rate at which heated air is flushed from the city. The artificial materials change the thermal properties of the land so that the surface reaches much higher temperatures with the absorption of solar radiation than it could with natural materials.

Of the heat generated in the urban landscape, most is released to the atmosphere as *sensible heat* rather than *latent heat*. Sensible heat, the heat of dry air, causes air temperatures to rise, whereas latent heat, the heat of moist air, does not. Latent heat is transferred into air with the vaporization of water from plants, soils, and water features. Because urban development largely destroys or covers these moisture sources, relatively little latent heat is produced from the urban landscape. Instead, the urban landscape generates mostly sensible heat, which drives up the air temperature.

In addition, the composition of the air over the city is altered by pollution so that its capacity to transmit radiation and heat is reduced. Airborne particles from various sources of combustion form an envelope of dirty air, called a **dust dome**, over the city that backscatters solar radiation, reducing the total amount of solar energy at times by 50 percent or more (see Figure 11.5). At the same time, heat and radiation flowing skyward are slowed by the polluted air. Added to this is a large amount of fugitive heat released into the atmosphere near ground level from heated buildings, industry, and automobiles.

The net effect of all these changes is a warmer climate over urban areas. On a thermal map of regions containing metropolitan landscapes, urban centers stand out as **heat islands** against the cooler rural environments surrounding them (Figure 10.13a). The size and intensity of

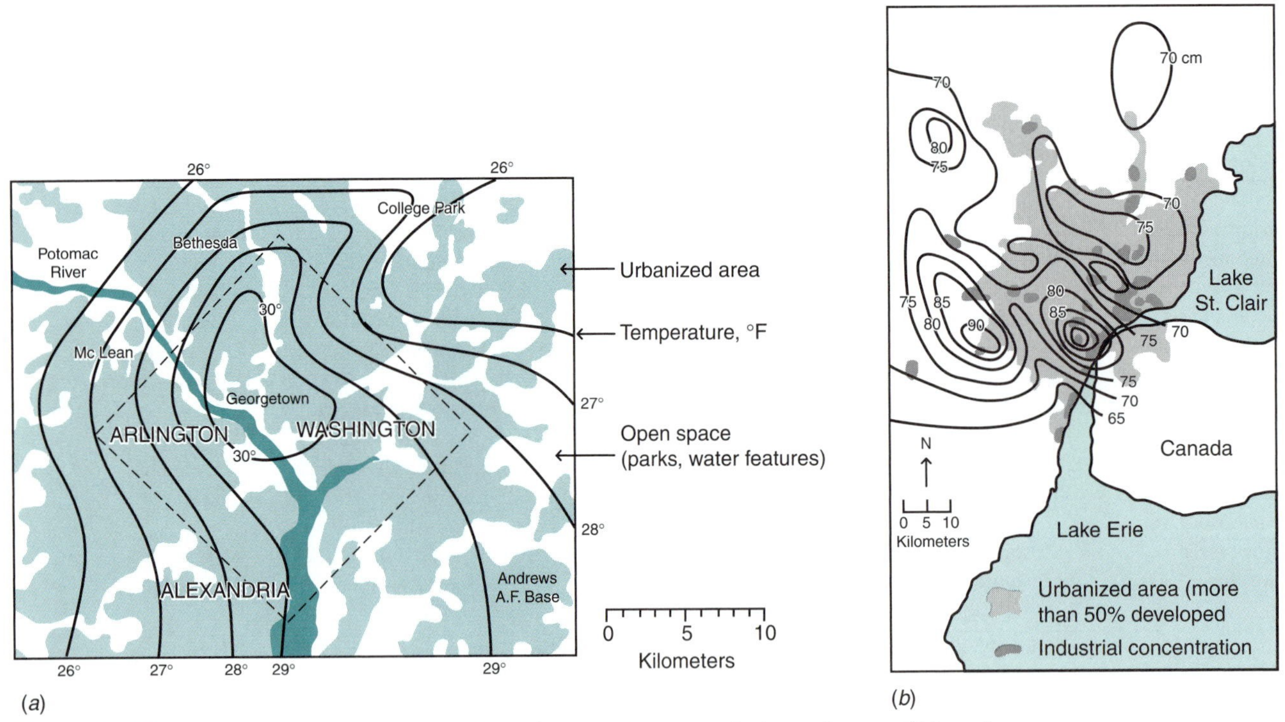

FIGURE 10.13 *(a) Heat island associated with Washington, D.C., a typical transient condition of large cities; (b) Precipitation patterns in the Detroit/Windsor (Canada) metropolitan region, also typical of many large cities.*

urban heat islands depends on city size (population), the geographic setting, and regional weather conditions at the time of measurement. The heat island is most pronounced during periods of calm weather when the interiors of large cities may be 5–10°C (10–18°F) warmer than surrounding regions. The most extreme conditions occur during summer heat waves, when the inner city may reach daily high temperatures of 40–45°C (105–115°F).

Precipitation Trends

Besides the higher temperatures, reduced solar radiation, and lower windiness, the urban climate is also characterized by greater precipitation (Figure 10.13b). The explanation for this has nothing to do with sources of moisture—indeed, the city is a much drier landscape than the rural landscape—but with air pollution. The particulates that make up the urban dust dome also provide condensation nuclei for precipitation droplets. Such nuclei are a necessary ingredient in precipitation formation, and pollution particles seem to work as well as natural particles in this process. In addition, the heat island over the city contributes to atmospheric instability (rising air), which enhances the precipitation processes.

The urban atmosphere, therefore, facilitates cloud development, fog formation, and precipitation. The frequency of fog is more than 100 percent greater in some urban areas than in comparable rural areas. Average annual precipitation is significantly greater over some large city centers and adjoining industrial zones compared to surrounding suburban and rural areas. For some cities, the zone of increased precipitation does not fall over the city itself but many kilometers downwind as the urban dust dome is blown into a large plume by prevailing winds.

Growth of Megacities

Since the 1930s there has been a massive shift in population from rural to urban areas. Today about 75 percent of the people and the vast majority of industrial and commercial land uses in the United States are located in and around urban centers. Cities have sprawled over such vast areas that many have merged together, forming **megacities** (Figure 10.14). The areas dominated by urban climate are no longer limited to small, isolated spots, but have spread to relatively large areas. This has happened, for example, on the East Coast in the New York City region and on the West Coast around both Los Angeles and San Francisco. Together these three metropolitan regions are populated by more than 30 million people, about 12 percent of the United States population.

While cities in the United States continue to grow through suburban sprawl, extending along highways to cover larger areas, the cities in developing countries are experiencing the fastest population growth. In these countries, the population trend is inward toward urban

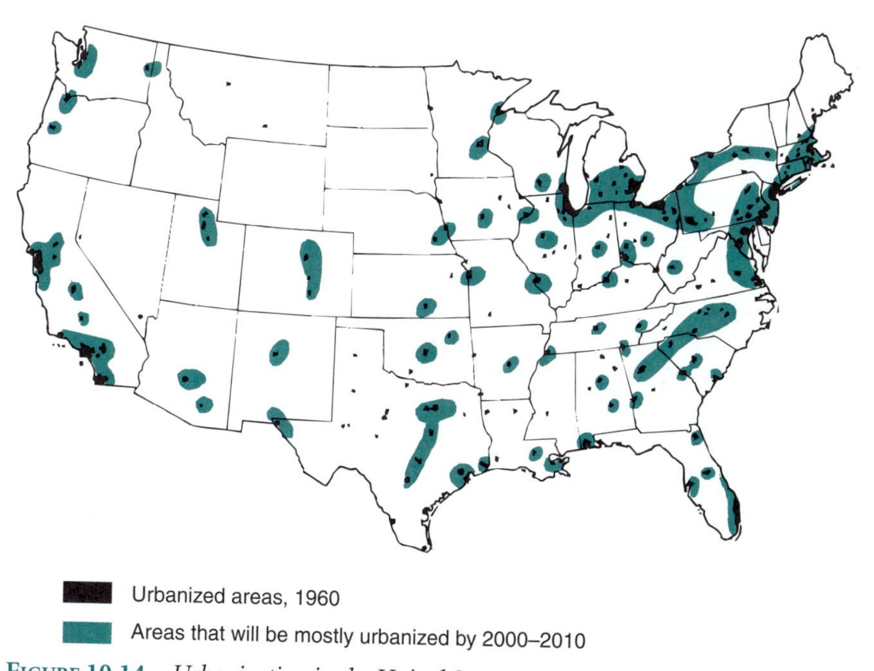

FIGURE 10.14 *Urbanization in the United States. Since 1960, cities have spread along highway corridors merging together to form great urban belts.*

centers—that is, from rural areas to the city—and the rate of growth is dramatic. Most of the people settle in shantytowns on the perimeter, where the original landscape of stream valleys, wetlands, and farmlands is replaced by makeshift structures, debris, and denuded surfaces (see Figure 7.11). Air pollution increases as well. Although automobile emissions are not greatly increased, air pollution from domestic fires, waste burning, and small industrial operations increases significantly. On balance, the urban climate and its various manifestations are destined to intensify in poor countries as cities such as Mexico City expand to accommodate more than 30 million people in this decade.

Heat Syndrome

In the past 30 years more than 10,000 people in the United States have died from heat syndrome. **Heat syndrome** is the term applied to several clinically recognized disorders in the heat regulation system of the human body. It can be caused by a number of contributing factors: excessive air temperatures, heating from extreme exposure to solar radiation, physical exertion, and various physiological disorders including low body fluids, salt imbalances, and respiratory problems.

Two environmental conditions contribute heavily to heat syndrome: regional heat waves and urban climate. Heat waves are usually associated with increased mortality, especially among the elderly, and the hazard increases in the more heavily built-up inner cities where the heat island is strongest. This is illustrated by the distribution of deaths in St. Louis, Missouri, during the July 1966 heat wave (Figure 10.15). These deaths were attributed directly to the heat; how many more were indirectly related to the heat is difficult to say, but undoubtedly the number would be significant.

Air pollution also has a major impact on the health and well being of the urban population. Respiratory disorders such as asthma and emphysema are high on the list of pollution-related diseases. Figure 11.10 in the next chapter compares representative air pollution levels in the inner city with those in suburban and rural locations. Inner-city levels are significantly higher, and data show that medical visits and costs for respiratory diseases are also higher there. Chapter 11 deals with the effects of air pollution on human health in greater detail.

Improving the Urban Climate

Improvement of climate-related health conditions in cities requires two major changes: (1) reductions in air pollution, and (2) modifications in the way we design and build cities. Reduced air pollution will not only improve public health, but reduce the impact of urban climate on the global climate, including atmospheric warming and the production of acid rain. The design of cities could be changed to bring about not only better air quality but less heat stress. This would require integrating belts of open space into built-up areas to improve ground-level airflow and also improving the balance between natural and artificial materials to reduce surface temperatures. Both pollution and thermal conditions

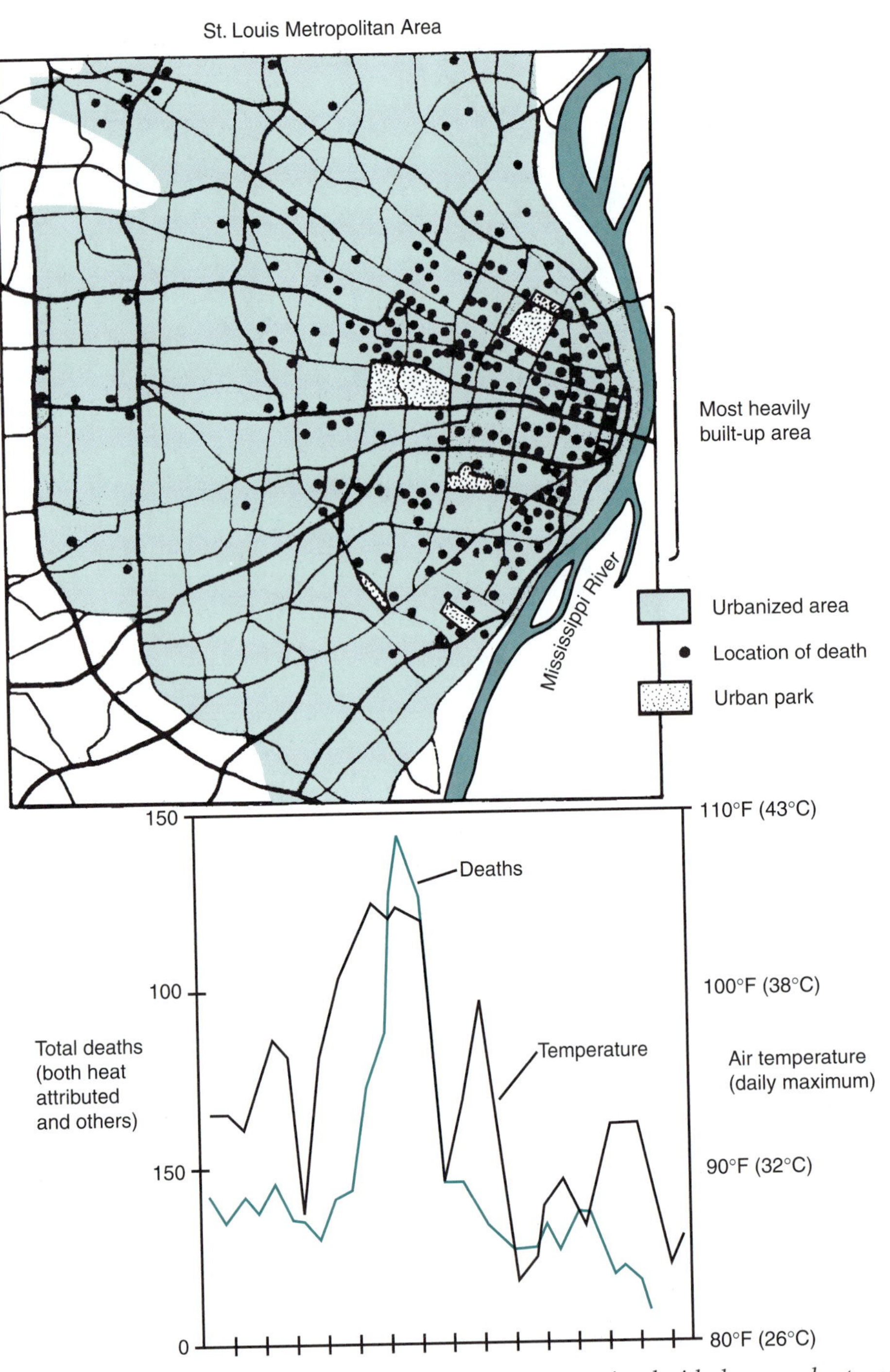

FIGURE 10.15 *Pattern of deaths in St. Louis, Missouri, associated with the severe heat wave of July 1966. The incidence of death was higher in the more heavily built-up urban area.*

would be improved by reducing automobile activity in the interiors of most cities.

10.10 MANAGING THE GLOBAL CLIMATE

Although climate and atmospheric processes have an enormous effect on life—shaping the distribution of plants, animals, and land use over large reaches of the Earth—until the 1980s little attention was given to managing this part of the global environment. There were several reasons for this, including doubts that human activities had much influence on the general condition and workings of the atmosphere and doubts that anything as large and complex as the atmosphere could be influenced by managing human activity on the land. The atmosphere is indeed complex and the relationships to human activities are not easily defined. Understandably, the scientific community has not enjoyed widespread consensus on what the effects of air pollution and land-use change and other factors are and should be on climatic conditions. But this is the nature of science. Complacent national and world leaders have pointed to this lack of agreement as cause for inaction in managing the factors known to alter the thermal balance, radiation, and composition of the atmosphere.

In the 1980s the international political scale began to tip as global leaders took a serious look at the trends in ozone destruction by CFCs and the buildup of carbon dioxide from air pollution. Resolutions passed as a part of international meetings, first the Montreal Protocol in 1987 and later in the 1992 United Nations Conference on Environment and Development (UNCED) at Rio de Janeiro, agreed to formulate programs for limiting the discharge of these pollutants into the atmosphere.

While these actions are encouraging, the general causes of climatic change related to human activity loom larger every year. The single greatest root cause is global population growth. Each year more than 80 million people are added to the planet. Most are added to poor countries, where they will clear land, burn wood, drain wetlands, and enlarge cities. These actions, in turn, will contribute to further alteration of the atmosphere both locally and regionally. Population growth in the rich countries is slower, but this trend is offset by higher levels of consumption: More automobiles, roads, sprawling cities, and energy use.

Overall, these activities, through both the effects of airborne pollutants and landscape alteration, are leading to significant changes in the atmosphere. In addition, in both poor and wealthy countries population is pushing into marginal environments that are naturally subject to severe droughts, heavy flooding, and intensive storms such as hurricanes. Coupled with the climatic changes that seem likely to occur in this century, people living in these marginal environments will be increasingly susceptible to larger and larger disasters from atmospheric events and related processes.

The impending conflict with the sea bears repeating. More than half the world's population is located within 100 kilometers (60 miles) of the sea. The world's largest cities and the most densely populated countries are found on the sea coast. Population trends reveal that there is a major shift of people toward the sea. The United Nations estimates that 20 to 30 million in the world's poor countries migrate to cities each year, many of which are coastal cities situated only 3 to 6 meters (10 to 20 feet) above sea level. Even if sea level remained constant, more people will be prone to storms, flooding, and other coastal disasters than ever before. If sea level rises as forecast with global warming, the disasters will be significantly larger and more frequent.

Air pollution control is a major route to managing the atmospheric environment, and it must include a substantial reduction in emission levels. The strategies already in place focus mainly on cleaner fuels, greater energy efficiencies for cars and factories, and filter devices on exhausts. It should also include a population management strategy, because much of the progress made in pollution control is offset by annual population growth and increased consumption. For example, between 1970 and 1989, when great improvements were made in automobile pollution emission rates, the number of miles traveled increased by 62 percent in the United States. A similar trend occurred in other countries.

It is also necessary to manage landscape change, especially the replacement of vegetated surfaces with hard, dry surfaces in the form of bare soil, roads, buildings, and other land-use facilities. The urban climates we now associate with large cities may be modest in magnitude compared to those of urban centers of the future (e.g., Mexico City, with a population expected to reach 30 million in this decade). The uncontrolled spread of cities and the related degradation of the surrounding environment are, by definition, a form of desertification, significantly drier and ecologically poorer than the landscape they replaced.

10.11 SUMMARY

The atmosphere is a mixture of gases held to the Earth by gravity. Nitrogen and oxygen are the two principal gases in air, making up nearly 99 percent by volume. Many minor gases constitute the remainder of the atmosphere, and two of the most important are carbon dioxide and ozone. Carbon dioxide is an important heat absorber and is thus critical to the maintenance of the Earth's greenhouse effect. Ozone, which is concentrated in the upper atmosphere, absorbs ultraviolet radiation, which in large doses is lethal to humans and many other lifeforms. Both carbon dioxide and ozone are being affected by air pollution resulting in a stronger greenhouse effect and a weaker ultraviolet screen over the Earth.

Water and airborne particles are also important minor constituents of the atmosphere. Water occurs in the form of vapor, an invisible gas, and droplets that make up clouds and fog. The water vapor content of the atmosphere is controlled by air temperature and moisture availability from surface sources. Polar air is dry mainly because it is cold, and desert air is dry mainly because of limited moisture supplies. Particles (dirt) from both natural and pollution sources are abundant in the atmosphere, and both affect the transparency of air and serve as nuclei in the formation of precipitation droplets.

The structure of the atmosphere can be described according to the vertical distribution of air pressure and temperature. Surface pressure at sea level averages 1013.25 millibars and decreases sharply with altitude. At 5.5 kilometers (3.3 miles) altitude, a height exceeded by hundreds of mountains, pressure is about 500 millibars. The thermal structure of the atmosphere is characterized by alternating zones in which temperature either increases or decreases with altitude. The lowermost zone is the troposphere, the zone of water vapor, active weather, and life.

Solar radiation is virtually the sole supply of energy to Earth. This energy heats the atmosphere and the Earth's surface and drives most of the essential processes such as winds, currents, evaporation, and plant growth. For the atmosphere to maintain its heat balance, it must release as much heat as it gains. Release takes place when heat is converted into thermal infrared radiation and transmitted through the atmosphere into space. The rate of release is controlled principally by carbon dioxide and water vapor, the main components of the Earth's greenhouse effect. As the carbon dioxide content of the atmosphere increases because of air pollution from a rapidly expanding world population, the greenhouse effect is likely to intensify. If the trend continues through the mid-twenty-first century, the result will be global atmospheric warming of at least 3°C (5.5°F). The environmental consequences could be severe, including changes in storminess and precipitation, shifts in the climatic belts, increased regional aridity, rising sea level, and changes in the distribution of infectious diseases.

Although scientists forecast global warming with carbon dioxide loading of the atmosphere, there is much uncertainty about the magnitude of warming. The uncertainty stems from evidence of countereffects (i.e., cooling effects) from increased aerosols and from possible variations in the extraction of carbon dioxide by the oceans and land vegetation. There is also uncertainty related to the ability of the developed countries to implement pollution-control programs.

Modern urban development alters the landscape and the atmosphere over it so dramatically that climatic change is the result. Compared to the surrounding rural landscape, urban climate is characterized by higher temperatures, less solar radiation, greater precipitation and fog, and lower wind speeds at ground level. These conditions threaten the health of urban residents, especially those of the inner cities, and call for alternative ways of designing and building today's cities.

10.12 KEY TERMS AND CONCEPTS

Atmosphere as buffer
Composition of air
- nitrogen
- oxygen
- carbon dioxide
- ozone

Water vapor and airborne particles
Thermal structure
- troposphere
- stratosphere
- mesosphere
- thermosphere

Atmospheric pressure
- changes with altitude

Wind and pressure systems
- pressure gradient
- global pressure belts
- prevailing wind systems
- Coriolis effect
- global air pollution

Solar radiation
- solar constant
- absorption and reflection
- electromagnetic spectrum

Breakdown of the solar beam
- diffuse solar radiation
- beam solar radiation
- albedo
- cloud reflection

Earth's heat balance
- equilibrium surface temperature

Greenhouse effect
- heat-absorbing gases
- secondary greenhouse gases
- global warming
- aerosols
- circulation changes
- hydrologic changes
- disease vectors

Urban climate
- dust dome
- heat island
- precipitation
- megacities
- heat syndrome

Managing the atmospheric environment
- international agreements
- population trends
- environmental conflict

10.13 QUESTIONS FOR REVIEW

1. Describe the composition of the atmosphere and role of carbon dioxide and ozone in the atmosphere's heat and radiation system.
2. What are the two forms of water in the atmosphere, and what factors control the amount of water vapor in the atmosphere?
3. Name the four main divisions of the atmosphere. What is the thermal characteristic of each, and what features make the troposphere distinctive?
4. What are transient and semipermanent pressure systems, and what are the two classes of the latter based on cause of origin? Name the four great belts of global pressure and draw a diagram showing the corresponding prevailing wind systems. What role do the global pressure and wind systems play in shaping air pollution patterns?
5. Describe what the radiation and thermal characteristics of the Earth's surface would be like without an atmosphere. What are the solar constant and the electromagnetic spectrum? What are the three main classes of solar radiation?
6. What is the role of the atmosphere in breaking down the solar beam? Name the key processes affecting the breakdown. What percentage finally reaches the surface as diffuse solar radiation and beam solar radiation?
7. What is *albedo*, and how does it vary with different surface materials? Can it be influenced by land use and cover changes? Can cloud cover and air pollution change atmospheric reflectivity and, in turn, the receipt of solar radiation?
8. What is meant by the concept of the Earth's heat balance? In which vertical division of the atmosphere and which climatic (latitude) zone does most of the heating take place? What is Earth's equilibrium surface temperature and how does the Earth maintain this temperature?
9. Describe the greenhouse model of the atmosphere. What major and minor gases affect the atmosphere's heat-absorbing capacity? What is the trend in atmospheric carbon dioxide since the eighteenth century, and what do you think about the likelihood of global warming and climatic change in the twenty-first century?
10. What factors (agencies) could play a part in the future carbon dioxide balance of the atmosphere? What is your evaluation of the kind of influence they will have (increase, decrease, or unchanged), and why? Describe the counterargument against global warming based on aerosol pollution.
11. Describe the major environmental changes that may result from global warming in the twenty-first century. Discuss the implications of accelerated desertification and rising sea levels for human population in light of a world population in 2050 at least twice that of today's.
12. What is urban climate, what conditions produce it, and what are its major manifestations? What are megacities, and what are some of the significant effects of urban climate on the quality of human life?

11

AIR POLLUTION: PATTERNS, TRENDS, AND IMPACTS

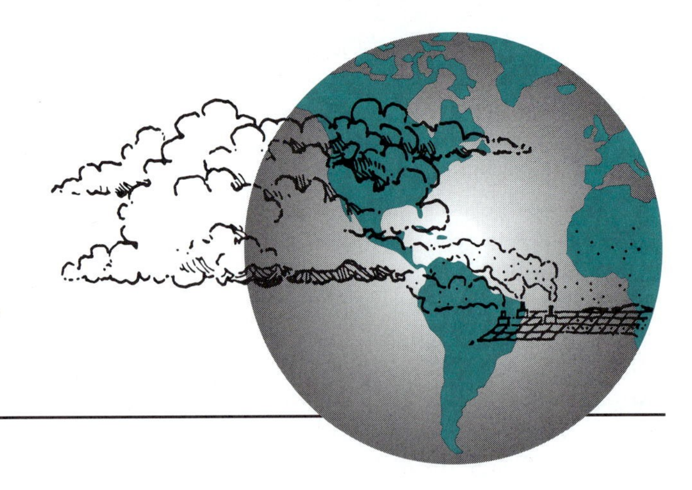

11.1 Introduction

Whenever society uses energy and matter there are both desirable and undesirable products. The desirable products are marketable goods and services. The undesirable products (or, more appropriately, the byproducts) are the discarded materials and energy given up in the course of performing work. Most environmental pollution comes from these byproducts. In olden times it was the heat, smoke, and ashes from the blacksmith's fires in his shop and home; in modern times it is, among other things, the heat, smoke, and related waste from one's home, workplace, school, and automobile and the dozens of support facilities and operations behind each.

There is no doubt that air pollution existed in ancient times, but it was much less severe and less widespread than today. Three factors account for this: First, human populations were much smaller then. Few people lived in cities; most people were tied to agriculture and they were lightly dispersed over the land. At the time of Christ, world population was only about 250 million, about 5 percent of today's population. As recently as 1900, most people in the United States and Canada lived in rural areas; today about 75 percent of these countries' more than 300 million people live in sprawling urban areas dependent on massive energy and material systems.

Second, before industrialization consumption per person was a small fraction of what it is today especially in terms of consumer goods. Energy and material use per capita was perhaps less than 1 percent today's rate for developed countries. Third, the types of materials used by ancient societies were very basic and did not include the specialized and exotic substances used by today's advanced technologies. Prior to 1900, mostly natural forms of raw materials—wood, glass, coal, iron, and lead, for example—were used in manufacturing. Today many synthetic materials such as plastics are developed in research laboratories from compounds that are not found in nature. Tens of thousands of these substances are used in manufacturing, and many end up as waste residues in the atmosphere, from which they enter respiratory systems, food chains, and ecosystems, causing a wide range of impacts that includes disease.

In this chapter we examine air pollution, beginning with the types of pollutants and their sources in developed and developing countries. This is followed by a discussion of the processes and patterns of atmospheric pollution, including the distribution of pollutants in the urban region. The latter part of the chapter deals with the impacts of air pollution on the environment and society and our efforts to control air pollution.

11.2 TYPES OF AIR POLLUTION

As we saw in the last chapter, 99 percent of the atmosphere is made up of two gases: nitrogen (78 percent) and oxygen (21 percent). The remainder is shared by a number of minor gases, including carbon dioxide and ozone, as well as particulate matter such as dust, salt particles, and water. Into this great mixture of gases and particles thousands of pollutants are discharged continuously from human activities and certain natural processes around the world. Those from natural sources, such as dust from soil erosion and carbon dioxide from volcanoes, combine with similar materials from human sources. Their impacts on the environment must ultimately be considered together, but our concern is chiefly with the fraction from human sources.

Pollutants enter the atmosphere as both gases and particles. Most pollution emissions, whether cigarette smoke, automobile exhausts, or forest fires, discharge gases and particles together. Once in the atmosphere, several different things may happen to these materials. They may settle to Earth, mix higher into the atmosphere, or undergo chemical alteration and change into other forms of pollution. Those that result from alteration of **primary pollutants** are called **secondary air pollutants.** Acid rain is a form of secondary pollution because it results from processes occurring within the atmosphere after primary pollutants such as sulfur dioxide are emitted from exhaust stacks and other sources.

The Major Pollutants

Of the great mass of pollutants discharged into the atmosphere we designate only a handful as major pollutants. They are singled out for two main reasons: (1) they are known to have a significant impact on environment and/or human health; and (2) they are produced in relatively large quantities by human activity over much of the Earth.

- **Sulfur dioxide.** This is a corrosive gas emitted in the burning of fossil fuels. It is harmful to human health and building materials and is a major contributor to acid rain. Coal-fired power plants are a major source of SO_2.
- **Nitrogen oxides.** Nitrogen oxide and dioxide are two gases produced from high-temperature combustion that are harmful to human health and also contribute heavily to acid rain. Nitrogen oxides (NO_x) are produced mainly by motor vehicles and power plants that burn fossil fuels.
- **Carbon monoxide.** This colorless, odorless gas is harmful to humans because when inhaled, it inhibits the absorption of oxygen by the blood. CO is also a byproduct of fossil-fuel combustion.
- **Carbon dioxide.** This gas absorbs longwave (thermal) radiation and in turn warms the atmosphere. It is released with the burning of fossil fuels, forests, and other vegetation.
- **Particulates.** Solid and liquid airborne particles of a wide range of sizes and compositions are known to cause a variety of illnesses, including lung cancer and respiratory disorders such as emphysema and asthma. Polluting particles include soot and ash from burning, lead from motor vehicle fuels, and pollen from plants.
- **Hydrocarbons.** A wide variety of carbon-based gases, including methane and benzene, are emitted in fossil-fuel combustion from automobiles, home heating, and industry when burning is incomplete. As a primary pollutant they are a major health hazard; they are also a source of secondary pollution because they contribute to the formation of oxidants such as ozone.
- **Oxidants.** These are part of a variable mixture of secondary pollutants commonly associated with photochemical smog in major urban areas, formed by the action of solar radiation on hydrocarbons and nitrogen oxides from automobiles and industry. They pose health risks ranging from eye irritation to cancer. Ozone is one of the chief oxidants.
- **Acid deposition.** Various oxides, led by sulfur dioxide and nitrogen oxides, combine with water in cloud droplets and precipitation to lower the pH (relative acidity) of atmospheric moisture. Precipitated on the land, oxides lower pH levels in freshwater, eliminating many species from ecosystems in lakes and streams. Motor-vehicle exhaust is a major source of NO_x and coal-fired power plants are a major source of SO_2.
- **Synthetic compounds.** A wide variety of manufactured organic compounds such as CFCs, DDT, PCB, and dioxins occur in both gaseous and particulate forms. They enter the atmosphere via combustion and wind erosion and are known causes of various disorders, including cancer in humans and other organisms.
- **Radioactive substances.** Particles such as ash from coal-burning, gases such as radon, and waste from nuclear operations emit harmful radiation. These cause cancer and genetic alterations in humans and other organisms.

- **Heat**. This is thermal loading of air as a result of direct heat loss from combustion sources such as power plants and automobiles and secondary heating as a result of other changes, such as decreased ventilation in cities or carbon dioxide pollution. See the previous chapter for a discussion of heat pollution in urban areas leading to an urban heat island.

Relative Human and Natural Contributions

The relative balance between **natural** and **human contributions** of pollutants is sometimes not easy to sort out. For instance, pollutants such as heat and harmful radiation are very difficult to separate into human and natural classes because the sources are extensive, mixed, and/or geographically complex in their distributions. On the other hand, synthetic organic compounds and nitrogen dioxide are produced solely by human activities, and when they are found in air samples, there is no doubt about their source. Sulfur oxides are intermediate: natural contributions total 55 percent and human 45 percent. Nature's contribution comes principally from the sea as a result of gas exchanges involving phytoplankton.

Of the remaining substances in the pollutant list, the largest share is contributed to the atmosphere from natural sources. The percentages from human sources are as follows: hydrocarbons, 16 percent; particulates, 11 percent; carbon monoxide, 9 percent; and carbon dioxide, about 3 percent (Figure 11.1a). Among the natural

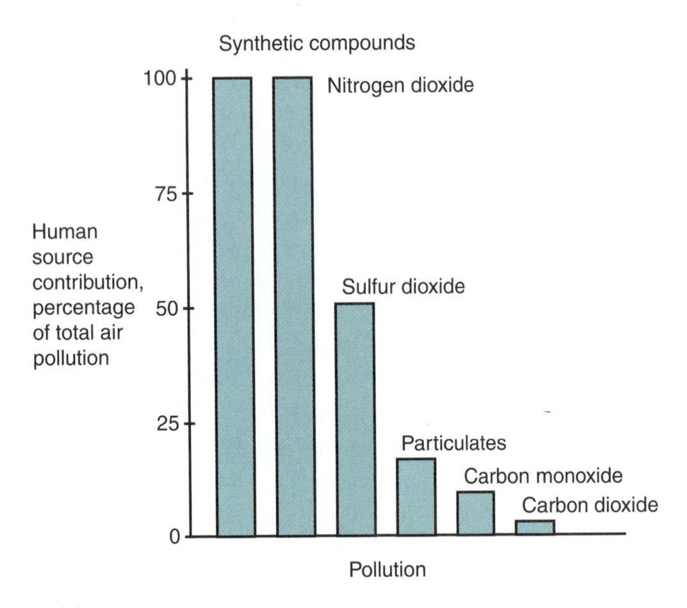

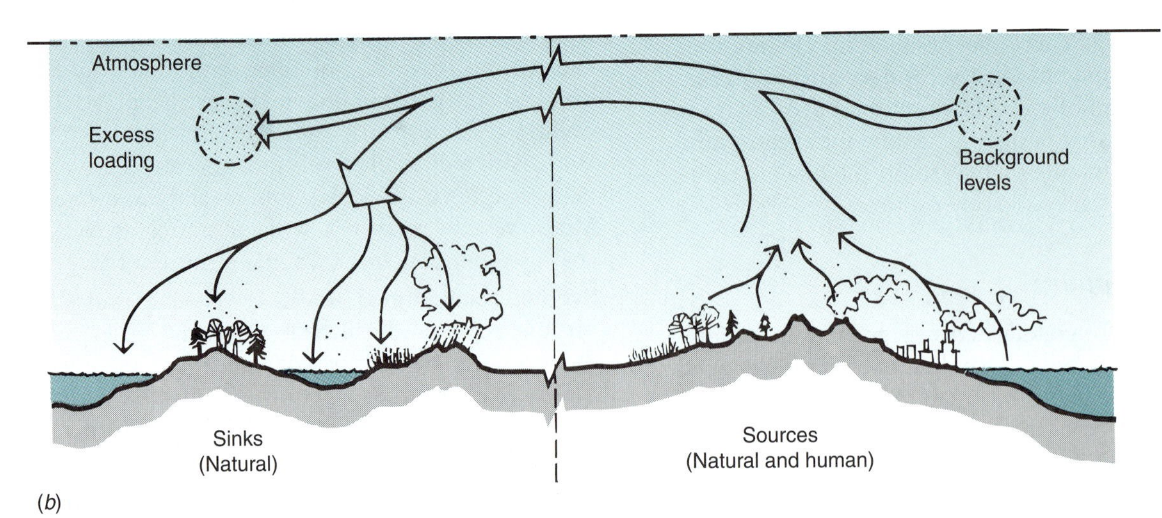

FIGURE 11.1 *(a) Percentages of air pollution from human sources by pollutant type. (b) The concept of an air pollution cycle beginning with multiple sources and ending with various sinks. Excess loading represents the fraction of pollution not removed by the sinks.*

contributors, volcanic activity is a significant source of particulates and respiration from living organisms is the major source of carbon dioxide.

To evaluate the role of pollutants from human sources relative to those from natural sources, we must bear in mind that these substances are cycled through the atmosphere. Each cycle is fed by natural sources (or inputs) from the surface and in turn is relieved by sinks (or outputs) back to the surface (Figure 11.1b). Under natural conditions, a balance is maintained that varies over different time periods in response to events such as volcanic activity. The key issue in air pollution is the influence of human contaminants as "unnatural," or excessive, events in the system.

When humans produce air pollution, it is added to the input side of the cycle. This quantity represents a surcharge on the atmosphere, or excess loading. On the other side of the ledger is the output rate from the atmosphere. If the natural extraction rate does not increase to compensate for the additional loading, then the pollution level in the atmosphere is going to increase. If, however, the system does increase the rate of extraction to offset the increased input, then a balance can be maintained at a natural or near natural level. For some pollutants, CO_2 in particular, increased natural extraction partially offsets rising inputs from human sources. However, for most pollutants this is not taking place in the atmosphere; measurements reveal that the atmosphere has been growing dirtier for more than a century.

11.3 SOURCES OF AIR POLLUTION

There is not a society on the planet that does not discharge debris into the atmosphere. The question, however, is how much and what types? Both wealthy countries and poor countries are major contributors to air pollution, but not equally so. Measured on a per capita basis, wealthy nations produce much more air pollution than poor nations. The wealthy also produce a greater variety of pollutants, including the technologically advanced ones associated with industrialized, consumer-oriented economies.

Pollution Sources in Developed Countries

Motor vehicles, power plants, and industrial plants are the principal sources of air pollution in wealthy, developed countries. Most pollution results from the combustion of fossil fuels, which emits carbon monoxide, sulfur dioxide, nitrogen oxides, particulates, and hydrocarbons (Figure 11.2). The amounts vary with the volume of fuel consumed, the quality of the fuel, and the efficiency of the burning operation. Less-efficient burning operations produce more particulates, carbon monoxide, and hydrocarbons.

Nations that are heavy consumers also tend to be heavy polluters. It follows that the United States, as the leading consumer nation in the world, is also a world leader in air pollution, especially when measured on a per-person basis. With less than 5 percent of the world population, the United States contributes 17 to 18 percent of the greenhouse pollutants (principally carbon dioxide). Comparison among nations is difficult, however, because there are no or few reliable data for most countries. We do not, for example, have reliable data for the former Communist-bloc countries, but indicators point to serious air pollution problems in Poland, the former USSR, and the former Czechoslovakia. On the other hand, reasonably reliable data are available for most

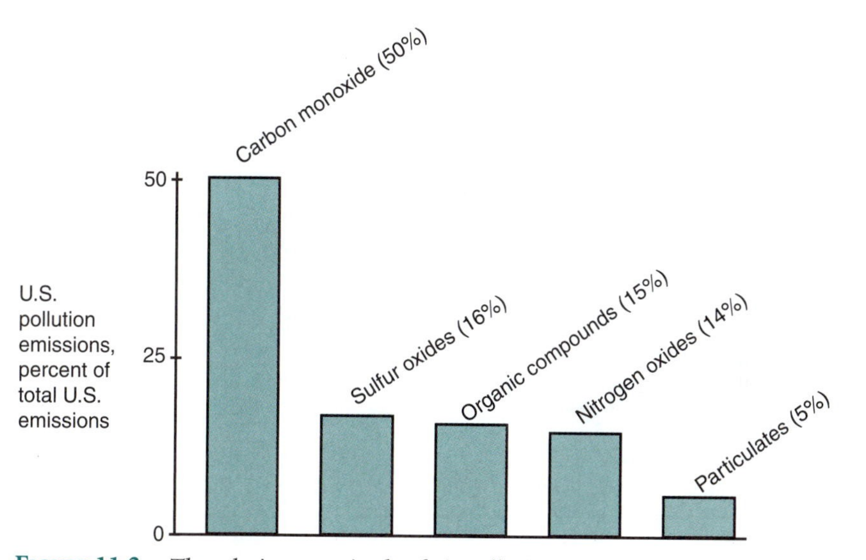

FIGURE 11.2 *The relative magnitude of air pollution emissions in the United States by pollution types.*

other developed countries, and they reveal that Canada, Australia, Germany, and the United Kingdom are also major contributors. Contributions from developing countries are small by comparison. The graph in Figure 11.3 shows just how striking the differences are. India's per capita output of carbon dioxide is about 5 percent that of the United States.

In the United States, automobiles are the primary source of air pollution, accounting for nearly half of the annual pollution load. The United States leads the world in the number of automobiles (approaching 200 million) and large, auto-based urban centers. In the Los Angeles region, for example, car and truck traffic daily exceeds 8 million vehicles and is the chief source of **photochemical smog**. This form of smog is a secondary pollutant resulting from three conditions: (1) heavy concentrations of hydrocarbons and nitrogen oxides; (2) abundant sunlight; and (3) frequent thermal inversions that hold the pollution near the ground. Smog forms when sunlight acts on the hydrocarbons and nitrogen oxides in the pollution layer, producing ozone and other oxidants. When this is added to many other pollutants, including particulates and carbon monoxide, they combine to make an obnoxious soup known as **Los Angeles - type air pollution** (Figure 11.4).

Photochemical air pollution has become a serious problem around the world. Worldwide, there are currently more than a half billion automobiles, and although most are found in developed countries, the number in poor countries is rising rapidly. In China, for example, there were only about 350,000 passenger cars in 1985, but by 1994 that number was approaching 1.5 million and today exceeds 2 millions. (The total number of trucks, cars, motorbikes, and other motor vehicles in China was about 8.5 million in 1994.) By 2010 China is expected to own a fleet of 20 million passenger cars with most concentrated in large cities.

Smog is common to all urban areas with massive automobile populations, though its severity depends on climate, city size, and weather conditions. Another type of smog is **industrial smog**, which is compositionally different from photochemical smog. Industrial smog is a mixture of sulfur dioxide, minute droplets of sulfuric acid, and other aerosols. Also found in urban regions, it smells different than photochemical smog and is produced mainly by manufacturing plants and power plants. It, too, varies with climate, weather conditions, city size, and land use.

Acid precipitation is another secondary pollutant common mainly to developed countries. In this case the primary pollutants are oxides of sulfur and nitrogen, which combine with moisture particles in clouds to form acids that lower the cloud's pH. (There is also a dry form of acid deposition, but it is poorly understood.) The term *pH* stands for the relative acidity/alkalinity of a substance. The pH is expressed in a numerical scale of 0 to 14 in which 7 is neutral; that is, neither acidic nor alkaline. Toward 0 conditions are increasingly acidic and toward 14 increasingly alkaline. Natural rainwater is slightly acidic at pH 6, whereas acid rain may go as low as 4 to 4.5 pH.

When the moisure of acid clouds or acid rain is deposited on the surface, it may damage vegetation and soil organisms or collect in streams and lakes, where it causes a chemical imbalance that many organisms, especially

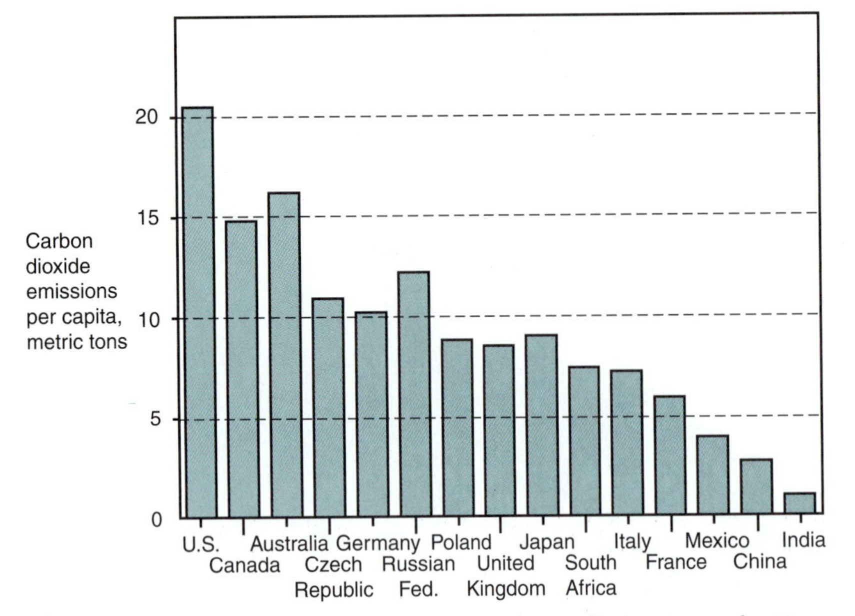

FIGURE 11.3 *Per capita carbon dioxide output from pollution sources for 15 countries in 1995. Note the differences between the developed and developing countries.*

(a)

(b)

FIGURE 11.4 *Contrasts in air quality over Los Angeles: a clear day and a smoggy day. Automobile traffic in the Los Angeles urban region exceeds 8 million vehicles per day.*

microorganisms, cannot tolerate. The environmental impact of acid deposition is serious and widespread in southeastern Canada, northeastern United States, northwestern Europe, and Japan (Figure 11.5). In North America the principal sources of acid-forming oxides are coal-burning power plants in the Midwest centering on Ohio and western Pennsylvania (see Figure 11.11). Power plants in this region draw on a ready supply of coal from the nearby Appalachian coal fields, which have relatively high sulfur content.

Ozone depletion also has its roots in the developed countries. Ozone is naturally concentrated in the stratosphere, 15 to 55 kilometers (10 to 35 miles) above the Earth, where it intercepts ultraviolet radiation (UV) in the incoming solar beam. Ultraviolet radiation is harmful to humans and other organisms in ways we will explore later in this chapter.

The loss of ozone is caused by a chemical reaction in which a variety of manufactured gases called **chlorofluorocarbons (CFCs)** produces a breakdown in the ozone

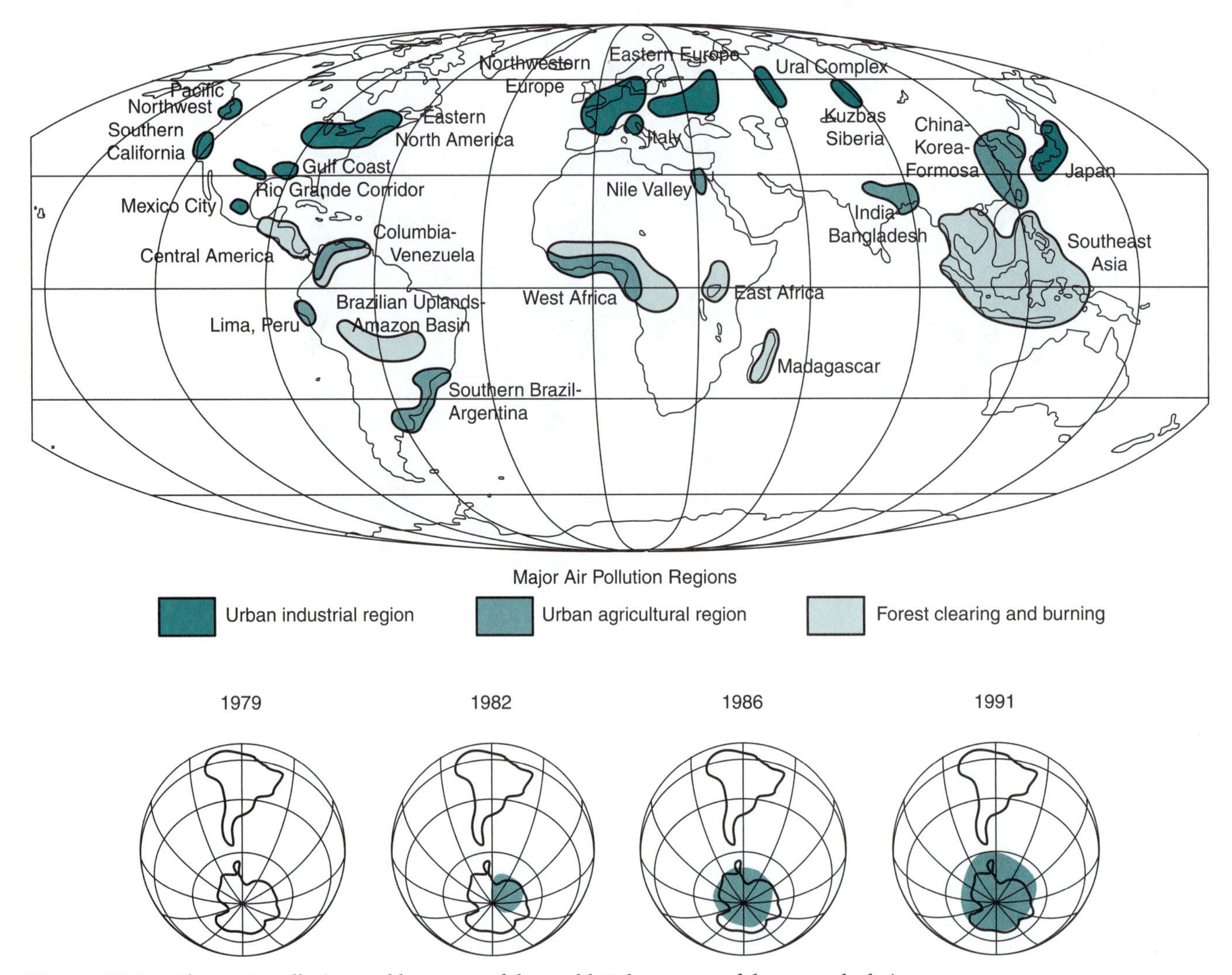

FIGURE 11.5 *Above, air pollution problem areas of the world. Below, maps of the ozone depletion "hole" over Antarctica, 1979–91.*

molecule. The process involves the release of a chlorine atom from the CFC molecule that breaks up the ozone molecule and destroys its capacity to absorb UV radiation. The individual chlorine atom does not do this just once, but many times, destroying as many as 100,000 ozone molecules before its reactive life is over.

CFCs are used in spray-can propellants such as spray paint, in solvents, and in cooling agents for refrigerator systems such as air conditioners. Atmospheric loading by CFCs increased dramatically between 1950 and 1985: In 1950 global emissions totaled 35,000 metric tons, whereas in 1985 they were up to 650,000 metric tons. When released to the atmosphere, CFCs rise into the upper troposphere and stratosphere where they may remain for 75 to 110 years. Because of the redistribution of CFCs by winds in the upper atmosphere, ozone depletion is tied to the seasonal variations in global circulation and therefore varies seasonally and geographically. Depletion is concentrated over the Earth's polar regions, especially in the Antarctic, where scientists have defined a so-called *hole* in the ozone layer that becomes most acute in September, October, and November (Figure 11.5). The polar regions are especially susceptible to ozone depletion because extreme cold is favorable to the conversion of chlorine into its ozone-destroying form.

Among the developed countries, air pollution in the lower atmosphere is significantly worse in some regions than others. Within the large nations such as the United States and Canada, the heavily urbanized regions such as the U.S. Northeast are the principal source areas of air pollution. In Europe a grievous picture of regional pollution has unfolded in the former Communist-bloc countries of eastern Europe. Although the details are not clear because of a general lack of air quality data, it is apparent that Poland, former East Germany, and parts of Russia and other countries had, and to a considerable degree still have, serious air pollution problems, especially in industrial areas. For example, in the period 1980–1985, when air quality in American, Western European, and Japanese cities generally improved, Warsaw's worsened significantly. Particulates increased by 58 percent as the government pushed industrial production in outmoded factories having few or no pollution controls (Figure 11.5).

Pollution Sources in Developing Countries

Developing countries are a significant and growing source of air pollution. Because of their large and rapidly expanding populations, these countries emit large amounts of traditional pollutants such as carbon dioxide and nitrogen oxides from forest burning as well as particulates from wind erosion of exposed farm fields. Increasingly, however, developing countries are becoming major centers of urban pollution and sources of technological pollutants such as insecticide residues used in agriculture and disease control. For example, DDT, which is banned in the United States and other developed countries, is widely used in developing countries. Unfortunately, developed countries continue to manufacture many harmful chemicals for the rest of the world.

As developing countries industrialize, pollution output in population centers is climbing rapidly. The rise is often especially dramatic because most countries do not have effective pollution control laws and if they do, they cannot enforce them. In Sao Paulo, Brazil, for example, which reports 45,000 industrial operations in the urban region, pollution control efforts are having little effect on emissions. In addition, massive shantytowns around many cities produce smoke from wood and debris fires. The result is some of the worst local cases of air pollution on earth. This is illustrated by the city of Cubatao near Sao Paulo, where industrialization and shantytown development have combined to produce exceptionally heavy air pollution; as many as one-third of the people in the inner city suffer from respiratory diseases.

The World Health Organization (WHO) reports that most of the world's worst polluted cities are found in developing countries. Of the 73 WHO monitoring sites in developing countries urban areas, 63 exceeded recommended WHO air quality standards for particulate levels for the period 1979–1985. Among the top ten cities were Kuwait City, Beijing, China, and Teheran, Iran. For sulfur dioxide, many of the same cities were listed; however, several cities in developed countries also fell in the top ten, including Milan, Italy, and Paris, France.

In the rural areas of developing countries, forest and grassland burning are major sources of air pollution. The destruction of tropical rainforest almost always involves burning, and given the enormous area of vegetation destroyed each year, the pollution load of carbon dioxide, carbon monoxide, nitrogen oxides, and particulates is considerable indeed (Figure 11.6).

FIGURE 11.6 *In 1998 forest fires in Southeast Asia produced some of the heaviest air pollution of the 20th century.*

11.4 PROCESSES, PATTERNS, AND SCALES OF AIR POLLUTION

The behavior of pollutants and the resultant pattern of their distribution in the atmosphere depends on many factors. Chief among these is the structure and motion of the receiving air and the densities of the pollutants themselves. In still air, airborne particles such as sooty debris from exhaust stacks tend to settle rapidly. The largest and heaviest particles fall out immediately around their discharge points (Figure 11.7). Near factory and power plant exhaust stacks, the fallout pattern can be traced in the distribution of dirt blanketing the neighboring landscape.

Intermediate-sized particles usually remain suspended for several days and spread over a much greater volume of air than large particles. In urban areas, they contribute to the formation of **dust domes**, large envelopes of dirty air over cities (Figure 11.7). The class of smallest particles, which are less than 1.0 micrometer (one-millionth of a meter) in diameter, also contribute to the dust dome. They may remain in the atmosphere for several weeks or months—the slightest motion of the air moves these particles upward and laterally. Most gaseous pollutants also have long residence times in the atmosphere. If they are lighter than air, such as methane and CFCs, they may float upward in the atmosphere, where they remain until they are altered chemically and/or recycled through the Earth's surface environments.

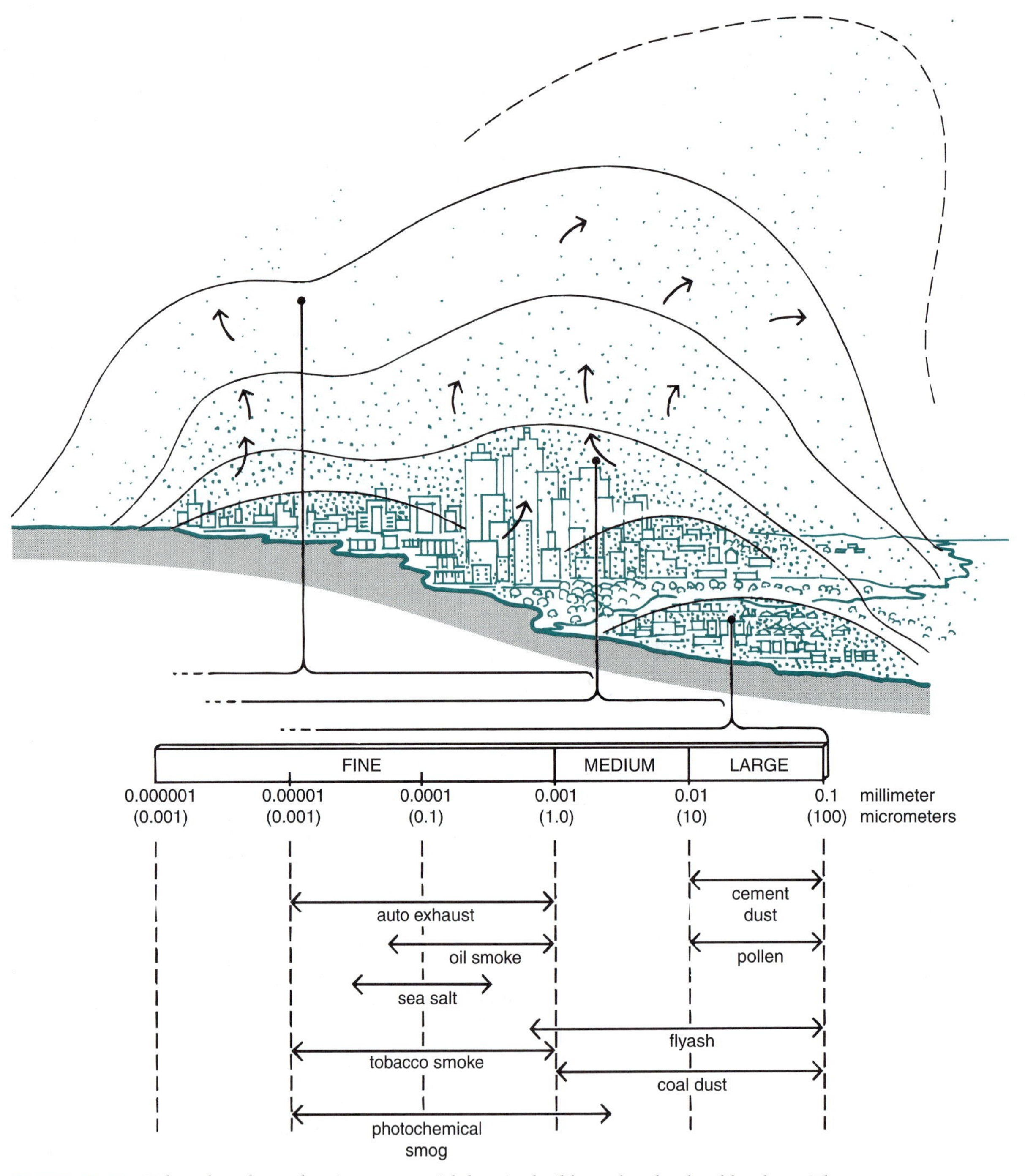

FIGURE 11.7 *Urban dust dome showing sectors with heavier buildup related to local land use. The scale relates particulate sizes to pollution sources.*

Plume Patterns

When the atmosphere is set into motion, the density of individual pollutants is less important because the force of the wind's turbulent motion tends to keep pollutants aloft. One of the critical questions in this regard is the behavior of the downwind trains, or **plumes**, of pollution emanating from a factory, power plant, forest fire, or a city. The behavior of a pollution plume depends not only on wind turbulence and direction but on the thermal structure of the receiving atmosphere. The thermal structure of the lower atmosphere controls the elevation and the general pattern of plume behavior. Many types of plume patterns can result. We describe three of them here (Figure 11.8).

In an unstable atmosphere—that is, one with warmer, lighter air at the surface and cooler, heavier air aloft—plumes move both vertically and laterally with the natural turbulence of the air. Indeed, their behavior in a crosswind may be erratic and take on a **looping plume** pattern before being dispersed into the atmosphere (Figure 11.8a). By contrast, if the atmosphere is stable, that is, thermally inverted with cool air at the surface and warm air aloft, the plume tends to be contained beneath the **inversion layer**. Under these conditions, which are usually associated with very light wind, the concentration of pollutants can become very great because they are boxed into a small volume of air and diffusion upward and laterally is retarded.

The most hazardous situations are caused by sinking air associated with low thermal inversions. This gives rise to **fumigation plumes** that settle to the ground and blanket the landscape (Figure 11.8b). **Fanning plumes** are somewhat less serious because the inversion layer is higher and the volume of air into which the pollutants are dispersed is greater (Figure 11.8c).

Thermal Inversions

Some of the worst episodes of traditional air pollution ever recorded were associated with thermal inversions. These episodes were especially prominent in industrial cities of the nineteenth and twentieth centuries when pollution from coal burning produced massive loadings of sulfur dioxide, particulates, carbon monoxide, and carbon dioxide in cities such as New York and London. One of the most frightening episodes occurred in December 1952 in London, when pollution held down by an inversion lasting nearly a week caused 4,000 deaths. The cloud cover associated with the stagnant air retarded the penetration of sunlight, causing surface temperatures to cool and thereby increasing the demand for home heating. This, in turn, led to increased air pollution, because

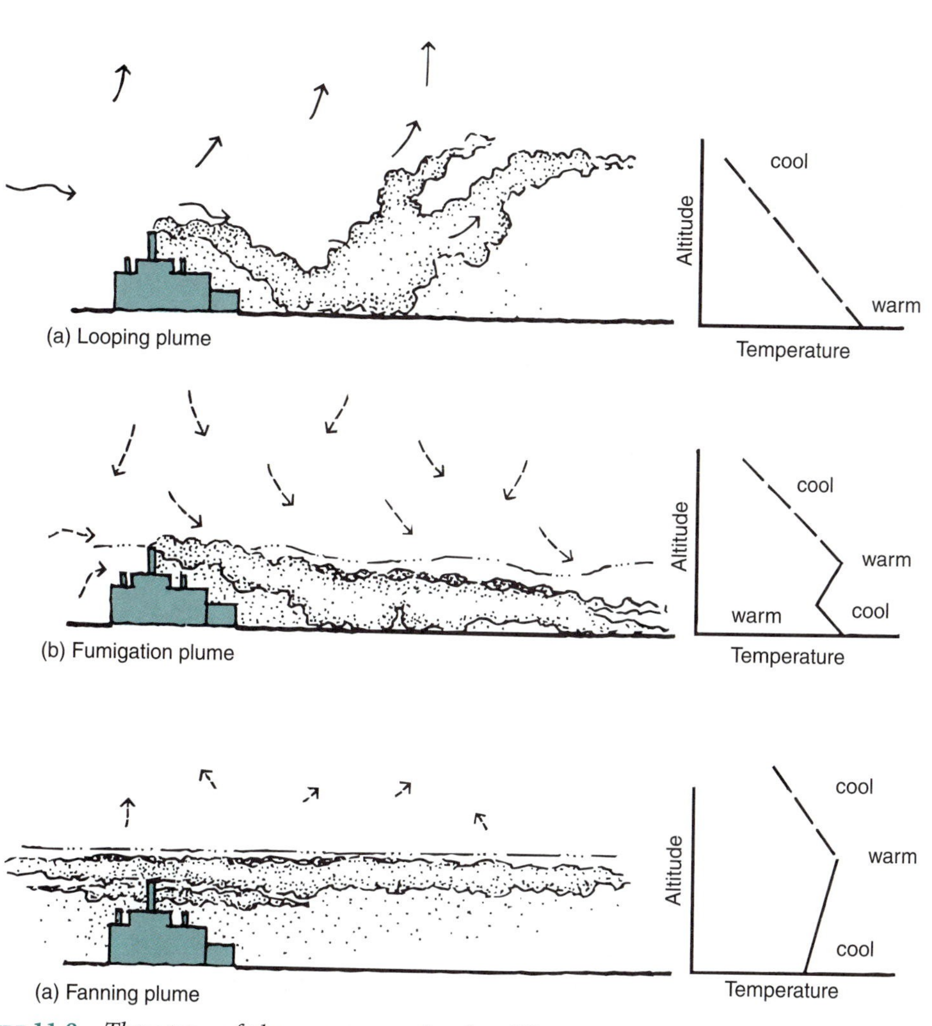

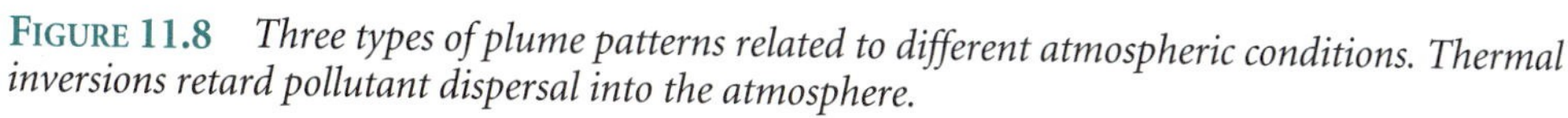
FIGURE 11.8 *Three types of plume patterns related to different atmospheric conditions. Thermal inversions retard pollutant dispersal into the atmosphere.*

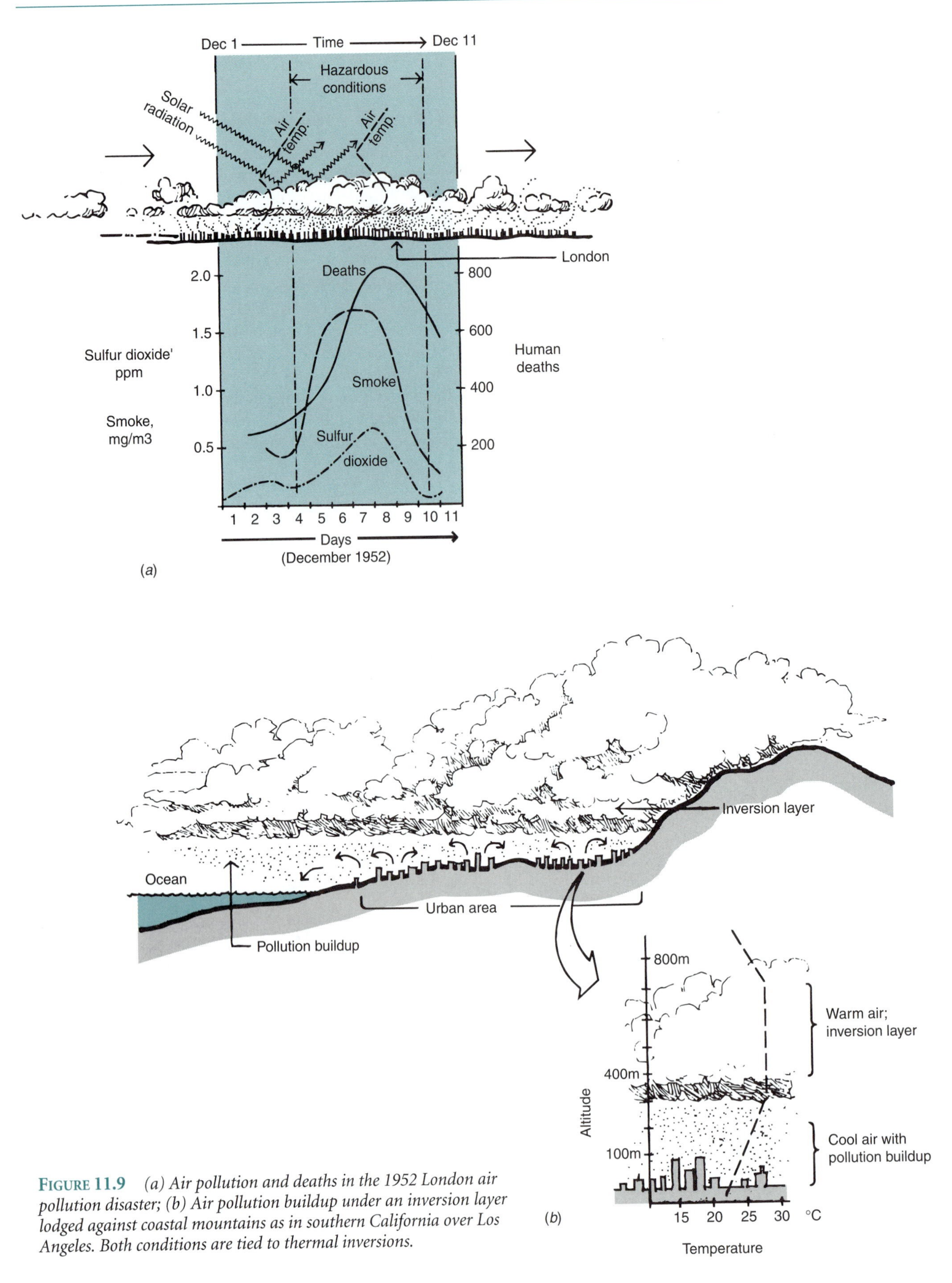

FIGURE 11.9 *(a) Air pollution and deaths in the 1952 London air pollution disaster; (b) Air pollution buildup under an inversion layer lodged against coastal mountains as in southern California over Los Angeles. Both conditions are tied to thermal inversions.*

most homes were heated with coal. Coupled with exhausts from vehicles and factories, pollution built up rapidly beneath the inversion layer (Figure 11.9a).

Thermal inversions are also associated with episodes of air pollution from photochemical smog. Los Angeles provides the classic example of photochemical smog associated with thermal inversions. In this case the city's coastal location and topographic setting in a broad, mountain valley contribute to both the magnitude and frequency of inversions. The inversion layer typically forms around 500 meters altitude as cooler air from the ocean slides under warmer urban air. Since the surrounding mountain ranges are higher than the inversion altitude, the valley tends to hold in the inversion layer and the pollution under it (Figure 11.9b).

Other valley settings are prone to even more serious (though less frequent) inversion episodes than Los Angeles. In the unfortunate Brazilian city of Cubatao, mentioned earlier, pollution flushing is limited by inversions that persist in the valley surrounding the city while pollution builds up under them. Closer to home, one of the worst pollution episodes ever recorded in eastern United States occurred in the Monongahela River Valley in Pennsylvania in 1948. A strong inversion fell over the industrial town of Donora that virtually precluded atmospheric mixing and the flushing of the industrial pollution. For several days pollution heavy in sulfur dioxide, particulates, and carbon monoxide built up, until half the town's 12,000 people became ill with respiratory difficulties.

Tropospheric Mixing of Pollutants

The troposphere is the only part of the atmosphere capable of vertical mixing. In most places thermal inversions in the lower atmosphere are the exceptional occurrences, and the atmosphere on most days is characterized by the mixing of surface air with air aloft. This process is extremely important in dispersing polluted surface air and thereby reducing pollution levels at the Earth's surface. But it also enables pollutants to be distributed regionally, even globally.

Mixing in the troposphere is driven by various types of storms and weather systems that draw in surface air and drive it into the upper parts of the troposphere, where it is recirculated by high-altitude winds. Studies show that this system draws up pollutants with the surface air and feeds them to upper-level winds such as the jet stream. These winds are faster than surface winds and sweep around the world over broad geographic zones (Figure 11.10).

Once in the zonal circulation system, pollutants can be quickly redistributed. We see it with most major volcanic eruptions, such as Mt. Pinatubo in the Philippines in 1991, whose ash was carried around the globe in a matter of weeks (see Figure 2.9). Similarly, pollution from the massive Kuwaiti oil well fires, also in 1991, was measured over western United States 12,000 miles away. Remarkably, it took only about 10 days to make the trip. The same scale of transport occurs in the midlatitudes, where developed countries are the sources. Japan, eastern United States, and Europe, for example, contribute to pollution problems thousands of miles beyond their regional borders.

One such midlatitude problem is acid precipitation. Since it is closely tied to the atmospheric moisture system, the distribution of acid-laden moisture is strongly influenced by the patterns of air mass and cyclonic storm movement. In the midlatitudes the prevailing wind and weather systems move from west to east. In North America, the oxides produced in the Ohio Valley, for example, are transported eastward and northeastward (Figure 11.11). In a matter of hours or days, a large part of the resultant acidic moisture is precipitated over southeastern Canada and New England. The remainder is carried over the North Atlantic, where most is probably precipitated onto the ocean. Some undoubtedly reaches northwestern Europe, where it combines with acid moisture emanating from European sources. This moisture, in turn, is carried by the Prevailing Westerly winds into northern Europe, and with additional contributions from eastern Europe, it is carried into the adjoining region of Asia.

Global Dispersal of Air Pollution

We must also recognize that the atmosphere is integrated into a global circulation system connecting one climatic zone to the next and that certain pollutants are passed along by this system. The decline of ozone as a result of CFC pollution over Antarctica is a case in point. Virtually no pollution is released from this continent and there is no way of accounting for the loss of ozone from CFC pollution than by broadly based global transfer mainly from developed countries in the Northern Hemisphere. As we noted earlier, ozone destruction from CFC pollution is favored, among other things, by extreme cold.

Evidence of the atmosphere's capacity for long-distance transport is also provided by certain pesticides. For example, the pesticide lindane, which is severely restricted in the United States and Canada but used in China and India, is showing up in North American waters. The most likely source is Asia, the only possible transporting agent is the atmosphere, and distances are in the range of 10,000 to 12,000 miles.

The global circulation system is also responsible for worldwide dispersal of carbon dioxide and other pollutants (e.g., methane, CFCs, and nitrous oxide) contributing to atmospheric warming. Most CO_2 from human sources is produced by developed countries in the midlatitudes of the Northern Hemisphere, from which it is thoroughly mixed across the globe. As the CO_2 is discharged into the lower atmosphere it is rapidly mixed into the troposphere as a whole, thereby increasing the potential for global warming over the entire world.

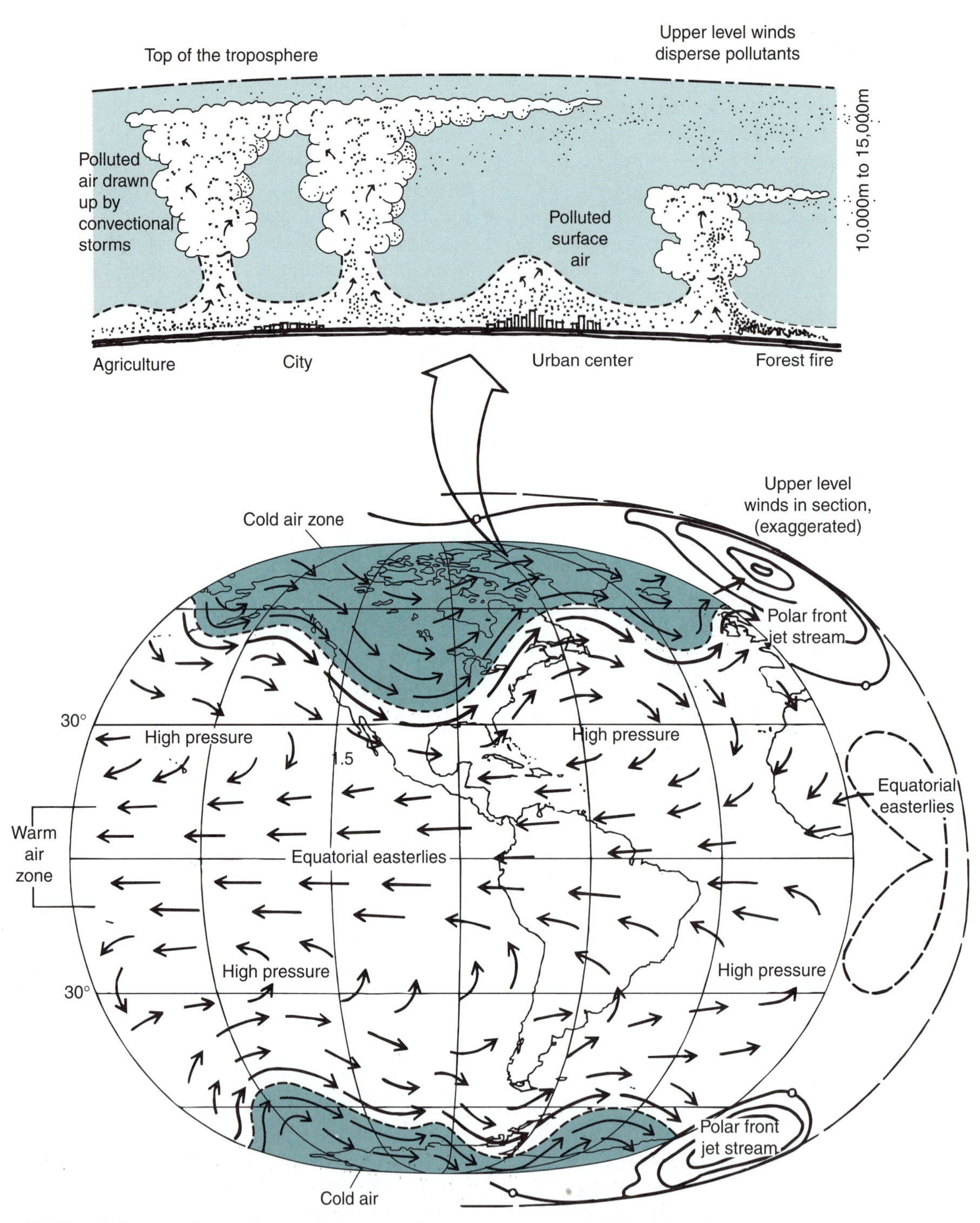

FIGURE 11.10 *Pollutants drawn from surface sources by storm systems and delivered to the upper troposphere, where they are redistributed by fast upper-level winds. Below, the polar front jet stream which meanders around the world at wind speeds typically greater than 100 mph (180 km/hr).*

According to current scientific thought, the global pattern of heating in the lower atmosphere related to carbon dioxide buildup will not be geographically uniform. The picture is complexly related to differences in atmospheric mass, cloud cover, solar heating, air circulation, and variations in the distribution of land and water. Computer simulation models, which take many of these factors into account, indicate that warming will be regional—that is, certain zones will get warmer while other regions will show no change and still other regions may actually cool a little. These thermal changes will be associated with changes in other aspects of climate, such

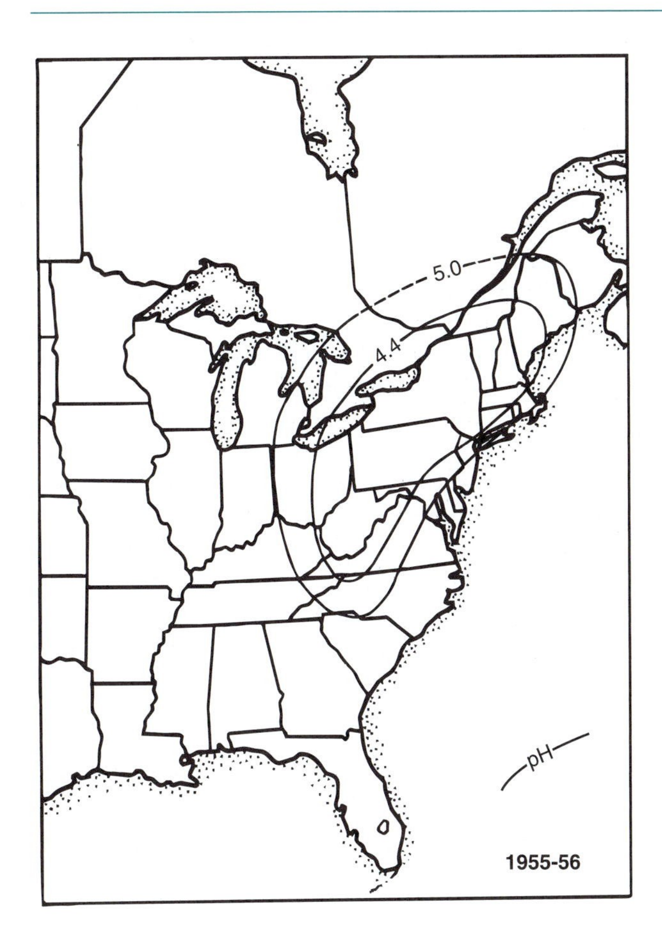

as precipitation, growing season, and storm patterns. The magnitude and geographic extent of these changes are uncertain, however. The one change that is predicted in all the forecasts is a rise in sea level, as we discussed in the previous chapter.

11.5 Patterns of Pollution in the Urban Region

In the previous chapter we also discussed the influence of urbanization on the atmosphere and the formation of a climatic variant called the urban climate. We have already noted in this chapter that urban centers are a source of pollution plumes and regional pollution such as acid rain. They are also major contributors to global air pollution related to ozone depletion and CO_2 warming of the atmosphere. Let us now take a look inside the urban region at pollution patterns in and around the environments where most of us live.

Figure 11.11 *Acid precipitation patterns in the United States and southern Canada between 1955 and 1986. Prevailing airflow is from west to east. The values represent the pH of rainwater.*

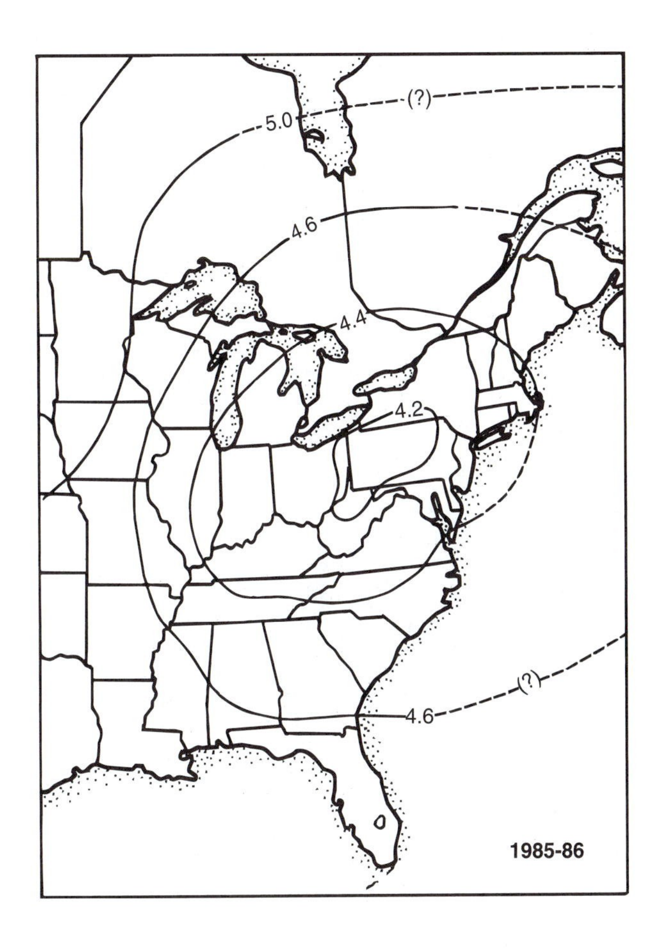

Urban Land-Use Patterns

Most cities are comprised of a complex system of closely packed land uses linked together by transportation systems. The pollution sources are both stationary (e.g., power plants) and mobile (e.g., vehicles), and for most large cities, the emission rates vary in different parts of the city. In general, the land-use density increases toward the city center, and the level of pollution generally increases with land-use density. For example, particulate levels in the inner city are typically two to three times higher than in suburban areas and five times higher than in rural areas (Figure 11.12).

If we look at most cities today, however, they are more complex in the geographic arrangement of land uses than the preceding statement suggests. They tend to be more *multinucleated*, meaning that there are many different centers of concentrated land use within the urban region rather than a single center in the middle. Each of these centers is characterized by a different land-use activity (e.g., power generation, industrial, or commercial) and different levels and types of air pollution. Interspersed among these centers is a broad matrix of residential and institutional land uses as well as parks, cemeteries, and other open space (Figure 11.13).

Local Pollution Patterns in Cities

Highway corridors are notorious for severe levels of air pollution, especially when traffic is backed up and the atmosphere is relatively still. This is related not only to the massive volumes of vehicles on urban highways (typically 200,000 to 300,000 vehicles a day on large, interstate segments), but to the fact that automobile engines are less efficient combustors at lower speeds. Carbon monoxide levels in traffic jams may exceed 100 parts per million (ppm), more than 1,000 times the atmospheric background level. If you are a smoker waiting for traffic to clear, carbon monoxide levels in your car can exceed 400 ppm.

Particulate levels are also high at slow points in the system, such as in central business districts or in shopping centers. Studies show that the particles are actually of both primary and secondary origins. Secondary particulates are formed by coagulation of gaseous pollutants shortly after their emission from vehicles, with the larger particles concentrated along streets and intersections.

Particles of all sizes, of course, are dependent on wind to flush away the surface air in which they are concentrated. Because of impaired air circulation in cities, especially the built-up inner cities, flushing of polluted surface air is often slow and incomplete. This is explained by the physical structure of the city: where buildings are high and closely spaced, surface wind is displaced upward, leaving calmer air near the ground. Pollution builds up in this layer of surface air precisely where most of the pedestrians move about. Highways and parking garages constructed below ground level and under buildings are also poorly flushed; tunnels are usually the most polluted segments in the urban highway system.

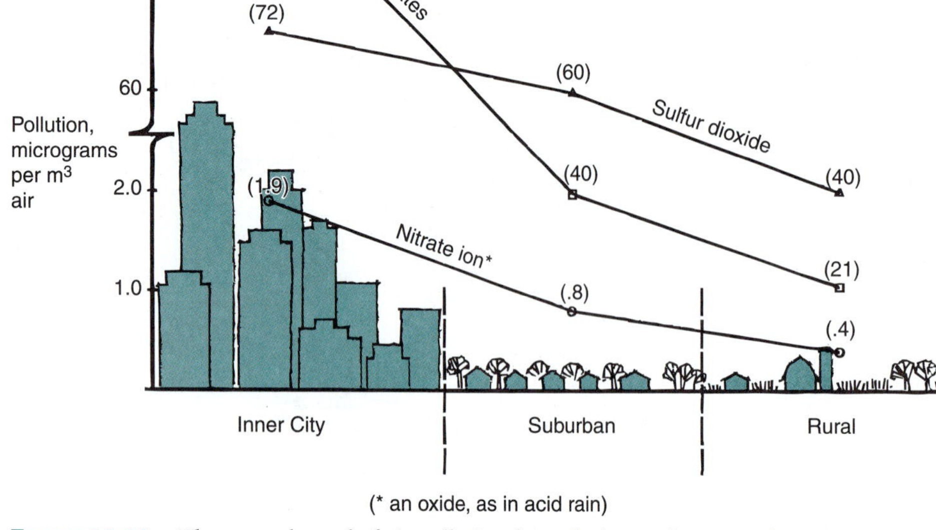

FIGURE 11.12 *The general trend of air pollution from the inner city to rural areas.*

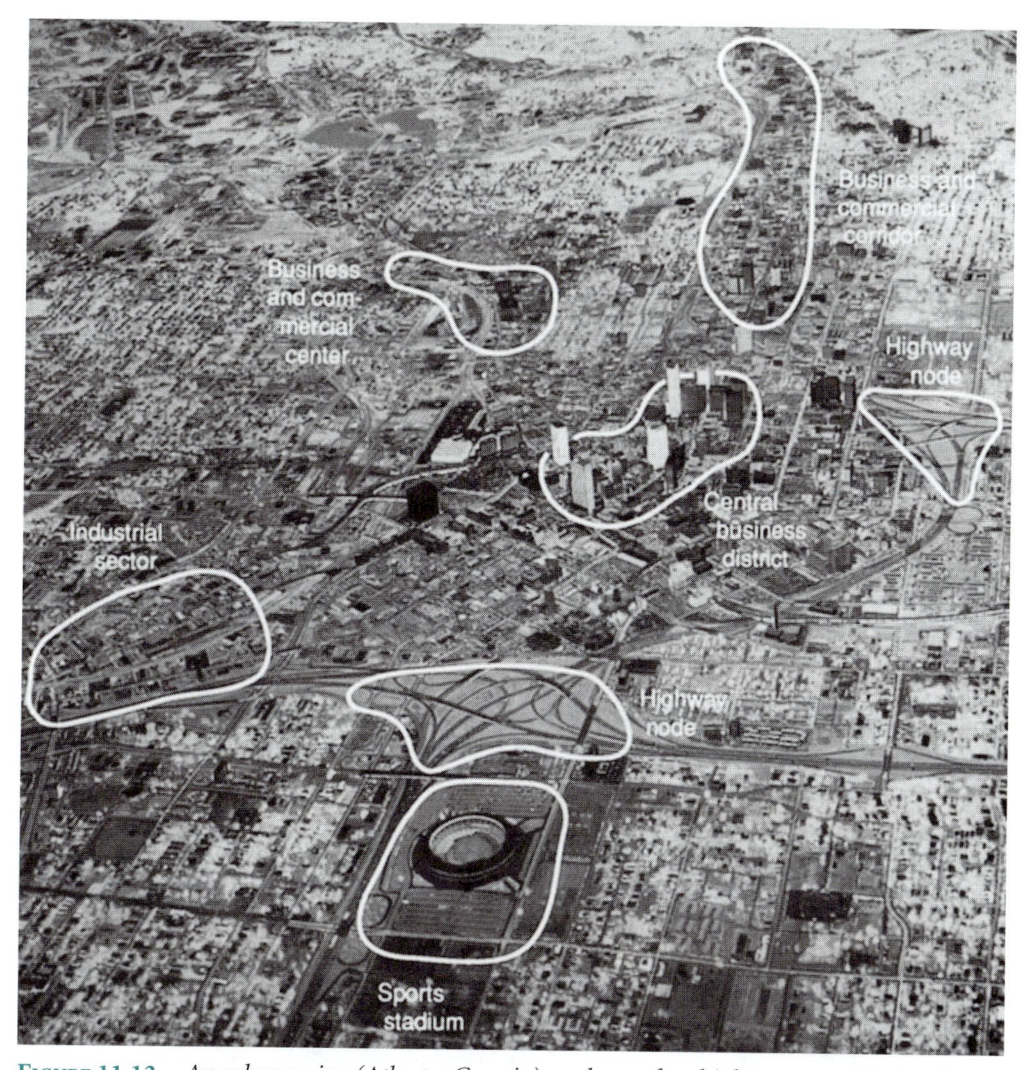

FIGURE 11.13 *An urban region (Atlanta, Georgia) made up of multiple centers or sectors of different land uses associated with different levels and types of air pollution*

Air Pollution Within Buildings

Generally the smallest scale of consideration in air pollution studies is the interior of homes, factories, and other buildings. In general the air inside buildings in the inner city, industrial areas, and similar settings is cleaner than the outside air. However, many buildings have their own peculiar air quality problems. For example, over the past two decades there is increased evidence of a phenomenon known as **new building sickness**. Some people become ill or develop disorders such as allergic reactions related to fumes from glues and other synthetic substances and from dust and residues from machinery.

In factories, internal air pollution has long been recognized as a serious problem. The pollutant types and levels depend on the types of industrial operations, the design and maintenance of the facilities, and the precautions, such as air filtering, taken to minimize pollution loading. In facilities with poor ventilation, such as old-fashioned factories in developing countries or former Communist-bloc countries, the levels of particulates and carbon monoxide can reach concentrations higher than the most polluted outdoor environments. Some of the worst indoor pollution is associated with heavy industry such as foundries, steel mills, and blast furnaces. From the standpoint of health, some of the most dangerous indoor pollution is associated with chemical processing and painting operations where aerosols are rich in synthetic compounds.

Although cleaner than industrial buildings, residential buildings are not free of indoor air pollution. Synthetic compounds from solvents and sprays and carbon monoxide and particulates from fireplaces and cigarettes are common household pollutants. Another pollutant is the radioactive gas **radon**, a known source of lung cancer. In the mid-1980s, it came to public attention that radon, a natural gas released from decaying radioactive elements in the Earth's crust, seeps into 6 percent (6 million) of the homes in the United States. The gas enters houses

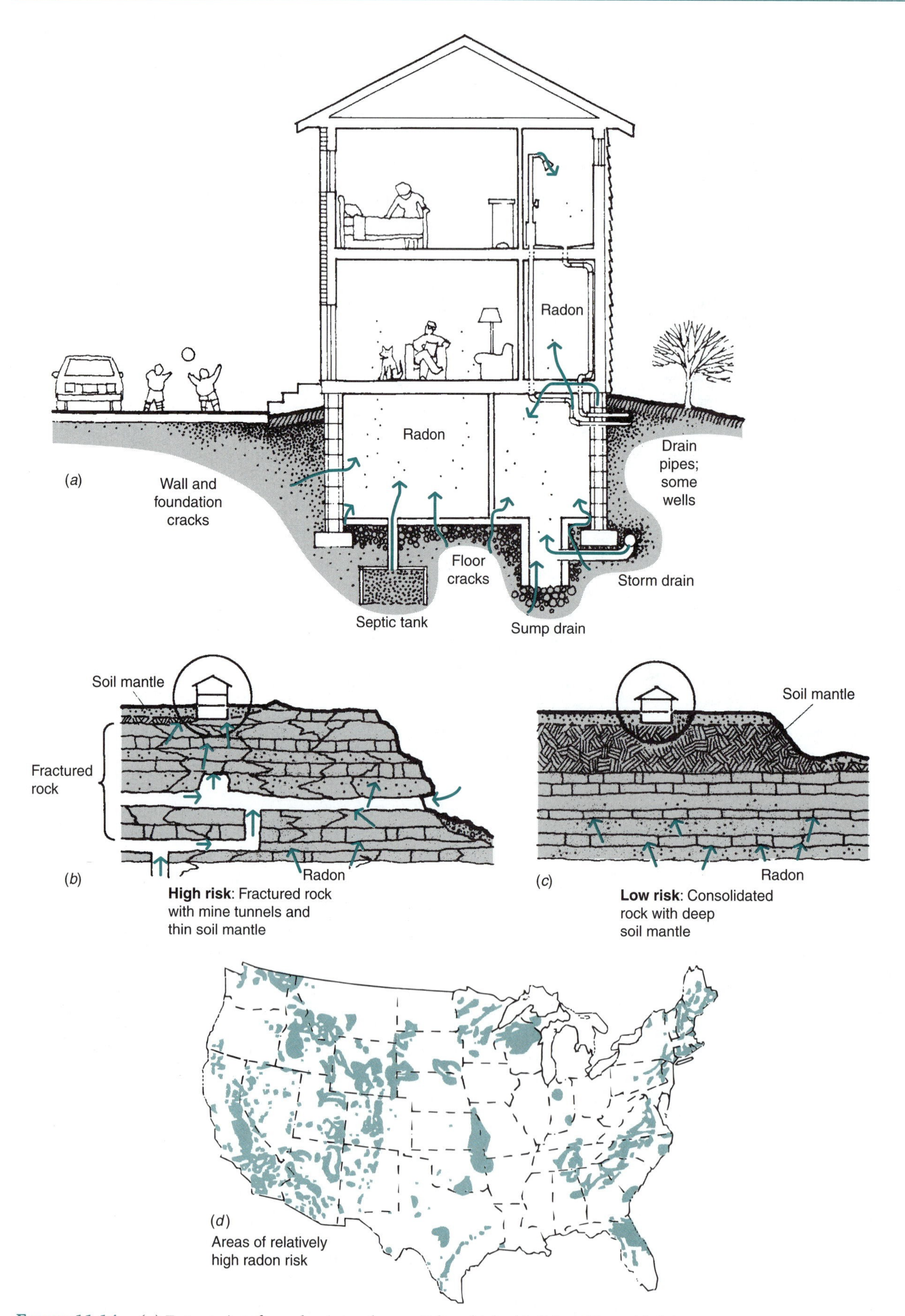

FIGURE 11.14 *(a) Entry points for radon into a house. Below, high-risk (b) and low-risk (c) geological conditions for radon discharge to the surface related to mine shafts, crack systems in bedrock, and the thickness of the soil mantle. Below (d), areas of relatively high radon risk in the United States.*

through foundation cracks and pipes. Once in the house the radon can break down into other radioactive elements that can lodge in the lungs and can cause lung cancer (Figure 11.14). Radon susceptibility is higher in areas underlain by coal deposits, especially where they are vented with mine shafts. Where bedrock is sealed by thick mantles of clayey soils, radon seepage to the surface tends to be retarded and is less a risk.

The Urban Thermal Landscape

Heat is another form of pollution in the urban region. As we noted in the discussion of urban climate in Chapter 10, cities tend to average several degrees warmer than surrounding areas. On closer inspection, there is a good deal of thermal variation within the urban region related to different land uses, surface materials, and flushing rates.

Where artificial materials have replaced natural surface materials (vegetation, soil, and water features), the urban heat balance changes dramatically. Heat released to the atmosphere as water vapor, called **latent heat**, is reduced, while **sensible heat**, the heat that raises the temperature of air, is increased. While the ratio of these two heat forms is often reversed with the conversion of land to hard materials, two other changes are also occurring:

1. The flushing capacity of the air at ground level is reduced with the construction of tall, closely spaced buildings.
2. Heat emission from building heating, lighting, and air conditioning systems dramatically increases.

It is difficult to believe, but in mid-winter New York City generates significantly more heat from artificial sources than it receives from solar heating. In combination, then, ventilation decreases while heat generation increases, thereby raising local temperatures by as much as 5 to 10°C in heavily built-up parts of inner cities. During heat waves the increase is even greater and poses a serious threat to human health (see Figure 10.15). In contrast, park areas located only blocks away may be significantly cooler.

Urban Planning for Air Quality

There are many reasons that the problem of building environmentally safe cities remains largely unsolved. Perhaps the chief one is the rapidly changing nature of the problem. Driven by accelerated urban growth and technological change, the patterns and magnitudes of urban air pollution have changed faster than planners have come up with solutions. The massively built-up inner city and its sluggish atmosphere, for example, have evolved rapidly on the urban landscape, with buildings growing higher and spreading farther out from the city each decade of the twentieth century. The first skyscrapers, built around 1900, were tiny by today's standards, only 30 to 60 meters (100 to 200 feet) high. The tallest buildings now exceed 360 meters (1200 feet) and the mass of buildings greater than 60 meters (200 feet) tall is enormous in cities such as New York, Chicago, Houston, and Philadelphia.

Perhaps more serious in the regional context is **urban sprawl**. With sprawl, residential development and services are spread over vast areas requiring extensive and frequent automobile travel (Figure 11.13). As a result, the amount of air pollution from cars and trucks is excessive. Overall, air pollution is 40 to 50 percent higher with urban sprawl compared to cluster development. In **cluster development** residences and services are grouped together, requiring less travel to schools, stores, and work and therefore less automobile pollution. However, land-use development in the United States and Canada, driven by automobile access, affluence, and a general aversion to cities by many people, has favored low-density sprawl, and in turn, greater air pollution.

11.6 AIR POLLUTION IMPACTS ON PEOPLE AND ENVIRONMENT

The impact of air pollutants on humans and the environment is highly variable, depending on the type of pollution, its concentration, and the part of the environment or human life affected. As a whole the impacts are enormous and it is difficult, given the magnitude and complexity of the problem, to assess them accurately. Among the impacts are many aspects of human health as well as damage to forests, crops, building materials, and ecosystems. Many of these are measurable in monetary (economic) terms, but others, such as climatic change and ecosystem decline, cannot be measured in this way with much accuracy.

Air pollution impacts fall roughly into two classes: direct and indirect. Direct impacts would be, for example, damage to vegetation in urban areas from a primary pollutant such as sulfur dioxide, whereas indirect impacts would be damage to ecosystems via the hydrologic system from acid rain, a secondary pollutant derived from sulfur dioxide. Impacts on human health are both direct and indirect.

Air Pollution and Human Health

Air pollution affects both healthy and infirm persons, although the infirm are far more susceptible to it. Particulates are especially serious, particularly small ones. Recent studies show a very close relationship between daily nonaccidental death rates and the daily rise and fall of particulate concentrations in American cites. People suffering from respiratory diseases such as asthma and emphysema, for example, are acutely affected by both solid and liquid **particulates**, especially when physical exertion is involved. Particulates composed of synthetic compounds and oxidants are also causes of lung cancer, as are the radioactive products of radon.

Other pollutants are also serious health threats. Ozone and sulfuric acid damage lung tissue and are a source of respiratory problems in both healthy and infirm persons. **Carbon monoxide** inhibits the absorption of oxygen by the blood, causing headaches, nausea, and reduced mental capability. Excessive heat in urban areas leads to many of the same maladies, and in extreme cases, such as St. Louis in 1966 and Chicago in 1995, to collapse and death.

Allergic reactions to a wide variety of pollutants are also a major source of illness in human populations. The **allergens** not only include noxious materials such as cigarette smoke, carbon monoxide, and smog, but also biologic pollutants, especially pollen from weeds, which are spread and nurtured by urban and agricultural land uses. Among the most serious weed pollutants is pollen from ragweed and junipers, both a major cause of hay fever, bronchitis, and more serious respiratory ailments such as asthma.

As damage to the ozone layer increases, health problems related to **ultraviolet** (UV) **radiation**, such as skin cancer and eye cataracts, are also expected to increase. Skin cancer caused by exposure to UV-B, the harmful form of ultraviolet radiation, will be especially pronounced in fair-skinned people. It is estimated that this disease will increase 2 percent for each 1 percent decline in stratospheric ozone.

The **health costs** of air pollution are significant. An accurate estimate expressed in dollar costs, however, is difficult to obtain because so many of the effects are secondary. That is, they are added to existing diseases from other causes; for example, as asthma is exacerbated by air pollution it often leads to emphysema. Some cancers appear to be caused by the combined effect of many contaminants in the air, water, and food environments, and it is difficult to define the fraction attributable to air pollution. In any case, realistic health cost estimates related to air pollution must include direct expenditures for medical care and treatment, as well as losses for job time and economic productivity. In the United States, direct health costs alone amount of billions of dollars yearly. In other countries such as Russia the costs would be even greater if health problems related to air pollution were more widely recognized and treated.

Impacts on Crops, Forests, and Ecosystems

Plants are damaged directly by several pollutants including ozone, photochemical oxidants, and acid rain. The damages vary with plant species and pollutant types but commonly include tissue loss, reduced photosynthesis, and lowered resistance to disease and environmental stress. In the United States crop yield losses due to air pollution are as high as 10 percent annually.

Reports of air pollution damage to forests are widespread in North America and Europe. Although ozone, oxidants, and acid rain are cited as the main sources, acid rain damage appears to be most widespread (Figure 11.15). Foliage may be damaged directly by contact with acidic fog, raindrops, and cloud moisture, and in the soil acidic moisture reduces fertility and promotes leaching, which in turn drives mercury, which naturally occurs in soil, into groundwater, lakes, and streams. Added to this are direct contributions of mercury from air pollution. In the form of methylmercury, the most dangerous form of the element, mercury readily enters food chains and bioaccumulates in organisms at higher trophic levels—including humans, where it is known to cause kidney and brain damage, among other disorders.

On a much broader scale, climatic change related to global warming is also expected to have a profound influence on ecosystems and crops. The most direct impact will be in coastal areas, where the rise in sea level will cause a landward shift of ecological zones and cropland. Some areas will be flooded, others damaged by storms, eroded shorelines, and increased salinity. In Bangladesh, 28 percent of the land would be lost to the sea by a 2-meter rise in sea level, flooding out tens of millions of the country's 130 million people (165 million projected by 2010) (see Figure 10.12).

Climatic change will also cause a latitudinal shift in ecological zones. Increased temperatures in North America, for example, will induce a northward shift in the ranges of many plants and the animals associated with them. Undoubtedly, some plants will not be able to make the shift and may suffer losses, even extinction. Crops may suffer too, especially in marginal agricultural lands where moisture supply is the limiting factor. As the dry zones expand with global warming, the combination of increased drought and human land-use pressure will likely accelerate desertification. According to some estimates, desertification is now overtaking about 200,000 square kilometers (75,000 square miles) of land each year.

Impacts on Water Quality

We have already established that water pollution in the form of acid rain is caused by air pollution, mainly from coal-burning power plants and industrial plants, and that air pollution contributes to eutrophication of inland waters. There are many additional contributions to water pollution from atmospheric sources. For example, coal-burning facilities also discharge two **heavy metal** contaminants (cadmium and arsenic) to surface waters via atmospheric emissions. Other pollutants, including DDT and PCBs, also enter surface waters through the atmosphere. In the Great Lakes atmospheric fallout is the principal source of DDT and PCBs. Banned in the United States in 1972, DDT continues to be transported through

(a)

(b)

FIGURE 11.15 *The effects of acidic moisture (rain, fog, and clouds) on North American forests: (a) Vancouver Island, Canada, and (b) the southern Appalachians.*

the atmosphere to the Great Lakes (and other surface waters, of course) from Mexico and other Central American countries where it is still used.

Acid precipitation also damages aquatic ecosystems in lakes and ponds as they lose their natural ability to neutralize acids after decades of acidic loading. As the pH drops, water bodies may lose their capacity to support some forms of aquatic life. Food chains are broken as insects and other lower organisms in the ecosystem disappear. Subsequently, dependent organisms in the food chain, such as fish, decline. As many as half the lakes in eastern Canada (including Ontario and Quebec)—more than 50,000 water bodies in all—have suffered increased acidity, some severely. Thousands of lakes in northeastern United States and northwestern Europe have also been affected.

Other lakes are threatened with nutrient loading from surface and atmospheric sources. The nutrients of greatest concern are nitrogen and phosphorus, because together they can induce accelerated growth in aquatic plants that advances **eutrophication**, the overgrowth and infilling process of water bodies. In Lake Michigan 20 percent of the annual phosphorus loading, the most critical of the two nutrients, comes directly from atmospheric deposition; in Chesapeake Bay, as much as 40 percent of the nitrogen loading comes from atmospheric deposition.

Water pollution from atmospheric sources is an especially serious management problem. The atmosphere is extremely mobile. Its wind and moisture system is capable of moving pollutants over thousands of miles in a matter of days. In addition, the exchange rate for atmospheric moisture is very fast, about once every nine days. Because of the atmosphere's rapid motion and enormous geographic range, pollution of surface water by the atmosphere is very difficult to control. In the Great Lakes, for example, management of terrestrial sources of some pollutants such as DDT and lead has been very successful in the past two decades. Management of atmospheric inputs, however, has been much less successful. Airborne contaminants continue to pollute the lakes with a large share coming from regions outside the Great Lakes' watershed.

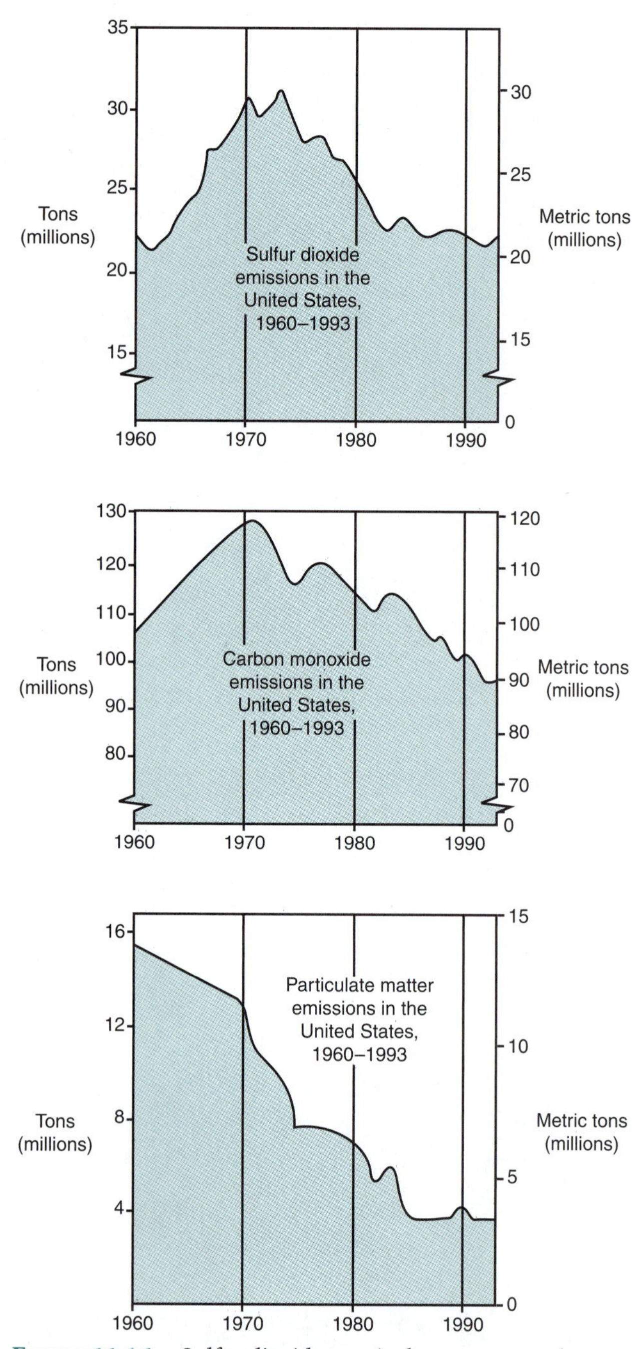

FIGURE 11.16 *Sulfur dioxide, particulate matter, and carbon monoxide emission trends for the United States since 1960. Concentrations of these pollutants in the lower atmosphere of the United States have also declined in response to pollution abatement efforts by the public, industry, and government.*

11.7 AIR POLLUTION CONTROL: HOW WELL ARE WE DOING

Air pollution control is pursued more or less at all scales from the global to local including the home and workplace. However, the effort is very spotty geographically with programs limited mostly to developed countries. Within the developed countries, progress has been made on certain fronts, especially with two major pollution sources, automobiles and power plants. On the other hand, acid rain is increasing and carbon dioxide buildup and ozone depletion continue to worsen over much of the industrial world in response to rising fuel and material consumption. Air quality remains deplorable in and around many industrial cities of eastern Europe, and in the United States, where air pollution control has been actively pursued for more than 20 years, seven major metropolitan areas are classified by the EPA as severely polluted, and one, Los Angeles, is classified as extremely polluted.

There are some signs of hope, however. Sulfur dioxide emissions, for example, which have been rising worldwide for more than a century, began to decline in several countries in the 1970s (Figure 11.16). These were mainly developed countries including Canada, the United Kingdom, France, and the United States, where reductions are attributed to a combination of factors, including economic change (e.g., reduction in heavy industry), stricter emission regulations, installation of exhaust-filtering equipment, and a shift to cleaner fuels. In the United

States, for example, most power plants have shifted to low-sulfur coal. Between 1975 and 1985 this resulted in a 44 percent reduction in SO_2 emissions per unit of power generated. Cleaner power plant emissions, together with reduced automobile and industrial pollution, also produced significant declines in particulate and carbon monoxide levels in the United States since the early 1970s (Figure 11.16). And in the winter of 1994–1995, cleaner-burning gasolines were required for automobiles in several major American cities. In Eastern Europe, the fall of communist power has improved the prospects of instituting pollution-control programs.

Prospects for Clean Air

Worldwide air-pollution control remains a great uphill battle and is certain to remain so for many decades. The reasons are compelling. First, world population is increasing by more than 80 million people a year. By 2020 it is projected to exceed 8 billion with the vast majority of the growth in the developing countries, where the prospects for implementing pollution-control programs are not good. Second, many of these people will be massed in sprawling super cities of 30 to 40 million people, where the sheer magnitude of air pollution problems may exceed reasonable management remedies. Third, the advantages of air pollution control in poor countries as a benefit/cost consideration are not encouraging when weighed against the overwhelming need for economic development in these countries. This will also be a consideration in former communist countries as they attempt to convert to free-market economies. Fourth, as developing nations strive to build consumer economies, per capita air pollution output will rise with the goods, energy, and services consumed. For example, China's drive toward industrialization will depend overwhelmingly on coal for energy, which is sure to add significantly to the global pollution load.

Building Pollution-Control Programs

Air-quality planning and management programs are founded in political systems. In order to establish regulatory policy and implement plans, several preliminary steps must be taken. Governments must first place air quality on their national agendas. This is often not politically feasible because many countries, by their own estimates, face more immediate national problems and simply do not find a place for air pollution in their political agendas. The more immediate problems vary from nation to nation but most deal with political unrest, economic development, military conflict, agricultural productivity, or health problems such as AIDS.

As the global implications of air pollution become more and more apparent, the circle of international dialogue has grown. Both developed and developing countries are being pressed by the global community to address the problem. The United Nations has played an important role in the process by promoting international dialogue, setting up air-quality monitoring programs, and making comparative evaluations among cities and nations worldwide. The United Nations has also promoted several international resolutions addressing global air pollution issues including carbon dioxide buildup and ozone depletion. One resolution established in 1987, called the **Montreal Protocol**, was designed to protect the ozone layer by cutting total CFC emissions by 50 percent by the year 2000.

Another resolution, the **Kyoto Protocol**, established in 1997, called for an overall cut in greenhouse gas emissions of about 5 percent from the world's 1990 emission levels. Developed countries were asked to make the largest cuts: Canada and Japan 6 percent, the United States 7 percent, and countries of the European Union 8 percent. But agreements have been slow in coming and negotiations between developed countries and developing countries continue over cutback quotas. International ratification is needed to make the protocol legally binding. The United States has been generally slow to back most international efforts in air quality management. In 2001, President George W. Bush rejected the U.S. allocation under the Kyoto Protocol on the grounds that it would hurt the U.S. economy.

Air Pollution Control in the United States

In the United States, the responsibility for setting and enforcing national air quality standards belongs to the U.S. Environmental Protection Agency. Under the **Clean Air Act** (originally enacted in 1970 and last amended in 1992), the EPA was directed to set air quality standards to protect human health with an adequate margin of safety. The resultant standards, called the **National Ambient Air Quality Standards (NAAQS)**, defined two levels of air quality standards: (1) primary standards designed to protect the health of people; and (2) secondary standards designed to protect against environmental degradation. The Clean Air Act amendments also set air quality standards for certain land-use areas. Three classes of areas were defined for the purpose of preventing air quality deterioration beyond certain base performance levels:

- **Class I**. Areas such as national parks and wilderness areas have the highest standards (that is, where least deterioration is allowed).
- **Class II**. These are broadly defined as intermediate areas and have intermediate standards.
- **Class III**. Industrial areas have the lowest standards.

In practice, all areas are defined as Class II unless they are moved into Class I or Class III by the EPA.

In the 1970s and 1980s, pollution control in the United States focused on the least expensive measures and largest sources of pollution. Two major areas of activity, large industrial facilities and motor vehicles, were pressed with most of the responsibility for cleanup. Emission standards were implemented for automobiles that included exhaust system devices called catalytic converters on all new vehicles. In addition, lead was banned from use as a gasoline additive in new automobiles. Fuel economy standards, called **CAFE standards** (corporate average fuel economy standards) were mandated (Figure 11.17a). As a result of these actions, pollution rates per automobile dropped significantly and lead levels in air and water declined over most of the United States. The Great Lakes, for example, saw lead concentrations drop dramatically between 1970 and 1980. On the other hand, the number of automobiles in use increased significantly between 1970 and 1995 and the total miles traveled by Americans annually increased from about 1,000 billion to 2,200 billion, both of which offset much of the gains made by reduced emissions and better mileage (Figure 11.17b).

In the late 1980s, further amendments were added to the Clean Air Act. These amendments address three major areas:

- Reductions in sulfur dioxide emissions from power plants with a target of 10 million tons reduction per year over the 1990s.
- Reductions in hazardous air pollutants (mainly synthetic compounds such as benzene, and PCBs).
- Improved air quality in selected areas, mostly urban areas, where national ambient air quality standards are consistently being violated.

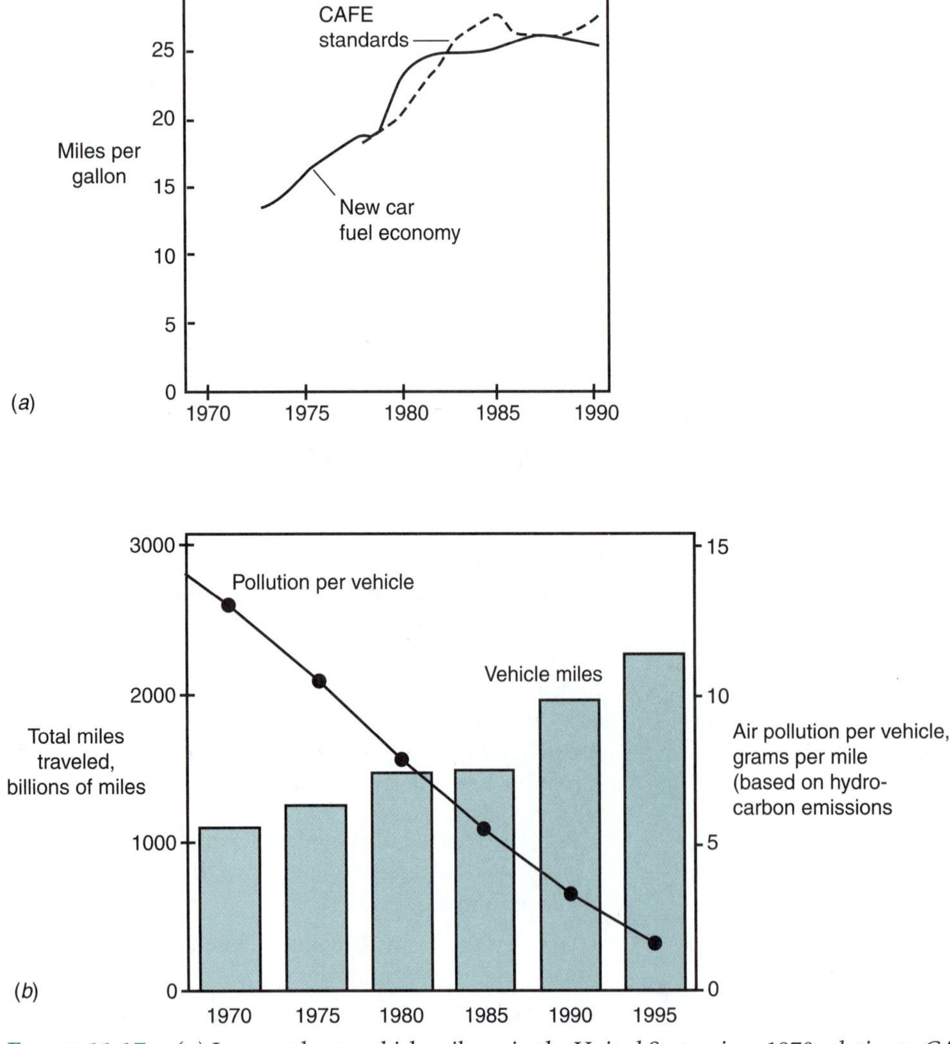

FIGURE 11.17 *(a) Improved new vehicle mileage in the United States since 1970 relative to CAFE standards; (b) the increase in vehicle miles in the United States relative to pollution emissions per vehicle since 1970.*

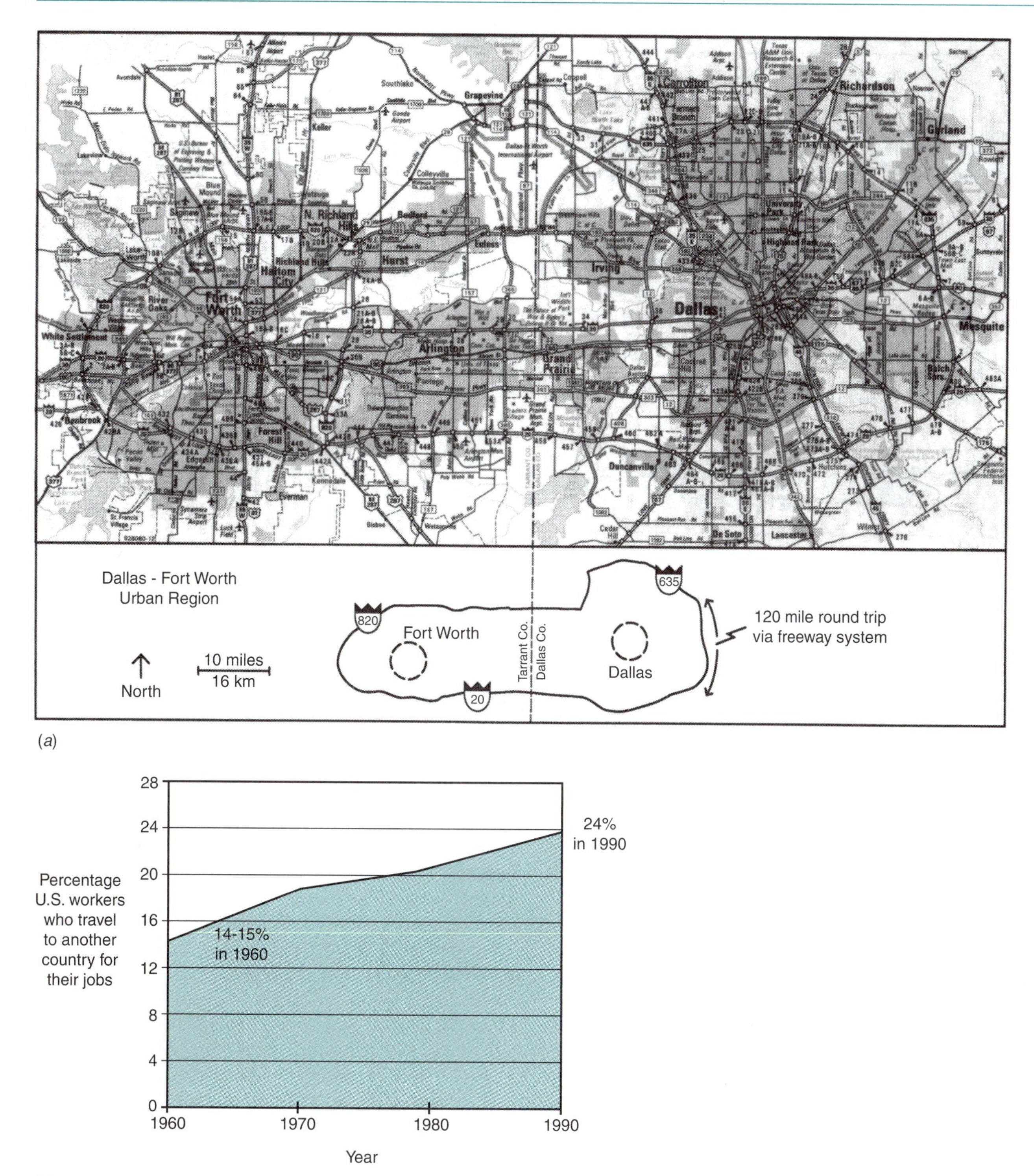

Figure 11.18 *(a) Highway system and edge city development in the Dallas-Fort Worth metropolitan region. (b) Graph showing the increase in U.S. daily commuter traffic across county lines, that is, from one city or suburb to another.*

In **southern California**, where regional air pollution is extremely critical, various new management concepts are being considered. The Los Angeles Air Management District is implementing a program based on a market incentives concept. Under this program, polluters are allocated a pollution allowance, a fixed quantity of allowable emissions. The total allowance for the district is designed to meet air quality standards set by the EPA. The novelty of the program is that it allows polluters to buy and sell allocations. This rewards those who are able to achieve lower emissions rates by allowing them to sell allocations for profit to less-efficient operations. The over-all balance, given both more-efficient and less-efficient operations, is designed to lower pollution levels over the district.

The Los Angeles situation is an extreme example of urban air pollution faced by virtually every metropolitan center in the United States. The principal source of the problem is increased automobile traffic brought on by population growth, increased use of cars, and longer travel distances. Many cities have sought various means of reducing the traffic load, including the use of ride sharing and mass transit systems. At the same time, however, the states and the federal government continue to build expressways in and around urban areas, which encourages more automobile travel over greater distances. And unlike most European countries, the United States imposses light taxes on gasoline, thereby maintaining relatively low fuel costs for vehicle operations.

In addition, the United States lacks a comprehensive national program for transportation and energy that could reduce car traffic, truck traffic, overall energy use, and pollution emissions. A national transportation plan, for example, could encourage rail transportation for heavy commercial and industrial goods as a means of reducing trucking, which is less energy-efficient with greater pollution emissions per ton of material transported. At the local level, urban sprawl has continued with the development of the expressway system, as we noted earlier. In fact, it is now giving rise to a new urban form, the **edge city**, at key interchanges around and between urban regions (Figure 11.18a). Served exclusively by automobile and truck traffic, edge cities foster even greater travel, especially among commuters. Daily one-way travel distances between home and workplace of 50 miles or 75 miles are not uncommon today.

On balance, we in the United States and Canada have developed a system that is very expensive for the citizen, local units of government, and the environment. It relies overwhelmingly on cars and trucks to move people and materials and encourages long daily travel distances. This, in turn, not only promoters massive fuel consumption and air pollution, but necessitates a huge and expensive highway infrastructure, sprawling land-use patterns, and highly inefficient communities. The 1990 U.S. census revealed that the trend toward long commuting distances continues to rise; nearly 25 percent of the workers in the United States actually drive to another county to work (Figure 11.18b). With the development of sprawling, urban areas and edge cities tied together by systems of expressways, a trend toward travel from suburb to edge city, more or less around the urban area, is emerging. The traditional travel pattern of the past three decades from suburb to city center is declining according to census data.

11.8 SUMMARY

Air pollution has been a part of the human environment since prehistoric times. Today, however, it has reached massive proportions and threatens to change the global environment. The sources of air pollution are many and are often complexly related to the environment. However, there is no doubt about the root cause: a world population of more than 6.1 billion growing by more than 80 million people a year, which is projected to exceed 8 billion people by 2025. Most of these people will be in developing countries where air pollution will likely reach unprecedented levels with higher levels of fossil-fuel burning related to consumption of goods from factories, home heating, cooking and transportation.

The developed countries are currently the major contributors to air pollution. Most pollutants in these countries come from motor vehicles, power plants, and industrial operations and result mainly from fossil-fuel combustion. The principal types of pollutants are carbon monoxide, carbon dioxide, nitrogen oxides, particulates, and hydrocarbons. In the developing countries, where motor vehicles are a less significant but rising source of air pollution, forest fires, residential fires, and industry are the primary sources.

Air pollution varies not only with emission rates but with the behavior of the atmosphere at different geographic scales. The heaviest levels of outdoor pollution are found locally within urban areas during work times, when the lower atmosphere is calm and thermally inverted. Crosswinds disperse pollutants from urban source areas, giving rise to great plumes extending sometimes hundreds of miles downwind. At a broader scale, polluted air is dispersed regionally by prevailing wind systems such as the Westerlies, which in the midlatitudes distribute acid precipitation across eastern North America and northwestern Europe. Tropospheric mixing by storm systems delivers pollution to the stream of upper level winds where CO_2, CFCs, particulates, and other pollutants are dispersed over all or much of the globe.

The effects of air pollution on humans and the environment are enormous. Human health is directly affected by the inhalation of particulates, various gaseous pollutants, and radioactive materials, resulting in allergic reactions, asthma, emphysema, and lung cancer. As a whole, cities are less healthy places for humans in both rich and poor countries because of higher air pollution levels. The environment is affected both directly and indirectly; the impacts include crop damage, forest damage, decline in ecosystem productivity, increased eutrophication, water pollution, and global climatic change. Climatic change related to atmospheric warming from CO_2 buildup is widely forecast by scientists. Among the anticipated effects of global warming is sea-level rise, which could have devastating effects on human populations and ecological conditions in coastal areas.

Attempts to control and reduce air pollution must incorporate sweeping changes from the local to the global scale. At the national scale for developed nations, energy and transportation plans are sorely needed, especially in the United States. In addition, more effective energy and material use in manufacturing and consumer habits are needed. At the international scale, economic change, political cooperation, emission abatement, and most importantly, population control are needed in order to manage air quality. It is widely agreed that if any one global measure were to be implemented it should be population control. Among the most threatened air quality environments of the next several decades are the large cities of the developing countries, such as Mexico City and Rio de Janeiro, which are growing rapidly toward populations of tens of millions.

11.9 KEY TERMS AND CONCEPTS

Primary air pollutants
Secondary air pollutants
Major air pollutants
Air pollution sources in developed countries
- photochemical smog
- Los Angeles-type air pollution
- industrial smog
- acid precipitation
- CFCs
- ozone depletion

Pollution sources in developing countries
Processes, patterns, and scale
- dust dome
- plumes
- inversion layer
- thermal inversion
- tropospheric mixing
- global dispersal

Urban air pollution
- land-use relation
- new building sickness
- radon
- latent heat
- sensible heat
- urban sprawl
- cluster development

Human health impacts
- particulates
- carbon monoxide
- allergens
- ultraviolet radiation
- health costs

Crop, forest, and ecosystem damage
Water quality
- eutrophication
- heavy metals

Pollution control
- Montreal Protocol
- Kyoto Protocol
- U.S. Clean Air Act
- National Ambient Air Quality Standards (NAAQS)
- CAFÉ standards
- southern California
- edge cities

11.10 QUESTIONS FOR REVIEW

1. Is air pollution unique to modern times? How was air pollution different in ancient times?
2. Name the major air pollutants and briefly describe their character, origins, and harmful effects. Distinguish between primary and secondary pollutants.
3. Name some major natural sources of air pollution. Compared to human sources, does nature produce more or less of pollutants such as carbon dioxide, carbon monoxide, nitrogen dioxide, sulfur dioxide, particulates, and hydrocarbons? How do you account for the concern over carbon dioxide pollution from human sources when it represents only about 3 percent of the annual input to the atmosphere?
4. Describe the character of the air pollution associated with developed countries. Is there a difference in the types, sources, and rates of air pollution between developed countries and developing countries? What trends are emerging in the developing countries?
5. Describe the sources and particular conditions associated with photochemical smog, acid precipitation, and ozone depletion. What is the CFC residence time in the atmosphere, and where geographically do CFCs have the most profound effect?
6. Describe the factors influencing the distribution of pollutants in the urban atmosphere. What conditions give rise to strong dust domes? How do thermal

inversions induce extreme levels of air pollution in urban areas?

7. How does the troposphere incorporate pollutants from surface sources and disperse them regionally and globally? Is there evidence that zonal circulation associated with prevailing wind systems plays a role in air pollution patterns?
8. Describe the typical pattern of land uses in large cities (urban areas) and the related distribution of air pollution, especially that associated with the transportation network. Is there any concern about air pollution in buildings? Explain.
9. Outline several important impacts that air pollution has on (a) human health, (b) crop, forest, and aquatic ecosystems, and (c) water quality.
10. Identify some recent trends in air pollution in developed countries. What are the prospects for reducing air pollution in developing countries faced with economic development and population growth in the next several decades? Discuss some of the key obstacles.
11. Describe the general character of the federal program for air pollution control in the United States. How has the program changed from the 1970s and 1980s to the present?
12. Argue a case for or against the trend toward expressway development, urban sprawl, and increased automotive travel in the United States on the basis of air pollution, energy use, and the cost of urban infrastructure (e.g., water, sewer, street systems). Can you recommend an alternative plan?

12

THE HYDROLOGIC ENVIRONMENT AND LAND USE

12.1 INTRODUCTION

Earth and Mars both contain vast reservoirs of water, but the contrasting thermal conditions on the two planets have created sharply different water regimes. On Mars, where the surface temperatures rarely rise above freezing, virtually all the water is frozen in the ground and in polar ice caps. On Earth, where the global surface temperature averages 15°C (60°F), less than 2 percent of the water is frozen and liquid water covers more than 70 percent of the surface.

The hydrologic condition of Earth is essential to understanding life and the life support system. The seas nurtured the first life billions of years ago, and modern life, both aquatic and terrestrial, remains water-dependent. Outside the Earth's cold regions, water is the most important limiting factor to life in terrestrial environments, not only in terms of availability but in quality as well.

All water is delivered to the land by the atmosphere as a part of weather and climatic systems. The rate of delivery is typically irregular and this poses a perennial problem for humans throughout most of the world. Not only is the water supply highly uneven geographically, but it also varies over time, so that most places rarely get the expected or desired amount of water. This has led people to manipulate the natural water system by building dams, aqueducts, irrigation systems, storm drains, and flood control structures in an effort to make the system more dependable and less threatening.

In this chapter we are interested in learning about the Earth's water system and the role of water in the environment. The hydrologic cycle is taken up first, followed by an examination of precipitation processes and their influence on the environment. Streamflow, the hazard of floods, and their relationship to human land use follows. The next two topics are groundwater, one of the most threatened parts of the modern hydrologic environment, and human water supplies and uses in North America.

The final topic centers on management and environmental problems associated with water use.

12.2 THE HYDROLOGIC CYCLE

The Earth's water is held in five **natural reservoirs**: the oceans, glacial ice caps, groundwater, surface water (lakes, streams, and wetlands), and the atmosphere (Table 12.1). The oceans hold more than 97 percent of the Earth's total water supply and serve as the central source of water for the rest of the world. Glacial ice, the largest reservoir of freshwater, contains nearly 2 percent of the Earth's water, held principally in the ice caps of Antarctica and Greenland. The last three reservoirs—surface water, groundwater, and atmospheric moisture—together contain about 0.6 percent of the total. These are the sources humans and other terrestrial organisms depend on most directly for their water supplies. Agriculture, industry, power plants, and homes use surface water and groundwater almost exclusively.

Components of the Cycle

The **hydrologic cycle** is a great flow system characterized by exchanges of water among the five major reservoirs (Figure 12.1). The cycle is critical to our understanding of Earth's life support system for, among other things, it is the only natural means of (1) supplying water to the land masses and (2) exchanging the saltwater of the oceans for freshwater. In addition, the hydrologic cycle is a major vehicle for the transport of many life-sustaining nutrients to terrestrial and aquatic ecosystems.

The hydrologic cycle is driven by surface heat derived from the Earth's absorption of solar radiation. The cycle begins with the evaporation of seawater, which supplies the atmosphere with water vapor. The atmospheric vapor condenses to form droplets of freshwater, which are precipitated to Earth. About 22 percent of this precipitation falls on the land masses, where it soaks into the ground, is stored on the surface (e.g., as snow or in lakes), or runs off into streams. The remainder of the Earth's precipitation (78 percent) falls on the oceans (Figure 12.1).

More than half (58 percent) of the water precipitated on the world's land masses is returned directly to the atmosphere by evaporation and transpiration. The rest goes into groundwater, soil water, streams, lakes, glaciers, and so on. Groundwater is Earth's largest and slowest reservoir of fresh, liquid water. The exchange time or replacement time for groundwater is as long as thousands of years (Table 12.2). This has grave implications for polluted groundwater because the rate of contaminant flushing is exceedingly slow. Ultimately, most groundwater is released from the land by seepage into streams and rivers that discharge into the sea. All told, 42 percent of water delivered to the land by precipitation is ultimately returned to the oceans as runoff.

TABLE 12.1

MAJOR NATURAL RESERVOIRS OF WATER ON EARTH

Reservoir	Proportion of Water (Percentage)
Oceans	97.410
Ice caps and glaciers	1.984
Groundwater and soil water	0.599
Streams, lakes, ponds	0.007
Atmosphere	0.001

Source: Speidel, D. H., and A. F. Agnew. 1988. The World Water Budget. In Perspectives on Water Uses and Abuses. New York: Oxford University Press.

Water Vapor, Condensation, and Precipitation

The total amount of water vapor in the atmosphere at any moment is surprisingly small (Table 12.1). If all of it could be precipitated onto Earth's surface, it would form a layer of water only 2.5 centimeters (1 inch) deep over the entire planet. Why, then, is there so much precipitation over much of the planet, more than 250 centimeters

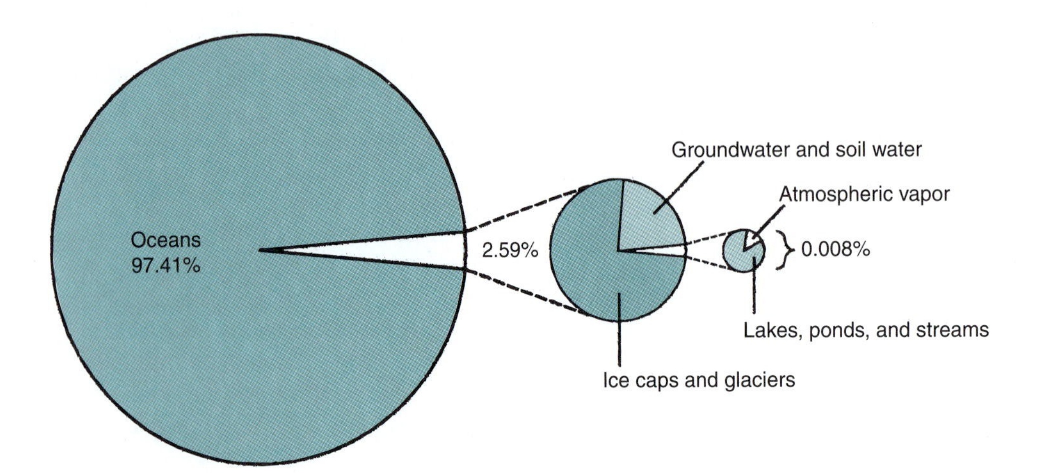

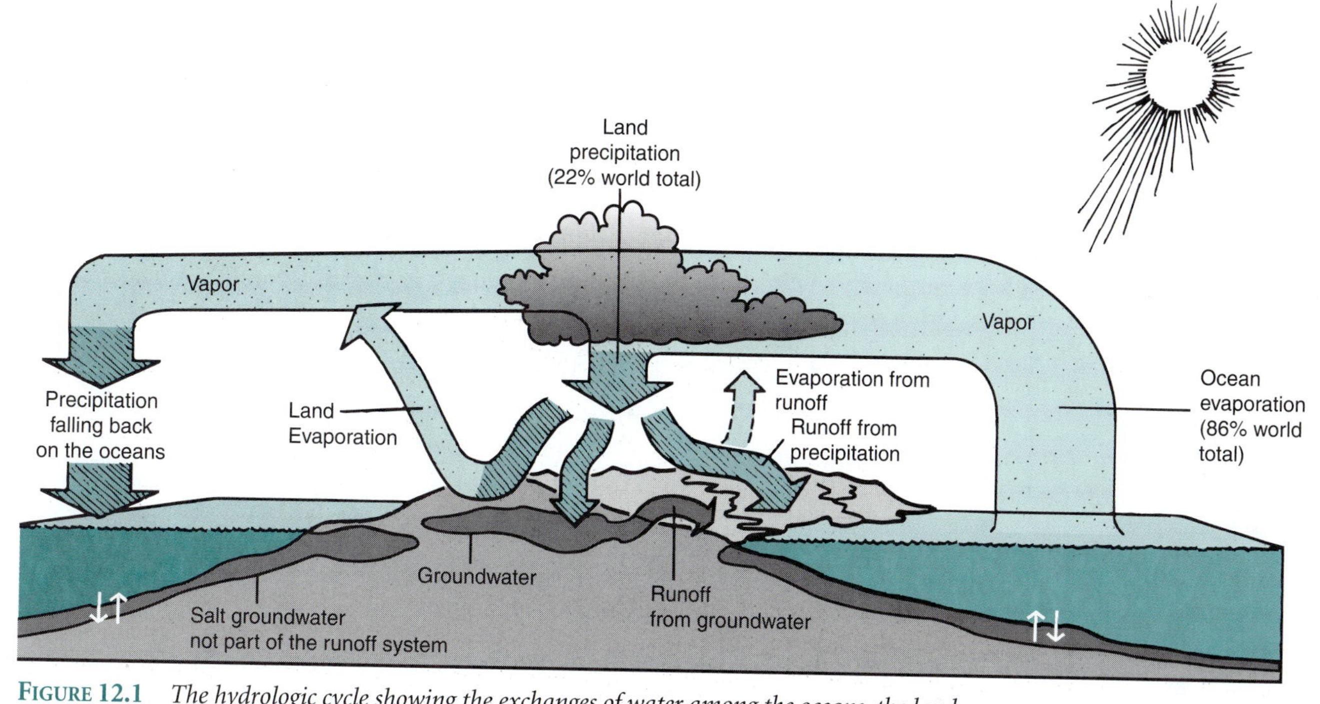

FIGURE 12.1 *The hydrologic cycle showing the exchanges of water among the oceans, the land, and the atmosphere.*

(100 inches) a year over large areas of the tropics and more than 50 centimeters (20 inches) a year over most of the Earth?

Moisture Exchange and Delivery

The answer lies in the rapid exchange of water between the atmosphere and the Earth's surface, that is, in the rapid recycling between precipitation and evaporation. On the average the atmosphere exchanges its total moisture content with the Earth's surface once every nine days (Table 12.2). This amounts to about 40 exchanges of water a year, which, at 1 inch per exchange, equals close to 40 inches of precipitation, the average annual precipitation for the Earth as a whole.

The **precipitation process** begins with the condensation of water vapor in response to atmospheric cooling.

TABLE 12.2

EXCHANGE TIMES IN THE HYDROLOGIC CYCLE

Reservoir	Exchange Time
Atmosphere	9 days
Streams and rivers	10 to 15 days
Large lakes	10 to 20 years
Oceans	3,000 years
Groundwater	100s to 10,000 years
Ice cap	10,000 years

Source: Modified from Ehrlich, Ehrlich, and Holdren. 1977. *Ecoscience: Population, Resources, Environment.* San Francisco: Freeman.

Minute water particles form first. As they grow, they coalesce to form droplets with enough mass to fall through the atmosphere. Cleaner than most surface water and decidedly free of most ocean salts, precipitation is nevertheless hardly pure freshwater when it arrives at the Earth's surface. Most droplets contain a nucleus of particulate matter or a **condensation nucleus**, a tiny airborne particle from natural or human sources around which condensation and droplet formation take place. Where debris such as volcanic dust or soot from industrial exhausts is heavy in the atmosphere, precipitation may incorporate a large number of particles, and in some instances may be quite dirty by the time it reaches the ground.

Gaseous pollutants also combine with precipitation water to alter its chemical makeup. As we revealed in the previous chapter on air pollution, oxides of sulfur and nitrogen combine with cloud and precipitation droplets, forming acids that lower the pH of the moisture. The result is **acid rain**, **acid clouds**, and **acid fog**. Fog is simply cloud particles at or near the surface. Where fog or low clouds containing acid moisture come into contact with the landscape, as is very common on mountain slopes, the moisture collects on vegetation and soil with the same effect as acid rain. The decline or loss of large tracts of forests in Europe and North America is attributed to the combined effects of acid deposition from clouds, fog, and rain (see Figure 11.15).

The meteorological conditions that produce precipitation always involve cooling of air containing water vapor and condensation nuclei. The principal means of cooling is air rising into the atmosphere. As temperature drops, the capacity of the air to hold water vapor declines until condensation sets in. The temperature at which air is fully saturated, that is, has reached 100 percent relative humidity, is called the **dew point**. Condensation increases rapidly (1) at temperatures below the dew point when the air is supersaturated (beyond 100 percent relative humidity) and (2) when condensation nuclei are abundant. One of the reasons for higher-than-average precipitation rates and incidences of fog in and around urban areas is the large supply of airborne particles from pollution sources such as factories, automobiles, and power plants.

12.3 PRECIPITATION, STORMS, AND THEIR IMPACT ON ENVIRONMENT

Four basic mechanisms or processes of cooling produce precipitation. The precipitation that results from each is termed a **type of precipitation**. From an environmental standpoint, precipitation processes are extremely important. They not only determine the amounts and distributions of freshwater received by the land, but they are also formative physical agents in shaping the landscape and its organisms. For example, violent weather such as tornadoes, hurricanes, and thunderstorms are associated with precipitation processes. In the following pages the essential processes, characteristics, and impacts of the four precipitation types are described.

Convectional Precipitation

One of the most common and powerful storms in nature is the **convectional storm**. In convectional storms a body of air (called a *parcel*) is heated in the lower atmosphere, which, in turn, forces it to expand. Expansion reduces its density, causing it to become buoyant just as a hot-air balloon becomes buoyant when heated. When the heated parcel of air ascends into the atmosphere, it expands further in response to the declining air pressure around it at higher altitudes. Beneath it the rising parcel entrains an upward flow of air currents from the surface, called **thermals** (Figure 12.2).

The expansion of the parcel caused by its decompression results in the decline of internal air temperature. Temperature within the rising parcel falls at a rate of −1°C per 100 meters of altitude (called the *adiabatic lapse rate*) until condensation and cloud formation begin. Condensation results in the release of stored heat, called **latent heat of condensation**, which was originally taken up from the Earth's surface in the evaporation of water from soil, water surfaces, and vegetation. With the release of latent heat, the rising parcel of air is given an extra charge of energy that induces a faster and stronger upward flow of air, driving the parcel even higher into the atmosphere. The temperature within the parcel now falls at a lower adiabatic rate, typically between −0.3°C and 0.8°C per 100 meters of altitude. With less energy lost in cooling, more energy is available to drive the storm.

Thus, the young convectional storm works like a great chimney, drawing huge drafts of air containing heat, moisture, and pollutants off the Earth's surface. Above the base of cloud formation the storm becomes highly turbulent with the release of latent heat, the formation of strong updrafts, and the release of precipitation with downdrafts falling earthward (Figure 12.2). The designation *thunderstorm* comes with further development, which includes violent turbulence sometimes accompanied by tornadoes, lightning, thunder, and towering vertical cloud formations typically reaching the upper levels of the troposphere.

Environmental Impacts

Precipitation from the thunderstorm may include hail, but most falls as rain. The intensity of the rainfall, measured in centimeters per hour, is typically great, ranging from 2 to 10 centimeters (1–4 inches) per hour. The combined effects of a thunderstorm—the hard, gusty winds, the large hail stones, the lightning, and the torrential rain—can be very damaging to the landscape. Each year Earth is hit by an estimated 3 billion lightning strikes that are a major cause of forest and grassland fires. Torrential rains cause extreme rates of runoff, which in small watersheds often produce severe flooding. The rise of flood waters from convectional rainfall is usually sudden and massive, sometimes resulting in dangerous **flash floods**, which are destructive to the landscape and certainly hazardous to humans and other organisms. Flash floods are often most severe in arid landscapes where there is little vegetation to slow the rate of runoff (Figure 12.3).

Since most thunderstorms are driven by moist air heated over land surfaces, the geographic distribution of convectional precipitation is reasonably easy to approximate. Most originate over land masses because land surfaces are capable of generating much higher air temperatures from solar heating than water surfaces. This is a function of the difference in a basic thermal property of earth materials, called *specific heat*, which reveals that water can take on large amounts of heat energy with only minimal rise in temperature. Land, on the other hand, has a much lower specific heat, that is, rises to a high temperature with the input of heat energy. Thus, for a given input of solar radiation, land surfaces generate much higher temperatures than water surfaces. Beyond this factor, the main controls on convectional precipitation are geographic differences in solar heating and moisture availability. The difference between southwestern and southeastern United States is a function of moisture availability (see map in Figure 12.2).

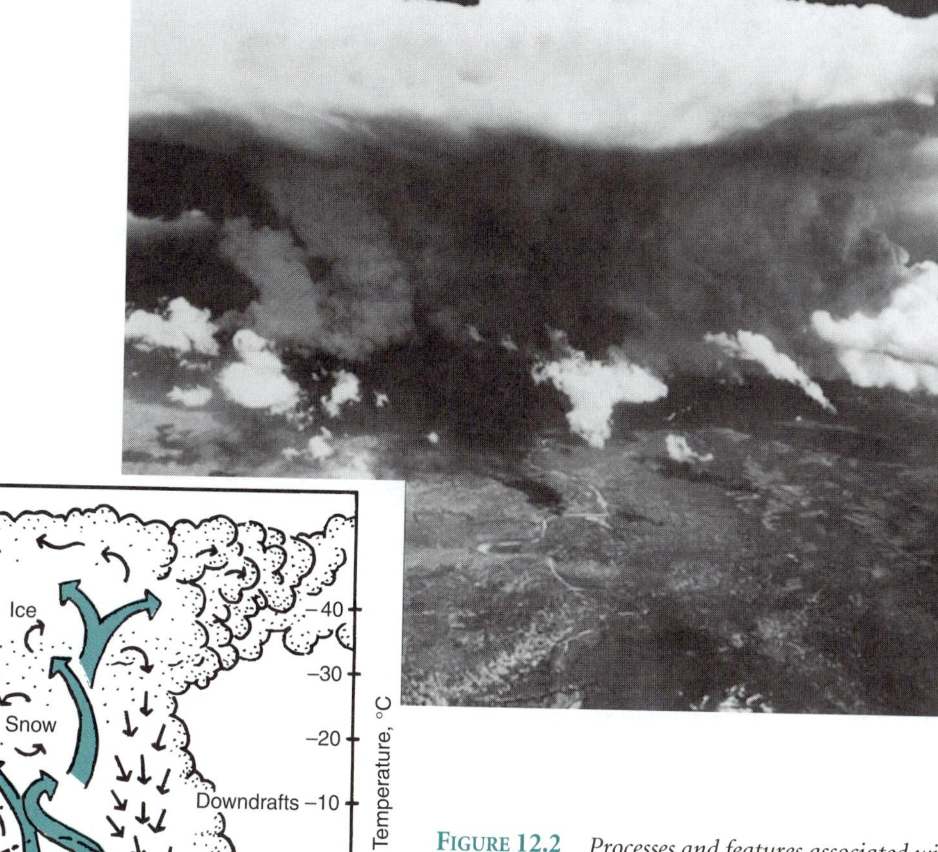

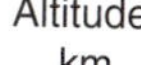

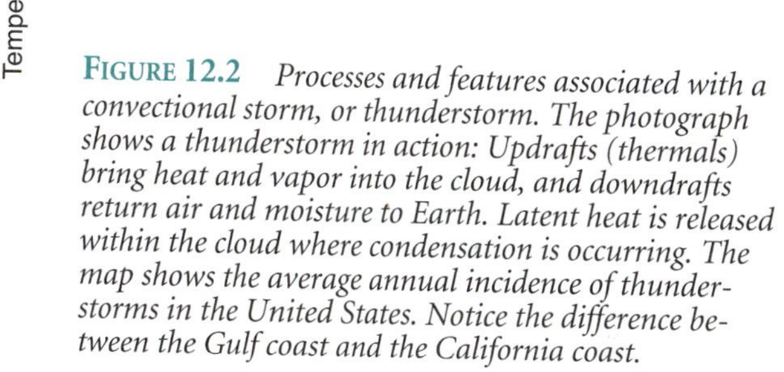

FIGURE 12.2 *Processes and features associated with a convectional storm, or thunderstorm. The photograph shows a thunderstorm in action: Updrafts (thermals) bring heat and vapor into the cloud, and downdrafts return air and moisture to Earth. Latent heat is released within the cloud where condensation is occurring. The map shows the average annual incidence of thunderstorms in the United States. Notice the difference between the Gulf coast and the California coast.*

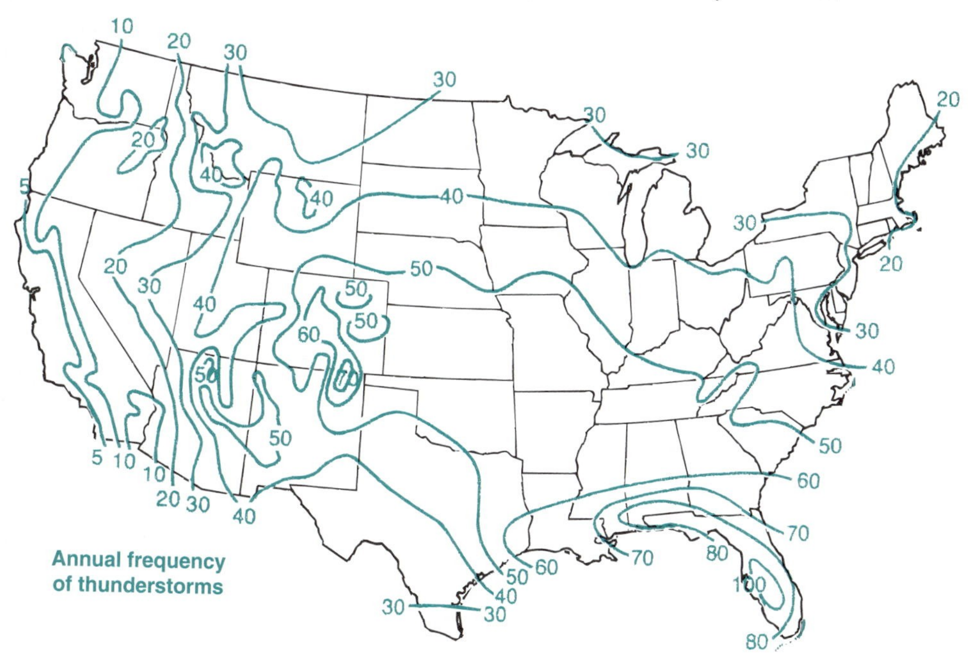

FIGURE 12.3 *A road washout from flash flooding in the American Southwest, the result of intensive rainfall and runoff from a thunderstorm.*

Within land areas, the greatest incidence of convectional storms occurs in wet tropical and humid subtropical climatic zones. Storm frequency in both is greatest during the high sun season, which near the equator occurs twice a year, because that is when solar energy is at maximum intensity. Thunderstorms are also common in humid midlatitude land areas in the summer. Most are caused by solar heating, but some are also caused by frontal activity associated with cyclonic storms, which are discussed shortly.

Orographic Precipitation

Orographic storms occur as air is cooled in passing over mountains, and they are much less violent than thunderstorms. The ideal geographic setup for this type of storm involves moist oceanic air blowing against a mountain range that runs along the coastline, such as the Prevailing Westerlies blowing against the ranges of the Pacific Coast of North America (Figure 12.4). Cloud formation and precipitation are concentrated on the windward side of the mountains. Precipitation rates are typically very high, including the highest annual precipitation values in the world, more than 1,000 centimeters (400 inches). Orographic environments generally produce heavy cloud covers that may be associated with temperature inversions in the lower atmosphere. A **temperature inversion** is a reversal from the normal thermal structure of the troposphere; that is, the atmosphere warms rather than cools with altitude. In orographic situations, the warming may be caused by the release of latent heat with condensation and cloud formation.

Environmental Impacts

Temperature inversions result in highly stable conditions in the lower atmosphere—that is, ones with no vertical mixing, because light, warm air aloft overlies heavier, cooler air near the ground. Such inversion layers may lodge against the mountains, and if regional winds and migrating air masses do not dislodge them, the inversions may last for several days, retarding the upward flow of surface air. If an inversion overlies an urban area, such as Seattle, Washington, or Vancouver, British Columbia, pollutants are held down and they build up in the surface layer often producing health-threatening conditions (see Figure 11.9). In addition, oxides from **air pollution** combine with the moisture to produce acid fog, clouds, and rain that damage vegetation and aquatic habitats on mountain slopes. See Chapter 11 for additional information on temperature inversions and air pollution.

The global distribution of orographic precipitation is based on a simple model of moist air driven by prevailing winds against a mountain range running more or less crossways to its path. Two belts of winds, the Prevailing Westerlies and the Easterly Tradewinds, are widely associated with this geographic arrangement (see Figure 10.5). Orographic precipitation associated with the Prevailing Westerlies is most pronounced in midlatitudes along the North American and northern European

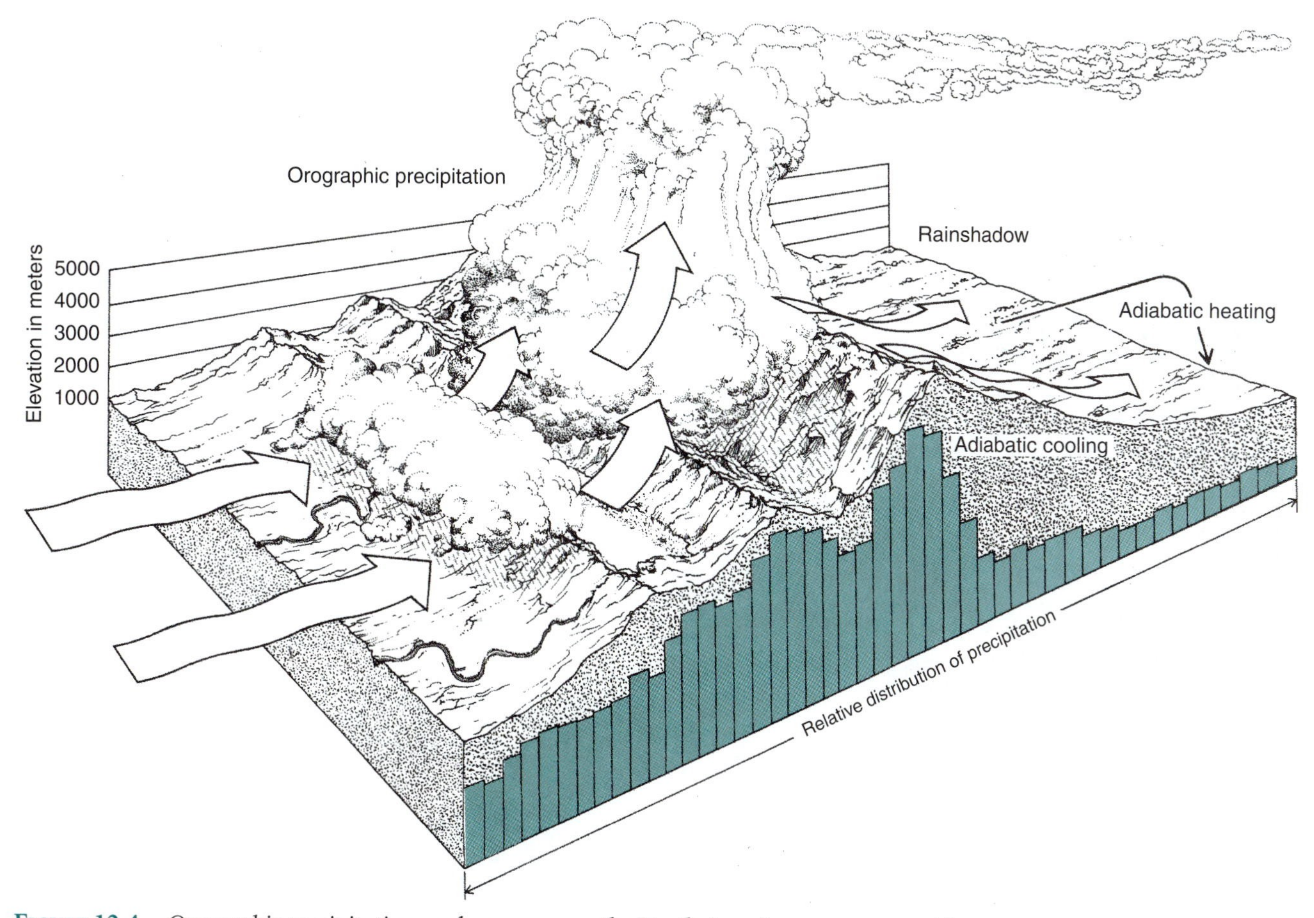

FIGURE 12.4 *Orographic precipitation, such as occurs on the North American west coast, with heavy rainfall and snowfall on windward slopes. The leeward slope is very dry by comparison.*

coasts. (In the Southern Hemisphere there is so little land in the midlatitudes that the Westerlies cause little orographic precipitation.) In the tropics, the moisture-laden Easterly Tradewinds bring heavy orographic precipitation to eastern-facing slopes of mountain ranges. The highest average precipitation on Earth, nearly 1,200 centimeters (486 inches) a year, occurs in the Tradewind belt on one of the Hawaiian Islands.

Frontal/Cyclonic Precipitation

Frontal (or cyclonic) **storms** are brought on by the interplay of cold and warm air masses in the middle latitudes. Within this broad zone of latitude the **polar front** forms a line of contact that separates the cold air of the high latitudes from the warm air of the low latitudes (Figure 12.5a). Along the polar front atmospheric disturbances develop causing a kink, or wave, in the front. The wave represents a small area of low atmospheric pressure that draws in surface air from both the warm and cold sides of the polar front (Figure 12.5b).

As the wave grows and pressure declines, air is pulled in from increasingly greater distances and the entire system is set into a great swirling motion. This swirling motion is caused by the *Coriolis effect.* The Coriolis effect is a deflection of moving airborne objects (wind, birds, aircraft) off a straight line of travel over a long distance. It is related to the Earth's rotation and latitude on the planet's surface. The direction of deflection is always to the right in the Northern Hemisphere and to the left in the Southern; however, there is no Coriolis effect along the equator.

Fronts form at the contact between the contrasting air types—a *cold front* at the leading edge of the cold air and a *warm front* along the leading edge of warm air (Figure 12.5c). The structure of the fronts is based on air density. Since cold air is denser than warm, cold air assumes the lower position and warm air the upper position along the frontal contact. As warm air is driven up along the fronts, cooling and precipitation take place. Both fronts produce precipitation, but in most other respects they are quite different. The warm front is characterized by a broad layer of clouds and widespread, usually gentle, rainfall or snowfall as the warm air gradually overrides the cold, denser air and is forced aloft. Cold fronts, by contrast, force warm air to rise rapidly and often produce strong winds and towering clouds, thunderstorms, heavy

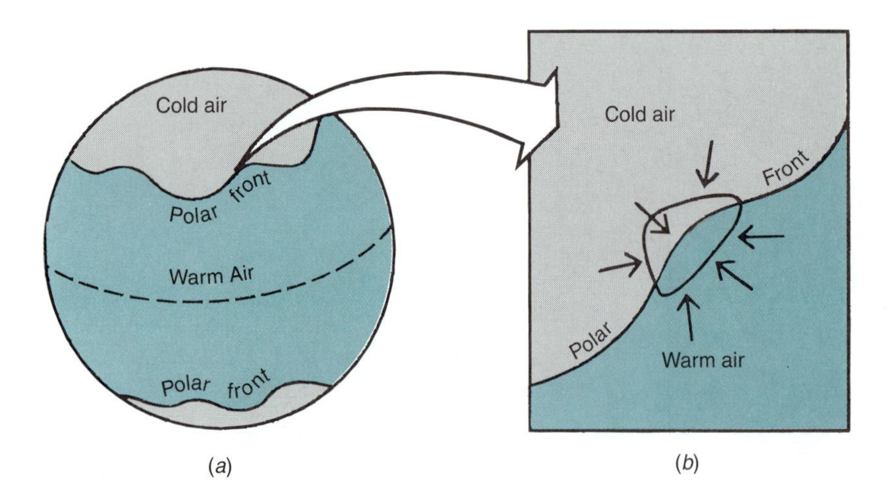

FIGURE 12.5 *(a) The polar front, a wavy boundary separating global regions of cold and warm air; (b) the formation of a disturbance along the polar front leading to a cyclonic storm; (c) a fully developed midlatitude cyclone showing the fronts, winds, and cloud cover; (d) a diagram of a cold front and a tornado.*

precipitation, and occasional tornadoes (Figure 12.5d). **Tornadoes** are small, violent windstorms characterized by air spiraling inward at tremendous speeds estimated up to 300 miles per hour or more. The spiraling air forms a funnel cloud that may sweep over the ground, sucking in air and destroying virtually everything in its path.

Environmental Impacts

Most tornadoes are spawned by severe weather conditions associated with cold fronts, isolated convectional cells, or clusters of convectional cells. The meteorological conditions that produce tornadoes are not well understood. They tend to develop where moist, unstable surface air rises rapidly into an overlying layer of dry air with strong winds aloft. Unfortunately, it is very difficult for meteorologists to forecast where tornadoes will occur under threatening weather conditions, and they therefore remain one of the most feared storms of the midlatitudes. Based on records of tornado occurrence and damage in North America, we do know that (1) they are most common in the United States east of the Rockies, especially in the Great Plains between northern Texas and northern Kansas in April, May, June and July; and (2) most are spawned by large thunderstorms called mesocyclones.

Between 1990 and 2000, 13,026 tornadoes were reported in the United States, and they caused 619 deaths and hundreds of millions of dollars in property damage. In the twentieth century more than 10,000 Americans were killed by tornadoes. The world's worst tornado disaster occurred in Bangladesh in 1989 where 1,109 people were killed and 15,000 injured.

Besides tornadoes, there are many other important environmental consequences associated with frontal precipitation. Most significantly, fronts are a major source of precipitation to lands in the midlatitudes. In contrast to convectional precipitation, which is usually localized, frontal precipitation is often extensive, stretching over distances of 800 kilometers (500 miles) to 1,600 kilometers (1,000 miles). In addition, the entire storm system migrates eastward (more or less) with the Prevailing Westerly winds of the midlatitudes; therefore, a single storm may affect fully half of the United States or Canada before it drifts over the Atlantic Ocean (Figure 12.6). When rainfall is heavy and prolonged, regionwide flooding may result, affecting large watersheds such as that of the Missouri and upper Mississippi in 1993 and 2001 (see Figure 12.12).

Frontal precipitation occurs with the greatest magnitude and frequency in the fall and winter. Occasionally, debilitating frontal storms known as *blizzards* generate heavy snows and strong winds. During a blizzard, drifting snow and frigid windchill disrupts communication and transportation, often kills unsheltered livestock, and causes significant human suffering. For example, the incidences of frostbite, hypothermia, and heart attacks, as well as accidents, increase dramatically as people fight the snow and press to maintain daily routines.

Cyclonic storms and frontal precipitation range over a broad belt of latitude centering on the midlatitudes in both hemispheres. They are common from the Arctic/Antarctic Circles to the subtropics and persist over both land and water. Since the occurrence of cyclonic storms is closely tied to the polar front, frontal precipitation tends to wane in the summer when the polar front weakens. In the midlatitudes, it is largely replaced by convectional precipitation in the summer but is never truly absent from the weather scene.

Convergent Precipitation

Convergent precipitation is produced when warm, moist air in the tropics is drawn into areas of low pressure. As opposing winds meet or converge, they are driven up-

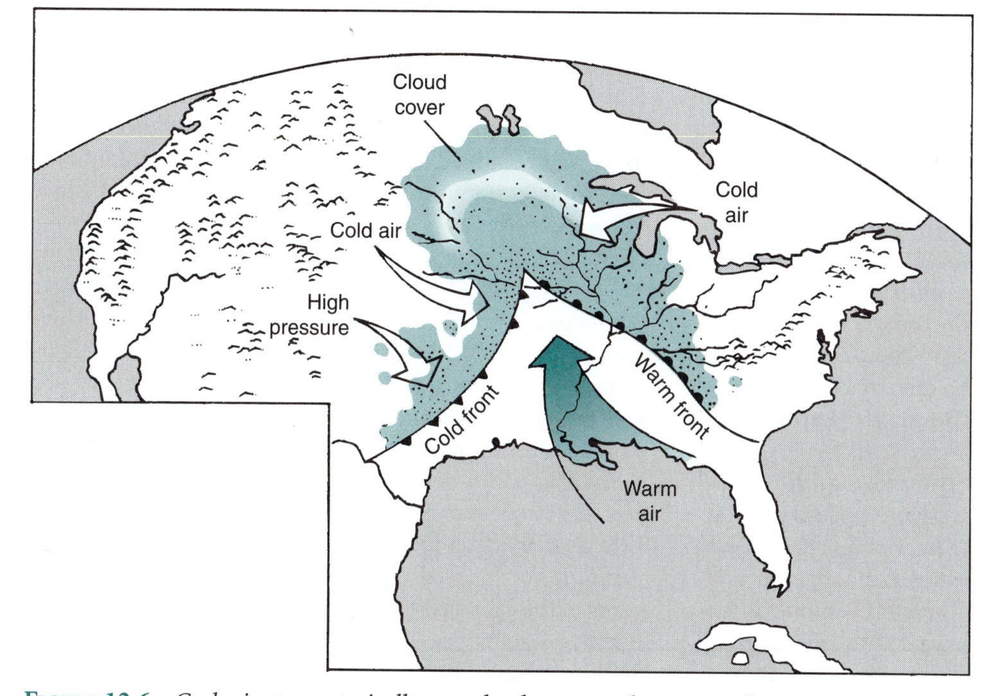

FIGURE 12.6 *Cyclonic storms typically cover hugh areas and may contribute vast amounts of water to large watersheds. Regional flooding such as the 1993 and 2001 Mississippi floods may result.*

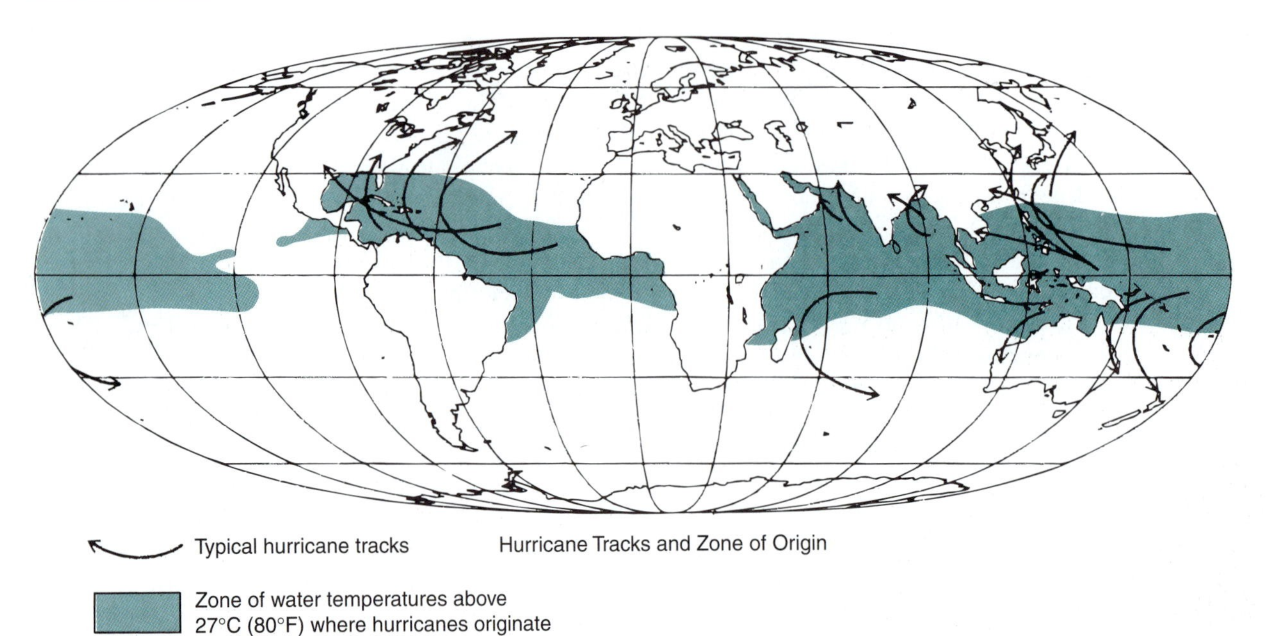

FIGURE 12.7 *The generalized global pattern of hurricane tracks and associated warm-water zone (above 27°C) where they originate.*

ward, causing cooling, condensation, and precipitation. This mechanism produces much of the precipitation in the tropics, such as the monsoon rains that spread over India and much of Southeast Asia. *Tropical cyclones* which are also known as **hurricanes** or **typhoons**, are the most powerful storms produced by convergence. These seasonal storms originate over tropical oceans where the late-summer sea surface temperature is greater than 27°C (80°F). A hurricane may begin as an atmospheric disturbance near the equator. As the disturbance develops into a small low-pressure cell, moist air is drawn into it and it may grow into a *tropical storm.*

Most tropical storms dwindle away over the oceans, but some expand and migrate poleward. When they reach latitude 5 to 10 degrees, their circulation may intensify as they are set into a broad swirling motion by the Coriolis effect. In addition, their path or track of migration, which until now had been mostly eastward under the influence of the Easterly Trade Winds, often takes a strong poleward bend in response to the Coriolis effect. This track takes them into the subtropics and the midlatitudes in both the Northern and Southern Hemispheres (Figure 12.7).

As the tropical storm grows and massive quantities of moist air are drawn into it, the converging winds are forced up, causing rapid condensation and heavy rainfall. Massive amounts of latent heat are released in condensation and this energy powers the hurricane. Therefore, as long as the hurricane remains over water it has a source of energy in the moist oceanic air on which the storm system can be sustained. Over land, by contrast, the supply of moisture is insufficient to maintain the storm and it declines.

Environmental Impacts

Widely regarded as the deadliest storm on the planet, hurricanes cause devastating **coastal damage**. Winds reach speeds of 120 to 240 kilometers per hour (75 to 150 miles per hour), strong enough to knock down trees, break windows, and blow away roofs. At maturity, large hurricanes may reach 1,600 kilometers (1,000 miles) in diameter. The powerful winds generate huge waves and the storm produces an elevated water surface known as a storm surge. In a surge, extreme-low air pressure allows the ocean surface to bulge up as much as 5 to 6 meters (15–20 feet). When the hurricane strikes land, it floods lowlands, forcing streams to back up and flood their lower valleys, and allows storm waves to advance inland.

Countries such as Bangladesh, where much of the land lies within 10 meters of sea level and is laced with stream channels, are especially vulnerable to surges (see Figure 12.13). During the twentieth century, tens of thousands of people were killed in Bangladesh by such events. There and in many other countries including the United States, immeasurable damage has been delivered to agriculture, fisheries, wetlands, and coral reefs by hurricanes during storm surges. Hurricane Andrew, which swept across south Florida in 1992, caused $8 billion damage and was the costliest hurricane in U.S. history (Figure 12.8). Along the U.S. Gulf and Atlantic coasts, the federal and state governments have required communities to prepare evacuation plans in the case of severe hurricanes. Despite the hurricane threat, residential development continues to push into low-lying shorelands throughout the U.S. Southeast. By the year 2010, it is estimated that 75 percent of the U.S. population will live within a one-hour drive of a coastline.

FIGURE 12.8 *Examples of damage to southern Florida from Hurricane Andrew in 1992.*

The next several decades may be especially critical for coastal land uses in hurricane-prone areas. Not only will population growth push more people into coastal lowlands, but the prospect of a sea-level rise associated with global warming raises the likelihood of greater storm damage. A sea-level rise of only 1 meter (3 feet) can significantly increase the penetrating distance and force of hurricane storm surges. In addition, global warming may increase ocean surface temperatures, thereby increasing the magnitude and frequency of hurricanes. The effects of these changes on Bangladesh and similar poor countries with large, subsistence agricultural populations in coastal lowlands may be devastating. In developed countries such as the United States, beach property damage from hurricanes might become so severe that laws will be passed prohibiting individuals and communities from rebuilding storm-damaged facilities. Today, the National Flood Insurance Program, which is underwritten by the federal government, provides funding to rebuild houses on ocean shorelines destroyed by hurricanes, thereby encouraging continued development.

12.4 STREAMFLOW AND FLOODS

The precipitation received by the land is disposed of in a variety of ways. For the Earth's land area as a whole, nearly 60 percent goes back to the atmosphere as vapor via evaporation and transpiration. The remainder goes to surface runoff and groundwater, both of which feed streams that carry the water to the oceans. An important question is how this water gets into streams and why most of the time it makes for smooth streamflows and only occasionally makes for highly erratic flows and destructive floods. In addition, streams are the principal source of water supplies for settlements and agriculture, which has led to widespread damming as a means of regulating and conserving their flows.

Sources of Streamflow

Streamflow is derived from two primary sources: (1) underground seepage of groundwater into the stream channel, and (2) surface runoff from rainfall and melting snow. Groundwater is the major source, especially for large streams, and it is characterized by a steady supply of water called **baseflow** that fluctuates only gradually over periods of weeks and months. Groundwater feeds streams during dry and rainy weather alike, which accounts for the fact that large streams can maintain steady flows in the absence of precipitation and surface runoff (Figure 12.9a).

Surface runoff, by contrast, is sporadic in nature, contributing water to streams only during heavy rainfall and melting snow. This water literally pours off the land, and when it reaches the stream, it is superimposed on the existing baseflow in the channel (Figure 12.9b). Because of its brief period of occurrence, this part of the stream's discharge is called *quickflow* or *stormflow*. It is the principal source of flooding in most stream valleys and the main vehicle for the delivery of pollutants to streams. These pollutants consist of sediments, chemicals, and microorganisms that are swept off farmlands, highways, parking lots, and the like.

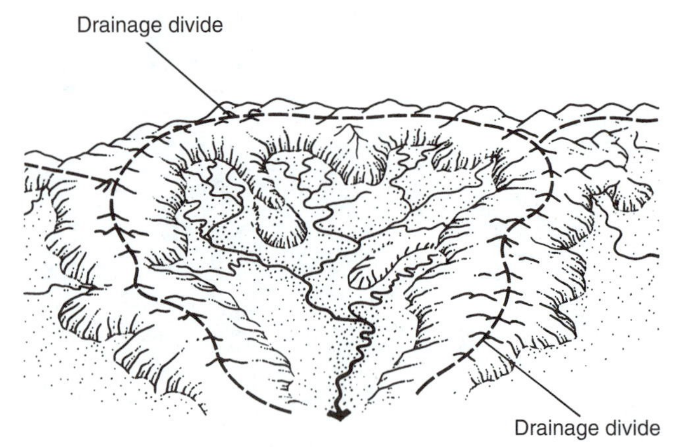

FIGURE 12.10 *The concept of a watershed and its network of streams.*

Drainage Basins

Streams drain an area of land called a **watershed** or *drainage basin*. All land surfaces, no matter how dry they may be, belong to a drainage basin of some size. A drainage basin functions like half a funnel placed on its side: Water runs toward the center and concentrates in deeper and faster flows toward the outlet (Figure 12.10). The bigger the basin, the more water it collects and the larger the streamflows it is capable of generating. Also, the faster the basin conducts water toward the outlet, the larger the resultant streamflow. Based on these simple observations, it should be apparent that geographic location within a drainage basin can be of paramount importance in determining the availability of water supplies and the potential risk from flooding.

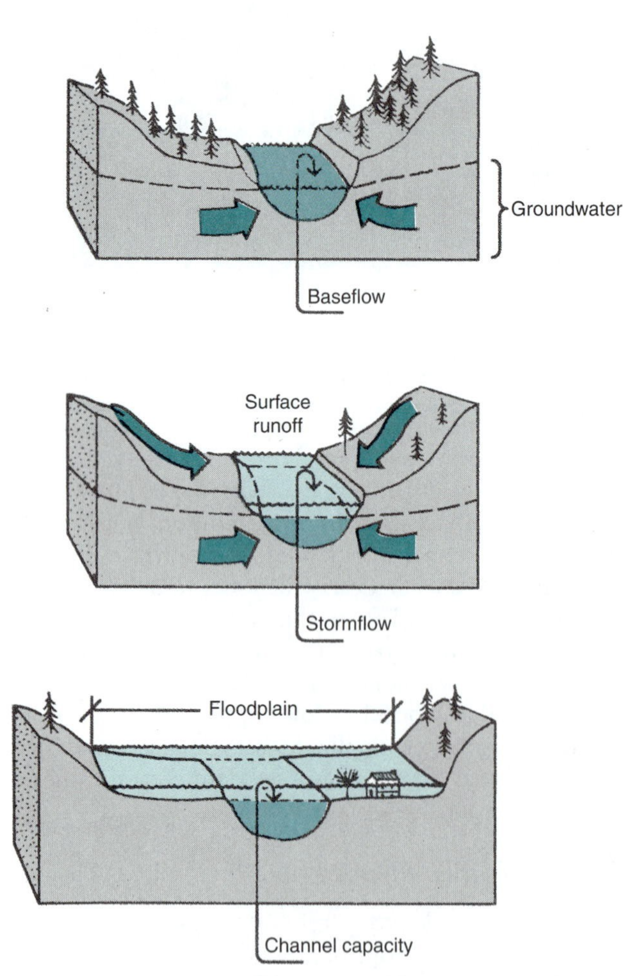

FIGURE 12.9 *(a) Baseflow generated by the seepage of groundwater into the stream channel; (b) stormflow generated by surface runoff in response to rainfall or melting snow; (c) channel capacity, flood flow, and the floodplain.*

The runoff produced within a drainage basin is carried away by an internal network of stream channels called a **drainage network**. The drainage net is organized like the branches on a tree; smaller ones, mainly on the perimeter, lead to larger ones toward the center. Each channel in the network has more or less an upper limit on the size of the flow that it is able to accommodate. Flows in excess of this limit over-top the stream bank and flood lands adjacent to the channel, called the **floodplain** (Figure 12.9c). Flood flows may be brought on by various means, including heavy surface runoff following a rainstorm, natural or human-made obstacles in the channel, or hydrologic changes from various human actions in the drainage basin such as increased stormwater discharge related to urbanization.

12.5 FLOOD HAZARD AND LAND USE

Flood magnitude and frequency have been increasing over the past century or so because of environmental changes related to land use. This is a source of great concern, not only because land-use facilities and human life are threatened, but because it indicates we are not managing our stream systems and related land uses very effectively. This is especially alarming in light of the extensive engineering projects to control streamflows and reduce flood damage in developed countries. There are at least four major factors behind the increase in the magnitude and frequency of floods.

1. As vegetation is cleared and land is covered with hard surface materials, called **impervious cover**, the amount of water soaking into the soil, called **infiltration**, declines. This, in turn, leaves more water to runoff. Infiltration diminishes with both land clearing and the addition of impervious cover; therefore, runoff increases with virtually all types of land development, including agriculture, residential, and urban (Figure 12.11). Hard surfaces are most critical;

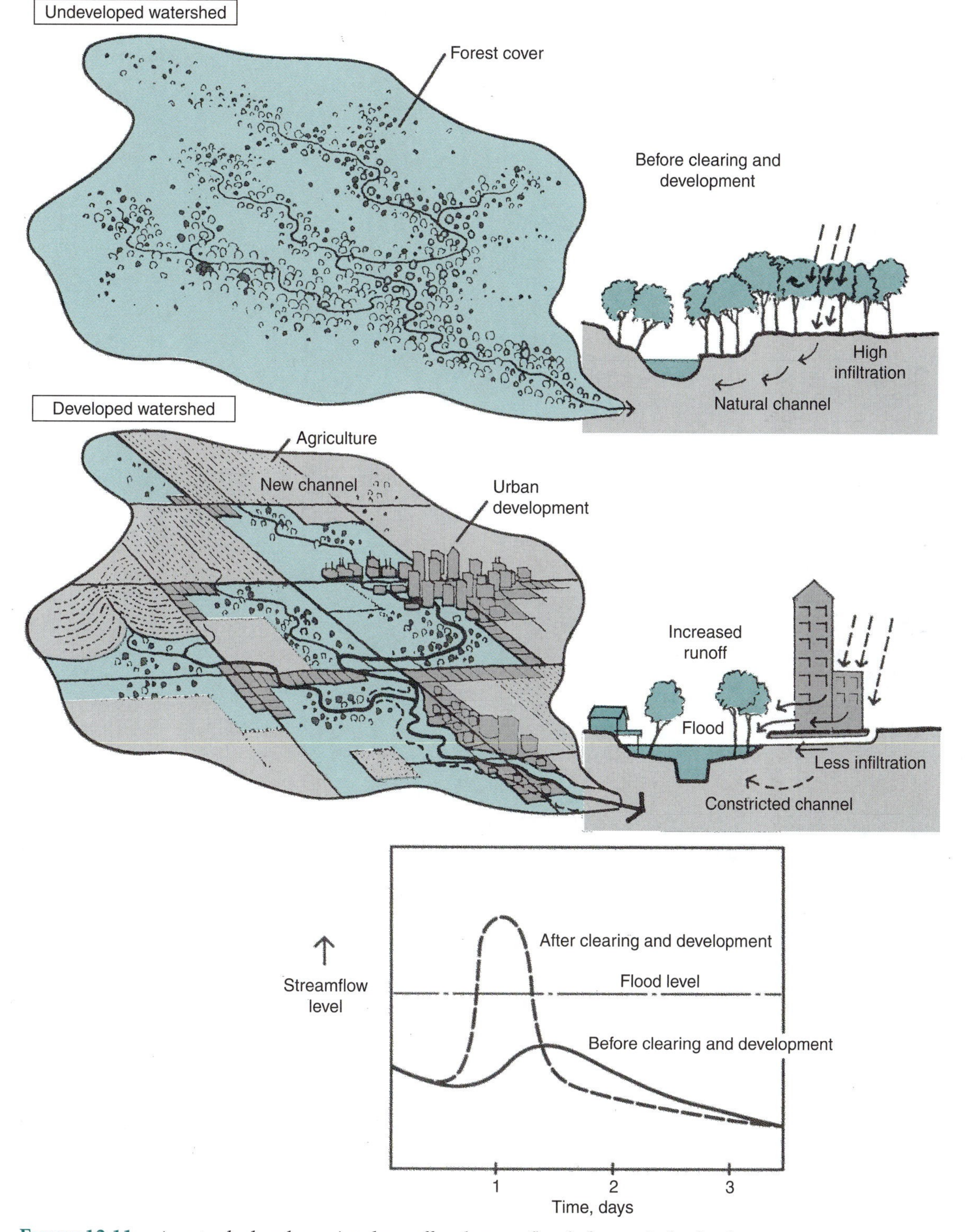

FIGURE 12.11 *A watershed and associated runoff and streamflow before and after land use development. The graph shows the corresponding change in streamflow from a given-sized rainstorm.*

in the United States hard surface cover averages about 600 square meters per person.

2. Land clearing and development increase the rate of runoff because pipes and ditches are constructed to speed the delivery of stormwater to streams. Pipes and ditches increase the rate of water movement by 2 to 3 times.
3. Rainfall from thunderstorms appears to have increased with large-scale urban development, resulting in larger volumes of runoff when it rains (see Figure 10.13).
4. Land development alters the stream channels themselves so they carry water away more rapidly in the upper tributaries (because of ditching, straightening, and piping) and reduce flow rates in the trunk streams (because of bridges, dams, landfills, and navigation facilities). On balance, modern streams typically receive more stormflow at a faster rate, but are often limited in their capacity to transmit the flows downstream (Figure 12.11).

FIGURE 12.12 *Severe flooding on the Mississippi River in the summer of 1993, the most destructive flood ever recorded in the United States, was due in part to constricting effects of channel-control facilities.*

FIGURE 12.13 *Villages and farm fields in one of the many floodplains of Bangladesh. Notice that little land is high enough to escape even modest floods.*

The severity of the 1993 Mississippi flood is explained in part by the presence of engineered works, both navigation and flood-control facilities, that constricted the river's flow at flood stage. In St. Louis, Missouri, engineers began tampering with the river as early as 1837. Between 1849 and 1969 the river was narrowed from 4,300 feet to 1,900 feet. At the same time the banks were built up with levees so that large flows could be contained at higher levels. These and similar actions along the Mississippi tended to hold back the river's massive 1993 discharge, thereby increasing the extent and duration of flooding upstream (Figure 12.12). In some places where constructed levees kept flood water out of natural floodplains, engineers actually broke the levees down to release water into the floodplain as a last ditch effort to lower the level of flooding.

Not surprisingly, property damage from flooding increased during the twentieth century despite huge expenditures to control flood flows through engineering. Only part of the explanation, however, can be based on the increased magnitude and frequency of flood flows. A large part of the blame also rests with our persistence in placing communities, farms, and industries in flood-prone locations, especially floodplains, which represent about 7 percent of the land area of the United States. There are several reasons for this state of affairs in the United States:

1. Americans have a tradition of protecting the rights of individuals to choose their own place to live, irrespective of the environmental hazards that might be involved.
2. Consumers and landowners are unwilling to recognize what parts of the environment are probably prone to flooding.
3. The United States relies on government programs to "fix" flooding problems with engineered facilities such as dams, levees, and deepened channels.
4. Society is reluctant to impose land-use controls on itself at national, state, and community levels.

In the developing countries, population pressure is also a factor. Because of the need for agricultural land in countries such as China, India, and Bangladesh, people are forced to occupy hazardous floodplains despite the evident risk (Figure 12.13). As development increases in floodplains, the magnitude and frequency of flooding also increase in response to greater storm-water runoff, channel constrictions, and other factors.

12.6 GROUNDWATER

Groundwater is an accumulation of liquid water in rock and soil materials underground. It can be found at depths ranging from several meters to more than a thousand meters beneath the surface. Groundwater is readily accessible as a source of water for humans. In the United States, about 50 percent of the population uses it as a source of drinking water. In rural areas groundwater provides more than 90 percent of the household drinking water and 40 percent of the water used in irrigation. Protected by a blanket of soil and rock, groundwater historically presented few environmental problems to

North Americans, other than finding enough of it in desired locations. In the past several decades supplies have begun to decline in some locations because of excessive pumping. More important, however, is the widespread emergence of groundwater pollution. This problem is taken up in the next chapter.

Aquifers and Their Functions

All sorts of materials are capable of holding groundwater; for example, loose sediments such as sand and gravel near the surface or bedrock formations such as sandstone and limestone deeper down. Materials containing large amounts of usable groundwater are called **aquifers**. The term *usable* is defined by the quality of the water; among the measures of quality for drinking water are total dissolved solids and bacteria content.

The amount of groundwater that can be held in any material is determined by its porosity. **Porosity** is defined as the total void, or pore, space available among particles or in cracks where water can reside. Aquifers have relatively high porosities, typically 20 to 30 percent, meaning that they are able to hold up to 20 to 30 percent of their volume in water. In addition, most aquifers have high yield capacities—that is, they give up water readily when subjected to pumping.

An aquifer functions as a basic water system whose reserve (supply) grows or shrinks with changes in the balance of inflows and outflows of water. The inflows, which originate mainly as surface water percolating down from the soil, are termed **recharge** (Figure 12.14). The outflows, represented by seepage into streams, evaporation, pumping withdrawals, and losses to other aquifers, can generally be termed **discharge**, or *depletion*. In order for an aquifer to maintain its reserve of water, depletion must not exceed recharge over some extended period of time, usually decades.

Among the most prized water resources in the United States is the Ogallala aquifer. This massive body of groundwater is one of the largest continuous aquifers in the world, stretching under the Great Plains from South Dakota to Texas (Figure 12.15). It provides highly usable water, and in the past five decades has become a principal source of irrigation water. Also prized are much smaller aquifers that support large urban populations. On Long Island, New York, for example, a large population is supported by an isolated aquifer formed in sandy deposits, surrounded by salty seawater and subject to both contamination and drawdown from intensive land use.

Aquifer Use and Decline

In many parts of the world aquifers are declining because of excessive pumping for agricultural and urban uses. Particularly alarming is the rapid decline of aquifers in dry lands where the recharge rate is exceedingly slow. In the American Southwest, for example, the water withdrawn from many aquifers, including parts of the Ogallala, represents ancient water, that is, water gained during wetter climatic conditions thousands of years ago. At the current pumping rates, a number of these aquifers are

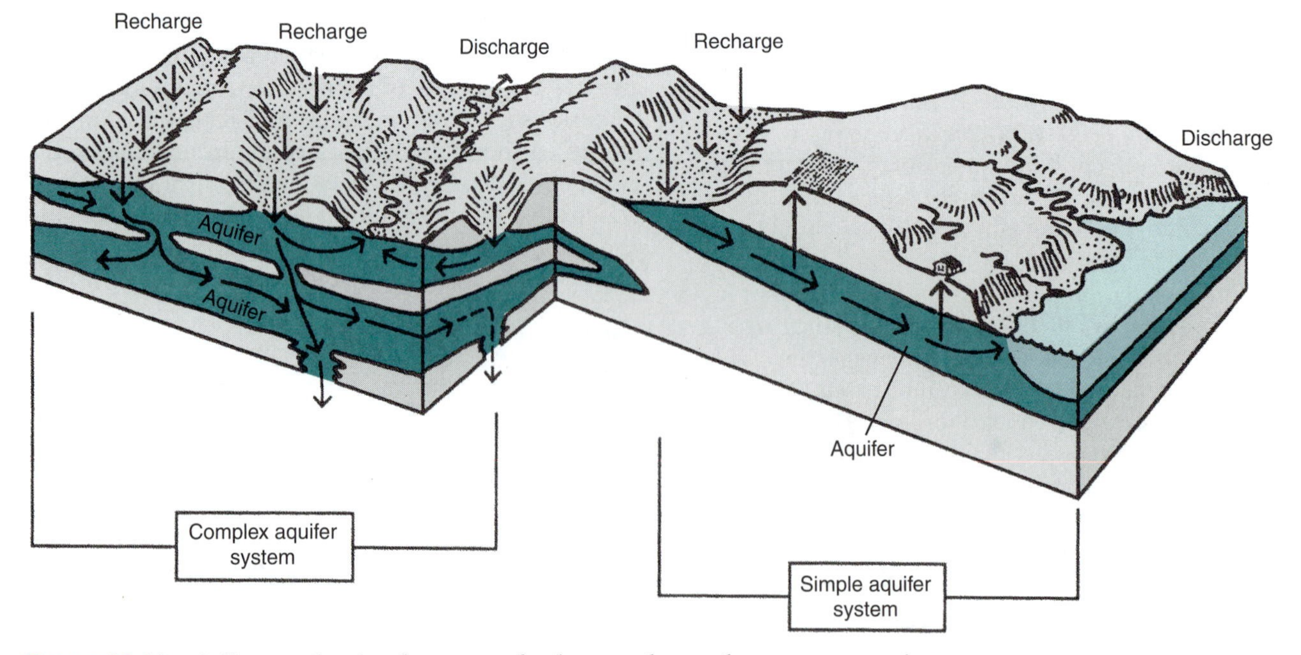

FIGURE 12.14 *A diagram showing the sources of recharge and groundwater movement in two aquifer systems, one simple and the other complex.*

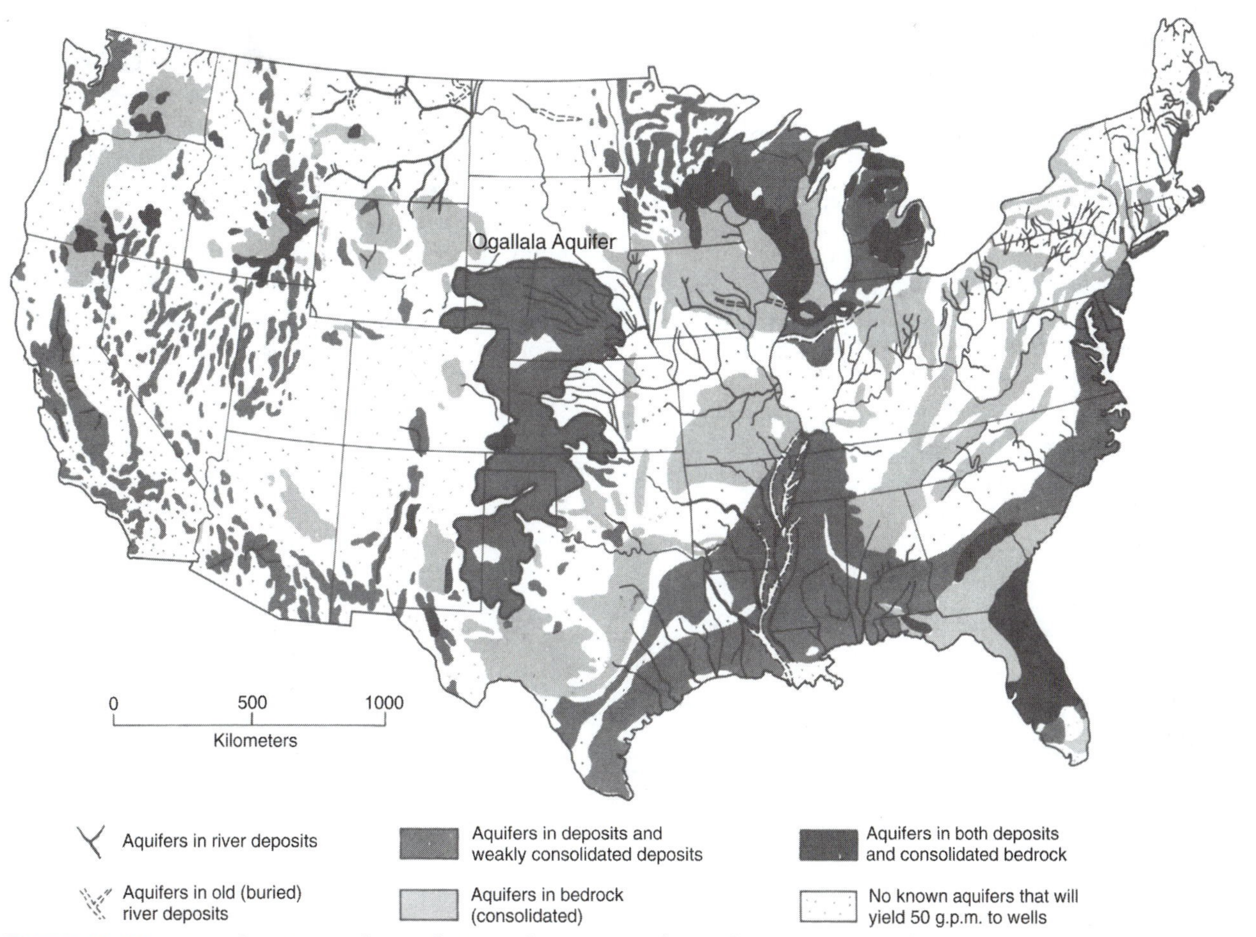

FIGURE 12.15 *Map of major aquifers in the United States. Aquifers in deposits are generally shallower whereas those in bedrock are generally deeper. Only those yielding 50 gallons per minute or more are shown.*

being drawn down several meters a year, which is sure to exhaust them in several decades. Once depleted, this water supply, is lost as a resource. Groundwater use is increasing substantially in the United States, especially in the West (including the Great Plains), where in the 1980s it is estimated that total pumping exceeded recharge by 7,200 billion gallons annually.

Management of groundwater supplies requires application of a concept called **safe aquifer yield.** This concept calls for setting pumping rates at levels that will not cause significant decline in the aquifer. Instead of pumping at the rate demanded by water users, pumping rates would be scaled down to a point where the aquifer is able to sustain a long-term balance. This concept is also applicable to many urban areas in humid regions where, as the cities grew, aquifers were lowered by exceptionally high pumping rates, forcing the city to shift to more expensive water supplies at greater depths or greater distances from the city.

In some coastal cities aquifer drawdown has led to **saltwater intrusion** into freshwater aquifers. Where the two types of water meet in coastal areas, fresh groundwater normally overlies salt groundwater because saltwater is denser than fresh. When part of the fresh groundwater is removed by pumping, the saltwater rises into the space vacated. If it rises to well depth, then the wells suck up saltwater, contaminating the water supply and rendering it useless for municipal purposes. One alternative course of action in cases of aquifer drawdown—which many cities have chosen to ignore or have found politically unacceptable—is to limit urban growth. Some cities, such as New York, have raised the price of water, and others, such as Los Angeles, have imposed use quotas. Both of these alternatives have resulted in lowered use rates.

Although the idea of an aquifer as a water system is simple, the dynamics of most aquifers are, in reality, complex. The recharge system may resemble a mazelike network in which connections to the surface and with other aquifers are often intricate and uncertain (Figure 12.14). Coupled with the enormous size of many aquifers, frequently covering hundreds of square miles, it is easy to see why the management and protection of groundwater is an especially difficult enviromental problem. Of greatest concern today is groundwater pollution. Suffice it to point out here that the ground is our favorite place to dispose of waste, and for decades we have been burying, dumping, and spraying everything imaginable in the same material—soil—that conducts recharge water to aquifers. Groundwater pollution is covered in the next chapter.

12.7 LAKES, PONDS, AND WETLANDS

The uppermost surface of the zone of groundwater is called the **water table**. The water table in humid regions is typically found 5 to 10 meters below ground level, except where the land surface breaks abruptly to a lower elevation, such as along a stream valley or mountain slope. In these locations, the water table often intersects the surface and the groundwater seeps out (see Figure 12.16c). In stream valleys seepage water is carried away as streamflow. But in low spots where an outlet to a stream is limited or nonexistent, seepage water accumulates, or *ponds*, forming a **natural impoundment** resulting in a lake, pond, or wetland. In addition to groundwater, impoundments also collect surface runoff, precipitation, and flood water. As the water builds up, mechanisms for the release of surplus water develop through overflows to

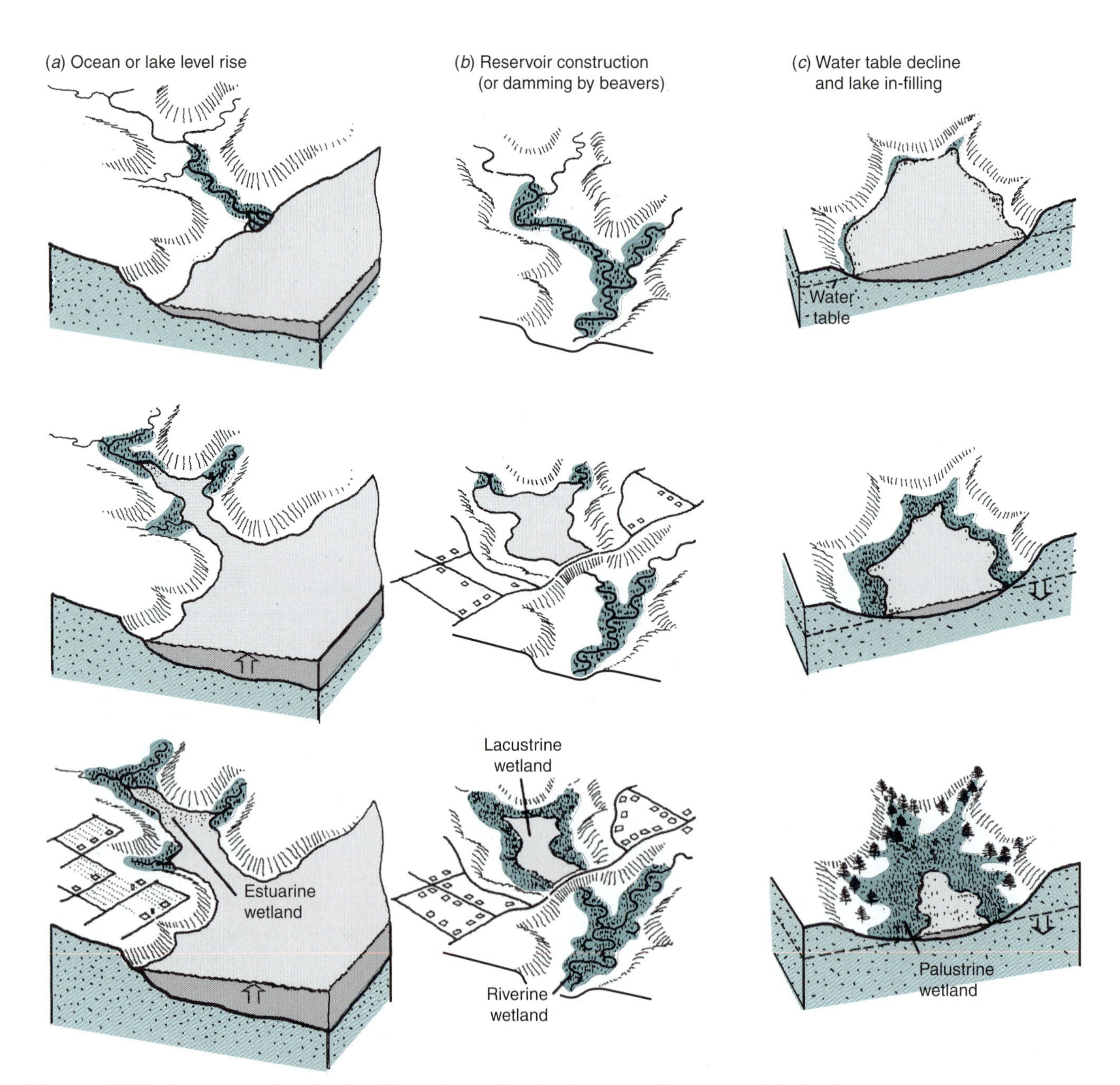

FIGURE 12.16 *Some examples of wetland formation and wetland systems: (a) ocean- or lake-level rise leading to an estuarine wetland; (b) reservoir construction leading to a lacustrine wetland; (c) groundwater rise leading to a palustrine wetland.*

streams, evaporation, transpiration, and as seepage back into the ground. In this way the impoundment maintains a **water balance** characterized by a relatively permanent body of water that fluctuates seasonally or from year to year with changes in inputs and losses.

Natural impoundments vary enormously in size. Among the freshwater bodies, Lake Superior, with a surface area of 82,477 square kilometers (31,869 square miles) is the largest, whereas Lake Baikal in Russia, with a depth of 1,741 meters (5,711 feet), is the largest in volume. Most of the millions of impoundments found on Earth are small, however, less than 1 square kilometer in area. Although their permanency is generally related to size, with small impoundments being most susceptible to infilling and losses of water supply, even large water bodies can be surprisingly susceptible to environmental change. As we noted in Chapter 8, for example, the Aral Sea of the former Soviet Union, which until 1970 measured about 67,000 square kilometers (26,000 square miles) in area, has been reduced by 40 to 50 percent because its water supply has been diverted for massive agricultural irrigation projects (see Figure 8.14).

Wetlands

Wetlands are characterized by shallow water (which may be relatively permanent or seasonal) with heavy growths of aquatic or semiaquatic plants and relatively thick organic deposits. In the landscape they occur as both independent water features and as parts of lakes, ponds, and seashores where they usually occupy the shallow waters fringing the shoreline (Figure 12.16). Inland, most wetlands are freshwater, but those along the oceans are both salty and fresh, with many *brackish*, or of intermediate salinity, from the mixing of freshwater and saltwater.

In the wetland classification scheme used by the U.S. Fish and Wildlife Service and other federal agencies, five **wetland systems** are recognized: marine, estuarine, riverine, lacustrine, and palustrine. The *marine* and *estuarine systems* are saltwater and brackish wetlands of the continental shelf and connecting bays and tidal marshes (Figure 12.16a). The remaining three are freshwater systems with *riverine wetlands* associated with stream channels (Figure 12.16b), the *lacustrine* with lakes and ponds, and the *palustrine* with all other inland wetlands. The palustrine wetland system includes swamps (wooded wetlands), marshes (grassy wetlands), and bogs (northern mixed wetlands), and is the largest system in terms of total area of coverage in North America (Figure 12.16c).

Wetland Trends

Wetlands are among the richest biological **habitats** on Earth. However, because they are a physical constraint to most land uses, wetlands have for centuries been drained and destroyed to make way for farming, towns, and navigational facilities. Wetland eradication can be traced back to the ancient civilizations, but some of the most dramatic examples come from Holland and England, where massive "land reclamation" projects have been active for centuries. The practice is still widespread in the world. As a countermeasure, many nations have recently adopted wetland protection laws in order to reduce further loss of wildlife habitat, endangered species, and water resources. In the United States wetland laws have been actively enforced since the 1980s, but they are currently the subject of widespread controversy, mainly because of the limitations they impose on local land-use development.

Prior to European settlement, wetlands in the contiguous United States were estimated to have covered 600,000 square kilometers (232,000 square miles), an area nearly equal to that of Texas. Since then, perhaps 40 to 50 percent of these wetlands have been eliminated. As in the rest of the world, wetland **eradication** in the United States has been caused primarily by agriculture. It is noteworthy that with the exception of land under certain federal programs, agriculture, as well as forestry and mining, are currently exempt from U.S. wetland protection laws.

On the other side of the ledger is the recent addition of literally millions of ponds and reservoirs to the North American landscape. Most are small, less than five acres, built for farms and rural residences. They have helped offset losses of palustrine wetland, although we are uncertain just how much. The uncertainty is made greater because many ponds created on farms were actually excavated from original wetland areas. In the United States the construction of ponds and reservoirs was promoted by the U.S. Department of Agriculture as a part of water management and erosion control programs between the 1930s and 1970s.

Lakes and Ponds

Lakes and ponds[1] are characterized by sizable areas of open water at depths great enough to inhibit heavy covers of rooted aquatic plants. The depressions, or basins, that contain lakes and ponds are of diverse origins; for example, in Canada, the northern United States, and the Rocky Mountains, they are formed by depressions in glacial terrain; whereas in Florida, they are formed in limestone sinkholes. Elsewhere in North America, most lakes and ponds have been formed artificially where streams have been dammed to create reservoirs. Lakes and ponds are also created from abandoned mining pits, animal watering ponds, stormwater basins, and residential landscaping and recreation ponds (Figure 12.17). Along with wetland eradication,

[1]A pond can generally be considered a small version of a lake. Use of the terms *pond* and *lake* varies regionally in North America and generally carries no scientific meaning.

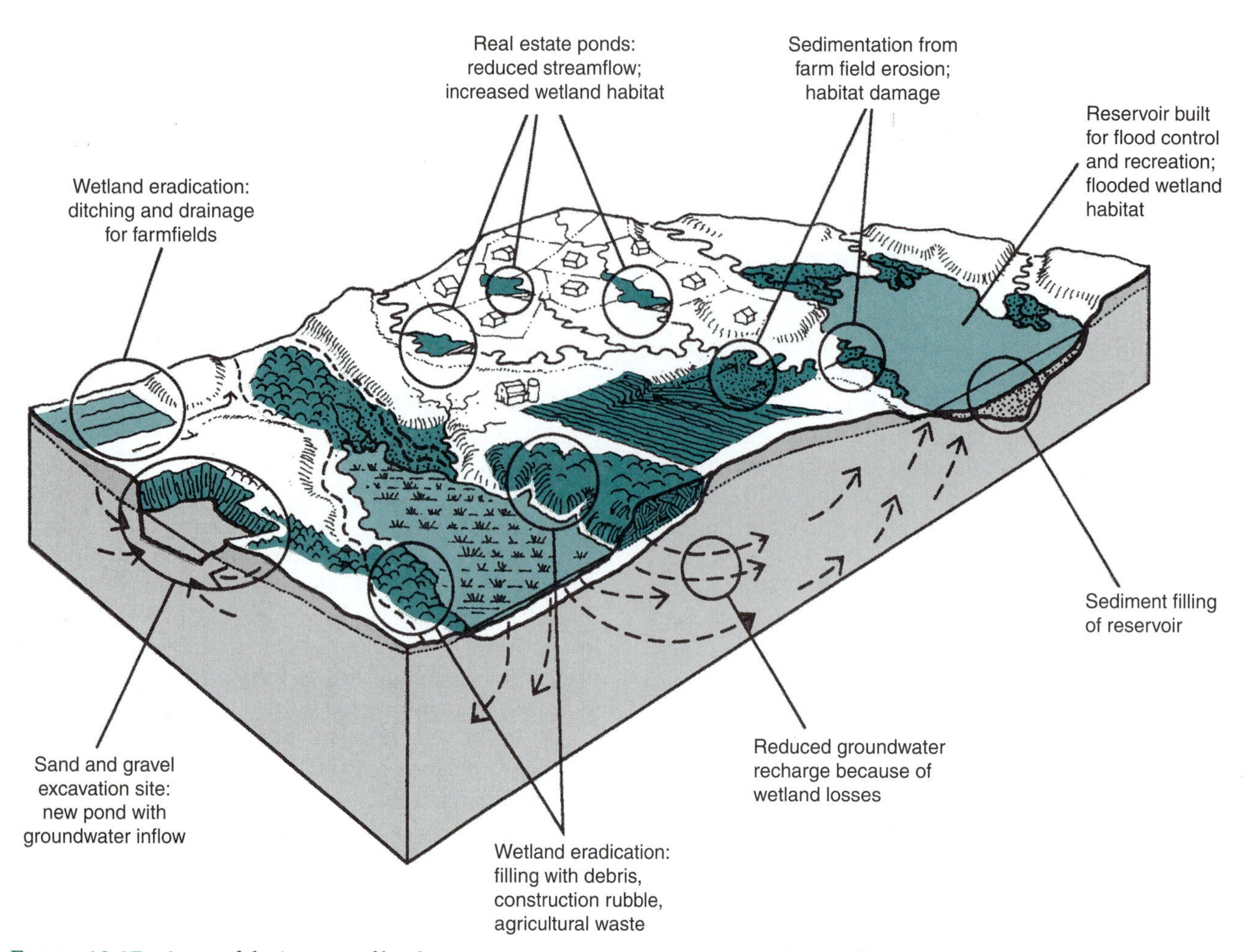

FIGURE 12.17 *Some of the impacts of land-use practices on streams, wetlands, and related habitats in the twentieth-century landscape: filling, excavation, drainage, damming, and sedimentation.*

dams, reservoirs, and other artificial water bodies have been major sources of disturbance to North American streams and freshwater habitats during the past century.

No matter their origin, the basins of all impoundments will eventually fill with organic debris and sediments. Most reservoirs; for example, have an expected lifetime of only 50 to 100 years before they are completely filled with sediment (sand, silt, clay) carried in by streams. Thousands of small reservoirs and ponds in the United States have already filled, leaving behind degraded land, altered drainage systems, and damaged habitat (Figure 12.17).

Another filling process is *eutrophication*, which we first mentioned in Chapter 6. In this process a water body becomes overgrown with aquatic plants fed by nutrients carried into them with runoff from natural sources and pollution sources. As the water body fills with decaying organic debris, it becomes smelly and unsightly, and the fish species change as the oxygen in the water declines. Furthermore, land-use development accelerates the eutrophication process as nutrient-rich runoff from farmlands, urban areas, and highways is discharged into lakes and ponds (also see Figure 6.17).

12.8 HUMAN USE OF WATER

The modern world uses huge amounts of water. It is estimated that total annual freshwater use is approximately 4,000 cubic kilometers worldwide. This is double the total annual use of 1950 and 35 times what it was in 1700. Populous countries such as China use more water, of course, than less populous countries such as Australia. However, there are two important differences. First, per capita water use is much higher in developed than developing countries, and second, a greater share of water in developed countries is used for industry and power generation. In the United States, the most extreme case, agriculture uses 34 percent and industry and power generation together use 57 percent. The relatively low percentage for agriculture is *not* explained by relatively small water use in that sector; rather, it is explained by

TABLE 12.3

WATER USE IN THE UNITED STATES TODAY (1990) AND IN THE RECENT PAST

User	Daily Total Use	Notes
Domestic/municipal	156 million cubic meters (41 billion gallons)	600 liters (160 gal) per person per day in 1980; in 1940 per person daily use was 285 liters (75 gal)
Industrial and related uses	120 million cubic meters (32 billion gallons)	declined from 174 million cubic meters (46 billion gal) in 1965 and 178 million cubic meters (47 billion gal) in 1970
Agriculture	530 million cubic meters (142 billion gallons)	mostly irrigation water; per person rate has more than doubled since 1940
Power generation	740 million cubic meters (196 billion gallons)	has increased substantially since 1940 with increased population and per capita use of electricity

Source: U.S. Geological Survey. 1990.

the comparatively massive water uses in the industry and power sectors, especially power.

In Canada and the United States, daily water use in homes and settlements (excluding industry and power plants) averages nearly 600 liters (160 gallons) of water per person, twice as much water as each person used in 1940. In the United States today this amounts to about 156 million cubic meters (41 billion gallons) a day (Table 12.3). Although this is a staggering quantity, it is small compared to the amounts of water used by the other three major sectors of water use.

Power, agriculture, and industry together account for more than 90 percent of the daily water use in the United States. This water and the water used in homes and settlements is drawn from both surface water and groundwater reserves. Considering all the freshwater used in the United States, about 80 percent comes from surface sources and 20 percent from groundwater.

Power generation is the single largest water user at nearly 740 million cubic meters (196 billion gallons) a day (Table 12.3). This water is used for cooling and then released as heated water to streams, lakes, and reservoirs. Since 1940 the water used in power generation has increased tenfold in response to increased population and increased per capita energy use. Currently the United States uses about 25 percent of the world's electrical energy.

Agriculture in the United States withdraws an average of more than 530 million cubic meters (142 billion gallons) of water each day. Virtually all this water is used in irrigation, and more than half is lost to the atmosphere via evaporation and plant transpiration. Thus, agriculture is the largest water consumer among the four user

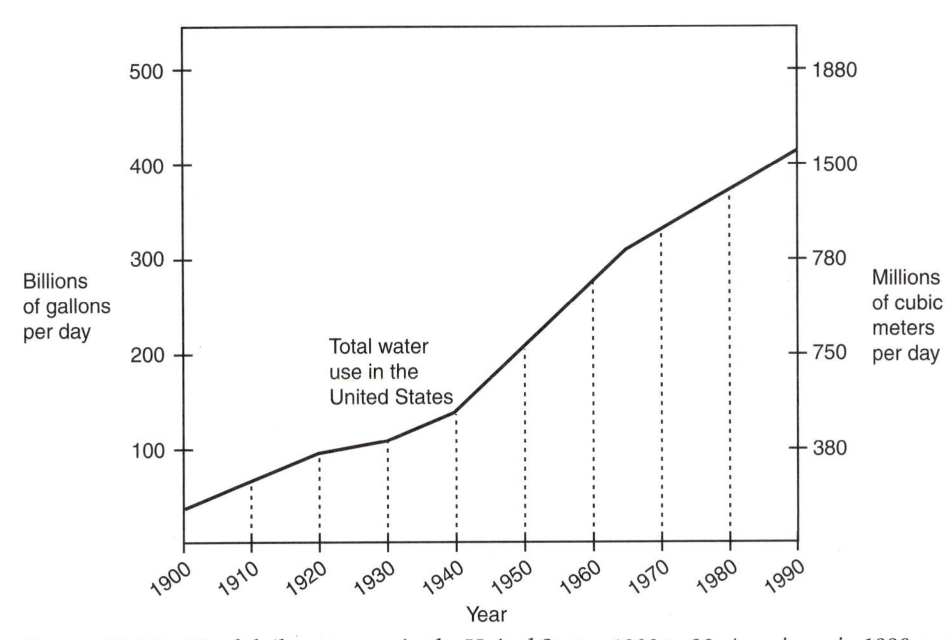

FIGURE 12.18 *Total daily water use in the United States, 1900 to 90. Americans in 1990 used more than twice as much water per person (based on all uses combined) as they did in 1940.*

groups. In California, where water supply has become a crisis issue and the urban population approaches 35 million people, nearly 85 percent of the water is used by agriculture. Manufacturing is the third-ranked water user in the United States at 100 million cubic meters (41 billion gallons) daily. Most of the water is returned to surface waters, although the quality is usually degraded by the various industrial processes in which it is used.

All told, the United States uses 1.55 billion cubic meters (411 billion gallons) of water daily (Figure 12.18). Despite ample water supplies in some regions of the country, the overall national water balance is negative. That is, more water is being extracted from the environment than is available from dependable supplies. As a result the national water supply is dwindling, but more for dry regions such as the Southwest. As population shifts toward southern states, water supplies there are declining, leading politicians to call for larger and more expensive extraction and distribution systems. Importing water from Canada has become a serious consideration for some regions.

Global Supplies and Problems

Global supplies of freshwater are under increasing demand for human use. The principal source of the increase is expanding world population and the increasing demand for food and consumer goods, none of which is possible without large water supplies for irrigation, power generation, and industry. The supply of usable water is unevenly distributed over the Earth, and human societies have long worked at redistributing it. In many parts of the world, however, population growth has outpaced water supplies. Despite redistribution projects, we are now facing serious regional shortages (Table 12.4).

Among these regions, the Middle East is one of the most seriously threatened, not only because it is arid and populous, but also because international political disputes threaten regional water planning and redistribution. The various conflicts among Turkey, Iraq, Iran, Kuwait, Syria, Jordan, and Israel illustrate this. An example is Turkey's project to build reservoirs on the headwaters of the Tigris and Euphrates, the primary sources of water for Iraq. Other threatened areas include the following:

- Northern China, where freshwater supplies are falling behind demands;
- The United States Southwest, where agriculture and urban development have dangerously stressed existing surface and groundwater supplies;
- Parts of Russia and Eastern Europe, where water supplies have become so fouled by pollution that they may be unusable for decades or longer;
- Many parts of sub Saharan Africa, where water demands greatly exceed supplies;
- Most of North Africa where ancient aquifers are being depleted and surface water is scarce;
- South Asia (India and Pakistan), where most water is used for agricultural irrigation, and groundwater pumping rates exceed recharge rates.

Table 12.4

Serious Water Shortages, 1990–2000

Location	Estimated Annual Shortage*
India	104.0
China	30.0
United States	13.6
North Africa	10.0
Saudi Arabia	6.0

* billions cubic meters

Source: Global Water Policy Project and World Watch Institute

In Pakistan, for example, it is likely that population growth will outrun water supplies within the next several decades. Pakistan is mostly arid and semiarid, and its population, which grew from 31 million in 1947 to 150 million in 2000, is expected to reach 225 million or more by the year 2025. In India, groundwater is being pumped at twice the recharge rate and India's population is expected to reach 1.5 billion by 2050. Both countries are overwhelmingly dependent on agriculture, and agriculture overwhelmingly dependent on irrigation. Thus, serious water shortages are sure to emerge as agricultural expansion attempts to meet rising food demands. Pakistan and India are not unique among developing countries in dry regions. A number of African and Middle Eastern countries face similar prospects.

Environmental Effects of Water Use

Water use is classified as **consumptive** or **nonconsumptive**, depending on whether the water is lost to the atmosphere as vapor or returned to surface waters as liquid. Agriculture is overwhelmingly consumptive because a large share of irrigation water is lost to evaporation and transpiration. The other major user classes are only marginally consumptive, consuming in the range of 10 to 20 percent of total use. However, most water uses can be defined as **degrading** because they usually pollute the water in some way before being released back into the environment. The sewer water from homes and settlements is heavy in sediment, nutrients, and biological contaminants; agricultural runoff is typically contaminated with sediment, fertilizer residues, and insecticides; water from manufacturing often contains heavy metals and petroleum residues; and the water from power plants is heated (thermal pollution) to a level far above that of receiving waters by the time it is returned to the environment.

Another major impact of water use is the effect of diversion on stream systems. The extraction and rerouting of water from streams often greatly reduces flows. With reduced flows, streams are less able to carry their sediment loads and channels often become clogged with sediments. This disrupts stream ecology, such as fish habitat and spawning environments, and reduces the stream's capacity to conduct flood flows. In addition, lakes and wetlands may be deprived of water by stream diversion. The Aral Sea is perhaps the most dramatic example in this century (see Figure 8.14).

Because of **diversions** and **dams** in the United States, little and sometimes none of the Colorado River reaches its mouth at the ocean in Mexico. This has deprived Mexican farmers of important irrigation water. In the Central Valley of California, the diversion and rerouting of streams has ruined most salmon migration routes and spawning areas. This process began with the gold-mining operations of the nineteenth century, which in only 30 years excavated more than 1 billion cubic yards of sediment from streams channels and rerouted hundreds of miles of streams. Today disturbance of streams continues with extensive agricultural irrigation systems that not only change routes and flow rates, but also change dramatically the stream bed environment and the quality of water flowing over it. In many areas only the imprint of former stream channels remain in the landscape (Figure 12.19).

FIGURE 12.19 *Only the ghost of natural stream channels often remains where water has been diverted by wetland drainage or irrigation projects. Stream water is now conducted in canals that run along roads and farmfields.*

Overall, dams and reservoirs probably have the greatest effect on stream environments. Not only do extensive reservoirs flood entire valleys and obliterate channel environments, but the channel downstream of the dam is severely altered as well (Figure 12.20). Reservoirs trap the natural flow of sediment, so that the water that emerges from the spillway is without its normal sediment load. As a result, the stream erodes its channel, picking up a new sediment load. This gives rise to deep-cut channels with degraded plant, insect, reptile, and fish habitats. Eventually (usually within 50 years for smaller reservoirs) the basin fills with sediment and the dam is abandoned. In the United States, more than 2,500 dams and reservoirs are classed as *retired* and typically are looked upon as environmental liabilities and safety hazards. Bigger reservoirs last longer, but they cover much larger areas, of course.

The United States, Canada, China, and the former Soviet Union have built thousands of dams in the twentieth century. Today about 80 percent of the major streams in these countries are interrupted by dams and reservoirs. The transformation of stream valleys is often extreme. For example, in the United States where the Missouri River crosses the Great Plains is today a continuous string of huge reservoirs. The stream channel that Lewis and Clark traveled in their famous trip to explore the Louisiana Purchase no longer exists in most places.

While dam construction in developed countries has declined significantly since 1970, in developing countries it has increased rapidly. As we noted in Chapter 6, China has already built 18,000 large dams (greater than 50 meters high) and is pushing ahead with many more projects. One project on the Yangtze River is being touted as the world's largest. Brazil is devel-oping massive dams and reservoirs that threaten huge tracts of rainforest. This trend is sure to continue as the demand for power, irrigation, water, and flood control rise throughout the developing world in the twenty-first century.

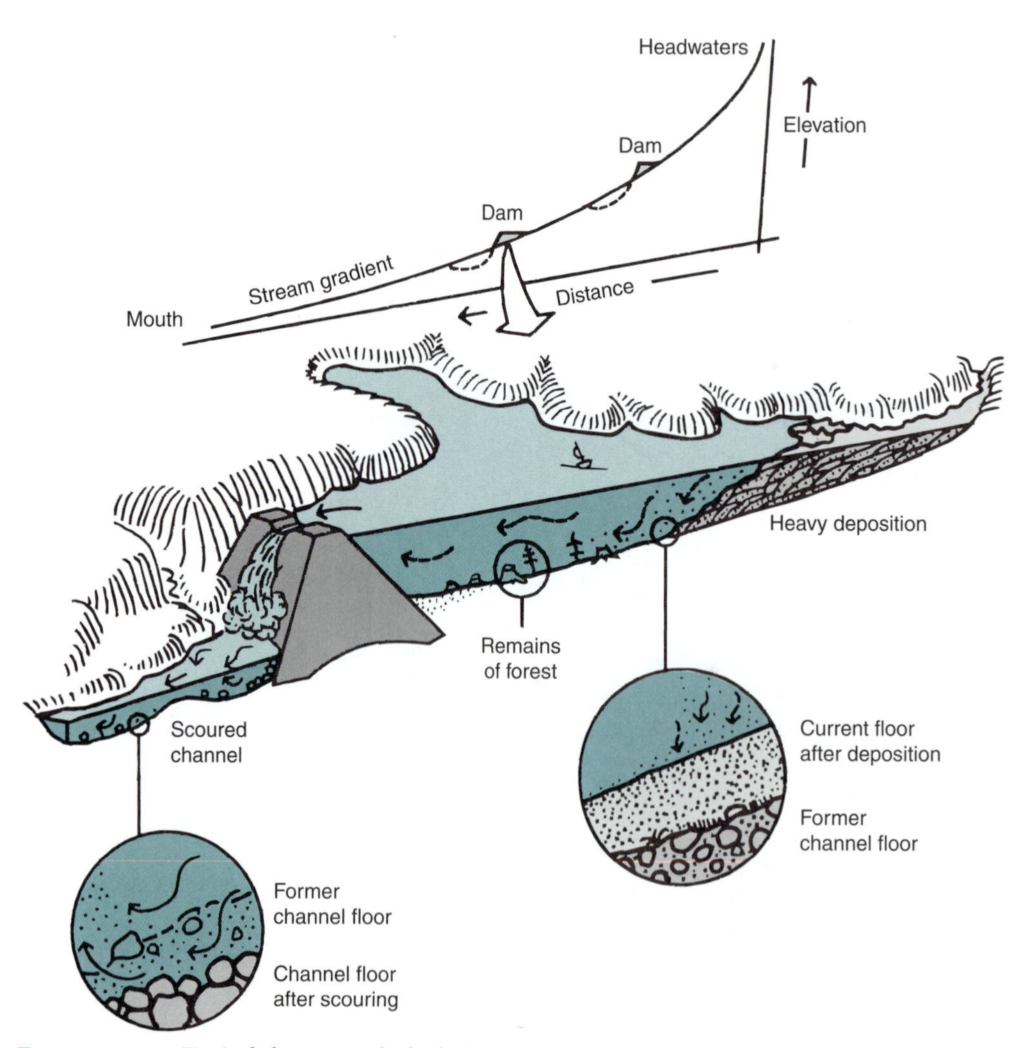

FIGURE 12.20 *Typical changes in the hydrologic environment of a stream valley as a result of reservoir construction.*

12.9 MANAGING WATER RESOURCES FOR SUSTAINED USE

Water management is one of the oldest organized environmental activities. The basic objectives of water management—to redistribute water and reduce the natural variability of its flow—have not changed in 10,000 years. The ancient civilizations of the Middle East, such as the Sumerian and Egyptian civilizations, practiced water management to provide irrigation for agriculture. The Romans built aqueduct systems to bring water to cities and stormsewer systems to drain excess rainwater from cities. There is a long list of examples from ancient times—some successful, some not—illustrating a rich log of human efforts to reorganize the natural distribution of water.

Today the need to manage water is stronger than ever. As population increases, there is a greater demand on existing supplies and new demands on untapped supplies, such as deep aquifers, ancient aquifers with little or no recharge, and distant streams. Redistribution systems in dry lands have become increasingly elaborate, extensive, and expensive worldwide. In Libya, for example, $25 billion was spent on a project called the Great Man-Made River Project to pump water 1500 km (930 miles) through a concrete pipe. Elsewhere, irrigated cropland has reached massive proportions; for example, the percentage of cropland currently irrigated is 70 percent in Pakistan, 63 percent in Japan, 48 percent in China, and 33 percent in India. In addition, there is growing competition between agriculture and cities for available water.

In the American Southwest, aqueducts transfer water from the Colorado River to the urban regions of Los Angeles and Phoenix. Coupled with the withdrawals for irrigation, 92 percent of the annual runoff in the Colorado River Basin is taken for human uses. In central California, an extensive network of aqueducts (hundreds of miles long) collects water from mountain streams and delivers it to agriculture in the arid San Joaquin Valley and to cities such as Los Angeles.

How is water allocated in the United States? In the West, where water resources are limited over wide areas, it is allocated according to a system called **prior appropriation**, in which water is assigned to users on a predetermined basis. Users have rights to water based on early allocation agreement, and they maintain these rights as long as they continue to use their allocation, whether they need it or not. To avoid losing their allocation, water is often wasted. In some of the driest parts of California's Central Valley, for example, where evaporation exceeds 1.5 meters (5 feet) of water a year, some farmers grow rice (a wetland crop) in flooded fields, sacrificing massive amounts of water to the atmosphere. In most humid regions such as eastern North America, water is allocated by **riparian rights**, which means that those who own land bordering on water have the rights to use it.

As with any environmental system utilized by humans, the desired long-term goal should be one of **sustained use.** Sustainability requires balancing human use with the flow of resources through the system. This means that maintenance of the system, both the quantity and quality of reserves, is given highest priority. Of course this is not the case with groundwater systems in many parts of the world, especially in dry regions such as the American West, where reserves are being used faster than nature is able to replace them. In addition, a depletion in flow and reserves beyond certain levels or thresholds in surface water systems damages aquatic habitats and reduces the quality of the overall system in terms of this biological diversity and productivity.

Broadly speaking, water management will require continued attention and improved planning and cooperation as world population grows. In developing countries, regional management plans based on the geography of watershed and aquifer systems rather than on the geography of political borders will have to be formulated for areas such as the Middle East and large parts of Africa. These plans must be built at appropriate scales for the water supplies and land uses involved and should avoid "quick fix" projects such as huge dams. Even with the best water management planning, however, the forecast is not good for countries in these regions, because population and agricultural demands will soon outgrow water supplies. In some areas it already has.

In developed countries new approaches will have to be taken to limit population growth and development in dry areas and in areas prone to extreme flooding. These approaches will have to include reductions in our reliance on technological schemes such as aqueducts, dams, and levees and rely more on guiding and managing land uses toward more environmentally sustainable distributions and activities. In other words, we will have to pay more attention in land development to the distribution of water resources and potential hazards if we are to avoid costly disasters like the great Mississippi floods of 1993 and 2001.

12.10 SUMMARY

More than 97 percent of the Earth's water is held in the oceans. From the oceans, water is cycled through the atmosphere to the land masses, where it goes into streams and related surface waters, soil, and groundwater. The rate of delivery of precipitation by the atmosphere is very rapid, with a complete exchange every nine days. Precipitation is the sole source of water for terrestrial environments.

The precipitation process begins with condensation of water vapor in the atmosphere. Precipitation particles form around condensation nuclei and may incorporate pollutants that can change their chemistry. One such change lowers the pH, yielding acid rain.

Condensation leads to cloud formation, and precipitation requires cooling of air containing vapor. There are four basic mechanisms of cooling and each produces a different type of precipitation: convectional, orographic, frontal, and convergent. The different types of precipitation are associated with different environmental settings and conditions. Each produces a number of impacts on environment and land use. Among these are lightning and fires, flooding, temperature inversions, tornadoes, blizzards, hurricanes, and storm surges.

Streams derive their flow from groundwater and surface runoff. These two sources produce distinctly different flows: that from groundwater is steady and long-term, whereas that from surface runoff is sporadic and short-term. Surface runoff, also called stormflow, produces most of the flooding in stream valleys. The magnitude and frequency of flooding has increased in this century with the clearing of land, the growth of cities, and the alteration of channels. Flood damage has also increased, although for different reasons in developed countries such as the United States and Canada and poor countries such as Bangladesh and India.

Excluding the glacial ice caps, groundwater is the largest reservoir of water on the land. Groundwater is held at various depths in porous sediments and bedrock. Those materials containing large amounts of usable groundwater are called aquifers. They function as water systems, receiving recharge water from the surface and giving up water to stream channels, pumping, and evaporation. Pumping in excess of the recharge rate can lead to the depletion of aquifers, a common problem in dry regions with growing populations (such as the American Southwest) and in large urban centers under any climatic conditions.

Shallow aquifers may lose water to topographic depressions in the land, giving rise to lakes, ponds, and wetlands. These impoundments also function as water systems, exchanging water through stream outflow, evaporation, and seepage back into the ground. Wetlands are usually shallow impoundments with heavy growths of plants and relatively thick organic deposits. Lakes and ponds are deeper water bodies that are not overgrown by aquatic plants. Both wetlands and lakes are seriously altered by land-use activity. Wetlands have for centuries been destroyed by draining and filling, resulting in, among other things, the loss of rich habitat. Only recently have some countries begun to protect wetlands.

Water use in the United States totals about 1.55 billion cubic meters (411 billion gallons) of water daily. This water is drawn from both surface supplies and groundwater. The three greatest users of water are power generation, agriculture, and manufacturing. Agriculture is the principal consumptive user because of the high rates of evaporative loss in irrigation. All the major water users, including homes and settlements, are degrading because they pollute the water in some way before releasing it back into the environment.

Humans have been involved in water management for thousands of years and have become very efficient at redistributing water and regulating flows. However, with freshwater in limited supply and world population growing by more than 80 million people a year, stress and conflict over water supplies is rising in regions such as the Middle East, southern Asia, and northern China. Sustained management of water supplies—if it is at all possible—will require limits on population growth and agricultural expansion in water-limited countries such as Pakistan; international water management plans; and water allocation programs based on most appropriate use rather than prior rights.

12.11 KEY TERMS AND CONCEPTS

Planetary water
- Earth
- Mars

The hydrologic cycle
- natural reservoirs
- breakdown and distribution

Precipitation processes
- condensation
- acid rain, acid clouds, and acid fog
- dew point

Types of precipitation

Convectional precipitation
- thermals
- latent heat of condensation
- thunderstorms
- flash floods
- geographic controls

Orographic precipitation
- prevailing winds
- temperature inversion
- air pollution

Frontal/cyclonic precipitation
- polar front
- cold and warm fronts
- tornadoes
- blizzards

Convergent precipitation
- hurricanes
- origin and distribution
- storm surges
- coastal damage

Streamflow
- baseflow
- surface runoff

stormflow
watershed
drainage network
floodplain
Flood hazard
causes and trends
flow constructions
the American dilemma
Groundwater
aquifer
porosity
recharge/discharge
safe aquifer yield
saltwater intrusion
Wetlands, lakes, and ponds
water table
natural impoundment
water balance
riverine wetland system
habitat
eradication
infilling and eutrophication
Water use
trends and rates
consumptive
nonconsumptive
supplies and problems
degrading
dams and diversions
Water management
redistribution tradition
prior allocation
riparian rights
sustained use

12.12 QUESTIONS FOR REVIEW

1. Briefly describe the hydrologic cycle, including the percentage of the Earth's water supply held in the atmosphere, oceans, groundwater, ice caps, and streams and lakes. What percentage of global precipitation falls on the land masses, and how fast is water recycled by the atmosphere?
2. Describe the processes and conditions leading to the formation of precipitation particles? Where do pollutants enter the process? What are some of the effects of acid precipitation?
3. What are the four main types of precipitation, and what geographic circumstances and environmental impacts are associated with each? Define the following terms: latent heat of condensation, specific heat, temperature inversion, and polar front. (Also see glossary.)
4. What are the two main sources of streamflow, and what is the nature of their delivery to streams? Which causes most flooding and pollution of streams?
5. What are watersheds and drainage networks? Name several ways in which land use and engineering projects change watersheds and drainage nets and in turn alter streamflow and flooding. Is there any connection here to the 1993 and 2001 Mississippi floods? Why has property damage from flooding continued to rise in recent decades?
6. What is groundwater, what is an aquifer, and what are the key environmental concerns about groundwater? Describe the Ogallala aquifer and the concept of safe aquifer yield.
7. What are the main types of wetland systems? What has been the traditional nature of the relationship between land use and wetlands? What is the trend in the development of ponds and reservoirs in the United States, and what is the ultimate fate of such water features?
8. Describe some of the historical and geographical trends in water use. What are the water use rates of the major land uses in the United States? Describe some of the impending water supply problems in the world.
9. Water use is not without its environmental impacts. Argue pro or con, and provide rationale for your positions. In particular, what are some benefits and costs of dams and stream diversions?
10. What is meant by the concept of sustainable water use, and what evidence is there that we are not achieving the objectives of this concept today? Can you build a set of guidelines for improving national and regional water management?

13

WATER POLLUTION: PATTERNS, TRENDS, AND IMPACTS

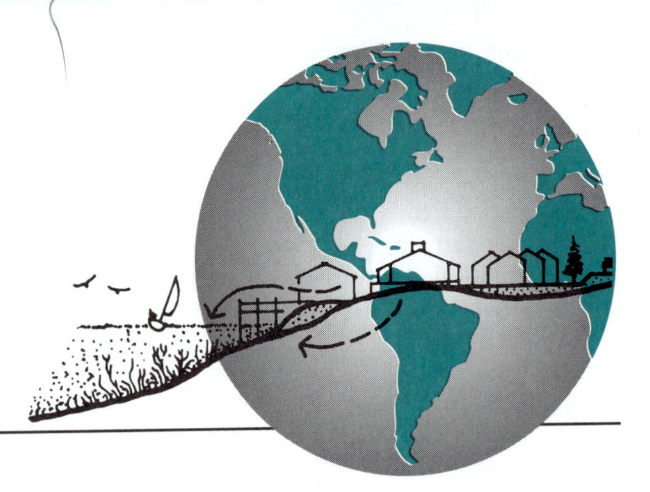

13.1 INTRODUCTION

During the twentieth century human activities introduced pollution into nearly all segments of the Earth's hydrologic cycle. Pollution has reached high into the atmosphere, seriously altering the chemical balance of airborne moisture. Rainfall over much of the Earth is actually polluted before it hits the ground (as acid rain, for example). On the ground runoff from urban and agricultural areas washes contaminants into streams from streets, buildings, and fields. Below ground, water percolating through the soil picks up additional pollutants from buried wastes, leaking tanks, and pipes and delivers them to groundwater aquifers. In selected spots even deep bodies of groundwater, thousands of meters down, have been contaminated by mechanical injections of hazardous wastes.

The general problem of water pollution revolves around three colossal issues: (1) the massive quantity of pollutants produced by more than 6 billion humans and their machines, plants, and animals; (2) the limited supply of fresh liquid water, about 0.6 percent of the world's total water supply, into which most water-destined pollutants are discharged; and (3) the growing number of "technological" pollutants released to the environment—namely, manufactured synthetic materials, which are new to nature and the aquatic environment. Water quality is declining worldwide and the costs are enormous, not only in terms of human diseases, waste-water treatment, and drinking water purification, but in terms of the degradation of aquatic ecosystems and the loss of irreplaceable habitats.

In this chapter, we investigate water pollution, beginning with a description of the different classes of pollution and how they are measured. Next, the sources of the major pollutants are examined. This is followed by a discussion of the distribution of pollutants in the hydrologic cycle and of pollution controls and the effectiveness of pollution control programs. The chapter concludes with some observations on water pollution trends.

13.2 TYPES OF WATER POLLUTION

Water pollutants can be classified in different ways according to several different criteria, including environmental effects, influence on human health, types of sources, and pollutant composition. For our purposes, pollutants can be grouped into eight classes according to their influences on the environment and human health.

- **Oxygen-demanding wastes.** As organic compounds in sewage and other organic wastes are decomposed by chemical and biological processes, they use up available oxygen in water that is essential to fish and other aquatic animals.
- **Plant nutrients.** Water-soluble nutrients such as phosphorus and nitrogen accelerate the growth of aquatic plants, causing the buildup of organic debris and the elimination of certain organisms in streams, lakes, and other water features.
- **Sediments.** Particles of soil and dirt eroded from agricultural, urban, and other land uses cloud water, cover bottom organisms, eliminate certain aquatic life forms, and clog stream channels.
- **Disease-causing organisms.** Parasites such as bacteria, viruses, protozoa, and worms associated mainly with animal and human wastes enter drinking water and cause diseases such as dysentery, hepatitis, and cholera.
- **Toxic minerals and inorganic compounds.** Substances such as heavy metals (e.g., lead and mercury), fibers (e.g. asbestos), and acids from industry or various technological processes are harmful to aquatic animals and cause many diseases in humans, including some forms of cancer.
- **Synthetic organic compounds.** Substances, some water-soluble (e.g., cleaning compounds and insecticides), some insoluble (e.g., plastics and petroleum residues manufactured from organic chemicals), cause various disorders in animals and humans, including kidney disorders, birth defects, and probably cancer.
- **Radioactive wastes.** Byproducts of nuclear energy manufacturing from both commercial and military sources emit toxic radiation, a cause of cancer.
- **Thermal discharges.** Heated water discharged mainly from power plants and some industrial facilities causes changes in species and increased growth rates in many aquatic organisms.

13.3 MEASURING WATER POLLUTION

In order to describe and analyze water pollution problems and recommend appropriate pollution control laws, we must be able to measure pollution levels. For pollutants such as thermal discharges, this is a relatively simple matter of measuring the rise in water temperature over the natural temperature of the receiving water. For most other pollutants, such as sediments and plant nutrients, measurement involves separating the pollution from the water and measuring the total units or parts of pollutant per parts of water, usually given in parts per million (ppm). A more widely used measure is milligrams per liter, which is the weight of the pollutants in a liter of water and is denoted by mg/l. Although they are different units, ppm and mg/l are equivalent and are standard measures of pollution concentration.

Pollution Concentration and Loading

Important to the understanding of pollution data is the distinction between **pollution concentration** and total **pollution load** or discharge. Pollution concentration is the amount of contaminants in water at a single moment such as during a pollution spill. It typically varies with the volume of water in response to changes in streamflow and changes in the amount of pollutants discharged. When there is a lot of water, pollution is diluted and the concentration is low. When there is a lot of pollution and little water, concentration is high. The pollutant load, on the other hand, is measured by adding up the total pollution discharge into a stream over an extended period of time, such as a year. Unlike pollution concentration, load is measured irrespective of the amount of water involved (Figure 13.1a).

Setting Standards for Water Pollution

Now we come to the important question of **safe limits** of pollution—or how much is too much. This is a source of endless debate and is at the very heart of the pollution regulatory process. Defining the actual level at which water can be called polluted depends on one's point of view. Among the many points of view held in society today, let us single out three. The first argues that any measurable increase above background levels, no matter how small, constitutes pollution. The second argues that concentrations must be great enough to exceed the water system's natural capacity (biological, chemical, and physical) to break down and/or disperse of the contaminants within a reasonable time and area. The third argues that the contaminants must threaten human health to constitute pollution. For example, drinking-

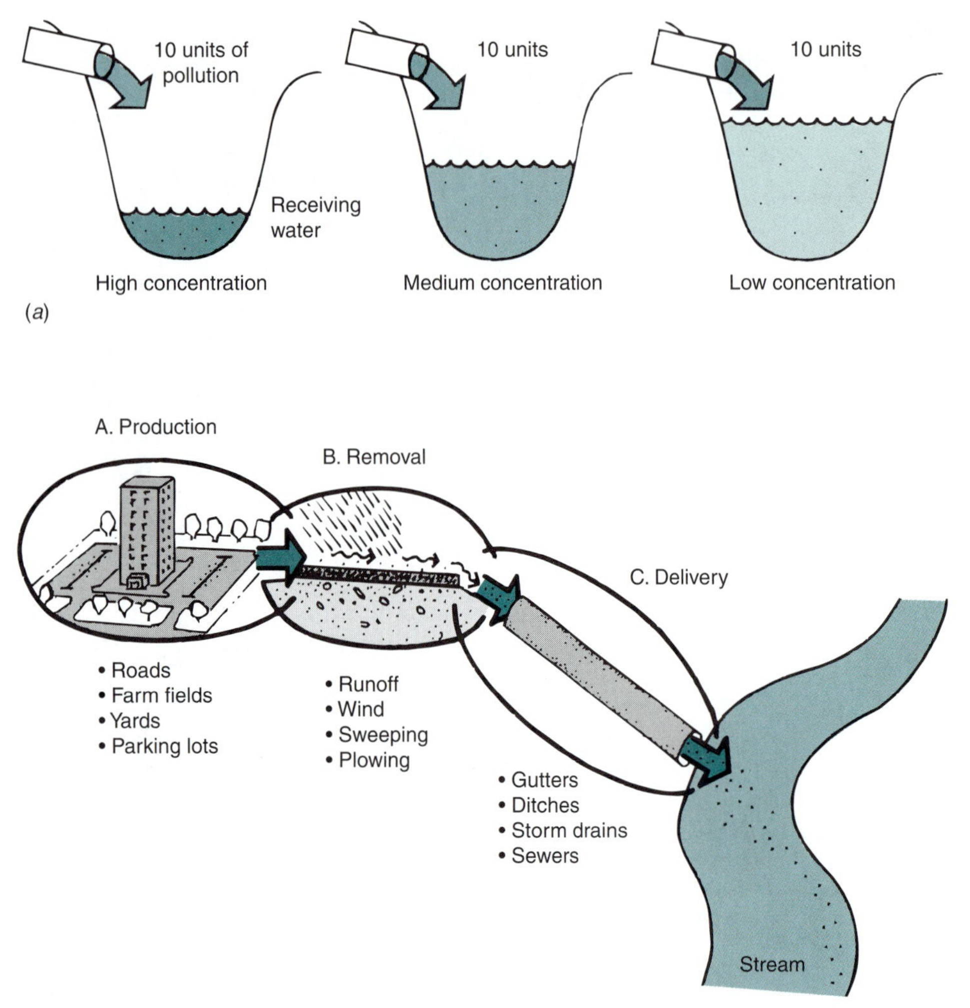

FIGURE 13.1 *(a) The concepts of pollution concentration and pollution load; and (b) the three components of a water pollution system: production, removal, and delivery.*

water standards define the limits of water pollution based on disease potential from ingestion. Another health standard, which is less stringent, is based on disease potential from bodily contact with contaminated water.

In most cases, the problem of measuring water pollution usually involves first determining background levels of the contaminant in question, because many substances labeled as pollutants (such as sediment, nitrogen, and certain heavy metals) are found naturally in water. For example, in a stream where sediment is the issue, the natural levels of sediment in the water would first have to be measured or estimated. If we define pollution according to the first point of view, then sediment loads above the background level would constitute pollution.

The more widely used definition is based on the second point of view—that is, the water system's capacity to break down and/or disperse contaminants. For this scientists often use indicators such as changes in the abundance or health of certain plant or animal species in aquatic ecosystems to define significant or critical pollutant concentrations or loads. However, such indicators rarely provide consistent readings of the safe limits of pollution, because, among other things, tolerances vary among organisms, even members of the same species.

In addition, for some pollutants the effects on indicator organisms are not immediate (or acute) but chronic and show up only in the long run. This makes it difficult to determine in complex systems what is actually causing some change. And, of course, some species are more sensitive than others, so which ones are used as indicators can make a big difference. Therefore, there is always room for debate. The standards that we see in environmental laws usually represent a best approximation based on several indicators as well as compromise among scientists, other experts, and lawmakers.

13.4 Water Pollution Sources and Distribution Processes

There are two conventional ways of describing water pollution sources. One is based on the activity that produces the pollutant, and the other is based on the way the pollution is discharged into the environment. The latter uses only two categories, generally called point sources and nonpoint sources, whereas description based on activities uses many categories, typically, land uses (such as agriculture, industrial, residential, and transportation), various technologies (such as petrochemical and electroplating), and disposal systems (such as sanitary sewers and landfills). Different activities not only produce different types of pollutants but strikingly different amounts as well. Moreover, both the types and amounts of pollutants change over time with economic, technological, and land use changes.

Point source pollution is that from a specific source, usually a facility, and is released at a known discharge point or outfall, usually a pipe or ditch. Chief among point sources are municipal sewer systems, industry, and power plants. While point source pollutants are very important, the majority of stream pollution comes from nonpoint sources.

Nonpoint source pollution represents spatially dispersed, usually nonspecific, sources that are released in various ways at many points in the environment. Storm water in both urban and rural areas is the principal source of nonpoint pollution. It is generated from various land uses over large areas and is discharged into streams, lakes, and other water features at countless places along channels and shores. In most countries, agriculture is the chief contributor of nonpoint source pollution and sediment is the main contaminant. However, the most geographically extensive source of water pollution is fallout (in both wet and dry forms) of pollutants from the atmosphere.

In general, nonpoint water pollution can be characterized as a three-part system consisting of on-site production followed by removal from the production site and ending with delivery to a stream, lake, or wetland. *Production* represents the amounts and types of contaminants generated by a land use. For example, on streets and highways cars and trucks are the principal contaminant producers (e.g., oil, gasoline, and tire particles) and the rate of output is directly tied to traffic levels. *Export* or *removal* is accomplished mainly by runoff. Roads are built with crowns in the middle so that runoff moves efficiently to the shoulders and collects with its load of contaminants in the drains along the road edge. *Delivery* entails routing the stormwater in a pipe or ditch from the road edge to a stream or water body (Figure 13.1b).

In the next several pages we describe the main sources of water pollution. Most of the discussion focuses on land uses, but attention is also given to processes such as soil erosion and atmospheric fallout.

Traditional Rural Landscape

Most of the world's population lives in small towns and villages scattered across agricultural landscapes. Water supplies are drawn from local sources, mainly streams and wells. Human and animal wastes are disposed of in the same locale—on the soil, in pits, or in the streams themselves. In countries such as India and China, sewage treatment is very rare, even in cities. Large cities such as Bombay and Calcutta, India, and Shanghai, China, have either grossly inadequate treatment systems or none at all (Figure 13.2). In the countryside pollution from expanding populations and crowded villages frequently contaminates water supplies and spreads disease. The United Nations estimates that 80 percent of all sickness and death among children in developing countries is related to unsafe drinking water. Among the diseases traditionally associated with impure water are dysentery, typhoid fever, infectious hepatitis, and cholera.

Whereas these are the most serious water pollution problems of traditional societies—because they directly affect human health—they are not the only ones. Beyond the cities and villages, agriculture is the chief source of pollution. Sediment loading of streams as a result of cropland erosion is serious in most parts of the world, and it commonly disrupts aquatic ecosytems, interferes with streamflow, enhances flooding, and impedes irrigation. Furthermore, wastes from livestock picked up by runoff represent another significant source of water pollution, especially nutrient-rich and oxygen-demanding wastes.

Added to the traditional rural pollutants in developing countries are increasing loads of chemical pesticides. For example, chlorinated organic compounds such as DDT, long banned in most developed countries, are widely used on cropland and to protect stored grains from vermin. Many of these substances are heavy in persistent (long-lived) organic compounds, which readily enter the water system with runoff or soil water and remain toxic for long periods. In addition, DDT is widely used in tropical areas for mosquito control to limit malaria.

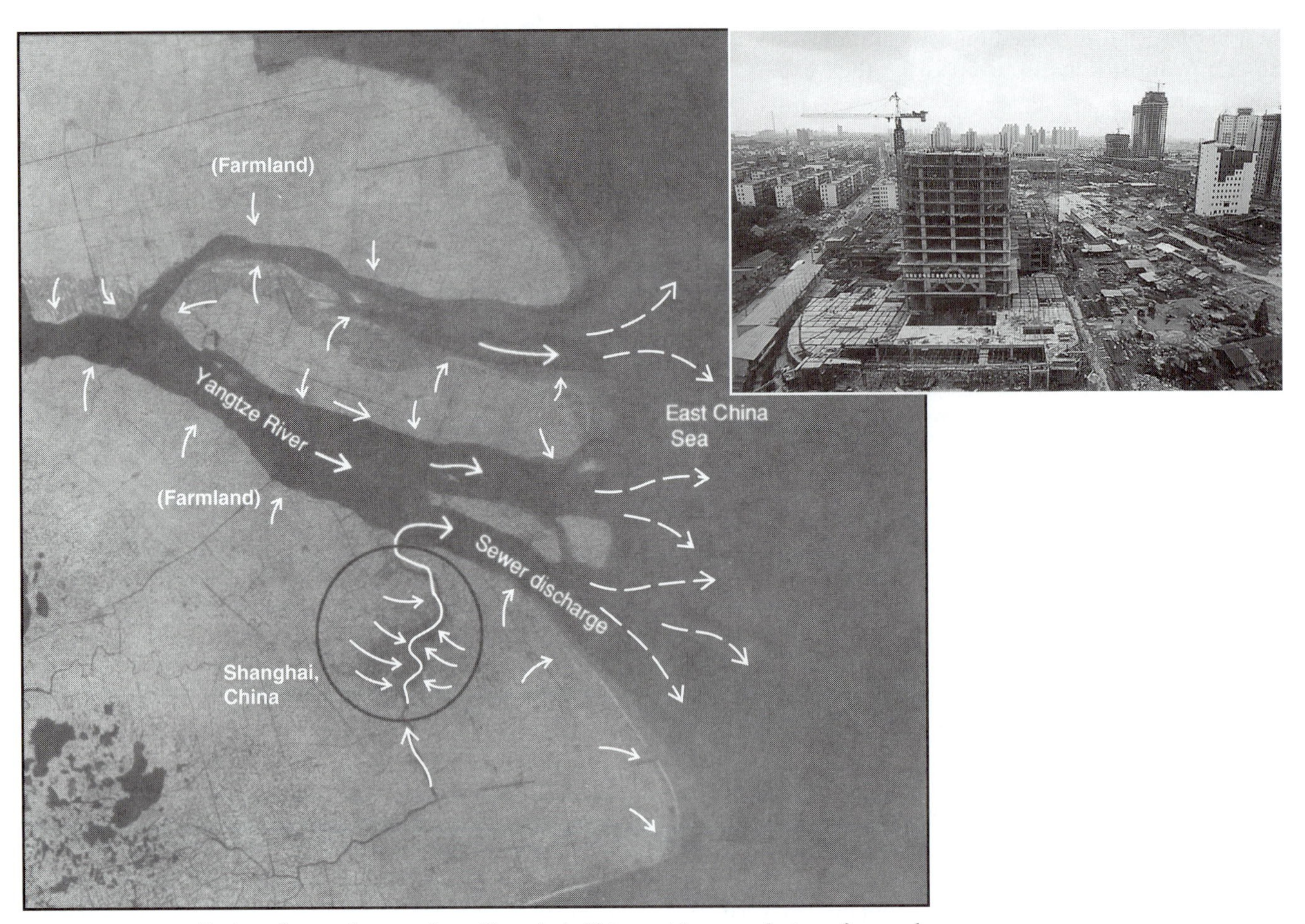

FIGURE 13.2 *Pollution of coastal waters from Shanghai, China, with a population of more than 11,000,000. Like cities in many other developing countries, Shanghai releases untreated sewage to streams and the ocean. The city is growing rapidly (inset) as China pushes toward a consumer economy.*

Modern Urban Landscape

In developed countries, 70 to 80 percent of the people live in urban areas. Most urban areas are complex amalgams of land uses, such as old city centers, industrial sectors, and sprawling suburbs, all tied together by an infrastructure of roads, power lines, water lines, and sewers. There are actually two systems of sewers—storm and sanitary—both of which consist of networks of underground pipes.

Sanitary sewers were developed in the late 1800s as a replacement for the outdoor (pit-style) toilet commonly used in rural areas. Initially sanitary sewer systems consisted only of networks of pipes to carry sewage beyond the city. Later, first-level or primary treatment plants were added to reduce pollution in receiving waters. But with massive urban development by 1950 or so, primary treatment systems proved inadequate. By 1980 more advanced systems were employed. Despite huge investments in the United States, Canada, and other developed countries in community and industrial waste-water treatment, water pollution from cities continues to rise because of large increases in stormwater due to land development and industrial growth, and because many cities continue to use outdated and inadequate sewage-treatment facilities.

In the United States, where federal law has mandated that communities build advanced sewage-treatment facilities, more than 1,000 communities, including New York, Detroit, Philadelphia, and San Francisco, still rely on inadequate systems involving *combined sewer overflows.* These systems allow untreated sewage to temporarily bypass treatment plants when the plants are overloaded. The overflow enters storm sewers and is discharged directly into lakes, streams, and harbors. Bypassing is most common during wet weather, typically in winter or after heavy rainstorms. In many southern European cities, sewage treatment is only practiced in summer when the climate is driest; otherwise raw sewage is bypassed directly to streams and/or the Mediterranean Sea.

According to one standard measure of water pollution, called the **biological oxygen demand (BOD)** index, about 66 percent of point source water pollution in the United States comes from cities and 33 percent from industry. Treatment facilities at municipal sewer systems serve 70 to 75 percent of the U.S. population, and although these facilities greatly reduce pollution levels, they still produce nutrient-rich effluent water, sediment, and sludge (solid organic waste). The remainder of the U.S. population, mostly in rural areas, is served by some form of **soil absorption system** such as a septic tank and drainfield, which is designed to serve individual households.

Industrial pollutants vary according to the type of industry and products manufactured. The power industry, for example, is a major contributor to air pollution, which, in turn, contributes to water pollution (Figure 13.3). For example, emissions from coal-burning power plants are a principal source of acid rain, which in the past several decades has resulted in a serious decline in the pH levels of lakes and streams in southeastern Canada, the northeastern United States, and northwestern Europe (see Figures 11.11 and 11.15). In both conventional (fossil fuel) and nuclear power plants, discharges of hot water from cooling operations are often a source of local thermal pollution of streams and lakes. Coal-burning power plants also contribute two heavy-metal pollutants (cadmium and arsenic) to surface waters through atmospheric fallout. Pulp and paper mills, which use large amounts of water, have traditionally been heavy polluters of streams and lakes. One of the most toxic pollutants in mill effluent is dioxin, a cancer-causing organic compound.

The list of pollution from industry is long and is not limited to large industrial operations. Small-scale industries can also be serious polluters; for example, metal-plating operations are a major source of heavy-metal pollution, and plastics manufacturing is a recognized source of synthetic organic compounds and paint residues. Such operations are commonplace around industrial centers, especially automobile manufacturing complexes. Because they are numerous and widely scattered, regulating them is often difficult. In the past several decades significant strides have been made in developed countries to reduce industrial water pollution. Treatment systems are required by law to reduce BOD, sediment, heavy metals, and other contaminants to specified outfall concentrations. However, the number of industrial facilities has increased with economic growth, so total pollution loading has risen in many natural water systems in North America and western Europe.

In other parts of the world, the picture can be very gloomy. During the communist regimes in Eastern Europe

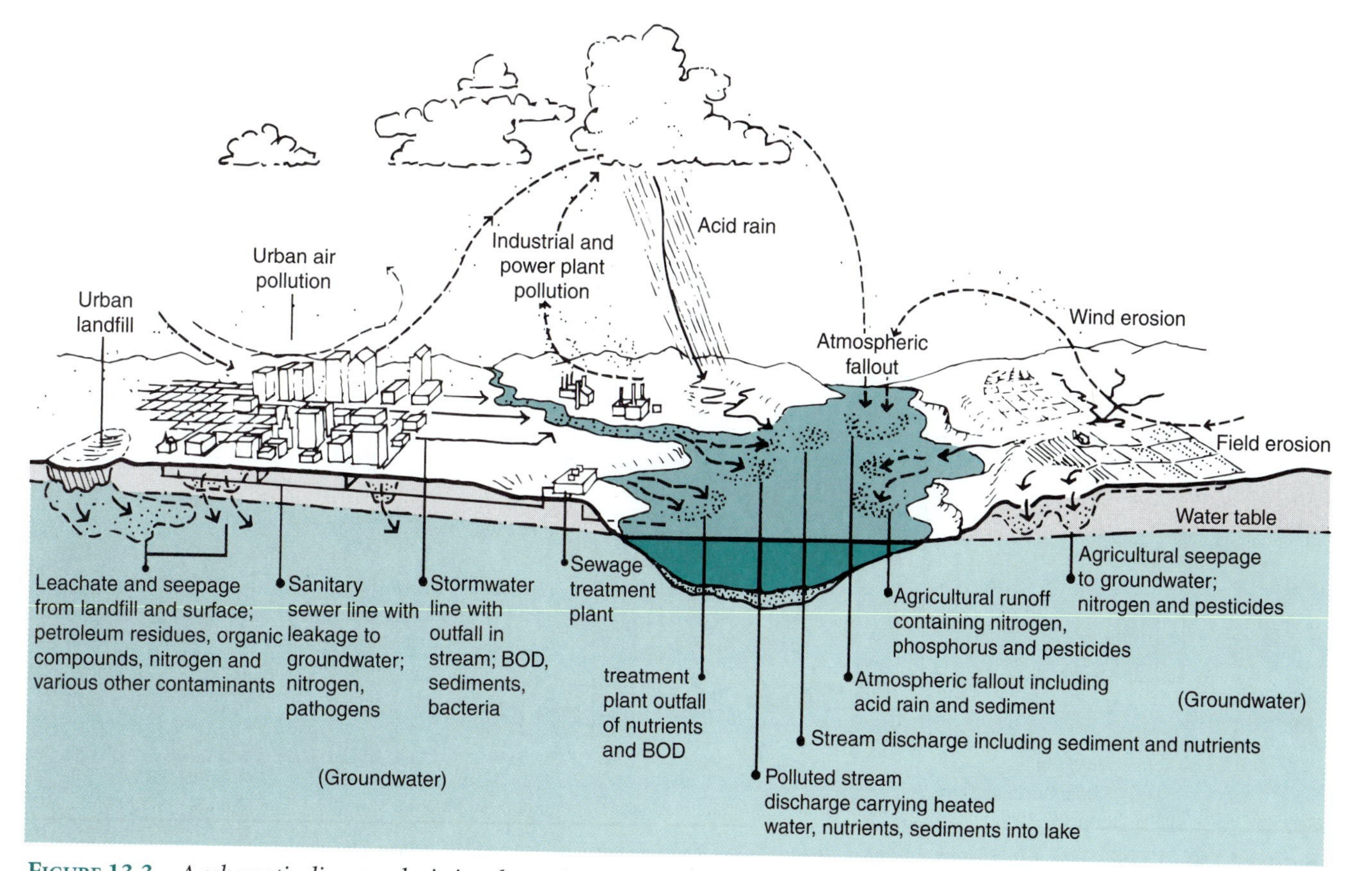

FIGURE 13.3 *A schematic diagram depicting the various sources and processes of water pollution associated with the complex system of land uses in and around a modern city.*

and the former Soviet Union, pollution regulation was ineffective and often ignored. Water-treatment facilities were either very primitive or were not required for industrial operations and for many cities. Direct dumping of industrial and municipal wastes into natural waters was common, resulting in some of the most deplorable and dangerous environmental conditions on earth in this century. These conditions persist today in many areas. In Siberia, Russian pipelines are so poorly maintained that it is estimated that millions of barrels of oil are being leaked at thousands of locations into soils, streams, and groundwater yearly.

Stormwater

The most significant nonpoint source of water pollution in most places is **stormwater**, the runoff from rainfall and melting snow. Because it impairs transportation, farming, and other land-use activities, stormwater is drained from the land as quickly as possible via ditches and pipes. As a result, all sorts of debris are flushed into streams, lakes, and oceans: sediment, oil, fertilizer, and pesticide residues, as well as organic residues from vegetation, animal droppings, and garbage (Figure 13.3).

The rise of urban stormwater pollution more or less parallels the development of the automobile industry and urban sprawl. In the past 90 years the number of automobiles in the world has grown from a few thousand to more than 500 million. In the United States there are now more than a million miles of paved roads and countless acres of parking lots, driveways, garages, and other hard surfaces for automobiles. As impervious surfaces, they generate huge amounts of stormwater (see Figure 12.11) containing a host of automobile contaminants—oil, paint, lead, organic compounds, and many other residues. Today the quantity of petroleum residues washed off streets, highways, parking lots, and industrial sites each year exceeds the total spillage from oil tankers and barges worldwide. In addition, to improve road safety in northern cities, salt is used to deice roads and it also becomes part of the stormwater pollution load. The salt content in expressway drains during spring runoff may be a hundred times greater than natural levels in freshwater. Unfortunately, it remains standard engineering practice to release contaminated stormwater from roads directly to streams in the quickest and cheapest way (see Figure 13.1b).

In urban areas, the loading of stormwater with pollutants increases with land-use density (Figure 13.4). Land-use density is measured by the percentage of land covered by impervious materials (mainly asphalt, concrete, and roofs), and it reflects the level of pollution-producing ac-

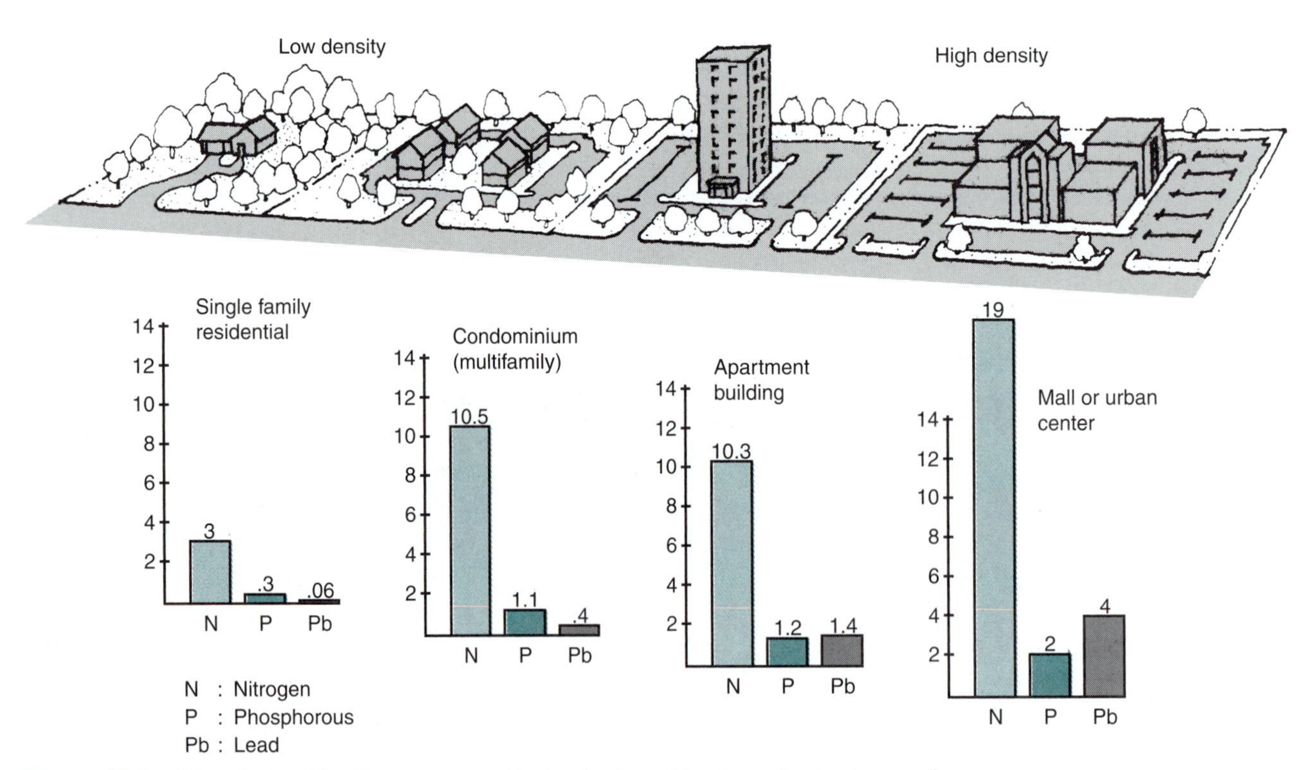

FIGURE 13.4 *The relationship of stormwater pollution levels and land-use density in an urban region. Pollution loading increases with density from the suburban fringe toward the city center. The quantities for nitrogen, phosphorus, and lead are pounds per acre per year.*

tivities such as automobile traffic, spills, leakage, atmospheric fallout, and garbage production per acre or hectare. It also reflects the efficiency of surface flushing by stormwater runoff, which is virtually 100 percent in heavily built-up areas. The heaviest pollution loads are typically carried by the so-called **first-flush flow**, representing the first 1 to 2 centimeters of a rainstorm.

The total BOD contribution from stormwater is difficult to gauge accurately, but it is very large in both urban and agricultural lands. It has been estimated that (1) in 35 percent of the watersheds in the United States, urban nonpoint sources (principally stormwater) are greater contributors than point sources, and (2) in 54 percent of the watersheds, agricultural nonpoint sources (also principally stormwater) are greater contributors than point sources. The concentration of BOD pollutants in stormwater is generally equal to or greater than that in effluents discharged from modern sewage treatment facilities, but the total loading from stormwater may be significantly greater than that from point sources because of the huge volumes of stormwater discharged from the land.

Agriculture

Among the world's land uses, agriculture is the single largest source of water pollution. It contributes sediment, fertilizer, pesticide residues, salts, bacteria, and manures to water systems. The manures, crop residues, and fertilizers increase BOD in surface waters. Chemical fertilizers are not a direct source of BOD pollution, but they greatly increase aquatic weed growth, which adds oxygen-demanding organic residues to streams and lakes.

In addition, nitrogen, a major constituent of all fertilizers, moves rapidly through most soils and spreads into groundwater supplies, where it poses a human health risk. In the presence of certain bacteria in the human digestive tract, nitrates can be converted into nitrites, which are toxic to the body. Disorders linked to nitrogen contamination of groundwater have become a problem in certain areas of the Corn Belt in the U.S. Midwest where groundwater is the main source of drinking water (Figure 13.5a).

The use of chemical fertilizers has increased substantially in the past 50 years in order to maximize crop yields. Among the major crops, corn is one of the heaviest fertilized; more than 90 percent of the U.S. corn acreage receives applications. Other crops, such as wheat and soybeans, are less widely fertilized, but the acreage receiving application is increasing (Figure 13.5b). Commercial fertilizers are composed of macronutrients, principally nitrogen, phosphorus, and potash. In the United States most of the phosphorus and organic nitrogen pollution of surface water and groundwater comes from agricultural sources. Nitrogen is especially abundant in the water system because soil and related materials have limited capacity to adsorb and retain it. Concentrations of nitrogen in groundwater are highest where soils are permeable or well-drained (Figure 13.6).

Herbicides (weed killers) and insecticides are also widely used in modern agriculture (Figure 13.5b). The residues of these chemicals enter streams with stormwater and enter groundwater with seepage through soil. Many are filtered out by the soil, where they may be broken down by microorganisms. However, significant amounts, especially of the more persistent ones, escape to groundwater and streams. Once in the water system, they infiltrate aquatic and terrestrial food chains and human water supply sources. The most controversial of the insecticides is DDT, which caused serious damage to many animals, especially birds of prey, before being banned in the United States in the early 1970s. Since DDT is a persistent chemical, having a long life in the environment, it can accumulate to toxic levels at the top of the food chain. It bears repeating that DDT is still widely used in developing countries.

Today other insecticides with various commercial names are used in the United States and other developed countries in place of DDT, with questionable results in terms of impact on the environment. Many insecticides, such as simazine, cyanazine, and aldicarb have been detected in groundwater supplies and have been restricted or locally banned. In California, America's principal agricultural state, as many as 22 organic compounds from both insecticides and herbicides have shown up in groundwater. While the concentration of most is very small and the decomposition times for many are relatively short (often days or weeks), some of these compounds persist for months and years (Table 13.1). The latter are called **persistent organic pesticides**, or POPs. Many chemical pesticides have been linked to human health disorders including cancer. At highest risk of pesticide-related illness are youngsters (because of their small body masses) and people who have been exposed to these compounds over long periods.

Worldwide, pesticide application is rising as farmers push to improve yields. In the United States the extent of groundwater pollution by pesticides is largely unknown, but the potential is high in many areas, as is shown by the map in Figure 13.7. Soil, which is the main recipient of

TABLE 13.1

PERSISTENCE OF SELECTED PESTICIDES

Substance	Persistence (half-life)*
Chlorinated hydrocarbons (e.g., DDT, aldrin, mirex, and kepane)	2 to 5 years
Organophosphates (e.g., Malathion, miazinon, and Diazinon)	1 to 10 weeks
Carbamates (e.g., Sevin, maneb, and Tenik)	1 week

*time taken for the quantity to be reduced by 50 percent

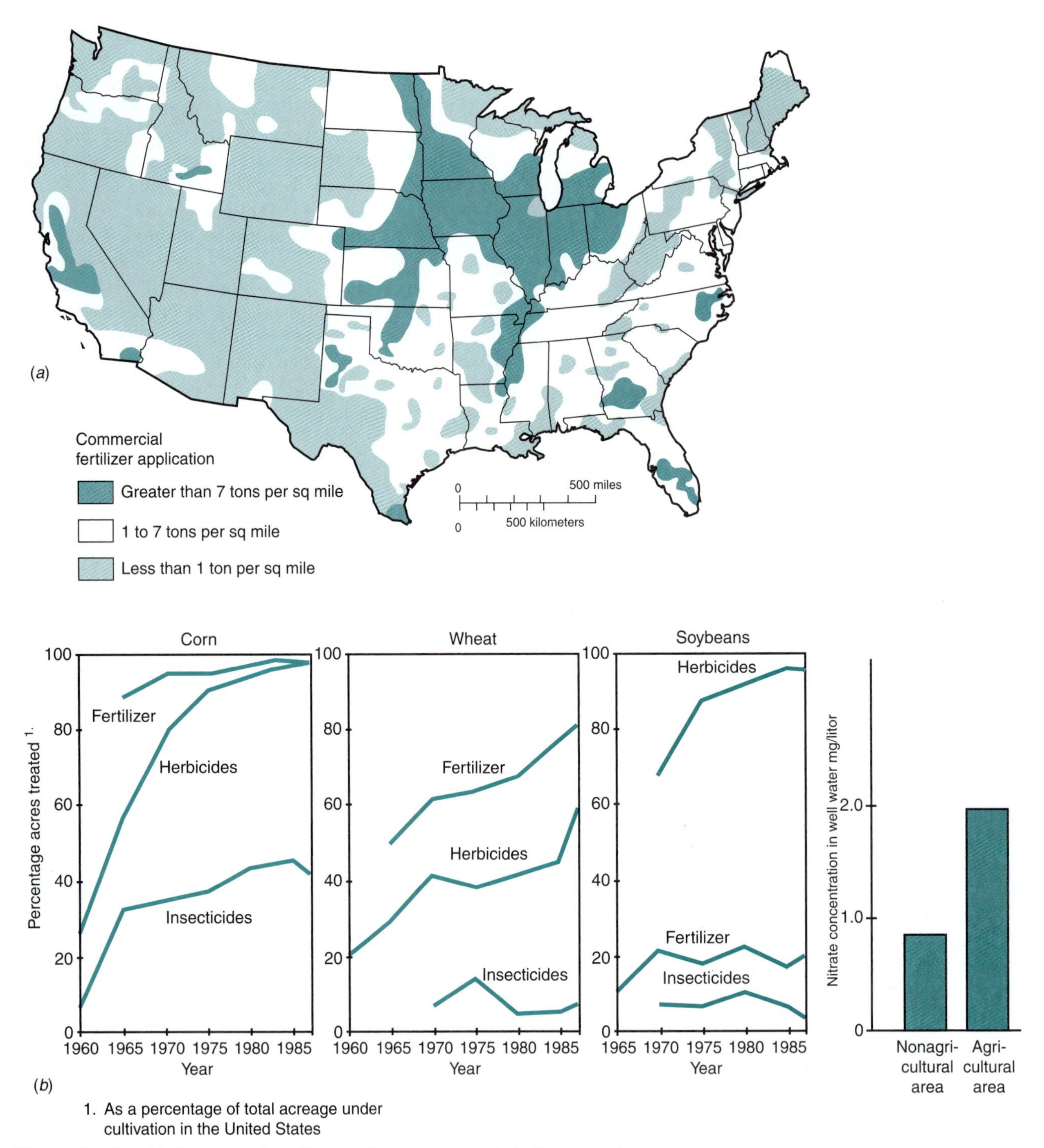

FIGURE 13.5 *(a) Commercial fertilizer applications in the United States. Nitrogen is a major component in fertilizers and risk of nitrogen contamination of groundwater is highest in major farming areas as the graph shows. (b) Trends in fertilizer, herbicide, and insecticide use in the United States for three crops.*

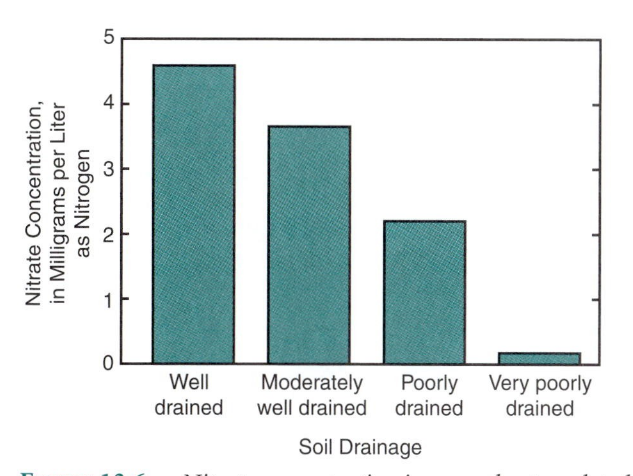

FIGURE 13.6: *Nitrate concentration in groundwater related to soil drainage. Well-drained soils are generally sandy; poorly drained soils tend to be clayey.*

pesticide residues, is actually an important line of defense against water pollution because of its capacity to hold many of these contaminants in place, bonded to soil particles. But POPs are a growing concern and a source of much debate. Many developing countries argue against bans on DDT, for example, because it is used in disease control, especially malaria.

Soil Erosion

For the world as a whole, **sediment** is the principal nonpoint source pollutant, and most of this sediment comes from soil eroded from cropland. Worldwide the annual soil loss rate is estimated at more than 20 billion tons, and the loss is increasing as more land is deforested and opened to agriculture. Since preagricultural times soil loss to erosion appears to have more than doubled worldwide, based on measurements of the sediment loads deposited in the ocean by the world's streams and rivers.

The burden of heavy soil erosion and sediment pollution is shared by both developed countries and developing countries, although the rates are lower in the developed countries. The United States and the former Soviet Union together account for nearly 4 billion tons of soil loss annually, whereas India and China—with much greater populations and less land under cultivation—together account for 8 billion tons. These four countries are responsible for more than half the world's soil loss and, in turn, a large share of the excess sediment loading in wetlands, streams, lakes, and ocean harbors.

Most of the soil lost from the land is **topsoil**, the organic-rich upper layer of the soil. As a result, the composition of the sediment that ends up in surface waters is

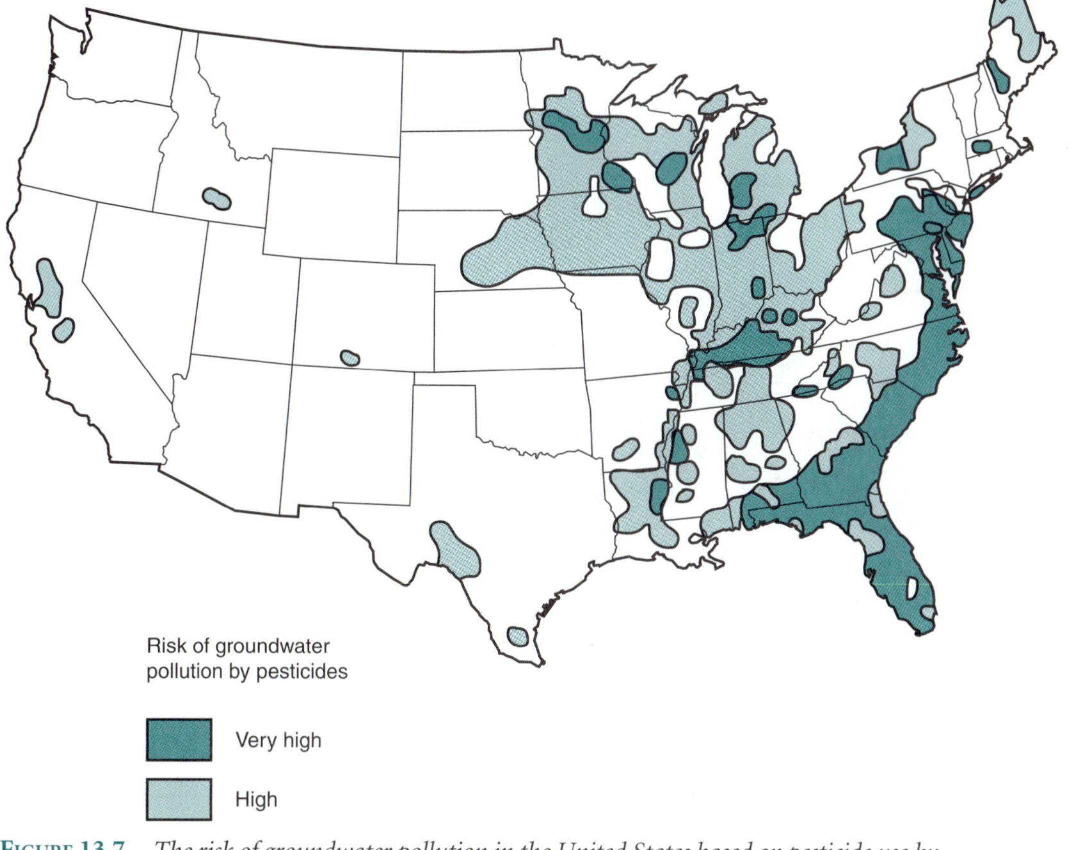

FIGURE 13.7 *The risk of groundwater pollution in the United States based on pesticide use by agriculture and groundwater vulnerability to pollution. (Compare with Figure 12.15.)*

organic as well as mineral (inorganic). The organic matter in topsoil is not only highly valuable to soil fertility, but is the part of the soil most effective in locking up organic compounds, heavy metals, and other pollutants, keeping them out of the water system while they are broken down. When topsoil is disturbed and subjected to erosion, these contaminants are released to runoff. In 1993, when massive floodwaters overwashed extensive areas of cropland in the upper Mississippi watershed, a great surge of sediment and chemical contaminants was discharged to the Mississippi Delta and the Gulf of Mexico (Figure 13.8).

Urban areas are not without sediment pollution problems, although they are far less severe than those in agricultural areas. The most serious sediment loading is associated with land-use change and construction activity (see Figure 15.18). When land is torn open for development, the soil, without protective vegetation, is very vulnerable to erosion. Sediment washed into streams often clogs channels, hinders flow, and damages organisms. Many other pollutants are also washed into streams along with sediment in urban and suburban areas, including nitrogen and phosphorus, petro-

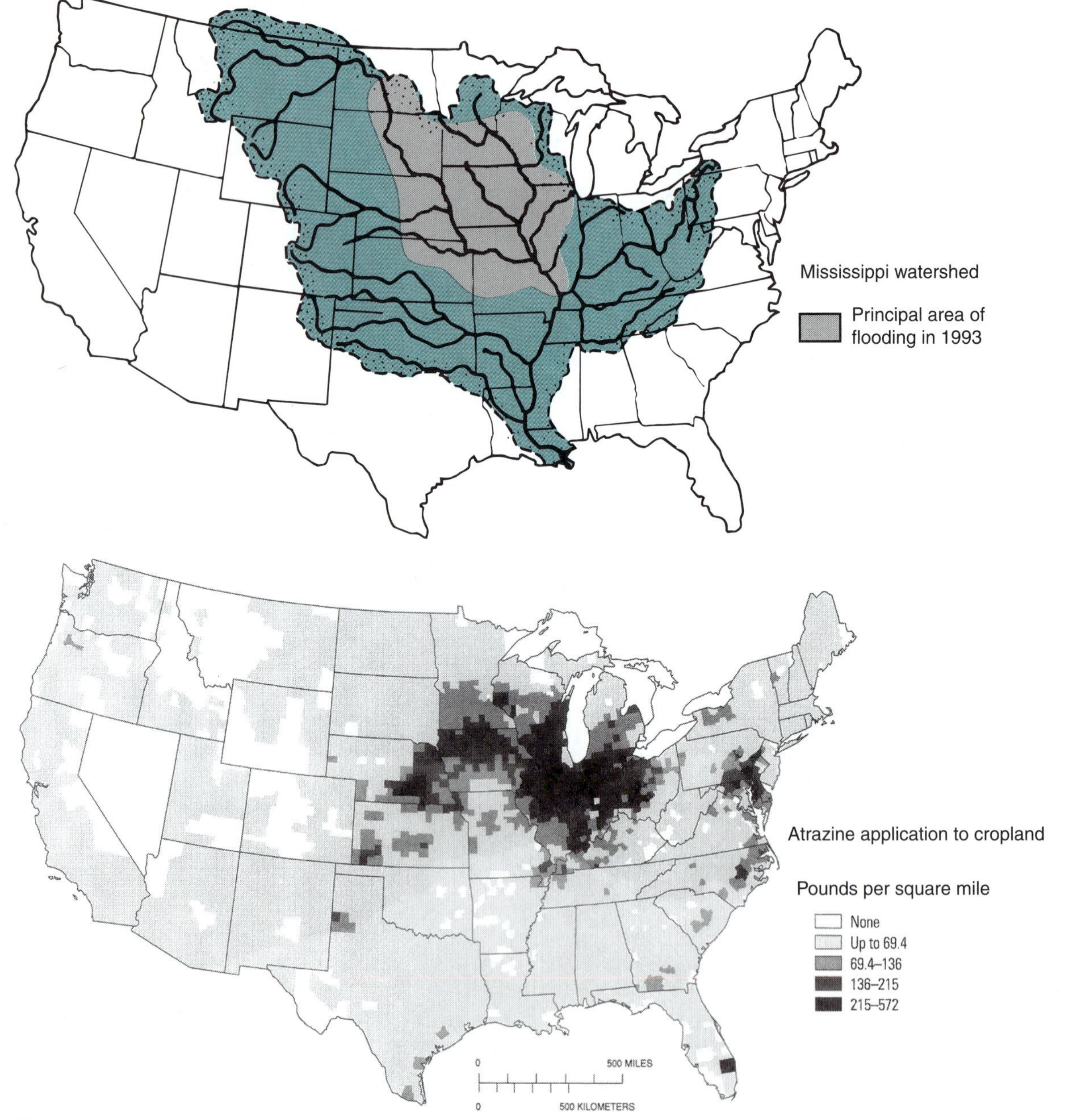

FIGURE 13.8 *The primary area affected by the 1993 Mississippi flood and the distribution of the agricultural pesticide atrazine. The flood released a surge of agricultural chemicals down the Mississippi.*

leum residues, and heavy metals such as lead and zinc (Figure 13.4).

Air Pollution

Because of air pollution the atmosphere has become a significant source of water pollution (Figure 13.3). **Airborne contaminants** that end up in the water system include organic compounds (such as PCBs, dioxin, and DDT), heavy metals, acid-forming gases, and particulate matter. They enter the atmosphere from a host of sources, including power plants, industry, automobiles, incinerators, and croplands and come back to the surface with precipitation (wet deposition) or as fallout (dry deposition). On the surface the pollutants enter the soil or are picked up by runoff and delivered to lakes and streams. The concentrations are usually greatest near the source areas, but the geographic diffusion of many pollutants such as organic compounds, small particulates, and carbon dioxide is transcontinental and circumglobal. (See Chapter 11 for additional information on the delivery of pollutants to water by the atmosphere.)

The atmosphere represents the major source for many toxic contaminants in the Great Lakes. The loading has become so serious that residents are warned to limit consumption of certain Great Lakes' fish species. After being banned by the United States and Canada, DDT in the Great Lakes declined steadily between 1970 and 1980, but since then has not declined to expected negligible levels. The explanation for the continued presence of DDT is that it is being blown into the Great Lakes from sources in Central America where the chemical is still widely used. Paradoxically, many insecticides banned against use in the United States are being manufactured by U.S. firms and sold to countries such as Mexico. The United States exports up to 150 million pounds of blacklisted agricultural chemicals each year.

13.5 Groundwater Pollution

Groundwater is polluted by many of the same sources that pollute surface waters, including fertilizers, pesticides, and stormwater. There are, however, some significant differences in surface water and groundwater pollution. First, most groundwater is separated from pollution sources by a layer of soil and subsoil materials that function as a buffer against direct contamination from the surface. Second, the soil has long been a favorite place to spread and bury wastes, leaving us a legacy of hidden pollution sources from which contaminants may discharge for decades. Chapter 14 presents a detailed discussion on land use and waste disposal.

Groundwater vulnerability to pollution depends on (1) how effective the soil buffer is in retarding the downward movement of pollutants, and (2) the types and concentrations of pollutants generated on the surface by land uses. Thin and permeable soil layers with low organic contents, such as sand and gravel beds less than 15 meters deep, offer poor protection for groundwater. Porous bedrock, such as cavernous limestone, also offers poor protection from surface water. If such soils and geological conditions are associated with high-risk land uses such as waste disposal or chemical storage facilities, then groundwater vulnerability to contamination is high.

Hazardous Waste Pollution

Although a great deal of attention is given to hazardous waste management in the United States, Canada, and other developed countries, most of the world handles hazardous waste haphazardly. In the developed countries, a great backlog of hazardous wastes that were once dumped in the environment continue to pose a threat to surface and groundwater today. **Hazardous waste** covers a broad range of potentially harmful substances from industry, agriculture, transportation, and consumers. This includes heavy metals, fibers, acids, synthetic compounds, petroleum residues, and radioactive substances.

Hazardous wastes enter the ground from accidental spills such as tanker trucks and train accidents; midnight dumping; direct applications such as agricultural pesticides; and burial in **landfills**. The U.S. Environmental Protection Agency in 1980 estimated that there are 32,000 to 50,000 landfills (planned disposal sites for solid waste) containing hazardous wastes in the United States. Current hazardous waste production in the United States is equivalent to more than 900 kilograms (2,000 pounds) a year for each citizen. Thus, there is an ongoing disposal problem related to current production, in addition to the hazardous waste already in the ground. The locations of most of the old sites are unknown but the types of land-use activities that created them are known for the most part (Figure 13.9).

There are several land uses most responsible for hazardous waste production and dumping (both accidental and purposeful):

1. **Industry**, especially petroleum refining and chemical manufacturing, which for decades disposed of wastes on-site or on lands near their operations;
2. **Agriculture**, through pesticide and chemical fertilizer applications to croplands;
3. **Transportation**, including highway and railroad corridors and their depots, storage, and fueling areas, where spills and leaking storage facilities are common;
4. **Urban centers**, where garbage and other wastes from residential, commercial, and industrial sources were buried on the urban fringe;
5. **Military installations**, where wastes at local, national, and international levels—ranging from garbage to fuel, old munitions, and lethal chemicals—were buried on-site.

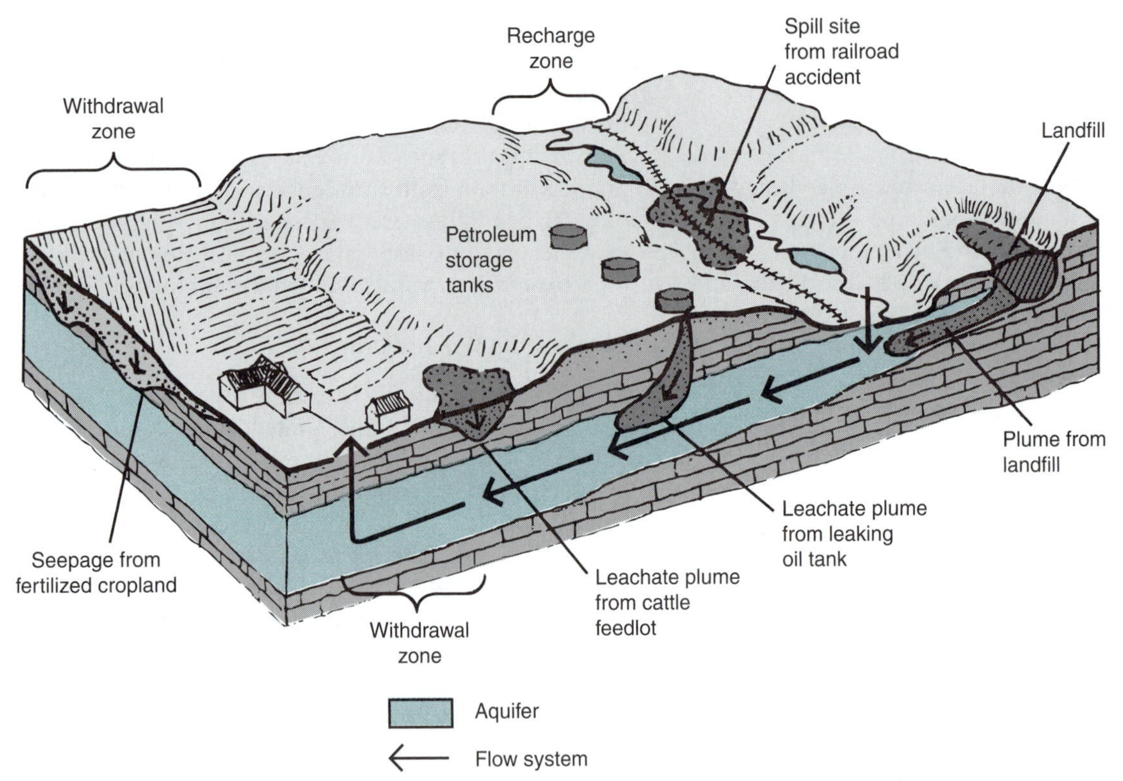

FIGURE 13.9 *Sources of groundwater pollution and their relationship to a typical groundwater system. The relative vulnerability of groundwater to pollution is also illustrated.*

Whereas industry, transportation, and urban areas have long been identified as major sources of groundwater pollution, agriculture and the military have been in the shadow of public attention for years. But the picture is changing; both are now recognized threats to the groundwater environment. It has come to light that a trail of environmental degradation follows the military of most nations, but especially the military giants such as the United States and the former USSR, who built huge installations within which they privately disposed of untold wastes. For U.S. installations, the story of these practices is unfolding as military bases are abandoned in Europe, Asia, and North America. In Germany, for example, the U.S. military itself has identified 309 known or suspected pollution sites associated with its installations. Within the United States and Russia there are undoubtedly thousands of such sites.

Solid Waste

Solid waste is the nonliquid and nongaseous debris discarded by cities, industry, agriculture, and mining. Industrial societies with high rates of consumption produce massive amounts of solid waste, and the United States is the leading producer. If we combine household garbage and solid waste from industry and other activities, the solid waste output in the United States averages 18,000 kilograms (20 tons) per person a year. For a family of four living in a three-bedroom home, it would take only five years to completely fill the house with refuse (including the family's share of solid waste from industrial, agricultural, and mining sources). It would take about twenty years to fill the house with the family's garbage alone.

The pollution of groundwater by buried waste takes place when contaminated liquid, called leachate, seeps into groundwater. **Leachate** is water that has percolated through waste and become contaminated with, among other things, acids from decomposing organic matter, heavy metals such as lead from discarded paint, and organic compounds from residues of cleaning agents (Figure 13.9). If the landfill is situated over permeable material, the leachate can migrate through a great volume of material in a matter of years. Once contaminated, the slow-moving groundwater, which is the largest source of fresh liquid water on Earth and the second leading source of domestic water supply, remains contaminated for decades or centuries.

Other Sources of Groundwater Pollution

There are also many other sources of groundwater pollution.

- **Lawns.** Since about half the American population lives in suburbs, lawns are a significant source of fertilizer and pesticide contaminants in groundwater. There are 55 million lawns in the United States and average fertilizer application rates are about 10 times higher than on cropland. This is a source of nitrogen contamination of drinking water.

- **Urban stormwater.** Especially in areas of permeable soils such as central Florida, nutrients, heavy metals, and petroleum residues are readily transferred underground.
- **Septic drainfields.** Septic systems are the favored sewage disposal method in rural areas, and are a source of nutrient-rich water in the ground.
- **Household materials.** Spills and leakages from yard and household equipment are a source of groundwater contamination. Products include paints, solvents, and used motor oil.
- **Pipes and tanks.** This source of contaminants includes gasoline, oil, and a host of chemicals. The number of underground pipe and tank leaks in the United States is unknown and most probably go undetected, but the U.S. EPA has confirmed 170,000 leaky storage tanks in the country. Other sources estimate at least 250,000 leaking above-ground facilities, mainly tanks and pipes (see Figure 14.4). In Russia and other Eastern European countries, the problem appears to be even worse.

Groundwater Protection

Guarding groundwater from pollution has become a serious concern for many nations. In the United States, it is approached at four levels:

1. Landfill planning and management aimed at proper siting and construction of waste disposal pits, including lining them to keep leachate from escaping;
2. Cleaning up wastes that were improperly buried, using funding from state and federal programs such as Superfund;

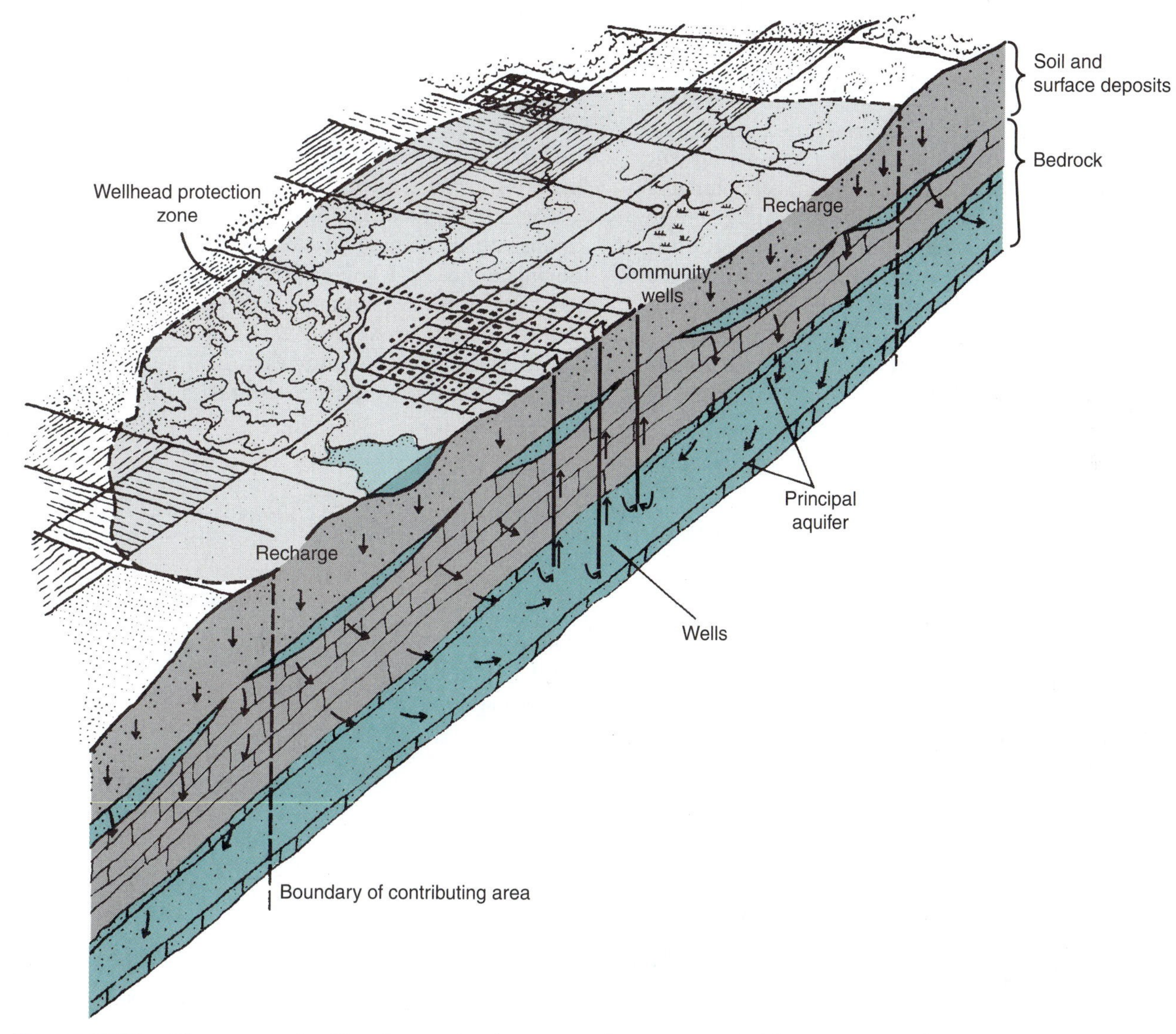

FIGURE 13.10 *The concept of wellhead protection for community water supply. The area protected supplies principal aquifers with recharge water.*

3. Proper design and installation of water-supply wells, including the use of sealed casings to keep surface water from trickling into wells;
4. Wellhead protection that involves managing the land-use activities and waste-disposal practices within the area contributing water to the aquifers that supply wells.

Wellhead protection is a particular concern for communities with municipal wells because the area contributing to their water supply may extend well beyond the community's geographic limits. Generally speaking, the larger the community, the larger and deeper the wells and the larger the contributing (recharge) area. Although often difficult to define, this area, called the wellhead area, needs to be managed to protect groundwater quality over the long term (Figure 13.10). This usually involves monitoring waste-producing land uses and providing means for safe contaminant disposal.

13.6 Pathways and Sinks in the Hydrologic Cycle

The impact of pollutants on the environment depends not only on the types and amounts of pollutants produced and released by society, but also on the nature of the receiving waters. One important consideration is the rate at which water is exchanged or flushed in the various segments of the hydrologic cycle. This influences the rate at which pollutants are dispersed in the system; for example, pollutants are dispersed rapidly in a fast-flowing stream (Figure 13.11).

Another consideration is the volume of water in each segment of the cycle. The volume of water, as we noted earlier, influences the **dilution** of pollutants. Where the volume is large, the potential for dilution is great. In alignment addition, **breakdown** and **storage** of pollutants must be considered when assessing the disposition of pollutants in the water system. Breakdown can result from chemical, physical, and biological processes in the soil, in wetlands, and in streams, lakes, and the sea. Storage can take place with deposition of sediments and the uptake of contaminants by plants.

Movement Through the Atmosphere and Streams

The atmosphere is the fastest segment of the hydrologic cycle, with an average water exchange time of only nine days. Therefore, for airborne pollutants that combine with water vapor and cloud droplets, (e.g., sulfur dioxide and nitrous oxides), the return to the Earth's surface is almost immediate. This helps explain why acid rain is so heavily concentrated immediately downwind of the major sources of oxide emissions. In the eastern United States the areas of heaviest acid rain lie only several hundred miles northeast of the major pollution sources (see Figure 11.11 in Chapter 11).

The atmosphere is also the most geographically expansive part of the hydrologic cycle. Being both fast-moving and of large volume, the atmosphere has the capacity to disperse pollutants over great distances and rapidly dilute them as well. As we noted earlier, in the Great Lakes the primary sources of DDT are now Mexico and the Central American countries. Long-distance

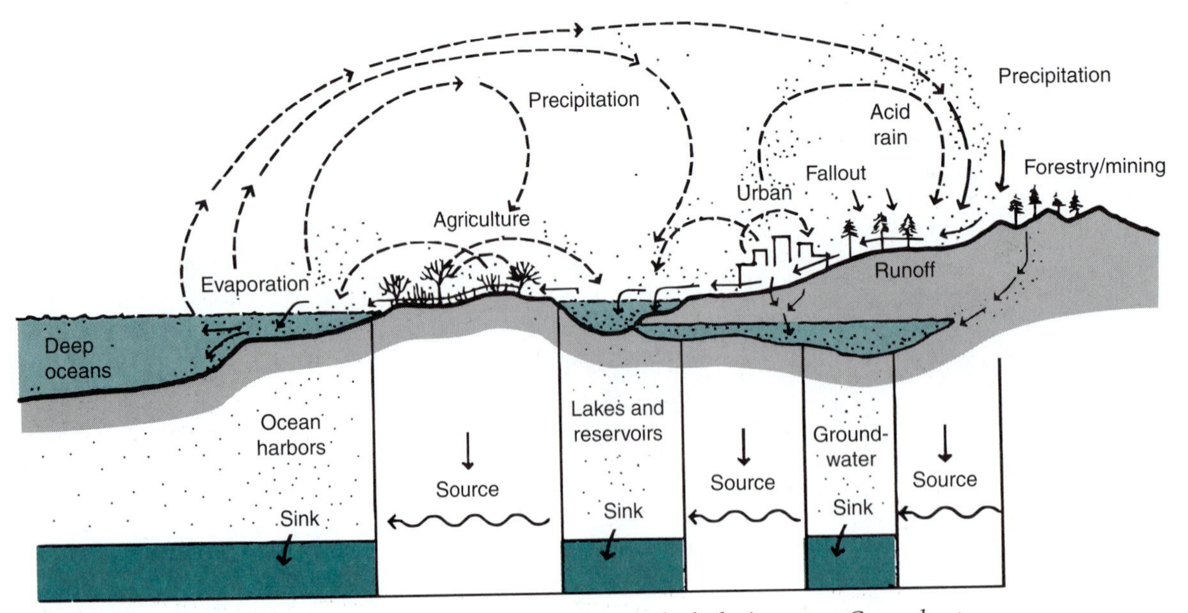

Figure 13.11 *The concept of pollution sources and sinks in the hydrologic system. Groundwater, lakes, and the oceans are important sinks, that is, places where pollutants accumulate.*

transport by the atmosphere favors contaminants such as DDT and PCBs, which are moved as gases rather than particles. Of the PCB pollution in Lake Superior, 70 percent enters via the atmosphere.

For streams and rivers the **exchange rate** is also fast, on the order of 10 to 15 days. Therefore, pollutants are rapidly dispersed downstream from an outfall point. Although streams have good flushing potential, the rapid transfer of pollutants makes containment and cleanup very difficult. Unfortunately, streams are ideal dumps for polluters because contaminated discharges are quickly carried away in the streamflow instead of building up at the discharge points. The adage "out of sight, out of mind," is frequently used to describe the attitude of stream polluters toward their problem. The heaviest stream pollution occurs in the urban segments of drainage systems and in segments where agricultural cropland is abundant.

Pollution Sinks

Any substance introduced to the environment has to end up somewhere, and so it is with pollutants. Generally speaking, pollutants tend to accumulate in slow segments of the hydrologic cycle. These segments are called **sinks**, and they function as the dust bins of the environment (Figure 13.11). On the continents, the first and most widespread sink is the *soil*, where many of the contaminants carried by infiltrating water are taken up. As we noted earlier, topsoil is especially effective in adsorbing heavy metals, organic compounds, and other contaminants.

For water that runs off the land, lakes and reservoirs are major sinks for many pollutants, especially sediment. *Inland lakes* have exchange times typically in the range of one to 20 years, meaning that it takes as long as 20 years to replace old water with new water. Lake Erie, the smallest and shallowest of the Great Lakes, has an exchange time of two years; whereas Lake Superior, the largest and deepest, has an exchange rate (or flushing time) of more than 100 years. In any water body, however, flushing is beneficial only if the new water is cleaner than the old water; in this century that has rarely been the case for most lakes and reservoirs.

In *ocean harbors*, such as those at New York and Boston, tides and currents are relied upon to flush polluted water to sea. Recent studies have shown that for New York, which generates 5.7 billion liters (1.5 billion gallons) of sewage effluent water per day, and Boston, which generates about one-third as much, harbor flushing is not as efficient as once thought. As a result, water quality is declining in many coastal areas near major cities, so that opportunities for swimming, fishing, and other activities are reduced or lost where recreation activities are sorely needed.

What happens to pollutants when they enter a sink? Within a soil, lake, reservoir, or harbor, pollutants take various paths depending on their chemical and physical forms and the environmental conditions of the sink. Some are taken up by plants, some in the water bodies settle to the bottom, some remain in the water in a dissolved state, and some undergo chemical change and are transformed into other forms. Increasingly, it is coming to light that microbes in soil and sediment are surprisingly effective in breaking down some of the most threatening contaminants such as organic compounds. On the other hand, some contaminants are merely locked up in sediment. Recent analysis of bottom sediments excavated from Boston Harbor, for example, revealed such high concentrations of heavy metals (lead, arsenic, chromium, and others) that the sediment must be handled as a hazardous waste.

In most water bodies a variety of pollutant paths are involved, making cleanup efforts difficult even with the most advanced remediation technologies. For example, to clean up and restore polluted reservoirs or lakes, it is usually not enough to simply reduce or stop the inflow of pollution and allow polluted water to be exchanged with cleaner water, because the bottom sediments usually hold large stocks of old pollutants. These pollutants are released into the overlying water by various means, such as plant root uptake, the activities of bottom organisms, and turbulence from waves and boats. They can continue to degrade the water for many more years.

The groundwater system moves even slower than the water in lakes and reservoirs. Shallow aquifers, located at depths from 10 meters (33 feet) to 1,000 meters (3,300 feet) or so, require up to 300 years on the average to exchange old water for new. For deep aquifers (at depths greater than 1,000 meters or so), the exchange time is several thousand years. Because of their great depth, deep aquifers are not as prone to contamination as shallow aquifers. In both cases, however, the time required to flush a polluted aquifer by natural means is excessive by human time standards.

Humans and Other Organisms as Pollution Sinks

Because ecosystems are tied directly to pollutant pathways in air and water systems, many organisms including humans have also become sinks for contaminants. Two systems are involved in creating biological sinks: (1) the physiological systems of individual organisms; and (2) the food chains of ecosystems. At the physiological level, we are dealing with a four-stage system, beginning with the entry of chemicals into the body by ingestion of food and water, inhalation of air, and penetration through the skin. This is followed by absorption of the chemical into the blood system and distribution to tissues and organs by the bloodstream. The last stage involves the final disposition of the chemicals in the body. Some chemicals

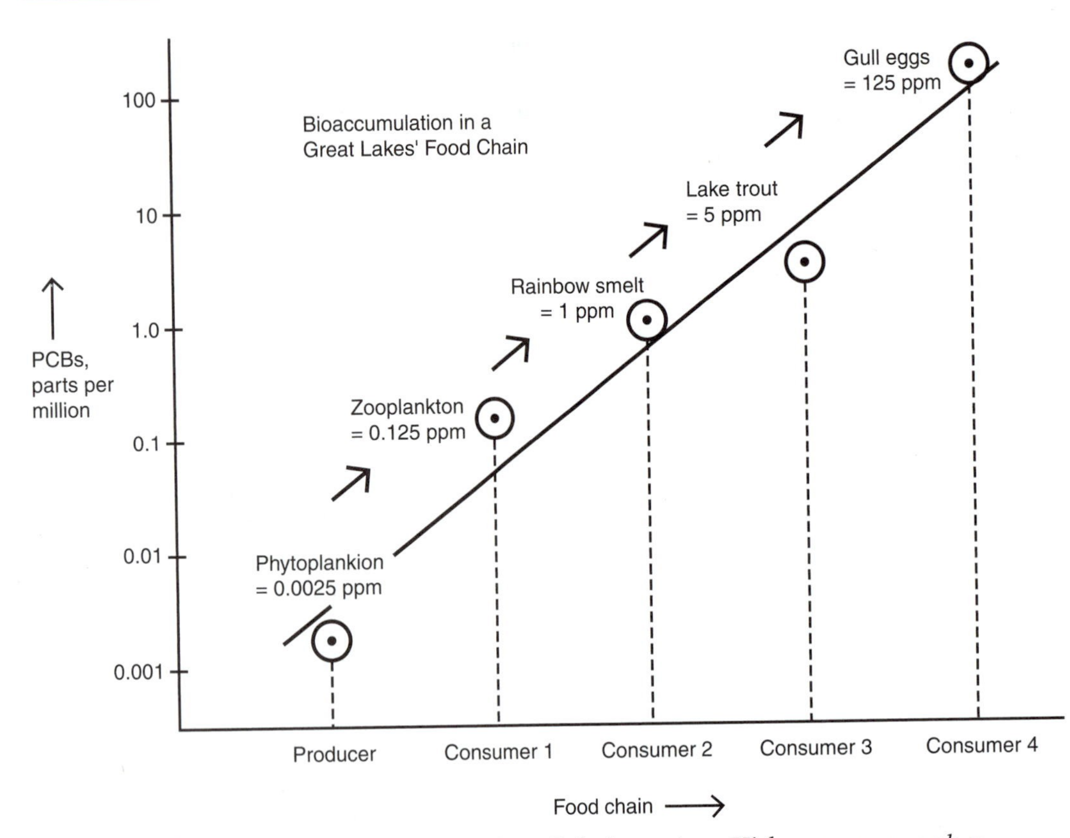

FIGURE 13.12 *Bioaccumulation of PCBs in Great Lakes' organisms. Higher consumers such as lake trout and gulls become sinks for organic compounds and heavy metals.*

are altered by metabolic processes and transformed into other chemical forms, some are eliminated with bodily wastes, and some are stored in organs and tissue.

Storage is limited to selected chemicals such as dioxin, PCBs, and methylmercury. These chemicals have low solubility in water and are resistant to breakdown by metabolic processes. Storage is concentrated in certain tissue and organs, notably membranes of lungs and gills, fatty tissue in muscle and other organs, and fatty material in the blood and in the liver. When a fish, for example, is consumed by another fish or by a gull, eagle, or human, contaminants in the fish's body are passed on to these organisms.

The transfer of contaminants in aquatic food chains begins with phytoplankton. When these tiny plants take in the nutrients phosphorus and nitrogen as part of growth processes, they can also ingest a variety of contaminants. Phytoplankton feed zooplankton, which in turn feed small fish and other organisms. Stored contaminants are not only retained in the food chain but accumulate in higher concentrations at higher trophic levels (see Figure 6.16). In addition, **bioaccumulation** (or *biomagnification*) is supplemented by intake of additional contaminants via respiration or skin absorption. The end result is a dramatic increase in contaminant levels in higher consumers in the ecosystem. In the Great Lakes, for example, PCBs are 50,000 times more concentrated in gull eggs (a fatty substance) than in phytoplankton (Figure 13.12).

13.7 POLLUTION OF THE OCEANS

The Earth's largest sink for pollutants is, of course, the oceans. Most pollutants enter the oceans from nonpoint sources via streams and the atmosphere. Additional pollutants such as garbage, dredge spoils, and oil are spilled or dumped directly into the oceans. As a result the most polluted ocean waters are bays, estuaries, and the shallow waters of the continental shelves, where most marine life is concentrated and the majority of the world's fishing grounds are located. Figure 4.14 in Chapter 4 shows the global distribution of offshore oil and gas operations, major tanker routes, and areas of major marine ecosystems.

More than 75 percent of the pollutants in the sea are delivered by runoff and atmospheric deposition (Figure 13.13). Runoff, which accounts for 44 percent of the total, contributes billions of tons of sediment from farm fields, rangelands, city streets, and other land uses, as well as many other pollutants such as nutrients, heavy metals, and oil. More oil is discharged into the oceans from *stormwater* than from accidental spills. The oil from stormwater, however, is less apparent in the ocean because it is more widely and evenly distributed than

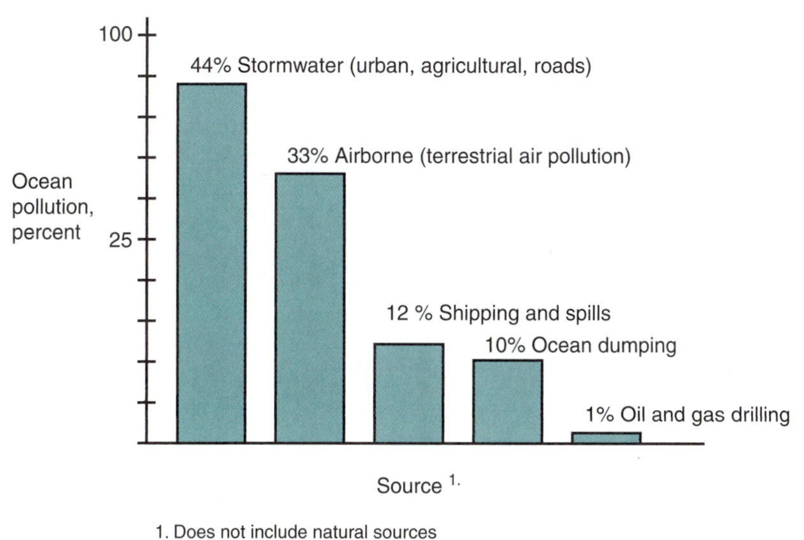

FIGURE 13.13 *General sources of ocean pollution worldwide. Runoff and the atmosphere together contribute 77 percent of the pollution to the oceans.*

the oil from spills. In the 1989 *Exxon Valdez* spill, for example, the oil was mainly limited to Prince William Sound, a large bay in the Alaska Coast, and to nearby islands and peninsulas. That year far more oil was discharged with stormwater from highways, settlements, and industry along the West Coast of North America; however, its environmental impacts were less profound and measurable.

Most of the sediment discharged into the oceans is deposited in shallow waters close to the shore in and around large river deltas. The sediment loads from rivers like the Ganges, Mekong and Yellow, which drain huge watersheds occupied by millions of people and extensive agricultural and urban land uses, are massive and growing. Added to this are smaller, more dispersed sediment inputs from thousands of smaller watersheds influenced by farming, forest clearing, and construction activity along coastlines. In the tropics, for example, deforestation near the coast is leading to soil erosion and heavy sedimentation of aquatic habitats. Coral reefs are especially susceptible to sediment damage because many coral species cannot survive under sediment in muddy water. Entire reef communities are being eradicated in Puerto Rico, Hispaniola, the Philippines, and many other tropical coastal areas throughout the world.

In urban areas, partially treated or raw sewage is discharged into the sea. As we noted earlier, even in developed countries where most cities have modern treatment facilities, wet-weather runoff can force sewage bypassing, resulting in direct discharge of raw sewage into the sea. In the developing countries, there is virtually no municipal sewage treatment, and where there is, treatment facilities are outmoded and overloaded so that raw sewage is being discharged most of the time. It is noteworthy that disease-causing organisms in the sewage can survive up to two and a half weeks in the seawater.

The sea also serves as a place to dump municipal garbage, sludge from sewage treatment plants, sediment (spoils) dredged from harbors, and radioactive materials. Large cities such as Boston and New York have long practiced garbage dumping at sea. Although the practice has been curtailed in the United States, it is still widespread in cities in the developing world. Sediment dredged from harbors and river mouths, which is often rich in contaminants, is the single-most abundant waste dumped in the sea through the world. Radioactive waste was dumped in the sea until 1982, when it was ended by an international agreement. However, it has come to light that the Soviet/Russian military continued the practice until 1993. After several decades of dumping, thousand of containers of radioactive waste remain on the sea floor off North America, Western Europe, and the former Soviet Union (see Figure 14.10).

Oil and other pollutants are routinely discharged into the oceans as tankers and other ships empty ballast and clean tanks—usually well beyond sight of land. Together with accidental spills from tankers and barges, these sources account for less than half of the oil pollution of the seas. However, there is growing evidence that ships also dump considerable amounts of other debris, such as plastics and paper, into the ocean.

13.8 POLLUTION CONTROL WITH TREATMENT SYSTEMS

Efforts to control water pollution were introduced in the nineteenth century when it became apparent that drinking water contaminated with sewage was the source of diseases such as cholera, typhoid fever, and dysentery. The first attempts at pollution control were designed to physically separate sources of drinking water from waters polluted by sewage. This usually involved collecting the sewage and piping it away from drinking water sources. This system, which came to be known as a **sanitary sewer system**, resulted in improved public health; however, the impact on the aquatic environment was usually severe because no treatment was applied to the sewage before it was released to streams and lakes.

In the twentieth century treatment facilities were added to municipal sewage systems to improve public health conditions. **Treatment plants** were built near the end of the sewer system where the raw sewage could be processed before being discharged into the environment. In most municipalities sewage is very consumptive of clean water, which is used to transport the waste through the sewer system. For example, in the United States and Canada, 100 or more gallons of clean water are used daily to transport the wastes of each person; consequently, the total volume of sewage (waste plus water) needing treatment is great. Put another way, this method sacrifices more than 360 kilograms (800 pounds) of clean water to dispose of less than a kilogram of solid waste for each person each day (Figure 13.14). For example, in New Mexico, the discharge from the Albuquerque treatment plant represents the fifth largest tributary of the Rio Grande.

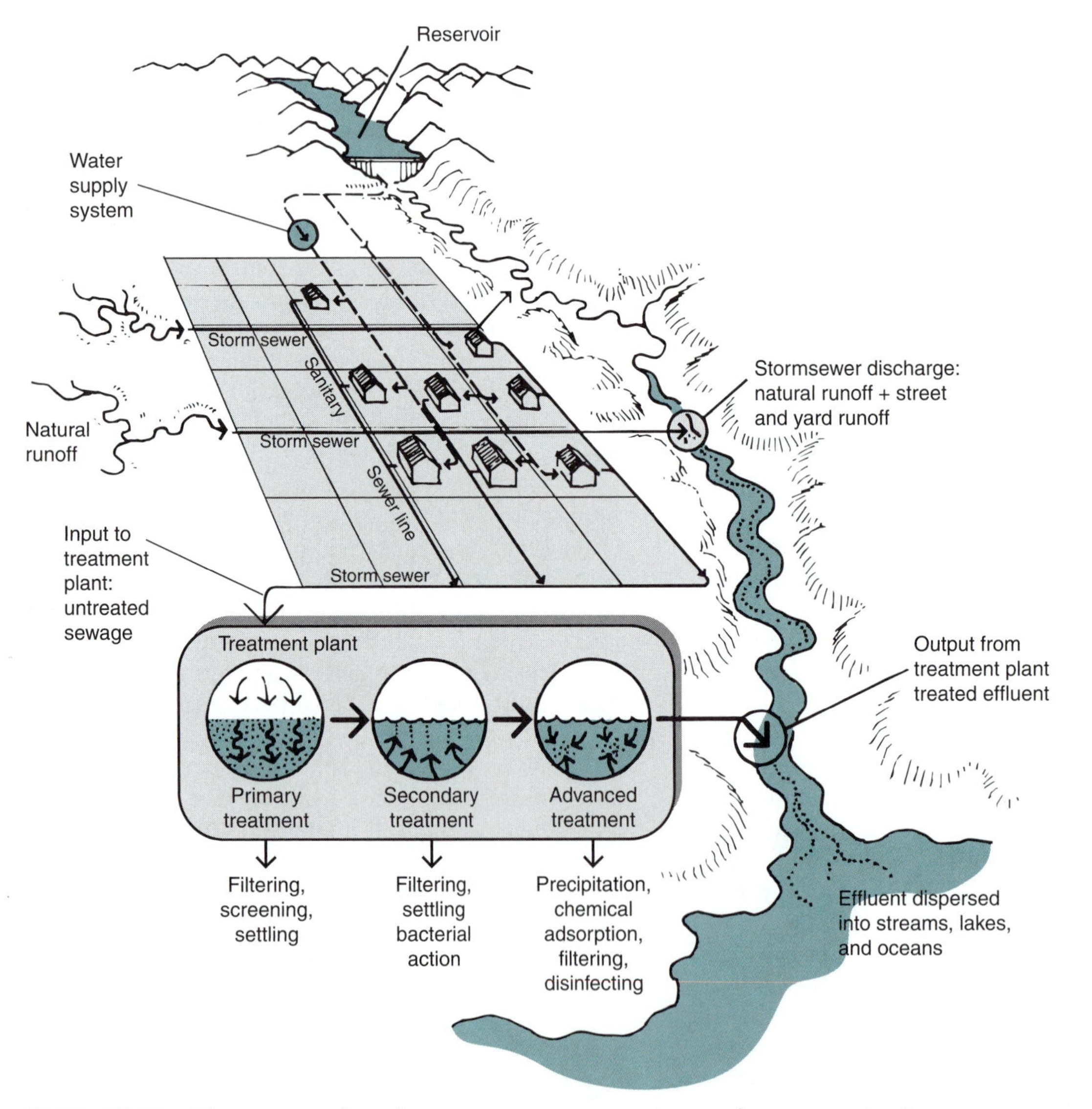

FIGURE 13.14 *The water supply and sewage treatment systems in a modern community. Large amounts of clean water come into homes and other buildings and most leave via sanitary sewer lines, carrying a load of pollutants that then require treatment before being discharged back into the environment. Stormsewers are a separate system in most, but not all, communities.*

Primary and Secondary Sewage Treatment

In **primary treatment** the objective is to reduce the amount of organic solids in the water through filtering, screening, and settling. This lowers the total organic load and thereby reduces the oxygen demand and nutrient content of the effluent. Primary treatment can remove 30 to 60 percent of the solid particles and 20 to 40 percent of the oxygen-demanding waste.

As cities grew in the 1950s and 1960s, the total mass of organic-enriched effluent released from sewer systems, both those with and without primary treatment, climbed significantly. Coupled with increased pollution from storm sewers, the impact on streams, lakes, and coastal waters in the form of oxygen-demanding waste became so great that aquatic ecosystems underwent radical deterioration. Among other things, the effluent was rich in nutrients that accelerated eutrophication, filling in lakes, ponds, and streams with aquatic plants and sediment and eliminating certain fish species.

Public pressure for improved water quality became a leading issue in the late 1960s and early 1970s. This pressure helped bring about federal legislation for improved water pollution control from municipal and industrial sources. The U.S. Clean Water Act of 1977 provided federal assistance to communities in the United States for the construction of additional treatment facilities. In the next fifteen years more than $100 billion were spent in the United States to improve sewage treatment by constructing more advanced facilities for second and third level treatment. By 1982, more than 60 percent of Americans were served by secondary or secondary plus advanced treatment systems (Figure 13.14).

Secondary treatment involves more filtering, further settling, and aeration to accelerate bacterial breakdown of the waste. The filtering process involves trickling the water through gravel beds. The bacterial processes are nurtured by injecting the sludge with pure oxygen and/or mixing it in basins. When coupled with primary treatment, the secondary processes can remove up to 90 percent of the oxygen-demanding waste. However, most secondary systems remove only 50 percent of the nitrogen, 30 percent of the phosphorus, and less than 10 percent of the heavy metals and certain organic compounds such as pesticides.

Advanced Treatment

To remove the pollutants remaining after secondary treatment, a series of processes, called **advanced** or **tertiary treatment**, can be added. They include the following:

1. Precipitation based on coagulation and settling to remove phosphorus and suspended solids;
2. Chemical adsorption in which organic compounds are reduced by bonding with a filter medium such as activated carbon;
3. Advanced filtering such as reverse osmosis to remove dissolved organic and inorganic substances;
4. Application of a disinfectant such as chlorine to kill disease-transmitting bacteria and some viruses.

Although effective, these processes are very expensive; advanced treatment costs about four times more than primary and secondary together.

This three-phase treatment system represents the conventional approach to municipal sewage treatment in the industrial world. It can be referred to as a high-tech approach because it demands relatively elaborate mechanical and chemical systems. There are, of course, drawbacks to this approach, including high costs of construction and operation, limited capacities to treat large volumes of water where sanitary sewers are combined with stormsewers, and some serious secondary impacts on the environment. Among the environmental impacts are periodic discharges of raw sewage from cities with combined sewer overflows, discharges of chlorinated, nutrient-rich effluent waters from treatment facilities, and the release of polluting gases such as methane and chlorine to the atmosphere. This approach also generates large amounts of solid organic waste, called *sludge* which is usually disposed of in landfills (Figure 13.14).

Low-Tech Treatment

Critics of the high-tech approach point to low-tech alternatives as an option to conventional treatment systems. Low-tech systems are mainly nonmechanical and use little or no chemical treatments. The traditional low-tech approach is the **soil absorption system** cited earlier in the chapter. This system utilizes-the soil as a filter medium for waste water, which is introduced to the ground from a septic tank via subsurface pipes called a drainfield. The soil absorption system is limited to individual households for the most part. In the United States and Canada it currently serves as many as 50 million people in rural areas and small communities (Figure 13.15).

At the community level, low-tech sewage treatment is a more difficult and controversial issue because the systems are slower than high-tech systems and are limited mainly to warm-season use. Most low-tech systems involve three steps:

1. A settling and filtering process;
2. Natural processing of effluent water in ponds where bacteria and algae break down organic material;
3. Application of tertiary water to fish ponds, cropland, and/or wetlands.

In the United States, the use of low-tech treatment by communities is still considered experimental, but it is showing success for some small communities with available wetlands and croplands (Figure 13.16).

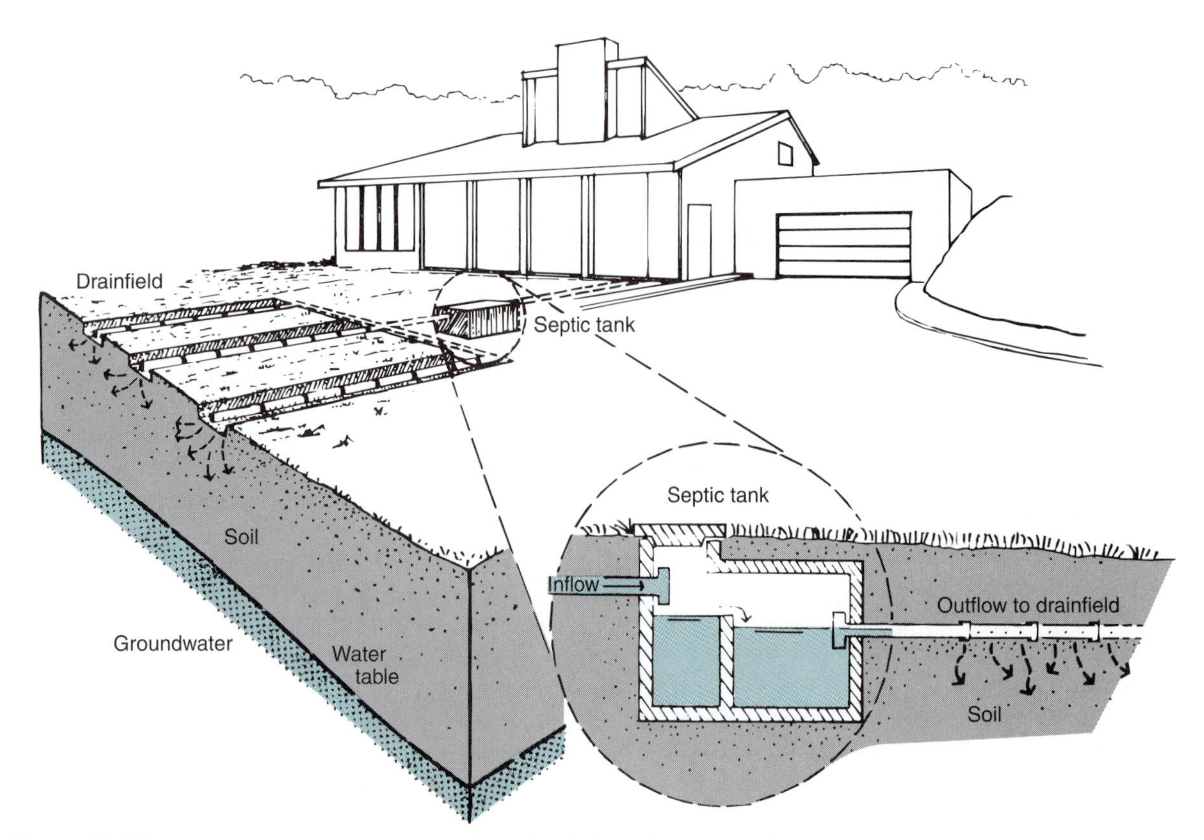

FIGURE 13.15 *A soil absorption system used for individual residents in rural aras. Given proper siting, design, and maintenance, such systems can be highly effective means of waste-water treatment.*

FIGURE 13.16 *Low-tech waste-water treatment systems using ponds and wetland are showing success in small communities.*

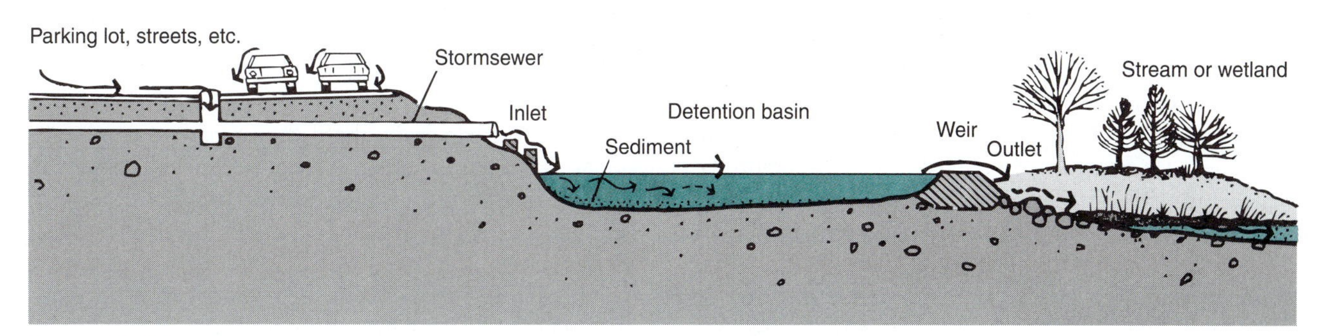

FIGURE 13.17 *An illustration of a stormwater detention basin. Such basins can provide basic water treatment functions, such as settling of sediment.*

Although stormwater is a major source of pollution, it is rarely treated. The reasons for this are that (1) polluted stormwater is not as dangerous to human health as sanitary sewer effluent, and (2) the volume is usually so large and the outfalls so widely dispersed in the landscape that treatment is technically very difficult. In the past several decades, urban stormwater has been subject to some treatment, mainly settling and BOD reduction, where it is held in storage basins built for flood control purposes (Figure 13.17). In addition, the U.S. EPA is now giving increased attention to urban stormwater management and will soon

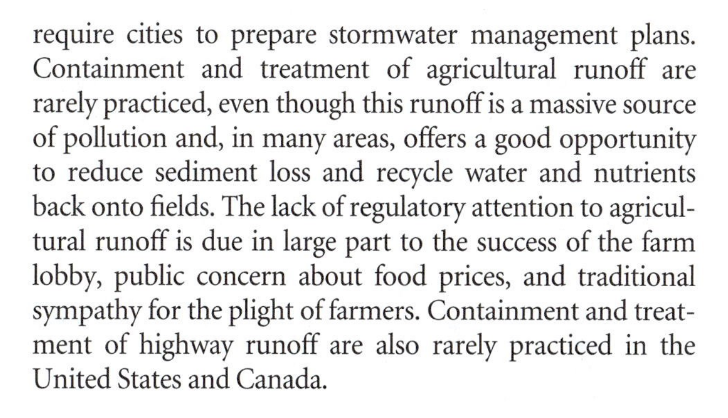

require cities to prepare stormwater management plans. Containment and treatment of agricultural runoff are rarely practiced, even though this runoff is a massive source of pollution and, in many areas, offers a good opportunity to reduce sediment loss and recycle water and nutrients back onto fields. The lack of regulatory attention to agricultural runoff is due in large part to the success of the farm lobby, public concern about food prices, and traditional sympathy for the plight of farmers. Containment and treatment of highway runoff are also rarely practiced in the United States and Canada.

13.9 SUMMARY

Despite improvements in some aspects of water quality in selected countries, water pollution is on the rise worldwide. This goes for most types of pollution in fresh surface water, groundwater, sea water, and atmospheric moisture. The sources include sediment and nutrients from nonpoint agricultural sources, oxygen-demanding wastes from human sewage, heavy metals from industry and city streets, and organic compounds from industry and agricultural pesticides via both the atmosphere and runoff. Broadly speaking, the causes are expanding world population and rising human consumption, both of which fuel growth in agriculture, industry, commerce, and cities. Cities are associated with the greatest water pollution concentrations, whereas agriculture and rural populations are associated with the greatest water pollution loadings worldwide.

For the developing countries the most serious water pollution problems are related to agriculture and uncontrolled urban growth. In Brazil, Malaysia, the Philippines, and Madagascar, for example, the push for new farmland and increased lumbering and mining is leading to widespread deforestation. This is resulting in accelerated soil erosion and sedimentation of streams and coastal waters. In coastal waters the sediment coats coral reefs and other valuable ecological communities, eventually destroying or greatly reducing them.

In active farming areas of developing and deprived countries, the use of pesticides, including DDT, is largely uncontrolled and has led to both water and air pollution. In the atmosphere, these contaminants are blown into other regions and precipitated into watersheds, lakes, and streams. Thus the world of poor farmers, like the industrial world, is the source of water pollution problems extending far beyond the borders of individual countries.

Poor countries are also plagued by urban sprawl. Economically destitute people abandon rural areas and move to cities, where water supply and sewage treatment facilities are already grossly inadequate. These people settle in shanty-towns on the urban fringe where there are no water and sewer services. Each year as many as 30 million poor people move to cities in poor countries throughout the world. For large cities such as Rio de Janeiro and Mexico City, shanty people number in the millions. They draw their water from local streams, ditches, and shallow wells, and there are no organized means of sewage disposal. Not surprisingly, the incidence of waterborne diseases such as cholera and typhoid fever has been increasing over the last decade. Outbreaks of cholera, a deadly disease transmitted in water contaminated by infected human feces, have been rising in South America and India. Worldwide it is estimated that as many as 30,000 people die every day from diseases linked to contaminated water.

In the developed world, water pollution is increasing in most regions and declining in only a few. Acute health problems related to impure drinking water is relatively uncommon in developed countries. However, several chronic health problems are tied to water impurities such as nitrogen in groundwater. Rising costs of water treatment are a source of concern, as are damages to the aquatic environment from agricultural and urban runoff, combined sewer overflows, and treatment plant outfalls.

Eastern Europe appears to have suffered one of the most dramatic increases in water pollution in modern times. The fall of the Iron Curtain has revealed a picture in Poland, former East Germany, and the former Soviet Union of environmental degradation caused by poorly managed cities and outmoded industrial operations. In Poland, for example, half the cities and one-third of the industry dump totally untreated waste into local waters. Although the true extent of the damage is currently unknown, it appears to be nearly catastrophic in places, affecting streams, lakes, and groundwater over broad regions around industrial centers and threatening the health of millions of people. And in some countries such as Bulgaria and former Yugoslavia, the prospects for change in the foreseeable future are remote because of a lack of governmental reform and continuing political strife.

In the United States the past 20 years have produced several significant trends. The Water Pollution Control Act of 1972 (now amended and known as the Clean Water Act) provided federal funding for new and improved sewage treatment facilities, which resulted in reduced fecal bacteria and nutrient loading and decreased BOD pollution in the nation's streams and lakes. Lead levels also declined with the change to unleaded gasoline for automobiles.

Despite these strides, the overall quality of the nation's surface waters has not met forecasts based on the programs implemented under the Clean Water Act. Sewage from many municipalities is still inadequately treated. More than 1,000 communities, rely on combined sewer overflow system that periodically discharge raw sewage directly into streams, lakes, and the ocean via stormsewers. These cities are reluctant to rebuild old combined overflow systems because of the high costs involved, estimated around $20 million per square mile.

Little progress has been made on the stormwater pollution problem although we understand this problem better today them we did a decade or so ago. Stormwater contributions have increased significantly with economic development and population growth throughout the developed world. BOD loadings from both agriculture and urban nonpoint sources have risen in most parts of the country, offsetting the gains made through improved municipal and industrial treatment facilities. Nonpoint sources are also responsible for increases in nitrogen, phosphorus, sediment, and chloride. The nitrogen increase is attributed primarily to a substantial rise in fertilizer applications from agriculture and secondarily to increased atmospheric fallout and urban runoff.

Agriculture is responsible for most of the increase in sediment loading. However, the pattern varies geographically as land is taken into and out of production with changes in market conditions and governement programs. Between 1950 and 1980, highway salt applications increased by a factor of twelve in the United States. Application rates remain high but little is being done by highway departments to reduce salt or other contaminants in road runoff.

However, most municipalities in the U.S. and Canada now require stormwater detention basins for new development, which help reduce sediment and some other contaminants in runoff. The U.S. EPA has implemented stormwater management rules and guidelines including a permitting requirement under the National Pollutant Discharge Elimination System (NPDES). This requirement includes communities with populations over 100,000, mandating them to obtain an NPDES permit for their stormwater management systems. Unfortunately, agriculture remains untouched by stormwater regulations despite its enormous contributions to the pollution of the nation's waters.

Although lead has declined, two other heavy metals, arsenic and cadmium, have increased in United States waters. Sources of these toxic metals include metal-processing industries, pesticides, herbicides, and phosphate-rich substances such as fertilizers and detergents, but the principal source is coal-burning power plants, which discharge both arsenic and cadmium into the atmosphere, from which they are released into surface water.

Finally, there is the issue of groundwater quality. Although data are inadequate to define a nationwide trend in groundwater quality, there seems to be little doubt that it is declining in the United States. In the Midwest and other agricultural regions, increasing nitrogen pollution from fertilizer is posing a health threat in some areas. In Canada there is rising concern about contamination of domestic wells from stromwater and septic drainfield seepage. The U.S. EPA estimates that 1.5 million Americans served by rural domestic wells are exposed to nitrogen above safe levels. Judging from the rising occurrence of pollution from hazardous waste, stormwater, fertilizers, insecticides, spills, and leaking petroleum tanks, groundwater quality is almost sure to decline in the next several decades in the U.S. and Canada. It is estimated that as much as 2 percent of the vast supply of groundwater in the United States is already contaminated.

13.10 KEY TERMS AND CONCEPTS

Pollutant classes
- effects
- sources
- composition

Measuring water pollution
- pollution concentration
- pollution load
- safe limits

Sources and Distribution
- point source pollution
- nonpoint source pollution
- traditional rural sources
- modern urban sources
- biological oxygen demand (BOD)
- soil absorption system
- industrial pollutants

Stormwater
- automobiles and roads
- first-flush flow
- pesticides and fertilizers
- persistent organic pesticides

Soil erosion and sediment
- cropland soil loss
- role of topsoil
- atmospheric contaminants

Groundwater pollution
- vulnerability
- hazardous waste
- landfill and leachate
- other sources

Groundwater protection
- waste disposal and clean up
- wellhead protection

Pollution pathways
- storage, dilution, and breakdown
- exchange rates

Pollution sinks
- soil
- inland waters
- organisms
- bioaccumulation

Ocean pollution
- stormwater
- ship wastes and spills
- dumping

Treatment systems
- sanitary sewer system
- primary treatment
- secondary treatment
- advanced treatment
- low-tech treatment
- stormwater

13.11 QUESTIONS FOR REVIEW

1. What are the three "colossal" issues concerning water pollution today? Name the eight classes of water pollutants and describe their basic characteristics.
2. Distinguish between pollution concentration and pollution load. Discuss the issues surrounding the concept of safe limits of water pollution.
3. Define point source pollution and nonpoint source pollution and give some examples of each.
4. Discuss the water pollution problems of the traditional land uses of developing countries. What are some of the health and environmental impacts of water pollution?
5. Describe the changes in water pollution accompanying urban growth and the development of the automobile industry in developed countries such as the United States.
6. What has been the trend in the development of municipal sewage treatment plants in the United States and Canada versus that in cities in poor countries? What are industry's contributions to urban water pollution? List some industries and examples of the pollutants they produce.
7. What are the trends in fertilizer use in agriculture, and what pollutants are produced by fertilizers? How do pesticides enter the water system? How long do they last before breaking down? What role does soil play in keeping pesticide residues out of the water system?
8. Sediment is widely identified as the principal nonpoint source pollutant. What land uses are most responsible for sedimentation of surface waters, what is the global trend in sediment loss from the land, and is soil loss related to any other pollution problems? Summarize the role of the atmosphere in water pollution.
9. What are the main sources of hazardous and solid waste, and why do these materials pose a threat to groundwater? What factors influence groundwater vulnerability, and what are four levels of groundwater protection in the United States?
10. What is meant by the terms *pathways*, *sources*, and *sinks* in the hydrologic cycle? What are water exchange rates, and in this context, what are the fastest and slowest segments in the hydrologic cycle? Where do pollutants end up?
11. Can you characterize the sources of ocean pollution? What is the role of oil spills? How do they compare with other forms of pollution and their impacts?
12. Briefly describe the three levels of sewage treatment employed today in treatment plants. What are some of the advantages and disadvantages of each? What is meant by "low-tech" sewage treatment?
13. Characterize some of the trends in water pollution in the developing countries and developed countries. What are the best arguments for and against declining water quality in the next decade or two?

14 HAZARDOUS WASTE PRODUCTION AND DISPOSAL

14.1 INTRODUCTION

Waste production has increased at an alarming rate over the past century in direct response to rising consumerism and population growth. Developed countries in particular have developed massive economies founded on "quick to use and quick to discard" consumer habits. These practices not only demand huge amounts of natural resources, but produce an endless stream of liquid and solid waste that enters the land and aquatic environment in various forms called solid and hazardous waste.

Solid waste is produced at every phase in the economic system, including raw material extraction, processing, and transportation; manufacturing, packing, and shipping; and advertising and sales, as well as with the products themselves once they are spent and/or discarded. In addition, research, experimentation, and development yield various technological materials including thousands of hazardous substances. About 35,000 of the chemicals currently approved for use in the United States are classed as hazardous or potentially hazardous to humans. Many are used in manufacturing, and in some form, they also become part of the waste stream. It is estimated that hazardous waste production in the United States is equivalent to as much as a ton or more per person per year.

All countries produce solid and hazardous wastes. Although developed countries produce the most waste per capita, they also exercise the best management practices in waste handling and disposal. Developing countries produce less waste, but generally lack adequate means to handle and dispose of many wastes, especially hazardous wastes, in an environmentally safe manner. As Western technology and consumer habits are adopted by developing countries, the problem of hazardous waste is certain to become more serious. To make matters worse, some poor countries are accepting imports of hazardous waste from industrial nations in exchange for payments or manufactured goods. Thus, hazardous waste has become the first form of pollution traded on the international market—with potentially grave implications for the receiving nations.

This chapter examines the processes and practices of waste production and disposal. The focus is on hazardous waste, but attention will also be given to solid waste. We open with a brief description of the types and sources of waste and their land-use relationships, and then examine why traditional disposal methods are inadequate in today's world. The chapter goes on to explore modern disposal and treatment methods and the opportunities presented through recycling programs.

14.2 TYPES AND SOURCES OF DISPOSABLE WASTE

There are three basic classes of waste requiring disposal in the environment: (1) *sewage*, which we examined in Chapter 13; (2) *solid waste*, also examined in Chapter 13 in the context of groundwater pollution; and (3) *hazardous waste*. Next to the liquid mass associated with sewage, solid waste is the most abundant form of human waste. **Solid waste** is assorted, discarded materials variously described as trash, garbage, refuse, and litter from urban and rural land uses. In most countries the vast majority of solid waste is produced by mining and agriculture mainly at extraction and production sites. In the United States these two activities account for about 85 percent of the solid waste, with the remainder coming from households, commerce, transportation, industry, and construction, mostly in urban areas.

The output of urban solid waste is massive, about 700 million tons annually in the United States, of which 150 to 200 million tons are municipal solid waste. *Municipal solid waste* is refuse from households, businesses, and institutions, that is, nonindustrial waste, and it is disposed of in landfills. **Landfills** are managed disposal sites. They are familiar features in the landscape around most American and Canadian cities and are commonly associated with unsavory environmental conditions and land-use problems.

In poor countries most solid waste is disposed of by *open dumping*, unmanaged and often unregulated piling of waste on marginal land. Dumps are usually more unsavory and environmentally threatening than landfills, and on the fringe of many African and Latin American cities, dump sites have become part of the living environment of the masses of displaced poor (Figure 14.2a). Water pollution, pests, and disease are common in these places and are vivid reminders of the consequences of improper waste disposal.

On the other hand, developed countries produce significantly more waste than developing countries. With only 5 percent of the global population, the United States has the unfortunate distinction of leading the world in waste output. Canada and Australia produce much less total waste (because of much smaller populations), but are comparable to the United States in per capita waste output. Most European countries produce considerably less than these countries. The United Kingdom, Sweden, and Japan produce about half as much waste per person as the United States, whereas poor countries like India produce only a small fraction of the person waste of the United States.

Solid waste is not classified as hazardous, yet it can emit environmentally harmful contaminants when the substances in it break down. Based on samples extracted from urban landfills in the United States, municipal solid waste is composed of eight basic classes of substances (Figure 14.1). Paper (including cardboard) is decidedly the primary constituent, representing 30 to 40 percent of

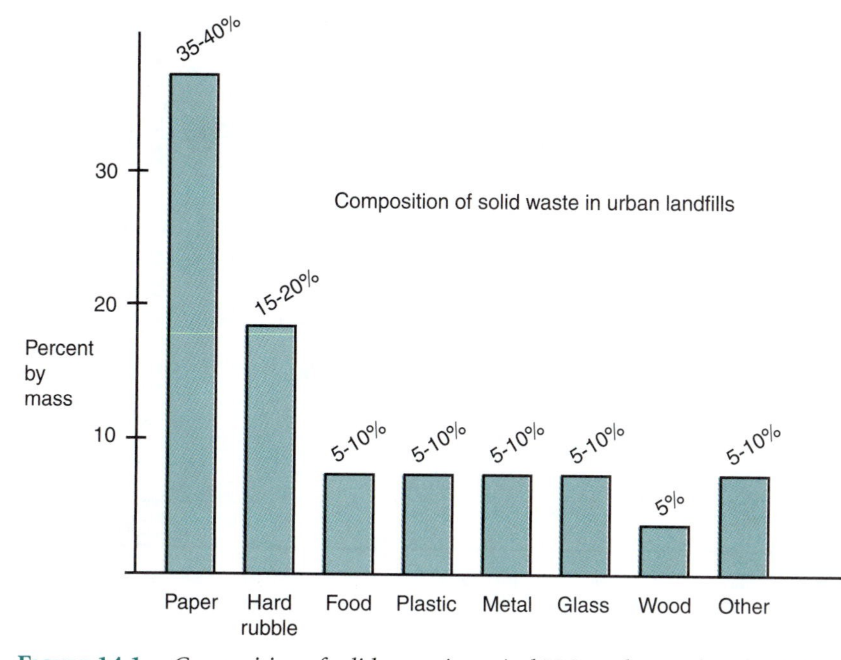

FIGURE 14.1 *Composition of solid waste in typical U.S. and Canadian landfills.*

the mass in most landfills. Hard substances, represented by materials like concrete and asphalt, constitute 15 to 20 percent. Glass, plastics, metal, and food each constitute 5 to 10 percent of the landfill mass, whereas wood represents about 5 percent. The material in each primary class also contains a host of secondary substances; for example, inks and glues in paper; oil in asphalt and spent containers; and paint residues in plastics, metal, and glass.

As the primary and secondary substances in solid waste break down, they yield a chemically concentrated liquid called **leachate**. If this foul material is allowed to seep from a dump or landfill, through the subsoil, and into the groundwater system, aquifer pollution can result. (See Chapter 13 for discussion of groundwater contamination from buried waste.) Solid waste can also be a source of pollution in surface water (via runoff) and air (via windblown debris), and solid waste landfills are a source of countless land-use problems, including traffic congestion, obnoxious odors, depressed land values, and political and legal disputes (Figure 14.2).

Hazardous wastes

Hazardous substances are chemicals that can be harmful to humans and other organisms. Most of these substances are *toxic materials* (such as DDT, solvents, and lead) from both manufactured sources and natural sources. A toxic substance is classed as **hazardous waste** if it is handled and disposed of in such a way that it poses a threat to human health and the environment. In other words, a toxic substance is hazardous if there is some risk of exposure and harm.

Throughout the world most hazardous waste can be traced to three industrial sources: *chemical manufacturing, primary metal manufacturing,* and *petroleum refining.* Together they produce about 75 percent of today's hazardous waste. From these industries there are three main paths to the hazardous waste stream:

1. The waste produced directly from raw material processing in the chemical, metal, and petroleum plants themselves;

FIGURE 14.2 *Plowing and compacting garbage in a conventional North American landfill operation.*

2. The waste produced from secondary manufacturing operations such as in plastics and paper plants that use the chemical and petroleum products;
3. The commercial products themselves, such as pesticides, plastics, and paints, that are finally discarded in the environment (Figure 14.3).

The final list is enormous, virtually incomprehensible, and changing and growing day by day. More than 70,000 chemicals are used daily worldwide, and we as consumers are ultimately responsible for a great many of them in our daily lives. It bears repeating that in the United States, hazardous waste production from all sources is equivalent to a ton or more a year for every man, woman, and child.

Generally speaking, there are at least six main classes of hazardous waste based on waste composition:

- Heavy metals (e.g., lead, zinc, and arsenic)
- Synthetic organic compounds (e.g., PCB, DDT, and dioxin)
- Petroleum products (e.g., grease, oil, and gasoline)
- Acids (e.g., hydrochloric acid and sulfuric acid)
- Biological substances (e.g., bacteria and plant toxins)
- Radioactive materials (e.g., nuclear fuel rods and nuclear medical materials)

Distribution of Hazardous Wastes

These wastes are distributed to the environment by virtually all major land uses. However, two land uses—industry and agriculture—are responsible for the delivery of most hazardous wastes to the environment. These land uses are major contributors of organic compounds, heavy metals, and acids to soil, air, and water systems.

Transportation is also an important hazardous waste delivery agent, especially of petroleum products, mainly through spills and leaky storage tanks. Nearly 60 percent of hazardous materials shipped in the United States are moved by truck, and the nation's highway system is

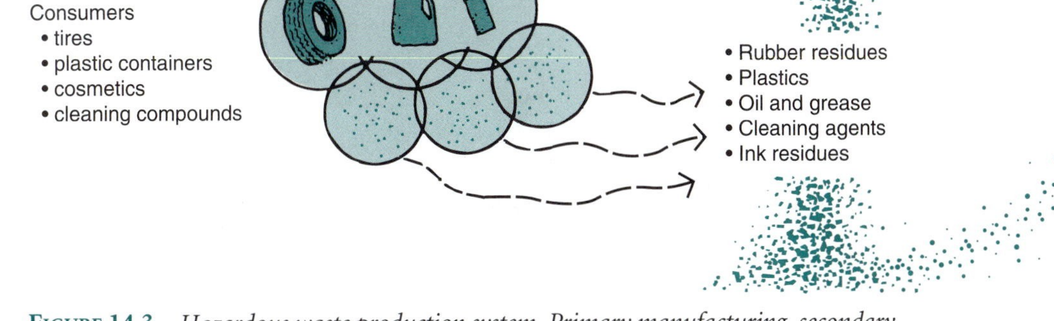

FIGURE 14.3 *Hazardous waste production system. Primary manufacturing, secondary manufacturing and consumers all contribute to the hazardous waste stream.*

becoming a network of contamination sites (Figure 14.4). Residential land use also contributes a substantial load of various chemicals used in and around our homes. For example, solvents, gasoline, oil, cleaning compounds, paints, and lawn pesticides enter the environment through yard applications, spills, and in garbage that becomes solid waste (Table 14.1). As many as 50,000 landfills (both active and inactive in the United States could contain significant amounts of hazardous waste.

This brings us to a major difference in the geography of hazardous and solid waste systems. Solid waste tends to be concentrated at points, mainly dump or landfill sites near its sources, whereas a large share of hazardous waste is widely dispersed into air, water, soil, and biotic systems. The latter is illustrated by pesticides applied to croplands. With more than 1 million square miles of cropland in the world receiving pesticide treatments, this is clearly the most widespread means of hazardous waste distribution to the land. In addition, some of the pesticide residues

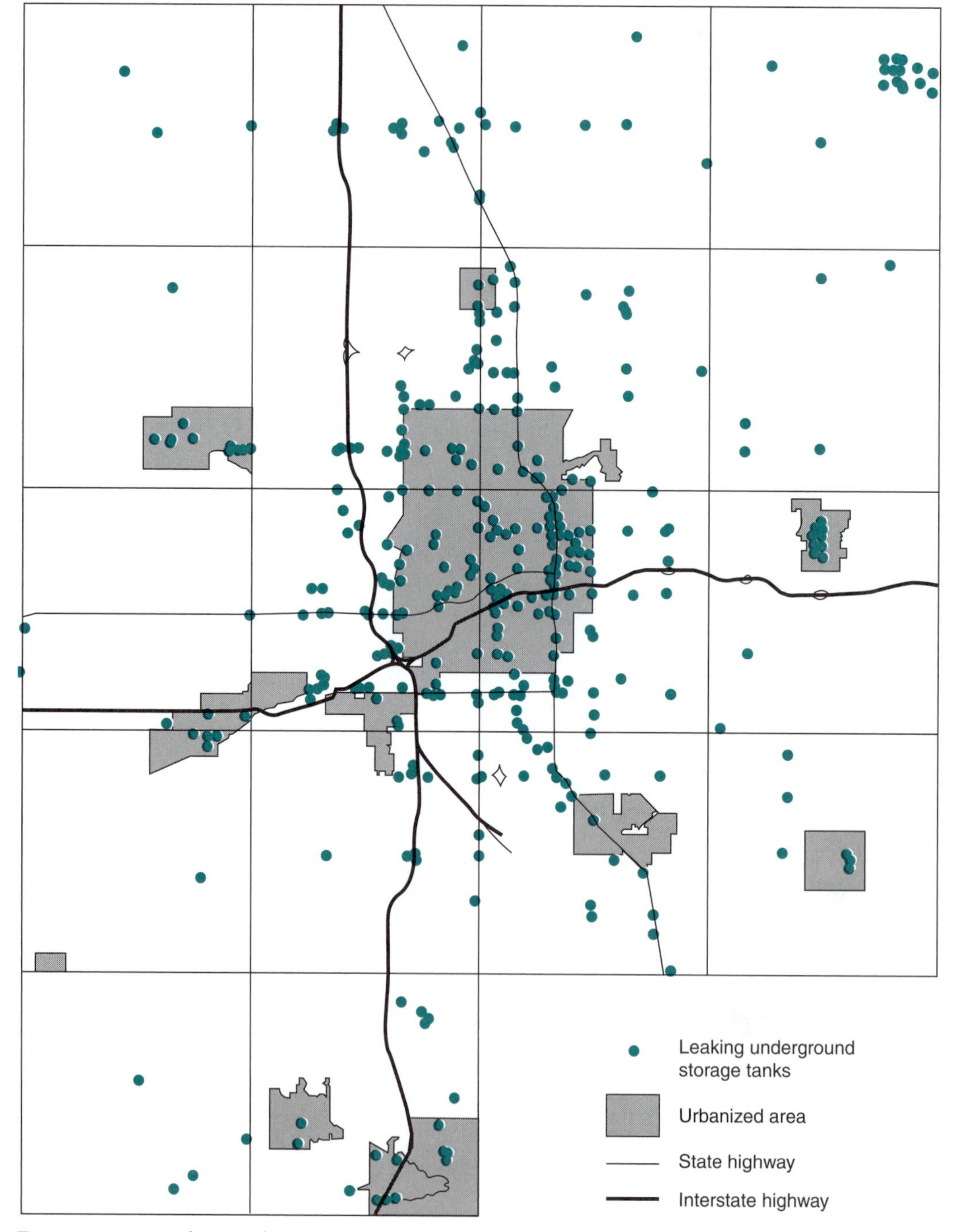

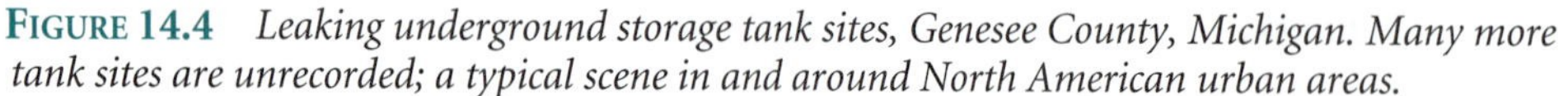
FIGURE 14.4 *Leaking underground storage tank sites, Genesee County, Michigan. Many more tank sites are unrecorded; a typical scene in and around North American urban areas.*

TABLE 14.1

COMMON HOUSEHOLD HAZARDOUS WASTES

Kitchen	
Floor-care products	Metal polish
Furniture polish	Bug sprays
Bathroom	
Disinfectants	Tub and tile cleaners
Toilet bowl cleaner	Nail polish remover
Garage	
Antifreeze	Car wax
Battery acid (or batteries)	Gasoline
Brake fluid	Motor oil
Workshop	
Paint (oil based)	Turpentine
Paint stripper	Paint thinner
Wood preservative	Rust remover
Yard	
Fertilizer	Rat poison
Fungicide	Weed killer
Insecticide	

from farmfields are further dispersed by the atmosphere. Unlike point distributions of waste in spills or burials, however, pesticide applications in agriculture are relatively light. Nonetheless, they are applied repeatedly, and in time they show up in soil, groundwater, and food chains over broad areas (see Figures 13.5 and 13.6).

14.3 TRADITIONAL METHODS OF WASTE DISPOSAL

Waste disposal became a serious problem with the Industrial Revolution and the growth of large cities such as London, Berlin, and New York. Early disposal methods by and large were grossly inadequate: Most solid waste was transported to nearby *open dumps* on the outskirts of settlements (Figure 14.5). Dump sites were typically low-value land such as swamps, floodplains, and abandoned mine pits. Unfortunately, many of these sites were the very worst choices for waste disposal from an environmental standpoint, because they brought the waste into direct contact with natural drainage systems, including streams, wetlands, and groundwater. In addition, they became habitats for pests and fostered disease and generally distasteful landscapes.

As towns and cities expanded and one land use gave way to another, dump sites were eventually abandoned and replaced by new ones farther out. With increased

FIGURE 14.5 *New York City garbage collectors around 1900 dumping garbage onto barges for transport to nearby dump sites such as the one shown in the inset photograph.*

pressure for land development in the late nineteenth and early twentieth centuries, urban land uses (including residential) often pressed around and pushed over old dump sites (Figure 14.6). The material in most dumps consisted of garbage, construction rubble, animal carcasses, and industrial waste. Hazardous wastes existed, but were much less common and far less complex than today. Consider that few petroleum products and fuels were manufactured before 1900 and the types and volumes of chemicals were minuscule by today's standards. Also bear in mind that cities were relatively few in number and small as population centers. In North America the majority of people lived in rural areas at the beginning of the 20th century.

By the 1930s municipal waste had become a serious problem, and a new method, called the **sanitary landfill**, was introduced as an alternative to open dumping. With a goal of minimizing nuisance or hazard to public health or safety, the sanitary landfill was designed to place solid waste in a planned disposal site, usually an excavated pit, where it could be contained and covered with soil. By 1960, more than 1,200 communities in the United States used sanitary landfills. However, the problem of waste disposal was far from solved. Indeed, it was growing.

In addition to the thousands of settlements still using open dumping, hazardous waste production was increasing rapidly in the 1940s, 1950s, and 1960s. Moreover, many sanitary landfills proved to be ineffective, because leachate leaked into the groundwater beneath them. On Long Island, New York, for example, two landfills dating from the 1930s and 1940s produced huge pollution plumes that contaminated groundwater and threatened public health. Such problems often coincided geographically with patterns of urban sprawl and the development of suburbs dependent on local groundwater for water supplies (Figure 14.6).

Hazardous Waste Dumping

Less apparent to the suburban or rural resident was the hazardous waste problem. After World War II, North American industry expanded enormously, as did the output and variety of hazardous wastes. For example, between 1945 and 1985 the production of synthetic organic compounds in the United States increased from less than 10 tons a year to more than 100 million tons a year. By and large, disposal methods for hazardous wastes were inadequate, not only in North America, but in virtually every industrial region in the world.

Companies either handled their own waste by dumping or burying it on company land or on nearby rented land, or they contracted haulers to truck it to "destinations unknown." Both strategies resulted in serious problems, but the latter was particularly frightening because it encouraged clandestine or **midnight dumping** in streams, lakes, wetlands, gravel pits, abandoned farms, and unsecured public properties (even parks in some places). It also led to the development of private depots where the waste from various sources was dropped and abandoned. One such site near Elizabeth, New Jersey, contained

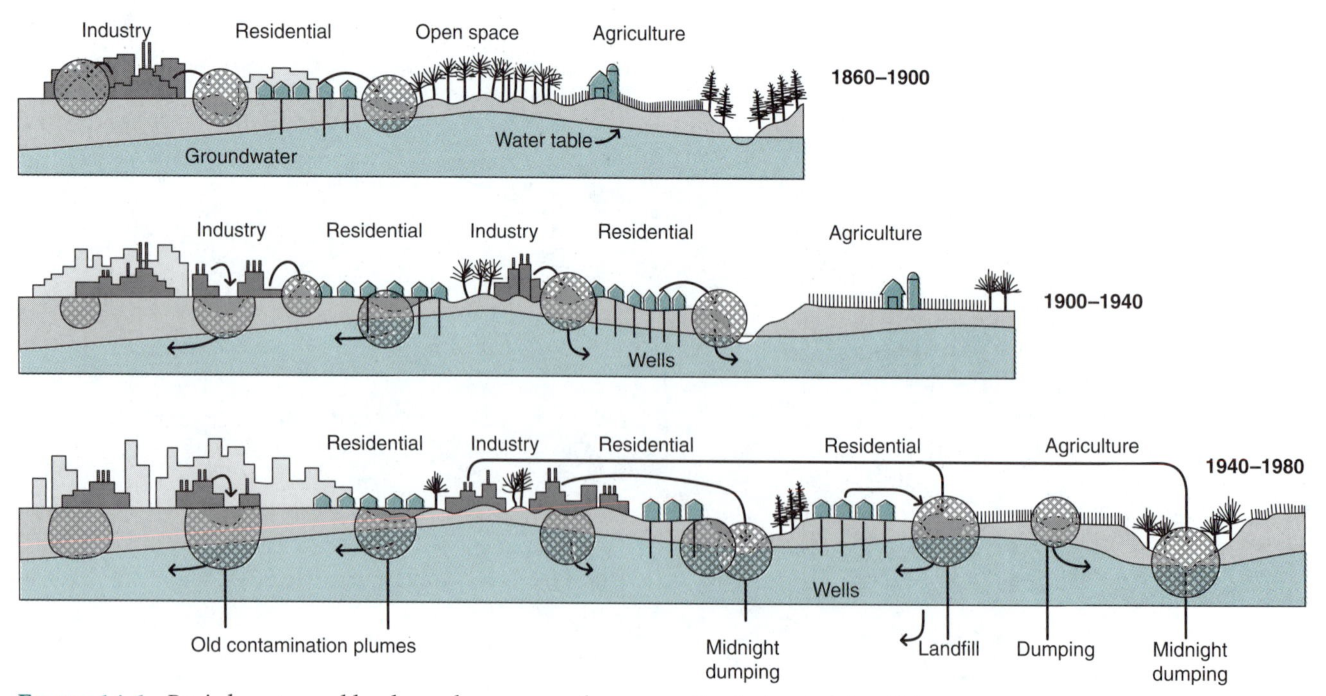

FIGURE 14.6 *Buried wastes and land-use change around an expanding industrial city. This diagram is designed to show the effect of urban growth and sequential land use change resulting in residential development over and around old hazardous waste sites.*

50,000 drums of chemical waste left to posterity by a bankrupt chemical company. These sorts of operations were common around every industrial region, and as urban land use spread and highway systems became more efficient in the 1960s and 1970s, dumping tactics expanded over larger geographic areas, often to other states. Meanwhile, changing land-use patterns in the urban region often led to new development on old hazardous waste sites (Figure 14.6). In many cases the waste was obscured by burial or ignored. But in some cases it came to public attention because of the illnesses it caused in local residents.

The most infamous of such hazardous waste sites in the United States was the Love Canal near Niagara Falls, New York, where many tons of nearly a hundred different types of chemicals from a factory were dumped in a large trench between the 1920s and the 1950s. The trench was covered over and secured from surface exposure, but eventually the land changed hands, and a school and several hundred homes were constructed over the trench. Construction involved surface grading and excavation, which exposed the toxic materials and contaminated the human living environment. Subsequently many residents developed chronic illnesses and serious diseases.

Over the past hundred years or more, four environmentally damaging methods were commonly used to get rid of hazardous waste. Although evidence is scarce, *midnight* dumping was common, and undoubtedly very serious in the urban hinterlands, especially where varied terrain, water features, and vacant land were accessible to trucks. Although greatly reduced, this illegal and destructive practice continues today. The practice of dumping liquid waste into uncovered and unlined pits, called *open lagoons*, was used in the United States as late as the 1970s. Although still practiced in some places, open lagoon disposal presents a serious hazard to both the surface and the subsurface environments. The third method consisted of *surface storage* of liquid waste in drums or barrels and open dumping of solid hazardous waste. The fourth method relied on *burial* of hazardous waste in containers. The main problem was that the containers usually deteriorated, leaking waste and contaminating the soil and groundwater (Figure 14.7).

By the 1980s, there were tens of thousands of known and suspected unsecured hazardous waste (dump) sites in the United States. How could this have happened? A key factor was the rapid rate of technological and industrial development after WWII. Most industries were unprepared to manage the types and volumes of waste they produced. Likewise, government regulatory programs lagged far behind; laws requiring monitoring and secure handling and disposal did not begin to emerge until the 1970s. Science also lagged behind in analyzing the effects of hazardous wastes on the human health and the environment. Given a climate of misunderstanding, the absence of incentives, and weak regulations, the handling and disposal of hazardous waste was usually relegated to second parties—middlemen who dumped the stuff wherever they could.

What about other countries? The postwar industrial nations of Europe faced many of the same problems as the United States and Canada. However, because much of the industrial base in Europe had been destroyed in the war, countries like Germany and the United Kingdom had an opportunity to build somewhat more efficient waste management systems. Nevertheless, the political atmosphere in Europe was one of economic growth as well as Cold War military buildup, and as a result hazardous waste regulation lagged behind a growing list of problems. But nowhere was the hazardous waste problem as severe as in the Communist-bloc countries.

Central and Eastern European Experience

With the emergence of the Communist bloc after WWII, Eastern European nations pressed hard to develop industrial economies. For nearly fifty years the true character of this development was not fully known to the rest of the world, but with the crumbling of the Iron Curtain a horror story of pollution came to light. All forms of pollution—air, water, and ground—were abundant around cities and industrial complexes in Poland, former East Germany, former Czechoslovakia, the former Soviet Union, Bulgaria, and other communist countries. Blatant discharge of untreated wastes with disregard for environment and human health occurred in thousands of places (Figure 14.8). In southern Poland, near the towns of Olkusz and Slawkow, soil contamination by lead and cadmium is the highest ever recorded in the world. Most experts agree that the problem is far worse than is currently known and decades will be required for cleanup, if cleanup is indeed possible. In Poland alone, the cleanup costs are estimated at $260 billion over the next 20 to 25 years.

How did this tragedy happen? Most observers point to the socialist economies and authoritarian governments of these countries as the root of the problem. In the struggle to demonstrate economic power, leaders were bent on increasing industrial output while ignoring responsibilities for waste management. The government-controlled pricing system for energy also contributed. The prices of fuels were set artificially low, leading to overconsumption, inefficiency, and high rates of pollution. In addition, pricing systems did not take into account the costs of controlling pollution. And since industrial enterprises were state-owned, the system of checks and balances, which would include environmental monitoring, enforcement of pollution laws, and levying of fines, was essentially useless. The government was not inclined to fine itself for pollution violations.

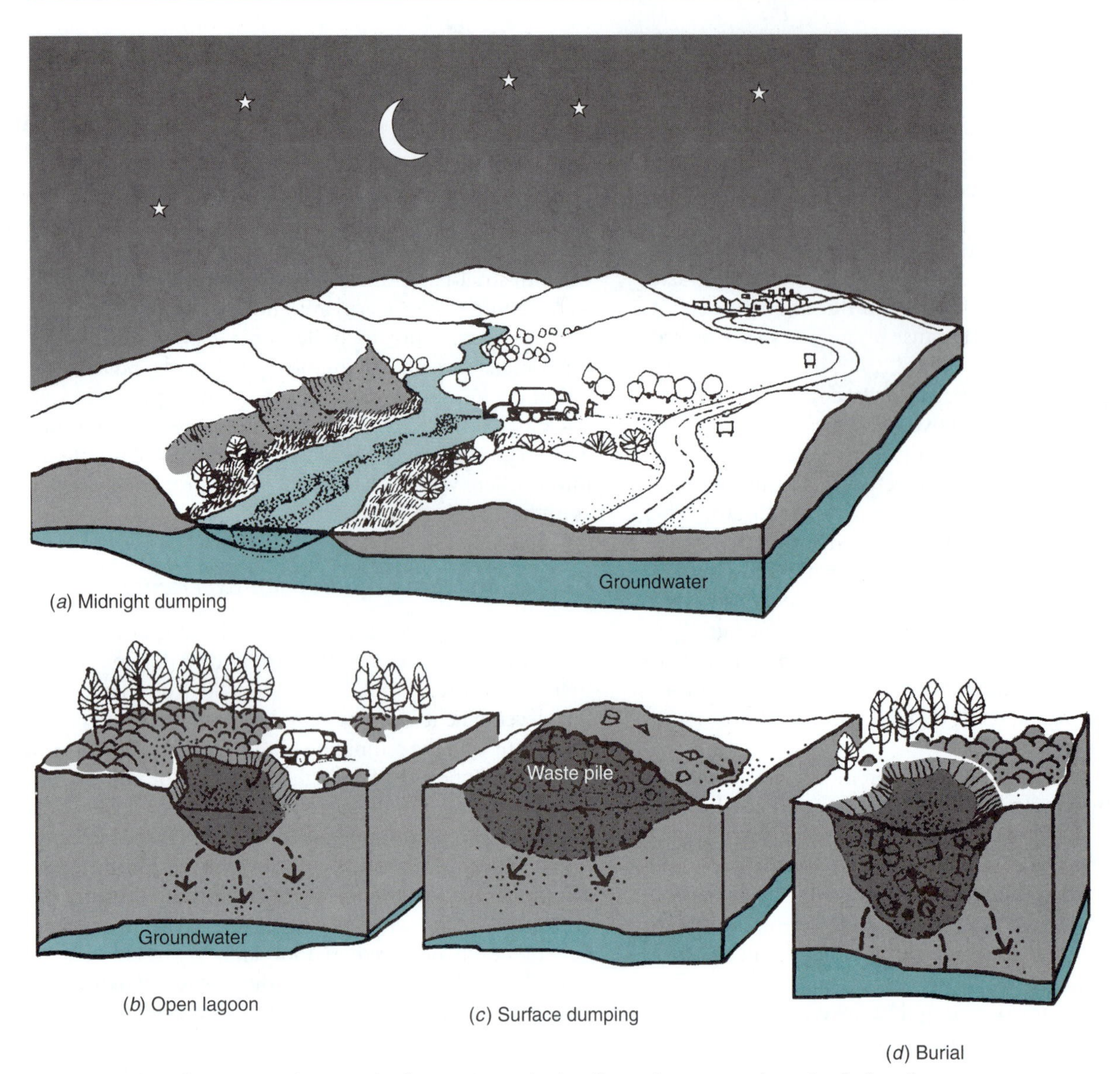

FIGURE 14.7 *Illustrations showing the four main methods of hazardous waste dumping before the advent of modern management techniques.*

The Evolving Geography of Hazardous Waste

There is little mystery to the traditional pattern of hazardous waste production in the world: Most operations are concentrated in urban, industrial regions of the developed nations. In North America four major regions stand out: the Eastern Seaboard, the Southern Great Lakes, Southern California, and the Gulf Coast where petroleum refining and chemical manufacturing are located. Because of transportation costs and associated liabilities, most hazardous waste disposal occurs within and around the regions where waste is produced. There are exceptions to this pattern, of course, including export to neighboring nonindustrial regions, dispersal of agricultural pesticides and chemical fertilizers in rural areas, and disposal associated with government weapons installations and military bases, most of which are scattered across the continent.

In the past several decades, many industrial operations producing hazardous wastes have relocated outside the traditional urban, industrial regions. This is in part a response to the general trend toward industrial decentralization, but in some cases it is clearly an effort to avoid the more stringent regulations of some states. In the extreme, some chemical manufacturers have moved their operations abroad to developing countries, where there are few and often weak regulations on hazardous waste production and disposal. As noted earlier, waste export to poor countries has also emerged as a trend of the modern era. Several developed countries, including the United States, Italy, France, the Netherlands, and Belgium, have purchased rights to export hazardous wastes to poorer nations such as Morocco and Senegal. These trends, coupled with widespread economic development in the world, have resulted in increasingly broader distributions of haz-

FIGURE 14.8 *An industrial operation in former East Germany showing excessive pollution from hazardous wastes, raw materials, and air pollution. The contamination from such operations will be a serious problem for decades.*

ardous waste pollution and more complex management arrangements.

14.4 Some Consequences of Improper Waste Management

Until recently most hazardous waste has been handled and disposed of in ways that pose real or potential threats to human health and environmental quality. With the exception of a dozen or so developed nations, hazardous waste is still improperly managed by most countries. Even for those countries such as the United States and Canada with modern management programs, the legacy of past errors will continue to haunt us for many years. At the same time, the development of new substances, new manufacturing processes, and new global trade and distribution patterns produce new problems.

Elizabeth, New Jersey

For nearly 10 years, 50,000 drums of chemical wastes lay corroding and leaking on a site near Elizabeth, New Jersey. Some of their labels had burned away, so that their contents could not be determined without testing. The chemicals from leaking drums seeped into the soil and entered runoff, which discharged into a local stream that flowed into the Hudson River. When cleanup efforts began, the drums were found to contain, among other things, solvents, heavy metals, pesticides, biological agents, and nitroglycerine. Cleanup efforts took months, and the waste and contaminated soil had to be trucked to a secure disposal site. Unfortunately, the true extent of the environmental damage will never be known, because much of it took place years ago as chemicals were carried by runoff to streams and the ocean.

Love Canal, Niagara Falls, New York

Here the problem literally emerged in the late 1970s when residents began to complain about chemical smells, signs of chemicals seeping into basements, yards, and playgrounds, and an unusually large number of ailments ranging from rashes and headaches to cancer, nerve disorders, miscarriages, and birth defects. Complaints to public officials were at first ignored, but testing, when finally done, revealed serious toxic contamination of soil, air, and water in the canal area. So serious was the health threat that 239 families had to be permanently evacuated. In 1980 the area was declared a federal disaster area. The cleanup program took 12 years and cost $250 million (Figure 14.9). The final assessment in terms of the effects on human health cannot, of course, be accurately measured in dollars, and for a number of people these effects will persist the rest of their lives.

Bhopal, India

In 1984, an accident at an American-owned chemical plant that manufactured the pesticide Sevin cost thousands of lives. The accident was caused by leakage of a contaminant, probably water, into a tank of methyl isocyanate (MIC), an extremely toxic chemical. The result was a violent chemical reaction that ruptured the tank's valves and released a poisonous gas into the air. The gas spread over a populated area of 25 square miles, instantly killing more than 1,000 people and injuring thousands of others. The final count by the Indian government was 3,000 people dead, 30,000 permanently disabled, and 25,000 severely injured, making Bhopal one of the two worst hazardous waste disasters ever recorded. The other was Chernobyl in 1986.

Health Effects of Hazardous Wastes

The effects of hazardous substances on human health are complex and diverse. As a whole, little is known about this area of public health, especially about chronic exposure to low levels of toxic chemicals. More is known about acute exposure to various toxic chemicals, particularly substances for which laboratory testing has been conducted. **Acute toxicity** is defined as the occurrence of serious symptoms, possibly death, after a single exposure to a substance. **Chronic toxicity**, on the other hand, is the delayed occurrence of symptoms after repeated exposures to a substance. The appearance of symptoms takes place after the body's threshold of resistance to the substance has been reached—this may not occur until months or years after the initial exposure. Moreover, the symptoms may be difficult to distinguish from those of other illnesses. Lead is a chronically toxic substance whose effects on human health are easily confused with other illnesses, but we now know that long-term exposure to lead can cause brain damage, behavioral disorders, and even death.

For most hazardous substances, however, the effects on human health are unknown. Where combinations of harmful substances are involved, the problem is even more perplexing. Consider that there are more than 10,000 types of chemicals in the chlorinated organic compound group, and we are exposed to dozens of them on a daily basis, including dioxin in white paper, benzene in solvents, and PCBs in electrical equipment. Many, such as dioxin, are known carcinogens and virtually all the rest are suspected carcinogens. **Carcinogens** cause cancer in humans and animals. They represent just one class of hazardous substances that cause specific illnesses. Others include *mutagens*, most of which are also carcinogenic and cause genetic changes that affect future generations; *teratogens*, which can damage fetuses and cause spontaneous abortions; and *infected substances* that transmit disease-causing organisms.

FIGURE 14.9 *Final stages in the remediation project on the Love Canal toxic waste site.*

Your risk of cancer from exposure to carcinogens is difficult to determine. Different countries use different forms of **risk assessment**, ranging from near indifference in many poor countries to detailed public information programs on risk and uncertainty associated with the use of suspected substances in many developed countries. In the United States, the federal government attempts to provide a line of defense for the public by labeling risky substances and banning some. Based on scientific evidence, such as laboratory tests, case studies, and related statistical approximations, a labeling statement warns the consumer of the potential risk of cancer or other illness associated with the use of the product at some level. The level may be defined in terms of the amount of a particular pesticide ingested with food as a part of a normal diet.

Among the problems with this approach is (1) the uncertainty of laboratory and case study data representing low dosage conditions, and (2) the emergence of new chemicals faster than evaluation can be performed. Some experts argue that evaluation of individual substances on a case by case basis should be replaced by group evaluation. In particular, it is argued that the chlorinated organic compounds should be banned as a group because they are consistently linked to cancer in general. However, the argument used by manufacturers, and one apparently accepted by regulatory agencies, is that each chemical is "innocent until proven guilty"; therefore, many synthetic compounds continue to be used while evidence for and against them is building up. In the end, the problem boils down to an argument over the health and environmental costs versus the economic benefits of a particular chemical, such as an agricultural pesticide. For most chemical substances, both costs and benefits are very difficult to measure and evaluate effectively, especially in cases where the health or environmental impacts appear only over the long term.

14.5 Hazardous Waste Regulation in the United States

The management and control of hazardous substances has become one of the most pressing environmental issues of our time. As with so many environmental issues, the national policies we adopt are as much a product of public opinion and politics as of science, and not surprisingly, opinion varies widely. Almost no one subscribes to indiscriminant, unregulated use of chemicals anymore, and most people agree that strict regulation is indeed needed. However, faced with the prospects of giving up certain modern conveniences provided by chemical inventions and products, few people favor wholesale outlawing of groups of chemicals such as chlorinated organic compounds. The favored approach seems to be one that combines monitoring and regulation of the entire stream of harmful and potentially harmful substances: production, distribution, use, and disposal—the "cradle to grave" approach.

Resource Conservation and Recovery Act

In 1976 the U.S. Congress enacted the **Resource Conservation and Recovery Act (RCRA)**, which provided a legal definition of hazardous wastes and established guidelines for managing, storing, and disposing of hazardous materials in an environmentally sound manner. The U.S. Environmental Protection Agency (EPA) was authorized to establish which substances were to be regulated (labeled) as hazardous. According to the RCRA, various criteria could be used to define a hazardous substance, including any substance found to be fatal to humans at low doses or shown through animal studies to be dangerous to humans. There are four main criteria defining hazardous materials:

- **Ignitability**. Products with this property include combustible substances such as gasoline and solvents.
- **Corrosivity**. Substances that dissolve steel or burn skin, such as rust remover and battery acid are corrosive.
- **Reactivity**. Unstable substances, such as bleaches and cyanide-plating substances, that undergo rapid or violent chemical reactions when they interact with water or other substances.
- **Toxicity**. This is based on tests to determine whether toxic substances are released from waste after leaching processes.

If a waste material meets one of these four criteria, the EPA can place it on the list of hazardous waste. The criteria are adequate for classifying most industrial waste; however, they are considered inadequate for defining biologically hazardous substances such as carcinogens.

Superfund

The most widely known regulatory program for hazardous waste management in the United States is the **Superfund** program. It was created in 1980 as an act of Congress (under the formal title Comprehensive Environmental Response Compensation and Liability Act) for the purpose of cleaning up hazardous waste sites and responding to the uncontrolled release of hazardous materials into the environment. Central to Superfund is that it empowers the federal government to provide funding for the cleanup of severely contaminated hazardous waste sites.

To date about 2,000 sites have been given priority status and the EPA estimates that it will cost more than $30 billion to clean them up. More than 10,000 waste sites are awaiting evaluation for possible priority status, and although most will not gain Superfund status, many will, so the list of Superfund sites is bound to grow. Since

1980 Congress has appropriated $13.5 billion to the program, but only about 10 percent of the money has actually been used for cleanup operations and only a handful of sites have been remediated. The lion's share of the funding has, unfortunately, gone to legal and administrative fees, principally for legal action to get responsible parties to help pay for cleanups. Because of the legal entanglement related to this polluter-pays provision, the Superfund program has become bogged down. In addition, there are serious criticisms of EPA's solutions to several of the sites already remediated. Understandably, the Superfund program is being seriously reevaluated by both public and private groups, and it appears that significant changes will be recommended to Congress.

Polluter-Pays Laws

Most hazardous waste sites do not qualify for government assistance in remediation. In response to growing concern by property buyers about the liability of hidden wastes on a site, several states have enacted **polluter-pays** laws. These laws place the responsibility for site cleanup on past property owners under whose ownership the waste was buried. Such laws are proving to be effective because the cost of remediation is added to real estate, thereby returning pollution costs to the market system rather than passing them along to the state or federal government.

14.6 Modern Hazardous Waste Disposal Methods

Several strategies are available for managing hazardous waste, though none is without drawbacks. In general they fall into two classes, disposal and treatment. **Disposal** involves collecting, transporting, and storing waste with no processing or treatment. **Treatment** involves submitting the waste to some sort of processing to make it less harmful. Because of increasingly stringent health and safety requirements, as well as endless local land-use and public opinion problems, land disposal has become very costly. For this reason and others, the trend in hazardous waste management is toward on-site treatment, but off-site disposal is still the most common management method.

Land Disposal Methods

The most widely used land disposal method is the **secure landfill**. The objective of this method is to confine the waste and prevent it from escaping the disposal site. The first step in planning a secure landfill is selection of an appropriate site, one that minimizes the risk of groundwater and surface water contamination. Normally this is a very lengthy and complex process involving a host of field tests, analyses, reviews, public hearings, and permit applications. Site selection is followed by landfill design, which must include: (1) a clay liner backed up (underlain) by a plastic liner; (2) a clay dike and clay cap; (3) a leachate collection and drainage system; and (4) monitoring wells around the site to check that leachate is not leaking into the soil and groundwater. It is argued that despite these measures, secure landfills can still leak. Therefore, it is extremely important that only the very best sites are used; namely, those with deep, dense-clay soils, little topographic relief (elevation variation), and water tables at great depths (Figure 14.10).

Another, less acceptable, disposal method is **surface impoundment**. Impoundments are natural or excavated pockets (pits) in the land that can be lined and used to store hazardous waste. They were fairly common to industrial sites in the past, but under today's more stringent storage and disposal laws, they are generally considered to be too risky. In addition to the potential of seepage, they can also be a source of air pollution, as the more volatile chemicals, such as chlorine, evaporate from the pond surface. Although open storage is now uncommon in countries such as the United States and Canada, it is very common to industrial areas of Eastern Europe and a growing number of developing countries.

Deep-well injection is another hazardous waste disposal method. With this method waste is pumped into the bedrock below the active groundwater zone, usually deeper than 2,000 feet. Deep-well injection is widely used today; in the United States the EPA estimates about 11 percent of hazardous wastes are disposed of in this fashion. The environmental risks are relatively minor if planning, construction, and operations are done properly. The main prerequisite is selection of geologically stable sites with few bedrock fracture systems to conduct liquid waste, which is normally injected under high pressure. But malfunctions do occur occasionally. These include *blow-outs* in which pressurized waste migrates along fractures and breaks out of the bedrock and discharges onto the surface; increased earthquake activity related to rock slippage under the influence of high-pressure waste injection; and groundwater contamination.

Ocean dumping is the only nonterrestrial disposal method. Though widely criticized, this practice is used by many countries for a broad range of wastes, including sludge from sewage treatment plants, spoils from harbor dredging municipal garbage, and hazardous materials (Figure 14.11). Although ocean dumping has been reduced in the United States since 1968 when 45 million tons of waste were dumped into the ocean, the practice continues today. Dumping municipal garbage and sludge in near-shore waters is now banned; however, the ban is ignored by some cities and some deep-water dumping is still allowed. In shallow coastal water near shore, extensive dumping of dredge spoils still takes place, and landfilling along shorelines as a part of urban development, navigational, and highway projects is common practice in the United States and most other developed countries.

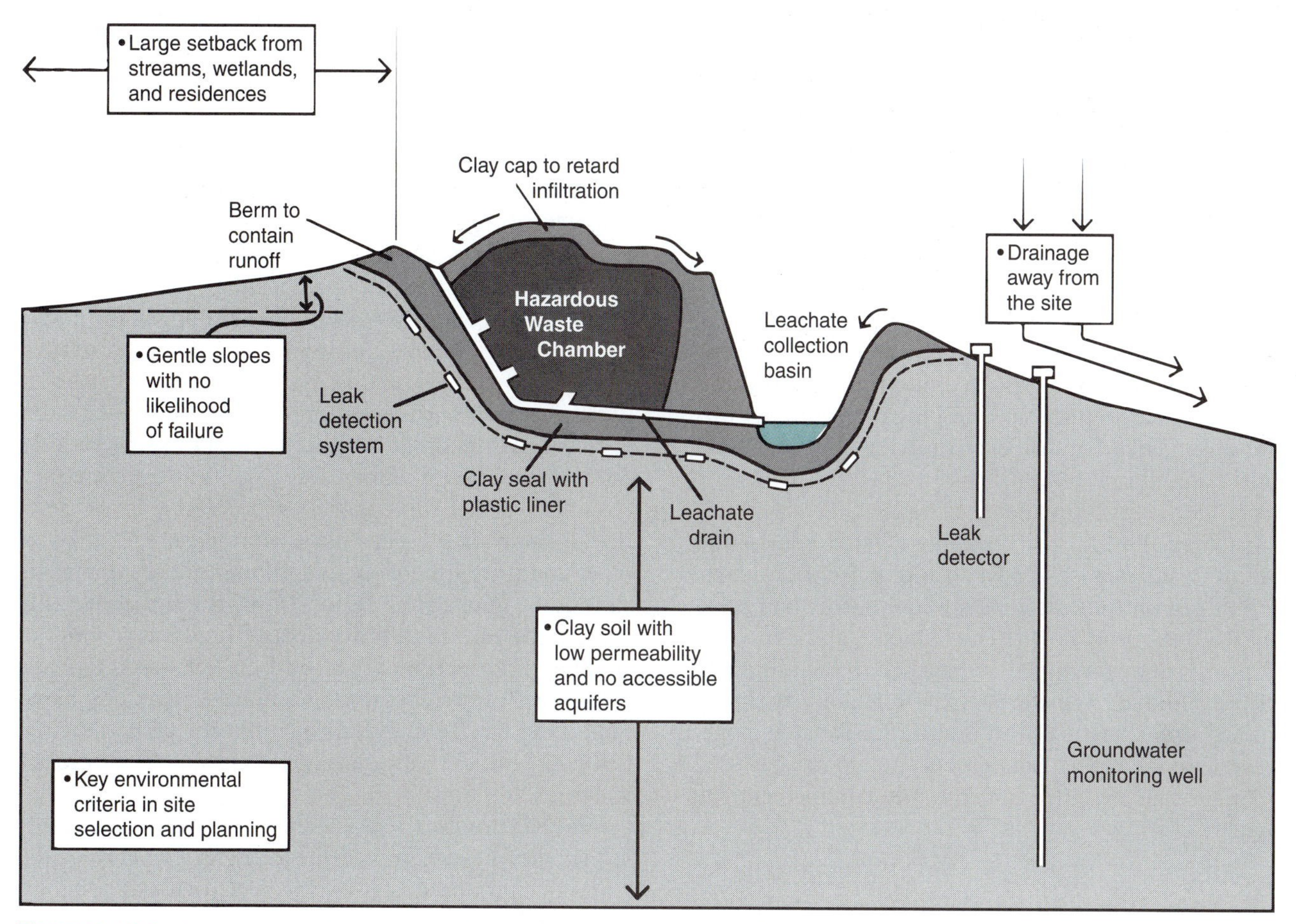

FIGURE 14.10 *Environmental criteria and design features of a secure landfill site according to current standards in the United States. Hazardous waste is sealed in a chamber.*

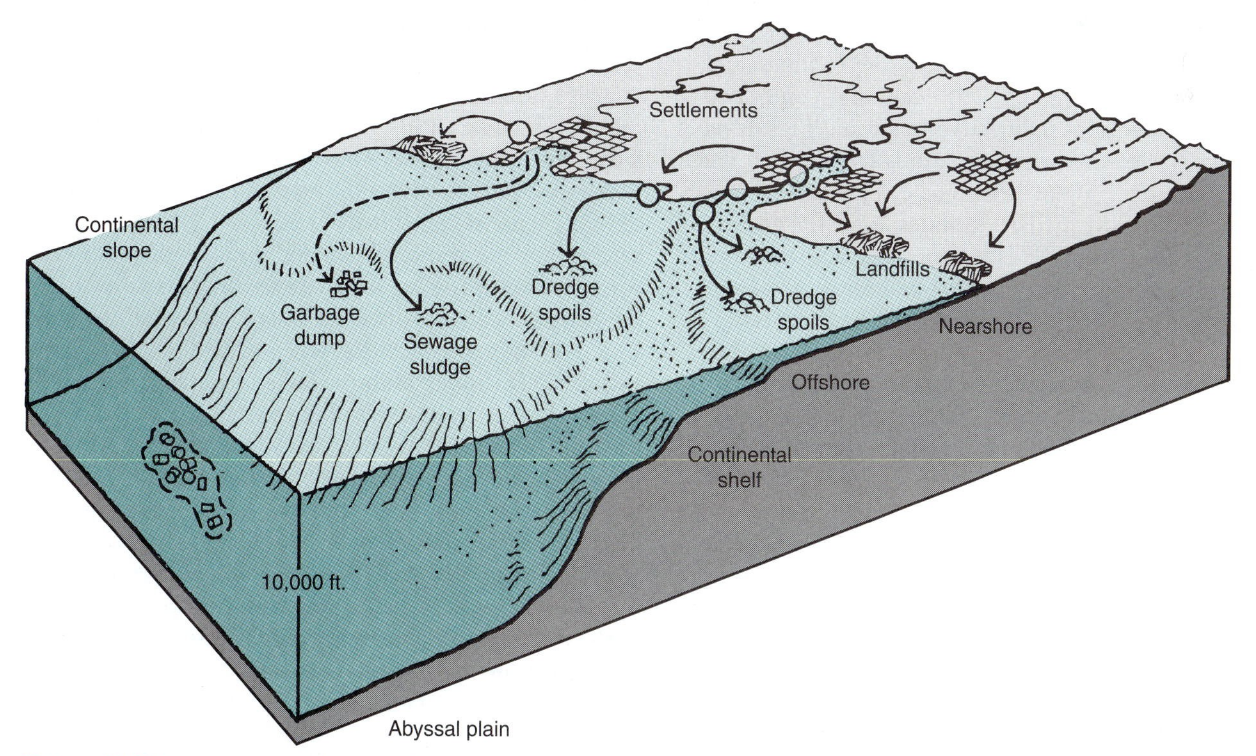

FIGURE 14.11 *A schematic diagram showing the location of dump sites for various types of wastes in coastal waters.*

Spoil materials (bottom sediments) from harbor and river dredging are the principal wastes dumped in coastal waters today. Most major coastal cities, including Tokyo, Singapore, Hong Kong, and San Francisco, have engaged extensively in such practices to improve navigation and provide additional land for growth. With the advent of huge ships such as supertankers, dredging activity has increased significantly for many port cities in recent years. Much of the material dredged up contains polluted sediments, especially in urban harbors, and where it is dumped into shallow waters (less than 100 feet deep or so) the pollutants can be remixed into the water column by wave and current action. Dumping in deep water offshore is less risky because ocean mixing is negligible at depths of thousands of feet (Figure 14.11).

For many years, containerized radioactive waste was dumped into the ocean by the United States, France, and other nuclear nations. The waste was sealed in concrete and dumped onto the deep ocean floor, more than 3,000 meters (10,000 feet) down. The United States and most European nations stopped radioactive waste dumping in 1982, although the former Soviet Union continued dumping in the Arctic Ocean until 1993. Today radioactive waste in the United States is being temporarily stored until a specially designed underground confinement site is prepared.

Radioactive Waste Disposal

Radioactive wastes are produced by a variety of land-use activities: nuclear-power generation facilities, nuclear weapons research and manufacturing installations, chemical manufacturing industries, medical care facilities, and uranium mining operations. The waste falls into two broad classes: **low-level radioactive waste** and **high-level radioactive waste**. The two designations are based on the level of radiation emitted from waste. Low-level waste, which is generated by medical facilities, industry, and research laboratories, as well as by nuclear reactors, is not considered a serious environmental or health threat unless it is improperly managed. Disposal is usually accomplished by burial. In the United States, low-level waste has been buried at 15 principal sites around the country. Some of these sites are questionable because of their proximity to groundwater and the possibility of radioactive leachate seeping into shallow aquifers, but most are considered environmentally safe at this time. However, the potential of environmental contamination may persist for several centuries; therefore, monitoring must be conducted over the very long term (Figure 14.12).

High-level radioactive waste, which is produced mainly by nuclear power facilities, is a much more serious problem. Currently more than 420 nuclear power plants are operating worldwide, with about 100 more under construction. If improperly managed, high-level waste from nuclear facilities poses a serious threat to human health, as the Chernobyl accident of 1986 demonstrated. At Chernobyl more than 30 people were killed and several hundred people were immediately injured from massive exposure to radiation. Thousands more are expected to be affected over the next several decades. By the late-1990s, approximately 9,000 people died prematurely from radiation-related illnesses. Radioactive fallout from the Chernobyl accident was recorded in Sweden and Norway more than 2,000 kilometers (1,200 miles) away.

In nuclear reactors, high-level wastes are produced when fuel materials become contaminated with radioactive byproducts such as kryton 85, strontium 90, cesium 137, and plutonium 239. The spent fuel must be periodically removed and either disposed of or reprocessed (recycled). In the United States, **reprocessing** is not practiced—although the technology is available and used in other countries—therefore, hazardous radioactive materials must be shipped to a disposal site and sealed away as the waste undergoes **radioactive decay**. Hundreds of years of confinement are required for most nuclear wastes to reach safe levels; however, because plutonium 239 has a very long radioactive life, 250,000 years of confinement are required before the nuclear waste assemblages produced in reactors will reach safe levels.

Another problem associated with nuclear power plants is that they have a limited life expectancy. Many of the plants that produced high-level radioactive wastes will soon have to be decommissioned. This requires either dismantlement or entombment (encasement in concrete), both difficult and very expensive processes. Within the next decade or so, more than 200 nuclear plants will face decommissioning (also see "Nuclear Power" in Chapter 9).

In addition to the waste from nuclear power facilities, high-level waste has also been produced by nuclear **weapons installations**. These are governmental operations involved in plutonium production and weapons development. In the United States, the U.S. Department of Energy has the responsibility for disposal of radioactive waste from government installations. Records show that at the three repositories used for many years, both production and storage facilities have tended to be poorly managed. Documentation of leaks of liquid waste at the Hanford, Washington, site between 1958 and 1973, for example, prompted the Department of Energy to make improvements in its management program, including double-sealing tanks and solidifying liquid wastes. Because of the secrecy traditionally surrounding government weapons programs in the United States and other developed countries with nuclear capabilities, hazardous waste problems have probably not been fully disclosed.

In 1982 the U.S. Congress enacted the **Nuclear Waste Policy Act**, which launched a comprehensive program for the disposal of high-level nuclear waste and charged the Department of Energy with selecting the first perma-

FIGURE 14.12 *Locations of low-level radioactive waste disposal sites in the United States. Some have been shut down.*

nent safe disposal site. In 1987 a site in Nevada, the **Yucca Mountain** site, was selected as the national repository. The site has been subject to controversy because of its geology (volcanic rock), but the climate is dry and the likelihood of drainage problems is low. By the time it is completed (around the year 2010), large amounts of contaminated fuel materials from both public and private nuclear facilities are expected to be awaiting disposal at Yucca Mountain. When and if reprocessing is implemented, however, the volume will decline significantly, making disposal a less serious problem in general. Until reprocessing is implemented, the volume of disposable radioactive waste will remain relatively high.

14.7 HAZARDOUS WASTE TREATMENT

Several hazardous waste treatment methods are used today, and as a whole, they are getting favorable attention, especially those that can be accomplished on-site. Treatment at the production site avoids both the risky problem of waste transportation by truck or rail and the expensive and often controversial problem of establishing and managing secure landfills.

Land Application

Among the more widely used treatment methods is **land application**. This method involves the spreading of waste over biologically active soil. Also known as *land farming*, this method is usable only with biodegradable wastes, such as oily materials and some organic chemical substances. The key criterion is the *persistence* or **biopersistence** of the waste material; only those wastes with low persistence (that break down quickly), such as grease, oil, gasoline, and certain organic compounds, are suitable for this method of disposal. The principal agents of soil biodegradation are the microflora that naturally inhabit the soil. These are mainly bacteria, algae, fungi, that have the capacity to decompose most organic-based wastes.

Land application has some distinct advantages in waste disposal. It is based on a natural ecosystem, it is technologically simple, and it is fairly inexpensive. However, it also has several limitations, including the susceptibility of treatment sites to erosion by runoff and wind, the inability of soil microflora to process inorganic wastes such as most heavy metals, and the necessity to manage carefully application rates in order not to overload the soil system. Where soils receive large doses of nonbiodegradable materials, the soil microflora can be damaged and waste breakdown is slowed significantly. In the United States, about half of all oily wastes are treated by land application methods.

Processing and Incineration

On-site treatment can also utilize various methods of physical and chemical processing to separate waste material into less hazardous components and/or reduce their volume. For example, heavy metals can be separated from oily residues by settling or filtering, thereby making the oils more suitable for land application. For some

FIGURE 14.13 *A toxic waste incineration facility (background).*

materials the logical corollary to this process is waste **recycling**, in which marketable substances such as oil, acids, and solvents are removed and reused.

Incineration is an effective but controversial disposal method. At high-temperature burning (900°C or greater), hazardous organic wastes such as PCBs can be broken down to safe residues and much smaller volumes (Figure 14.13). The byproducts are ash, which requires land disposal, and air-borne pollutants, which require pollution-reduction devices on exhaust stacks. In addition to organic compounds, incineration is particularly effective for medical byproducts, especially if they can be treated on-site with minimal handling. Incineration is ineffective in treating heavy metals, however, which remain unchanged with burning.

14.8 Waste Reduction and Recycling

To reduce the costs and environmental impacts associated with solid and hazardous waste, broadly based, multifaceted management programs are required. The first component in any such program is **source reduction** or **minimization** in waste production. This can be achieved in many ways, including resource recovery systems in manufacturing; reduced rates of public consumption; improved packaging design; bulk shipping and sales; and reduced use of disposable products such as baby diapers, plastics, glass, and newspapers. As we move further into the era of electronic communications, the opportunities for reducing paper waste from newspapers, catalogs, and telephone books, for example, are improving. On the other hand, gross consumer consumption and waste production are increasing with population growth. Clearly more is needed than simply reducing waste production. Recycling is also needed.

Why Recycle?

Currently the opportunities for recycling favor municipal solid waste over hazardous waste for several reasons. First, handling and management procedures are simpler and less risky for solid waste. Second, less processing, especially that requiring advanced technology, is required for solid waste, and third, municipal garbage collection systems provide ready access to large volumes of solid waste in accessible local concentrations. In some developed countries as much as 30 to 40 percent of solid waste is recycled, whereas less than 5 percent of the hazardous waste is recycled in most industrial nations, according to the United Nations.

Besides the technical problems, hazardous waste recycling is also limited by inadequate communications among waste producers. In general, little information is shared by producers on substances and rates of production; therefore, little can be exchanged about common problems and opportunities. However, with improved information flow among producers, manufacturers are finding that one industry's waste is another industry's resource. Networking services are helping find compatible parties and linking them together for the exchange of material. Thus, **waste exchanges** are emerging that provide brokering services at regional, national, and international levels. The U.S. EPA estimates that it is now possible to recycle 15 to 30 percent of hazardous waste, using available technology. Currently Japan has the best recycling record for hazardous waste among the industrial nations. As the risks, costs, and regulations associated with hazardous waste disposal escalate, the incentives for recycling also rise, and most industrial countries will have to follow Japan's lead.

What are the incentives for recycling solid waste? Clearly, there is a major energy saving represented in recycled material. For paper, aluminum, glass, and steel, recycling represents an estimated energy savings of more than 90 percent compared to extracting and processing raw materials. In addition, the burden to landfills is reduced, and, very important, the environmental impacts from fugitive pollutants, including leachates in groundwater, air pollution, and plastic debris in streams and lakes, are also reduced. For aluminum, steel, and paper production, air pollution is reduced by 70 to 95 percent through recycling.

The first successful recycling efforts were carried out long before the modern environmental era, and they were motivated purely by economic opportunities. In the first half of the 20th century and before, scrap metal, paper, rags, and bottles were commonly recycled in urban areas (Figure 14.14a). Although little factual information is available, these entrepreneurial activities appear to have waxed and waned over the years with the national economic climate. During hard times recycling increased, both at the market and personal levels, and during good times it tended to decline. With the spreading affluence of the late 20th century, small-scale recycling—with the possible exception of used auto parts—declined and manufacturers marketed increasing amounts of lavishly-packaged throwaway products.

Modern Recycling Efforts

The original incentive for recycling programs in the United States was driven not by economics or overflowing landfills, but by public revulsion to the sight of debris along roads, in waterways, and on beaches. For many of us it is hard to imagine North American highways and country roads in the 1960s and 1970s lined with beer bottles, soda cans, plastic containers, and assorted paper debris. These revolting scenes were instrumental in mounting public pressure to pass **container recycling laws**, or *bottle bills.*

The first bottle bill was passed by Oregon in 1971, and over the next two decades, a number of other states and Canadian provinces passed similar ordinances. These laws require a cash deposit on certain drink containers that can be reclaimed at a store or recycling center. Despite complaints by drink manufacturers, distributors, and stores, most states claim marked success with container recycling. For example, in New York State, where 80 percent of the drink containers are recycled, solid waste has been reduced by 8 percent, the landscape is much cleaner, and less has to be spent in roadway cleanup programs. Ontario's bottle

(a)

(b)

(c)

FIGURE 14.14 *Photographs of (a) old-time recycling; garbage being sorted and glass, rags, metal, and other materials selected for recycling (circa 1900); and (b) waste separation in a modern residential neighborhood; and (c) plastic being sorted at a modern collection center, Monterey County, California.*

program is recognized as the most successful in North America, with greater than a 90 percent recovery rate for alcoholic beverage containers.

Most other recycling efforts at the community level focus on paper, glass, plastic, and aluminum. The principal requirement for success in such programs is development of an efficient and inexpensive collection and processing system. The first step is **waste separation**—that is, segregating recyclable materials from other wastes and from each other. The most efficient way of doing this is at the source of the waste, at homes, businesses, schools, and factories (Figure 14.14b). Programs that rely on separation of mixed or comingled waste at a central processing installation are generally more expensive.

The next step is to gather the material at a collection center and prepare it for transportation to reprocessing installations. Transportation is often expensive because reprocessing installations may be located great distances from many communities, especially small communities. In addition, some materials may require handling and preparation for shipment, such as bailing of waste paper. In any case, handling and shipment significantly add to costs, so it is extremely important that recycling programs are well planned and carefully managed (Figure 14.14c).

How successful is recycling in the United States? Aluminum recycling has shown the greatest success: about 50 percent of all aluminum cans are now recycled. **Paper recycling** is also showing success. Waste paper is the major material in landfills, and the bulk of it comes from newspaper. About 33 percent of newsprint is now recycled, most of which is mixed with virgin fiber and reprocessed into cardboard, home insulation, and similar products. On the average, new cars now contain about 60 pounds of recycled paper, which is used mainly in interiors. About 50 percent of cardboard (corrugated) boxes are recycled, mostly from supermarkets and other large stores. Most recycled bottles are used to make new bottles. According to U.S. bottle manufacturers, about 25 percent of new bottles are made up of recycled glass. This represents an energy savings (because the recycled glass can be fired at lower temperatures), as well as a reduction in air pollution and solid waste.

Plastics are recycled, but here the success story is less encouraging. The amount of waste plastic is increasing and is expected to make up about 10 percent of the municipal waste load in this decade. Of the many types of plastics manufactured, only a few types are actually recycled in the United States. High-density plastics and drink containers show the best potential, and today about 20 percent of this material is recycled. Most goes into new products such as carpet backing, paint, and floor tiles. Most other plastics are not recycled, and most end up in landfills—including many plastic items collected by households for recycling. A major obstacle in the final disposition of American plastics is that they are not biodegradable within a reasonable time frame. Although biodegradable plastics are successfully used in Europe, they are currently not widely used by U.S. manufacturers. Most plastics manufactured in the United States take 200 years or more to decompose.

Recycling the Automobile

Throughout this book we have encountered the automobile at or near the heart of many serious environmental problems, including air pollution, water pollution, energy consumption, and urban sprawl. How does the automobile rank in terms of waste recycling? Huge amounts of steel, iron, aluminum, plastic, glass, and other materials are committed to vehicles. Typically 2,000 pounds or more of metal and as much as 200

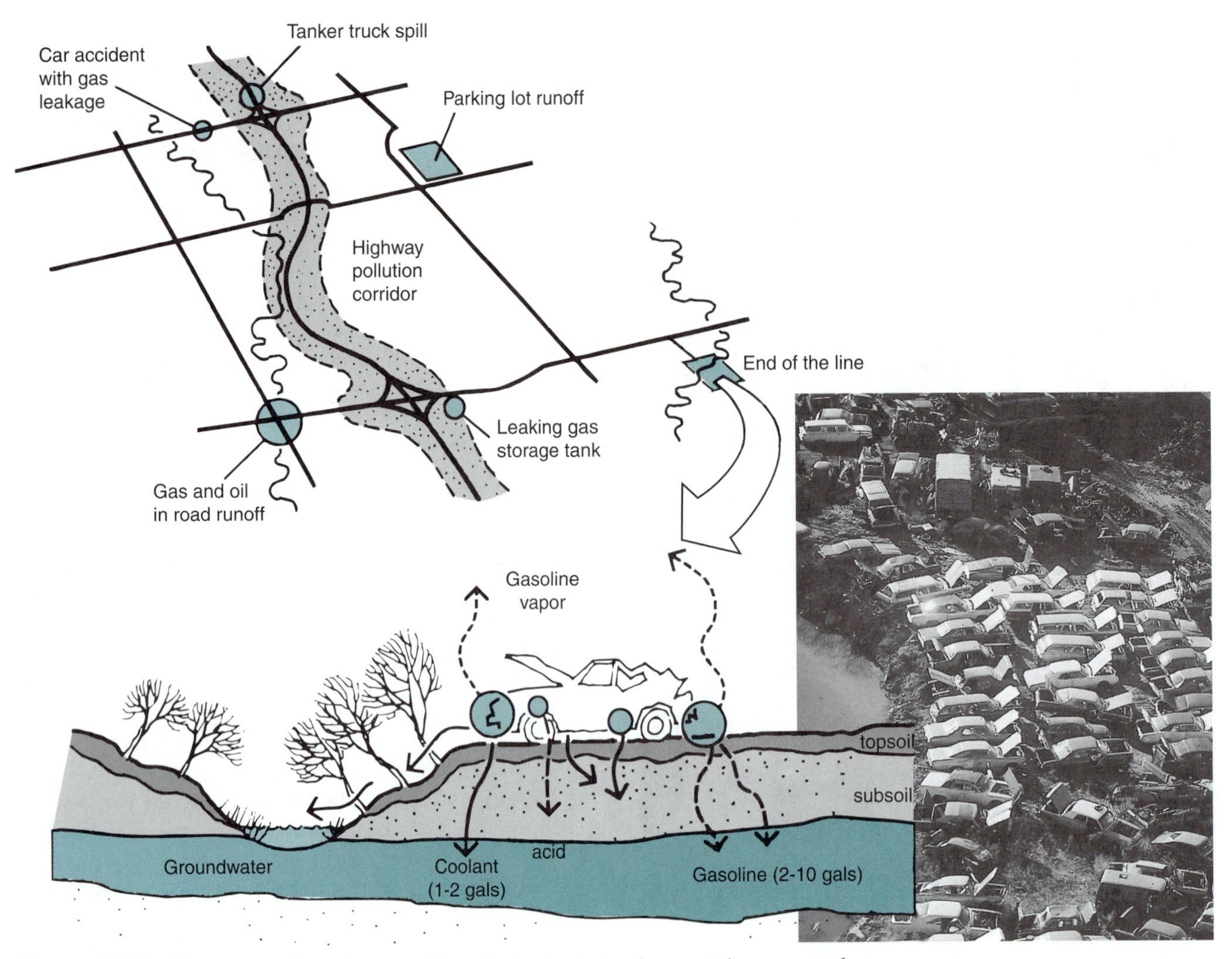

FIGURE 14.15 *The various sites of automobile pollution in the landscape. Below, wastes that emanate from a junked car. Whether in junk yards or in the back forty, these wastes are substantial and can be a serious threat to water quality.*

pounds of plastic are used in passenger cars. In the United States, about 9 million cars and 2 million other vehicles (trucks, buses, etc.) are retired (junked) every year, resulting in about 12 million metric tons (13 million tons) of waste. How much is recycled?

Most of the metal is recycled either as used parts or melted down and reprocessed. This is a well-established practice in the United States, but it is by no means universal. In particular, it is less common in rural areas and small communities because of the cost of hauling to processing centers located in distant urban areas and the absence of local ordinances governing the disposal of junked vehicles. As a result, hundreds of thousands of vehicle carcasses are abandoned in yards, in fields, and in dumps. There they slowly disintegrate, releasing oil, grease, gasoline, cooling fluid, and various other contaminants to the ground (Figure 14.15).

Although most of the metal from junked vehicles is recycled, a sizable amount of other material such as plastic, glass, and cardboard is not. Most of this material must be separated from the metal before recycling. In the United States, most of the plastic (in bumpers, dashboards, moldings, and many other parts) is not recyclable and must be disposed of in landfills. This is needlessly wasteful, of course, and several American automotive manufacturers are now developing systems using recyclable plastics. European manufacturers, on the other hand, are well into plastic recycling and are now mandated by law to achieve 80 percent recycling efficiency.

Tires are also a major automobile waste management problem. With nearly 200 million vehicles on U.S. roads, tens of millions of tires are discarded each year. In the Los Angeles region, where there are 8 million vehicles, 250 tons of tires are discarded every day. Some are retreaded, some are ground up and used as asphalt and playground mulch, but most are discarded to landfills, stored in empty lots, or used as fuel for power generation. The latter option produces serious air pollution unless proper emission-control devices are used by power plants. Other options for recycling old tires include use as an additive in asphalt and in erosion control. Placement of tires in gullies and along shorelines to reduce erosion is a questionable practice and often causes more problems than it solves.

14.9 SUMMARY

The trends toward consumer economies, population growth, and technological advances in material development and manufacturing indicate conclusively that virtually every nation must now or will shortly have to wrestle with serious waste management problems. There are three factors driving the problem:

- **Cost.** The cost of managing waste includes handling, transportation, disposal, and treatment.
- **Economic loss.** This is represented by the resources (energy, raw materials, and labor) spent in the substances discarded.
- **Impacts.** Waste affects both the environment and human health.

As developing nations make the transition to consumer economies, their level of waste production rises with economic development. Old methods of disposal, such as dumping and unregulated landfilling, are no longer acceptable, especially for hazardous wastes. Modern methods of safe disposal, on-site treatment, reprocessing, and recycling are necessary. As the developed countries have demonstrated, there is a need to make vast improvements in waste management at all levels. The need for these improvements is critical for both hazardous waste, where progress in reduction and recycling has been modest at best, and solid waste, where recycling programs have been more successful at the municipal level.

But as the Eastern European countries illustrated under communism, there are many obstacles to the development of safe management programs. Authoritarian governments, for example, generally show an unwillingness to manage hazardous wastes in environmentally safe ways. On the other hand, free-market economies without the guiding hand of governmental regulation do not seem to do so well either, as the legacy of hazardous waste problems from the 1950s, 1960s, and 1970s in the United States, Canada, and Europe illustrates. The developing countries face the entire range of obstacles in managing hazardous waste. For some it is unregulated market economies, for others it is the lack of technical and management capabilities and/or irresponsible central governments. For many it is national agendas that give environmental issues low priority despite political lip service in the media.

In the last several decades another facet of the problem has unfolded as some poor countries have begun to make a business of accepting other nations' wastes. Under a recent United Nations agreement addressing this problem, waste can be exported only to countries that have signed a contract to accept it and only if it has been properly packaged and handled. Further, waste cannot be accepted by nations that do not have environmentally appropriate means of disposal. The agreement is necessary and commendable, but monitoring and enforcement will be difficult and not all nations have subscribed to the agreement.

Integrated waste management programs involving reduction, recycling, and reprocessing are the ways of the future. The energy costs of extracting raw materials and manufacturing throwaway goods currently exceed the costs of material recovery and reuse for many resources. In addition, there is the cost of pollution from discarded waste, as well as the basic issue of society's values about the scenic quality of the landscape and the morality of excessive resource consumption and the loss of natural resources.

Recycling is firmly established in most developed countries, especially in European countries, where it is widely mandated. Many states, provinces, counties, communities, industries, and institutions in the United States and Canada have established recycling programs. Their successes are serving as models for other organizations and communities. Most community recycling programs in North America are marginally profitable as businesses, but that will change as the cost of conventional waste collection and disposal rise in the next decade or two. In 1990 the U.S. Environmental Protection Agency (EPA) estimated that 73 percent of municipal solid waste is buried in landfills, 14 percent is burned, and 13 percent is recycled. EPA expects these numbers to improve, driven by a combination of factors, including the changing economics of waste management, public pressure, public image considerations, liability and insurance costs, and new and stiffer environmental laws. In the latter area, additional regulations are needed to push manufacturers to produce more environmentally compatible products such as biodegradable plastics and to push consumers to accept greater responsibility for waste production and management.

Integrated waste management involves multiple operations linked together in comprehensive programs. Recycling is central to integrated waste management, as are reduced consumption of expendable goods and new means of packaging, transportation, advertising, and sales. Understandably not all wastes are recyclable, and therefore disposal practices will remain with us for a long time. But the amounts of waste materials, land requirements, environmental impacts, and health risks should be reduced significantly for most developed countries as they achieve stable or near stable populations and stable rates of material consumption. On the other hand, for many developing countries with rapidly growing populations and rising consumerism, integrated waste management will lag behind and unchecked dumping will continue for decades.

Finally, it is necessary to consider the costs of waste management, especially hazardous waste, relative to the health benefits. The argument can be made that, compared to other health risks such as poor diet, smoking, and air pollution, hazardous waste sites are relatively minor. For example, it is estimated that less than 5 percent of all cancers can be related to living near hazardous waste sites, whereas as much as 30 percent can be related to smoking. Should the huge expenditures for the Superfund program be devoted to this relatively small problem, or should these funds be directed elsewhere—for example, to antismoking and air pollution control programs?

Three qualifying observations are noteworthy: (1) planning and management programs (and their budgets) are a response to public policy that is driven by public attitudes and politics. Neither is necessarily rational, but they do represent the will of society, which is critical in the planning process; (2) the true impacts of buried wastes may not be felt for decades because of the slowness of the soil and groundwater systems they affect. For example, groundwater recharge and exchange rates commonly require several decades to hundreds of years. Thus, contaminants released into the ground in this century may not appear in the water supply system in significant quantities until the next century; and (3) the full extent of the effects of hazardous wastes may not be known until their combined effects, with the influences of synergism, are assessed together, especially in combination with contaminants from other sources such as smoking and air pollution.

14.10 KEY TERMS AND CONCEPTS

Disposable waste
- classes of waste
- solid waste and landfills
- waste composition
- leachate

Hazardous waste
- toxic materials
- principal sources
- distribution

Waste disposal
- historical trends and practices
- midnight dumping
- open lagoons
- surface storage
- burial
- communist bloc practices

Some consequences
- Elizabeth, New Jersey
- Love Canal, New York
- Bhopal, India

Health effects
- acute toxicity
- chronic toxicity
- carcinogens
- risk assessment

Waste regulation
- Resource Conservation and Recovery Act
- Superfund
- polluter-pays laws

Modern disposal methods
- disposal vs. treatment
- secure landfill
- surface impoundment
- deep-well injection
- ocean dumping

Radioactive waste
- low-level waste
- high-level waste
- reprocessing
- radioactive decay
- weapons installation
- Nuclear Waste Policy Act
- Yucca Mountain

Hazardous waste treatment
- land application
- biopersistence
- recycling
- incineration

Waste reduction
- source reduction, minimization

Recycling
- traditional activities
- waste exchanges
- container recycling laws
- waste separation
- paper recycling
- automobiles

Integrated waste management

14.11 QUESTIONS FOR REVIEW

1. Name the three main types of waste requiring disposal. What is solid waste? In the United States what rural and urban land-use activities are responsible for most solid waste? What are the main ingredients in a U.S. landfill?
2. What countries have the highest waste production rates in the world? Why are rates in developing countries so much lower than in developed countries? Does this mean that waste disposal is not a problem in developing countries?
3. What is the difference between a dump and a landfill, and why is there concern about environmental impacts from such disposal sites?
4. What is hazardous waste? What industries are most responsible for hazardous waste production? What are the three main paths in the hazardous waste stream? Name the six main classes of hazardous waste.
5. Outline the relationship between urban expansion, waste disposal practices, and land-use change associated with an industrial city. How do you characterize the legacy of past disposal practices? How could these sorts of practices have occurred in countries such as the United States? What circumstances encouraged irresponsible waste management in the former communist countries of central and Eastern Europe and the former Soviet Union?
6. What are acute toxicity and chronic toxicity? Which appears to apply to the Love Canal, New York, and Bhopal, India, incidents? Name the four main classes of health problems associated with hazardous substances. What is the basis for the health risk statement provided by the U.S. government on various substances?
7. What are the two main regulatory policies for hazardous waste in the United States, and what are their targets? What is the main difficulty with the Superfund program?
8. Distinguish between disposal and treatment. Briefly describe the main disposal methods for hazardous waste. Expand on the situation for nuclear waste.
9. What is the favored trend in the siting (location) of hazardous waste treatment facilities? Outline the chief features of the three main treatment methods.
10. What are the benefits of waste recycling? What were the incentives for recycling before the modern era? What were the first modern recycling programs in the United States, and what was the motivation behind them?
11. Describe the main components of a waste recycling operation. What sorts of wastes are recycled, and how successful are recycling programs in the United States? Is there room for improvement in solid waste recycling? What about hazardous waste?
12. What are the destinies of junked automobiles in North America? Is recycling universally practiced in urban and rural areas? Which materials in cars are actually recycled and which are not?

15

SOIL, LAND, AND LAND USE

15.1 INTRODUCTION

The total area of land on Earth is less than half that of the oceans, about 30 percent of the planet's surface. This is an important statistic because in the extreme this area represents the habitable limits of the planet for terrestrial organisms. But we must also recognize that the Earth's land surface is changing, both in area and habitat quality. In the past 20,000 years, for example, total land area has changed significantly as sea level rose and fell in response to changes in the volume of glacial ice masses on the land. Early populations of humans living in coastal areas shifted with these changes.

As a consequence of global warming over the next fifty years or so, sea level might rise 2 to 3 meters. If so, hundreds of millions of people who now live in low-lying coastal lands could be displaced, and valuable agricultural lands on the fertile river deltas and coastal plains could be inundated. For example, Bangladesh, a crowded lowland nation on the Indian Ocean, would lose about 28 percent of its land area, including much of its productive farmland.

Change in land area is only one of our concerns about the habitability of Earth. Perhaps more important is the capacity of terrestrial environments to support life. Much of the Earth is limited by harsh climates. The dry zones and cold zones take up more than 60 percent of the available land. The area of dry land is expanding because of desertification brought on by misuse of the desert margins and semiarid lands. In addition, a small but significant area is limited by rugged mountain terrain.

The remaining one-third of the Earth's surface has landforms, climates, and soils that are most suitable to human occupancy. These are the lands where most humans live, so they suffer the heaviest environmental impact. Our objective in this chapter is to examine the land and soil environment. We are especially interested in the formation of soil and how it is used and misused. The loss of soil as a result of land use is one of the most serious problems facing human survival and the quality of life on the heavily populated land masses.

15.2 Geographic Organization of the Land

Running water is the principal natural agent shaping the land surface. Most of the landforms we see around us, no matter where we live, are directly or indirectly the work of running water. The most apparent examples are the valleys and streams that are everywhere in the landscape, even in arid lands. With very few exceptions, the valleys that streams flow in have been carved by the streams themselves. It follows that there is a relationship between the size of valleys and the size of streams—large streams flow in large valleys, for example.

Streams and their valleys are linked together in large networks called **drainage nets**, which transfer water from higher ground in the continental interiors to lower ground on the continental margins and hence to the oceans. Most drainage nets have patterns similar to the branching pattern in a tree, with small branches joining to form larger branches and these, in turn, joining to form yet larger ones (Figure 15.1). The largest drainage nets, such as that of the Mississippi River, cover millions of square miles. This area is the **watershed**, or **drainage basin**, and it contributes water to the system of streams that make up the drainage net and feed the trunk streams.

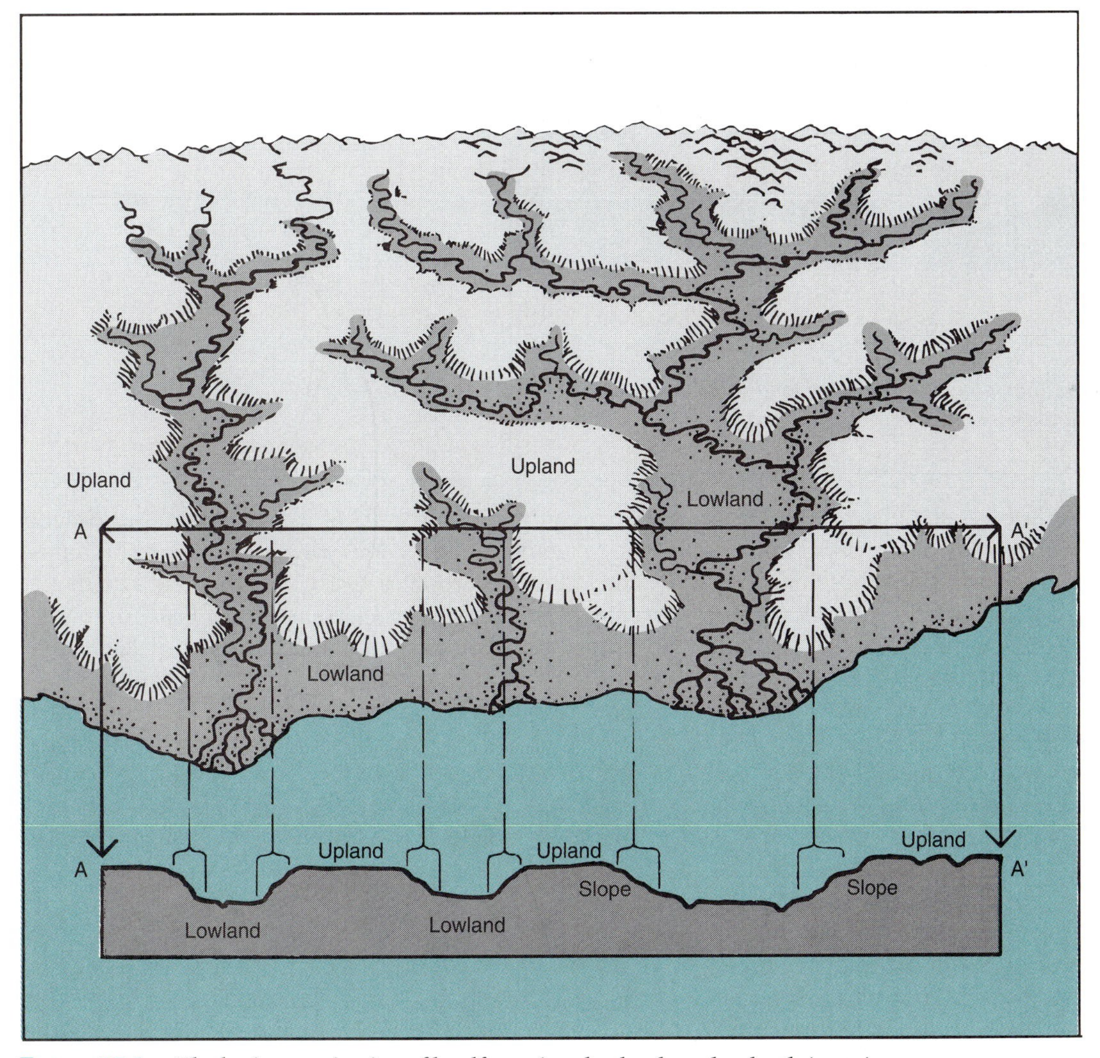

Figure 15.1 *The basic organization of landforms into lowlands and uplands in regions dominated by stream systems. These two classes make up most land masses. Slopes, which separate uplands and lowlands, are a third class of land.*

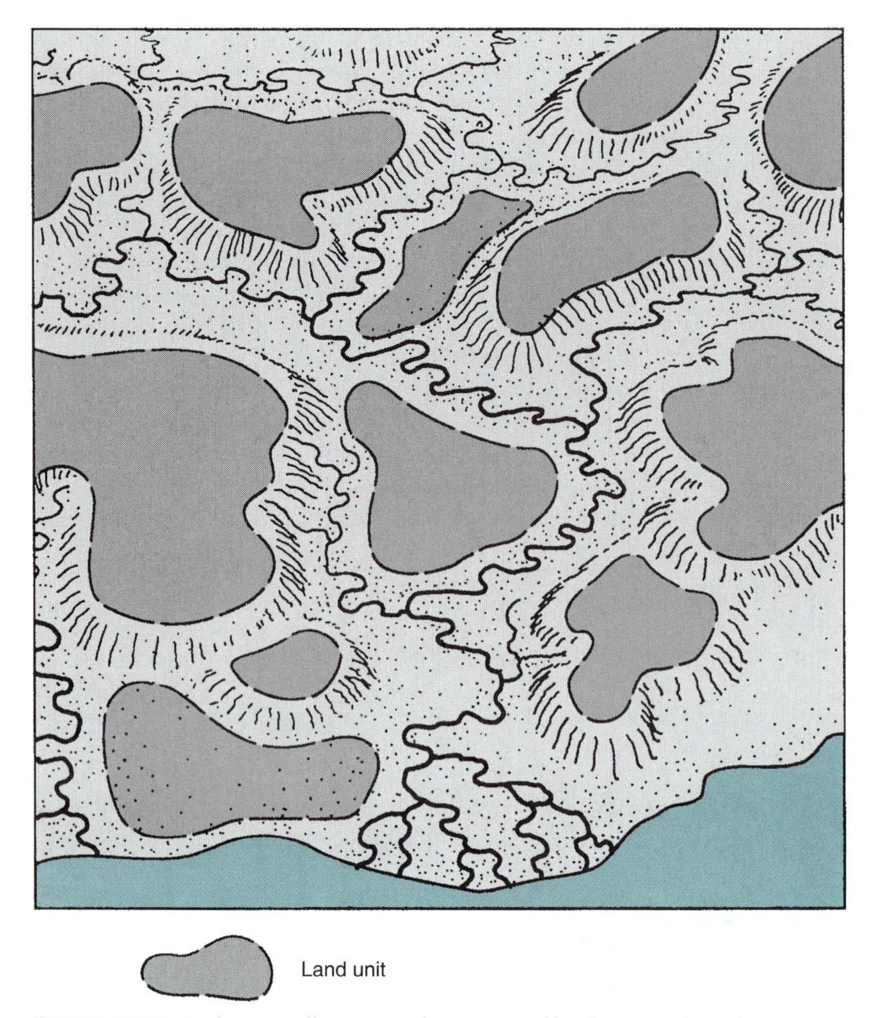

FIGURE 15.2 *A diagram illustrating the concept of land units, relatively homogeneous parcels within uplands and lowlands with the potential to support land use.*

Between the stream valleys lies elevated ground, called **upland**. This land is different in many ways from the lowlands of the stream valleys. Among other things, it is drier and not subject to flooding, and the soils may be thinner and less diverse. In the **lowlands**, on the other hand, the soils are made up of diverse streams deposits, moisture is ample, and flooding is common. Uplands, and lowlands, along with a third land-type represented by the hills and slopes separating them, dominate most of the continents at scales ranging from the local to the regional landscape (Figure 15.1).

Within uplands and lowlands, it is possible to define geographic parcels, or **land units**, of various sizes. These are relatively homogeneous areas that lie on upland surfaces or within large stream valleys. Each land unit has a capacity to support various land uses based on its size, soil characteristics, water supplies, topography, location, and other factors. In the lowlands of a stream valley, for instance, the land units are defined by the position of the trunk stream in the valley, the curvature of the valley, and the location of tributary streams. Large parcels that are well drained and accessible are attractive to commercial agriculture, whereas small, less accessible strips of land or poorly drained parcels are unsuitable for mechanized agriculture, and in countries such as the United States and Canada, these are often left in forest cover (Figure 15.2). In areas of rugged terrian where traditional, nonmechanized agriculture is practiced, land units are articulated at a much smaller scale—commonly less than an acre per field among the potato farmers of the Peruvian highlands, for example.

15.3 TOPOGRAPHY AND LAND USE

Topography is the main determinant of the land's capacity to support land use, especially uses involving facilities (roads or houses, for example) and

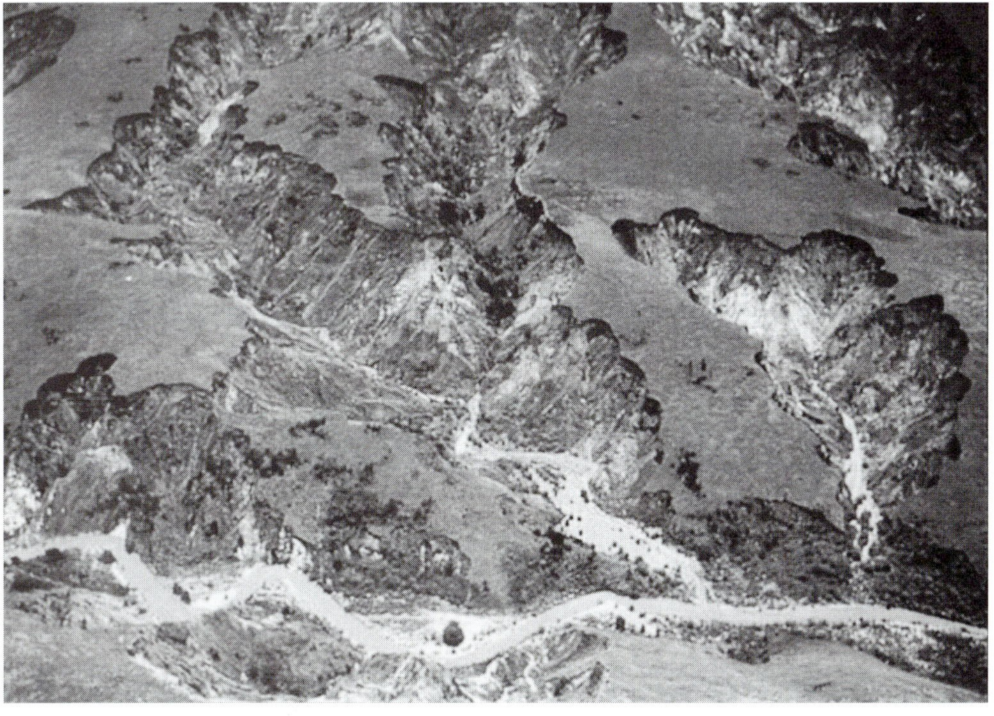

(a)

(b)

FIGURE 15.3 *Examples of slope failure and erosion related to land use: (a) slopes cleared for agriculture in Madagascar; (b) urban development on steep slopes in Los Angeles.*

crop farming. Construction and maintenance costs for buildings and transportation systems are far greater in rugged terrain than on flatter land, and mechanized agriculture is limited to land with gentle slopes. When land use spreads out of this topographic framework and onto **hillslopes** and mountain sides, environmental degradation often results. Only in selected places, such as in the ancient landscapes of south China, has agriculture been successful on steep slopes. With massive and skillful effort, the Chinese have transformed hillslopes into sequences of terraces in order to increase land available for rice farming. The success of such terrace systems depends on careful management by farmers. But hillslopes are tricky environments, and even after centuries of practice, failure of terraces is not uncommon.

Why are rugged topography and steep slopes limiting to land uses such as houses, roads, and most modern agriculture? The reasons are related to a number of factors, including the actual stability of the ground itself. **Slope failure** such as landslides and heavy erosion is likely on steep slopes where the vegetation has been removed and the ground has been disturbed by construction or plowing. Even in dry lands, slopes are prone to erosion and rapid degradation under heavy grazing or plowing. In addition, it is very difficult for equipment, people, and animals to maneuver safely on steep slopes. For these and other reasons, including greater susceptibility to earthquake damage, most land uses do much better on flatter ground with less damage to the environment (Figure 15.3).

In many parts of the world the rational relationship between land use and terrain is breaking down. In the developing countries the principal cause of this change is population pressure and the need to use hillslopes for agriculture. This is vividly illustrated in Madagascar, a mountainous, tropical country off the southeast coast of Africa. Most of Madagascar's arable land on valley floors and coastal plains has been taken by farms and settlements. In the past several decades population has increased substantially, driving farmers onto the precarious mountain slopes. Once deforested, the slopes are subject to high rates of runoff, accelerated soil erosion, and massive slope failures (Figure 15.3a). Not only are forests and farmland lost, but streams in the valleys are loaded with excessive runoff and sediment, resulting in damage to traditional farmland on the valley floors. The same sort of trend has been suggested regarding the mountainous country of Nepal; however, there the story is not so clear nor conclusive, especially beyond the fringes of major settlements.

In developed countries, hillslopes have also been subject to misuse, although the causes often extend beyond forest clearing and agriculture. In the United States and other affluent developed countries urban development has spread onto steep slopes. Much of the pressure for hillslope development has come from elevated real estate values on the urban fringe and from the desire of homeowners to live on steep, remote terrain because of the views and environmental amenities it affords. Nowhere is this more evident than in cities such as Los Angeles and Hong Kong where the environmental consequences include slope failure and mudflows during wet periods (Figure 15.3b).

Although the failure of hillslopes occasionally extends tens of feet below ground, most failure and erosion occurs in the surface layer and affects only soil. (Faults and earthquakes, on the other hand, which often cause slope failure, occur in the bedrock well below the soil layer.) When soil is lost from the land, the capacity of the land to support life, including agriculture, is greatly diminished. Throughout the world soil erosion is a persistent and serious problem. Year by year it reduces Earth's carrying capacity, driving up the costs of food production and adding greatly to water pollution in streams, lakes, wetlands, and the sea itself. The problem has reached crisis levels in some developing countries, but it is also serious among many developed countries, including the United States, where soil loss is a gradual but widespread problem on agricultural lands. In short, the capacity of nature to build and sustain soil is being exceeded over large areas of the world because of land-use practices. This raises the question of what we need to do to bring about a sustainable balance in the world's soil resources.

15.4 The Soil Mantle: Sources of Plant Material

Soil is the mixed assemblage of materials—particles of minerals and organic matter, water, and air—that blankets the land. It is the material that plants sink their roots into and from which they draw water and nutrients. The life-supporting soil, called the **solum**, is a relatively thin surface layer, less than 6 feet deep in most places. Beneath it lies the subsoil, the inorganic part of the soil that ranges from several feet to several hundred feet deep and rests on the bedrock. Together the solum and the subsoil comprise the **soil mantle**. The soil mantle forms nearly a continuous cover on the land and is especially deep in the lowlands of stream valleys and coastal areas. In mountainous upland areas, on the other hand, the soil mantle is thin, and in some places, absent altogether (Figure 15.4).

The soil mantle is made up principally of mineral particles. These particles are of different compositions and

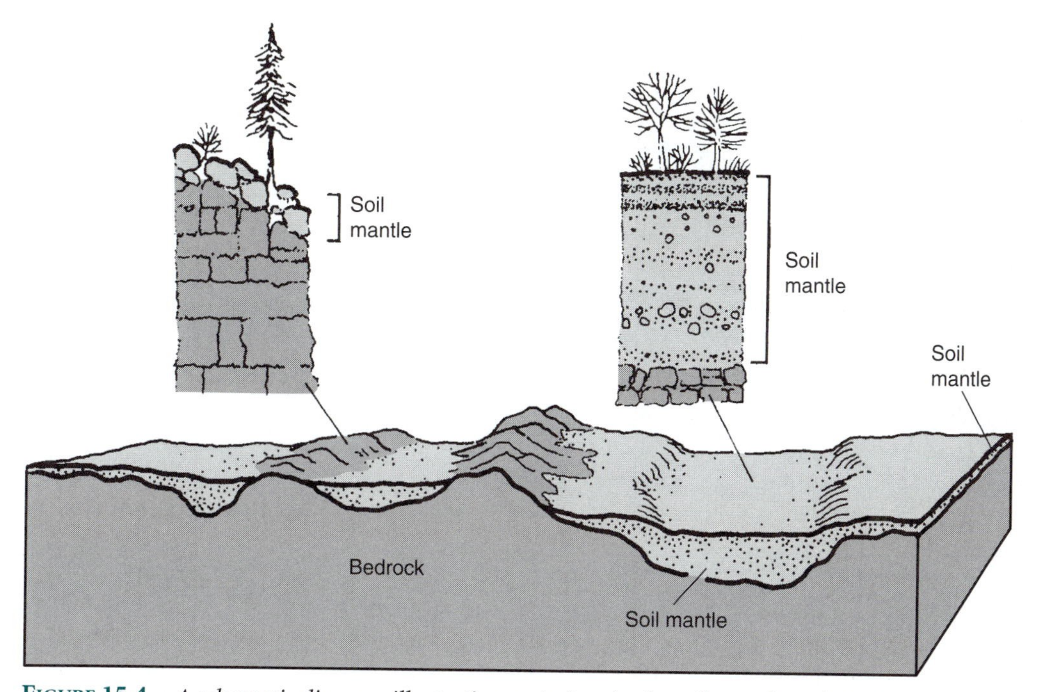

FIGURE 15.4 *A schematic diagram illustrating variations in the soil mantle with landforms.*

sizes, ranging from the microscopic to large boulders. In most places the different-sized particles are mixed together to form a relatively stable body, or *soil skeleton*, capable of supporting the overlying soil mass as well as surface objects such as forests, water bodies, roads, and buildings. Within the soil skeleton, air and water occupy the empty spaces between the particles, and near the surface in the solum, the mineral particles are mixed with organic matter.

If we could trace soil particles to their origins, we would find that most came from bedrock, the solid rock underlying all land. When bedrock is exposed to conditions near the Earth's surface, it undergoes *weathering*. Weathering produces physical and chemical breakdown of rock into smaller and smaller fragments. In general, weathering is dependent on heat and moisture; therefore, it varies with climate. In moist climates, for example, the more soluble minerals, such as salts, are dissolved and carried away in runoff, leaving behind the less soluble minerals, such as quartz.

The remains of these undissolved minerals form residues, of particles—fragments of various sizes and shapes—over the bedrock surface. As the bedrock is lowered by weathering, the layer of particles may grow thicker. Soil material formed in this manner is termed **residual parent material**. In order for residual material to form, however, particles freed by weathering must not be carried off by erosional processes such as running water and wind.

In most places, weathered particles are indeed eroded and transported away from their place of origin. Most are moved by running water, but wind, glaciers, waves, and currents also move particles. In any case, this material is eventually deposited somewhere on land or in the oceans. Those deposited on land are called **surface deposits**. Examination of the soil mantle in most places will usually reveal several different types of deposits representing different episodes of deposition associated with different processes. Soil material formed in this manner is termed **transported parent material**, and it is the principal material of the soil on both uplands and lowlands.

Looking at large geographic regions, we find that the coverage of certain surface deposits is extremely vast. Figure 15.5 shows three examples for the United States and southern Canada. **Alluvium**, which is river-deposited sediment, extends along the floor of every stream valley regardless of size. Most are too narrow to show up in the map in Figure 15.5 but there are thousands of alluvial-floored valleys in the area shown. The largest single area is the Mississippi River Valley south of Illinois, where a great belt of alluvium stretches to the Gulf of Mexico. **Glacial drift**, the material deposited by glaciers or by the meltwater draining from a glacier, covers a huge region from the Missouri and Ohio rivers northward to Hudson Bay and beyond. **Loess**, which is wind-deposited silt, covers large tracts of the prairie states, the Great Plains, and the lower Mississippi Valley.

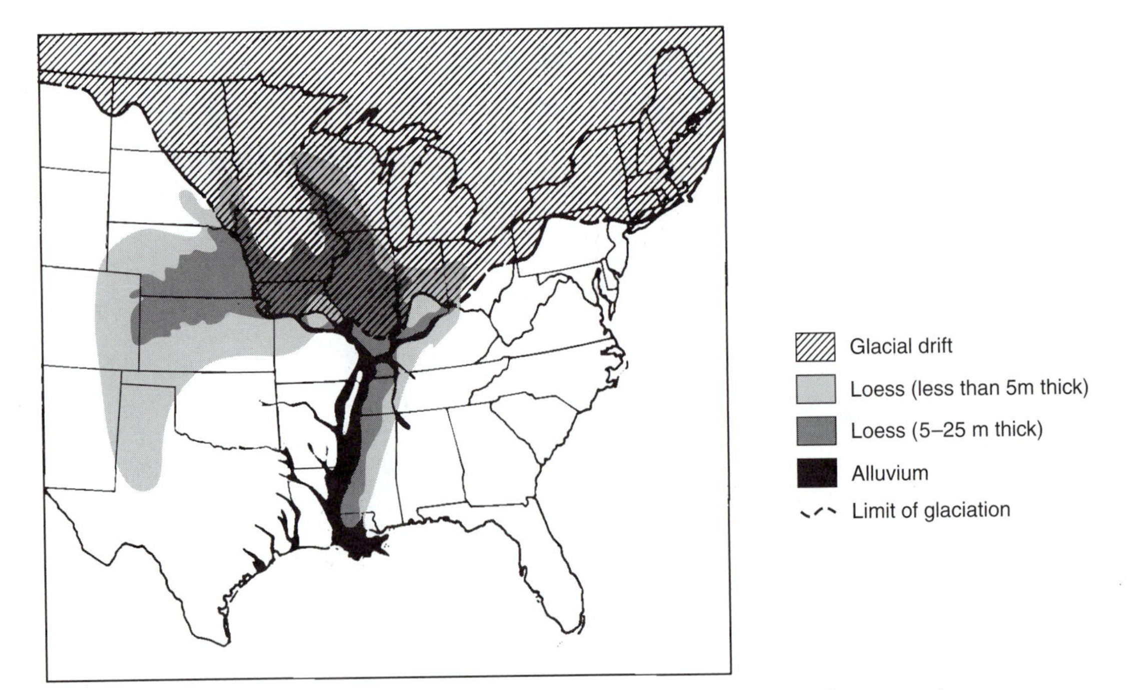

FIGURE 15.5 *Major areas of surface deposits in the central and eastern United States and southern Canada: alluvium, glacial drift, and loess.*

15.5 SOIL PROPERTIES: KEY SOIL TRAITS AND COMPONENTS

In order to understand the varied nature of soils, we must first identify some of their important traits. These traits, called *soil properties*, include composition, texture, and soil water.

Composition

Composition is generally classed as either mineral or organic. Mineral composition is represented by inorganic substances such as particles of quartz and calcium, whereas organic composition is represented by largely dead plant tissue in various states of decomposition. In environments where decomposition is slow, such as swamps and marshes, soil may consist of massive accumulations of organic matter; however, most soils contain only about 3 to 5 percent organic matter. Most is concentrated in a 10- to 50-centimeter-deep surface layer called **topsoil**.

Organic matter has many important roles in the soil. It is a source of plant nutrients and other chemicals that are released from plant tissue as it decomposes. Organic matter has a high capacity for moisture absorption and retention of dissolved chemicals important to soil fertility. Soil organic matter is part of a very active **organic energy system** and, under most conditions at the soil surface, is subject to relatively rapid decomposition and replacement.

Raw organic matter is supplied by the plant cover. This material is decomposed by microflora such as algae and bacteria, eaten by worms and other consumers, and lost to runoff and water percolating through the soil (Figure 15.6). Because heat and moisture speed the decomposition process, the wet tropics have the highest rates of breakdown of any major bioclimatic zone. The wet tropics also have the highest rates of plant productivity, but because organic matter breaks down so rapidly, topsoil is very thin, often only a few centimeters deep. In contrast, semiarid grasslands have much deeper topsoil because the dry climate limits the rate of organic matter breakdown.

When the plant cover is reduced or eliminated by land use, the soil's organic system is thrown out of balance. In the wet tropics, destruction of rainforest curtails the input of organic matter to the topsoil. In a few years, the topsoil itself may disappear and soil fertility declines, thereby breaking the critical system of nutrient recycling that supported the original forests. In the grasslands, agriculture breaks the balance of the organic system by eliminating native grasses and promoting topsoil loss to erosion by runoff and wind. With reduced inputs and increased outputs, the soil's reserve of organic matter dwindles, rendering it less fertile and less productive.

Texture

Average diameter is used to classify the size of individual soil particles and three main size classes of particles make up most soil: sand, silt, and clay (Figure 15.7a). The composite of particle sizes in a representative soil sample, say, several handfuls, is termed **texture**. Soil texture is defined

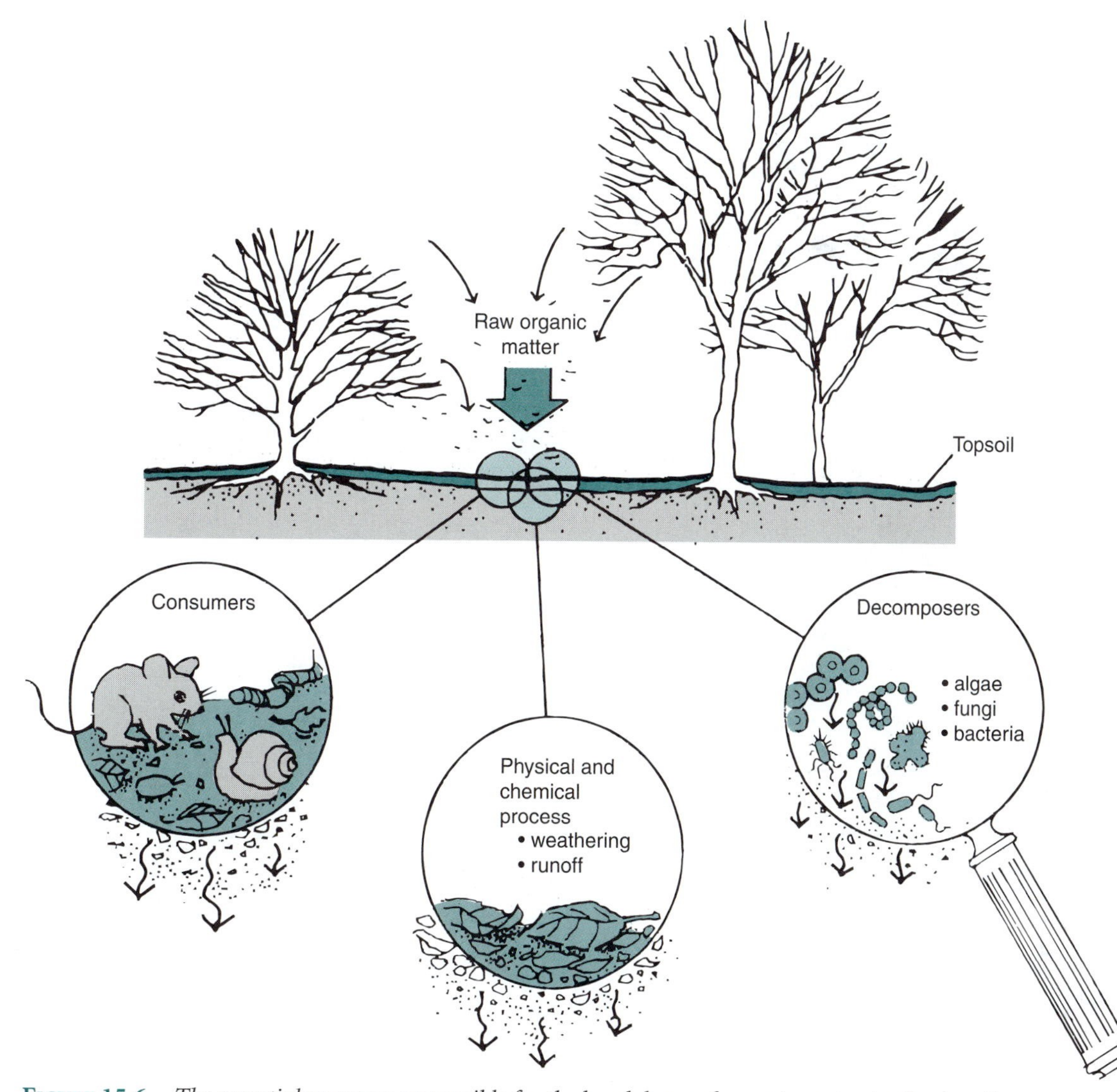

FIGURE 15.6 *The essential processes responsible for the breakdown of organic matter in the topsoil.*

in terms of the percentage of weight of particles in various size classes. Since it is unlikely that all the particles in a soil will fall within one textural class, texture is usually expressed by a set of terms. At the center of this scale is the textural class **loam**, which is a mixture of sand, silt, and clay. Loam is composed of 40 percent sand, 40 percent silt, and 20 percent clay. If there is a slightly heavier concentration of sand, say, 50 percent, with 10 percent clay and 40 percent silt, the soil is called sandy loam. The textural names and related percentages are given in the form of a triangular graph (Figure 15.7b).

Because most soils are formed in surface deposits, a correlation can often be found between certain landforms and soil texture. Such correlations between topography and soil are termed **toposequences**. In mountainous terrain, for example, streams deposit their sediment loads in the form of alluvial fans along the foot of the mountain ranges (Figure 15.8). Coarse sediments are laid down in the upper part of the fan and the finer ones are deposited farther downslope. The finest sediments (usually clays) may be carried into shallow lakes on the basin floor, where they settle out. In addition, the availability of abundant sandy sediment around the alluvial fan often leads to wind erosion and sand dune formation. A sand dune is formed purely of sand and contrasts sharply with the alluvial fan, which contains all sizes of sediment (Figure 15.8).

Soil Water

Water is central to virtually every facet of soil. Erosion and deposition by runoff are important determinants of soil thickness and composition. Within the soil, water movement and associated chemical processes are responsible for the dissolution and precipitation of minerals, and, at depth, the weathering of the bedrock. Soil mois-

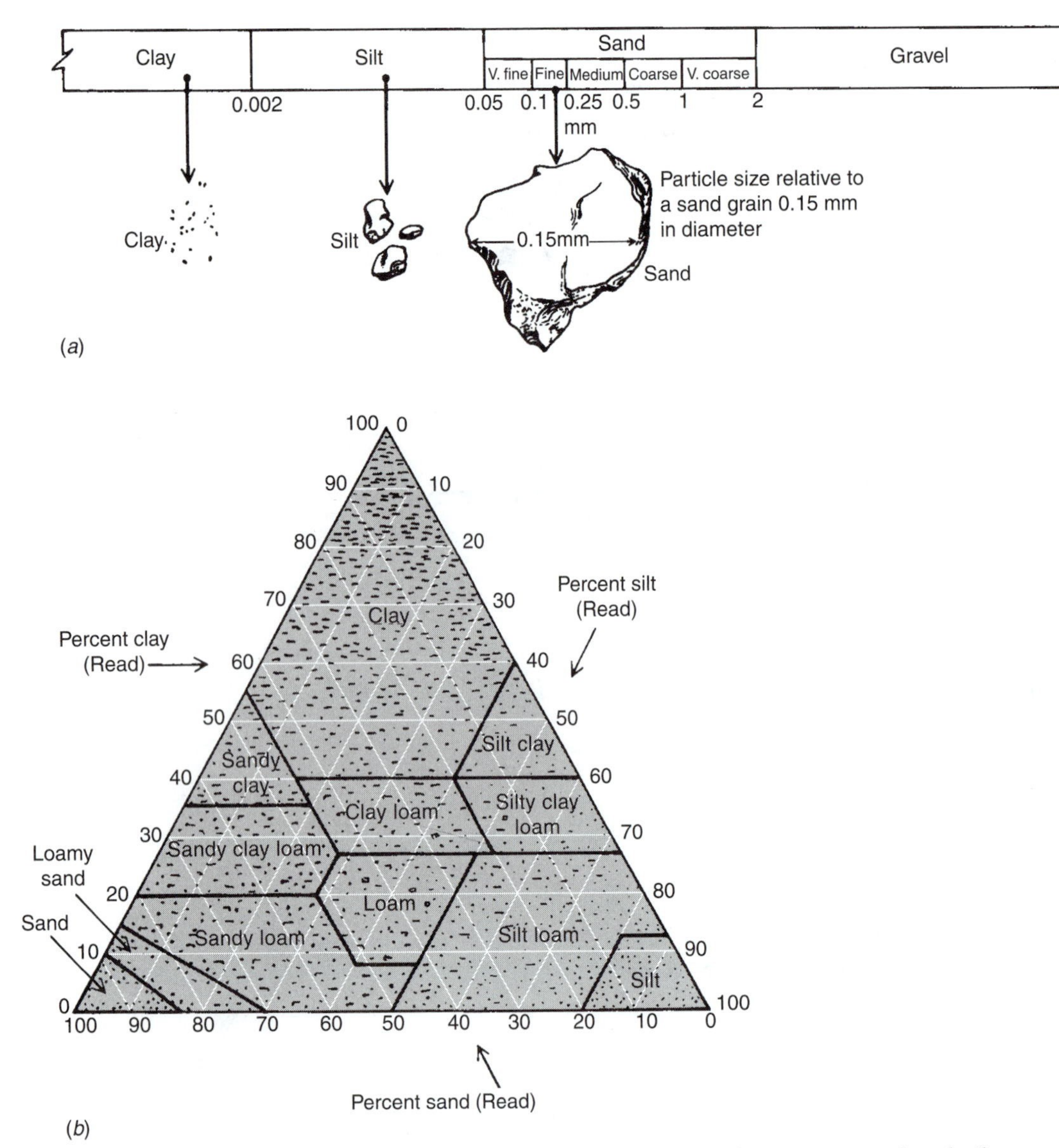

FIGURE 15.7 *(a) Scale of soil particle sizes; (b) textural triangle giving the percentages of sand, silt, and clay for different texture classes.*

ture is a key control on plant growth and soil organisms and thus has an important influence on the organic content of the topsoil. These generalizations appear to be basically sound for arid and humid climates alike, and so it is important that we examine the origin, movement, and dissipation of water in the soil.

Soil hydrology begins on the soil surface, where precipitation penetrates the soil mass. The rate of penetration is called the **infiltration capacity** and is expressed in millimeters or centimeters of surface water lost to the soil per hour. The infiltration capacity is controlled by many factors, including soil texture, plant cover, existing soil moisture, ground frost, and land use. Sandy soils with abundant plant covers and low moisture contents normally have high infiltration capacities, usually several centimeters per hour. In contrast, infiltration capacities are low in bare, frozen, wet, and fine-textured soils.

Within the soil, water is moved along three paths (Figure 15.9). The first is **capillary water**, or soil water, which is held around and between soil particles by cohesion among the water molecules. Capillary water makes soil feel damp, and it moves slowly among soil particles by a wicking process. Most terrestrial plants draw on capillary water for their moisture supply. The second path is that represented by fluid water, called **gravity water**, that moves downward under the force of gravity, eventually becoming groundwater at depth. The third path is water, represented mainly by capillary water, lost in **evapotranspiration** (the combination of evaporation and transpiration) near the surface (Figure 15.9).

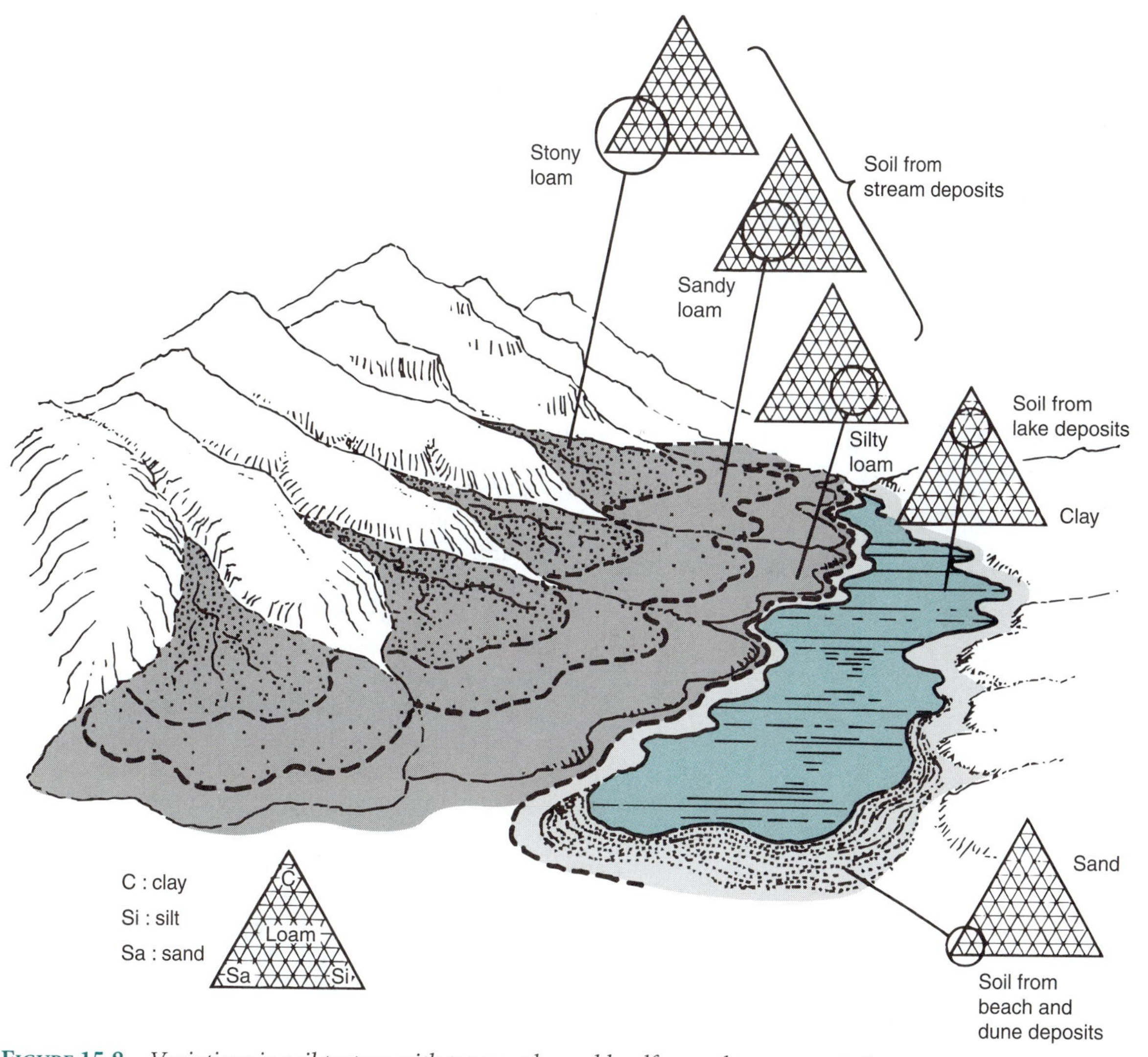

FIGURE 15.8 *Variations in soil texture with topography and landforms along a mountain range and connecting valley.*

The amounts of water that go to each path depend on soil composition and climate. In humid climates much water infiltrates the soil. Once the soil soaks up its capacity of capillary water, called its **field capacity**, any additional water moves on through the soil as gravity water to become groundwater. In dry climates, on the other hand, infiltration water rarely equals field capacity, and thus little or no water is available to move through the soil as gravity water. Most is taken up near the surface and is quickly lost to evaporation and plant transpiration.

Based on the relative balance of moisture received from the atmosphere and moisture lost to the atmosphere, we can define the soil's **moisture budget**. In humid lands the soil moisture budget is *positive* (more water received than given up) and excess water drains through the soil. In arid lands, not only is all the moisture received by the soil lost to the atmosphere, but the high surface temperatures provide ample thermal energy to evaporate even more water if it were available. Thus, arid land soils are classified as having *negative* or potentially negative moisture balances.

Where climate is intermediate between humid and arid, as in the world's semiarid grasslands, the moisture balance is also *intermediate*. Characteristically, the cool season tends to favor a positive balance when soil moisture is recharged and the warm season favors a negative balance when soil moisture is depleted. As a whole, the exchange of water in these intermediate soils takes place over a greater depth than in arid lands but not as deep as in humid lands. Not surprisingly, the soils in humid, semiarid, and arid lands are strikingly different. The soil moisture balance also has a profound effect on ecosystems, influencing both productivity and species diversity. Likewise, it is very important to agriculture because it is a key determinant of crop yields and irrigation needs. It is also important in soil formation because as the water moves up and down in the soil, it transports and alters chemicals and particles.

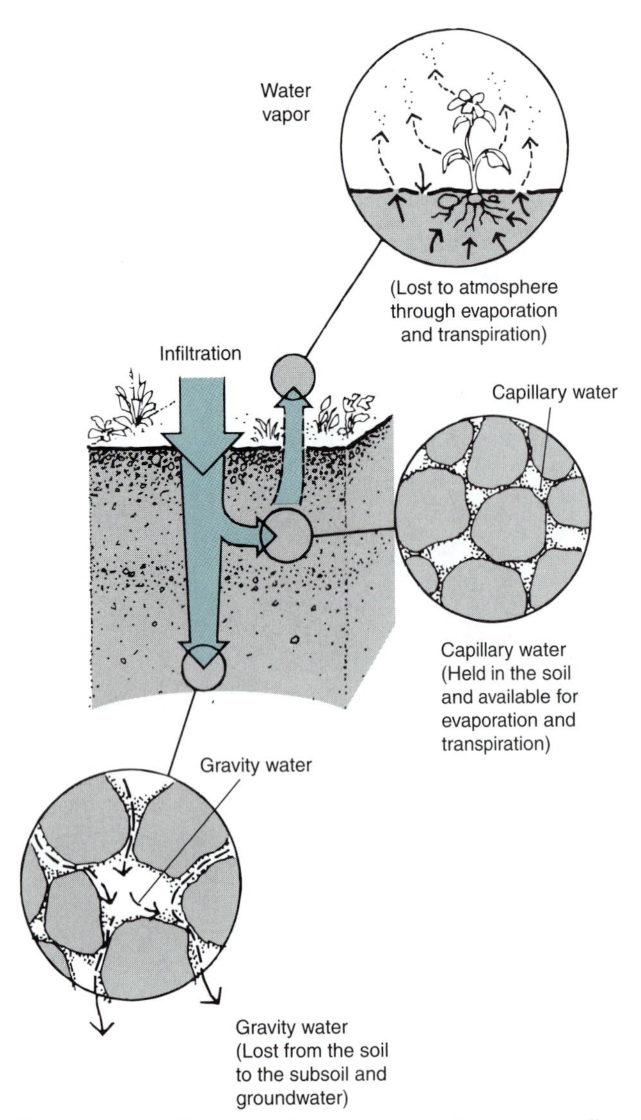

FIGURE 15.9 *The soil moisture system: infiltration, capillary water, gravity water, and evapotranspiration.*

15.6 SOIL-FORMING PROCESSES: SOLUM AS A CHEMICAL SYSTEM

Not only is the solum a highly active system of water, particles, and organic matter, but a dynamic chemical system as well. This chemical system is fed with minerals in different forms from four primary sources: bedrock, vegetation, runoff, and atmosphere. The relative amounts and types of minerals from these sources determine much of the soil's character and fertility (Figure 15.10).

Sources of Chemical Matter

In some soils bedrock plays a prominent role, contributing large amounts of weathered material to the solum. This material is principally in the form of particles, but the particles break down chemically and release their minerals into the soil. This is often the case where the bedrock lies close to the soil base so that weathered residues can be readily incorporated into the solum. For instance, where limestone bedrock lies near the surface soils tend to be chemically similar to the rock itself—that is, high in calcium content.

In contrast, where bedrock is deeply buried, it may play no role at all in the soil's chemical system. In stream valleys, for example, thick deposits overlie the bedrock, precluding any influence from bedrock. The solum's mineral input is dominated by contributions from runoff such as floods and channel flows. The minerals are delivered as both dissolved load and as particle load, and in some places may include contaminants from pollution sources such as phosphorus and nitrogen from urban stormwater and herbicide residues from agriculture. In and around cities, industrial areas, and cropland, fertilizers, road salts, sewage, and oil are also part of the mineral load deposited on the solum. They may be either direct deposits or carried to the solum by runoff (Figure 15.10).

Minerals also reach the solum from the atmosphere in precipitation and fallout. In the deserts, wind-blown dust is a major source of salts such as calcium on the soil surface. In the humid regions, such as Europe and eastern North America, precipitation contributes phosphorus and nitrogen from urban air pollution and other sources, as well as synthetic compounds such as insecticide residues from agriculture (Figure 15.10).

Vegetation is also an important source of soil minerals, including phosphorus, nitrogen, carbon, calcium, and many other elements. The types and amounts of minerals contributed by plants vary widely with the composition and productivity of the plant cover. Most plants draw minerals, including various contaminants in some cases, directly from the soil, but a few plants are able to draw large supplies of mineral nutrients from the atmosphere itself.

One group of plants, called **nitrogen fixing** plants, facilitate the transfer of nitrogen from air to the soil. These plants attract certain bacteria to their roots, which in turn extract nitrogen from the atmosphere and fix it in a solid form in the soil. In this way these plants, many of which are members of the pea family, actually increase soil fertility. Other plants, such as those in many of the world's great forests including the tropical rainforest, do not increase soil fertility, but rather help maintain it by recycling mineral nutrients between the soil and the plant cover. As we noted earlier, the recycling process is dependent on a critical balance between the rate of productivity by the plant cover and the rate of breakdown in organic matter in the soil.

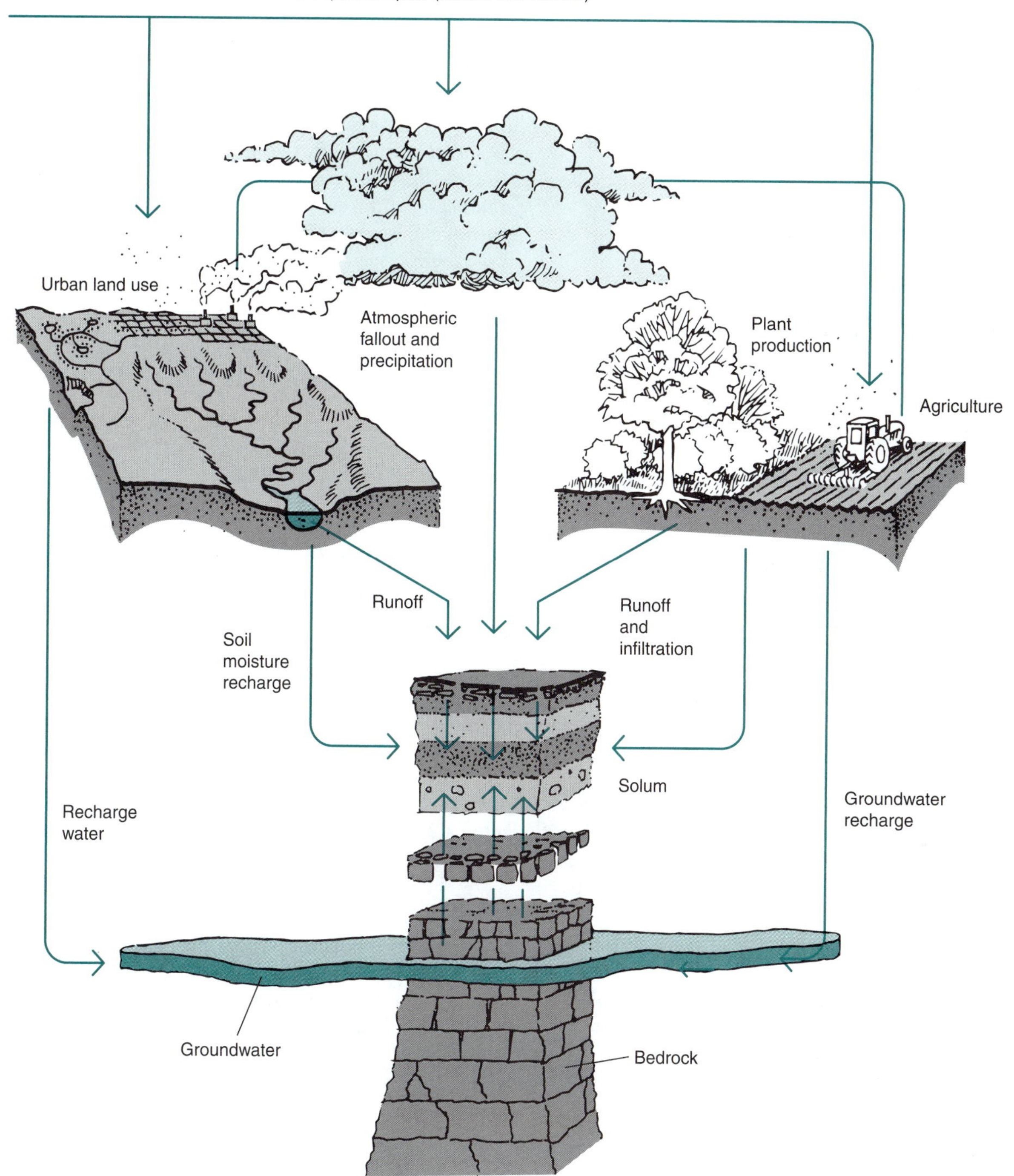

FIGURE 15.10 *Principal sources of minerals in the soil chemical system. This system can include pollutants from the atmosphere, cropland, urban runoff, and buried wastes.*

Chemical Process of the Solum

How are all these minerals involved in the soil chemical system? First, it is important to understand that they must first enter into solution as ions in water before chemical processes can take place. Chemical **ions** are atoms or groups of atoms that carry electrical charges that facilitate bonding with other ions or with minute clay particles. They are the main ingredients of the soil's chemical system and the system's operation depends, among other things, on the types and amounts of ions, the availability of water, the soil temperature, and the soil composition.

Ions are subject to three fundamental processes in the soil:

1. They may be taken up by plants.
2. They may be moved with soil water to other locations in the soil, including being washed from the solum and into groundwater at depth.
3. They may become adsorbed by tiny clay particles, called **colloids**, and held in the soil.

Both colloids and ions carry small charges or attracting forces that drive the adsorption process. *Adsorption* involves bonding of ions to colloids and it is particularly effective among organic colloids and ions such as calcium, phosphorus, and sodium, which are important plant nutrients. Therefore, soils with abundant colloids, mainly clayey and organic-rich soils, have high **nutrient retention capacity** and thus tend to be the soils with higher fertilities. Sandy soils, on the other hand, are poor in colloids and thus tend to have low nutrient retention capacities.

The ion retention capacity of soils has important implications for the behavior of contaminants from pollution sources. Especially significant is the capacity of organic colloids to retain synthetic compounds. These compounds, which include some toxic pollutants such as PCBs, DDT, and dioxin, are themselves made up of organic ions, which are readily adsorbed by the organic matter in the soil. As a result, the topsoil can function as an important **sink** for such contaminants, helping to keep them out of runoff and the groundwater system. On the other hand, if the soil is disturbed by plowing or erosion, these chemicals can reenter the hydrologic system and move into streams, lakes, wetlands, and the food chain.

Soil Formation and Features

As water moves up and down within the soil, it can move ions and colloids. The removal of mineral ions from a zone within the soil is known as **leaching**. This process is characterized by the relocation, or *washing out*, of ions to lower levels in the soil. If the washing out involves colloids as well as ions, the term *eluviation* may be used to describe it. The layer or zone in the soil that loses the materials is called the **zone of eluviation** (Figure 15.11).

If the downward movement of ions and colloids ceases at some depth in the solum, the material may be deposited, or *illuviated*, forming a **zone of illuviation**. Illuviation can take place at depths ranging from several centimeters to 5 meters or more, depending mainly on the soil moisture balance. In time, the zone of illuviation can collect such a mass of colloids and minerals that the interparticle spaces become clogged, cementing the parent particles together. In heavily leached soils, the zone of illuviation takes the form of a concretelike layer or horizon called *hardpan* or *duripan*.

In very rainy climates, such as the wet tropics, oxides of iron and aluminum are illuviated in the lower soil, forming a thick layer called **laterite**. Most other minerals, however, including many mineral nutrients such as nitrogen and calcium, are carried through the soil to groundwater and then to streams and rivers (Figure 15.11a). In humid agricultural areas such as the U.S. Midwest, where farmers employ heavy fertilizer applications for corn crops, the nitrogen leached from the fertilizer also ends up in groundwater, often contaminating domestic water supplies (see Figure 15.5b). For example, in parts of Illinois and Iowa the concentrations of nitrogen compounds in well water now threaten public health.

Where the moisture balance is negative, as in the arid southwestern United States, mineral-rich water does not pass through the soil but penetrates only the first 10 to 30 centimeters below the surface (Figure 15.11b). As the water evaporates, its ion load of calcium carbonate and sodium is deposited in the upper soil. Eventually a hard crust, traditionally known by the Spanish word *caliche*, develops.

Where irrigation water is applied to soils in arid regions, leaching is increased and the salt is washed downward. However, if too much water is applied, salt may accumulate in the solum and ruin its agriculture potential. This process, known as **salinization**, is caused by the buildup of groundwater from excessive irrigation. As the water table rises, the salts in the soil column are displaced upward and eventually are deposited in the solum or on the surface when the water evaporates (see Figure 4.8).

The Soil Profile

Working in combination, the various chemical, biological, and physical processes of the soil environment are capable of producing horizons within the upper 5 meters or so of soils. **Horizons** are horizontal or nearly horizontal zones or layers, which are generally distinguishable on the basis of color, texture, and chemical composition. They are the "fingerprints" of the soil's formation and relation to the environment. A sequence of horizons in a section of soil is called a **soil profile** and several common profiles are shown in Figure 15.12.

In humid environments where the upper soil is stable, well drained, and plant-covered, we can usually expect to see four horizons in the profile as illustrated in Figure 15.11a:

1. An organic, rich upper layer (O);
2. A zone of eluviation under the organic layer (A);
3. A zone of illuviation beneath the zone of eluviation (B)
4. A lower horizon (C), which is transitional to the subsoil.

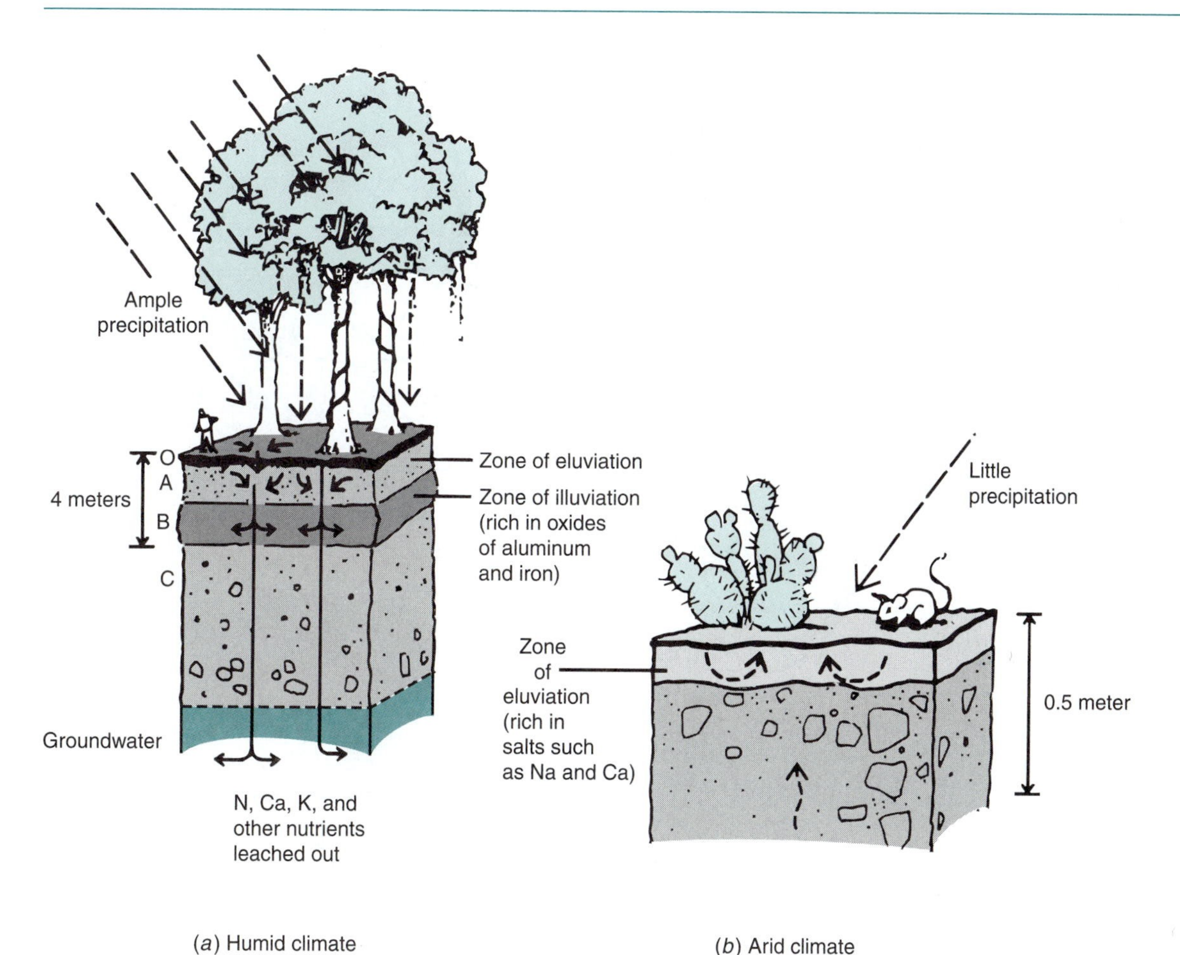

FIGURE 15.11 *Soil formation under (a) humid conditions and (b) arid conditions. The amount and depth of water penetration are important determinants of the types of minerals in the soil.*

In arid regions, only one horizon may be apparent: a calcic layer (B horizon) at or near the surface, as is shown in Figure 15.11b. Where parent material is newly deposited (as on floodplains) or continually shifting (as in sand dunes) or regularly mixed in cultivation soil, horizons may be absent. As a result, it is possible to find soils of three classes based on profile development: (1) those with horizons, (2) those without horizons, and (3) those with weak or partially formed horizons.

15.7 INTEGRATED MODELS OF SOIL, LAND USE, AND ENVIRONMENT

Within the broad, global bioclimatic framework, we can identify many combinations of natural and human conditions, or *soil-forming factors*, that contribute to soil formation and related environmental conditions. These combinations are called **soil-forming regimes**. Although more than a dozen regimes could be defined, we will highlight seven key ones here: wet tropical; desert and grassland; midlatitude forest; permafrost tundra; wetland; floodplain; and cropland.

Wet Tropical Regime

In the wet tropics rainfall is abundant in all seasons and soils are intensively leached. Most nutrient minerals are removed and oxides of iron and aluminum are illuviated in the B horizon. This horizon may become massive, forming a laterite layer, with concentrations of iron and aluminum great enough to make mining profitable in some places. As we noted earlier, despite the high productivity of the tropical rainforest, topsoil is light because organic decomposition and consumption rates on the forest floor are so high (Figure 15.12). With both heavy leaching and weak topsoil, the soils of the wet tropics are generally infertile, making the occurrence of the lush rainforest seem enigmatic. The explanation for the rainforest centers on the ability of the plant cover to recycle nutrients from the soil before they are lost to leaching.

Sustainable agriculture is difficult because once forests are cleared, soil fertility declines rapidly without a source of nutrients. As a result native peoples employ a system of **shifting agriculture** in which fields are used only for a few seasons before being abandoned. After several

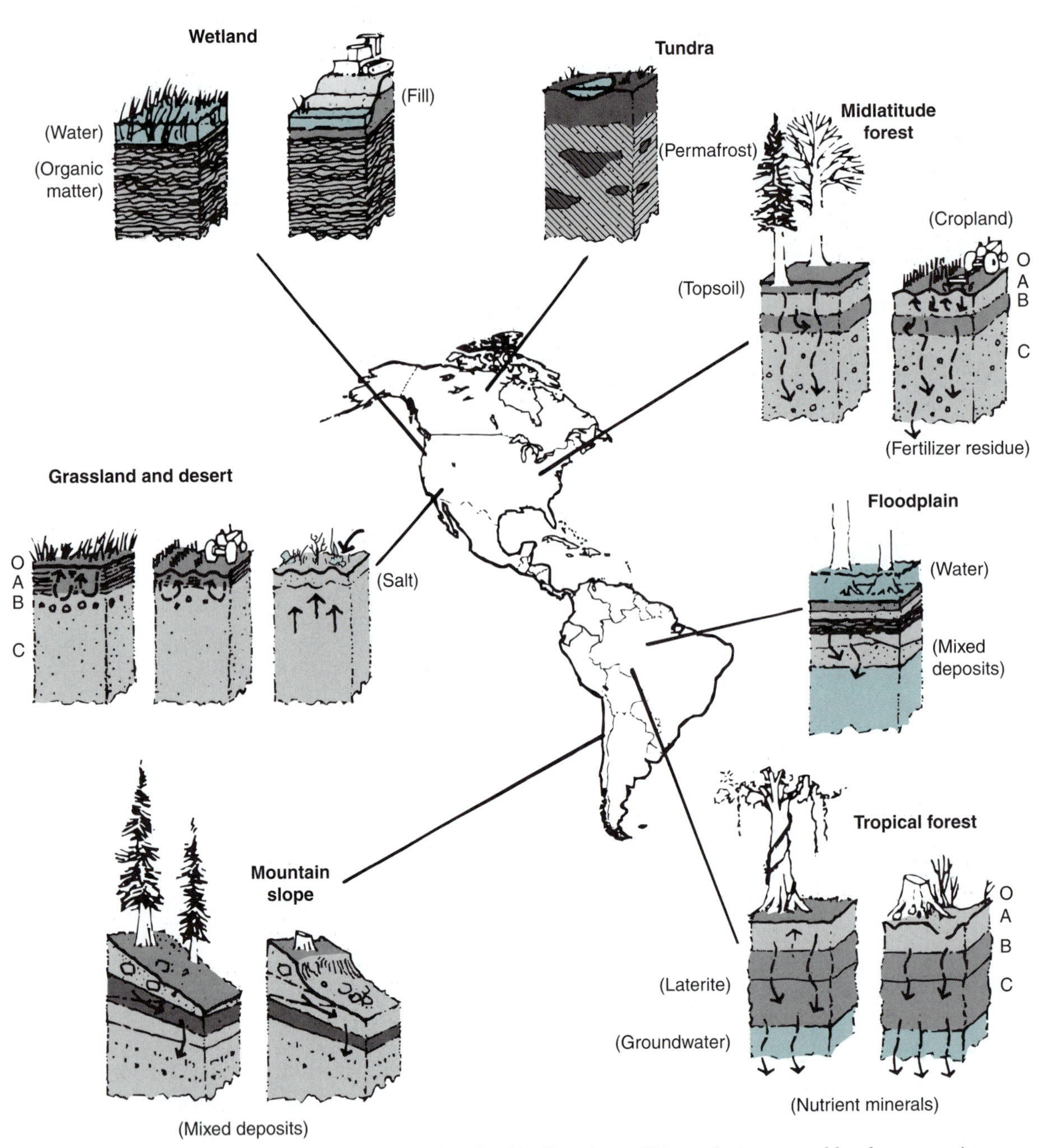

FIGURE 15.12 *Basic models of soil formation related to bioclimatic conditions, drainage, and land-use practices.*

decades the forest becomes re-established and soil nutrients are once again restored through organic recycling. (see Figure 16.10). Soil scientists use the term *oxisol* and *ultisol* for the soils of the wet tropics. With the massive destruction of the tropical rainforests in this century, vast areas of these fragile soils may no longer be able to support forest or crop production.

Desert and Grassland Regime

At the opposite end of the spectrum are the dry lands where soil moisture is slight to modest and plant productivity is generally low. The severely arid soils are characterized by light topsoils—hardly noticeable in many cases—and heavy concentrations of salts. The salts range from sodium chloride, which is toxic to agricultural plants, to calcium carbonate, which at proper levels is a nutrient. With salinization in areas of excessive irrigation, these salts are brought to the upper soil where they become concentrated at such high levels that the soil is toxic to most plants (Figure 15.12). The soils of the deserts are classed as *aridosols.*

The semiarid grasslands are climatically less severe than the arid lands. Moisture is sufficient to support a moderately productive grass cover that builds a substantial topsoil, often 20 to 30 centimeters deep. The grassland soils retain calcium and other nutrients that are not leached away by percolating water delivered during winter or occasional rainy periods (Figure 15.12). Together the organic matter, fertility, and seasonal moisture make for some of the richest soils in the world. Since they support abundant natural grasses, they are ideally suited to agricultural grasses, notably wheat. Today a large share of the world's grain is produced on the grassland soils of North America, Russia, Argentina, and other regions.

Soil scientists call grassland soils *mollisols*. By and large, the mollisols are a widely threatened resource because of their susceptibility to erosion by wind and water when placed under cultivation. The North American Dust Bowl of the 1930s is vivid testimony of the susceptibility of the mollisols to erosion when mismanaged. Even without events such as the Dust Bowl, these soils suffer incremental erosion year by year under the widely accepted agricultural practices used by modern grain farmers in the Great Plains and the prairies of North America.

Midlatitude Forest Regime

In the midlatitudes under humid climatic conditions, soils are leached by infiltrating water. Profiles typically show well-developed horizons, with the A horizon taking on a buff-to-grayish color and the B horizon a reddish color because of iron oxide concentrations (Figure 15.12). Topsoil is far more substantial in this regime than in the wet tropics, not because productivity is higher, but because decomposition rates are lower. This is attributed to the colder ground temperatures of the winter season when biological processes are largely dormant.

The soils of the midlatitude forests are among the most disturbed in the world. In North America, Europe, and Asia, the great deciduous forests of maples, oaks, and beeches have been cleared for crop and livestock farming. Today nearly 40 percent of the world's population lives in this vast zone. Since most of these soils, called *alfisols*, support agriculture, they are subject to heavy plowing, fertilization, and erosion. Only to the north, under the subarctic climate, are the forests and soils largely intact. Here the soils, known as *spodosols*, which developed under the cold, damp conditions of the boreal forests, are acidic and heavily leached.

Permafrost Tundra Regime

North of the boreal forests in North America and Eurasia lie the vast expanses of tundra. Here the climate is too cold for trees and the dominant plant cover is shrubs and herbs. In many places the subsoil is permanently frozen, which retards drainage and keeps the topsoil cool and saturated in summer (Figure 15.12). As a result, decomposition rates are slow and tundra soils often have high organic contents. Soil profiles are poorly developed and subject to intensive frost heaving and thrusting with freeze/thaw cycles. Because of these soil conditions and the harsh climate, land use is severely limited in the tundra.

Wetland Regime

Whenever the ground is saturated all or most of the time, there is a tendency toward the development of organic-rich soils called *histosols*. Abundant water ensures both high plant productivity and slow organic decomposition rates, which can lead to the buildup of thick organic deposits over time. Histosols can form in almost any bioclimatic zone because impaired drainage, rather than climatic conditions or parent material, is the primary soil-forming factor (Figure 15.12).

Wetland soils are common in floodplains, coastal plains, lake shores, the tundra, and local depressions of both natural and human origins. Locally, they may form anywhere drainage is impeded—in quarries, gravel pits, and along highways and railroads. Wetland soils are among the most maligned soils in the world because they are too wet for most agriculture, too unstable for buildings, and are often associated with pests and diseases (Figure 15.12). The traditional response to wetlands has been to drain and fill them.

Floodplain Regime

Floodplains, seashores, and mountain slopes are among the most changeable environments in the landscape. They are subject to frequent episodes of erosion and deposition and as a result are usually too unstable for soils to develop horizons and "trademark" profiles. Soils in these environments are given the name *inceptisols*, meaning new soils. In floodplains, soil materials are frequently rearranged by erosion and deposition. The deposits are usually mixtures of clay, silt, sand, and organic debris with high moisture contents. Mountain soils, which are also illustrated in Figure 15.12, are similar in that they are also mixed assemblages of eroded materials and deposits.

The land-use potential for floodplain soils may be very good, and in many of the great river valleys such as the Nile, Mississippi, and Ganges they are prized agricultural resources. Where the floodplain is too wet or is subject to frequent flooding, agriculture and settlements are risky and the soils are left to wetland or forest. The major human influence on floodplains soils occurs when dams and reservoirs are built. In the twentieth century, thousands of reservoirs were constructed, inundating tens of thousands of square miles of valley floor.

Cropland Regime

Through agriculture humans have created one of the major forces shaping soil environments, especially in the midlatitudes and subtropics where farming has displaced forests, grasslands, and wetlands. Crop agriculture is as important as climate, parent material, drainage, and erosional forces as a soil-forming factor for 10 to 12 percent of the Earth's land area. Virtually all physical, biological, and chemical aspects of soil are influenced by agriculture. The chief physical influence is plowing, which is essential to most crop farming practices.

Plowing mixes the upper 20 to 60 centimeters (8 to 24 inches) of soil, inverting the topsoil, and altering the soil's moisture system, organisms, and chemical activity. The roots of natural plants are destroyed and the input of organic matter from natural vegetation is truncated. Only where farmers bring in crop debris and/or manure or where floods bring in organic debris is the soil resupplied with organic matter. Otherwise, the topsoil declines as crops extract nutrients and weathering and leaching processes break down organic matter and remove organic and mineral particles (Figure 15.12). With commercial agriculture, the loss of natural organic nutrients is usually made up by fertilizer applications, and where insects are a problem, pesticides are added, both of which alter the soil chemistry.

Without a permanent cover of plants to protect the soil surface and roots to bind the topsoil, erosion increases significantly with cropping. Erosion takes place both when crops are on the fields and when fields are barren, but the rates vary widely with crop management practices, soil composition and texture, climatic conditions and slope, a topic we take up in the next section. Suffice it to say that the agricultural regime gives rise to a distinctly different soil than the one it replaced, usually with less organic matter, less moisture, reduced fertility, and lower total mass. In addition, most agricultural soil is biochemically different, with residues from pesticides, fertilizers, and salts and reduced populations of microorganisms that are critical to soil fertility.

15.8 Soil Loss by Erosion

Erosion is the single most degrading factor to soil worldwide. The principal erosional agents are runoff and wind; of the two, runoff is the most significant. **Erosion** involves the dislodgment of particles from the soil. Sometimes it leads to movement of only a few feet, especially with coarse particles, but often it leads to movement of hundreds or thousands of miles via streams or strong winds. Where sediment from eroded soil ends up is also important. These places are called **sediment sinks**, and they include floodplains, wetlands, lakes, and the ocean. With few exceptions, once mineral and organic material is lost from the soil, the material is not recoverable; that is, it is taken out of the soil/food production system.

Erosional Controls and Processes

Erosion is a natural process, but before humans entered the global scene, soil loss rates were much lower as a whole. Soil erosion has increased worldwide over the past 12,000 years with the spread of agriculture as farmers cleared forests and grasslands and plowed the soil. The main cause has been crop farming, but grazing (cattle, sheep, and goats) has also contributed to soil loss by causing extensive damage to rangeland. Broadly speaking, we can say that soil loss from erosion has followed the trend of world population. Today it is estimated that 11.5 million square kilometers (4.5 million square miles) of land, mostly cropland and pastureland, have been seriously degraded by erosion, deforestation, overgrazing, and other factors (Figure 15.13). This is an area larger than all of Canada.

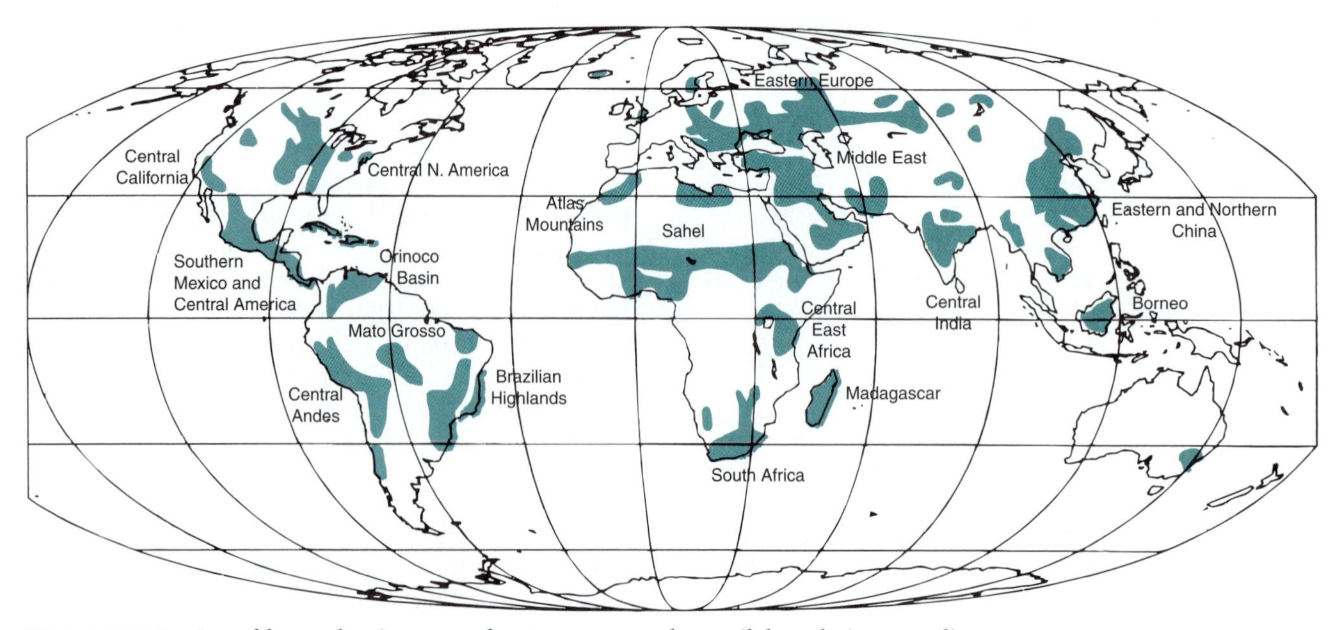

FIGURE 15.13 *A world map showing areas of greatest concern about soil degradation according to the United Nations.*

The costs of erosion are enormous. Scientists estimate that the annual costs of soil erosion are $44 billion in the United States and $400 billion worldwide. These estimates include direct costs represented by loss in soil fertility (productivity) and indirect costs represented by damage to waterways, infrastructure, and human health. The latter includes sediment clogging of channels which increases flooding and facility and land-use damage, and reduction of water quality. Direct costs are calculated according to the value of fertilizers needed to replace soil fertility losses.

Vegetative cover is the most important control on erosion. When vegetation is disturbed or removed, no matter whether it is grass or trees, erosion increases. Other factors influencing erosion by runoff include rainfall energy (based on the intensity and total annual rainfall), soil erodibility (based mainly on texture), slope of the land, and crop management (plowing and planting techniques). These five factors are combined in a formula called the **universal soil loss equation**, which can be used to compute expected soil loss under different combinations of factors.

The most vulnerable soils are those of high erodibility (usually loams and silts) on sloping ground without a protective plant cover. If this scene is coupled with heavy rainfall and no contour plowing, then soil loss may be as high as 5 to 10 centimeters a year. This is equivalent to 200 to 400 metric tons of soil loss per acre. At this rate it is only a matter of a few years before the topsoil is entirely wiped out, leaving only the underlying mineral soil (Figure 15.14).

The erosional process itself begins with *rainsplash*, or the impact of raindrops. When the drops hit the ground, they explode, driving small soil particles downslope. If the rainfall rate (intensity) exceeds the soil's infiltration capacity, then runoff also results. This form of runoff, called *overland flow*, moves in thin sheets and tiny threads, causing additional soil erosion known as *rainwash*. Further downslope, the overland flow begins to concentrate in small channels capable of eroding gullies into the soil. Known as *gullying*, this process is a severe form of erosion. In erodible soils such as exposed farm fields, gullying may carve channels 2 meters or more deep and tens of meters long in a single year (Figure 15.15). Entire hillsides may be gutted in a matter of several years (also see Figure 15.3a).

FIGURE 15.14 *Soil erosion on grasslands in Australia. Grasses protect the topsoil until they are weakened by overgrazing, deforestation, and plowing and then dislodged by runoff.*

FIGURE 15.15 *Gullying on farmland in Asia. Gullying is the most blatant and damaging form of soil erosion, literally consuming soil and land as it advances.*

Soil Erosion and Sediment Loss Rates

The rates of soil erosion are difficult to determine for large geographic areas. The rates usually given in books and reports are based on sediment accumulations measured in reservoirs and on sediment carried to the ocean by streams. These are actually export rates or **sediment yield rates** and they are decidedly lower than soil erosion rates themselves. In fact, only 10 to 30 percent of the soil eroded in most watersheds is actually delivered to streams for export to reservoirs and/or the oceans. Most eroded materials are deposited in various sinks within the watershed itself. These sinks are wetlands, swales, lakes, ponds, floodplains, and vegetated areas (Figure 15.16).

Estimates of global sediment yields are based on the load of sediment delivered to the oceans by the world's large rivers. According to these estimates, yield rates have more than doubled since the spread of agriculture. Before agriculture, the world's rivers delivered about 9 billion tons of sediment to the oceans each year. Today, the sediment export rate to the oceans is estimated at more than 20 billion tons a year, with most coming from the world's farm-lands. The main contributors are the large agricultural nations, both developed and developing countries.

The United States, the former USSR, India, and China are responsible for more than half the estimated world total sediment yield. Prorated according to the area of cropland in these countries the per-acre yield rates are close to 3.6 tons for the United States and the former USSR and close to 13 tons for China and India. Translated into soil erosion rates (as opposed to the rate of export to the sea), the values would be perhaps three to four times greater.

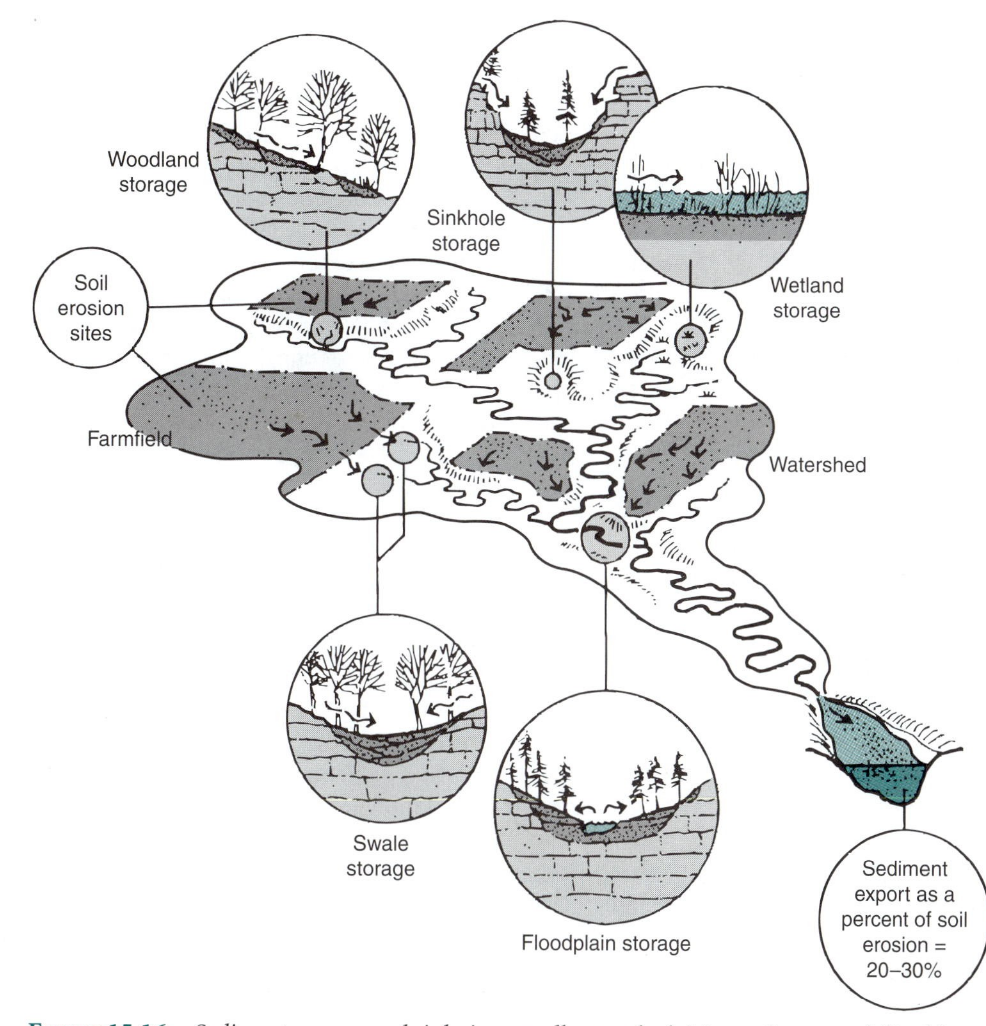

FIGURE 15.16 *Sediment sources and sinks in a small watershed. Most sediment mobilized by erosion is deposited in various places before leaving the watershed.*

15.9 ENVIRONMENTAL IMPACTS OF SOIL EROSION

The principal impact of soil erosion is reduction in the quality of the world's soils. Because erosion is concentrated on the surface, topsoil—the most fertile part of the soil—is most severely affected by erosion. The most sobering consequence of *topsoil loss* is reduced agricultural productivity. It is estimated that for each inch of soil eroded from wheat fields in the United States, wheat yields are reduced by 6 percent. For each inch of cornfield erosion, productivity is reduced by 5 to 6 bushels per acre per year. To make up for such fertility loss, farmers in developed countries apply commercial fertilizers. This adds significantly to the cost of food production. The quality of the soil as a medium for crops declines as topsoil grows thinner (Figure 15.17).

Where gullying occurs, there are multiple environmental impacts. In addition to the soil loss, workable land units are reduced in area and are often fragmented into smaller parcels. *Fragmentation* not only makes for less efficient land use, both in mechanized and nonmechanized farming, but also breaks up the biogeographical environment. Ecosystems linked by hedge rows, woodlots, and fallow land can be fractured by gullying into smaller biogeographical entities that are less efficient ecologically (see Chapter 16 for discussion of habitat fragmentation).

Since most of the sediment eroded from farm fields is stored in the landscape, sediment accumulation is another prominent impact of soil erosion. Known as *sedimentation*, the deposition of excess sediment in wooded areas, swales, wetlands, ponds, lakes, and reservoirs has become a serious problem throughout most agricultural regions (Figure 15.16). Wetlands, for example, are naturally adjusted to slow rates of organic accumulation and decomposition. When a surge of inorganic sediment is deposited in a wetland, among other things, biochemical processes and nutrient availability in the underlying soil will be altered and many aquatic organisms will be buried. The outcome is reduced wetland species diversity and productivity. Where sedimentation is heavy and frequent, the wetland may be transformed into a monoculture comprised of only a few hardy plant species.

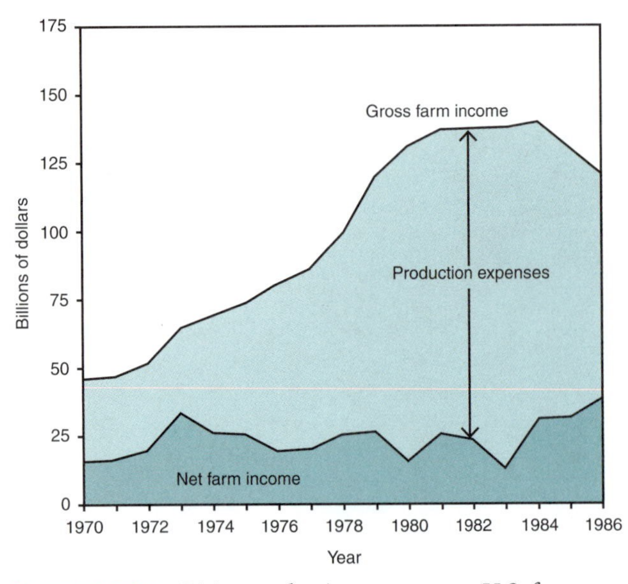

FIGURE 15.17 *Rising production expenses on U.S. farms reflects, in part, the declining quality of soil.*

Another impact of sedimentation is *clogged stream channels* and *turbid water*. Fine particles, mainly clay and fine organic material, are carried in suspension in streams and lakes, making the water muddy or turbid. This is a major type of water pollution that impedes biological activity (sunlight is reduced, among other things) and damages drinking water supplies.

When heavier particles, such as sand and gravel, sink to the bottom, stream channels become glutted and stream ecology is seriously damaged because bottom organisms are buried. The deposits reduce the channel's flow capacity, which may increase the magnitude and frequency of flooding. Figure 15.18 shows the trends in sediment yield and stream channel formation with a typical sequence of land-use change from woodland to urban. Note the large increase in sedimentation as a result of construction processes associated with suburban development. Although this increase is short-lived, the glut of sediment results in environmental degradation. In addition, sediment-choked channels increase flooding and damage to the built system of drains, irrigation channels, bridges, and reservoirs in and around both urban and agricultural regions.

In many areas where irrigation is practiced there is a tendency toward excessive water application resulting in the saturation of the subsoil. In time the zone of saturation can rise into the solum, *waterlogging* the soil root zone, and reducing its capacity to grow crops. Waterlogging can take place in any climatic zone where irrigation is practiced. In arid lands it is often accompanied by *salinization* as salt is displaced upward when the water table rises with waterlogging. At or near the surface, the salt-rich water evaporates, leaving a salt residue behind that reduces or even eliminates the soil's capacity to grow crops.

Some of the world's most productive cropland has been damaged or destroyed by waterlogging and salinity, including parts of the rich Imperial Valley of southern California. In Pakistan, which is overwhelmingly dependent on irrigation, about 75 percent of the cropland has been damaged by salt saturation, and in China it is estimated that nearly 50 percent of the irrigated cropland is salt-damaged. In the extreme, salt-saturated land must be abandoned. It is recoverable in time, but only after the water table has receded and the salt has been washed back down with heavy applications of fresh water.

Wetland eradication, which we touched on earlier, is another serious environmental impact. For centuries agricul-

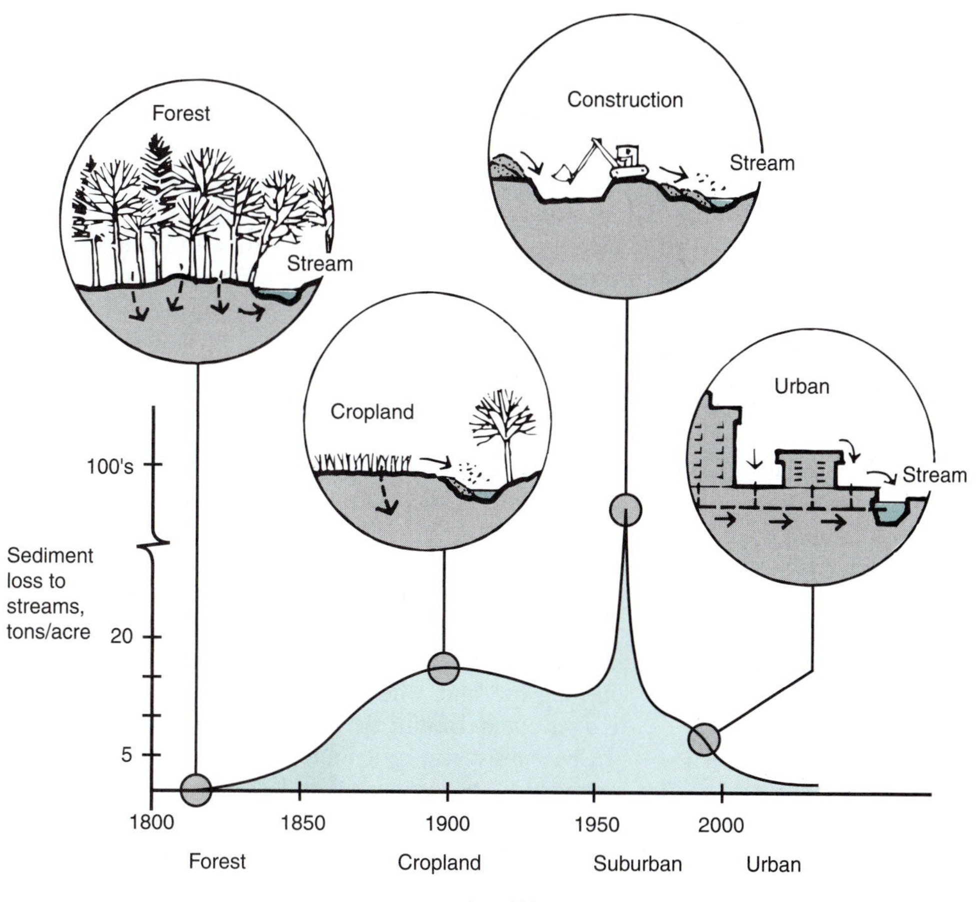

FIGURE 15.18 *Sediment loss to streams as a result of land-use change and construction leading to urban development.*

ture has deforested, drained, and cultivated wetlands. Without groundwater to saturate and preserve the organic material, the soil mass breaks down and is eventually lost. In the United States, as much as 50 percent of the wetlands in the lower 48 states have been destroyed. In many European countries, the percentage may be even larger.

15.10 SUMMARY

The most basic consideration in assessing the status of Earth's land and soils is global population. With an annual increase of more than 80 million people, the pressure to use more land for food production is inevitable. Likewise, the pressure to use existing farmland more intensively is also inevitable. Fallow periods are shorter or nonexistent and fertility losses from overcropping and erosion are, where the technology and resources are available, made up by heavy fertilizer applications. Not surprisingly, the system is probably not sustainable in the long run because the fundamental resource, soil, is not being maintained while a rapidly growing world population expects substantial increases in the food supply.

The urgency to bring more land under cultivation has pushed farming beyond the reasonable capacity of the land and its soil base in many parts of the world. Increasingly, marginal and submarginal lands are being placed under cultivation and grazing. **Marginal lands** are those where successful crop farming is possible about 50 percent of the time. **Submarginal lands** are those where less than half the years will produce profitable yields. The nonprofitable years are those plagued by drought, floods, cold, or some other setback.

Attempts to beat such odds have proven to be not only economically disastrous but often environmentally disastrous as well because farming is overextended. This means that when cultivation is applied, the threshold of soil stability is lowered, as was illustrated earlier in this chapter for the mountain-slope farming in Madagascar. The same can be argued for the North American Dust Bowl of the 1930s, where the threshold was drought-related. In the Dust Bowl, farms were abandoned and plowed fields were left unprotected from eroding winds. The result was massive wind erosion of silty grassland soils.

Although crop farming is the leading cause of soil loss worldwide, grazing and deforestation are also significant causes of soil loss. Cattle, sheep, and goats are the principal grazing animals. Where their numbers are too great to sustain a healthy grass cover, overgrazing is the result. With overgrazing, grasses and near-surface roots are destroyed, leaving the soil vulnerable to erosion by runoff and wind. Much of the world's grassland is currently overgrazed, resulting not only in high rates of soil loss, but permanent damage to the grass cover. The consequence is an irreversible reduction in the land's life support capacity, or *desertification.* As we noted in Chapter 4 according to some estimates, as much as 200,000 square kilometers of land are being lost yearly to desertification.

Deforestation followed by crop agriculture also gives rise to a sharp increase in soil loss. It is especially so in regions of heavy precipitation and steep slopes such as mountain rainforests. If agriculture follows, then forest regrowth is prohibited, and runoff and soil loss rates soar. Lumbering of mountain slopes in the midlatitudes, such as the American Northwest also causes soil instability, especially where logging roads are cut into slopes. Sediment loading of local streams usually increases, particularly where clear-cutting is practiced, but then decreases as second-growth trees become established.

In all of the situations previously described—intensive cultivation, submarginal farming, overgrazing, lumbering, and hill slope cultivation—gambler's luck and misfortune play a significant role. To begin with, these environments are inherently risky places, and when disturbed by land use, they are made even riskier. As a result, it is often the roll of nature's dice that determines if a disturbance such as forest clearing or cultivation on a mountain slope will lead to sudden destruction of soil and damage to land. The magnitude and frequency of natural events, including powerful wind storms, torrential rains, and prolonged droughts, follow a pattern in which large and destructive events occur infrequently and small, less damaging events occur frequently. The periods of small events tend to encourage risky land-use practices because the conditions they produce do not appear threatening. Large events, by contrast, are often hidden from short-term human experience (one or two generations) because of their infrequent occurrence. This is especially so where society does not have a long record of experience with a particular environment or geographic location.

Most of the environments that experience sudden, severe soil loss are those where the probability of powerful, potentially destructive events is relatively high. For example, semiarid grasslands are prone to greater magnitudes and frequencies of drought than midlatitude woodlands. In addition, when disruptive land uses are introduced to these environments, the landscape's resistance to powerful events is significantly lowered. This was the point illustrated earlier with the U.S. Dust Bowl. Had it not been for farming and the exposure of soil caused by breaking up the grass cover, the drought and strong winds of the 1930s would have had inconsequential effects on the Great Plains.

On balance, societies worldwide are not doing well in managing land resources, especially soil. Annual loss of arable land is presently estimated at 100,000 square kilometers a year. In developing countries agriculture is pushing into lands where both the risk of economic setback and serious and lasting environmental damage are high. Intensive subsistence agriculture, which is a way of life for more than 2 billion people in developing countries is not sustainable in the twenty-first century as it is currently practiced. And looming on the horizon are rapidly growing populations to feed. Of the 2.1 billion people that will be added to world population in the next 30 years, more than 96 percent will be added to developing countries.

In developed countries agriculture is more sophisticated technologically and far more productive, but it is not without serious problems. Soil loss per acre of cropland is about one-third that of developing countries, but it is nonetheless substantial. Moreover, much of the agriculture in developed countries is arguably not sustainable through the twenty-first century because of heavy dependence on exhaustible water supplies such as dryland aquifers and on expensive fertilizers, pesticides, and fuel. Some estimates claim that in commercial agriculture it now takes more calories of energy to produce a pound of food than there are calories of energy in the food itself.

15.11 KEY TERMS AND CONCEPTS

- Organization of land
  - drainage nets and watersheds
  - uplands and lowlands
  - land units
- Topography and land use
  - hillslope limitations
  - slope failure
  - environmental impact
- Soil mantle
  - solum
  - soil skeleton
  - weathering
- Parent material
  - residual
  - transported
  - alluvium
  - glacial drift
  - loess
- Soil properties
  - composition

- texture
- moisture

Toposequence
Topsoil
Organic energy system
Soil hydrology
- infiltration
- capillary water
- gravity water
- evapotranspiration
- field capacity

Soil moisture budget
Soil chemical system
- mineral sources
- nitrogen fixation
- ions
- adsorption
- nutrient retention capacity
- contaminant sink

Soil-forming processes
- leaching
- eluviation and illuviation
- moisture budget relations
- laterite
- salinization

Soil profile and horizons
Soil-forming regimes
Models of soil formation
- wet tropical
- desert and grassland
- midlatitude forest
- tundra
- wetland
- floodplain
- cropland

Soil erosion
- processes and factors
- universal soil loss equation
- sediment yield rates
- land-use relations
- environmental effects

Marginal and submarginal lands
Threshold of soil stability
Risk and natural events

15.12 QUESTIONS FOR REVIEW

1. Can you present an argument that the total area of land, the quality of land, and the habitability of Earth for humans has changed and is continuing to change?
2. Describe the general character of uplands and lowlands in watersheds. What are land units? Where are slopes found among uplands and lowlands?
3. Why is steeply sloping ground limiting for most land uses, and what are some of the consequences of misuse of rugged topography in developing and developed countries?
4. What is the soil mantle, and what are the two main sources of the particles found in soil. Define alluvium, glacial drift, and loess. What are the three main properties of soil?
5. What is topsoil? What is the principal source of this material, and why is it appropriate to view topsoil as an organic energy system? How is this system thrown out of balance by agriculture?
6. Describe the three main paths of water movement in the soil. What is meant by the term soil moisture budget? Where do we find negative, positive, and intermediate soil moisture budgets?
7. What are the main sources of chemical matter in the solum? How do contaminants from human sources fit into this picture? What are nitrogen fixing plants?
8. What are chemical ions, and what are the three fundamental processes affecting ions in the soil? What is meant by nutrient retention, and what role does the soil play in this regard as a sink for pollutants?
9. Define eluviation and illuviation. What is their role in soil formation? What are hardpan, laterite, caliche, and salinization? What is meant by the term horizon, and what is a soil profile?
10. Briefly describe the following models or regimes of soil formation, including the climate and land uses associated with each: wet tropical, desert and grassland, midlatitude forest, permafrost, tundra, wetland, floodplain, and cropland.
11. What is the definition of soil erosion? What factor has been most instrumental in accelerating worldwide soil erosion in the past 12,000 years? Name the principal controls on soil erosion. Define rainsplash, rainwash, and gullying.
12. Distinguish between soil erosion and sediment loss. In general, what percentage of eroded soil is lost to reservoirs and/or oceans? Where do sediment sinks occur in water-sheds?
13. Are there differences in the soil erosion rates on cropland in developing and developed countries? Describe a number of environmental impacts resulting from soil erosion.
14. Discuss the dilemma faced by societies pressed to use marginal and submarginal lands because of population growth and food production needs. What are the inevitable impacts on environment and society? What do you think needs to be done in order to achieve a sustainable balance between land use and land resources in developing countries?

16
BIOLOGICAL DIVERSITY AND LAND USE

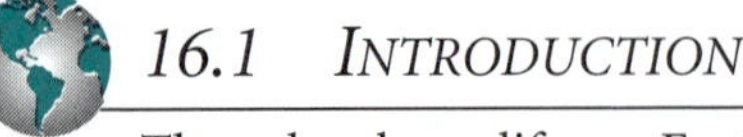

16.1 INTRODUCTION

There has been life on Earth for perhaps 3 billion years. During most of this time life has been limited to the sea and comprised mainly simple life forms such as bacteria and algae. About 550 million years ago, at the beginning of the Paleozoic era, many new and more complex forms of life emerged.

During the Paleozoic, life made the great transition from the sea to the land. Terrestrial plants evolved first. The ancestors of ferns and mosses were prominent among the early land plants. They were followed by the cone-bearing plants, the ancestors of modern conifers such as pines and junipers. About 170 million years ago the flowering plants appeared, flourished, and spread over the Earth, becoming the dominant organisms of the landscape.

With the appearance of new plant species, new animals, insects, and microorganisms also evolved. During the Paleozoic, Earth witnessed the evolution of tens of millions of new organisms. But it also witnessed the natural extinction of a similar number of species. Taken together at any moment, the net balance between the emergence of new species and the extinction of existing species constitutes the Earth's **biological diversity**.

Biological diversity has become one of the major environmental issues of our times. The basic problem is the increased rate of extinction as a result of expanding human populations, resource exploitation, land clearing, and land-use development, especially in tropical forest environments. The remaining tropical forests cover about 10 percent of Earth's land surface, but they contain more than 50 percent of all species. The current rate of extinction worldwide is estimated at more than 1,000 times the natural rate, and the vast majority of the species loss is taking place in tropical forests. In this chapter we examine the conditions and causes of Earth's declining biodiversity and why the rising rate of extinction should be a serious concern to humanity. We are especially interested in the impacts of land use on habitat and in those places where the conflict between land use and biodiversity is most acute.

16.2 THE CONCEPT OF A SPECIES

Biodiversity is defined as the number of species inhabiting a prescribed geographic area. The term species is used to define the basic unit of biological organization in nature. A **species** is defined as a group of organisms within which, under natural circumstances, there is free flow of *genes* among individuals. Individuals of the group typically look alike and are capable of interbreeding with each other, but not with members of other such groups.

Not all species, however, require interbreeding to transmit their genes to other individuals. Many plants, for example, are capable of reproducing by nonsexual means. One such means is vegetative regeneration, or cloning, in which a new plant is formed from a piece of living tissue from another plant. Nevertheless, interbreeding is the most common means by which members of a species share genes. Conversely, the inability of members of different species to interbreed is the most common means of keeping species separate and distinguishable from one another.

Each species represents a distinctive pool of genes, the genetic resources of all life. The total number of genes in the chromosomes of species varies widely, but for most organisms it is in the tens of thousands. Genes contain the genetic information of each organism, and this huge compendium of information is unique to each species. When a species is lost from Earth, its genes are forever gone and there is no way that nature can recreate this species.

Speciation

The process that results in the formation of new species is termed **speciation** (see Figure 16.1). Speciation takes place in nature at radically different rates depending on the organisms involved and the conditions of the environment. Broadly speaking, it appears that the frequency of speciation among plants, animals, and insects typically ranges from hundreds of thousands to several million

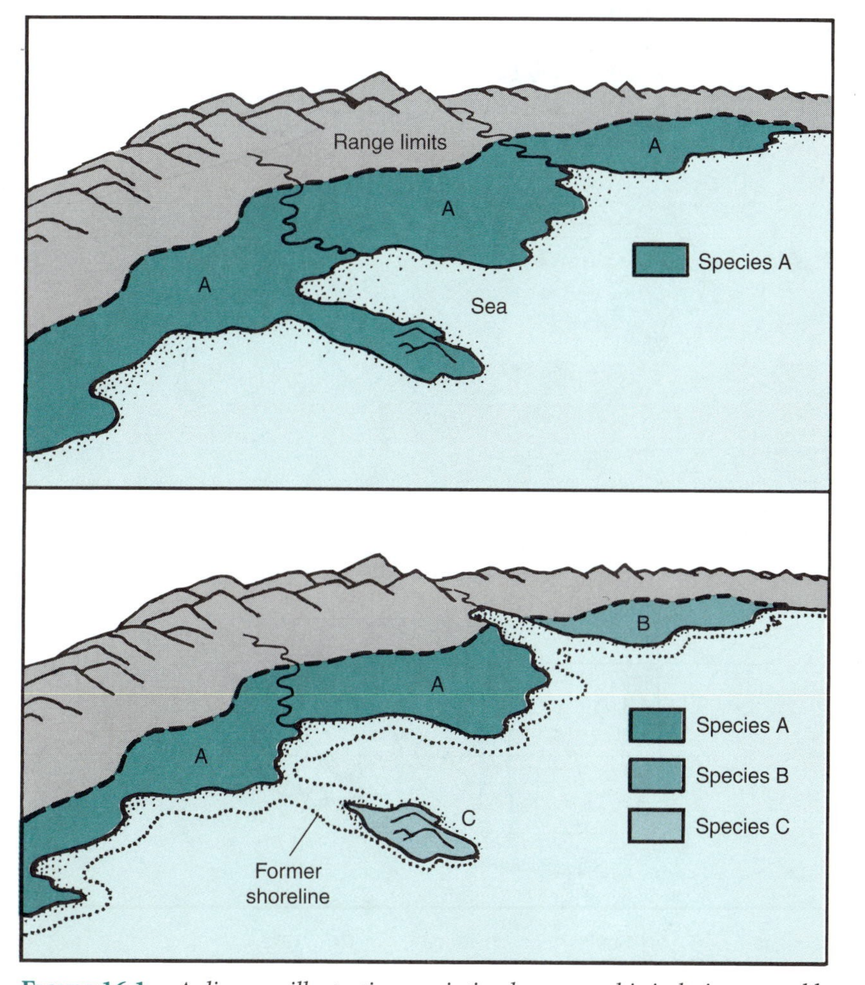

FIGURE 16.1 *A diagram illustrating speciation by geographic isolation caused by a rise in sea level.*

years. Extinction, of course, represents the opposing process to speciation.

Biologists define two basic mechanisms of speciation: geographic isolation and polyploidy. **Geographic isolation** involves the splitting apart of a population by a physical barrier of some kind. The barrier may be created, among other things, by the formation of a mountain range, the formation of a deep valley such as the Grand Canyon or the emergence of a water body (Figure 16.1). In any case, some members of a species are restricted geographically from interbreeding with the main population. As a separate population, these isolated individuals evolve independently, forming a new species. Crossbreeding or hybridizing with the original population is strictly limited by the barrier; thus, the two populations tend to diverge along separate lines.

Polyploidy involves the formation of mutations within the gene pool of a particular species. The mutation is maintained by interbreeding with individuals, such as parents, possessing the same gene traits. Given favorable conditions and time, a distinct population is formed that is genetically independent from the species of origin. Of the two means of speciation, polyploidy is least common and appears limited mainly to plants.

Species Distributions

The geographic area occupied by a species is called its **range**. The ranges of species vary enormously in size, shape (or pattern), and location. Over the life of a species, its range may change substantially as a result of environmental change such as the removal of a barrier to

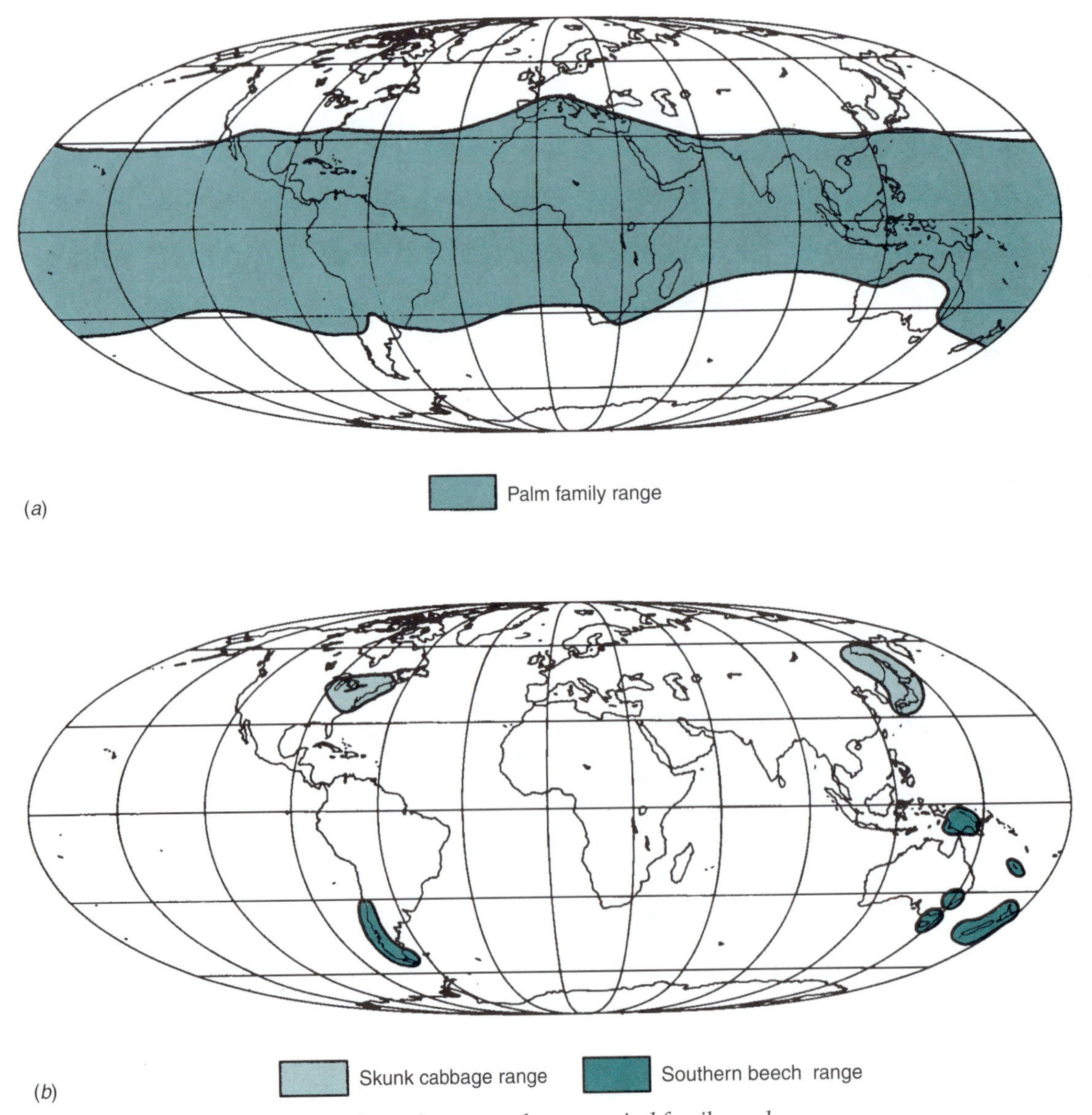

FIGURE 16.2 *Examples of intercontinental ranges: palms, a tropical family; and skunk cabbage and southern beech, two midlatitude species.*

migration or biological adaptation to new environmental conditions. Many agricultural plants (e.g., wheat, rice, and corn) and many weeds have undergone dramatic increases in ranges over the past 10,000 years with the spread of farming and transoceanic travel. In our times, however, the opposite trend is more common: The ranges of thousands of wild species have been truncated, reduced, fragmented, or eradicated.

Several types of ranges are common to terrestrial species. The largest are **intercontinental ranges**, which span two or more continents. These include *cosmopolitan* or worldwide distributions in which the species range is continuous over all land masses wherever its habitat type is found. Many of the weeds and vermin that have followed humans and the spread of agriculture and settlements are cosmopolitan species. Rats, cockroaches, and houseflies are good examples. Among the plants, many species of palms, for example, are intercontinental, but their ranges are strictly limited to the tropics. Within this broad zone, the palms are continuous, spanning the entire globe (Figure 16.2a). Most intercontinental species, however, have *discontinuous ranges* characterized by several or more independent subareas of various sizes (Figure 16.2b).

At the other extreme from intercontinental species are **endemic species**. These are species with small ranges usually limited to a single geographic locale or to a single habitat that may be distributed in small, local patches over a single geographic area. Endemic species are common through the world but especially so in the tropics.

In the tropics many endemic species are actually *microendemic species*. These are species with very small ranges; for example, the canopies of several trees or an island in a lake or river channel (Figure 16.3). Some of these, called **vicariads**, are genetically very similar species because they evolved from a common population. They have become different species because of geographic separation owing to a barrier of some sort. The barrier may be a geographic obstacle such as a large river or merely a certain distance that limited interbreeding among the populations of origin.

Extinction

Almost all the species that ever inhabited Earth are extinct, yet more species appear to inhabit Earth today than at any other time in the planet's history. How can this be

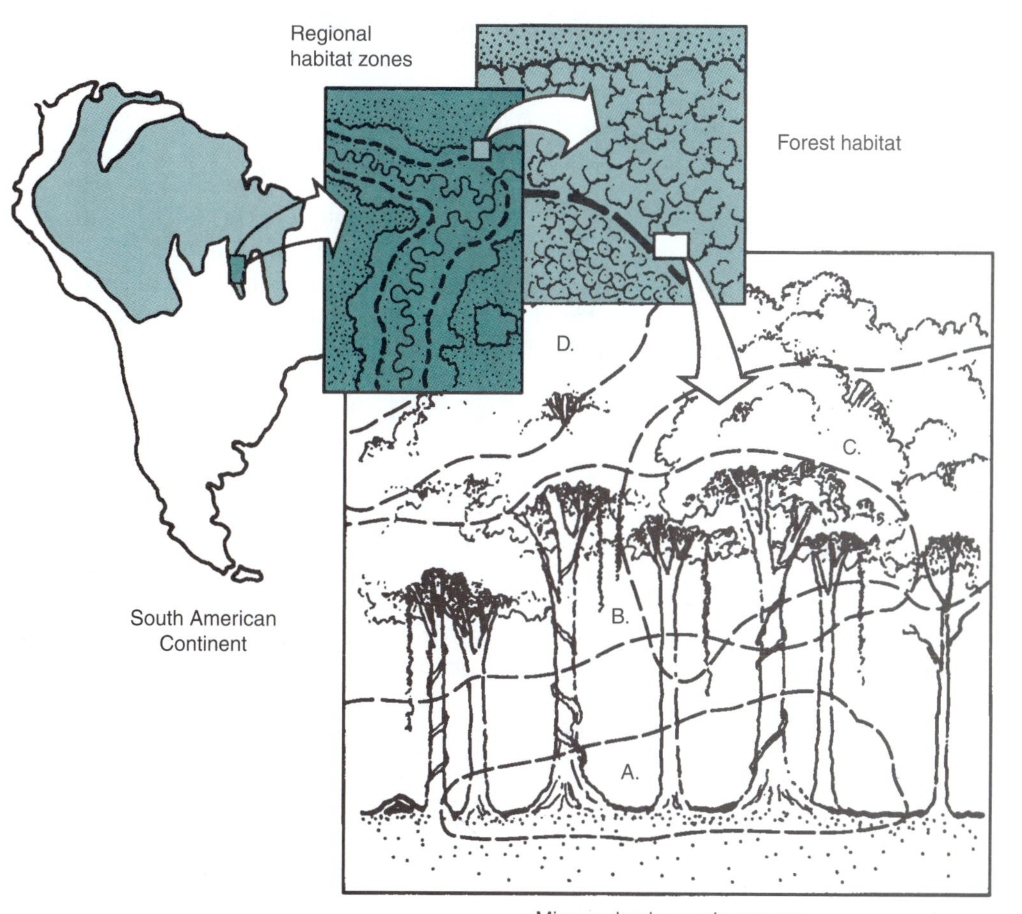

FIGURE 16.3 *An illustration of geographic scale considerations in the distribution of microendemic ranges of four species (A–D), such as those typical of many insects in the tropical rainforest.*

so? The answer is related to the great length of time organisms have been around, 3 billion years or more. During that time millions of species have evolved and disappeared. The length of an organism's tenure on Earth, however, is a matter of definition based on whether we measure extinction by the disappearance of a species or by the disappearance of its genes.

Biologists make a distinction between *chronospecies extinction*, or **pseudo extinction**, and **true extinction**. All species evolve from other species; most speciation is characterized by a gradual shift of a population into a new form closely resembling its parent species. The genetic makeup of the parent and offspring species is nearly identical; therefore, extinction of the parent species usually does not represent loss of the genetic package. In other words, the genetic lineage endures as one species evolves into another. The disappearance of older species, therefore, is defined as chronospecies or pseudo extinction because the original genes endure. A more meaningful measure of extinction is based on the life of a **clade**, defined as a species and all its descendants. A clade includes all the species of a genetic line from the time the first ancestral species split off to the moment the last species dies off. The life of clades appears to range from 1 to 10 million years for plants.

Extinction caused by humans is clearly different from chronospecies extinction because it does not allow for maintenance of genetic lineage through evolution of offspring species. Moreover, human-driven extinction may result in the elimination of clades as well, especially for closely related endemic species such as vicariads, which together occupy a small geographic area that is easily destroyed by land-clearing operations, reservoir floodings, or land use development.

16.3 GENETIC DIVERSITY

You do not have to venture far into the popular media to run across a statement something like this: "What's so serious about the loss of a few thousand species each year? The Earth's got millions of them, and most have no demonstrable benefit to humans." Such statements are horribly naive for several reasons. Putting aside the moral issue of whether one species, humans, has the right to exterminate other species, there are purely practical considerations such as the loss of undiscovered food and medicine sources.

Humans currently use a very small number of plants for food. Only 20 species provide 90 percent of the world's food, and only three plants, wheat, rice, and corn, provide the bulk of our food supply. There are literally thousands of plants—perhaps 30,000 or more—with promise as agricultural food crops, many with nutritional and caloric values equal to or greater than those of many common food crops. Many of these plants are known to indigenous cultures—especially tropical fruits, root plants, and grains such as amaranth, a favorite of the Indians in Central America. But thousands of food plants remain undiscovered or poorly understood in terms of their ecology and potential for domestication and food production. Many of these plants may disappear before we find them and define their value to humanity.

It might surprise you to learn that about 40 percent of our prescription drugs come from organic sources, principally plants. Many of the most promising sources of medicines for cancer are found in tropical plants. The *rosy periwinkle* of Madagascar, for example, produces chemicals that cure most victims of two deadly cancers, Hodgkin's disease and acute lymphcytic leukemia (Figure 16.4). Five other species of periwinkle are found in Madagascar, and one is known to be seriously threatened by rainforest clearing for agriculture. Countless plants with medicinal value will be lost in the next 50 years with the eradication of the tropical rainforests worldwide. Thousands more organisms with medicinal potential lie undiscovered in threatened environments. This includes plants known only to indigenous people, such as the various Indian tribes of the Amazon rainforest.

Recognizing the considerable resource represented by folk knowledge, some countries are sending scientists into rainforests to uncover the secrets of ancient tribes. One such team, working in the Amazon rainforest, learned about more than 1,000 plants with potential in medicine, agriculture, and industry. In Asia, for example, the *neem tree* (a relative of mahogany), long known among native peoples as a remedy for various pains, fevers, and infections, has recently come to the attention of scientists. We are slowly learning that the potential for new biological resources of value to humanity is not only greater than we imagined but probably greater than we can imagine.

In addition to losing organisms of potential value to human survival and the quality of life, it is also necessary to preserve the genes of the plants and animals upon which we currently depend for food, medicine, and other needs. All domesticated organisms originated from wild species: wheat, corn, rice, goats, cattle, and chickens, for example. In the wild, each of these species exhibited considerable gene diversity within its own ranks. Such **genetic diversity** was extremely important to survival of the species because it ensured the existence of strains with resistance to certain stresses in the environment such as disease and drought.

With domestication the natural diversity within a species is reduced as its breeding is refined for selected characteristics. In wheat, for example, early breeding selected for plants that, among other things, would hold the seed heads without breaking off as they did in the wild. After thousands of years as a crop, wheat lost much of its genetic diversity, including its ability to survive without the aid of humans to collect and broadcast its seed. Many other traits valuable to its future, including resistance to new diseases, have undoubtedly also been lost over the years.

FIGURE 16.4 *Rosy periwinkle from the tropical forest of Madagascar provides chemicals that are effective in the cure of two forms of cancer.*

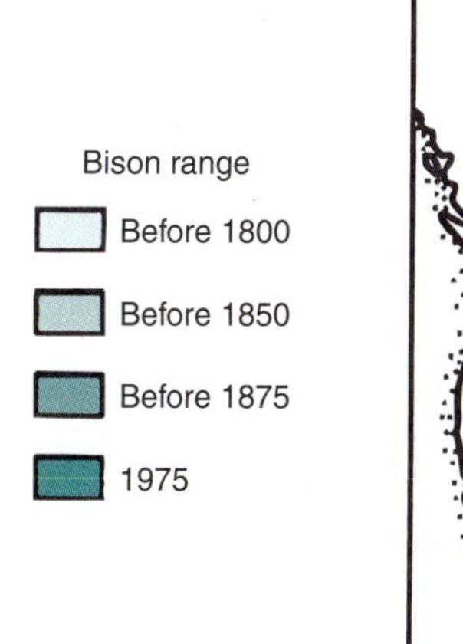

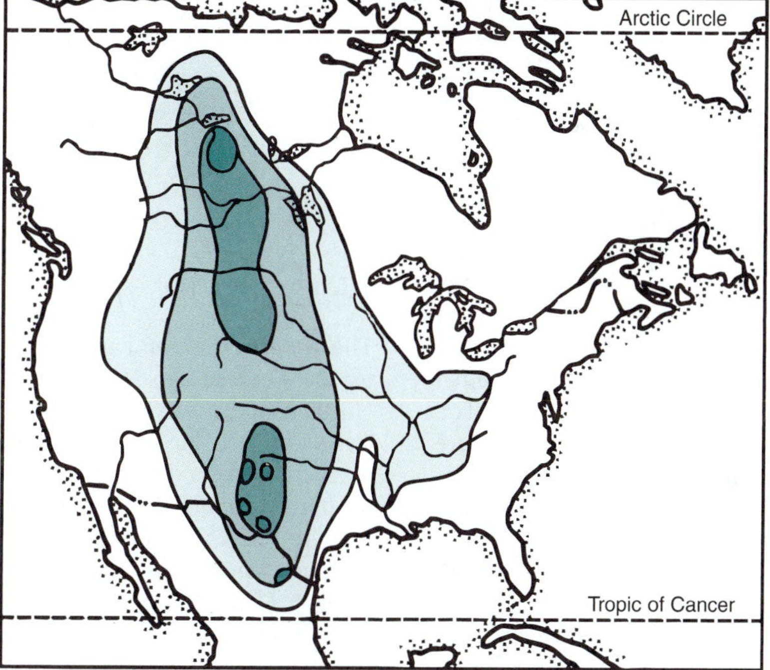

FIGURE 16.5 *The shrinking range of the North American bison from pre-1800 to 1975.*

Genetic diversity can be restored in many organisms by breeding them back with their ancestral wild stock. But what if the wild stock no longer exists? Since the domestication of plants such as wheat and corn, most of the natural landscapes of these plants have been taken up by agriculture, which has destroyed most of the original habitat and their populations.

To ensure that wild stock is preserved, efforts are being made to search out and protect these organisms against extinction. Some no longer exist, and for many others, their once large and diverse gene pools have been reduced to small and localized populations from which it is impossible to recreate the original diversity. This is probably the case with the American bison and North American wild rice. Both have been saved from extinction, but much of their genetic diversity has been lost because of their minuscule populations (Figure 16.5).

16.4 SPECIES COUNTS, ESTIMATES, AND EXTINCTION TRENDS

Scientists have been recording species since the 1700s. To date they have cataloged about 1.4 million species over the entire planet. Among these there are 750,000 insects, 250,000 plants (mainly higher or vascular plants), 350,000 invertebrates and microorganisms (including microflora such as algae and fungi), and 41,000 vertebrates. The larger organisms, represented principally by flowering plants, mammals, birds, reptiles, and fish, have received the most scientific attention. Here the record is most accurate. On the other hand, it is widely agreed that the count is grossly shy of the true number for insects, invertebrates, and microorganisms.

Insects, in particular, are extremely underrecorded. Sampling programs recently carried out in the tropics typically reveal dozens of unknown insects inhabiting areas no larger than several trees and their canopies. When these sample data are projected over large areas, such as the total area of the tropical rainforest, the estimate for insects alone may be 5 million species or more. Of these, beetles are decidedly the most abundant species. Based on similar projections, the estimates for the total species currently inhabiting Earth fall between 10 and 30 million.

Where are these undiscovered species? The principal source of unknown species is the tropical rainforest. The rainforest is a diverse and complex environment where the zone of life extends high into the forest canopy. Here in the treetops, species abound in numbers often exceeding our wildest expectations. But most of these organisms are small, and finding them in their lofts 100 to 150 feet above the ground is difficult.

Besides the tropical rainforest, three other environments hold many unrecorded species: coral reefs, the deep oceans, and soils. Again, each is relatively remote, making field research and specimen collecting difficult. In the case of soil, most of the organisms live in the upper 3 feet or so, and again it is tropical soils that hold the greatest promise for unrecorded species.

Extinction Trends

The geologic record reveals a very dynamic picture of Earth's biodiversity with new species emerging and old ones dying away, resulting in a broad continuum of biodiversity over periods of millions of years. On closer inspection the geologic record also shows distinct periods of accelerated extinction, termed **extinction events** or **episodes.** One such event occurred 65 million years ago and marked the end of the Mesozoic Era, the age of reptiles. At least half of Earth's, plant and animal species were wiped out. The dinosaurs, which were abundant over much of Earth, died off.

For years the cause of this event puzzled scientists. However, in the 1980s a thin layer of unique sediment containing the element *iridium* was discovered in sedimentary bedrock whose age coincided with this event. Research showed that this sediment was created by an explosion on the Earth's surface, caused by an asteroid impact. The explosion threw a huge mass of debris into the atmosphere, which, scientists reason, reduced incoming solar radiation, sending Earth into a deep freeze. Plants declined and the food sources for many herbivores apparently disappeared, followed shortly thereafter by a sequence of extinctions among predators.

Many other extinction events are also identifiable in the geologic record. Some seem to occur at 26-million-year intervals, and scientists speculate that they may be associated with extraterrestrial impacts similar to the one at the end of the Mesozoic Era. Between these events, many smaller episodes of extinction are also apparent. The influence of humans on extinction patterns, however, does not appear until about 10,000 to 15,000 years ago.

Extinction and the Human Scene

The last glaciation on Earth reached its maximum 15,000 to 20,000 years ago. At this time the northern half of North America was covered with great sheets of glacial ice. A similarly large mass of ice covered a large part of northern Eurasia. Humans lived in the zone south of the ice border, where they survived by hunting and gathering. Among the animals hunted were large mammals such as mammoths and mastodons, which are now extinct. The extinction process appears to have taken place over several thousand years in the presence of humans as the ice sheets were melting and the bioclimatic environment was undergoing change.

By 8,000 to 10,000 years ago, dozens of mammal species had disappeared. Scientists debate whether the extinctions were driven mainly by changing climate or mainly by human exploitation. No doubt these animals were under stress from a changing environment, but the evidence for population reduction by early hunters is also compelling. On balance, it appears certain that humans had a hand in the process. At the very least, hunting hastened the extinction process.

The next wave of extinction began about 10,000 years ago with the spread of agriculture and the growth of human populations. As we noted in earlier chapters, before 1500 A.D. or so, much of the area taken up by sedentary agriculture in Europe and Asia was in subtropical and midlatitude environments (see Figure 4.2). Many species were undoubtedly eradicated in these environments, but for most there is little or no record of their existence, especially for plants, insects, and smaller organisms. In the Mediterranean basin, on the other hand, where the early historical record is most complete, there is sound evidence of widespread mammal extinction associated with the spread of early agriculture.

Though less widely cultivated than the subtropics, the wet tropics were not exempt from early agriculture. Large areas of monsoon forest and rainforest were destroyed in Asia and Africa by both crop farmers and herders. In the Americas, tropical rainforest and wetlands were also lost to agriculture. Recent evidence indicates that considerable areas of rainforest were cleared by the Aztec and Mayan civilizations in Central America. All told, the first several thousand years of agriculture undoubtedly forced the extinction of thousands of organisms, most of which we can never know about because they left no measurable record of their existence.

After 1500 and European colonization of the Americas and Africa, the rate of global extinction increased again. In North America, most of the midlatitude forests were eradicated. In the tropics, plantation agriculture, mining, and expanding subsistence farming further reduced the tropical rainforest. Before the twentieth century, however, rainforest eradication was small by today's standards. Today's crisis began in 1950 or so as tropical forest destruction began to climb with rapidly increasing population in the developing countries. By 1975, it had reached epidemic proportions.

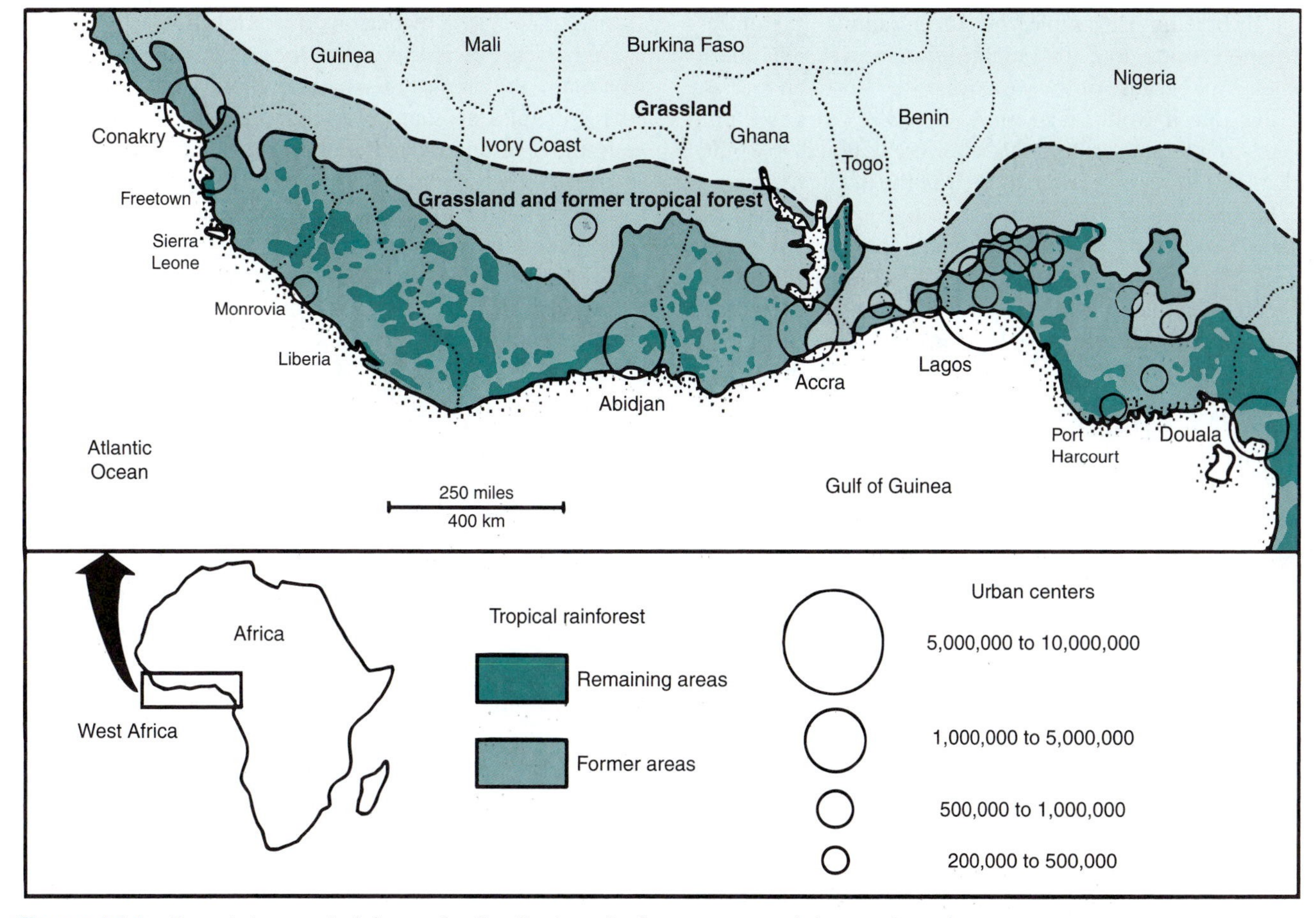

FIGURE 16.6 *Remaining tropical forest, the distribution of urban centers, and the past forest limits in West Africa.*

Tropical Forest Loss

In the past 20 years, tropical forest destruction has reached a rate estimated around 1 percent of the total remaining tropical forest per year. This is equivalent to an annual loss of 65,000 square miles, an area about the size of North Dakota or Washington. This number includes both wholesale eradication and significant disturbance (degradation) of all tropical forests. The resultant extinction rate is not known with certainty, but reliable estimates place it between 1,000 and 10,000 times the prehuman rate. By the year 2040, as much as 35 percent of the world's species could be in danger of extinction.

The rate and geographic distribution of tropical forest clearing needs some clarification before we move on. Estimates of the area currently covered by tropical forest are made with the aid of satellite imagery and are accurate within 5 to 10 percent of actual tree coverage. In the late 1950s, tropical forest coverage, which includes rainforests, monsoon forests, and various other types of tropical tree covers, were set at 17.6 million square kilometers (6.8 million square miles) for Earth as a whole. But the rate of tropical forest loss is highly variable from country to country, and for this reason, the gross global loss rate of about 1 percent annually can be somewhat misleading. In Brazil, for example, the current annual rate of destruction is close to 22,000 square kilometers (8,500 square miles), but because of the vast area of the Brazilian forests, at this rate it would take more than 300 years to completely obliterate them. If the rate were doubled, it would still take 150 years or more to destroy the Brazilian forest.

In contrast to Brazil are African countries such as Nigeria and the Ivory Coast, which have many small areas of rainforest and relatively high destruction rates (Figure 16.6). It is estimated that complete deforestation will take about twenty years in Nigeria and about fifteen years in the Ivory Coast. While these rates may change, the trends indicate that by the year 2020 or 2030, remaining rainforest will be highly variable from country to country. The only sizable tracts remaining will probably be in large countries such as Brazil, Zaire, and Indonesia.

16.5 GEOGRAPHICAL BIODIVERSITY

Much of our understanding about the relationship between biodiversity and environment has come from studies in **island biogeography**. These studies have revealed the powerful influence of two geographic factors, area and distance, on species diversity. The larger the area of an island, the larger the number of species it can support; and the greater the distance separating neighboring islands, the lower the genetic mixing and the lower the resultant biodiversity (Figure 16.7).

If you look at a map of a vast tract of tropical rainforest undergoing destruction, you will see that the pattern is usually very patchy (Figure 16.8). Patches of forest are typically left standing as islands in a great sea of cleared and burned land. As clearing progresses, many of these forest patches disappear, but some survive because they are protected by rough terrain, land ownership, or some other factor.

What chance is there that these islands of forest can maintain the rainforest's natural biodiversity? Although the principles of island biogeography are not strictly applicable to remnant patches of forest, they indicate that the prospects are not good. According to the area factor, these island remnants of forest, despite the fact that they look like the original forest, are incapable of performing biologically as the former rainforest once did. As a general rule, a 90 percent reduction in habitat area results in a 50 percent reduction in biodiversity. In addition, as more islands are

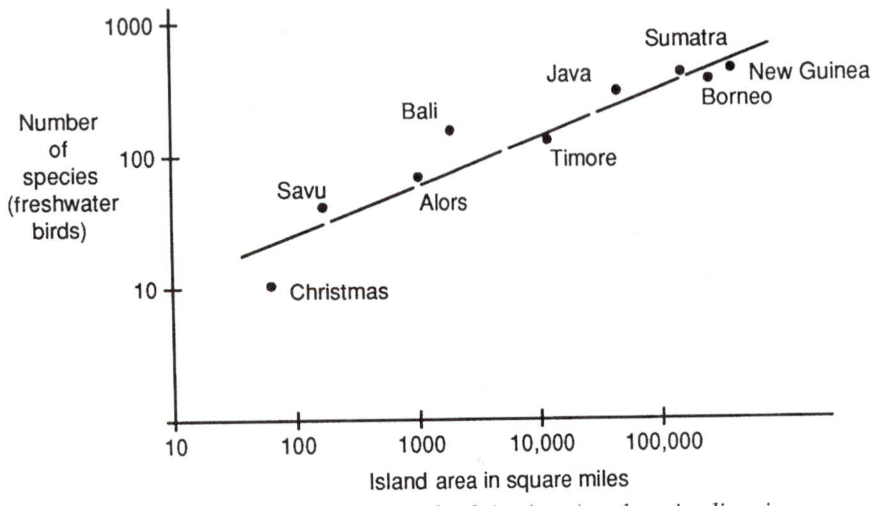

FIGURE 16.7 *The relationship between island size (area) and species diversity.*

FIGURE 16.8 *The pattern of deforestation in an area of southern Rodonia, Brazil, 1988. Notice the isolated patch of forest (which measures 2 by 9 miles) and the extensive cleared strips.*

taken down, the remaining forest islands are left farther and farther apart. This, of course, reduces mixing among isolated members of islands populations (Figure 16.8).

In the tropical rainforest the loss of species may actually exceed 50 percent with 90 percent habitat reduction. The reason is that the ranges of many rainforest species are endemic, often not more than a few acres or a few square miles. This contrasts with midlatitude species, which, although fewer in total number, are often distributed in much larger ranges. Thus, in the tropics eradication of large areas of forest is more likely to destroy the entire ranges of many endemic species resulting in greater than 50 percent species loss as forest loss approaches 90 percent.

In addition, studies of the short-term relationship between species decline and forest loss sometimes reveal lighter than expected losses. The reason for this, as other studies have shown, is that reduction in biodiversity does not occur suddenly with habitat reduction. Rather, species counts decline gradually over many years or many decades. Thus, short-term postclearing evaluations of biodiversity changes, based on indicator organisms such as birds, must be extended over the long-run in order to get a reliable picture of the impact of habitat loss on species numbers in general. Finally, we must recognize that our current understanding of these problems is very limited and much remains to be learned about biogeographical relations as a whole.

Relationship between Organisms and Their Habitat

Habitat is defined as the local environment of an organism. It is sometimes referred to as the environmental address of an organism, but more frequently it is thought of as a unit of space and its environmental features—principally microclimate, soil, topography, water available nutrition, and other organisms (Figure 16.9). Although different organisms may occupy the same or very similar habitats—for example, different birds in the same wetland—interaction with the habitat is different for each species.

The term **niche** is used to define an organism's way of life or what it does in its habitat. Insects occupying a tree canopy habitat, for instance, may survive by acquiring food in different ways: some eat leaves, some suck nectar, and others chew bark. *Niche* defines an organism's functional relationship with its habitat: through its niche, each organism "sees" its habitat as unique.

Because of the special relationship each organism has with its habitat, it is virtually impossible to trade one habitat for another. Unlike humans, who have a wide range of habitat versatility, most organisms displaced

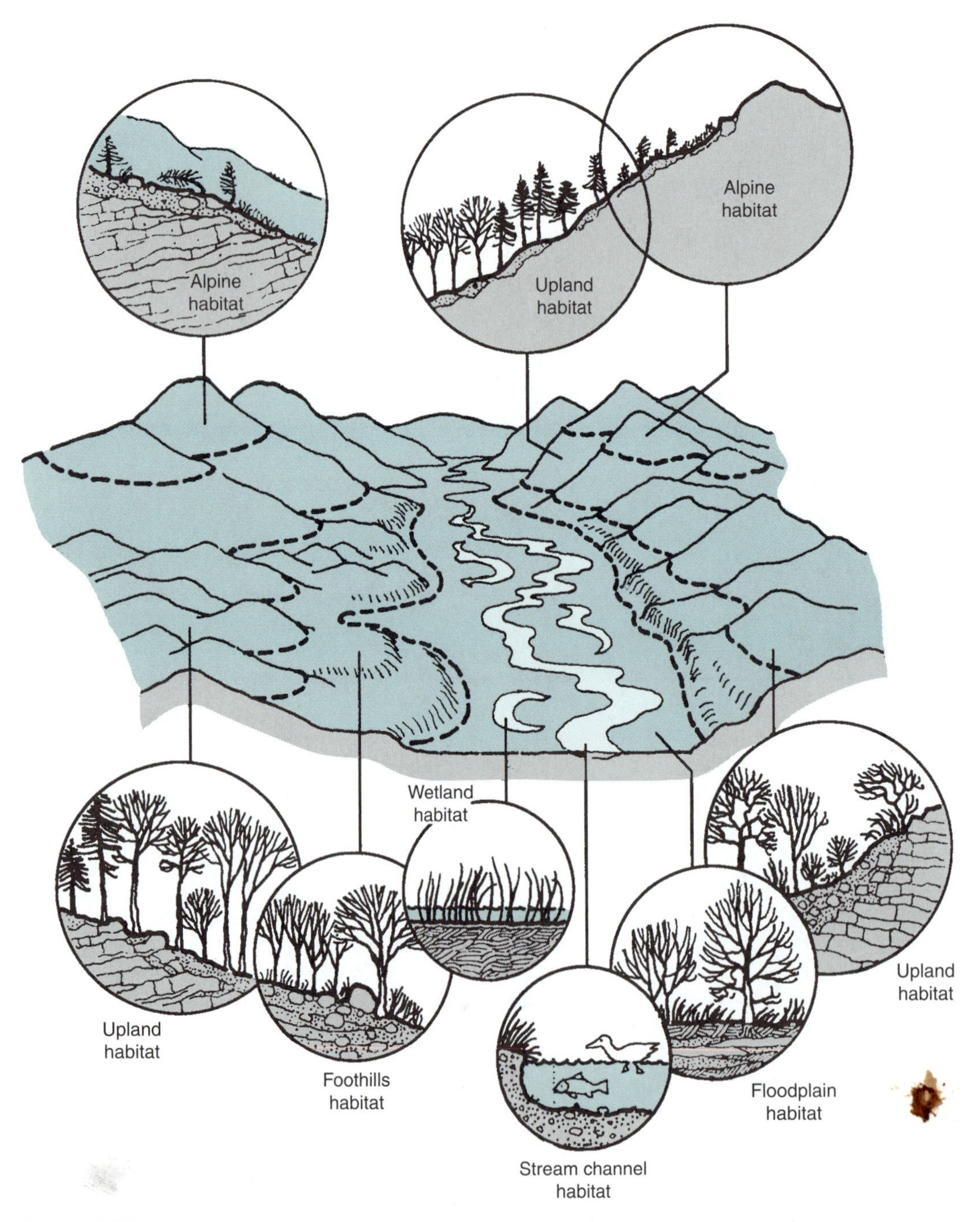

FIGURE 16.9 *Different habitats associated with a local geographic setting consisting of a stream valley and surrounding uplands.*

from one habitat cannot simply take up life in another. In addition, when habitats are destroyed, they cannot be recreated in a manner suitable to sustain most of the organisms that once occupied them. This is especially so for complex habitats, such as the tropical rainforest, where niches are formed by many and often subtle interrelationships with other organisms, such as insects and microflora. Herein lies the fundamental dilemma of preserving organisms in zoos or seed banks whose habitats have been destroyed. If their habitats are gone and cannot be resurrected, then there is no place to sustain the organism and the zoo becomes an artificial life support system.

Land Use and Habitat Loss

What do land uses do to habitat? The effects are highly variable, depending on the type and density of land use. The least disruptive land use is decidedly hunting-gathering because land clearing and agriculture are not involved and human populations are very light. In fact, it is not inappropriate to consider hunting–gathering societies as part of the natural landscape, because they, like other mammal populations, do little to manipulate the environment. When agriculture is added to hunting and gathering, such as in the land-use system known as **slash-and-burn** or **shifting agriculture**, the impact on habitat increases significantly (see Table 4.1).

In slash-and-burn farming, forest is cleared and burned, destroying canopy, understory, ground cover, and much of the topsoil (Figure 16.10). On the other hand, traditional shifting agriculture is limited to small, isolated patches, often widely scattered in the rainforest. The likelihood exists of destroying a species with very small ranges, but the overall effects of traditional shifting agriculture are not considered highly detrimental to biodiversity unless the cycle of land clearing and recovery are compressed in time and forced to reuse the same land repeatedly.

When agriculture is organized into large-scale systems, whether subsistence or commercially based, the balance between cleared land and natural habitat shifts in favor of cleared land. The actual ratio between the two depends on the suitability of the land for agriculture and the regional population pressure to use more land. The conversion of natural landscapes to farmland progresses more or less in stages, as illustrated in Figure 16.11.

In the early stages, land clearing follows a reasonably predictable pattern corresponding to the landforms in watersheds. Farmland occupies belts of arable land lying between corridors of less usable land stretching along valley floors, stream systems, and hilly terrain. As population increases in the latter stages of agricultural development, pressure to clear more land pushes farming beyond the reasonable limits of cropland into marginal lands, such as slopes and lowlands. Wetlands are drained and the remaining **habitat corridors** are broken up or *fragmented* into smaller segments. The remaining uncleared landscape is left in small, widely separated islands (Figure 16.12).

Within the agriculture sector of the landscape, **fragmentation** severely limits habitat potential. Trees and shrubs, the habitats of many birds and insects, are reduced to field margins or **edge habitats**. Some of the original plant and animal species are replaced by introduced species. Natural ground cover is largely eliminated and topsoil is reduced significantly by crop farming. Plowing greatly disrupts or destroys root, insect, and microorganism habitat and pesticides further limit the soil as habitat.

Under intensive cultivation, the landscape becomes biologically impoverished. Farm fields are reduced to biological monocultures dominated by several crop species, and the remnant islands of habitat are too small and widely spaced to support more than a fraction of the original species. As agriculture intensifies with commercialization and/or expansion into marginal lands, remnant habitat is chopped up into smaller fragments that are subject to severe damage from surrounding cropland and farm animals.

Opportunistic Species

Whereas many species are reduced or eliminated by land clearing, agriculture, and settlement, a number of species are favored by these changes. Some of these are native species, but many are aliens (Figure 16.13). A recent U.S. government report identified more than 6,500 alien plants, animals, and microorganisms in the United States. Many of these reduce biodiversity by displacing native species or whole communities or by killing or weakening native species with new diseases. Among the aliens are hundreds of pathogens brought in with imported ornamental plants. Among the plants, weed species, some native and some alien, such as sumac, ragweed, and Russian olive have increased their geographic coverage and populations with the spread of land use development. These plants have responded to (1) edge habitats such as fence lines and hedgerows along farm fields where the original species cannot sustain themselves, and (2) disturbed areas such as road margins, playgrounds, and construction sites. In extreme cases, alien plants such as kudzu, an Asian vine introduced to the American South to stabilize eroding farmland, have overrun and smothered local landscapes (Figure 16.13).

Many mammals and birds have also increased, taking up edge habitats and disturbed areas. Fragmentation of forested landscapes in the U.S. Midwest, for example, has favored cowbirds, blue jays, and crows. After being introduced into North America, European house sparrows and starlings have been very successful in agricultural areas as well as in cities and suburbs. Coyotes, rabbits, opossums, raccoons, and white-tailed deer, all North American natives, have shown a remarkable capacity to

FIGURE 16.10 *The slash-and-burn method of tropical forest clearing for shifting, subsistence farming in the Amazon. Land clearing involves (a) chopping down trees and vines, (b) burning the debris, and (c) planting the cleaned area. Plots are usually small and scattered; however, population growth and development may force larger, permanent plots and settlements and deny forest recovery.*

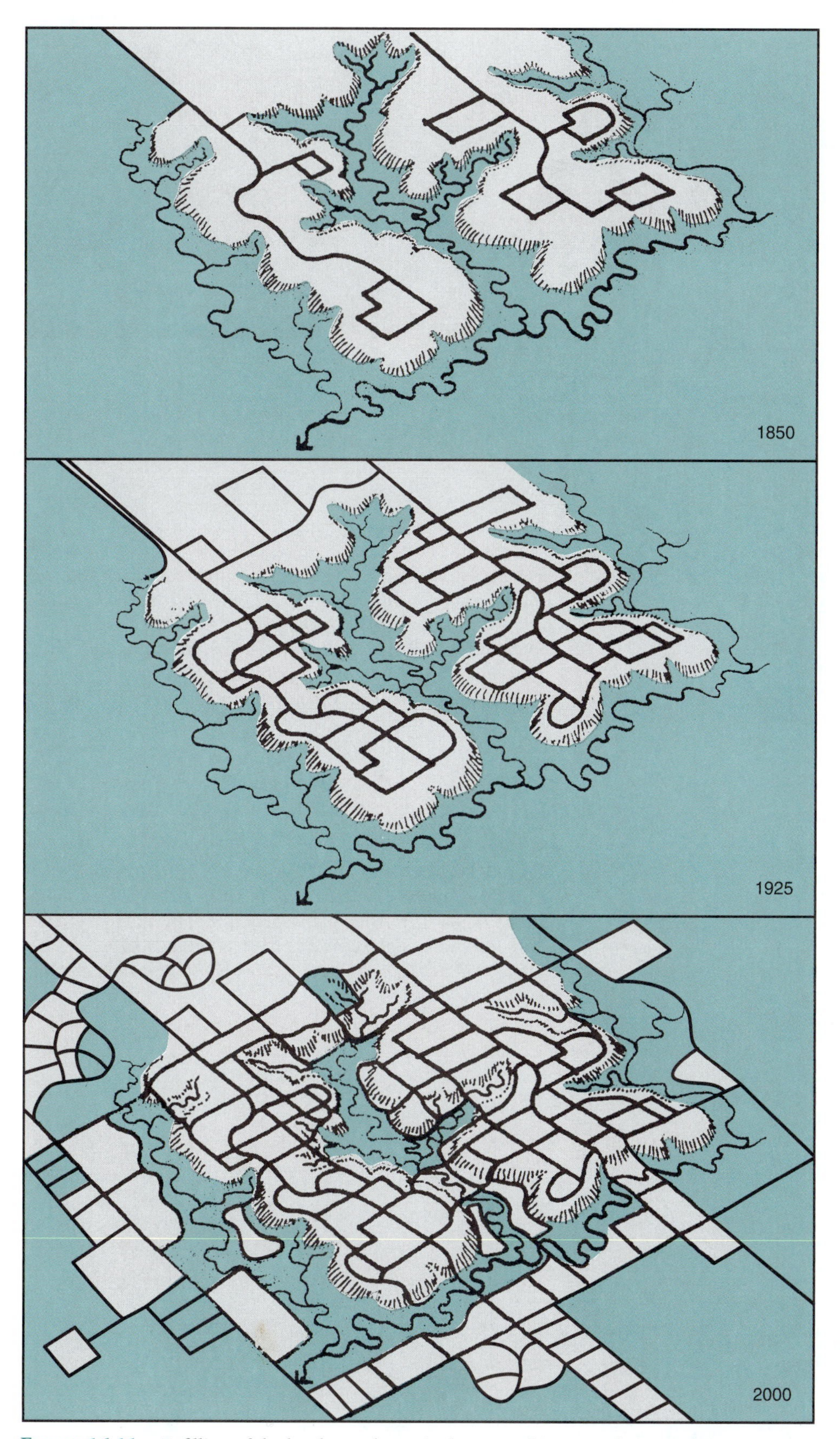

FIGURE 16.11 *Infilling of the landscape by agriculture resulting in reduction of open space and fragmentation of habitat corridors. Land use eventually spills over slopes and into stream valleys and wetlands.*

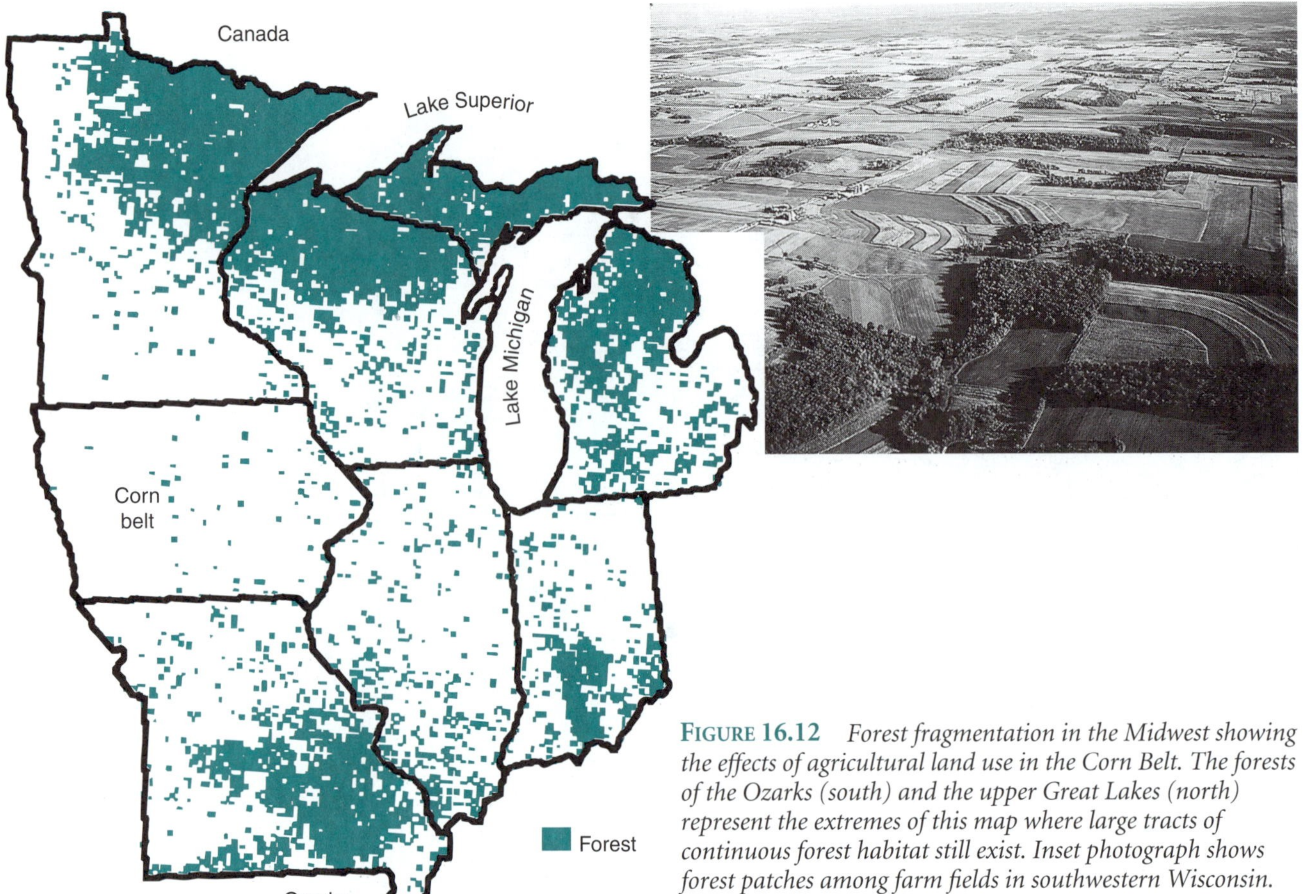

FIGURE 16.12 *Forest fragmentation in the Midwest showing the effects of agricultural land use in the Corn Belt. The forests of the Ozarks (south) and the upper Great Lakes (north) represent the extremes of this map where large tracts of continuous forest habitat still exist. Inset photograph shows forest patches among farm fields in southwestern Wisconsin.*

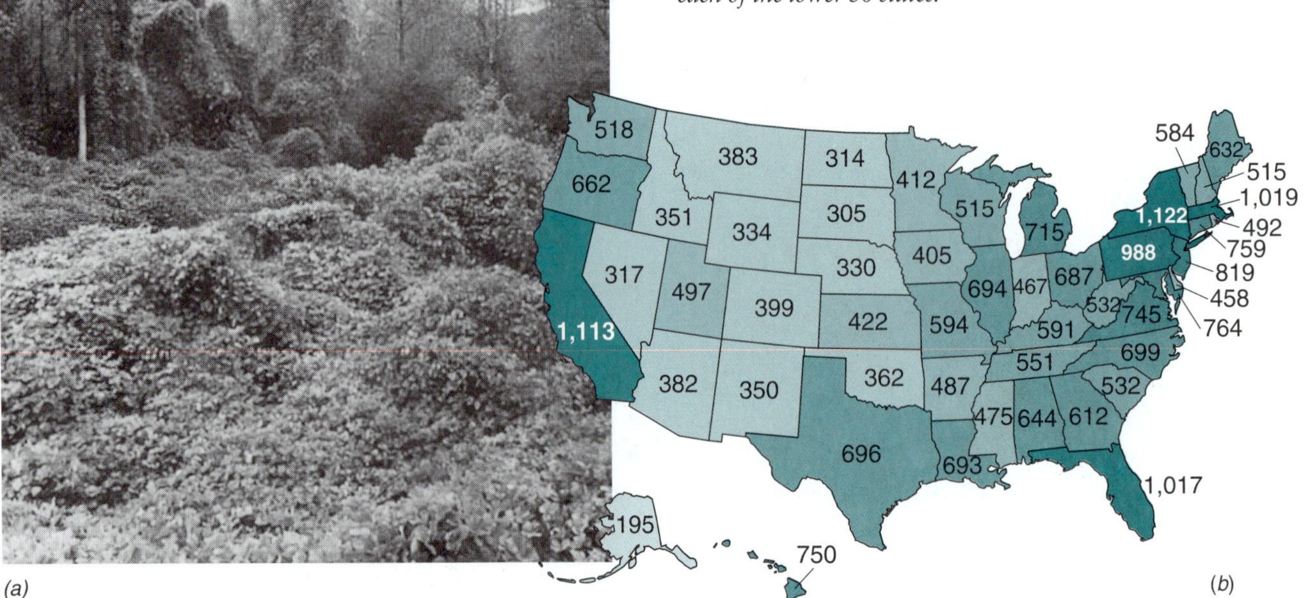

FIGURE 16.13 *(a) A landscape overrun by kudzu, a plant introduced to the American South to help arrest field erosion in the 1930s. This Asian plant spreads rapidly and overgrows both natural and planted vegetation, eventually killing most large plants. (b) The number of alien plants known to exist in each of the lower 50 states.*

adapt to a new and changed habitat. Such versatility has enabled them to reinhabit suburban areas from which they were displaced during the clearing and construction phases of the development process. The trend to reoccupy is explained in part by the observation that as the suburban landscape ages and tree and shrub covers increase, the landscape becomes more diverse and richer as habitat. In addition, the geographical linkage within and among suburban areas is often improved as habitat corridors are created with the establishement of parks and wooded areas around schools, in vacant lots, and along streets.

16.6 ENDANGERED, THREATENED, AND PROTECTED SPECIES

Many countries are giving attention to the species loss issue. In most developed countries and many developing countries, policies have been enacted designating special status to selected organisms as protected species. The protected organisms that have gained most international attention are isolated species with some special significance to society. These are often exotic organisms such as the black rhino or the African elephant, which are subject to overexploitation for economic purposes.

In 1973 the United States Congress passed the **Endangered Species Act**. This controversial law established two classes of protected species. **Endangered species** are defined as those in imminent danger of extinction in all or a significant portion of their ranges. **Threatened species** are those with rapidly declining populations that are likely to become endangered within the foreseeable future. While the act applies to all lands and marine environments (private, state, and federal), it has focused particularly on federal construction and land-use projects (e.g., military bases, interstate highways, and flood-control facilities) and on private projects using federal money or requiring federal permits. The act also makes it illegal to capture, kill, possess, buy, sell, transport, import, or export threatened or endangered species.

Although the importance of individual protected species should not be trivialized, it is not the central problem in species protection. The central problem is the much broader issue of biodiversity and habitat protection. The greatest loss of species is caused by habitat destruction, not by exploitation of individual organisms; therefore, the focus should be on the protection of entire ecosystems. In protecting ecosystems, habitat must automatically be incorporated because habitat, represented by soil, topography, water features, and so on, is part and parcel of the ecosystem. The passage and enforcement of wetland laws represent a move toward habitat protection and conservation of biodiversity.

On the other hand, the protection of individual organisms is not without some broader ecological benefits. By protecting individual species, their ecosystems must also be protected, and in turn a larger network of organisms and habitats also receive protection. The least prudent approach to species protection is the creation of zoolike preserves where special organisms are given showcase status without the appropriate habitat, area, and ecosystem arrangements for long-term viability.

The Most Vulnerable Species

Some species are decidedly more prone to extinction than others. At the top of the list are microendemic species, which are found mainly in the tropical rainforest. For some of these species, their range is smaller than the area of land cleared for a peasant farm. They, therefore, can be swept to extinction in one season of land clearing. Endemic species that are *relics* of once large populations and ranges are also vulnerable. Among other things, these species often have very narrow habitat requirements and are therefore subject to decline from subtle imbalances in the environment. The Chinese giant panda, for example, depends on only a few species of bamboo for virtually its entire food supply. Should these bamboos (which are already severely limited by land use) be destroyed by development or natural change, the pandas are likely to starve. Only about 1,000 pandas remain in the wild in China.

Species with small populations are also prone to extinction because loss of relatively few individuals may pull the population below the critical threshold of a breeding population. This is especially significant for animals, such as whales, accustomed to living and breeding in groups. If the group becomes too small, breeding may cease even though breeding adults are available. Small populations are also vulnerable for species, such as cheetahs, that reproduce slowly. These species produce one or fewer offspring a year and the young require a number of years to reach breeding age. The recovery time is dangerously long for such populations, because they are subject to other perturbations during recovery.

Another extinction-prone class is species that have evolved in an isolated environment that has protected them from competition and predation. Many island species, such as the giant tortoises on the Galapagos Islands, declined when goats were introduced to the island because goats consumed the tortoise's food sources. In other cases, island species were decimated by predators introduced by humans, or by humans themselves. Among the predators commonly associated with humans are pigs, dogs, chickens, and rats. Many bird populations in New Zealand and Hawaii, for example, evolved without serious natural predators, making them vulnerable to the new and vigorous predators introduced with early settlers. Introduced snakes have been particularly effective in decimating bird populations on a number of tropical islands. Witness that two-thirds of the birds in the Hawaiian

Islands became extinct over the two hundred years following the initial settlement of the islands by Polynesians. The World Conservation Union estimates that since 1600 nearly 40 percent of the animal extinctions in the world have been caused by introduced species.

16.7 WAYS OF CUTTING BIODIVERSITY LOSSES

There are several ways of minimizing species loss under the circumstances. National **land-use programs** hold promise as a means of guiding and limiting land-use development. But because of political obstacles they have proven difficult to formulate and even more difficult to implement, especially in developing countries. National land-use planning is a difficult concept to promote because it is widely viewed as limiting to economic development. Moreover, there are few major examples to follow because most of the large developed countries, including the United States, do not have national land-use plans. Nonetheless, national land-use planning is inevitable for many countries in the twenty-first century if they are to achieve both sustainable economic systems and conservation of valued environmental resources (including rangeland habitat, old forests, coastal lands, and prime agricultural land).

Conservation programs also hold promise. These programs focus on open space: parks, wildlife preserves, wilderness areas, forest reserves, and landscape quality. Although parks and preserves, which are the most popular measures, have drawbacks for biodiversity preservation, they are nonetheless one important line of defense. The largest areas of protected or preserved forests are found in North America, Europe, and the former Soviet Union (see the following chapter for more details). In Asia, Africa, and South America, parks and preserves currently make up about 5 percent of the total areas of tropical forest, savannas, and associated woodlands. Broadly speaking, a larger total area of preserves—say, 20 percent or more of the remaining tropical forests—would provide better insurance for species preservation.

Because most of the areas in need of conservation programs already contain local human populations, accommodations will have to be made for these people. This will not be easy, because many of the populations in Africa, South America and Asia are already stressed because of rapid growth, ethnic conflict and lack of farmland. **Integrated conservation—development projects** have been proposed as a way of reducing the human threat to protected areas such as parks and preserves. The objective is to provide these people with sustainable, income-generating opportunities. This could include (1) purchasing additional land or negotiating for the use of land for farming and (2) setting up buffers around protected lands where certain economic activities would be encouraged. Other approaches call for **integrated landscape management**, which involves coordinating government agencies, businesses, community leaders, landowners, and others in a region to ensure that biodiversity objectives are included in the overall planning and management process.

Although more land and aquatic environment must be set aside for habitat protection, not just any accumulation of land will do. The ecological benefits of preserved land can be maximized by selecting the right locations, sizes, and numbers of preserves. Many small, widely scattered tracts are generally not advisable. Large tracts are preferable, and their locations and geographic

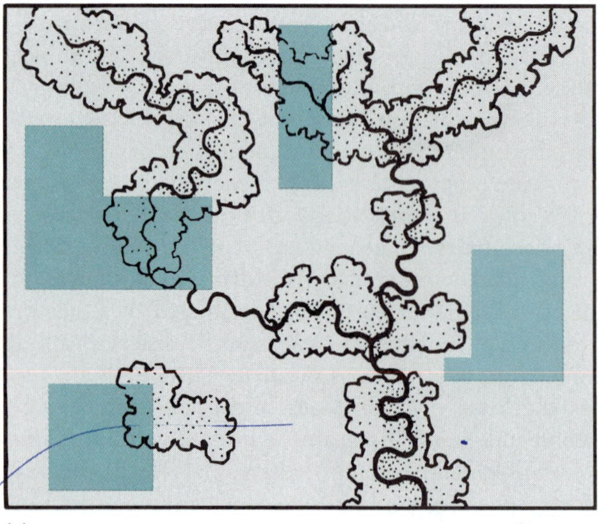

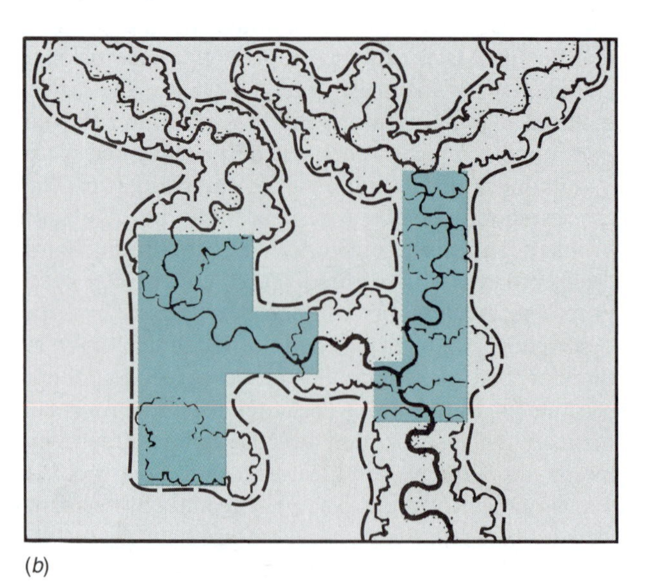

FIGURE 16.14 *Alternative designs of open space tracts for habitat conservation: (a) dispersed tracts and (b) linked tracts forming habitat corridors. The latter favors habitat continuity.*

shapes should conform to existing ecosystems. Where large areas are not possible, the tracts selected should be linked together in corridors that facilitate unimpaired movement among animals (Figure 16.14).

The concept of corridors is very important in habitat management. Not only do corridors minimize landscape fragmentation, but they are a means of linking larger preserves together. Watersheds often serve as models for integrated corridor plans because stream networks and associated floodplains are excellent sites along which to build the corridors. Not only are stream networks continuous over large areas, but they are ecologically rich and often among the last areas taken up by land use.

Repairing Damaged Landscapes

Most of the forest lands and grasslands in the midlatitudes and subtropics were cleared, farmed, and settled before the twentieth century. Extensive areas of forest habitat in Canada and the United States, for example, were reduced to little more than small woodlot patches among farm fields. The woodlots were used for a variety of purposes: firewood, windbreaks, and small-game hunting. Because of their small areas and isolation, the biodiversity of these islands was extremely limited. Not only that, but surrounding fields and roads posed barriers to linkage among neighboring woodlots (Figure 16.15).

The concept of networks and spatially continuous land-use systems is very much a part of planning tradition in rural areas. Small roads link to large roads; villages and towns are linked by railroad and highways; and small streams are linked to larger streams to form drainage networks. Thus, the notion of linking fragments of habitat into corridor systems is not foreign to our way of thinking about landscape organization. New opportunities to create landscape ecology corridors for improved biodiversity are emerging in the United States with the decline in family farming and the shift in rural land-use patterns. As small farms fall out of production, fields are abandoned, especially where the land tends to be marginal, such as in floodplains and hilly terrain. The fields fill in with woody vegetation, topsoil begins to rebuild, and habitat generally improves. This process, however, does not automatically build habitat corridors, because plowed fields, communities, and roads often remain as barriers. Thus, planning action that includes land acquisition and cooperation from private landowners is usually required to facilitate the linkage process.

Offsetting habitat recovery at the local level is the emergence of major ecological barriers in the American landscape over the past several decades. Chief among these is the system of interstate highways and similar expressways, especially where development has filled in the land along the highway. The resultant corridor is an imposing barrier, an ecological wasteland, that not only limits seasonal migrations and interbreeding but limits longer-term migrations as well (Figure 16.16). These longer-term migrations might be essential to the survival of some species as they shift their ranges in response to climate change in this century.

FIGURE 16.15 *Agricultural landscape in the Pacific Northwest where highways and farm fields form barriers between scattered woodlot habitats.*

FIGURE 16.16 *Expressway corridor across the Florida Everglades poses a formidable barrier to animal migrations. Few highways are designed to facilitate animal movement.*

16.8 SUMMARY

Species have been disappearing from Earth as a result of human actions for more than 10,000 years. Until 1500 or so, a large part of the human-induced extinction in Europe and Asia was concentrated in the subtropics and midlatitudes, where land clearing, crop farming, and herding destroyed or altered vast areas of habitat. In the Americas and Africa, the wet tropics were also subject to extinctions related to indigenous agriculture during the pre-Columbian period. We have some records on the disappearance of many birds and mammals since 1600 or so, but no information on the loss of thousands of soft, small organisms in the form of microflora, insects, invertebrates, and microorganisms.

In the last century, the rate of extinction accelerated with the spread of land use into tropical forests, the reservoirs of more than half the Earth's species. The driving force behind this trend is the rapid growth of world population. Between 1950 and 2000 global population increased from 2.9 billion to more than 6.1 billion. The fastest growing populations are in developing countries, and many of these developing countries contain the world's only remaining tropical rainforests. By 2030, world population is projected to increase by another 2.1 billion with more than 98 percent of that growth in developing countries.

The threat to biodiversity comes mainly from habitat loss and introduced species rather than economic exploitation of selected species. In the tropics the leading cause of habitat loss in the rainforest is the spread of farming, mainly the subsistence variety. Forest is cut, burned, and converted to small farms resulting in habitat loss, reduction, and fragmentation. Other land uses, driven both by foreign and domestic interests, are also responsible for forest destruction: lumbering, commercial farming, and mining. Some of the global "hot spots" in the tropical and subtropical forests are shown in Figure 16.17. These are especially rich and/or unique ecological areas that are being threatened by land-use development.

A leading factor in tropical forest eradication is improved access to remote areas. In the past several decades, road-building programs promoted by governmental policies and often supported by outside funding sources such as the World Bank have pushed roads deep into the tropical forest. In Brazil the roads have encouraged economic development such as logging and mining as well as clandestine forest clearing by peasant farmers (Figure 16.18). Many of these farmers, or hope-to-be-farmers, are from urban slums, and their tenure on the land after clearing is often short because the farming effort frequently fails. The Brazilian government is attempting to curb illegal land clearing, but it is extremely difficult because of the huge areas that must be policed.

In global perspective, given the population projections and the momentum of the wave of tropical forest destruction currently underway, the prospects for avoiding widespread habitat loss and massive extinction are very poor. By the year 2015, as much as 35 percent of the world's species could be in danger of extinction. Given even the most optimistic forecasts on curbing population growth in the developing countries, expanding land use into the world's tropical forests will continue for decades.

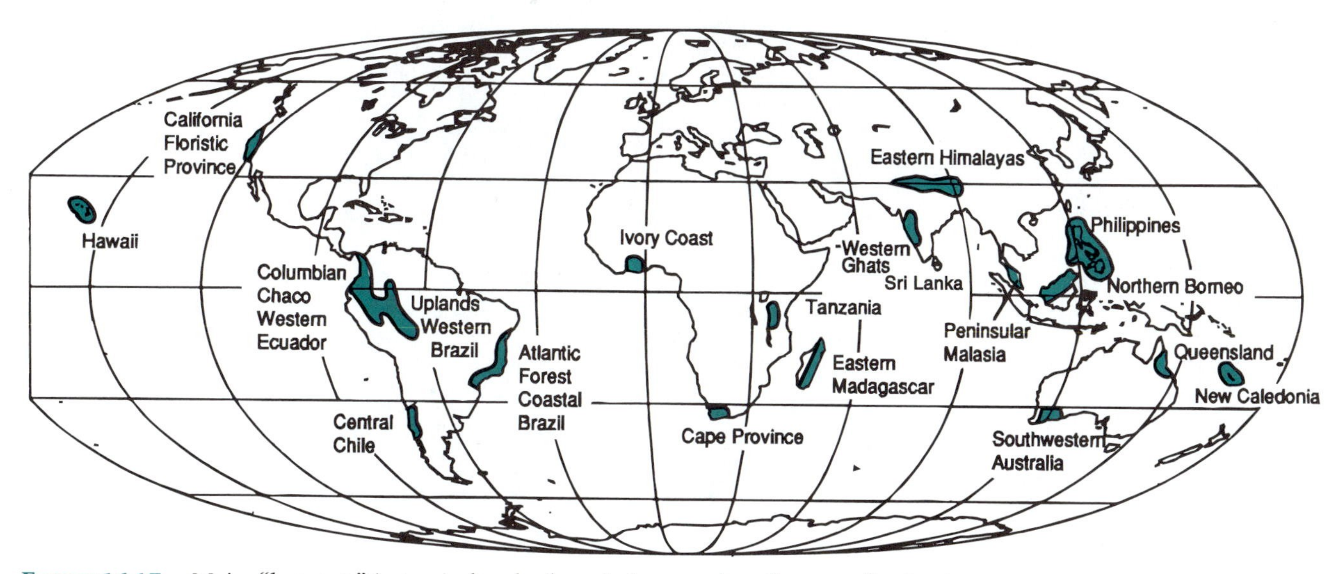

FIGURE 16.17 *Major "hot spots" in tropical and subtropical zones where forest eradication is sharply reducing biodiversity.*

(a)

(b)

FIGURE 16.18 *The development of roads into the Amazon Frontier has opened the region to forest clearing and settlement: (a) tropical forest clearing along roads (white lines) as revealed by satellite imagery; (b) a closer view of the clearing pattern.*

It appears inevitable that Earth is destined to be much poorer biologically in the next century.

With the projected eradication of most remaining tropical forest over the next 50 years, the loss of tropical species is estimated at 30 to 50 percent, or as many as 2.5 million species if we use 10 million as the Earth's species count. With the loss of biodiversity, the Earth's resource base in terms of future food, medicine, and industrial materials will be reduced immeasurably. But the reduction may be less severe if we promote and implement various management measures, including national land-use programs, international conservation programs with greatly expanded and integrated systems of preserves, and programs to repair damaged landscapes that reduce barriers and fragmentation.

16.9 KEY TERMS AND CONCEPTS

- Biological diversity
    - species
    - speciation
    - geographic isolation
    - polyploidy
- Species range
    - intercontinental
    - endemic species
    - vicariads
- Extinction
    - psuedo extinction
    - true extinction
    - clade
- Genetic diversity
    - food and medicine
    - wild stock
- Species count and estimates
- Extinction events
- Extinction and land use
    - historical trends
    - tropical forest loss
- Geographical biodiversity
    - island biogeography
    - habitat and niche
    - slash-and-burn/shifting agriculture
    - habitat corridors
    - fragmentation
    - edge habitats
    - opportunistic species
- Threatened and endangered species
    - Endangered Species Act
    - vulnerable species
- Cutting biodiversity losses
    - land-use programs
    - conservation programs
    - integrated conservation–development projects
    - integrated landscape management
    - repairing damaged landscapes

16.10 QUESTIONS FOR REVIEW

1. What is biodiversity, what geographic region of Earth possesses the greatest biodiversity, and what are humans doing to reduce this biodiversity?
2. What is a species, what is speciation, and can you describe the two principal mechanisms of speciation?
3. Define the term range and describe various types of ranges and their relative geographic scales. What influence does land use typically have on ranges? Why are endemic or microendemic species especially vulnerable to extinction related to land-use development?
4. Can you describe the difference between true extinction and pseudo extinction? Which tends to be associated with human action and why?
5. Build a good argument against this statement: The Earth has millions of species and the loss of several thousand per year can't make much of a difference to the quality of life and our future on the planet.
6. How many species do scientists estimate currently exist on Earth? How many have actually been cataloged? What accounts for the big gap between the recorded and estimated number? Where do these unrecorded species live, and what kinds of organisms are most of them?
7. Outline some of the major extinction trends or events that we are aware of on the Earth. How do humans fit into this in terms of time and geographic location? What is the current rate of tropical forest destruction, and which tropical countries will be the first to experience complete forest eradication?
8. How does the concept of island biogeography apply to tropical forest clearing, and why is it that leaving patches of rainforest intact is not a good way of ensuring preservation of biodiversity?
9. Define the terms habitat and niche. Discuss the effects of various types of agriculture on habitat. What

is meant by edge habitats and how do opportunistic species respond to habitat changes?

10. What are endangered and threatened species? Should species preservation programs focus on species or habitats? Describe some of the most vulnerable species and give some reasons for their vulnerability.
11. Describe some ways of cutting biodiversity losses. Outline a plan for improving biodiversity in the area of your neighborhood, town, or farm.
12. Summarize your position on biodiversity. How serious do you think this problem is? What principal force(s) drive global biodiversity trends? How would you respond to a Brazilian who says, "You midlatitude societies have already cut your forests away, and now you point an accusing finger at us when we harvest *our* forests."

17

OPEN-LAND RESOURCES: FORESTS, RANGELANDS, PARKS, AND PRESERVES

17.1 INTRODUCTION

Although most of the world's 6.1 billion people live in towns and cities or in rural communities supported by productive crop agriculture, these areas occupy only about one-sixth of Earth's total land area. Beyond these areas, the uninhabited icecaps, and some desert areas, the majority of the land is *open land.* This land is ill-suited for permanent agriculture and large-scale settlement because of rugged terrain, infertile or saturated soils, or climates that are too dry or too cold. Until recently, these difficult environments have supported only small, isolated populations of subsistence farmers, herders, or gatherers. However, as we noted in Chapter 4, over the last century many of Earth's open lands have become frontiers of commercial resource development and settlement. As logging, livestock ranching, and crop farming have replaced traditional subsistence activities, *deforestation, overgrazing,* and *soil erosion* have seriously damaged large parts of these environments. This chapter examines our changing perceptions, use, and management of open lands.

Not all open lands have been developed or used for commercial purposes. Today, in most countries some areas have been set aside as *parks, wilderness areas,* or *wildlife preserves.* In some large countries such as the United States and Canada, significant areas of open lands are administered by government agencies as *multiple-use resources.* For example, the U.S. national forests are supposed to be managed to protect watersheds and wildlife habitat and to support recreation opportunities, as well as to provide a sustainable timber supply. This chapter will also discuss the government's role in managing multiple-use lands, parks, wilderness areas, and wildlife preserves as well as the expanding role of nongovernmental organizations in land resource conservation.

17.2 DEVELOPMENT AND EXPLOITATION OF OPEN-LAND RESOURCES

At the dawn of the Industrial Revolution (1750), the majority of Earth's tropical forests, savannas, and grasslands were truly natural, wild landscapes, little modified by human activity. In general, the small populations who inhabited these areas practiced their subsistence ways of life with only slight modification or alteration of the environment. However, by 1750 the impetus for rapid and extensive change had already begun because the Europeans had established colonies and outposts in Africa, Asia, Australia, and the Americas. Due to the technological revolutions of the eighteenth and nineteenth centuries in agriculture, medicine, sanitation, transportation, and industry, population grew rapidly. The burgeoning populations in Europe prompted many people to emigrate to the Americas or colonial outposts worldwide, where settlement was encouraged and subsidized.

Settling a Continent: The American Experience

During the first half of the nineteenth century the newly independent United States had gained control of huge expanses of land that stretched from the Mississippi River to the Pacific coast (Figure 17.1). To promote settlement west of the Appalachians the federal government initiated land grants and sales. The vast wilderness of forests, plains, mountains, and deserts represented a seemingly inexhaustible fount of resources to support a rapidly growing population and economy.

Spurred by the railroads that penetrated the Midwest's prairies and plains and the new steel plows that could till the tough sod, these fertile grasslands quickly became America's agricultural heartland. After the Civil War, the transcontinental railroads and the Homestead Acts effectively opened the rest of the west to settlement. Dry-land and irrigated agriculture expanded rapidly in parts of the western plains and intermountain areas during the 1880s and 1890s. As a result, cropland acreage in the United States increased nearly fourfold between 1850 and 1900, with most of the expansion west of the Mississippi River.

Grazing expanded rapidly on public rangelands. Use of these lands was free, and this provided strong economic incentive for building large herds. Without controls against overgrazing, rangeland vegetation was depleted and soil erosion set in. By 1900 millions of acres of range had been severely depleted by sheep and cattle herds estimated to be 85 to 90 million head. Early in the twentieth century, special homesteading acts encouraged the transfer of nearly 100 million acres of the better

FIGURE 17.1 *Acquisition of territory by the United States. By 1853, the entire territory which makes up the contiguous 48 states was under U.S. control. Alaska was added in 1867 and Hawaii in 1898.*

grazing lands to private ownership as ranching replaced herding. Even so, this change in the organization of commercial grazing did not reduce overgrazing. The condition of the remaining open range continued to deteriorate.

Forest Use and Mining

Along with the settlement frontier, lumber production expanded westward during the nineteenth century. By the 1840s, New York had replaced Maine as the leading lumber producer and was succeeded by Pennsylvania in 1860 and Michigan in 1870. By the late 1870s, the forests surrounding the Great Lakes were supplying more than one-third of the nation's lumber (Figure 17.2a). However by the first decade of the twentieth century, the southern states had become the leading source, supplying nearly two-fifths of the nation's construction lumber. As the northern and southern pine forests were cut over, lumber production expanded to the west coast. By 1920, the Pacific forests were a leading source of the nation's lumber.

As lumber production declined in one region with the loss of old-growth forests, lumberjacks moved to more distant virgin forests. Declining old-growth forests and higher timber prices did not provide great enough economic incentive to conserve mature forests or to replant and nurture young forests. Natural forest regeneration was hampered by forest fires that repeatedly swept across the cut-over lands and by widespread loss caused by insects and diseases. No attempts were made to conserve the forest resource until national forest reserves began to be set aside toward the end of the nineteenth century.

Mining was also encouraged by the federal government. Mineral prospecting and mining on the public domain were open to anyone. The Mining Act of 1872 provided that free claims could be staked on public lands wherever minerals were discovered. These claims provided exclusive mineral and surface rights for indefinite periods, provided some assessment work was done each year. Once mining began, the land could be purchased at a nominal cost and the claim patented without royalty payments to the government. Over the decades, thousands of mining claims have been staked on public lands, with many made to simply gain free use of the land for grazing, timber, or other nonmining purposes (Figure 17.2b). The Mining Act of 1872 continues today as the law of the land!

Decline of Wildlife and Closing of the Frontier

Many species of wildlife were eradicated in the east as a result of habitat destruction and hunting. Similar decline and demise of species occurred as the frontier proceeded westward, the extensive forests were cut, and the vast grasslands were converted to cropland and range. By the late 1880s, the millions of bison that had roamed the plains decades earlier had been reduced by hunters to a few remnant herds (see Figure 16.5). The enormous flocks of passenger pigeons that once darkened the skies over the eastern forests were gone forever, and most of the large game animals and predators were greatly reduced or eliminated over much of their natural range.

With the closing of the American frontier, the natural resource base that seemed nearly unlimited just a century before now appeared overexploited. Future use would have to be tempered with both resource conservation and preservation, if production were to remain sustainable.

(a)

(b)

FIGURE 17.2 *(a) Great Lakes Region in the late 1800s. Shortly after the turn of the twentieth century this timber supply had been depleted and lumbermen moved on to the more distant forests in the south and west. (b) The Mining Act of 1872 allowed free mineral claims to be staked on most public domain. Most provisions of the Mining Act of 1872 continue today.*

Settlement and Resource Development Trends in Other Regions

In contrast to the extensive areas of North America that proved suitable for settlement during the nineteenth century, the vast deserts, savannas, and tropical forests of Africa, South America, and Australia proved formidable barriers to large-scale settlement by immigrants. Even as these continents were divided into colonies or countries, most of these areas remained wild and little affected by distant civilizations. Until two or three decades ago, many of these lands remained lightly populated by indigenous people practicing their subsistence ways of life. Since then, however, booming populations, economic development, and expanding international markets have created twentieth-century settlement frontiers, and some of these natural landscapes have been cleared and converted to human uses. For more details on the processes, causes, and environments, see Chapters 4 and 16.

17.3 CONSERVATION AND PRESERVATION OF OPEN-LAND RESOURCES

Near the end of the nineteenth century, after promoting settlement and resource exploitation for decades, the U.S. government began to initiate policies that would preserve some land and control the use of some resources on the public domain. It was clear that the continent's abundant resource base could no longer be considered inexhaustible. Furthermore, there was the growing perception that undeveloped wildlands were an integral part of America and that those with outstanding scenic resources should be preserved. Much earlier, this had been argued by intellectuals such as Henry David Thoreau and John J. Audubon, who extolled the beauty and inherent value of nature through their writings and paintings.

The role of government in protecting scenic lands and natural resources emerged gradually. In 1864, Congress granted the Yosemite Valley to the state of California to be used as a park. In 1872, when the United States designated more than 800,000 hectares (2 million acres) of spectacular mountain scenery, waterfalls, and thermal features in northwestern Wyoming, the Yellowstone region became the world's first national park. The concept of preserving open space was also evolving in other industrialized countries. In 1885, Canada established Rocky Mountain (later Banff) National Park, and a year later Ontario set up a provincial park at Niagara Falls.

The American Conservation Movement

While the first parks were being established, the **conservation movement** was beginning in the United States. Based on the concepts of sustained yield and multiple use, the progressive or **scientific conservation** movement was promoted actively by President Theodore Roosevelt during the first decade of the twentieth century. It called for natural resources to be managed "for the greatest good for the greatest number for the longest time."

Roosevelt established the first national wildlife refuge in Florida in 1903 and used the Antiquities Act of 1906 to reserve areas of scientific and historic value as national monuments. In 1905, the U.S. Forest Service was established as a branch of the U.S. Department of Agriculture, with more than 20 million hectares (50 million acres) of western forest reserves under its management. This would become the national forest system. Also in the west, federal irrigation projects were initiated, which led to massive federal dam projects during the next several decades. Although popular, the conservation movement was not without controversy. During Roosevelt's administration, it split into two camps: the *preservationist school* and the *wise-use school.* Preservationists opposed encroachment on parklands for dams and other utilitarian uses, whereas the wise-use proponents thought that other uses should be permitted for the greater good of the people. The argument rages on today.

The national parks program gained widespread popularity, and in 1916 the National Park Service (NPS) was established. To encourage visitors and to enhance their experiences, the NPS built roads, trails, campgrounds, and related facilities in parks. The U.S. Forest Service began a similar program to promote recreation in the national forests as a part of its multiple-use concept (Figure 17.3). By 1920, the idea of preserving tracts of wilderness within national forests had emerged, and in 1929 Forest Service policy was initiated to keep some of the remaining roadless areas in an undeveloped state.

New Deal Conservation

With the Great Depression of the 1930s, the role of the government in land management and conservation expanded rapidly. Under New Deal legislation, programs were established that put people to work building roads, campgrounds, and trails and planting trees in national forests and parks. New Deal conservation also initiated managed grazing programs on the vast federal rangelands. To combat the serious soil erosion problems such as the infamous Dust Bowl of the Great Plains, the U.S. Soil Conservation Service was established. Major dam projects in the west and the south were initiated to stimulate economic development by providing hydroelectric power, water supplies, navigational assistance, and flood control. At the same time, the government expanded its role in wildlife management with the establishment of the U.S. Fish and Wildlife Service.

Today, four **federal agencies**—the Forest Service, Park Service, Bureau of Land Management (formerly the

FIGURE 17.3 *By the 1920s, both the U.S. National Parks and National Forests had developed roads, trails, campgrounds, and other facilities to enhance the recreation experiences of the automobile tourist.*

Grazing Service), and the Fish and Wildlife Service—manage over 260 million hectares (650 million acres) of land—nearly 90 percent of the federal land in the United States. The vast majority of this land is in the west and Alaska. In contrast, most eastern states, settled long before the conservation movement was established, have relatively little federal land. On the other hand, most of these states have established conservation programs with state forests, parks, and wildlife preserves. The conservation movement is also active in most other countries, especially in Europe, where the Americans originally got the idea of conservation and resource management.

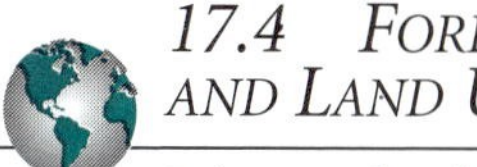

17.4 FORESTS, WOODLANDS, AND LAND USE

Prior to the development of agriculture, forests and wooded landscapes covered nearly 40 percent of the continents; today, it has dropped to about 30 percent (Figure 17.4). Net forest losses total more than 15 million square kilometers (6 million square miles), an area about the size of Russia, with most losses occurring over the past two centuries. Furthermore, most of the remaining forests have been heavily modified by humans, often fragmented into relatively small forest patches. According to a 1997 study conducted by the World Resources Institute, only about 20 percent of the world's original forests remain as large, relatively unbroken, natural ecosystems.

During the 1980s the world's forests and woodlands declined by almost 8 percent. Similar rates of deforestation continued through the 1990s with most concentrated in the tropical forests of the developing world. Data from a 1999 U.N. Food and Agriculture Organization study indicate that between 1990 and 1995 the average annual loss of tropical forests in South and Central America, Asia, and Africa was about 13.5 million hectares (33 million acres). At this rate, South America's forest cover is being reduced by about 0.7 percent annually, Asia's by about 1.1 percent, and Africa's by about 0.7 percent each year. In addition, the areas of degraded and fragmented tropical forests that threaten plant and animal diversity are likely to be significantly larger than the deforested area. In Latin America, government-sponsored programs designed to promote economic development, spur agricultural production, and gain more political control over

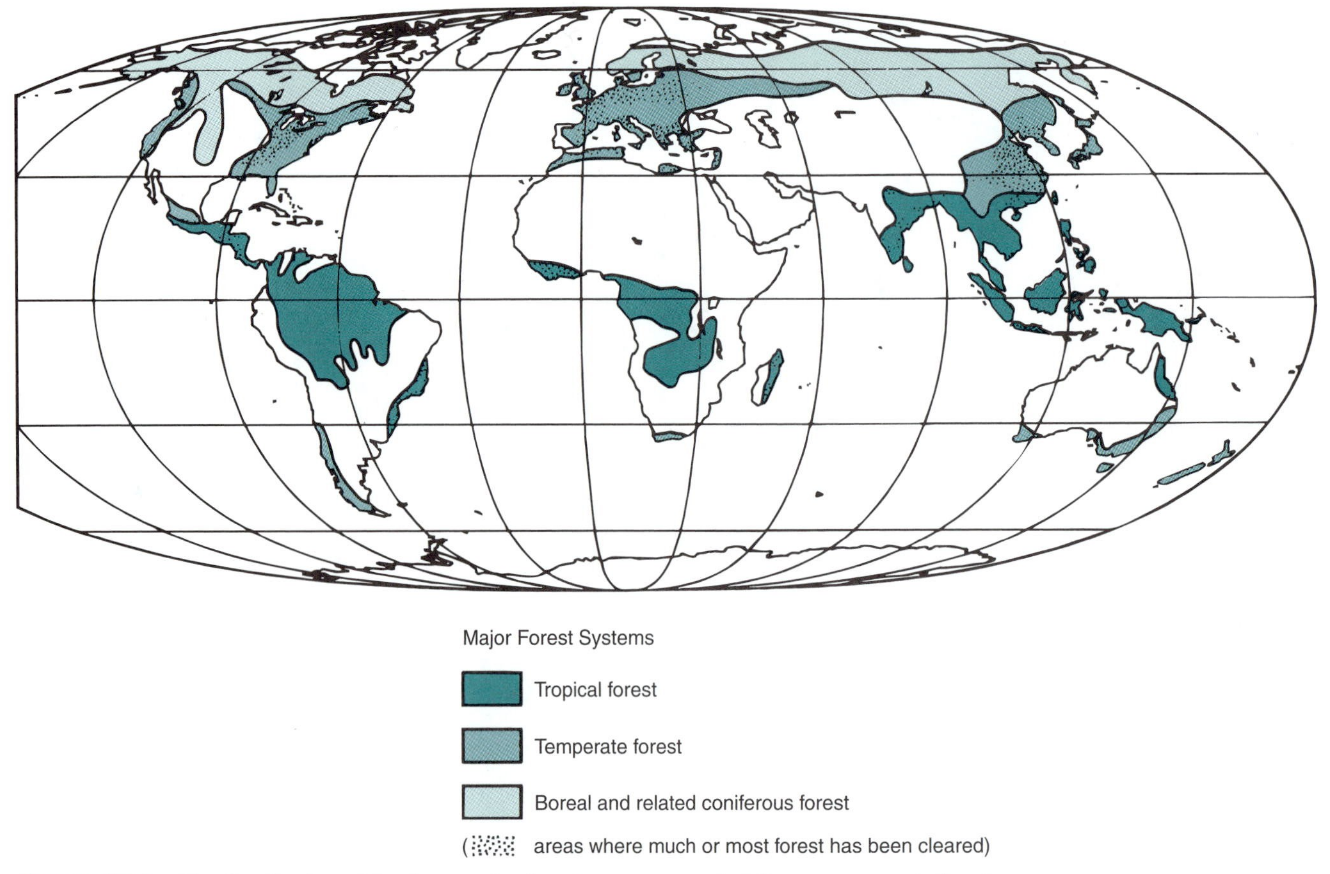

FIGURE 17.4 *The distribution of the three major forests systems; tropical forests, temperate forest, and boreal and related coniferous forests. In agricultural areas, such as Europe, China, and eastern North America, forest coverage is highly fragmented and often sparse. See also Figure 16.12.*

remote regions are responsible for most forest conversions and fragmentation while in Africa and Asia, expansion of subsistence agriculture accounts for most deforestation.

Until the twentieth century most of the deforestation had occurred in the middle latitudes and subtropics, where 25 to 35 percent of the forests had been cleared. In the United States, for example, nearly half of the land was forested when European settlement began. Today, about one-third of the United States is forested, with most made of relatively young *secondary forests* or plantations. Outside of Alaska, only about 5 percent of the frontier or primary forests in the United States remain.

Types of Forests and Woodlands

The world's major forest ecosystems form three great belts: boreal, temperate, and tropical. The **tropical forests** cover about 17.6 million square kilometers (6.8 million square miles) and lie astride the equator in South America, Africa, and parts of south Asia. They are the world's most massive forests in terms of coverage (40 percent of global forest land), volume of living matter, numbers of species, and rates of growth. Most tropical forests are closed-cover woodlands, meaning that tree crowns form a continuous canopy over the forest floor. Most are made up of **rainforest** and **monsoon forest** (deciduous tropical forests). Together they represent 75 percent of all tropical forests and cover about 13.1 million square kilometers (5 million square miles).

Over half the tropical forests are found in Latin America, about 18 percent in Asia, and 30 percent in Africa (Figure 17.4). With forests covering more than 5.7 million square kilometers (2.2 million square miles), Brazil has more than 30 percent of the world's tropical forests and nearly five times that of any other country (see Figure 7.12). However, programs to develop and settle portions of the Amazon basin over the last two decades have resulted in significant deforestation and fragmentation. Estimates from satellite imagery indicate that an average of about 20,000 square kilometers (7,800 square miles) of Brazilian forest were cleared annually over the past two decades.

At the other geographic extreme are the **northern coniferous forests**, which extend from the midlatitudes to the northern limits of forest growth in the subarctic. They are found only in the Northern Hemisphere and are made up of mostly hardy, coniferous softwood species such as firs, spruces, and pines (see Figure 17.4). Covering about 10.9 million square kilometers (4.2 million square miles), including mountain conifer forests, these forests supply the majority of the world's lumber and pulpwood. In the far north the vast **boreal forests** stretch from coast to coast across central and southern Canada and 11,000 kilometers (7,000 miles) across Eurasia from Scandinavia to eastern Siberia. In their most northern reaches, which extend to the Arctic Circle in Alaska and Eurasia, the boreal forests are too remote and slow-growing to be of much commercial value.

South of the northern coniferous forests are the **temperate forests**. They vary in composition, but deciduous hardwood species such as oak, beech, and maple dominate. Virtually all of the frontier stands of hardwoods in North America and Europe have been lumbered or cleared for agriculture. Today, secondary forests and tree plantations make up most of the forest cover throughout these areas. Temperate forests cover about 16 million square kilometers (6.2 million square miles). An inventory of world forests reveals that Russia has nearly 25 percent of the current forest cover, Canada has 9 percent, and the United States has 7 percent.

Forests as a Multiple-Use Resource

Coincident with two centuries of rapid population growth, large tracts of forest were cut and cleared as demand surged for lumber, fuel, pasture, and cropland. This trend continues today, and the world's forest cover continues to dwindle in both quantity and quality as the primary tropical forests are cleared or degraded and most of the remaining frontier forests in the midlatitudes are being logged away. These changes have taken place in the face of rising concern about forests as ecological and recreation resources. In addition, loss of forests, especially the tropical forests, is recognized as critical to the global environmental balance. These forests are a major component in the carbon dioxide and water cycles as well as the world's principal reservoir of species (Figure 17.5). Finally, in parts of the developing world forests are the homes of several million indigenous people who depend directly on the forest ecosystem for their livelihoods. When considering the benefits provided by forest ecosystems it can be shown that for some forests, the greatest long-term values are realized by leaving the forest intact rather than cutting it for timber or converting it to other uses such as grazing or crop agriculture.

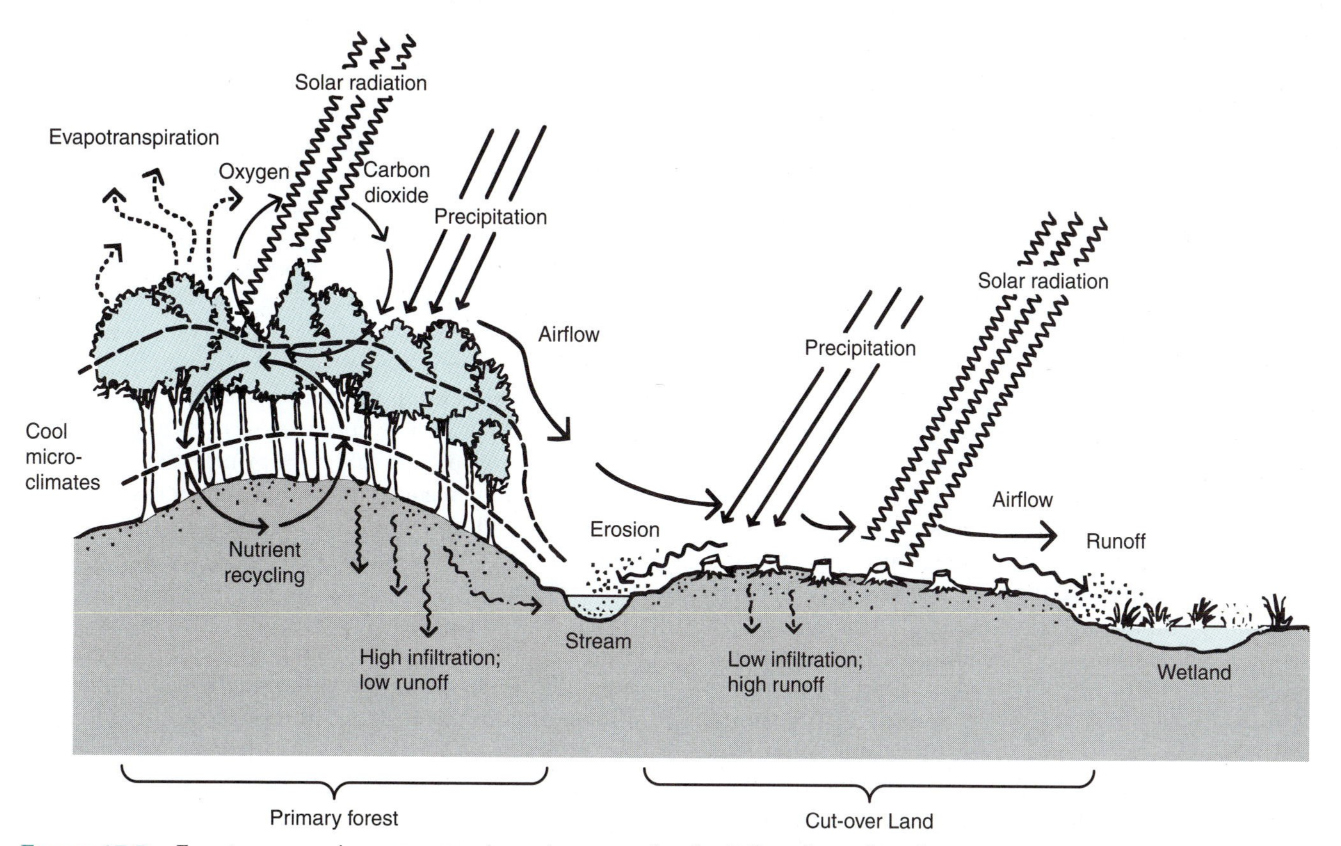

FIGURE 17.5 *Forests are a major component in environmental cycles. Where forests have been cleared nutrients are removed, runoff and erosion increase, while wind velocity and temperatures are more extreme.*

Forests in Parks, Wilderness, and Other Reserves

Since the United States established the first national parks and forest reserves, other countries have set aside parks, wilderness, wildlife reserves, or scientific study areas. Currently there are more than 2,000 of these legally protected areas in the world encompassing about 3 percent of the world's forests and woodlands. About 10 percent of the tropical forests and woodlands have been protected. Much of the protected tropical open woodlands are in Africa's extensive national parks, where illegal timber harvesting, settlement, and fires constantly threaten these areas. In Latin America only about 2.7 percent of the tropical forest is currently protected. About 12 percent of the forestland of Europe, North America, and the former Soviet Union is legally protected or reserved.

In the United States, about 10 percent of the forested area, 32 million hectares or (80 million acres) is protected in national parks. Another 20 percent is in the national forests, which are managed for multiple uses that include managed timber production and livestock grazing, along with noncommercial uses such as recreation, wildlife conservation, and watershed protection. Commercial timber production is permitted on about half of the acreage of the national forests. In Alaska, a considerable amount of forested land is administered for multiple uses by the Bureau of Land Management (BLM). All forests in the 425,000 hectares (105 million acres) of the National Wilderness Preservation System (NWPS) are protected from logging. The majority of the NWPS is located within national parks, national forests, or national wildlife refuges and is managed to preserve wilderness qualities.

Lumber, Fuelwood, and Other Forest Products

Figure 17.6 shows that over the past four decades the annual harvest from the world's forest and woodlands for fuel and for commercial products has increased by more than 50 percent. With rapid population growth in many developing countries, the demand for firewood, the main heating and cooking fuel for nearly 2 billion people, has increased dramatically. Fuelwood now accounts for more than half of the annual world harvest of wood, and it is being cut more rapidly than it can grow back in many developing countries. Today more than half of those depending on fuelwood are facing current or potential shortages.

In parts of Africa, Asia, and South America, nearly 100 million people cannot meet their minimum fuelwood needs even with overcutting. Another 1.5 billion people are cutting wood faster than it can grow back, so shortages are imminent if current trends continue. To combat this, many developing countries are expanding or initiating tree-planting programs that emphasize fast-growing fuelwood plantations or *agroforestry* programs where trees are interplanted with agricultural crops. As demand for the materials and products made from wood continues to expand, the annual cut for commercial purposes continues to grow by about 1 percent. More than half of the world's current production of wood for lumber, plywood, veneer, pulp, and other industrial uses is produced in the United States (25 percent), Canada (10 percent), and the former Soviet Union (20 percent).

Significant **logging** of tropical forests has taken place mainly since 1950. During the 1970s and 1980s, log production and exports rose rapidly and then decreased in several Asian and African countries as forests were depleted and controls were imposed on cutting. In 1985, Indonesia banned all log exports, making Malaysia the world's leading exporter of tropical hardwood. By 1989, the rate of cutting was more than twice the sustainable yield and Malaysia's forests were disappearing, just as they had earlier in the other tropical Asian nations. During the 1990s production of valuable tropical timber species expanded rapidly in the Amazonian rainforests. In the Brazilian Amazon, the annual forest loss during the 1990s rose from about 1.2 million hectares (3 million acres) to more than 1.9 million hectares (4.7 million acres).

Clear-cutting one forest and then moving on to exploit virgin forests elsewhere has been standard procedure for large-scale commercial lumbering worldwide for the last century and a half. Frontier forests in Washington, Oregon, and northern California are still being cut, even as efforts are made to preserve these centuries-old forests and their specialized ecosystems. Although some magnificent primary forest stands of redwood, fir, spruce, and cedar are protected in parks and designated wilderness areas, most remain vulnerable to logging. Currently about 30 percent of the U.S. supply of construction lumber and plywood is cut from Pacific Northwest forests.

As the world's leading producer, the U.S. forest industry now harvests more wood from second-growth forests and plantations than it cuts from primary forests. Today more than 50 percent of the nation's timber is harvested from the southeastern states where long growing seasons and plentiful rainfall are favorable for relatively rapid tree growth.

The recent rapid increase in demand for paper and other products made from wood pulp has brought about a significant change in both the geography and method of operation of this segment of the U.S. timber industry. Today much of the timber harvested for wood pulp is being processed in large chipping mills that grind entire trees into wood chips—the raw material for wood pulp from which paper products are made. Over the past decade more than 150 chipping mills have been placed into operation, primarily on private wooded land in the

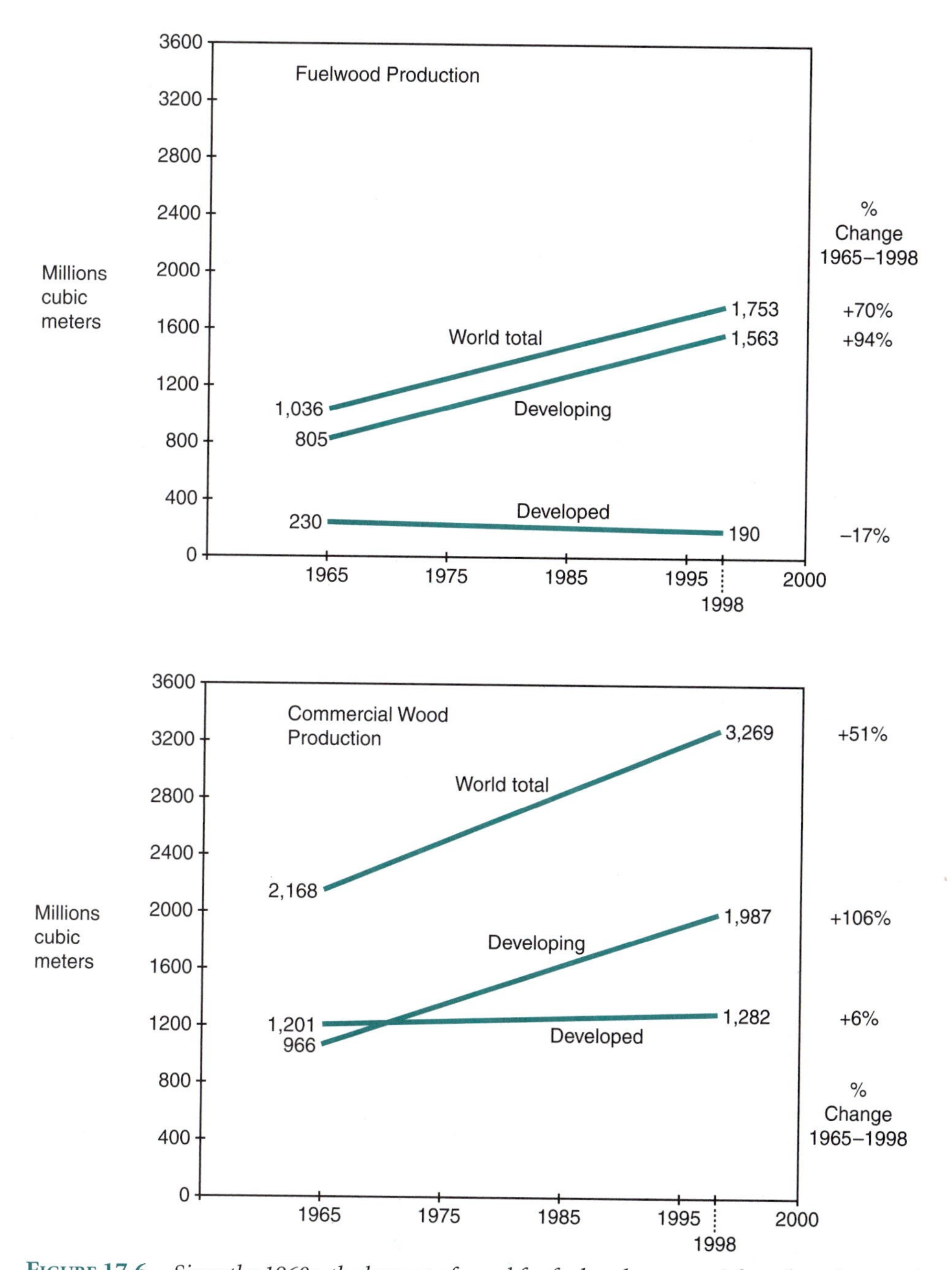

FIGURE 17.6 *Since the 1960s, the harvest of wood for fuel and commercial products has nearly doubled. Fuelwood accounts for over half of the world's annual cut and more than 2 billion people in the developing countries depend on it for heating and cooking. Most commercial wood is cut for lumber, plywood veneer, and pulp.*

South. By 2000 these chipping mills were consuming trees clear-cut from about 1.2 million acres, mostly in the southeastern states—a rate that studies indicate is not sustainable. There is growing opposition to the impact that these large-scale operations are having on the landscape, but the cutting goes unrestricted in most states. In 2000 Missouri became the first state to order a two-year moratorium on new chipping mill permits to allow for more detailed impact assessment.

Commercial Logging and Environmental Impact

The traditional logging practice of cutting the old-growth frontier forests and then moving on to exploit forests elsewhere is declining because most of the primary forests have been cleared. However, where forest frontiers remain, such as the tropical forests in Brazil, Indonesia, Papau New Guinea, Cambodia and Cameroon, the Russian and Canadian coniferous forests, and the North American western forests, this practice of **forest mining** continues to destroy these biologically diverse, old-growth forest ecosystems. For example, in the U.S. Pacific Northwest, where timber companies have cut more than 90 percent of the old-growth forests on private land, the industry is exerting pressure to allow increasing cuts in the national forests where ancient forest stands remain. Across the Canadian border, in British Columbia, the annual cut of old-growth forests reached nearly two-thirds of a million acres by 1990, and companies there hold licenses to harvest nearly all of these remaining forests except the 5 percent that are protected in parks.

Of the harvesting methods used by loggers today, **clear-cutting** is the most widely used and most controversial. Clear-cutting involves harvesting all trees in a given forest stand. It is favored by loggers because it usually produces the greatest volume of timber and requires less road building and transportation. Furthermore, clear-cut areas can be reforested to produce an even-aged, single-species stand that can be managed and subsequently harvested, much like an agricultural crop. Clear-cutting currently accounts for about two-thirds of the U.S. timber harvest, and it remains an approved harvesting method for certain species in most areas of the national forests (Figure 17.7).

Clear-cutting large blocks on steep slopes, leaves the soil unprotected and often leads to high erosion and runoff rates. With higher sediment loads and higher runoff volume, streams that drain clear-cut watersheds become sediment-laden and flood-prone. In the humid tropics, where most nutrients are stored in the trees and

(a)

(b)

FIGURE 17.7 *In the United States clear-cutting is the most widely used and most controversial timber harvesting method. It yields the greatest volume of timber at the least expense, but causes high runoff rates, soil erosion, and ecological disturbance. In (b) the light lines are roads.*

recycled to the soil, tree removal leaves a nutrient-deficient soil that might not sustain reforestation efforts. The practice at least temporarily destroys the forest habitat and biodiversity declines at all ecological levels, from large plants and animals through insects and microorganisms. In addition the stark landscape of recent clearcuts precludes most recreation activities and, for most passersby, creates an eyesore (Figure 17.7).

Selective cutting has less environmental impact. This method harvests only trees of certain species, leaving others in place. However, it is not without significant environmental impacts. The practice of '*creaming*' the few valuable trees from tropical forests often results in destruction of other trees and plants during road building and logging operations. As a result, selective cutting can reduce the biological diversity of the forest, and in some areas, has led to elimination of certain species. On the other hand, selective cutting that removes only mature trees or thins crowded stands may promote the growth of remaining trees. Since this harvesting practice maintains the forest cover, watersheds remain protected from erosion and heavy runoff, and streams from sedimentation and flooding. Forest-oriented recreation activities such as hiking and camping are compatible with selective cutting and the openings created within the forest improve the habitat for some wildlife species.

Tree farms and **industrial plantations** are common alternatives to managing native forests, but they have a number of drawbacks. When wild forests are cut and replaced by single-species plantations or tree farms, the diversity of plant and animal life is lost. Habitat is lost for animals that require uneven age stands, dying trees, or snags. In addition, the even-aged, genetically uniform plantations are prone to epidemic diseases and outbreaks of pests. Numerous tropical plantations—as well as many coniferous plantations in Europe, North America, and Asia—have been damaged or destroyed by pests and diseases. Forests with such a narrow genetic base may be more vulnerable than natural forest stands to climate change, air pollution, and acid rain.

In all forest management programs there is serious environmental impact related to harvesting operations. The network of roads built to harvest a forest stand not only clears as much as 10 percent of the forest but degrades or destroys large areas of habitat by carving large, isolated forested blocks into numerous, small, accessible islands. The roads are also the sites most prone to erosion and slope failures (Figure 17.7). Roads also open the forest to human incursion, and in Brazil and other tropical countries, have been a leading factor in tropical rainforest destruction by squatters and miners (see Figure 16.18). In addition, loss of forest reduces global carbon dioxide (CO_2) absorption, and clearing and burning of forests releases large amounts of CO_2 into the atmosphere.

Recent Trends in Forest Resource Use in the United States

Since the middle of the twentieth century there has been growing conflict over U.S. forest resources. At the same time that demand for lumber and other forest products has risen, there has been increased emphasis to preserve forests for recreation and environmental protection. Today much of the conflict and debate focuses on the use and management of the public forests. For example, most of the frontier forests that remain in the Pacific Northwest are in the government owned national forests. These virgin stands represent not only a valuable timber resource but also support a productive and unique ecosystem. The timber industry argues that these forests should be cut to maintain jobs and economic growth. On the other hand, those wishing to preserve these forests articulate the long-term economic and noneconomic benefits that accrue from forest-based recreation and tourism, ecological preservation, watershed protection, and wilderness values.

Over the last four decades, forest land has declined in the contiguous 48 states by about 8 percent. Most forest loss has resulted from urban and industrial development and some cropland expansion. Currently, the U.S. forest inventory is approximately 2.9 million square kilometers (1.1 million square miles) of commercial-class forest, with nearly 80 percent in the lower 48 states. Although Alaska's forests are vast, most are not classified as **commercial forest**,[1] because they are so slow-growing under harsh climatic conditions of the north. Commercial forest makes up 80 percent of the forest lands in the contiguous states. More than 70 percent of this commercial forest land is privately owned (Figure 17.8).

Two-thirds of government-owned forests are in the west. Some 75 percent of these western public forests are included in the national forest system managed by the U.S. Forest Service, and the most valuable of these are found in the Pacific Northwest. Because the great coniferous forests of this region have been logged from private lands, the forest industry has turned to the national forests. As a result of logging, during the 1980s and 1990s, there has been a significant reduction in the area of old-growth forest and corresponding expansion of dense, coniferous plantings and less valuable hardwood species.

The **national forest system**, which includes nearly 773,000 square kilometers (300,000 square miles) of forests and grasslands, comprises 25 percent of all federal

[1]Commercial forests include lands capable of producing at least 20 cubic feet of wood per acre that have not been withdrawn from timber use by law or administrative regulations.

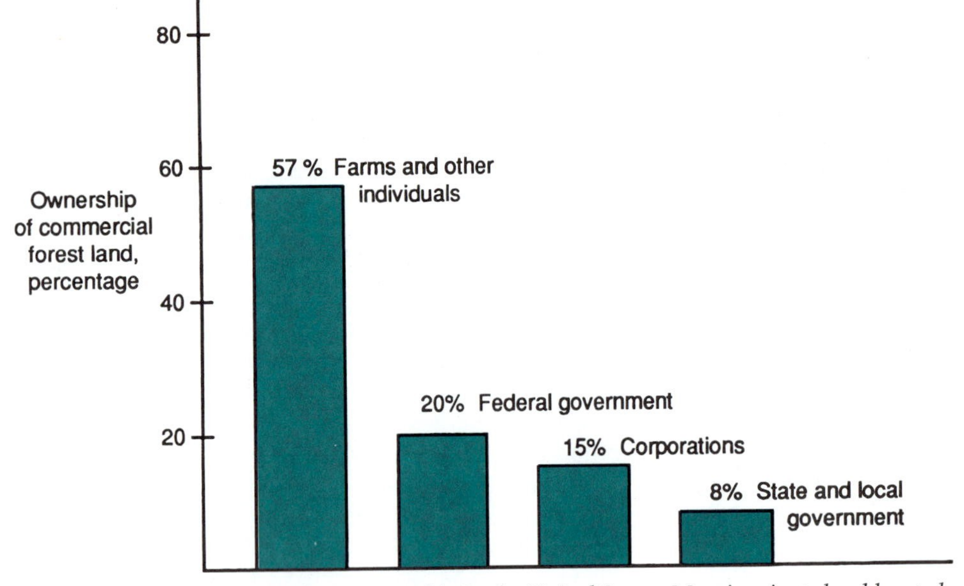

FIGURE 17.8 *Commercial forest ownership in the United States. Most is private land located in the eastern United States.*

lands and nearly 8 percent of the nation's total land area (Figure 17.9). In the west, coniferous forest ecosystems of fir, pine, spruce, or redwood are dominant, whereas in the north and east, northern hardwood and coniferous ecosystems dominate. In the Ozarks and Appalachians, hardwoods dominate. In the south, several national forests support southern pine ecosystems from the piney woods of east Texas to the subtropical forests of Florida. The system also includes about 1.6 million hectares (4 million acres) of *national grasslands*, with most units on the short-grass prairies of the western Great Plains.

The U.S. Forest Service oversees all resource use in the national forests. By law, these forests are managed for multiple resource uses. These include commercial activities such as logging, grazing, mining, and resort-based recreation. The national forests also support a variety of noncommercial activities that include watershed, wildlife habitat protection, and many types of recreation. The total activity associated with these uses is enormous. For example, about one-tenth of the nation's commercial timber harvest is from the national forests. More than 2.1 million head of cattle, horses, sheep, and goats graze under permit on the national forests. In the late 1990s there were nearly 25,000 active mineral leases and permits on the national forests. Finally, the national forests provide an important resource for recreation. Estimates from the late 1990s indicate that national forest recreation use, 859 million visits, is more than twice the number of recreation visits in National Park Service units.

In the past several decades, there has been increasingly contentious conflict between environmentalists and the Forest Service as to the appropriate balance of forest uses. Environmentalists argue that forest management is geared too heavily toward timber production with large-scale road building, fire protection, and insect and disease control programs that cost more than the receipts from the annual timber cut. Between 1992 and 1994, timber sales from the national forests lost an estimated $1 billion in direct costs such as road building. This figure does not include costs of reforestation, stream erosion, or lost recreation revenues from damage to fisheries or water resources. Furthermore, environmentalists maintain that the widespread clear-cutting practices are degrading the forest resource and precluding other uses prescribed by law. In response, the Forest Service points out that managed commercial activities such as logging that utilize the national forests contribute direct economic benefits to local and regional communities and to the thousands of people employed by these commercial activities. Furthermore, wood and other resources produced in the national forest provide a steady supply of materials and products to the national economy while generating revenue for the U.S. people.

Recently, the Forest Service has eliminated some of the unprofitable timber sales that required building expensive access roads to remote forest stands. As a result, the annual timber harvest from the national forests has declined significantly.

Over the years, various laws have been passed to help ensure balanced use of national forests. Among them is the *Wilderness Act* of 1964, which established the *National Wilderness Preservation System (NWPS)*. The *National Forest Management Act (NFMA)*, implemented in 1976, requires that every national forest develop a management plan that integrates each of the primary forest uses. This law also established specific policies relating

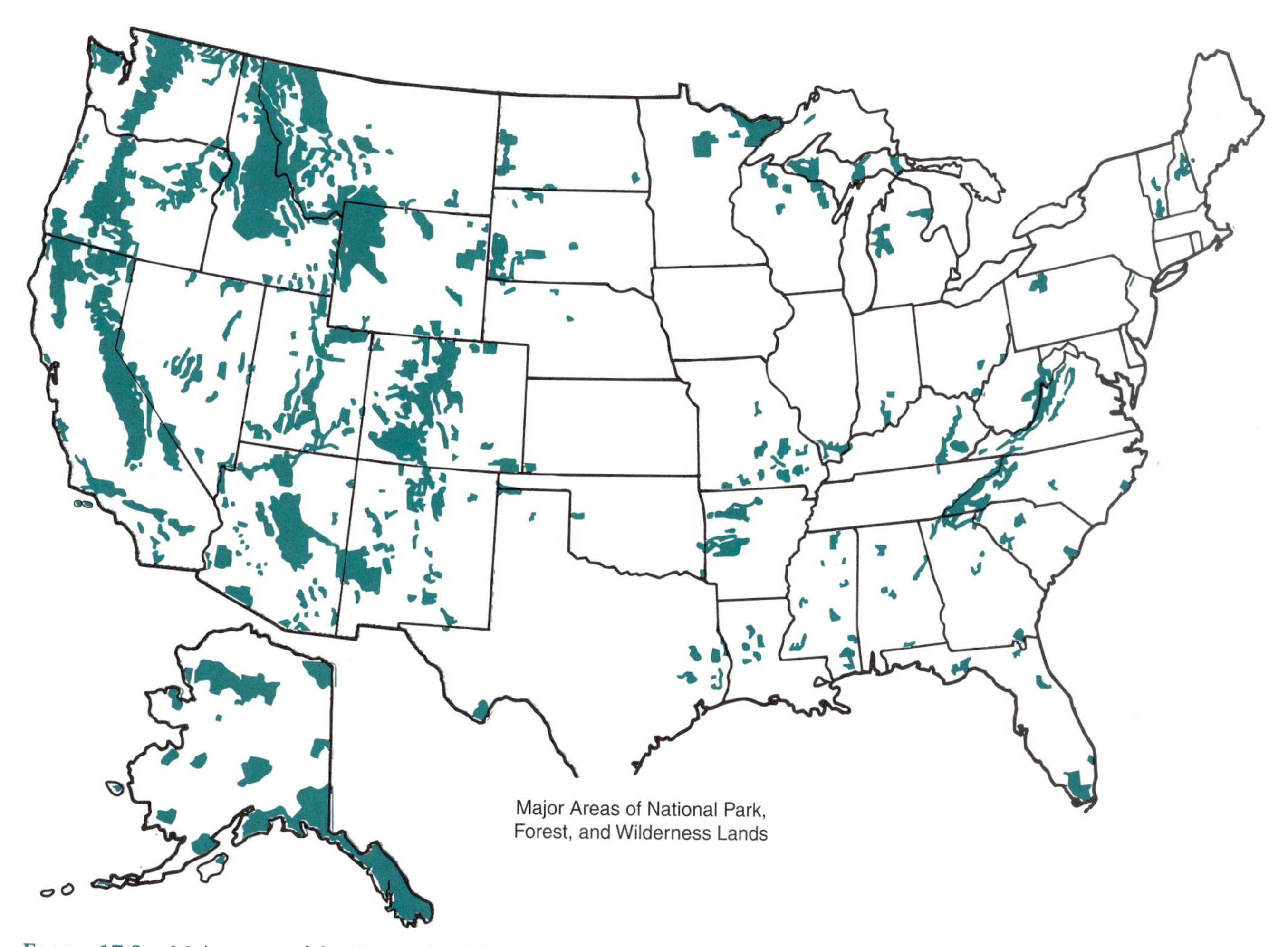

FIGURE 17.9 *Major areas of the U.S. national forest system, the U.S. national park system, and the U.S. Wilderness Preservation system.*

to clear-cutting practices, stream corridor (riparian) protection, and the rate of harvesting on the national forests. However, the controversy over the use of the national forests rages on, with environmental interests opposing additional road building and cutting of remaining old-growth forests in the west.

Sustainable Forest Management

How can we meet our increasing needs for timber, paper, and other forest products, and yet maintain the essential services and benefits of forest ecosystems? **Sustainable forest management (SFM)** considers the total range of forest resources and strives to manage the forest as a complete ecosystem. During the decade of the 1990s, progress has been made to develop SFM criteria for the world's various forest types. Today, the SFM concept continues to evolve based on the central theme that forests be managed to realize social and ecological needs as well as the economic needs of current and future generations. This means that biodiversity and other environmental services provided by the forest ecosystem are to be maintained and that tree harvesting strategies ensure the diversity in species, age, and size of the natural forest community.

To accomplish sustainable forest management, an international Forest Stewardship Council (FSC) was established in 1993 to develop SFM criteria. Composed of foresters, timber producers, environmental groups, and certification organizations, the FSC has set management standards that apply to all commercial forests. By the late 1990s, about 3 percent of the wood traded internationally was certified by the FSC. Consumer demand for wood products produced under SFM criteria was growing even faster than the supply.

This certification program is a voluntary market-based approach to promote sustainable forest management. Although it is a promising first step, for international SFM to progress it will require legislative and policy support from countries that are major producers and consumers of forest products.

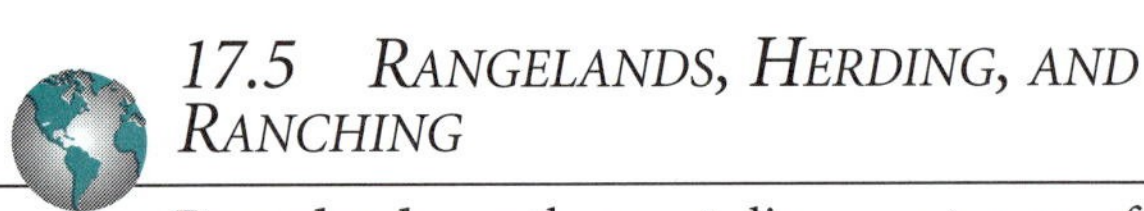

17.5 RANGELANDS, HERDING, AND RANCHING

Rangelands are the most diverse category of the world's open land resources. They cover more than 25 percent of Earth's land surface and are made up of prairie, steppes, savannas, shrubland, open woodlands, and even portions of the tundra (Figure 17.10). Most of these lands are too dry, too cold, or too steep or rocky for crop agriculture. Nevertheless, they are an important food-producing resource because the domestic livestock and wild animals that graze the range convert plant materials that cannot be used directly by humans into protein-rich foods and other animal products. Furthermore, some rangelands support many species of wildlife that provide a reservoir of genetic diversity. In Africa, parks and preserves that have been established on the savannas support the world's largest wild animal herds.

Rangeland provides forage for a significant but declining share of the world's 3 billion domestic animals, along with millions of wild grazing animals. This land provides about one-fourth of domestic livestock forage. However, in developing countries where the majority of the livestock forage from the range, there has been widespread environmental stress from overgrazing. Large tracts are undergoing **desertification**, which by some estimates is consuming 200,000 square kilometers (75,000 square miles) a year.

Traditional Use of Rangelands

Subsistence livestock herding or **pastoralism** has been a way of life since the beginning of agriculture. In the grasslands and deserts of Asia and Africa, small bands of herders move with their livestock along traditional migration routes where seasonal forage and water are available on commonly used grazing lands. Today, this way of life is threatened as migration patterns are disrupted by national boundaries and as grazing lands are converted to crop agriculture. As a result, this pastoral system and its distinctive societies have disappeared in some areas and are severely curtailed in others. Today perhaps 15 to 20 million nomadic pastoralists still range with their herds over Old World deserts, grasslands, and savannas (Figure 17.10).

The greatest concentration of subsistence herders remains in *sub-Saharan Africa.* Here, as elsewhere, the pastoral tradition has declined as national governments have tried to impose settlement programs and commercial livestock production among nomadic groups. At the same time, crop agriculture has expanded onto the communal seasonal pastures. These changes, coupled with increasing population, have resulted in the breakdown of the traditional pastoral system. With larger herds grazing less pasture land, degradation and even desertification have occurred in parts of Africa, the Middle East, and the dry interior of Asia. As a result, the

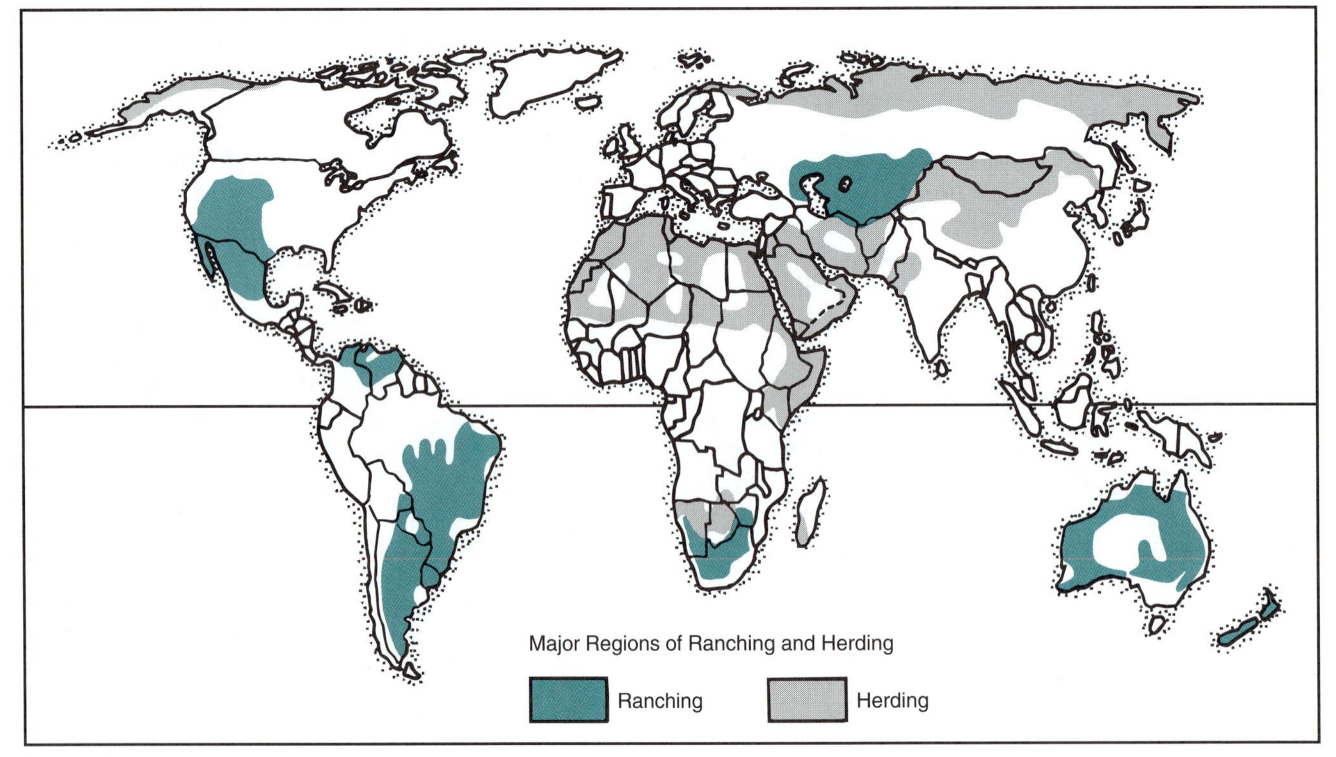

FIGURE 17.10 *World distribution of livestock ranching and grazing. Rangelands occupy more than 25 percent of the Earth's land surface.*

subsistence herders have declined, and in Africa, have suffered food shortages and even famine when the ranges withered and livestock starved during the prolonged droughts of the 1970s and 1980s.

Well adapted to the inevitable droughts that sear the grasslands, pastoralists build herds during wet years when the range can accommodate more animals. This surplus is drawn on during dry years when the forage base shrinks. This has been effective for generations, but such strategies have not been maintained because of imposed settlement programs and rapidly growing population. As a result, pastoralists have borne a disproportionate share of the suffering, privation, and even starvation that has become commonplace during the prolonged droughts of the last two decades in sub-Saharan Africa.

Simply transplanting practices and technologies from commercial livestock systems to pastoralists have not been successful. For example, programs designed to assist herders by establishing permanent water supplies have had negative impacts on the range. Where livestock have had free access to new wells or boreholes, the range has been severely overgrazed. Rings of barren land now extend for several kilometers around boreholes. Developing sustainable programs that increase the productivity of these rangelands will require a better understanding of the rangeland ecosystems—not only its plants and animals, but also the long-term occupancy and use by subsistence herders. Programs are now underway that incorporate aspects of the traditional pastoral system to ensure traditional grazing and water rights with innovative new approaches such as growing forage during the wet season to be fed to livestock in fenced enclosures during the dry season.

Commercial Ranching

Ranching began during the eighteenth and nineteenth centuries on the frontier of lands being settled by Europeans. With rapidly expanding domestic and international markets for meat and livestock products and improved transportation systems, ranching evolved as a commercially viable use of the arid and semiarid portions of these lands. In North America, ranching became established on the grasslands of the western Great Plains and the desert scrublands of the intermountain west. In South America, tropical grasslands (e.g., the *llanos* in Venezuela and the *campos* of Brazil) became ranching centers, as did the lush *pampas* of Uruguay and Argentina. Ranching is the dominant economic endeavor throughout much of Australia's arid interior and the highlands of New Zealand's South Island. Currently, only a few areas of the Old World rangelands support commercial ranching. Two examples include cattle ranching in South Africa and some commercial herding on the Russian steppes (Figure 17.10).

Unlike traditional pastoralism where herders migrate with their stock over commonly-used range, modern ranching practices restrict grazing to lands owned or leased by the ranch. These ranches are the largest operating units in commercial agriculture. In the western United States, ranches are often measured in square miles rather than acres. In addition, many ranchers purchase grazing permits on extensive areas of public rangeland. The world's largest ranches extend over hundreds or even thousands of square kilometers of "outback" in Australia's arid interior.

A major reason for the variation in the average size of commercial ranches relates to the carrying capacity of the rangeland. Grazing **carrying capacity** refers to the number of animals that can be sustained by the natural vegetation growing on the range. The quality and quantity of range forage is, of course, influenced by local climate, soil, and topographic conditions. However, the character of the rangeland vegetation also changes in response to other factors, such as fire and overgrazing. Fire suppresses the growth of woody vegetation while encouraging grasses and other herbaceous plants. Overgrazing, defined by animals in excess of carrying capacity, will reduce some preferred plant species while less desirable species increase. Furthermore, weedy plants not present before grazing become established after livestock are introduced (see Figure 6.13).

Carrying capacity is also affected by livestock species. For every large grazer like a cow, the range will usually support several smaller grazers such as sheep. On the public rangelands in the United States, for example, range-carrying capacity is stated in terms of **animal unit equivalents**, where one horse or cow is equated with five sheep or goats. Typical carrying capacity for healthy short-grass prairie in the western Great Plains and Intermountain West, for example, might range from 30 to 50 cattle per square mile.

Evolution of Rangeland Use in North America

The first livestock were introduced to North American rangeland by the Spanish in the early 1500s, but large-scale commercial grazing did not begin until the 1850s. After the Civil War, the railroads penetrated the western range, and domestic and foreign markets for meat, hides, and wool expanded rapidly. Between 1870 and 1890 the number of sheep and cattle on the open range grew from less than 10 million to 47 million head.

By the turn of the twentieth century, millions of acres of range, including vast tracts of public lands, had been seriously damaged as a result of uncontrolled grazing. As with all other common property resources where use is unrestricted, there is no incentive to conserve the resource or restore productivity. By withdrawing livestock from the overgrazed range, the rancher would simply forgo any benefit, while the range resource would continue to decline as grazing by others continued. Early in the twentieth century the Public Land Commission, the agency that oversaw federal lands, recommended that grazing districts and rangelands leases be established as a means to control overgrazing. Although this initiative failed, several special homestead acts were passed that resulted in the transfer of nearly 100 million acres of public rangeland to private ownership.

Managing the Public Range

By the 1930s, decades of uncontrolled grazing had left 84 percent of the 142 million acres of U.S. public open range in fair to poor condition. In 1934, the Taylor Grazing Act closed the public domain to further settlement and divided the range into grazing districts to be managed by the Interior Department's Grazing Service and local advisory boards of ranchers. Under this system, the range was managed in the interests of the ranchers and conditions did not improve. In 1976, Congress enacted the *Federal Land Policy and Management Act* (FLPMA) that authorized the federal agencies to plan and manage the public rangelands (now called the **national resource lands**) on the basis of multiple use and sustained yield.

Ranchers perceived the shift toward multiple-use management of the range as a threat to their traditional grazing practices. They joined with mining interests and others seeking greater access to resources of the western public lands in a concerted political effort dubbed the *Sagebrush Rebellion*, to transfer the federal rangelands to state control or private ownership. Although the movement received considerable political support in the early 1980s during the Reagan administration, pressures from conservation and environmental interests to maintain federal control and to manage the national resource lands for multiple uses prevailed.

The public lands used for commercial grazing are limited to the 17 western states (Figure 17.11). These states

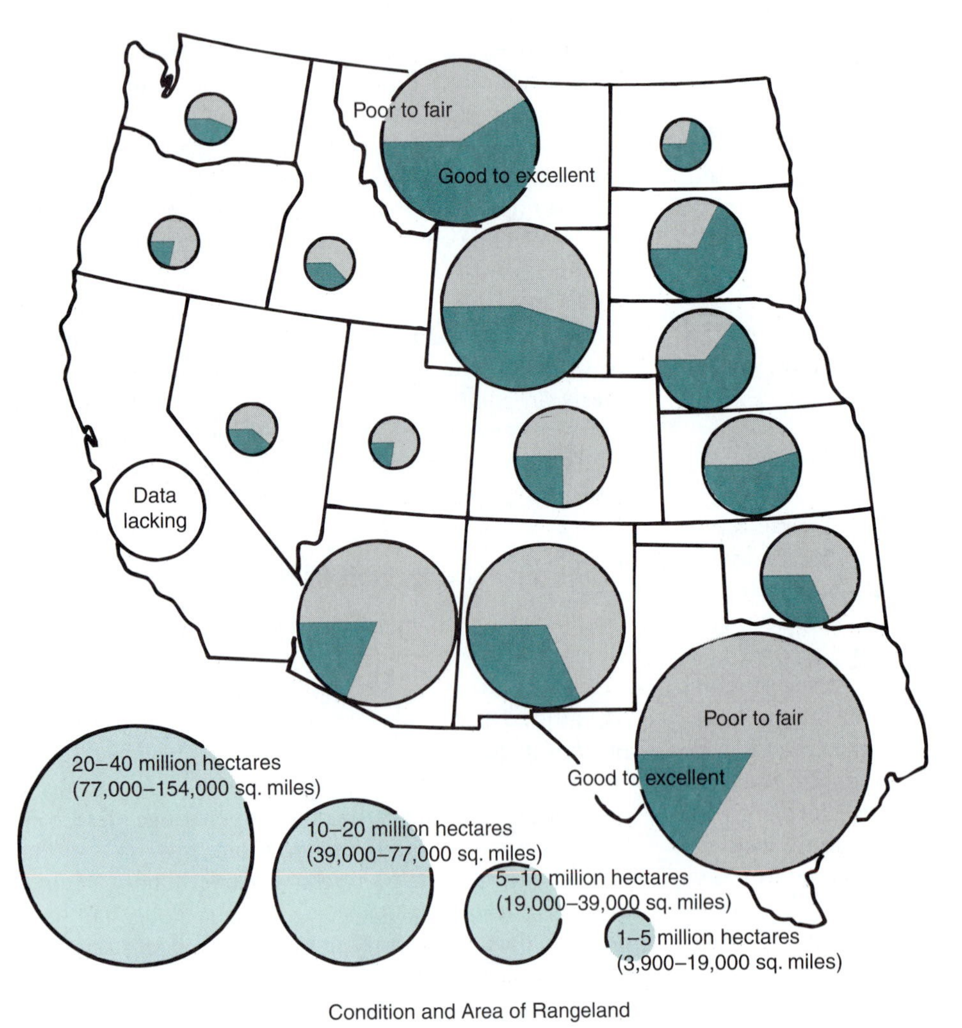

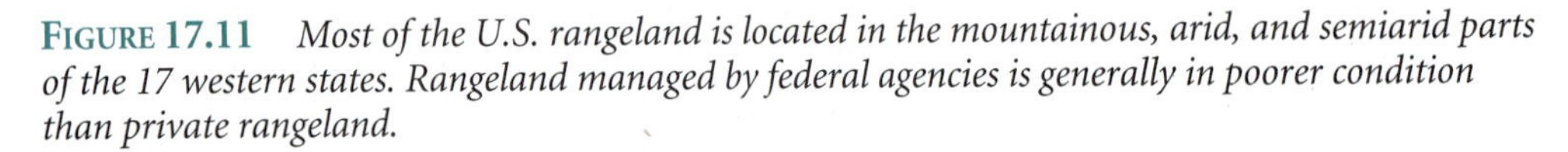

FIGURE 17.11 *Most of the U.S. rangeland is located in the mountainous, arid, and semiarid parts of the 17 western states. Rangeland managed by federal agencies is generally in poorer condition than private rangeland.*

contain 107 million hectares (264 million acres) of federal rangeland, which represents about one-third of the grazing lands in the United States. The remainder of the nation's rangeland is privately owned and is generally in better condition than the federal lands.

In recent years, livestock grazing on western public rangelands has come under increasing scrutiny. Critics point out that grazing is responsible for degradation of rangeland ecosystems and conflicts with the legal management mandate of sustained yield and multiple uses. Although commercial grazing occurs on more than two-thirds of the federal rangeland, it supports less than 2 percent of the nation's cattle and sheep. Furthermore, the costs of administering public range grazing far exceed the revenues from the grazing fees—they are less than half of those changed on comparable private land. Finally, this extensive and environmentally damaging use conflicts with virtually all other recognized public land uses and values.

Today an increasing number of range scientists and environmental groups are calling for a *no grazing policy* on most of the public range. They argue that by eliminating grazing, the rangeland ecosystem could be restored and managed primarily for nonconsumptive uses and aesthetic values that would provide enhanced benefits to the millions of recreational uses of these public lands. Although public support for a no grazing policy is growing and federal management policies recognize the value of restoring rangeland ecosystems, it is unlikely that grazing will be eliminated soon due to the strong political influence exerted by ranchers and livestock interests.

17.6 The Parks Movement

In the middle of the nineteenth century as urbanization and industrialization expanded in Europe and North America, a movement began to set aside land specifically for recreation and leisure pursuits. Central Park in New York City, one of the first urban parks, became a model for many public parks built in and around large cities during the latter part of the 1800s. At about the same time the concept of preserving wild, scenic landscapes as parks was evolving. Not only did these precedent-setting events protect these lands from private exploitation, but they established the first parks administered by the federal government. The **national park movement** in the United States established a model for parks in other countries. Today there are national parks and reserves in more than 100 countries.

The national park movement in Canada began in 1885 when Rocky Mountain National Park (now Banff) was established. Over the decades, the Canadian system has expanded to include 35 national parks, which preserve some of the nation's most spectacular landscapes, as well as numerous national historic parks (Figure 17.12a). In New Zealand the first park was established in 1894 the national park system now includes 10 parks occupying more than 5 million acres, or about 8 percent of the nation's area (Figure 17.12b).

National parks vary greatly from nation to nation. In countries where preserving wildland resources is new, national parks may be no more than boundaries drawn on a map. In Russia, large tracts of wild land are retained as natural reserves used primarily for scientific research, with little public use. The national parks of Africa, on the other hand, extend over vast open landscapes that support the world's remaining great concentrations of large animals. The best-known parks occupy the open woodlands and savannas in Kenya, Tanzania, and South Africa.

The National Park System in the United States

With the establishment of Yellowstone, Yosemite, Sequoia, and Mt. Rainier Parks before the end of the nineteenth century, the United States was the first country with a national park system. More than a century later, the system managed by the National Park Service (NPS) has expanded to include a wide array of cultural as well as natural resources, such as historic sites and battlefields, national monuments, lakeshores and seashores, and even urban-oriented recreation areas. Today, with more than 380 units covering a total area as large as New Mexico (80 million acres), the U.S. national park system is the largest and most diverse in the world. Some of the major NPS units are shown on Figure 17.9.

Most of the 54 national parks are large, predominantly undeveloped areas that contain outstanding, often unique natural resources that are part of the national heritage. Often called the nation's "crown jewels," many consist of a major landscape unit such as a canyon, valley, or mountain range. Several of Alaska's huge parks, such as Denali, Wrangell-St. Elias, and Gates of the Arctic, are composed of millions of acres of isolated mountain ranges. Many of the parks in the contiguous states like Grand Teton, Rocky Mountain, Grand Canyon, Canyonlands, Yosemite, and Death Valley take their names and their character from the landscape features they include (Figure 17.13a).

The 75 **national monuments** in the NPS consist of a diverse set of resources that range from small, monument-like units such as the Statue of Liberty to large, mostly undeveloped areas such as Great Sand Dunes (Figure 17.13b) that include natural resources similar to those in national parks. However, the majority of the national park units preserve historical or archeological sites of national significance, such as Gettysburg National Military Park (Figure 17.13c). The NPS also includes 17 **national recreation areas**. These are among the most heavily used facilities in the national park system and

(a)

(b)

FIGURE 17.12 *National parks and reserves have been established in more than 100 countries; for example, (a) Banff National Park in Canada, and (b) Fiordland National Park in New Zealand.*

include mostly ocean beaches, large reservoirs, or rivers, often located within or near major urban areas. In addition, there are 10 **national seashores** and 4 **national lakeshores**, which also provide water-oriented recreation while preserving stretches of ocean and Great Lakes' coastline.

The national park system has changed dramatically since the NPS was formed in 1916. At that time, the NPS had to manage only 15 parks and 21 national monuments. Use of the park system increased rapidly as American society became more affluent in the years following World War II. New recreational areas and urban parks that were added to the system during the 1960s and 1970s provided high-quality recreation opportunities close to large population centers. As a result, annual visits to national park areas grew from less than 50 million in 1950 to

(a)

FIGURE 17.13 *The U.S. national park system includes cultural as well as natural resources: (a) Great Sand Dunes National Monument in Colorado, a unique dune ecosystem; next page: (b) Grand Teton, one of the most spectacular national parks; and (c) Historic sites such as the Gettysburg National Military Park.*

(b)

(c)

FIGURE 17.13 *(continued)*

more than 436 million by the end of the 1990s. With many parks receiving more than a million visitors each year, traffic congestion, pollution, crowding, noise, and even crime have become problems. As budgets dwindle and use increases, the mission of the NPS to provide quality recreation experiences and to maintain the environmental integrity of the resource is beginning to fail.

State, Provincial, and Urban Parks in the United States and Canada

In addition to the extensive national park systems in both the United States and Canada, their states and provinces also sponsor extensive park systems and other recreation areas. Today estimated attendance at the 4,000 state parks is nearly double that of national parks (over 850 million visits). In Canada, park systems have been established by each of the provinces, and several have provincial parks that are comparable to national parks in size and natural features. For example, Ontario recently designated Polar Bear Provincial Park. Stretching along the south shore of Hudson Bay, more than 300 kilometers from the nearest highway, this 9,400 square kilometer (6 million acre) park protects prime habitat for polar bear, woodland caribou, and numerous species of nesting waterfowl.

The urban park movement developed more than a century ago as North American and European cities became larger and more industrialized. With stark, pol-

FIGURE 17.14 *Central Park in New York City, one of the world's first urban public parks. It includes 840 acres in the center of Manhattan.*

luted urban landscapes expanding rapidly, parks were established to set aside open space for the recreation and relaxation of the growing urban population. New York City's Central Park, which recreated a rural landscape with forests, meadows, and lakes, influenced the creation of numerous large landscaped parks in other North American cities during the late nineteenth century (Figure 17.14). Since then park systems and other recreation areas have been established in and around towns and cities throughout the industrialized world.

17.7 The Wilderness Movement

As the North American frontier closed and the United States emerged as an urban industrial nation, the perception of wilderness evolved from that of a desolate, inhospitable landscape to that of a valued resource. Today, the United States has a large national wilderness preservation system consisting of more than 630 individual areas and 425,000 square kilometers (105 million acres). A few other countries are also developing similar systems to protect and manage their wildland resources.

As society has come to endorse the concept of wilderness as a resource, three main themes have emerged.

1. The *experiential and spiritual values* of wilderness were espoused a century ago by John Muir, founder of the Sierra Club. To him, the solitude, beauty, and feeling of freedom afforded by wilderness were qualities to satisfy human needs.
2. Wilderness as a place to *build character and conduct natural scientific studies* was a view held by American wilderness proponents Aldo Leopold and Robert Marshall. They argued that the wilderness recreation experience could build both individual and national character. Educated as scientists, they also recognized that wilderness tracts were natural laboratories.
3. Today, scientists recognize the importance of wildlands as a *reservoir of genetic diversity* and as a place to study ongoing processes in undisturbed ecosystems. These pristine lands also afford baselines to gauge environmental changes occurring elsewhere and as sanctuaries for plants and animals that flourish only in truly wild habitats.

Inventory of the Earth's Wildlands

The Sierra Club conducted an inventory of the world's wilderness in the late 1980s and found that nearly one-third of the world's land surface remains a wilderness (Figure 17.15). Virtually all the world's wilderness is found in the frontier environments, with the majority in cold lands, dry lands, and mountains: the Antarctic and Greenland icecaps; the tundra and boreal forest of North America and Eurasia: the arid lands of North Africa, the Arabian Peninsula, Central Asia and Australia (see Figure 4.4).

In the tropics, more than one-quarter of Brazil remains a forest wilderness, and some forest wilderness remains in equatorial Africa and in southeast Asia. Elsewhere, most wildlands are associated with rugged mountain ranges. Most of the 810,000 square kilometers (200 million acres) of wildlands in the United States are in the mountainous west and Alaska. More than half of these lands have been legally designated and protected in the national wilderness preservation system. Among the continents, Europe has the least wilderness at 7 percent.

International Wilderness Programs

In most parts of the world, national and international programs are being established to protect some wildlands. It is important to remember, however, that the values associated with wildland resources differ markedly among cultures and nations. In the developed countries, wilderness has become a valued resource to be protected. When the wilderness preservation movement began in the nineteenth century, the eastern two-thirds of the United States was already settled. The remaining wilderness in the western mountains and deserts was recognized as a resource that should be protected. European countries had few if any wilderness areas, so they turned their attention to protecting wildland and wildlife in their colonies, especially in Africa, by establishing parks and preserves.

In contrast, many developing countries where wildlands are commonplace are poor, and they hope to improve their living standards through economic development. Here there may be little recognition of the wildland values shared by developed countries. Some of these lands continue to be used in their traditional manner by subsistence gatherers, herders, or shifting cultivators. Others are being exploited for timber, minerals, or commercial agricultural production. Only when wildlands become a significant source of income do they become a prized resource to these countries. For example, the national parks of Africa attract tourists from throughout the world with their spectacular animal herds. By simply keeping their ecosystems intact, these wildlands have become important sources of revenue for poor countries such as Tanzania and Kenya. Some Central and South American countries are beginning to realize revenues generated by a growing tourist trade—known as **ecotourism**—to the tropical forests.

Currently only Australia, Canada, New Zealand, South Africa, United States, and Zimbabwe have wilderness preservation programs. With the exception of Zimbabwe, all are affluent countries that still have sizable tracts of uninhabited land. Compared with the extensive wilderness preservation system in the United States that has evolved over four decades, these programs are relatively new and small. Australia has 15 designated wilderness tracts, New Zealand has 6, and South Africa has 12. In Canada, zoned areas within national parks are protected

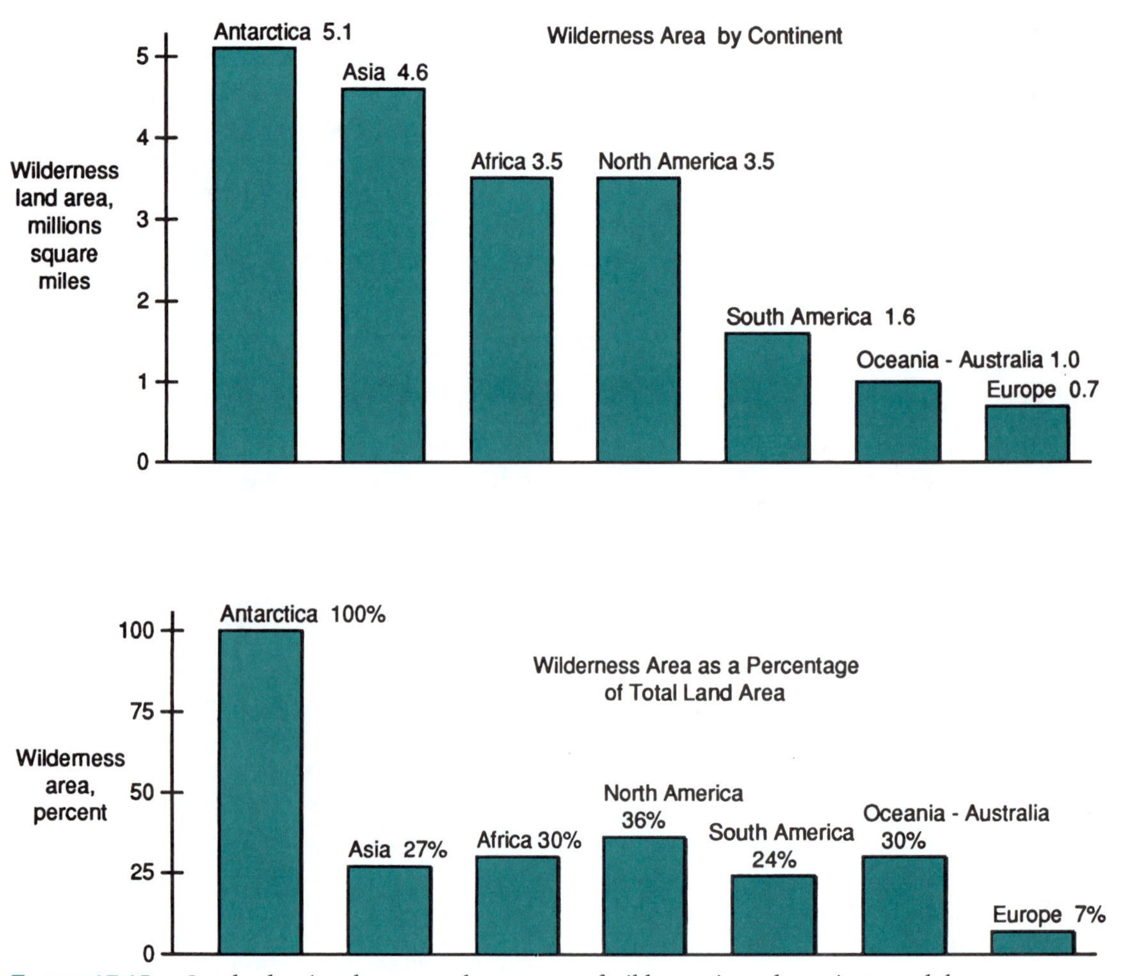

FIGURE 17.15 *Graphs showing the area and percentage of wilderness in each continent and the world.*

wilderness. Four provinces—British Columbia, Alberta, Ontario, and Newfoundland—have legally established wilderness. In 1989, Zimbabwe became the first developing country to designate a wilderness area, setting aside about 500 square kilometers (120,000 acres) of communal tribal land that will depend on hunting and other recreational activities to generate revenues.

The Role of Nongovernmental Organizations

In addition to government programs that manage open-land resources, private conservation organizations (e.g., The Nature Conservancy, Sierra Club, National Wildlife Federation, and Wilderness Society) have become politically and economically powerful advocates for protecting open-land resources throughout North America and the world. In the United States, **The Nature Conservancy** has directly purchased land or developed management programs to protect more than 28,000 square kilometers (7 million acres) of prime wildlife habitat. While the Conservancy actually manages some of the land that it has purchased, some is deeded to federal or state resource management agencies. Other tracts owned by private corporations are managed by the Conservancy to maintain natural ecosystems. At the international level, The Nature Conservancy sponsors initiatives to promote sustainable development strategies while protecting valuable open-land and wildlife resources. Finally, the Conservancy has begun cooperative programs with state and national agencies, both in the United States and abroad, that couple sustainable development with ecosystem protection.

Organizations such as the Sierra Club, Wilderness Society, and Audubon Society work to implement laws and policies that protect and preserve land resources. For example, the **Sierra Club**, one of the largest and most politically influential of these conservation organizations, has recently focused its environmental protection goals to promote policies and laws that transcend political boundaries in order to more effectively protect and restore some 21 regional ecosystems covering North America and its offshore marine environments.

In contrast to the large, well-organized, well-funded conservation organizations based in developed countries, numerous *grass roots* groups have organized in many developing countries to address local developmental and environmental issues. The goal of many of these organizations is to provide for long-term economic growth and development based on sustainable resource practices, such as that described in Chapter 3 for the Rishi Valley in India. More about the growing role of nongovernmental organizations in environmental management is presented in the next chapter.

17.8 SUMMARY

Forests, woodlands, and rangelands, which cover nearly three-quarters of Earth's habitable land, provide food, fuel, building materials, and pharmaceuticals. They represent vast and valuable stores of biological productivity and genetic diversity. Genetic resources, especially those of the tropical forest, may prove to be the most valuable resource remaining on the planet. Open lands also provide a host of environmental services and are critical links in temperature, and moisture budgets, and biogeochemical cycles.

Although forests and rangelands have supported humans in a sustainable fashion for millennia, the rapid worldwide population growth and affluent living standards in the developed countries, supported by the technological and medical revolutions of the past two centuries, have resulted in overuse, degradation, and even depletion of these resources. Because they are located within developing countries with rapidly growing populations, the tropical forests, woodlands, and grasslands are most seriously threatened. Decline and desertification of tropical grasslands from overgrazing is widespread. Deforestation and forest fragmentation is advancing at alarming rates in the tropics, and the losses of species and ecosystems are mounting. Prospects for change in the countries affected by these trends are not good, at least in the short run, as population grows and new lands and resources are needed.

Most developed countries have begun managing forests and rangelands for sustained yields and multiple consumptive and nonconsumptive uses, and reserves have been established. For instance, the national forests and national resource lands (mostly rangeland)—more than one-fifth of U.S. land—are managed by federal agencies primarily for multiple uses and sustained yield. More than 324,000 square kilometers (80 million acres) are now included in the U.S. national park system, which legally protects them from consumptive uses. In addition, the national wilderness preservation system provides legal standing to more than 425,000 square kilometers (105 million acres) of federal wilderness, mostly in national parks, forests, and wildlife refuges. Overall, the United States has set aside more than 6 percent of its land as federally protected parks, refuges, and wilderness. Worldwide, a relatively small fraction of open and wildland resources has been legally protected from consumptive uses as parks and preserves. Only about 5 percent of the world's woodlands are protected, and rangelands that are primarily used for grazing are afforded little or no protection as national reserves or wildlands.

To date, most professional management and preservation efforts have been implemented in developed countries. However, in recent years there has been increasing awareness and concern that the open-land and wildland resources remaining in the developing world must be managed and protected. Although many developing countries have set aside land as parks and other land preserves, most are not fully protected from consumptive uses such as livestock grazing, farming, poaching game, and cutting wood. In fact, extensive open-land reserves are perceived in many developing countries as frontier areas for new settlement, agriculture, and other economic development opportunities as their populations continue to grow rapidly. Only when undeveloped lands attract significant revenues, such as the national parks on the East African savanna, are they considered anything more than potential resources. International cooperation will be necessary to preserve wildland resources and intact ecosystems in many developing countries. Parks, wildland, and biodiversity preserves will need to be established as an integral part of development programs in these countries. The problems in doing this are daunting, but they must be addressed and accomplished for the benefit of all.

17.9 KEY TERMS AND CONCEPTS

North American frontier
- settlement
- lumbering
- mining

Conservation movement
- scientific conservation
- New Deal conservation
- federal agencies

Types of forests
- tropical forest
- rainforest
- monsoon forest
- northern coniferous forest
- boreal forest
- temperate forest

Multiple-use resources
Protected forests
Logging practices
- forest mining
- clear-cutting
- selective cutting
- tree farms
- industrial plantations
- commercial forests

National forest system
- sustainable forest management (SFM)

Rangelands
- desertification
- pastoralism
- commercial ranching
- carrying capacity
- animal unit equivalent
- grazing policy on public rangeland

Parks movement
- U.S. national parks
- national monuments
- national recreation areas
- national seashores
- national lakeshores
- parks in other countries
- state and urban parks

Wilderness movement
- perception
- world inventory
- preservation in United States
- ecotourism

Nongovernmental organizations
- Nature Conservancy
- Sierra Club

17.10 QUESTIONS FOR REVIEW

1. Describe how the perception and exploitation of open-land resources changed as North America was settled during the eighteenth and nineteenth centuries. Contrast the settlement patterns in Africa and South America during the same period that North America was undergoing its frontier period. Discuss why these tropical regions are currently experiencing their own frontier period.
2. In North America efforts to conserve and preserve open-land resources began more than a century ago. What prompted this conservation movement? Describe some of the major legislation that was enacted to conserve or preserve parts of the North American landscape. Why did the conservation movement soon split into the preservationist and the wise-use school? How did the U.S. government expand its conservation role during the Great Depression?
3. List the four (4) U.S. federal agencies that manage most of the public land resources. Where is most of this land? What specific land resources does each agency manage? Which lands are managed for multiple uses? Which lands have more restricted uses?
4. The global forest cover has been shrinking over the past 10,000 years as humans cut and cleared forest. Where did most forest clearing occur in the past? Where is most deforestation occurring now? Discuss these past and current trends. Why does tropical deforestation present an especially critical loss to the global resource base? What are the environmental impacts associated with practices such as forest "mining," clear-cutting, selective cutting, and tree farms and plantations?
5. Why do we call forests multiple-use resources? Which different forest uses are compatible, and which nearly always conflict? The United States and other countries manage public forest reserves for multiple uses. Describe the multiple uses recognized in the U.S. national forests. Discuss the reasons for the growing conflict over U.S. forest resource use during recent decades.
6. Why is subsistence herding no longer a viable way of life in Old World rangeland? What social and environmental impacts have occurred as a result of the disruption of this once-sustainable agricultural system?
7. As with many commonly held resources, public rangelands of western North America were seriously degraded from decades of unrestricted grazing. What steps have been taken to regulate and control grazing on the U.S. public range? Have these policies resulted in improved range conditions? Under current U.S. law, the national resource lands are to be managed using multiple-use-sustained-yield practices. Besides grazing, what other uses are recognized for the national resource lands? Why are the public ranges generally more degraded than private rangeland?
8. The national park movement coincided with growing public recognition that some unique examples of open-land resources should be preserved. When,

where, and why did this occur? Why do developed countries in the New World have well-established public national parks, whereas most European countries do not? Which developing region has the most successful and famous national parks? What is the unique attraction of most of the parks in this region? Why are parks in many developing countries simply "paper" parks?

9. Although wildlands include nearly 30 percent of the world's land area, they have only recently been considered a natural resource. In what types of environments are most wildlands? Where are wilderness resources declining most rapidly? Why are most protected wildland resources (parks, wildlife refuges, and wilderness areas) in developed countries? How does the perception of wilderness as a resource differ between developed and most developing countries? How can this perception change?
10. What is the role of nongovernmental conservation organizations in protecting open-land resources? Where have they proven to be most effective and politically influential?

18 MANAGING THE ENVIRONMENT

18.1 INTRODUCTION

In the preceding chapters, we have attempted to provide an overview of current environmental conditions and trends as well as efforts to control environmental degradation through sustainable development. After reprising the most challenging issues that threaten the environment and listing several emerging problems, our final chapter focuses on the management strategies that have been developed to protect the environment through sustainable development practices. We conclude with a short section on the use of two advanced information systems—remote sensing and geographic information systems (GIS)—that can be used to monitor and study global environmental processes and how human activities are affecting these processes.

18.2 GLOBAL ENVIRONMENTAL TRENDS AT THE BEGINNING OF THE NEW MILLENNIUM

By several measures, global trends with respect to sustainable development are worse today than when the United Nations Conference on Environment and Development (UNCED) and the Earth Summit convened in 1992. Conditions are worst in the world's poorest countries where the number of people living in poverty has increased as the income gap between the poorest countries and developed countries has widened considerably. Such trends are even more alarming given the progress that has been made in terms of institutional development, international consensus-building, and accords. Moreover, the trends persist despite the actions of government and private sectors to push for environmental protection through sustainable development.

A number of critical environmental situations exist that will be very difficult to ameliorate, and time is running out for us to make the transition toward sustainability (Figure 18.1). At the beginning of the new millennium some of Earth's most pressing environmental conditions include the following:

- **Global climate warming.** There is now strong evidence that increasing levels of greenhouse gas emissions are directly linked to global warming. Annual average global temperatures have risen to record levels each year since the early 1990s.
- **Tropical deforestation.** The vast tropical forests have been irreversibly damaged, and the annual rate of tropical deforestation continues at nearly 1 percent. Destruction of these original forests has caused the demise of native cultures and the extinction of innumerable species.
- **Biodiversity decline.** While data on species loss and endangerment as a result of ecosystem destruction or decline are meager, as many as half of the world's biological species may be at risk.

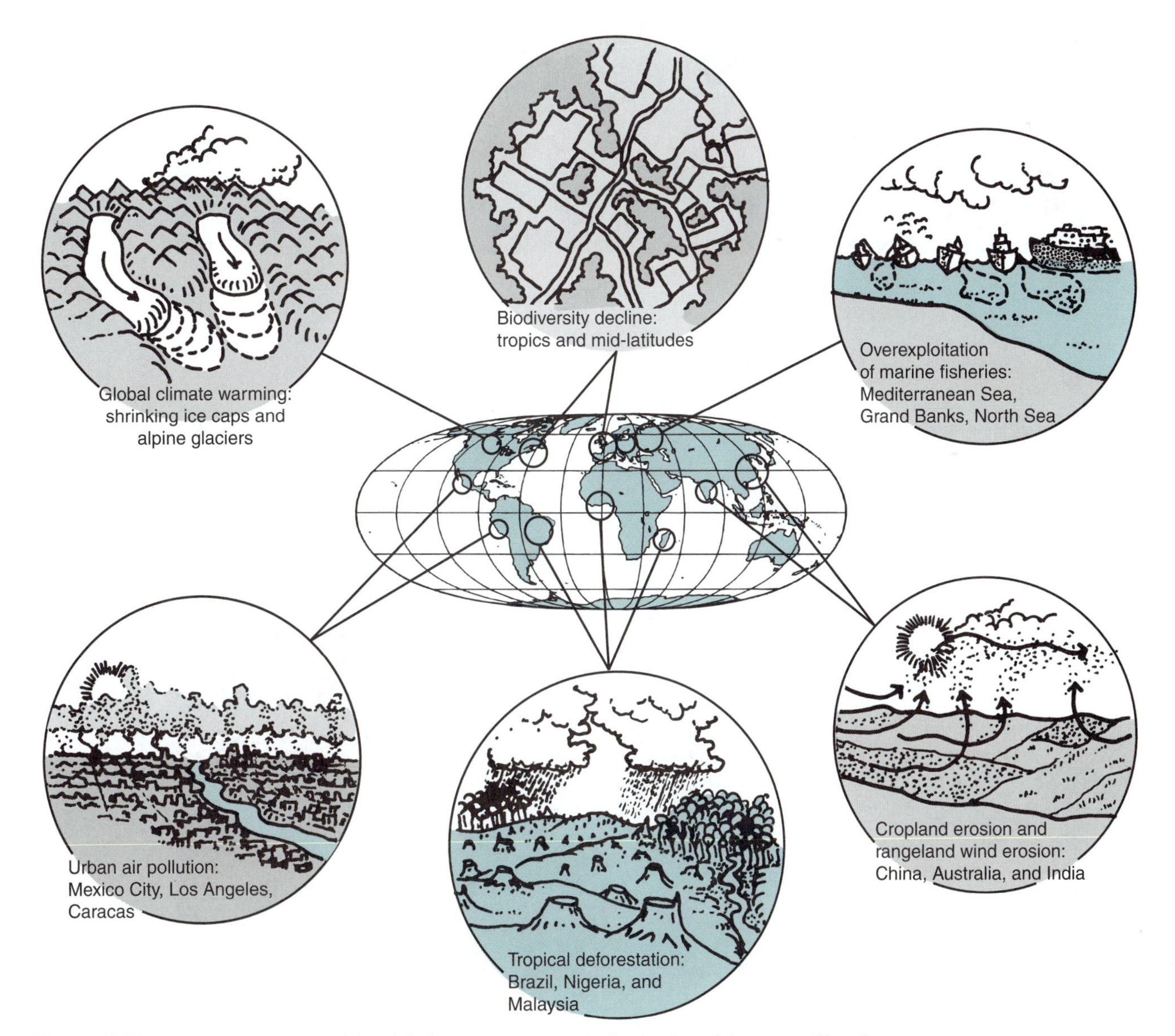

FIGURE 18.1 *Critical problems of the global environment at the beginning of the new millennium. Most are tied to poverty, population growth, and economic expansion.*

(a)

(b)

FIGURE 18.2 *(a) Refugees fleeing Kosovo for Macedonia in 1999; (b) Rainforest burning in Indonesia, fall 1999. In a region stricken by drought, fires set by humans advanced at alarming rates, (c) Massive flooding in Mozambique in 2000 which left thousands of people homeless and cropland and other land uses devastated, (d) Damaged coral reef in Man O'War Bay, Bahamas. An example of the spreading impact of land use on ocean aquatic habitat.*

(c)

(d)

- **Land degradation.** With expansion and intensification of agriculture, the resulting widespread and massive erosion of topsoil has reduced fertility and agricultural potential over significant areas of the world's arable land. These soil losses have negated many of the spectacular productivity advances made in recent decades.
- **Overexploitation of marine fisheries.** Several important stocks of ocean fish have collapsed and others have been greatly reduced as the result of over fishing.
- **Urban air pollution.** Although urban air quality has generally improved in most developed countries, many of the rapidly growing *megacities* in developing countries suffer acute air pollution problems. During the 1990s in Mexico City, New Delhi, and Beijing, air pollution exceeded international standards more than 75 percent of the time.

In just the last few years, *additional environmental threats* are emerging at the same time there seems to be declining governmental and media attention to environmental issues:

- **Armed conflict.** Numerous civil wars and other military actions not only devastate the environment in the area of conflict but in surrounding areas as well. Furthermore, the large numbers of refugees fleeing these conflicts exert significant environmental impact as they move *en masse* to temporary settlements (Figure 18.2a).
- **Forest fires.** More numerous and larger forest fires are caused by abnormal weather conditions and changes in land use and land cover (Figure 18.2b).
- **Natural disasters.** Increased frequency and severity of natural disasters such as floods and hurricanes injure and kill thousands of people each year and displace millions more (Figure 18.2c).
- **Coastal damage.** The rapid decline of coastal marine environments includes the loss and irreparable damage to more than half of the world's coral reefs (Figure 18.2d).

18.3 PROGRESS TOWARD SUSTAINABLE LAND USE SYSTEMS

On the positive side, there has been significant progress in environmental management, consensus building, and decision making at local, regional, national, and international levels. Especially in the developed countries, increased public concern over environmental issues has resulted in numerous regulations and laws governing environmental matters, as well as increased expenditures to remediate or improve environmental conditions. For example, in the United States the number of federal laws relating to environmental protection has doubled over the past 30 years. During the same period, the European Union has increased its environmental directives and recommendations by threefold to more than 300. And since the 1970s European countries have agreed to several pacts to protect or improve coastal and marine environments. Although resultant changes in the environment are often subtle the net effect is expected to produce a more sustainable system of land uses in Europe and North America.

In many areas, environmental regulations have mandated improved technologies in the industrial and transportation sectors that reduce pollution and improve land use efficiencies. For example, catalytic converters that reduce exhaust emissions on motor vehicles are now required in many developed countries, and coal and oil-fired power plants are required to remove sulfur dioxide from emission before releasing them into the atmosphere.

Governmental and nongovernmental institutions that focus on environmental protection and management have emerged in recent years. Today nearly all developed countries have environmental protection agencies. And thousands of nongovernmental organizations (NGOs)—large and small alike—operate in developed and developing countries to promote environmental issues at international, national, regional, and local levels. In Europe, the political influence of green parties with strong environmental agendas have ebbed and flowed over the past three decades. Under the auspices of the United Nations, international agencies have formed to develop and implement policies and programs that protect the environment and foster sustainable development. The following paragraphs discuss the evolution and growing influence of these institutions.

18.4 INTERNATIONAL INSTITUTIONS TO MANAGE THE ENVIRONMENT

When the **United Nations Conference on Environment and Development (UNCED)** convened in June 1992, delegates from more than 150 countries and representatives from more than 1,400 nongovernmental organizations (NGOs) adopted a massive action plan designed to tie together sustainable development and environmental protection as well as international agreements on climate change and biodiversity. The conference concluded with the **Earth Summit** as heads of state gathered for two days of meetings to reach final agreement on the principles, treaties, and programs produced by the conference.

The documents approved at the Earth Summit were the result of several years of planning and negotiating among the world's nations and representatives from environmental and development groups and other NGOs. In effect, they represented a management plan that was designed to guide environmental protection and economic development throughout the world well into this century. Have these broad-ranging set of principles, treaties, and action plans made progress toward sustainable development?

Many are skeptical that the UNCED agenda will not be effective, but this strategy of international cooperation and participation has made progress in dealing with several environmental problems during the past three decades.

Setting the Precedent: Establishing Institutions

In 1972 the UN Conference on the Human Environment was held in Stockholm to identify and study environmental problems that were international or worldwide in scope and could only be solved through cooperative efforts among nations. The conference established a precedent for using international diplomacy to address environmental issues by establishing the **United Nations Environment Programme (UNEP)**, a permanent UN institution that would advise and act on environmental issues. Perhaps the most significant impact of the Stockholm conference was that it drew worldwide attention and publicity to environmental degradation. As a result, it increased public awareness of environmental problems and it served notice to national governments of their responsibility to participate in international efforts to address these issues.

Since the Stockholm meeting, programs to mitigate a variety of environmental problems have been forged through international diplomacy.

- The *Long Range Transboundary Air Pollution (LRTAP) convention* was established in 1979 by European governments to reduce acid rain. Since then, in response to LRTAP protocols, measures have been taken by national governments and the European Union (EU) to reduce emissions of pollutants that cause acid precipitation. Rather than regulations, these protocols establish emission standards. By monitoring emissions and publicizing results, nations that fail to meet the standards are subject to public pressure to comply.
- International efforts among countries bordering the North and Baltic Seas to control marine pollution began in the 1970s. Following the establishment of the *North Sea and Baltic Sea Ministerial Conferences* during the 1980s, stringent measures to reduce pollution were enacted. By the late 1990s, emissions of dioxins, mercury, cadmium, and lead had been reduced by 70 percent, while emissions of 37 additional pollutants had been reduced by 50 percent.
- In 1990 international agreement was reached to phase out production of CFCs, the chemicals that threaten to deplete the stratospheric ozone layer. This successful example of environmental diplomacy began with the 1985 *Vienna Convention*, which established a nonbinding framework for controlling CFCs and was strengthened by the 1987 *Montreal Protocol*, which cut CFC production by 50 percent. The 1990 amendment to phase out CFCs came as a result of increased scientific evidence of the dangers of ozone depletion and the development of CFC alternatives. More than 70-countries have agreed to these phase-out goals. Financial assistance has been available to developing countries, and nations continue to share technical information for developing CFC alternatives.
- Joint efforts have been made by UNEP and the UN Food and Agriculture Organization to promote safe use of pesticides in the developing countries by registering pesticide imports and sales. Involvement of UNEP has led to tighter controls on international pesticide sales. Now prior informed consent (PIC) must be obtained before pesticides are sold to developing countries. Although compliance is voluntary, most governments with large agricultural chemical companies, including the United States, have endorsed the new registration policies.

Additional international institutions have been organized to manage ocean resources and reduce pollution. For example, stringent government proposals by the United States and European nations, backed by public concern over several major oil-tanker accidents, led to regulations that oil tankers install specific equipment in order to reduce intentional discharges of oil residues into the oceans. And although international fisheries management in the oceans has evolved over the past two decades to a system that gives nations the authority to allocate fish catches within 200 miles of their coasts where the most productive fishing grounds are found, there has been a significant decline in several important commercial species.

Sustainable Economic Development: The Key to Environmental Management in Developing Countries

Although several of these examples of international efforts to deal with environmental problems have proven to be more or less successful, most have involved only the developed countries. This is not surprising, because most developed countries have well-established environmental protection policies. Furthermore, large, well-organized nongovernmental organizations such as the Sierra Club, World Wildlife Fund, Greenpeace, and The Nature Conservancy, which are based in developed countries, have used their political and economic clout to involve themselves in national and international environmental issues and to influence environmental policy making.

In general, the developing countries have been less able to respond effectively to environmental degradation that is inexorably linked with rapid population growth and economic development strategies. With real economic growth[1] slowing or stagnant and with increasing poverty in many poor countries, environmental protection efforts have been ignored or minimized. It has become increasingly clear that environmental protection must be tied to programs that promote sustainable economic development. Recognizing this need to integrate environmental protection and sustainable development, the United Nations began planning the UN Conference on Environment and Development (UNCED) in 1989. Over the next two years, meetings were held to negotiate environmental protection treaties and to develop programs to promote sustainable economic development.

The United Nations Conference on Environment and Development (UNCED)

Held in Rio de Janeiro, Brazil, in June 1992, the 12-day UN Conference on Environment and Development brought together delegates from more than 150 nations and representatives from more than 1,400 nongovernmental organizations (NGOs) to reach final accord on several important issues, which included international agreements on climate change and biodiversity and a comprehensive work plan for sustainable development. These documents were then formally presented and signed by ministers and heads of state during the two day *Earth Summit* that concluded the Rio Conference. Highlights of the two international agreements and the action plan are briefly summarized:

- **Framework Convention on Climate Change (FCCC).** The objective is to stabilize atmospheric greenhouse gas concentrations in order to prevent or slow climate change. By the late 1990s the FCCC had established specific emissions limits on six greenhouse gasses for each of the developed countries and had set dates for meeting these emission standards. While no emissions levels have been established for developing countries there are provisions to assist developing countries to reduce greenhouse gas emissions. The FCCC agreements have been ratified by more than 150 countries and while the United States has been active in negotiations, as of 2001 it had not ratified any of the binding agreements.
- **Convention on Biological Diversity (CBD).** Initiated in 1988 by UNEP, negotiations for the CBD were concluded just two weeks before UNCED. The main goals of this agreement are to conserve biological diversity and to share sources and revenues of biotechnological products. More than 70 national signatories have agreed to develop habitat and species protection programs. Convening on a regular schedule, meetings of CBD have since focused on a variety of biodiversity issues and the individual nations' reports on their efforts to preserve biodiversity.
- **Agenda 21.** This is the UNCED action plan to guide environmental protection and development into the twenty-first century. The plan contains four major sections. The first section, *Social and Economic Dimensions*, emphasizes the need to accelerate sustainable development by improving physical and economic well-being of most people in developing countries and curbing excess consumption in the developed countries. Section two, *Conservation and Management of Resources for Development*, considers sustainable management strategies for atmospheric, land, water and biological resources, as well as safe management practices in transport, storage, and disposal of toxic and other waste materials. The role of specific nongovernmental sectors in implementing sustainable development is the focus of section three, *Strengthening the Role of Major Groups*. Each chapter suggests the contributions to be made toward sustainable development by such groups as business and industry, trade unions, science and technology, local authorities, women, children, and indigenous people, among others. The last section, *Means of Implementation*, proved most difficult to negotiate as it deals with financing procedures, scientific knowledge and technology transfer, training, and education, as well as legal issues relating to compliance and disputes.

Progress Since UNCED

The two international agreements and the action plan that were the products of the 1992 Earth Summit set ambitious plans to mitigate climate change by reducing greenhouse gas emissions, to conserve biological diversity and to plan for sustainable development. What has happened during the intervening years to implement these plans?

Most of the world's attention has turned to the Framework Convention on Climate Change. Since 1994 when this international agreement came into force after being signed by over 150 nations, annual meetings of the signatory nations called Conferences of the Parties (COP) have been held to discuss and develop specific strategies to reduce greenhouse gas emissions. At the 1997 COP held in Kyoto, Japan a draft document was negotiated that established emissions targets for the developed countries. The **Kyoto Protocol** set overall cuts of about 5 percent from the world's 1990 emission levels. Each developed country is required to reduce its greenhouse gas emissions by a certain percentage between 2008 and 2012. For example, Canada and Japan must cut emissions by 6 percent, the

[1]Real economic growth is tied to increases in per capita economic development as measured by national productivity, such as the Gross National Product (GNP). Countries with rapidly rising population and modest economic development are actually falling behind when measured on a per capita basis.

United States by 7 percent, and the countries of the European Union by 8 percent. As the protocol currently stands, developing countries need only to voluntarily reduce greenhouse gas emissions.

The Kyoto Protocol also includes mechanisms that afford some cost control measures and flexibility in meeting emissions standards, such as **emission trading** among the developed countries. The United States has been a strong proponent of trading greenhouse gas emissions rights, in part due to the successful experience with sulfur-dioxide trading among electric utility companies as part of the program to control acid rain. Critics point out that the United States would likely purchase a large quantity of low-cost emissions permits from Russia and other Eastern European countries where economic activities (and therefore, greenhouse gas emissions) have declined during the last decade.

To provide additional cost control and flexibility a **Clean Development Mechanism (CDM)** was implemented that could provide emissions credits to developed countries for projects that assist sustainable development in developing countries. For example, a CDM project might involve a company from a developed nation that plans to increase its domestic greenhouse gas emissions, building an efficient power plant in a developing country. The reduced emissions from the new facility would provide emissions credits for the firm in the developed country.

At the beginning of the twenty-first century, the Kyoto Protocol remains a work in progress. In the annual Conferences of the Parties since Kyoto, negotiations have continued. Amendments and compromises have been made, but a legally binding protocol has yet to go into effect. In fact, none of the countries that emit the most greenhouse gasses (including the United States) are likely to ratify the protocol until equity issues are settled between the developed countries whose greenhouse gas emissions are high and the developing countries whose emissions are low but rising rapidly. Another issue, of course, is the high costs involved in quickly reducing emissions to meet the somewhat demanding targets of the Kyoto Protocol. As a result, it seems unlikely that an effective international agreement will be reached until these major issues are resolved.

On the positive side, the goals of the FCCC and the Kyoto Protocol provide a sound basis for international climate change policy for the new century. Furthermore, the ongoing debate has made it eminently clear that the issue of climate change is among the most pressing issues of the twenty-first century. In 2001 the newly elected U.S. political administration indicated that it would not support ratification of the Kyoto Protocol.

With much less fanfare and publicity than the FCCC, both the Convention on Biological Diversity (CBD) and Agenda 21's **Commission on Sustainable Development (CSD)** have continued their work with Conference of the Parties (COP) meetings in most years. By 2000, the CBD had convened five COP meetings. Among the important themes that have been considered at these meetings include access to genetic resources, forest biodiversity, sustainable tourism, biodiversity of marine, coastal, and inland waters, and traditional knowledge. Most signatory countries have filed national reports detailing their efforts to preserve biodiversity (the United States is a notable exception).

The Commission on Sustainable Development is responsible for coordinating Agenda 21 programs. Each spring since its formation in 1992 it has met to discuss current and future programs and to promote and coordinate sustainable development activities worldwide. In 1997, **Earth Summit+5**, a special session of the United Nations General Assembly held on the fifth anniversary of UNCED and the Earth Summit, adopted the *Programme for Further Implementation of Agenda 21* including a comprehensive CSD workplan through 2002. The document emphasized that, while progress had been made in some areas, the state of the global environment had continued to decline. Despite five years of institutional development efforts and international consensus-building, public policy and private sector actions, overall environmental trends had worsened. It itemized the critical environmental issues and reiterated implementation strategies and mechanism similar to those of Agenda 21.

Nongovernmental Organizations (NGOs)

Outside of government there are many organizations that influence the direction of environment and development policy. Some of these nongovernmental organizations (NGOs) have large membership, are well-funded, and often deal directly with government officials in influencing and formulating environmental policy. Such well-known environmental organizations as Greenpeace, The Nature Conservancy, Sierra Club, National Audubon Society, and National Wildlife Federation often mount well-publicized campaigns in opposition to government programs or positions. On the other hand, they often work with government agencies to develop and implement environmental policies that they deem appropriate, such as "debt for environment" programs in which a small share of developing countries' international debt is paid in exchange for preservation of a special environment, such as a tract of tropical rainforest. Either way, these large, well-organized, activist organizations have become powerful political forces in some developed countries.

Unlike the well-known environmental groups, other NGOs are small, community-oriented organizations focusing on specific local needs. In developing countries, many NGOs are organized to solve local problems and meet community needs that poor governments cannot.

Greenpeace posting a warning sign for toxic waste in a Louisiana swamp.

Because of the evolving linkages between environmental and development issues in the developing world, such organizations usually support sustainable resource practices as the basis for long-term local growth and development. Additionally, coalitions and networks of NGOs are emerging as the interrelationships among sustainable economic development, environmental protection, and human well being are recognized. Furthermore, cooperation is expanding between large international NGOs and local NGOs in the developing countries.

As the numbers of these highly diverse NGOs continue to grow, their influence in creating and implementing environmental and development policy is also increasing. As mentioned earlier, more than 1,400 NGOs participated in the UNCED sessions, including business associations, trade unions, academic groups, and organizations representing women and indigenous people, as well as environmental and development organizations.

NGOs from around the world met at the loosely structured *Global Forum*, at the same time and place as the UNCED. The forum consisted of a series of technical, scientific, and policy meetings. It also issued a set of NGO treaties that paralleled those of UNCED. Perhaps the most important results of this international gathering were the alliances and transnational links formed during the forum. As permanent institutional linkages develop among NGOs, their role in influencing sustainable development and environmental protection policies will grow.

Skeptics observe that the measures for reducing global environmental problems and achieving sustainable land-use systems are "Too little, too late," and there are many good reasons for their skepticism. Among them are rapid population growth and poverty in the developing countries, which make any environmental programs exceedingly difficult to implement simply because of the need for people to consume resources and degrade the environment to ensure their own short-term survival. People in this situation do not plan for future generations.

Another reason for skepticism is political pressure for economic growth. In the United States and most other countries in the world, political platforms and governmental policies are predicated on the assumption that economic expansion is not only possible but necessary and good. Economic growth is achieved by expanding markets through population growth and increased consumerism. Economic sustainability might be a part of political rhetoric today, but economic growth is the political goal in most countries. In addition, geographic circumstances severely limit some countries' options. China and India, for example, possess small oil and gas reserves, but relatively large coal reserves. Massive populations and rapidly mounting political and consumer pressure for economic development guarantee that both countries will have to use their coal reserves for industry, power generation, and residential fuel. But coal is dirty compared to natural gas, and air pollution can be expected to rise dramatically, contributing to acid rain and carbon dioxide problems.

18.5 MONITORING ENVIRONMENTAL CHANGE WITH REMOTE SENSING

As the international agreements and programs for sustainable development and environmental protection signed during the Earth Summit are initiated, and as Earth's environments are put to increased use, there is a need to measure and monitor environmental change.

Because of the vastness of the environment and the rapid rate at which it is changed by human and natural forces, special surveillance systems are needed to measure and monitor the environment efficiently. Among the most widely used are **remote sensing systems.** These instruments, mounted on orbiting satellites or aircraft, produce maplike images and gather other nonpictorial data about the environment, including land use, vegetation cover, water temperature, and air pollution.

An Introduction to Some Remote Sensing Systems

The most widely used remote sensing systems include photographic cameras and electronic remote sensing instruments that record electromagnetic energy that has been reflected or radiated from Earth's surface or atmosphere.

Electromagnetic energy, which includes sunlight, radiant heat energy, and radar signals, travels via radiation waves. We can classify electromagnetic energy by its *wavelength*, the distance from one wave to the next. Figure 18.3 illustrates the range or *spectrum* of wavelengths that are recorded by most remote sensing systems. Note that photographic film used for remote sensing can record wavelengths from 0.3 to 0.9 micrometers, which includes visible light and some wavelengths of ultraviolet and infrared energy. A micrometer is a standard unit of length equal to one-millionth of a meter.

Aerial photography, which records reflected sunlight from the Earth's surface on photographic film, is the most widely used form of remote sensing imagery. In fact, until electronic imaging systems became available in the 1960s, aerial photography was the only remote sensing system used. Compared with aerial photography, electronic imaging systems record electromagnetic energy (radiation) emitted or reflected from Earth over a much broader spectral range (Figure 18.3). Electronic remote sensing systems include digital cameras, line scanners, and side-looking radars. Instead of photographic film, **digital cameras** use several million tiny light sensitive detectors to form electronic images. **Line scanners** record energy in narrow strips aligned either at right angles or parallel to the lines of aircraft or satellite flight, whereas **side-looking radars** send out an energy beam and record a returning signal reflected from the surface (Figure 18.4).

With the development of electronic sensors, longer wavelengths, including radiant heat or thermal infrared and microwaves (which, despite the term, are very long waves), could be used to gather environmental information. Because Earth continuously radiates most of its energy at infrared wavelengths between 3 and 14 micrometers, **thermal infrared sensors** record radiant energy from Earth's surface or atmosphere during darkness as well as daylight, revealing thermal conditions of the features (Figure 18.4a). **Multispectral line scanners** generate several images simultaneously by recording bands of electromagnetic energy from different portions of the spectrum. For example, multispectral scanners that are used on meteorology satellites for weather forecasting and to monitor storms, transmit visible and infrared images back to Earth during daylight and thermal images at night.

The long wavelengths generated by remote sensing radar systems are unaffected by clouds, so radar is an all-weather remote sensing system (Figure 18.4b). Among the first environmental applications of remote sensing radar were land-mapping projects in Panama and Brazil's Amazon basin, where persistent cloud cover precluded the use of aerial photography. When used from satellites,

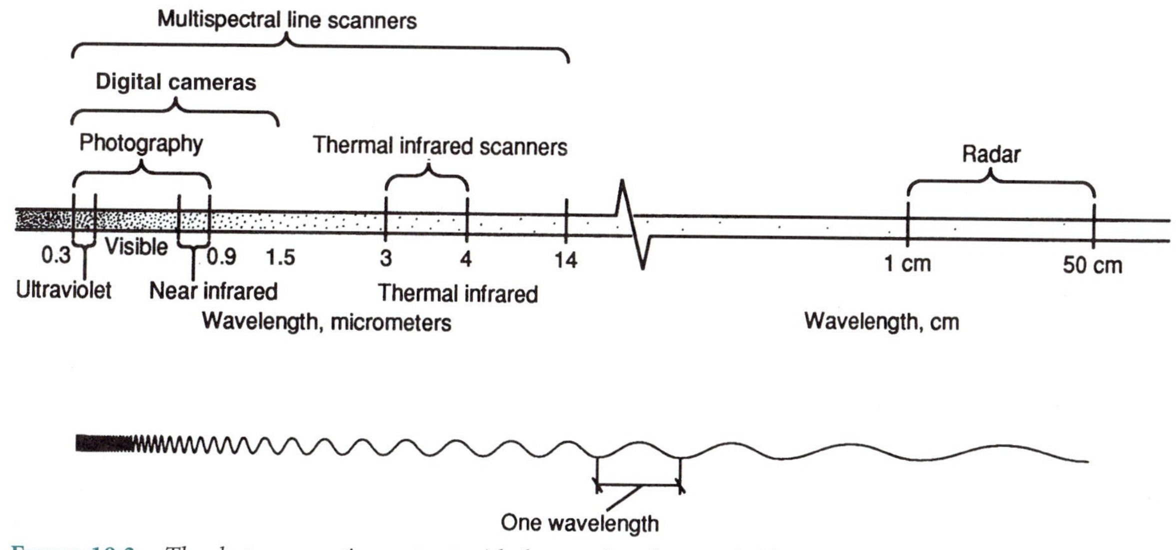

FIGURE 18.3 *The electromagnetic spectrum with the wavelengths recorded by various remote sensing systems.*

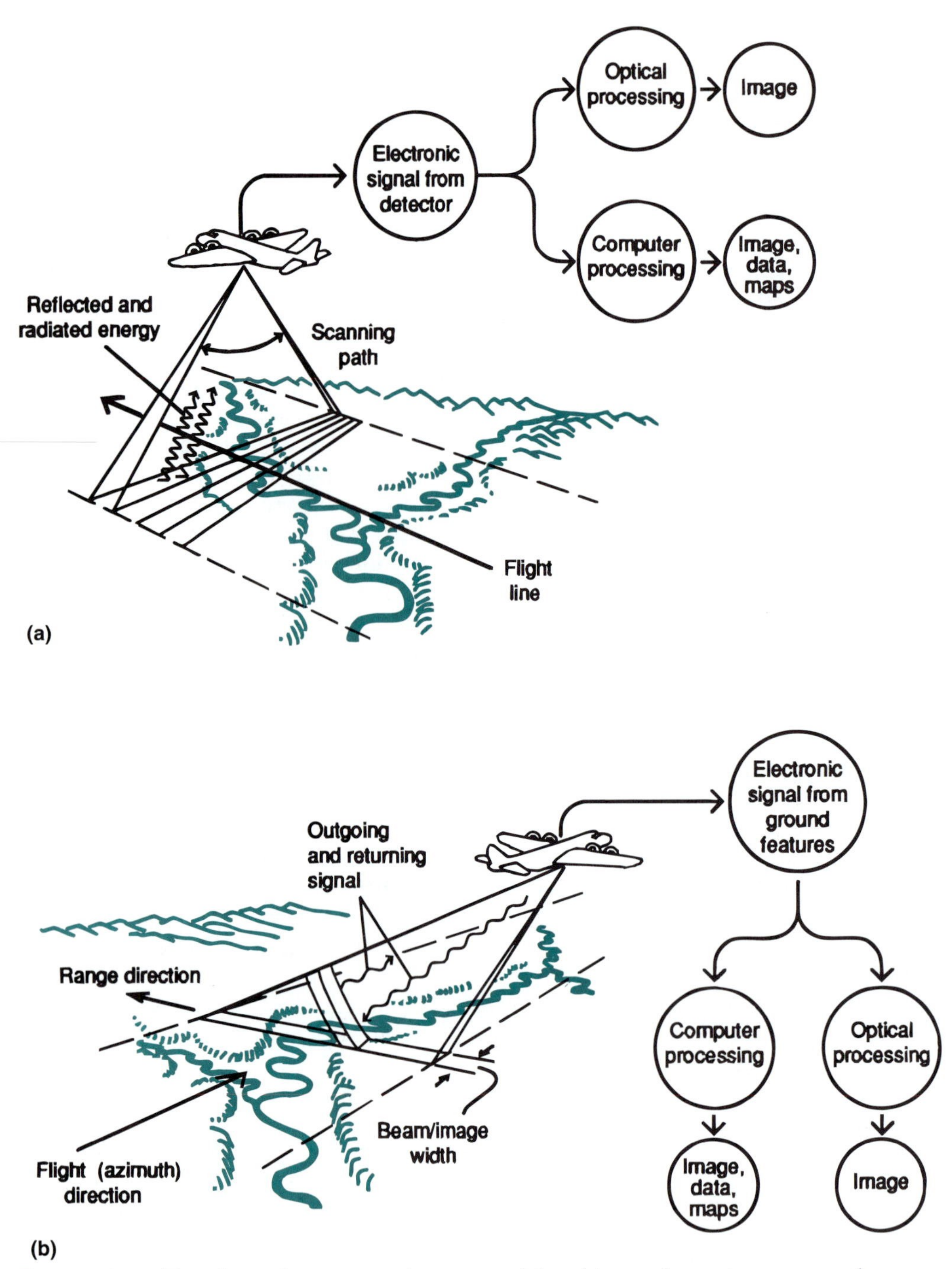

FIGURE 18.4 *Two electronic remote sensing systems: (a) multispectral scanning system and (b) side-looking radar. The first records natural electromagnetic energy reflected and emitted from the earth in different wavelengths, whereas radar beams energy to the surface and records the energy reflected back to the radar.*

remote sensing radar also generates unique imagery of the oceans that display wave patterns and currents.

In recent years, digital cameras have become an important new tool for electronic remote sensing. Digital cameras look like traditional photographic cameras but instead of film, a two-dimensional array of several million detectors converts reflected energy from the Earth to computer-compatible digital data. Currently digital cameras are producing the most detailed satellite images of the planet's surface.

Perhaps the greatest advantage of electronic remote sensing systems is that data can be recorded in digital format that can be processed and analyzed using computers. **Digital data** are created by converting the electronic signals of the remote sensing detectors into an array of numbers that can be processed by a computer. To create visible images, the digital data can be transformed into video signals that can be viewed on a computer monitor or recorded on photographic film. Another advantage of electronic remote sensing is that the data can be readily transmitted via telemetry from satellites to ground receiving stations. Today more than three dozen scientific remote sensing satellites are collecting and transmitting digital imagery and other data used to monitor environmental processes and change.

Aircraft Remote Sensing to Map and Inventory Resources

For nearly a century aerial photography has been used to produce accurate topographic maps and as a planning and management tool to inventory, assess, and monitor rural and urban land uses and forest, soil and wetland resources (Figure 18.5). Today with high resolution panchromatic (black and white), color, and color infrared films, aerial photography remains an indispensable tool for resource management (Table 18.1).

As a result, many countries now have national programs for acquiring aerial photographs. U.S. agencies such as the Geological Survey (USGS) and the Department of Agriculture (USDA), along with state and local agencies throughout the country, have been systematically acquiring photography for decades. Much of this photography has been cataloged and archived and is available through the USGS. Currently, the U.S. government sponsors the **National Aerial Photography Program (NAPP)** to meet federal and state agencies' needs for aerial photos. This program acquires color infrared (CIR) or panchromatic photography. NAPP photos are also available to the public through the USGS.

Electronic remote sensing systems such as digital cameras, line scanners and side-looking airborne radar (SLR) are also important tools for environmental management. Aerial imagery acquired with multispectral line scanners

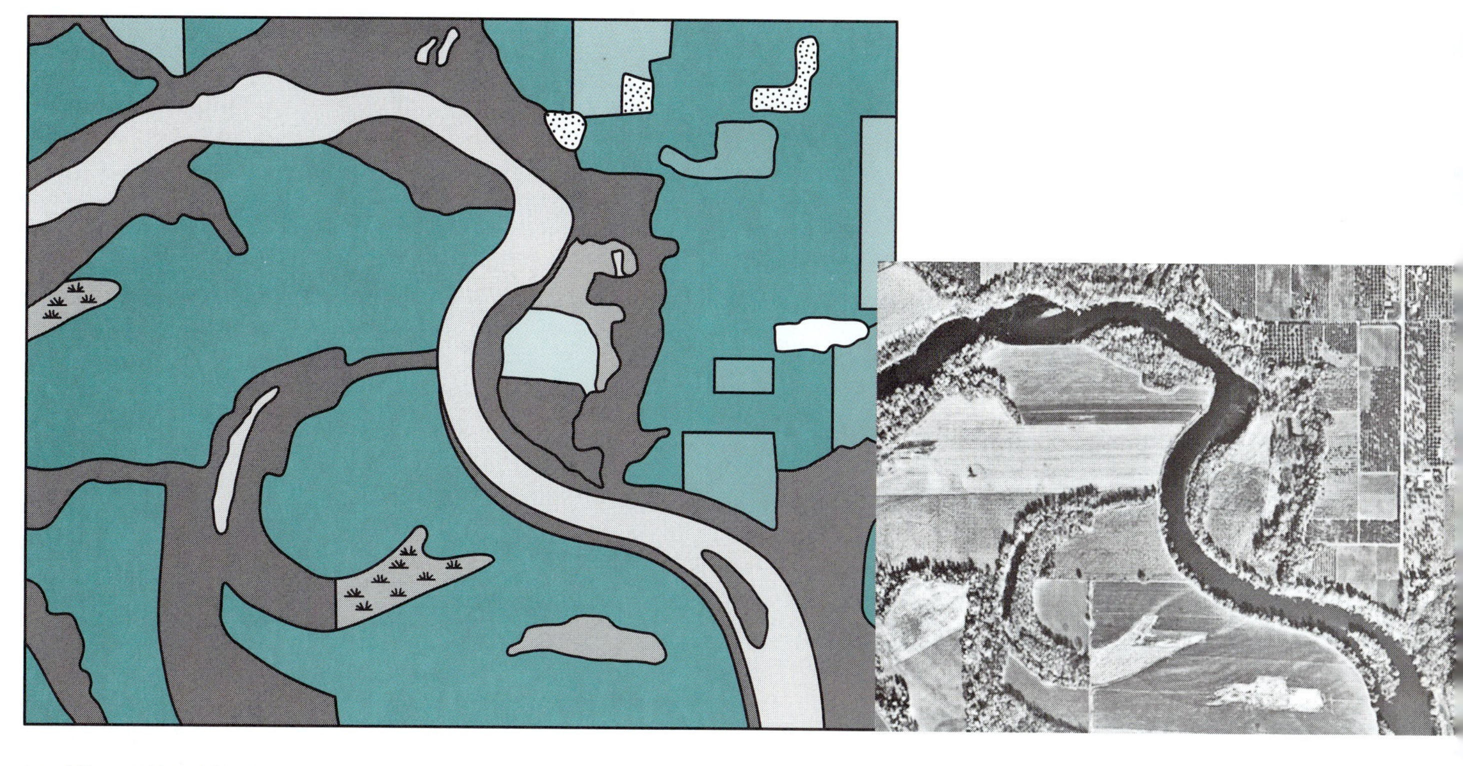

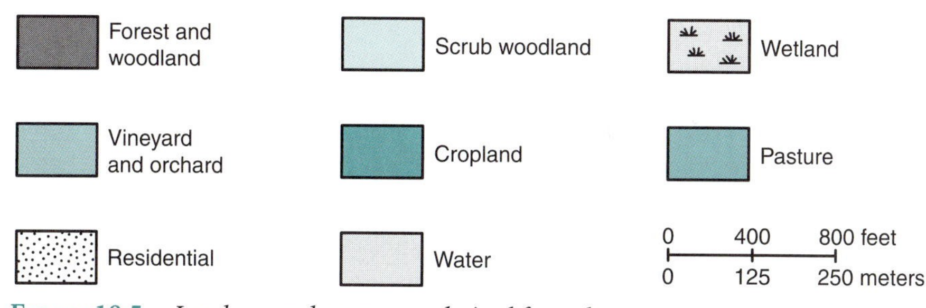

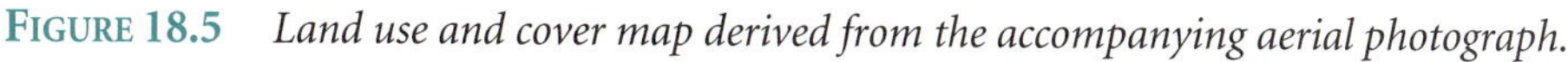

FIGURE 18.5 *Land use and cover map derived from the accompanying aerial photograph.*

TABLE 18.1

CHARACTERISTICS AND APPLICATIONS OF SELECTED REMOTE SENSING SYSTEMS

Remote Sensing System	Wavelength Range	Image Types	Representative Applications		Examples of Current Remote Sensing Programs
			Aircraft	Satellite	
Photographic Cameras	0.3-0.9 micrometers	Photographic film • Panchromatic (b+w) • Panchromatic infrared • Normal color • Color infrared (CIR)	• topographic mapping • land use mapping • resource monitoring and assessment	• Military reconnaissance	• National Aerial Photography Program (NAPP) acquires 1:40,000 scale panchromatic and color infrared photos for U.S. federal agencies and states • Various resource inventory and mapping photography programs for state, regional, and local agencies • Large-scale topographic mapping photography by commercial aerial survey films
Digital Cameras	0.3-1.5 micrometers	Digital panchromatic or color images	• (same as photographic cameras)	• Update land use • Resource monitoring	• Digital orthophotography • Commercial satellite high resolution imagery (IKONOS)
Line Scanners: • **Multispectral** • **Thermal Infrared** • **Hyperspectral**	0.3-1.4 micrometers	• Black and white or color images from digital or analog electronic signals recorded as photographs	• Monitoring heat losses from buildings, pipelines	• Monitoring and mapping earth resources • Weather forecasting and storm monitoring	• Earth resources satellites • Landsat (U.S.) • SPOT (France)
	3-14 micrometers	• Digital imagery processed and displayed on computers	• geothermal mapping • forest fire detection and mapping	• Mapping surface and atmospheric temperatures	• Operational meteorology satellites • NOAA (U.S.) • GOES (U.S.)
	0.3-2.5 micrometers			• Monitoring atmospheric water vapor and ozone concentrations	• Experimental satellites • UARS (U.S.) • TERRA (U.S.)
Radar	1-100 centimeters	• (same as line scanner imagery)	• All-weather mapping and resource monitoring	• Monitoring and mapping land resources with high resolution imagery • Monitoring waves, currents and other ocean surface characteristics	• Radar equipped satellites • ERS 1,2 (European Space Agency) • Radarsat (Canada) • Shuttle Radar Topographic Mapping Mission (U.S.)

and digital cameras provides an efficient way to monitor landscape change and to update landuse maps. Aircraft equipped with thermal infrared line scanners have been used by the USGS to map and monitor areas of geothermal and volcanic activity and by the U.S. Forest Service for forest-fire detection and mapping. Commercial firms use thermal infrared aerial surveys to detect heat loss from buildings, steam tunnels, and pipelines. As mentioned earlier, aircraft equipped with remote sensing radar have been used to map extensive areas in the wet tropics where persistent cloud cover makes traditional photographic mapping difficult (Table 18.1).

Satellite Remote Sensing to Monitor the Environment

For more than four decades orbiting remote sensing satellites have been recording information about Earth's environment and relaying these data, usually in digital form, to receiving stations on Earth.

Although manned satellites operated by the Russians and the United States have periodically acquired photographs and other remote sensing images of the Earth, the majority of remote sensing from space has been with unmanned Earth observation satellites. Along with the United States and Russia, several other nations, including Japan, France, Canada, India, and the European Union, currently operate environmental remote sensing satellite programs. There are two major types of these remote sensing satellites: **meteorology satellites** and **Earth resources satellites.**

Meteorology Satellites

Although specially designed to monitor regional and global weather conditions and to provide data primarily for weather forecasting, meteorology satellites or **metsats** also gather useful information about the conditions of the Earth's atmosphere and land surface. One type of metsat is launched into a geostationary orbit 38,500 kilometers (24,000 miles) above the Earth's equator. From this position the satellite orbits in the same direction and speed as Earth's rotation and therefore is essentially "parked" at a specific longitude over the equator (Figure 18.6). Currently, five *geosynchronous* (geostationary) *environmental satellites* operated by the United States, the European Union, Japan, and India make up the geostationary metsat network. Operated by the **National Oceanic and Atmospheric Administration (NOAA)**, the U.S. geosynchronous satellites are called **GOES** for **Geostationary Operational Environmental Satellites.** Producing both visible and thermal infrared imagery, each satellite generates an image of the facing hemisphere of the Earth every 30 minutes throughout the day and night. While geostationary satellites continuously monitor cloud patterns and other weather phenomena, their images can also be used for mapping regional snow and sea ice conditions and surface temperatures (Table 18.1).

Currently, the United States and Russia operate **polar orbiting environmental satellites (POES)**, metsat systems. Orbiting Earth at altitude of about 825 kilometers (520 miles), each satellite acquires images of all points on the earth, twice each day (Figure 18.6). The U.S. NOAA satellites monitor weather and surface conditions from contin-

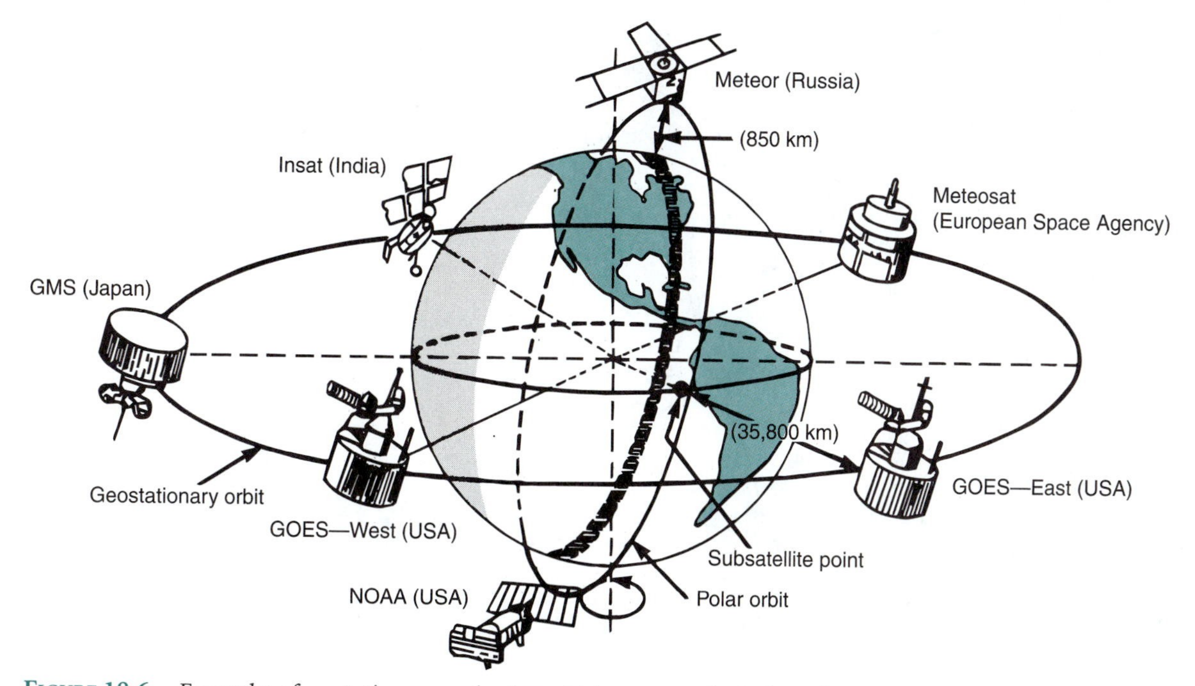

FIGURE 18.6 *Examples of geostationary and polar orbiting meteorological satellites.*

uous images covering a 2,400-kilometer (1,500-mile) swath. Figure 18.7 is a satellite image showing the position and details of a typhoon in the Indian Ocean off the coast of Bangladesh. The daylight orbits are recorded in visible, reflected infrared and thermal infrared bands, while night-time orbits produce thermal infrared images. Using a ratio of the visible and reflected infrared bands, computer-generated *vegetation index images* have been developed to monitor vegetation growth over large areas such as major agricultural regions and the tropical forests. These are useful indicators of the condition or presence of natural or planted vegetation. Besides providing worldwide imagery twice daily, each NOAA satellite receives data from automatic ground sensors, ocean buoys, weather balloons, and aircraft that measure local atmospheric and surface conditions, then relays these data to ground receiving stations.

In addition to the operational metsats, the United States has supported experimental remote sensing satellite programs designed to study environmental processes. For nearly three decades, experimental **Nimbus** satellites operated by NASA produced vast amounts of atmospheric and oceanographic data from a wide array of remote sensing instruments. For example, the *Coastal Zone Color Scanner (CZCS)* on board Nimbus 7 measured subtle color variation in coastal waters to estimate concentrations of phytoplankton (the microorganism at the base of the ocean food chains), suspended sediment, and dissolved organic matter. Reflected infrared bands mapped marine vegetation, and thermal infrared bands recorded temperatures at the sea's surface. During the eight years that the CZCS operated, scientists learned much about these most productive yet most polluted portions of the oceans. For the past decade the **Upper Atmosphere Research Satellite (UARS)**, launched by the United States, has studied the dynamics, chemistry, and energy balance of the upper atmosphere. One instrument measures concentrations of ozone and a chlorine molecule, which is a key reactant in ozone destruction (Table 18.1). With scientific operations beginning in 2000, the *Terra* satellite uses five state-of-the-art remote sensing instruments to study the Earth's atmosphere, oceans, and land surface Figure 18.8). Terra is an integral component of the international *Earth Observing System (EOS)* which will provide remote sensing satellites and instruments for the *International Geosphere-Biosphere Program (IGBP).*

Earth Resources Satellites

In contrast to the metsats, which produce frequent but low-resolution (low-detail) imagery, the **Earth resources satellites**, produce imagery with relatively high ground resolution but with less frequent repeat coverage. The first Earth resources satellite program, **Landsat**, has been providing the scientific community with worldwide digital data since 1972 (Table 18.1). The first five polar-orbiting Landsats were equipped with a *Multispectral Scanner (MSS)*, which produces digital imagery in four bands—green, red, and two reflected infrared bands. Each full MSS Landsat scene covers a ground area ap-

FIGURE 18.7 *A NOAA meteorology satellite image of a typhoon over the Indian Ocean off the coast of Bangladesh.*

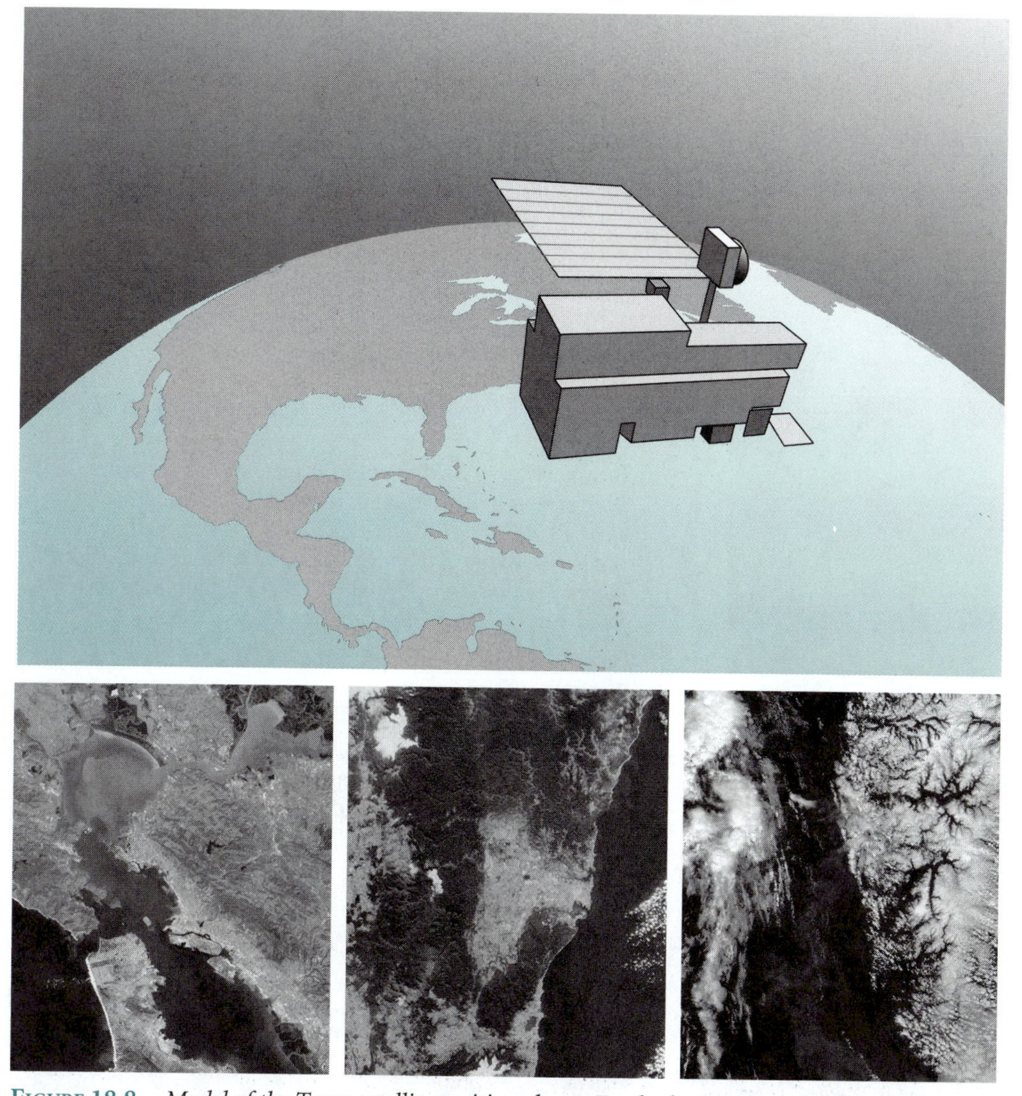

FIGURE 18.8 *Model of the Terra satellite positioned over Earth along with actual images from three of Terra's experimental remote sensing devices. From left to right are images of the San Francisco Bay area, Sydney and the southeastern Australia coast, and portions of the rugged Norwegian coast and an algae bloom in the Norwegian Sea.*

proximately 185 kilometers by 185 kilometers (115 miles by 115 miles), or more than 34,225 square kilometers (13,220 square miles). The MSS can detect features as small as 79 meters by 79 meters, (260 feet by 260 feet). The green band provides the best water penetration and is most sensitive to suspended sediment and pollution, while the red band usually provides good contrast between vegetated and nonvegetated areas and is therefore useful in detecting land-use change, desertification, and forest loss. Because healthy leaf-covered vegetation has high infrared reflectance, both of the infrared bands are useful for monitoring natural and planted vegetation and since water reflects little infrared, the dark water features usually contrast sharply with land and are more easily detected that on the visible bands (Figure 18.9). Color infrared composite images can be generated by combining the green and red bands with one infrared band.

A more advanced multispectral scanner, called the *Thematic Mapper (TM)*, was used along with the MSS aboard Landsats 4 and 5. The TM provides six bands of visible and reflected infrared data with 30 meters by 30 meters (100 feet by 100 feet) ground resolution and a thermal infrared band with 120 meters by 120 meters (400 feet by 400 feet) resolution. Besides the black-and-white images for each band, normal color composite images as well as many different false color composites can be produced by combining three different TM bands.

Although sponsored by the United States, the Landsat program provides data to the worldwide scientific community and the agencies of many national governments. The digital data collected onboard Landsat are transmitted to receiving stations in fifteen countries. Since the launch of the first Landsat in 1972, scientists from more than a hundred countries have studied environmental processes and monitored changes in land cover and land use using

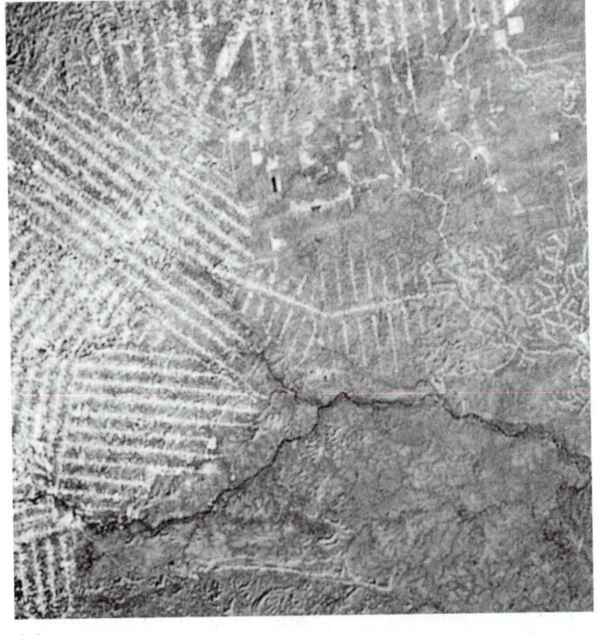

(a)

FIGURE 18.9 *Examples of Earth resources satellite imagery showing (a) Landsat image of tropical forest clearing in Brazil; (b) a SPOT image of the confluence of the Mississippi – Missouri Rivers, near St. Louis, Missouri before the massive 1993 flood and (c) a spot image after floodwaters were at their peak.*

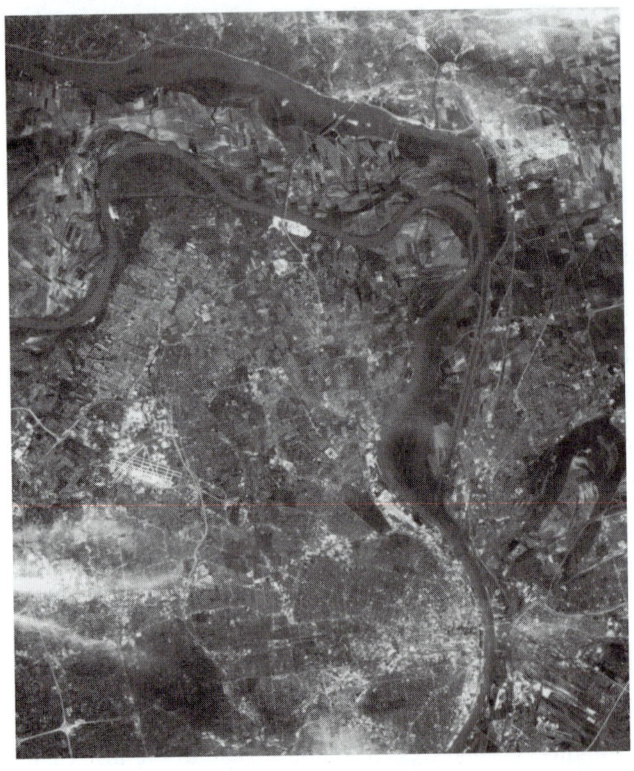

(b)

(c)

some of the several million Landsat images. After the failure of Landsat 6 in 1993, the program faced an uncertain future. But the aging Landsat 5 continues to generate Earth resources data even after Landsat 7 was successfully launched and became operational in 1999.

Landsat 7's *Enhanced Thematic Mapper Plus (ETM+)* produces the same six visible and reflected infrared bands with 30 meter × 30 meter resolution as the earlier TMs. As a result, these data are consistent with the earlier TM multispectral data. Perhaps ETM+'s most notable characteristic is a new panchromatic band that records visible and reflected infrared energy with ground resolution of 15 meters by 15 meters (50 feet × 50 feet) (Figure 18.10). The ETM+ will update the global archive of Landsat multispectral images and provide new data to monitor the terrestrial environment and assess land use and land cover change.

FIGURE 18.10 *Landsat 7's ETM+ produces a panchromatic band image with 15 meters by 15 meters ground resolution. In this image of the Washington, D.C. region, features as small as 50 feet by 50 feet can be detected.*

France's, **Systeme Pour l'Observation de la Terre (SPOT)** was the first commercial satellite remote sensing program. (Table 18.1). When the first SPOT satellite was launched in 1986, a worldwide network of ground receiving stations, as well as marketing and distribution centers for SPOT products, had been established. The polar-orbiting SPOT satellites employ two identical *High-Resolution Visible (HRV) scanners* that operate either in a wide single spectral band that records visible light in a digital format with 10 meters by 10 meters (33 feet by 33 feet) ground resolution, or in a three-band multispectral mode (green, red, and reflected infrared) that generates imagery with 20 meters by 20 meters (66 feet by 66 feet) resolution. SPOT 4, launched in 1998, has a mid-infrared band that is part of the multispectral array. Each HRV scanner image covers an area of 60 kilometers by 60 kilometers (36 miles by 36 miles, and the scanner's viewing angles can be adjusted to image either side of the satellite's path. This capability allows more frequent coverage of short-duration events such as floods, forest fires, and volcanic eruptions. Plans are to launch new SPOT satellites with improved imaging capabilities through 2007 and to archive more than 1 million images, making SPOT digital imagery a valuable environmental management tool well into this decade. The SPOT images in Figure 18.10b and c show the Mississippi and Missouri Rivers near St. Louis before and during the 1993 flood.

After the Cold War, remote sensing technologies that were developed for military reconnaissance satellites

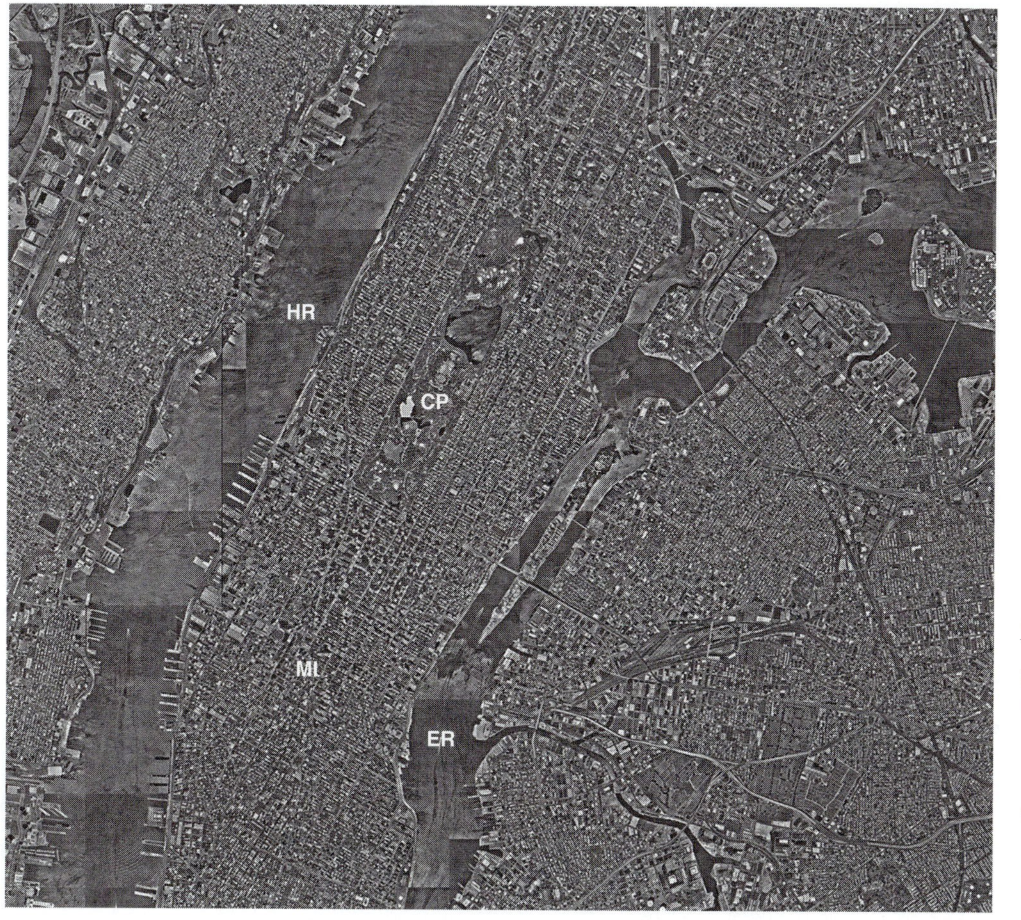

CP: Central Park
MI: Manhattan Island
HR: Hudson River
ER: East River

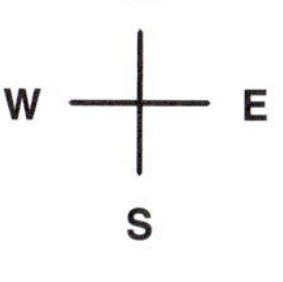

FIGURE 18.11 *Russian Spin – 2 satellite image of New York City featuring Central Park in the middle of Manhattan Island. For orientation, check the photograph of Central Park in Figure 17.14.*

became available to the scientific and commercial remote sensing communities. In 1993, the Russians began to market digital versions of military satellite photographs that have ground resolution capabilities from 2 meters to 20 meters (6.6 feet to 66 feet) (Figure 18.11).

During the late 1990's, three U.S. corporations were building remote sensing satellites capable of producing images with 1 meter by 1 meter to 3 meters by 3 meters (3 feet by 3 feet to 10 feet by 10 feet) ground resolution. In 1999, the corporate **IKONOS 2** satellite began commercial operations producing digital panchromatic images with 1 meter resolution and multispectral color images with 4 meters by 4 meters resolution (Figure 18.12). These high-resolution digital satellite images provide sufficient detail for many planning and Earth resources applications at lower costs than traditional sources of aerial imagery.

Satellites Using Remote Sensing Radar

In contrast to the satellite remote sensing systems that record energy reflected or radiated from the Earth's surface in the visible and infrared wavelengths, several

FIGURE 18.12 *IKONOS satellite image of the pyramids near Cairo, Egypt. The ground resolution of the original image is approximately 1 meter by 1 meter.*

satellites have been equipped with image-forming radar systems. Radar systems propagate their own electromagnetic energy in the microwave portion of the spectrum, with wavelengths ranging from less than 1 centimeter to 1 meter, and then record the energy that is reflected back to the radar antenna from the Earth's surface (Figure 18.4b). Because the radar generates its own energy and because clouds do not absorb or reflect radar energy, space-borne radar can always produce clear images of the ground, no matter the weather conditions.

Because sea surface conditions produced by waves, currents, or even bottom configuration are best detected by radar, the first civilian satellite designed primarily to monitor the oceans was equipped with remote sensing radar. Launched in 1978 by NASA, **Seasat** collected thousands of images of land as well as the sea (Figure 18.13). Several U.S. Space Shuttle missions during the 1980s further demonstrated the utility of space-borne remote sensing radar as a resource management tool. In 2000 the **Shuttle Radar Topographic Mapping Mission (SRTM)** generated data to produce the most accurate topographic (elevation) maps ever produced (Figure 18.14).

During the past decade several remote sensing radar satellite missions have been sponsored by Russia (ALMAZ), Japan (JERS) and the European Space Agency (ERS1 and 2). Launched in 1995, Canada's **Radarsat** was the first commercial radar satellite. Be-

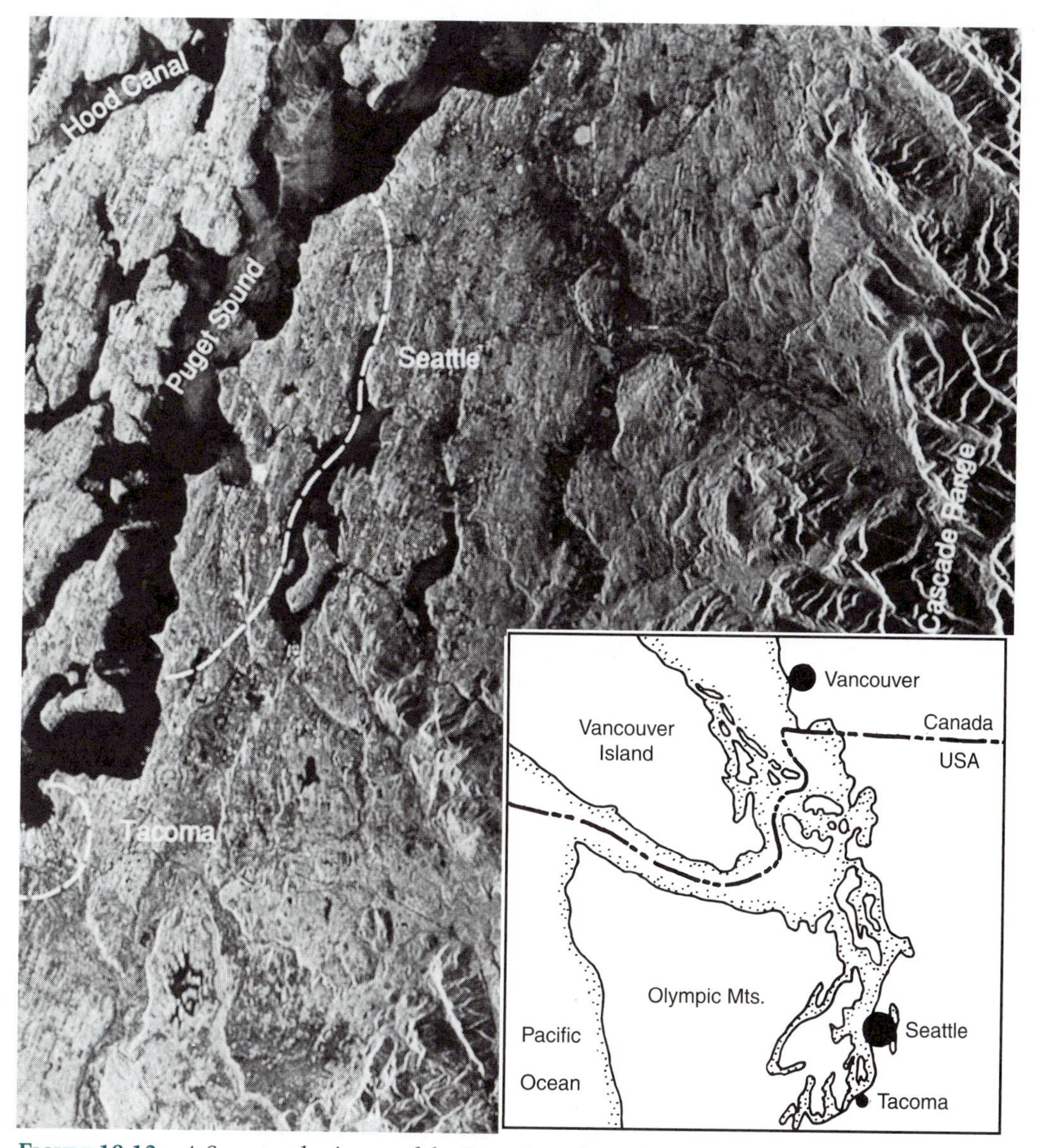

FIGURE 18.13 *A Seasat radar image of the Puget Sound area in Washington. This area is often cloudy but remote sensing radars are able to "see" through clouds and detect land use and topography as well as ocean waves.*

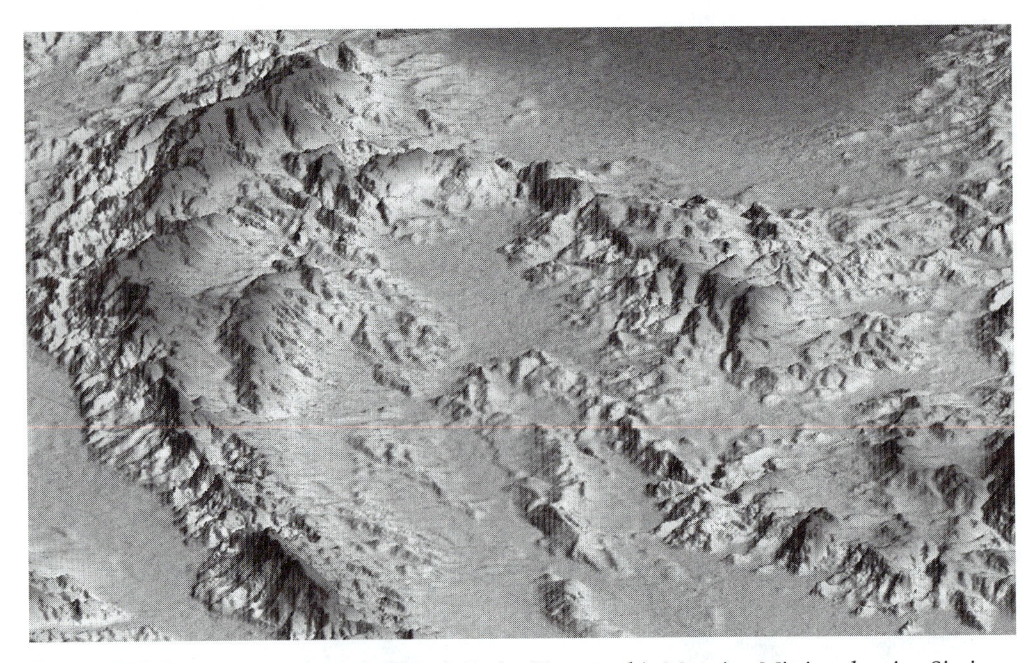

FIGURE 18.14 *An image from the Shuttle Radar Topographic Mapping Mission showing Simi Valley, California, in shaded relief.*

cause Radarsat has been able to provide images with a broad range of ground resolution capabilities (10 meters by 10 meters to 100 meters by 100 meters), area coverage (50 kilometers by 50 kilometers to 500 kilometers by 500 kilometers), and different viewing angle perspectives, it has been a commercial success (Figure 18.15). Radarsat II, which will provide even higher resolution images, will be launched in 2003. Scientists believe that data from such systems will provide new information about how variations in sea ice, snow cover, soil, and vegetation moisture may effect the world's hydrologic cycle and energy budget.

Mission to Planet Earth and Earth Science Enterprise

More than a decade ago, NASA proposed **Mission to Planet Earth,** an international program to study global change using data collected by remote sensing satellites operated by the United States and other countries. The long-term goal of the project now called **Earth Science Enterprise** is to understand global environmental processes well enough to predict environmental changes, such as rising sea level with global warming. The program relies on research conducted by teams of scientists utilizing data from satellite networks such as the Earth Observing System (EOS). This international remote sensing program to study the global environment will provide information to better manage resources and protect the environment.

In summary, both the meteorology satellites with their small scale (large area) but frequent patterns and the Earth resources satellites that provide larger scale (smaller area), more detailed data have unparalleled value for detecting environmental change and monitoring global systems. Both Landsat and SPOT data have been used to map rainforest clearing, desertification, land-use change, oil spills, and floods. Radar-equipped Earth resources satellites have been providing unique and timely high-resolution imagery of both terrestrial and marine features to the scientific community. For four decades metsat data have been used to develop runoff models based on snow cover measurements, as well as models that assess regional vegetation growth and vigor. The seasonal ozone "holes" that form over the Antarctic and Arctic were discovered and are now monitored by polar-orbiting metsats (see Figure 11.5) During the 1990s, UARS has collected information on chemistry, temperate wind, and energy flow. In the upper atmosphere and the U.S. and French sponsored TOPEX/Poseidon satellite precisely measures the sea-surface height to calculate the direction and velocity of ocean currents from which we have observed the El Niño and La Niña phenomena. In this decade experimental satellites such as Terra use new remote sensing technologies to investigate global-scale atmospheric oceanic and land-based processes and interactions. Only through the use of satellite remote sensing can we obtain the data that will enable us to better manage our terrestrial and oceanic resources while continuing to gain a more thorough understanding of the energy and material flows and interactions that drive interregional and global processes.

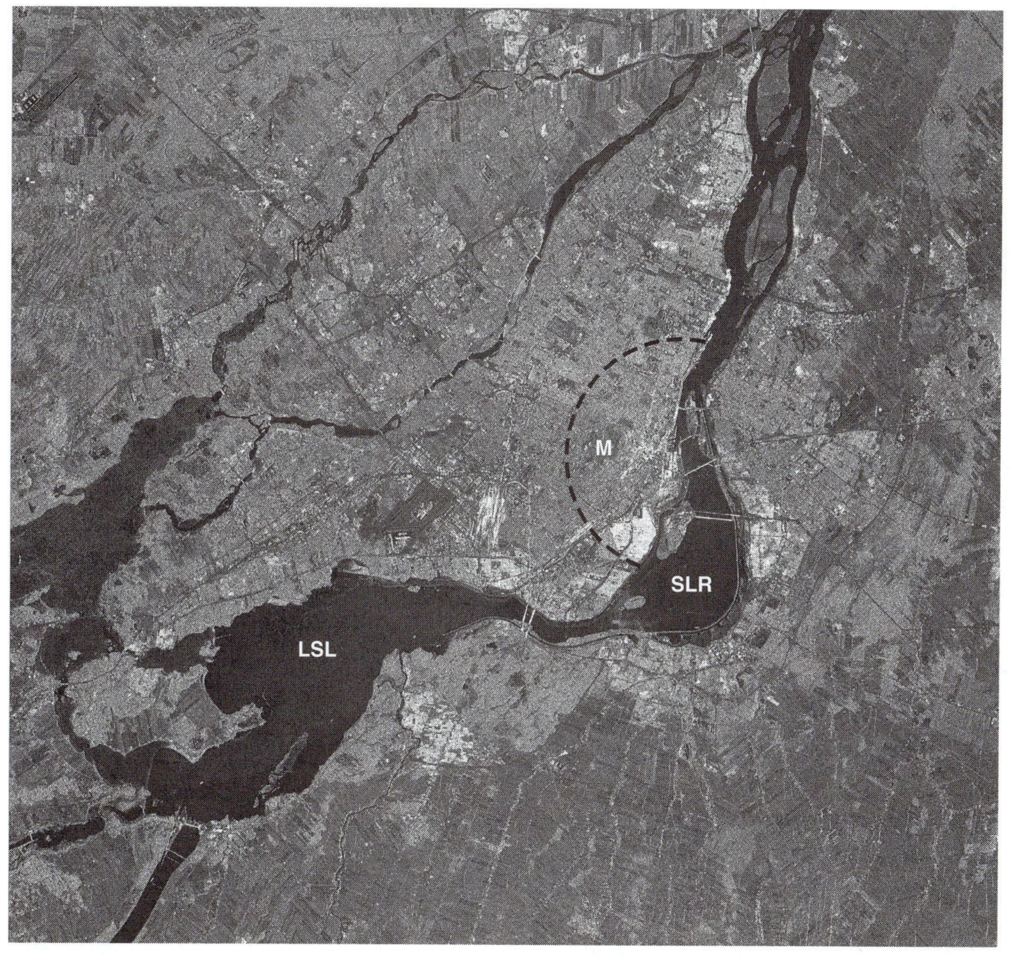

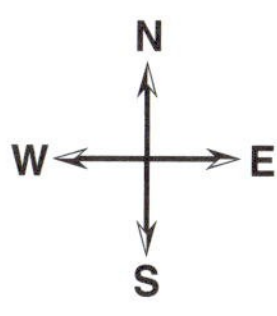

FIGURE 18.15 *Remote sensing satellite radar systems such as Canada's Radarsat can produce high quality images of the Earth's surface even in areas where cloudy weather is persistant as can be seen in the Radarsat image. This scene shows the Montreal metropolitan region.*

18.6 GEOGRAPHIC INFORMATION SYSTEMS

Today the amount of environmental data gathered from remote sensing satellites and many other sources is nearly overwhelming. And each year, as new systems go into operation, the data stream increases and becomes more complex. In order for these data to assist in environmental management decision making, they must be available to researchers in a timely and organized way. This means that efficient data and information management systems must be implemented to handle, store, manipulate, and disseminate these data. Over the past 30 years sophisticated computer-based information systems have evolved that are proving to be essential tools to help in developing management strategies for sustainable development and environmental protection. **Geographic Information Systems (GIS)** are designed to store, analyze, overlay, and map diverse types of digital data in a geographically referenced format. GIS have become an essential link in the global monitoring system between the data collection technologies, such as remote sensing, and the policy makers such as environmental agencies and organizations.

As the name implies, GIS are different from other computer mapping or data base systems because they organize data according to their location. Each bit of data on vegetation, soil, temperature, water quality, or other spatial variables contained in a GIS, is stored as a separate geographically referenced layer. Figure 18.16 illustrates GIS layers composed of georeferenced environmental, economic, and demographic variables. Because GIS presents and evaluates both the locational characteristics of variables as well as qualitative and quantitative attributes of the variables, such as soil characteristics and land value within agricultural areas, these systems provide unique and powerful tools for environmental management. Especially valuable is the capacity to add layers of new data to existing layers of data in exact locations, and to compare past and present conditions in precisely the

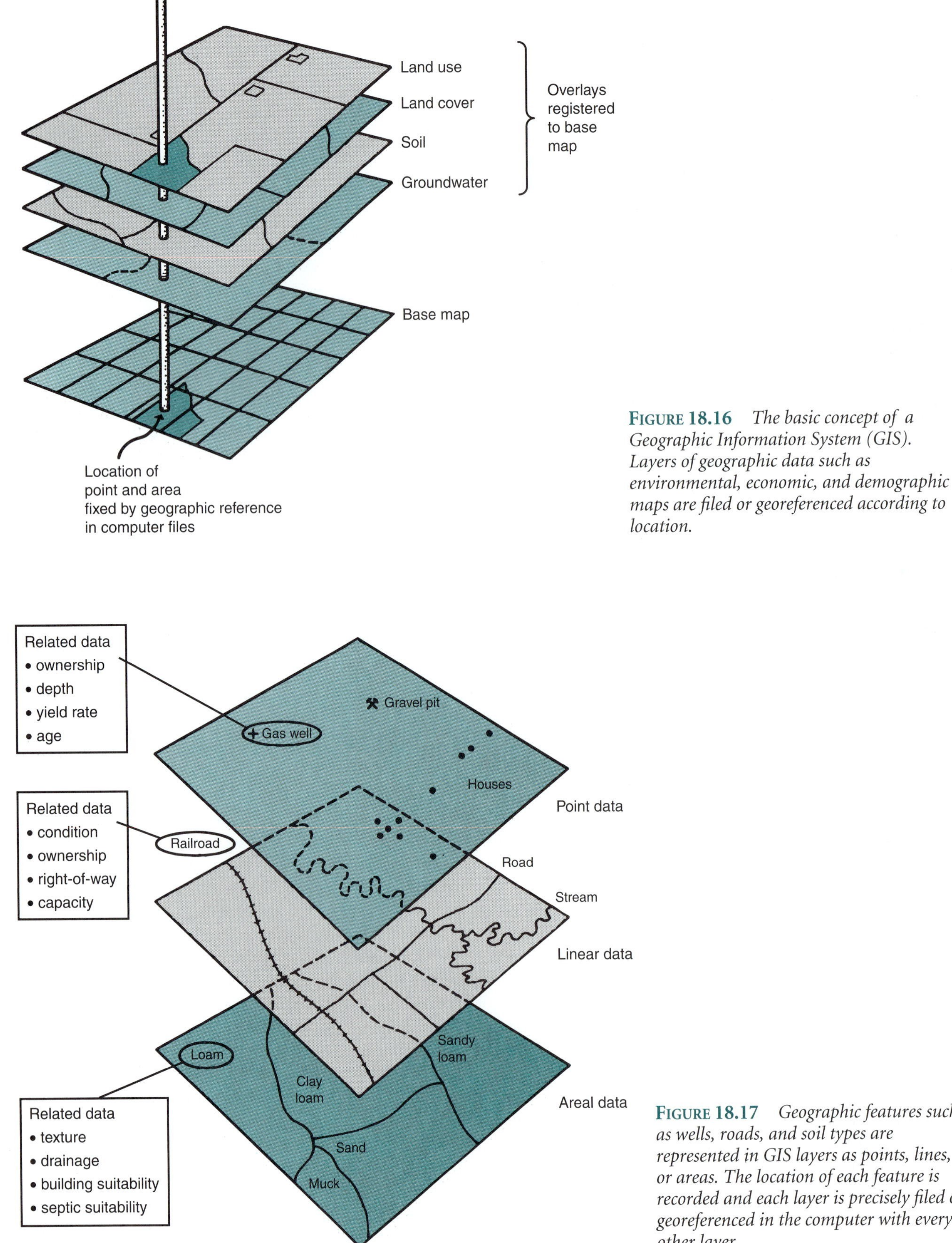

FIGURE 18.16 *The basic concept of a Geographic Information System (GIS). Layers of geographic data such as environmental, economic, and demographic maps are filed or georeferenced according to location.*

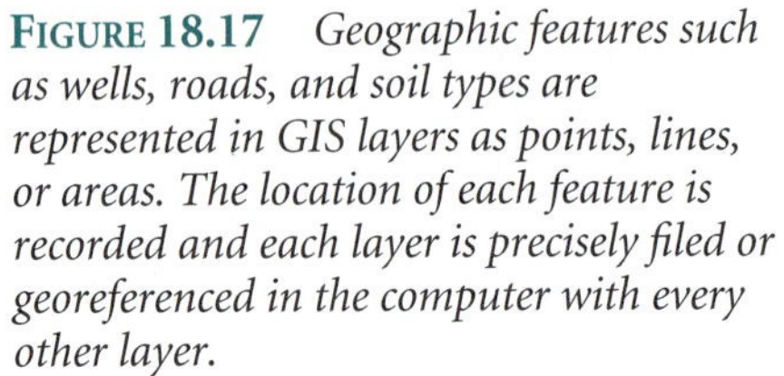

FIGURE 18.17 *Geographic features such as wells, roads, and soil types are represented in GIS layers as points, lines, or areas. The location of each feature is recorded and each layer is precisely filed or georeferenced in the computer with every other layer.*

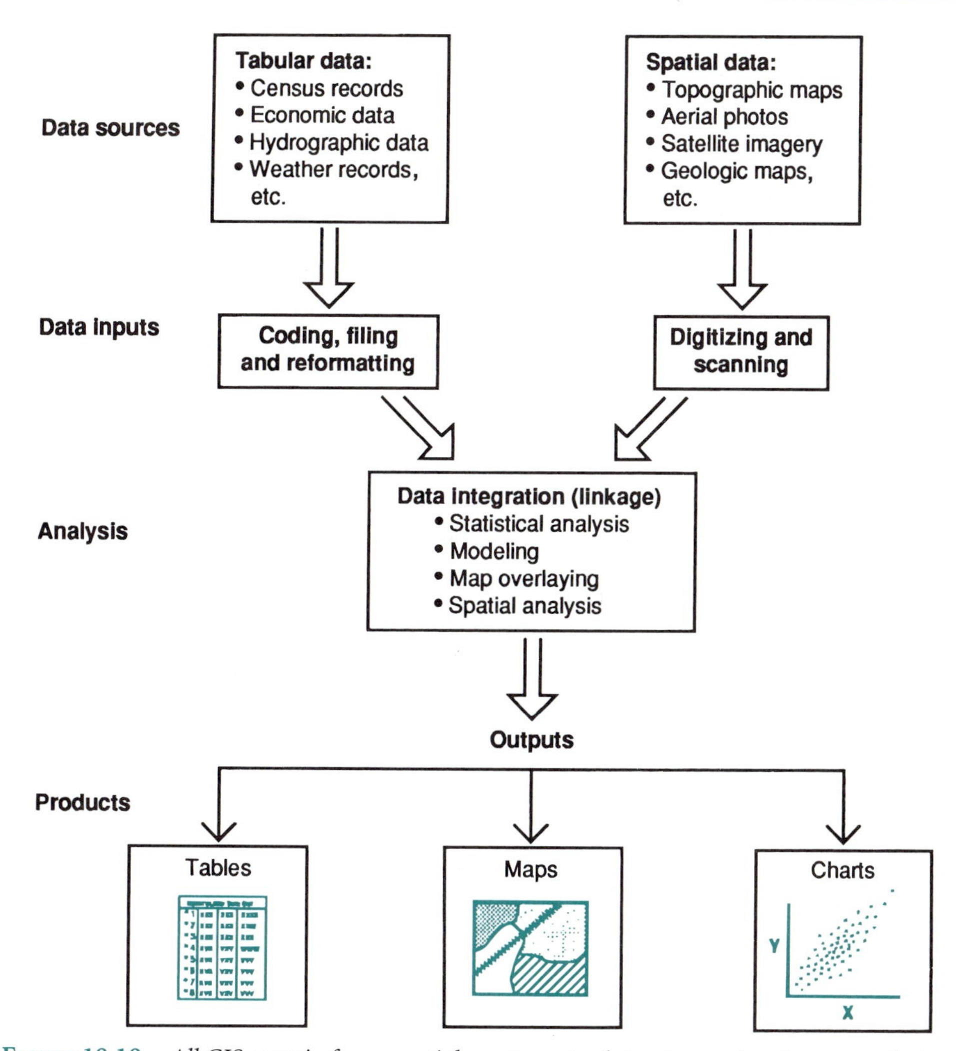

FIGURE 18.18 *All GIS contain four essential components: data sources, data inputs, analysis or data integration, and products such as maps, tables, and charts.*

same locations on Earth. Finding the same location in which to record data is incredibly difficult on the vast surface of the sea or great deserts, for example, and GIS eliminates this uncertainty.

Unlike other computerized information systems, GIS considers both the location as well as certain characteristics or attributes of the environmental variables in the database. For example. Figure 18.17 depicts GIS layers representing three different types of spatial data: point-like features (wells), linear features (transport routes), and areas (soil types) with examples of attributes of these features that might be included in a GIS. Figure 18.18 illustrates the essential components common to all GIS, namely, data input, analysis and manipulation, and information output.

Data Input

Data are entered into a GIS in a digital form. Common methods of data entry involve tracing the outline of features on maps with an *electronic coordinate digitizer* or using *drum* or *laser scanners* that systematically transform maps into digital data. After each variable has been digitally encoded, it is stored as a digital map or overlay as illustrated in Figure 18.17. Since each overlay has the same scale and map projection, it is spatially registered with every other overlay.

Data Manipulation and Analyses

These operations are performed using one or more of the variable maps stored in the database to generate new information, usually in the form of new maps or summary tables. There are two broad types of analytical and manipulative GIS operations: surface and overlay operations. *Surface operations* include a single GIS layer. For example, a map showing prime agriculture land could be derived from the land-use map layer or the area of lakes and the density of stream channels could be derived from a map of surface hydrology. Such types of surface operations are called *reclassification* functions. Most GIS also use *overlay operations* to manipulate two or more layers to create a new map. For example, GIS maps of tropical deforestation can be generated by overlaying a recent forest cover map onto an older map of forest cover for the same area.

Through the use of GIS's manipulative and analytical procedures, various types of models can be developed. *Evaluative models* are used for environmental assessment. For example, a tropical rainforest tract could be evaluated for its potential as an ecosystem preserve or for its current lumber value and related potential for crop, pasture, or wood pulp production. In contrast, *allocative models* are developed to select the optimum location for a specified land use. Such a model would incorporate several environmental, economic, and cultural variables that would be evaluated to select the "best" route for a new highway.

Information Output

Maps, charts, and other graphics generated through GIS analytical operations can be displayed on a computer monitor or printed. Most graphics are initially displayed on a computer monitor which can be edited and formatted into the desired map or graphics. Finally, these maps or other graphic output are printed on computer-driven plotters or printers.

18.7 THE ROLE OF GIS IN ENVIRONMENTAL MANAGEMENT

Because many different types of information can be integrated and mapped using GIS, these systems have become essential tools for environmental management. With these mapping and analytical systems, we are able to develop new information that will help us learn about the interrelationships between human activity and environmental processes. Some of the environmental management applications that currently use GIS are briefly noted as follows:

- Agencies that manage public land resources such as forest, parks, and wildlife refuges are employing GIS. These tools are used to manage and allocate multiple-use or restricted-use resources and then monitor the effects or impacts of these management programs. For example, GIS might be used to select the grazing or timber-cutting alternative that will have the least negative impact on wildlife habitat or recreation opportunities.

- Regional and local planning agencies use GIS to develop rigorous evaluation methods when formulating land-use policies for private land. For instance, what changes might be made in zoning regulations that will better serve current and future land-use needs? Or what is the optimum location for a highway considering both environmental and economic issues? Or where might conservation easements be used most effectively to provide wildlife habitat corridors?

- Land cover and land-use change can be monitored using GIS to determine the impact on biologically sensitive species or to measure attendant changes in erosion or air and water quality. For instance, a baseline land cover map in a GIS could be updated to show changes observed from satellite imagery, while ground surveys could provide GIS inputs regarding species enumeration and air and water quality. Furthermore, just as GIS can monitor and document habitat loss, it can be used to monitor habitat restoration.
- GIS can also be used to create models that depict the spread and influence of point and nonpoint pollution sources. For example, the diffusion of a contaminated groundwater plume spreading from a landfill can be predicted and mapped using GIS, just as atmosphere and surface water pollution plumes can be modeled and their potential impacts estimated with GIS.
- Because GIS can produce and update maps quickly and inexpensively, it is possible to provide current information in relatively large quantities. For example, during the Mississippi River flood in 1993, GIS maps showing the extent of the flood and the land uses affected were available during emergency relief efforts and for insurance adjustment estimates. Individual maps can also be inexpensively prepared to illustrate the role of each parcel of land when proposing easements or zoning changes. Finally, inexpensive GIS maps make it possible to provide a wide range of environmental information in easy-to-use map form. Such maps are useful for environmental education efforts and to promote ecotourism.

18.8 GLOBAL CHANGE, RESEARCH, AND GEOGRAPHIC INFORMATION SYSTEMS

Large computer based information systems, including GIS are needed to study the environmental processes producing global change. Today, global GIS are possible given the advanced technologies of data storage, management and analysis, and the expanding network of remote sensing satellites and ground stations that gather data. To learn more about complex environmental systems, scientists develop computer models that simulate these processes. GIS are designed to manage and analyze large georeferenced data sets, so they will be useful tools for developing and testing models and to assist in managing for global change. The GIS map overlays make it possible to integrate remote sensing data from meteorology or Earth resources satellites with maps of land use, soils, vegetation, and transportation.

GIS and Environmental Change

Research into global change utilizes GIS techniques in several significant ways. Most notable is research into global climatic and hydrologic change, as well as ecosystem analysis and the role of human interaction with environmental systems.

GIS and Climate Change

Atmospheric scientists build complex general circulation models of the atmosphere to describe energy and material flows. Such models are being developed to simulate climate modifications that could occur as atmospheric CO_2 concentrations build. The GIS factors used by atmospheric circulation models include surface reflectance, evapotranspiration, and surface roughness. The proportion of solar radiation reflected by Earth's surface (and conversely, the amount of energy absorbed) is affected by the surface cover. Seasonal and geographic variations in snow and ice cover, natural and planted vegetation, bare surfaces, and built-up areas that affect surface reflections or albedo are monitored with remote sensing satellites and incorporated into GIS. GIS data on global soil moisture, vegetation, land cover, evapotranspiration, and their projected changes can also be used in general circulation models.

GIS and Hydrologic Systems

Any change in regional or global climate would affect the hydrologic systems. Analyses of stream networks, cover, and surface characteristics within watersheds—as well as topographic, geologic, and soil characteristics—are all

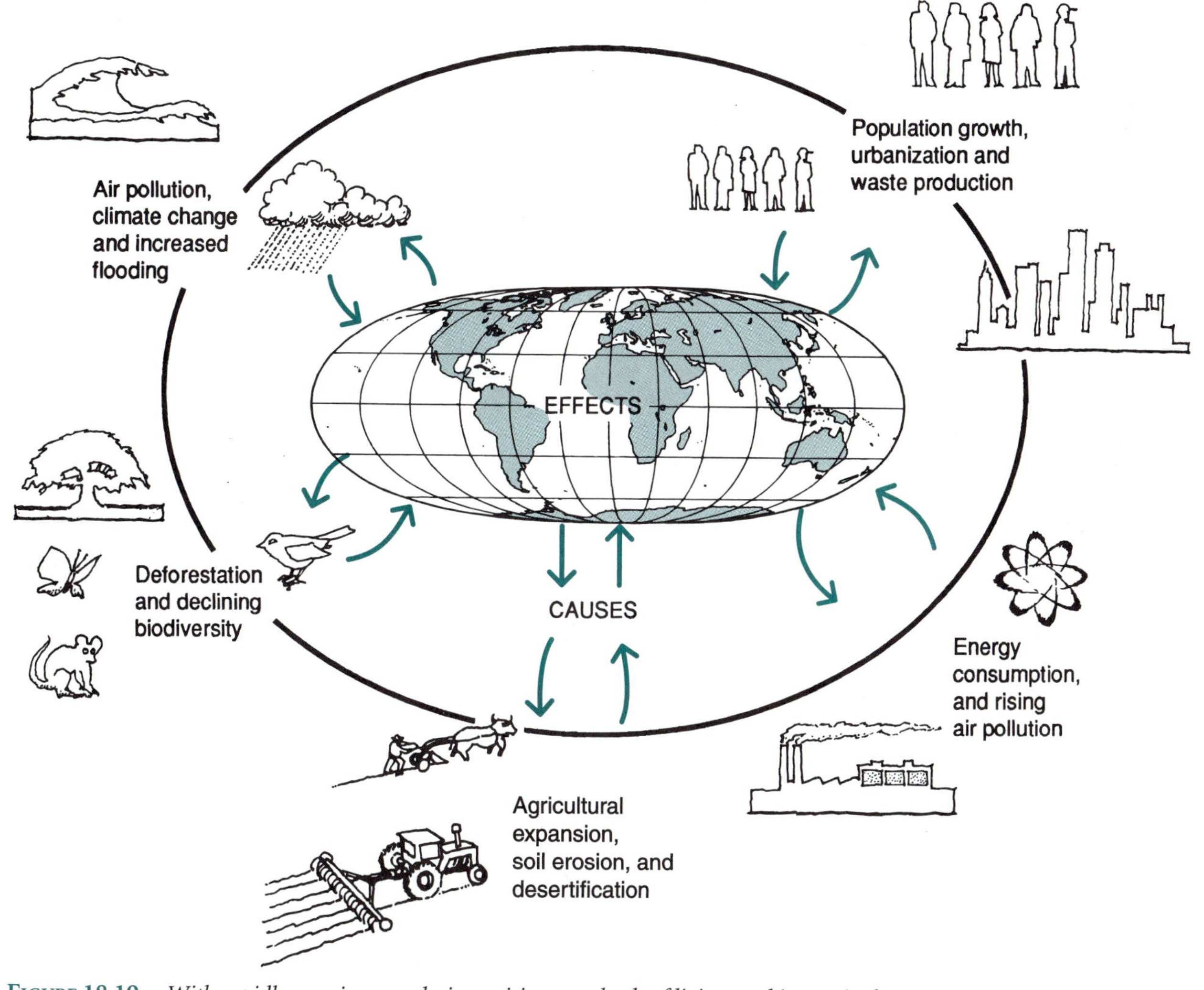

FIGURE 18.19 *With rapidly growing populations, rising standards of living, and increasingly sophisticated global economic and production technologies, human activity is producing global environmental changes and at the same time being affected by these changes. Both the environmental changes and the feedback they produce will increase substantially in the next several decades.*

spatial components associated with hydrologic systems that can be analyzed by GIS.

Ecosystem Change Modeling and GIS

Computer models are being used to simulate ecosystem response to climate and other environmental changes. Because so many variables affect ecosystem development and disruption, it is useful to inventory current environmental conditions and patterns that can be used as a baseline for the modeling process. GIS are used to quantitatively define many basic parameters such as soil, vegetation, and topography. Data from polar-orbiting metsats have been used to generate global vegetation indices with 4 kilometers by 4 kilometers resolution (Figure 18.6). Time series of these indices can be correlated with ecological system models of biomass production.

Human Interactions and GIS

As models are developed to forecast potential changes in climatic, hydrologic, ecologic, and other environmental conditions, GIS techniques will be used to assess their impacts on human settlement and economic activities. As scenarios of environmental change are developed, GIS maps and information will enable societies to evaluate the potential consequences of change.

Several national and international research programs are currently investigating global change. The International Council of Scientific Unions (ICSU), a nongovernmental organization that promotes international scientific research, administers a global change research program, the *International Geosphere-Biosphere Programme (IGBP)*. The IGBP studies how physical, chemical, and biological processes interact to regulate the global environment and monitors detectable changes. IGBP projects are investigating how biological processes regulate atmospheric composition; how vegetation controls energy and water flows; how global changes will affect marine and terrestrial ecosystems; and how processes in the open ocean and coastal zones affect global cycles. Currently, more than 50 countries have national IGBP projects. In the United States the IGBP project is the Global Change Research Program (USGCRP). This interdisciplinary program staffed by scientists from eleven federal agencies and departments has established several high-priority research areas, including climate and hydrologic systems, biogeochemical and ecosystem dynamics, and human interactions.

Complementing the IGBP is the *Human Dimensions of Global Environmental Change Program (HDGECP).* which has been established by the International Social Science Council (ISSC) to focus on environmental changes driven by human activities. Figure 18.19 illustrates how human activities contribute to and are affected by changes in the environment. Social scientists study how human activities influence natural processes and how social and economic behaviors might change in response to environmental changes. Additional research will develop through cooperative efforts between IGBP and HDGECP. For example, models might be developed that take into account the socioeconomic impacts of climate change or how activities such as logging and agriculture may affect greenhouse gas emissions.

18.9 SUMMARY

In this first decade of the new millennium, the global community is developing management objectives and establishing institutions that can monitor, forecast, and begin to change the destructive course of unsustainable economic growth and development that has spread rapidly over the globe during the last century. Such actions have been initiated in response to the growing perception and scientific evidence that the increasing scale and intensity of human activities are altering some basic environmental processes and will likely bring about undesirable regional or even global changes that reduce the planet's carrying capacity for humans and other life forms.

As world population quadrupled during the twentieth century from about 1.5 billion in 1900 to more than 6 billion in 2000, environmental degradation increased dramatically. Overgrazing and intensified crop production in semiarid lands have led to dust bowls and desertification. Pollution from expanding urban and industrial centers has sullied air and water resources as well as the land itself. Acid rain, smog, hazardous waste, and oxygen-depleted waterways were the unexpected, unwanted, but increasingly persistent byproducts of the urbanization and industrialization process. Moreover, many of these problems are not local or even regional in their geographical extent. They are increasingly intercontinental and global.

Beginning about three decades ago, most developed countries began to establish policies to reduce many of these forms of pollution. At about the same time, international accords and treaties were being negotiated among western European countries to protect common air and water resources. An international agreement was reached in which each coastal nation was responsible for managing fish stocks within 200 miles of its coast.

During this same period, evidence was mounting that the protective shield of stratospheric ozone was being depleted by chlorofluorocarbons (CFCs) and that increasing greenhouse gas emissions were likely to produce regional or global climate changes. As a result, binding international agreements have been signed by most nations that take steps to reduce ozone depletion and greenhouse gas emissions. Growing concern over the loss of wild plant and animal species and agricultural crop varieties has resulted in an international accord to preserve biodiversity that has been signed by most nations.

The most comprehensive attempt to develop strategies for economic development that do not deplete or damage the renewable resource base has come from the UN Conference on Environment and Development (UNCED). Agenda 21, the massive action plan promoting sustainable development practices unveiled at the Earth Summit in 1992, took more than two years to develop and involved input from representatives of most nations, as well as from numerous nongovernmental organizations. This nonbinding work plan details the priorities and goals regarding many development and environmental issues, as well as the related financial, legal, and institutional frameworks necessary for implementing the plan.

It is still too early to know if Agenda 21 will indeed provide the guidelines for worldwide sustainable economic development and environmental protection into the next century. However, certain activities that have taken place and international accords that have been reached since the Earth Summit indicate that governments, the international business community, and NGOs are beginning to work within the framework or guidelines of Agenda 21. To date several significant mileposts had been reached. For example, the Commission on Sustainable Development is beginning to review the specific sustainable development plans submitted by national governments and international organizations to initiate numerous Agenda 21 programs. The Framework Convention on Climate Change has reached an important agreement, the Kyoto Protocol, which will stabilize or reduce global greenhouse gas emissions over the next decade if ratified by the nations of the world.

International meetings sponsored by NGOs and UNEP have begun to develop strategies for implementing the Convention on Biodiversity, an international treaty unveiled at the Earth Summit. Furthermore, the United States, which had opposed part of the treaty in 1992, has now signed the convention. In the years since UNCED, there have been hundreds of international government meetings and many more NGO gatherings to address UNCED issues. Such activities may indicate that worldwide efforts continue toward developing and implementing sustainable environmental management strategies.

Finally, a wide range of international scientific programs are underway to study and monitor environmental processes and earth resources using satellites and geographic information systems (GIS). The vast quantities of data generated by the satellites' remote sensing systems are being made available to the worldwide scientific community via GIS and the Internet. For example, the United States and several other nations have committed to a series of up to 50 scientific investigations of the environment using remote sensing satellites over the next two decades. The program is under the aegis of the UN Committee on Earth Observation Satellites (CEOS). To date several state-of-the-art satellites have been launched to begin these scientific investigations.

In conclusion, important initiatives are being taken by the world community to promote environmentally sustainable economic practices. Fortunately, both the developed and developing countries now recognize the need for profound change. It is also fortunate that the world community recognizes the need to monitor the global environment, to produce timely information on land use, pollution, and other problems, and to share this information freely with all nations and concerned organizations. The global community has finally come to grips with the fact that human survival and the condition of the environment are inextricably linked and that sustainable land use, if it is possible, will require comprehensive, international programs rooted in a healthy global environment. However, many fundamental societal, political, and economic changes will have to be made in the developed and developing countries alike. Such changes will be difficult and very expensive, but they must succeed if the Earth is to remain a hospitable environment for life as we know it.

18.10 KEY TERMS AND CONCEPTS

- Current global environmental trends
  - global climate warming
  - tropical deforestation
  - biodiversity decline
  - land degradation
  - overexploitation of marine fisheries
  - urban air pollution
  - additional environmental threats
- Progress toward sustainable development
- International institutions to manage the environment
  - United Nations Conference on Environment and Development (UNCED)

- The Earth Summit
- United Nations Environment Programme (UNEP)
- Framework Convention on Climate Change (FCCC)
- Convention on Biological Diversity
- Agenda 21

Progress since UNCED
- Kyoto Protocol
- emissions trading
- Clean Development Mechanism (CDM)
- Commission on Sustainable Development (CSD)
- Earth Summit +5

Nongovernmental organizations (NGOs)

Monitoring environmental change with remote sensing
- remote sensing systems
- electromagnetic energy
- aerial photography
- digital cameras
- line scanners
- thermal infrared scanners
- side-looking radar
- digital data

Aircraft remote sensing
- National Aerial Photography Program

Satellite remote sensing
- Meteorology satellites (metsats)
  - National Oceanic and Atmospheric Administration (NOAA)
  - Geostationary Operational Environmental Satellites (GOES)
  - Nimbus
  - Upper Atmosphere Research Satellites (UARS)

Earth resources satellites
- Landsat
- SPOT
- IKONOS
- Terra

Radar satellites
- Seasat
- Shuttle Radar Topographic Mission (SRTM)
- Radarsat

Mission to Planet Earth

Earth Science Enterprise

Geographic Information Systems (GIS)
- data input
- data manipulation and analyses
- information output

Global change research and GIS
- GIS and climate change
- GIS and hydrologic systems
- ecosystem modeling and GIS
- human interactions and GIS
- organizations studying global environmental change

18.11 QUESTIONS FOR REVIEW

1. Unsustainable resource use has degraded or depleted basic resources such as soil, forests, water, and the atmosphere. List several of the most critical environmental situations that we face.
2. What are some of the environmental threats that have emerged during the last few years?
3. Describe some of the institutions and organizations that have been established to resolve environmental problems and manage sustainable economic development.
4. The United Nations conference on Environment and Development (UNCED) was held in 1992. In the years since UNCED how have its environmental initiatives fared? Are the treaties being honored? What, if any, parts of the monumental Agenda 21 are being carried out? Are any changes being made to implement resource conservation or sustainable development programs?
5. List three types of remote sensing systems that produce images of the Earth's surface. Which system is most widely used? List several advantages and disadvantages of each system.
6. What type of remote sensing systems is used on most Earth resources satellites such as Landsat, SPOT, and the meteorology satellites? Why are most of the data collected in digital format? How have these data been used in environmental research and management applications?
7. What are Geographic Information Systems (GIS)? Why are they becoming useful tools for environmental management?
8. Although research programs to study global change are just being established, both satellite remote sensing and geographic information systems will play important roles. Discuss the need for these tools in order to study climate and ecosystems changes.

REFERENCES FOR FURTHER READING

CHAPTER 1: REFERENCES FOR FURTHER READING

Byers, Bruce. 1991. *Ecoregions, State Sovereignty and Conflict. Bulletin of Peace Proposals.* London: Sage Publications.

Devall, Bill, and Sessions, George. 1985. *Deep Ecology.* Salt Lake City: Peregrine Smith Books.

Diamond, Jared. 1997. *Guns, Germs, and Steel: The Fates of Societies.* New York: St. Martins Press.

Erhlich, P. R., and Erhlich, A. H. 1991. *Healing the Planet.* Reading, Mass.: Addison Wesley.

Homer-Dixon, T. F., J. Boutewell, and G. Rathjens. 1993. Environmental Scarcity and Violent Conflict. *Scientific American.*

Kaplan, R. D. 1996. *The Ends of the Earth.* New York: Vintage.

Kaplan, R. D. 1994. The Coming Anarchy. *The Atlantic Monthly* (February).

Miller, Morris. 1991. *Debt and the Environment: Converging Crisis.* New York: United Nations Publications.

Schneiber, Allan, and K. A. Gould. 1994. *Environment and Society: The Enduring Conflict.* New York: St. Martins Press.

Silver, C. S., and R. S. Defries. 1990. *One Earth One Future.* Washington, D.C.: National Academy of Sciences Press.

CHAPTER 2: REFERENCES FOR FURTHER READING

Brunn, S. D., and J. F. Williams. 1993. *Cities of the World.* New York: Harper Collins.

Buck, S. J. 1991. *Understanding Environmental Administration and Law.* Washington, D.C.: Island Press.

Goudie, Andrew. 1994. *The Human Impact on the Natural Environment.* Cambridge, Mass.: MIT Press.

Jellicoe, Geoffrey, and Susan Jellicoe. 1995. *The Landscape of Man.* New York: Thames and Hudson.

Merritts, D., A. DeWet, and K. Menking. 1998. *Environmental Geology.* New York: W.H. Freeman.

Press, Frank, and Raymond Seiver. 1994. *Understanding Earth.* New York: W. H. Freeman.

Summerfeld, M. A. 1991. *Global Geomorphology.* Essex, England: Longman.

CHAPTER 3: REFERENCES FOR FURTHER READING

Brown, L. R., et al. 1991. *Saving the Planet: How to Shape an Environmentally Sustainable Global Economy.* New York: W. W. Norton.

de Steiguer, J. E. 1997. *The Age of Environmentalism.* New York: WCB/McGraw-Hill.

Hardin, Garrett. 1968. The Tragedy of the Commons. *Science* 162.

McHarg, Ian. 1995. *Design with Nature.* New York: Wiley.

Meadows, D. H., et al. 1972. *The Limits of Growth: A Report for the Club of Rome's Project on the Predicament of Mankind.* New York: Universe Books.

Naess, Arne. 1989. *Ecology, Community and Lifestyle.* Cambridge: Cambridge University Press.

Nash, R. F. 1989. *The Rights of Nature: A History of Environmental Ethics.* Madison: The University of Wisconsin Press.

Rosenberg, A. A., et al. 1993. Achieving Sustainable Use of Renewable Resources. *Science* 262.

Schmidhieny, S. 1992. *Changing Course: A Global Business Perspective on Development and the Environment.* Cambridge, Mass.: MIT Press.

World Resources Institute. 1992. Dimensions on Sustainable Development. In *World Resources 1992–1993.* New York: Oxford Press.

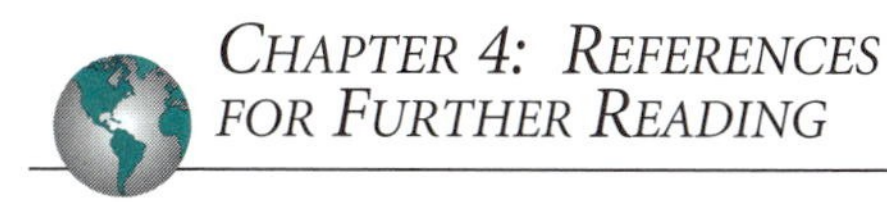

CHAPTER 4: REFERENCES FOR FURTHER READING

Deshmukh, Ian. 1986. *Ecology and Tropical Biology.* Palo Alto, Calif.: Blackwell Scientific.

Drew, D. 1983. *Man-Environmental Processes.* London: Allen and Unwin.

Glautz, M. H., ed. 1977. *Desertification: Environmental Degradation in and Around Arid Lands.* Boulder, Colo.: Westview.

Goudie, Andrew. 1994. *The Human Impact on the Natural Environment.* Cambridge, Mass.: MIT Press.

Grainger, A. 1983. "Improving the Monitoring of Deforestation in the Humid Tropics." In Sutton, S. L., et al. (eds). *Tropical Rainforest Ecology and Management.* Oxford: Blackwell Scientific.

Nitecki, M. H., ed. 1984. *Extinctions.* Chicago: University of Chicago Press.

Oldfield, M. L. 1984. *The Value of Conserving Genetic Resources.* Washington, D.C.: U.S. Department of Interior.

Roberts, N. 1989. *The Holocene: An Environmental History.* Oxford: Basil Blackwell.

Smith, B. D. 1995. *The Emergence of Agriculture.* New York: Scientific American Library.

Smith, N. J. H. 1982. *Rainforest Corridors: The Transamazon Colonization Scheme.* Berkeley and Los Angeles: University of California Press.

Stanley, S. M. 1998. *Children of the Ice Age.* New York: W.H. Freeman.

Tucker, C. J. 1991. Expansion and Contraction of the Sahara Desert from 1980 to 1990. *Science* 252.

Turner, B. L., and Butzer, K. W. 1992. The Columbian Encounter and Land-Use Change. *Environment* 34:8.

Webb, W. P. 1986. *The Great Frontier.* Lincoln: University of Nebraska Press.

Webb, W. P. 1981. *The Great Plains.* Lincoln: University of Nebraska Press.

Williams, P. J. 1979. *Pipelines and Permafrost.* New York: Longmans.

Wolman, M. G., and Fournier, F. G. A., eds. 1987. *Land Transformation in Agriculture.* SCOPE: New York: Wiley.

CHAPTER 5: REFERENCES FOR FURTHER READING

Berkner, L. V., and L. C. Marshall. 1965. On the Origin and Rise of Oxygen Concentrations in the Earth's Atmosphere. *Journal of Atmospheric Science* 22.

Bolin, B., and R. B. Cook, eds. 1983. *The Major Biogeochemical Cycles and Their Interactions.* Chichester, England: John Wiley.

Castro, Peter, and M. E. Huber, 1997. *Marine Biology.* Dubuque, Iowa: Wm. C. Brown.

Cook, R. B. 1984. "Man and the Biogeochemical Cycles: Interacting with the Elements." *Environment* 26:7.

Dixon, R. K., et al 1994. Carbon Pools and Flux of Forest Ecosystems. *Science* 263.

Ehrlich, P. R. 1986. *The Machinery of Nature.* New York: Simon and Schuster.

Gould, S. J. 1994. *The Evolution of Life on Earth,* 271.

Horgan, J. 1991. In the Beginning. *Scientific American,* 264.

Lovelock, James. 1988. *The Ages of Gaia.* New York: W. W. Norton.

Marsh, W. M. 1987. *Earthscape: A Physical Geography.* New York: John Wiley.

Odum, E. P. 1971. *Fundamentals of Ecology.* 3d ed. Philadelphia: W. B. Saunders.

Schneider, S. H. 1990. Debating Gaia. *Environment* 32:4.

Whittaker, R. K., and G. E. Likens. 1973. Carbon in the Biota. In *Carbon and the Biosphere* (G. Woodwell and E. Pecans, eds.) Wash., D.C.: U.S. Atomic Energy Commission.

CHAPTER 6: REFERENCES FOR FURTHER READING

Angel, M. V. 1993. Biodiversity of the Pelagic Ocean. *Conservation Biology* 7:4.

Brown, J. H., and A. C. Gibson. 1983. *Biogeography.* St. Louis: C. Y. Mosby.

Brown, J. H., and B. A. Maurer. 1989. Macroecology: The Division of Food and Space among Species of the Continents. *Science* 243.

Cairns-Smith, A. G. 1985. The First Organisms. *Scientific American* 252:6.

Deshmukh, Ian. 1986. *Ecology and Tropical Biology.* Palo Alto, Calif.: Blackwell Scientific Publications.

Drake, J. A. 1989. *Biological Invasion: A Global Perspective.* Chichester: Wiley.

Ehrlich, P. R., and A. H. Ehrlich. 1991. *Healing the Planet.* Reading, Mass.: Addison-Wesley.

Hutchinson, G. E. 1978. *An Introduction to Population Ecology.* New Haven, Conn.: Yale University Press.

Jarvis, P. H. 1979. The Ecology of Plant and Animal Introductions. *Progress in Physical Geography* 3.

McDonnell, M. J., and S. T. A. Pickett. 1993. *Humans as Components in Ecosystems.* New York: Springer-Verlag.

Odum, E. P. 1971. *Fundamentals of Ecology.* 3d ed. Philadelphia: W.B. Saunders Co.

Polunin, Nicholas. 1967. *Introduction to Plant Geography.* London: Longmans.

Primack, R. B. 1993. *Essentials of Conservation Biology.* Sutherland, Mass.: Sinauer Associates.

Sumich, J. L. 1980. *Biology of Marine Life.* 2d ed. Dubuque, Iowa: Wm. C. Brown.

Vitousek, P., et al. 1986. Human Appropriations of the Products of Photosynthesis. *Bioscience* 36.

Whittaker, R. H. 1975. *Communities and Ecosystems.* 2d ed. New York: Macmillan.

CHAPTER 7: REFERENCES FOR FURTHER READING

Ashford, Lori S., and J. A. Noble. 1996. Population Policy: Consensus and Challenges. *Consequences* 2:2.

Bongaart, John. 1994. Population Policy Options in the Developing World. *Science* 263.

Chen, L. C., W. M. Fitzgerald, and L. Bates. 1995. "Women, Politics, and Global Management," *Environment* 37:1.

Durning, Alan. 1991. Ending Poverty. *State of the World 1990.* Washington, D.C.: Worldwatch Institute.

Ehrlich, P. R., and A. H. Ehrlich. 1990. *The Population Explosion.* New York: Doubleday.

Jacobson, Jodi. 1988. *Environmental Refugees: A Yardstick of Habitability.* Washington, D.C.: Worldwatch Institute.

Kaplan, Robert, D. 1994. The Coming Anarchy. *The Atlantic Monthly* (February).

Kaplan, Robert D. 1996. *The Ends of the Earth.* New York: Random House.

Keyfitz, Nathan. 1990. The Growing Human Population. *Managing Planet Earth: Readings from Scientific American.* New York: W.H. Freeman & Co.

Mitchell, Jennifer D. 1998. Before the Next Doubling. *World Watch* (January/February).

Nelson, Toni. 1996. Russia's Population Sink. *World Watch* (January/February).

Newman, J. L., and G. E. Matzke. 1984. *Population: Patterns, Dynamics, and Prospects.* Englewood Cliffs, N.J.: Prentice-Hall.

Peterson, Peter G. 1999. Gray Dawn: The Global Aging Crisis. *Foreign Affairs* (January/February).

Population Reference Bureau. *World Population Data Sheet.* Washington, D.C.: Population Reference Bureau Annual Report.

Ridker, Ronald, G. 1992. Population Issues. *Resources.* Resources for the Future (Winter).

Robinson, Warren C. 1998. Global Population Trends. *Resources.* Resources for the Future (Spring).

Sen, Gita. 1995. The World Programme of Action: A New Paradigm for Population Policy. *Environment* 37:1.

Simon, Julian L. 1990. *Population Matters: People, Resources, Environment and Immigration.* New Brunswick, N.J.: Transaction Publishers.

World Bank annual. *World Development Report.* New York: Oxford University Press.

World Resources Institute, 1998. Population and Human Well Being. *World Resources 1998–1999.* New York: Oxford University Press.

CHAPTER 8: REFERENCES FOR FURTHER READING

Brown, L. R. 1987. Sustaining World Agriculture. *State of the World 1987.* New York: W.W. Norton & Co.

Brown, Lester. 1999. Feeding Nine Billion, *State of the World, 1999. The Worldwatch Institute,* Ch. 7, pp. 115–132.

Crosson, Pierre. 1997. Will Erosion Threaten Agricultural Productivity? *Environment* 39:8 (October): 4–9; 29–31.

Crosson, Pierre. 1992. Sustainable Agriculture. *Resources* 106 (Winter): 14–17.

Crosson, Pierre., and N. J. Rosenburg. 1990. Strategies for Agriculture. *Managing Planet Earth: Reading from Scientific American.* New York: W. H. Freeman and Co.

Ervin, David E., C. F. Runge, E. A. Graffy, W. E. Anthony, S. S. Batie, P. Faeth, T. Penny, and T. Warmon, 1998, Agriculture and the Environment: A New Strategic Vision, *Environment* 40:6 (July/August): 4–15; 35–40.

Faeth, Paul. 1994. Building the Case for Sustainable Agriculture. *Environment,* 36:1 (January–February): 16–20, 34–39.

Furuseth, Owen J. 1997. Restructuring of Hog Farming in North Carolina: Explosion and Implosion, *Professional Geographer* 49:4, pp. 391–403.

Gasser, C. S., and R. T. Fraley. 1992. Transgenic Crops. *Scientific American* (June): 62–69.

Grigg, D. B. 1974. *The Agricultural Systems of the World: An Environmental Approach.* London: Cambridge University Press.

Kates, Robert W. 1996. Ending Hunger: Current Status and Future Prospects *Consequences* 2 (November 2): 3–11.

Kromm, David J. 1992. Low Water in the American High Plains, *The World and I* (February): 312–319.

Lappe, F. M., and J. Collins. 1986. *World Hunger: Twelve Myths.* New York: Grove Press.

Mann, Charles. 1997. Reseeding the Green Revolution. *Science* (August 22): 1033–1043.

Meadows, D. H. 1986. *Systems Analysis and the World Food System: Natural Resources and People.* Boulder, Colo.: Westview Press.

Pimental, David. 1988. *Industrialized Agriculture and Natural Resources.* The Cassandra Conference: Resources and the Human Predicament. College Station: Texas A & M University Press.

Postel, Sandra. 1989. *Water for Agriculture: Facing the Limits.* Worldwatch Paper 93. Washington, D.C.: Worldwatch Institute.

Reichelderfer, Katherine. 1991. The Expanding Role of Environmental Interests in Agricultural Policy. *Resources* 102 (Winter): 4–7.

Sanchez, P. A., and J. R. Benites. 1987. Low Input Cropping for Acid Soils of the Humid Tropics. *Science* 238 (December 11): 1521–27.

Tarrant, John. 1990. World Food Prospects for the 1990s. *J. of Geography* 89:6 (November/December): 234–38.

Weintraub, Pamela. 1992. The Coming of the High-Tech Harvest. *Audubon* (July/August): 93–100.

Wolf, E. C. 1986. *Beyond the Green Revolution: New Approaches for Third World Agriculture.* Worldwatch Paper 73. Washington, D.C.: Worldwatch Institute.

World Resources Institute, 1998, Feeding the World, *World Resources 1998–99,* Oxford University Press, pp. 152–160.

CHAPTER 9: REFERENCES FOR FURTHER READING

Annual Energy Review. Washington, D.C.: Energy Information Administration, U.S. Department of Energy.

Barcott, Bruce. 1999. Beyond the Valley of the Dammed. *Outside.* (February).

Campbell, Colin, and Jean Laherrere. 1998. The End of Cheap Oil. *Scientific American.* (March).

Chiles, James. 2000. A Second Wind. *Smithsonian.* (March)

Dostrovsky, Israel. 1991. Chemical Fuels from the Sun. *Scientific American.* (December).

Dunn, Seth. 1999. King Coal's Weakening Grip on Power. *World Watch.* (September/October).

Energy for Plant Earth: Readings from Scientific American Magazine. 1991. New York: W.H. Freeman and Co.

Flavin, Christopher. 1999. Bull Market in Wind Energy. *World Watch.* (March/April).

Fowler, John M. 1984. *Energy and the Environment* (2d ed.). New York: McGraw-Hill.

Hoffert, Martin, and Seth Potter. 1997. Beam It Down. *Technology Review* (October).

Horgan, John. 1991. The Muddled Cleanup in the Persian Gulf. *Scientific American* (October).

International Energy Annual. Washington, D.C.: Energy Information Administration, U.S. Department of Energy.

Quinn, Randy. 1997. Sunlight Brightens our Energy Future. *The World and I* (March).

Tester, Jefferson (ed.). 1991. *Energy and the Environment in the 21st Century.* Cambridge, Mass.: MIT Press.

World Resources Institute. 1994. Energy. *World Resources 1994–95.* New York: Oxford University Press.

Yergin, Daniel. 1990. *The Prize: The Epic Quest for Oil, Money and Power.* New York: Simon and Schuster.

CHAPTER 10: REFERENCES FOR FURTHER READING

Budyko, M. I. 1982. *The Earth's Climates: Past and Future.* Orlando, Fla.: Academic Press.

Charlson, R. J., et al. 1992. Climate Forcing by Anthropogenic Aerosols. *Science* 255.

Hobbs, P. V., and M. P. McCormick, eds. 1988. *Aerosols and Climate.* Hampton, Va.: Deepak.

Kauppi, Pekka E., et al. 1992. Biomass and Carbon Budget of European Forests, 1971 to 1990. *Science* 256.

Kellog, W. W., and M. Mead. 1977. *The Atmosphere: Endangered and Endangering.* Washington, D.C.: Department of Health, Education and Welfare.

National Academy of Sciences. 1975. *Understanding Climatic Change,* Washington, D.C.: U.S. Government Printing Office.

Neiburger, Morris, et al. 1982. *Understanding Our Atmospheric Environment.* 2d ed. San Francisco: Freeman.

Oke, T. R. 1978. *Boundary Layer Climates.* London: Methuen. New York: Halsted.

Penner, J. E., et al. 1992. Effects of Aerosol from Biomass Burning on the Global Radiation Budget. *Science* 256.

Perry, A. H., and J. M. Walker. 1977. *The Ocean–Atmosphere System.* New York: Longman.

Quay, P. O., et al. 1992. Oceanic Uptake of Fossil Fuel CO_2: Carbon-13 Evidence. *Science* 256.

Roberts, Leslie. 1988. Is There Life after Climatic Change? *Science* 242.

CHAPTER 11: REFERENCES FOR FURTHER READING

Elsom, D. M. 1992. *Atmospheric Pollution: A Global Problem.* 2d ed. Cambridge, Mass.: Blackwell Scientific.

Garreau, Joel. 1991. *Edge City: Life on the New Frontier.* New York: Doubleday.

Gleason, J. F., et al. 1993. Record Low Global Ozone. *Science* 260.

Krupnick, A. J., and P. R. Portney. 1991. Controlling Urban Air Pollution: A Benefit–Cost Assessment. *Science* 252.

Likens, G. E., et al. 1979. Acid Rain. *Scientific American* 241:4.

Lin, G. Y., and W. R. Bland. 1980. Spatio-Temporal Variations in Photochemical Smog Concentrations in Los Angeles County. *California Geographer.* v. 20.

Mnatsakanian, R. A. 1992. *Environmental Legacy of the Former Soviet Republics.* Edinburgh: University of Edinburgh.

Nadis, Steve, and J. J. Mackenzie. 1993. *Car Trouble.* Boston: Beacon Press.

National Wildlife Federation. 1995. The 27th Environmental Quality Review. *National Wildlife.* v. 33:2.

Stern, A. C., et al. 1984. *Fundamentals of Air Pollution.* 2d ed. Orlando, Fla.: Academic Press.

U.S. Environmental Protection Agency. 1994. *Deposition of Air Pollutants to the Great Waters: First Report to Congress.* Washington, D.C.: U.S. EPA Office of Air Quality Planning and Standards.

U.S. Environmental Protection Agency. 1993. *National Air Pollutant Emission Trends, 1900–1992.* Washington, D.C.: U.S. EPA Office of Air Quality Planning and Standards.

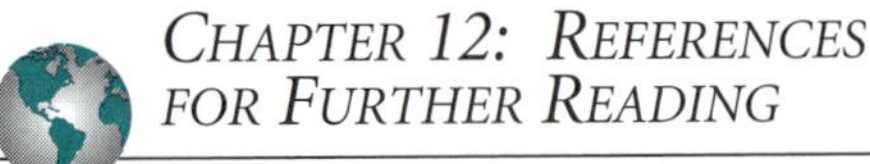

CHAPTER 12: REFERENCES FOR FURTHER READING

Ambroggi, R. P. 1966. Water Under the Sahara. *Scientific American* 214:5.

Black, P. E. 1991. *Watershed Hydrology.* Englewood Cliffs, N.J.: Prentice-Hall.

Boroden, Charles. 1977. *Killing the Hidden Waters.* Austin: University of Texas Press.

Dunne, T., and L. B. Leopold. 1978. *Water in Environmental Planning.* San Francisco: W. H. Freeman.

Ebert, C. H. V. 1988. *Disasters: Violence of Nature and Threats by Man.* Dubuque, Iowa: Kendall/Hunt.

Gilbert, G. K. 1917. *Hydraulic-Mining Debris in the Sierra Nevada.* U.S. Geological Survey Professional Paper 105. Washington, D.C.: U.S. Government Printing Office.

Gleick, P. H. 1993. *Water in Crisis.* New York: Oxford University Press.

Glynn, P. B. 1991. Western Ground Water Supplies Causing Concern. *Farmline.* 10.

Heath, Ralph C. 1991. *Basic Groundwater Hydrology.* U.S. Geological Survey Water-Supply Paper 2220. Washington D.C.: U.S. Government Printing Office.

Marsh, W. M. 1998. *Landscape Planning: Environmental Applications.* 3d ed. New York: John Wiley.

Postel, Sandra. 1999. "When the World's Wells Run Dry." *World Watch* 12:5, 30–38.

Pye, V. I., et al. 1983. *Groundwater Contamination in the United States.* Philadelphia: University of Pennsylvania Press.

Speidel, D. H., et al. eds. 1988. *Perspectives on Water Uses and Abuses.* New York: Oxford University Press.

Wipple, W., et al. 1983. *Stormwater Management in Urbanizing Areas.* Englewood Cliffs, N.J.: Prentice-Hall.

CHAPTER 13: REFERENCES FOR FURTHER READING

Chesters, G., and L. J. Schierow. 1985. A Primer on Nonpoint Pollution. *Journal of Soil and Water Conservation.*

Eisenreich, S. J. 1987. Toxic Fallout in the Great Lakes, *Issues in Science and Technology.*

Faulke, Judith. 1989. Low-Input Farming Faces Profitability Issue. *Farmline* 10.

Fetter, C. W. 1993. *Contaminant Hydrogeology.* New York: MacMillan.

Lape, J. L., and T. J. Dwyer, 1994. A New Policy on CSOs. *Water Environment and Technology* (June).

Marsh, William M., and R. Hill-Rowley. 1989. Water Quality, Stormwater Management and Development Planning on the Urban Fringe. *Journal of Urban and Contemporary Law* 35.

Moody, David W. 1990. Groundwater Contamination in the United States. *Journal of Soil and Water Conservation.*

Mueller, D. K., et al. 1996. Nutrients in the Nation's Waters—Too Much of a Good Thing? *U.S. Geological Survey Circular* 1136.

Nadis, Steve, and J. J. MacKenzie. 1993. *Car Trouble.* Boston: Beacon Press.

Rapaport, et al. 1985. New DDT Inputs to North America: Atmospheric Deposition. *Chemisphere* 14.

Smith, R. H., R. B. Alexander, and M. G. Wolman. 1987. Water-Quality Trends in Natoin's Rivers. *Science* 235 (March).

Sun, Marjorie. 1989. Mud-Slinging Over Sewage Technology. *Science* 246 (October).

Wolman, M. G., and C. E. Chamberlain. 1982. Nonpoint Sources. *Proceedings of the National Water Conference.* Philadelphia: Philadelphia Academy of Sciences.

CHAPTER 14: REFERENCES FOR FURTHER READING

Durning, A. T. 1992. *How Much is Enough?* New York: Norton.

Elliot, J. 1980. Lessons from Love Canal. *Journal of the American Medical Association* 240.

Gots, R. E. 1993. Toxic Risks: *Science, Regulation and Perception.* Boca Raton, FL: Lewis Publishers.

Hunt, C. B. 1983. How Safe are Nuclear Waste Sites? *Geotimes* 28:7.

Keebler, Jack. 1990. Popular, Durable Plastic Faces Dilemma on Recycling. *Automotive News.*

Keller, E. A. 1992. *Environmental Geology.* 6th ed. New York: MacMillan.

Marczewski, H. E., and Michael Kamrin. 1987. *Toxicology for the Citizen.* Michigan State University: Center for Environmental Toxicology.

Office of Technology Assessment. 1992. *Green Products by Design: Choices for a Cleaner Environment.* Washington, D.C.: U.S. Government Printing Office.

Rathje, W. L. 1989. Rubbish! *The Atlantic Monthly* (December).

Schmidt, Karen. 1994. Can Superfund Get on Track? *National Wildlife* 32:5.

Schnaiberg, A., and K. A. Gould. 1994. *Environment and Society: The Enduring Conflict.* New York: St. Martin's.

United Nations. 1990. *Global Outlook 2000.* New York: United Nations Publications.

CHAPTER 15: REFERENCES FOR FURTHER READING

Barry, Wendell. 1981. *The Gift of Good Land.* New York: North Point Press.

Brown, L. R., and E. C. Wolf. 1984. Soil Erosion: Quiet Crisis in the World Economy. *Paper No. 60.* Washington, D.C.: WorldWatch Institute.

Clark, E. H. 1985. The Off-Site Costs of Soil Erosion. *Journal of Soil and Water Conservation* 40.

Judson, S. 1981. What's Happening to Our Continents. In B. J. Skinner, ed. *Use and Misuse of Earth's Surface.* Los Altos, Calif.: William Kaufman.

Lal, R., ed. 1988. *Soil Erosion Research Methods.* Ankeny, Iowa: Soil and Water Conservation Society.

Larson, W. E., et al. 1983. The Threat of Soil Erosion to Long-Term Crop Production. *Science* 219.

Marsh, W. M. 1998. *Landscape Planning: Environmental Applications.* 3d ed. New York: John Wiley.

Morgan, R. P. C., ed. 1981. *Soil Conservation Problems and Prospects.* Chichester, England: John Wiley.

National Academy of Science. 1974. *Productive Agriculture and a Quality Environment,* Washington, D.C.

National Research Council. 1989. *Alternative Agriculture.* Washington, D.C.: National Academy Press.

Pimentel, David, et al. 1995. Environmental and Economic Costs of Soil Erosion and Conservation Benefits. *Science* 267.

Schmidt-Vogt, Dietrich. 1990. *High Altitude Forests in the Jugal Himal (Eastern Central Nepal): Forest Types and Human Impact.* Vol. 6. Franz Steiner Verlag: Stuttgart Geological Research.

United Nations. 1975. *Gross Sediment Transport to the Oceans.* U.N. Educational, Scientific, and Cultural Organization and

International Association of Hydrological Sciences. Paris, France.

U.S. Department of Agriculture. 1988. 1988 Agricultural Chartbook. *Agricultural Handbook No. 673.* Washington, D.C.: U.S. Government Printing Office.

Chapter 16: References for Further Reading

Andrewartha, H. G., and L. C. Birch. 1954. *The Distribution and Abundance of Animals.* Chicago: University of Chicago Press.

Deshmukh, Ian. 1986. *Ecology and Tropical Biology.* Palo Alto, Calif.: Blackwell Scientific.

Ehrlich, P. R. 1986. *The Machinery of Nature.* New York: Simon and Schuster.

Eisner, Thomas. 1990. Prospecting for Nature's Chemical Riches. *Issues in Science and Technology.* 6:2.

Kaufman, L., and K. Mallory. 1986. *The Last Extinction.* Cambridge, Mass.: MIT Press.

MacArthur, R. H. 1972. *Geographical Ecology.* New York: Harper and Row.

MacArthur, R. H., and E. O. Wilson. 1967. *The Theory of Island Biogeography.* Princeton, N.J.: Princeton University Press.

McNeely, J. A., et al. 1990. *Conserving the World's Biological Diversity.* Washington, D.C.: World Resources Institute and others.

Nisbet, E. G. 1991. *Leaving Eden to Protect and Manage the Earth.* Cambridge, Mass.: Cambridge University Press.

Oldfield, M. L. 1989. *The Value of Conserving Genetic Resources.* Sutherland, Mass.: Sinauer.

Polunin, Nicholas. 1967. *Introduction to Plant Geography.* London: Longmans.

Robinson, S. K., et al. 1995. Regional Forest Fragmentation and the Nesting Success of Migratory Birds. *Science* 267.

Turner II, B. L., and C. W. Butzer. 1992. The Columbia Encounter and Land Use Change. *Environment* 34:8.

Welhan, T. 1988. Central American Environmentalists Call for Action on the Environment. *Ambio* 17.

Wilson, E. O. 1985. The Biological Diversity Crisis: A Challenge to Science. *Issues in Science and Technology.*

Wilson, E. O. 1992. *The Diversity of Life.* New York: Norton.

Chapter 17: References for Further Reading

Abramovitz, Janet N. 1998. Sustaining the World's Forests. *State of the World 1998.* Worldwatch Institute.

Abramovitz, Janet N., and A. T. Mattoon. 1999. Reorienting the Forest Economy. *State of the World 1999.* Worldwatch Institute.

Chubb, Michael, and Holly Chubb. 1981. *One Third of Our Time?* New York: Wiley.

Donahue, Debra, L. 1999. *The Western Range Revisited.* Norman: University of Oklahoma Press.

Durning, Alan T. 1994. Redesigning the Forest Economy. *State of the World 1994.* Worldwatch Institute.

Durning, Alan, and Holly Brough. 1992. Reforming the Livestock Economy. *State of the World 1992.* New York: W. W. Norton Co.

Everhart, William C. 1983. *The National Park Service.* Boulder, Colo.: Westview Press.

Fedkin, John. 1989. *The Evolving Use and Management of the Nation's Forest, Grasslands, Croplands, and Related Resources.* USDA Forest Service, General Tech. Report RM-175, Fort Collins, Colo.

Grove, Richard H. 1992. Origins of Western Environmentalism. *Scientific American* 267 (July): 42–47.

Hendee, John C., G. H. Stankey, and R. C. Lucas. 1990. *Wilderness Management.* 2d ed. Golden, Colo: North American Press.

Joyce, Linda A. 1989. *An Analysis of the Range Forage Situation in the United States: 1989–2040.* Fort Collins, Colo.: USDA Forest Service. General Tech. Report RM-180.

Laurance, William F. 1998. Fragments of the Forest. *Natural History* (July/August).

Ledec, George, and Robert Goodland. 1988. *Wildlands: Their Protection and Management in Economic Development.* Washington, D.C.: The World Bank.

Myers, Norman. 1984. *The Primary Source.* New York: W.W. Norton and Co.

Nash, Roderick F. 1982. *Wilderness and the American Mind.* 3d ed. New Haven, Conn.: Yale University Press.

Norse, Elliot A. 1990. *Ancient Forests of the Pacific Northwest.* The Wilderness Society. Washington, D.C.: Island Press.

Postel, Sandra, and John C. Ryan. 1991. Reforming Forestry, *State of the World 1991.* New York: W. W. Norton Co.

Runte, Alfred. 1987. *National Parks: The American Experience.* 2d ed. Lincoln: University of Nebraska Press.

Wilson, E. O., ed. 1988. *Biodiversity.* Cambridge, Mass.: Harvard University Press.

Wolfe, Edward C. 1986. Managing Rangelands. *State of the World 1986.* New York: W.W. Norton Co.

World Bank. 1992. *World Development Report 1992: Development and The Environment.* New York: Oxford University Press.

World Resources Institute. Biennial. Forests and Rangelands. *World Resources.*

Wuerthner, George. 1998. High Stakes: The Legacy of Mining. *National Parks* (July–August).

Chapter 18: References for Further Reading

Anderson, J. W., R. D. Morgenstern, and M. A. Toman. 1999. At Buenos Aires and Beyond. *Resources* (Winter).

Ausubel, Jesse H., D. G. Victor, and I. K. Wernick. 1995. The Environment Since 1970. *Consequences* (Autumn).

Avery, T. E., and G. L. Berlin. 1992. *Fundamentals of Remote Sensing and Airphoto Interpretation.* 5th ed., New York: Macmillan.

Berry, J. K. 1987. Fundamental Operations in Computer-Assisted Map Analysis. *International Journal of Geographical Information Systems* 1:2 119–136.

Bliss, N. B. 1991. GIS Technology Benefits Global Change Research. *GIS World* 4:9 (December): 55–58.

Daly, H. E., and K. N. Townsend, eds. 1993. *Valuing the Earth.* Cambridge, Mass.: MIT Press.

Darmstadter, Joel, ed. 1992. Environment and Development. *Resources.* Resources for the Future. No. 106 (Winter).

Depledge, Joanna. 1999. Coming of Age at Buenos Aires. *Environment* (September).

Edmonds, James A. 1999. Beyond Kyoto: Toward a Technology Greenhouse Strategy. *Consequences* 5:1.

The Global Partnership for Environment and Development: A Guide to Agenda 21. 1992. New York: United Nations Publications.

Haas, P. M., M. A. Levy, and E. A. Parson. 1992. Appraising the Earth Summit. *Environment* 34:8 (October): 6–11; 26–33.

Kennel, C. F, Pierre Morel, and G. J. Williams. 1997. Keeping Watch on the Earth: An Integrated Global Observing Strategy. *Consequences.* 3:2.

Jensen, John. 2000. *Remote Sensing of the Environment.* Princeton: Prentice Hall.

Levy, M. A., P. M. Haas, and R. O. Keohane. 1992. Institutions for the Earth: Promoting International Environmental protection. *Environment* 34:4 (May): 12–17, 29–36.

Lillesand, T. M., and R. W. Keifer. 2000. *Remote Sensing and Image Interpretation* 4th ed. New York: Wiley.

Managing Planet Earth 1990. Readings from *Scientific American.* New York: W. H. Freeman & Co.

Nisbet, E. G. 1991. *Leaving Eden: To Protect and Manage the Earth.* New York: Cambridge University Press.

Ott, Herman E. 1998. The Kyoto Protocol: Unfinished Business. *Environment* (July/August).

Piel, Gerard. 1992. *Only One World: Our Own to Make and Keep.* New York: W. H. Freeman & Co.

Ripple, William J., ed. 1989. *Fundamentals of Geographic Information Systems: A Compendium.* Bethesda, Md.: American Society of Photogrammetry and Remote Sensing.

Schmidheiny, S. 1992. *Changing Course: A Global Business Perspective on Development and the Environment.* Cambridge, Mass.: MIT Press.

U. S. National Research Council, Committee on the Human Dimension of Global Change. 1992. *Global Environmental Change: Understanding the Human Dimensions.* Washington, D.C.: National Academy Press.

UNEP. 1997. *Global Environment Outlook.* Oxford University Press.

VanDeveer, Stacy D. 2000. Protecting Europe's Seas. *Environment* (July/August).

Wheeler, Douglas, J. 1993. Linking Environmental Models with Geographic Information Systems for Global Change Research. *Photogrammetric Engineering and Remote Sensing* 59:10 (October): 1497–1501.

APPENDIX

UNITS OF MEASUREMENT AND CONVERSION

CONVERSION FACTORS AND DECIMAL NOTATIONS

Energy, Power, Force, and Pressure

Energy Units and Their Equivalents

joule (abbreviation J): 1 joule = 1 unit of force (a newton) applied over a distance of 1 meter = 0.239 calorie

calorie (abbreviation cal): 1 calorie = heat needed to raise the temperature of 1 gram of water from 14.5°C to 15.5°C - 4.186 joules

British Thermal Unit (abbreviation BTU): 1 BTU = heat needed to raise the temperature of 1 pound of water 1°Fahrenheit from 39.4° to 40.4°F = 252 calories = 1,055 joules

Power

watt (abbreviation W): 1 watt = 1 joule per second

horsepower (abbreviation hp): 1 hp = 46 watts

Force and Pressure

newton (abbreviation N): 1 newton = force needed to accelerate a 1-kilogram mass over a distance of 1 meter in 1 second squared

bar (abbreviation b): 1 bar = pressure equivalent to 100,000 newtons on an area of 1 square meter

millibar (abbreviation mb): 1 millibar = one-thousandth (1/1,000) of a bar

pascal (abbreviation Pa): 1 pascal = force exerted by 1 newton on an area of 1 square meter

atmosphere (abbreviation Atmos): 1 atmosphere = 14.7 pounds of pressure per square inch = 1,103.2 millibars

Length, Area, and Volume

Length

1 micrometer (μm) = 0.000001 meter = 0.0001 centimeter
1 millimeter (mm) = 0.03937 inch = 0.1 centimeter
1 centimeter (cm) = 0.39 inch = 0.01 meter
1 inch (in) = 2.54 centimeters = 0.083 foot
1 foot (ft) = 0.3048 meter = 0.33 yard
1 yard (yd) = 0.9144 meter
1 meter (m) = 3.2808 feet = 1.0936 yards
1 kilometer (km) = 1,000 meters = 0.6214 mile (statute) = 3,281 feet
1 mile (statute) (mi) = 5,280 feet = 1,6093 kilometers
1 mile (nautical) (mi) = 6,076 feet = 1.8531 kilometers

Area

1 square centimeter (cm^2) = 0.0001 square meter = 0.15550 square inch
1 square inch (in^2) = 0.0069 square foot = 6.452 square centimeters
1 square foot (ft^2) = 144 square inches = 0.0929 square meter
1 square yard (yd^2) = 9 square feet = 0.8361 square meter
1 square meter (m^2) = 1.1960 square yards = 10.764 square feet
1 acre (ac) = 43,560 square feet = 4,046.95 square meters
1 hectare (ha) = 10,000 square meters = 2.471 acres
1 square kilometer (km^2) = 1,000,000 square meters = 0.3861 square mile
1 square mile (mi^2) = 640 acres = 2.590 square kilometers

Volume

1 cubic centimeter (cm^2) = 1,000 cubic millimeters = 0.610 cubic inch
1 cubic inch ($in.^3$) = 0.0069 cubic foot = 16.387 cubic centimeters
1 liter (1) = 1,000 cubic centimeters = 1.0567 quarts
1 gallon (gal) = 4 quarts = 3.785 liters
1 cubic foot ($ft.^3$) = 28.31 liters = 7.48 gallons = 0.02832 cubic meter
1 cubic yard (yd^3) = 27 cubic feet = 0.7646 cubic meter
1 cubic meter (m^3) = 35.314 cubic feet = 1.3079 cubic yards
1 acre-foot (ac-ft) = 43,560 cubic feet = 1,234 cubic meters

Mass and Velocity

Mass (Weight)

1 gram (g) = 0.3527 ounce* = 15.43 grains
1 ounce (oz) = 28.3495 grams = 437.5 grains
1 pound (lb) = 16 ounces = 0.4536 kilogram
1 kilogram (kg) = 1,000 grams = 2.205 pounds
1 ton* (ton) = 2,000 pounds = 907 kilograms
1 tonne = 1,000 kilograms = 2,205 pounds

Velocity

1 meter per second (m/sec) = 2.237 miles per hour
1 km per hour (km/hr) = 27.78 centimeters per second
1 mile per hour (mph) = 0.4470 meter per second
1 knot (k) = 1.151 miles per hour = 0.5144 meter/second

*Avoirdupois, i.e., the customary system of weights and measures in most English-speaking countries.

QUANTITIES, DECIMAL EQUIVILENTS, AND SCIENTIFIC NOTATION

Quantity	Decimal Notation	Scientific Notation		Prefix
One trillion (U.S.)	1,000,000,000,000	10^{12}	T	tera-
One billion (U.S.)	1,000,000,000	10^9	G	giga-
One million	1,000,000	10^6	M	mega-
One thousand	1,000	10^3	k	kilo-
One hundred	100	10^2	h	hecto-
Ten	10	10	da	deka-
One tenth	0.1	10^{-1}	d	deci-
One hundredth	0.01	10^{-2}	c	centi
One thousandth	0.001	10^{-3}	m	milli-
One millionth	0.000001	10^{-6}	μ	micro-
One billionth (U.S.)	0.000000001	10^{-9}	n	nano-
One trillionth (U.S.)	0.000000000001	10^{-12}	p	pico-

GLOSSARY

Abiotic: The nonbiological factors in ecosystems, such as solar radiation and water.

Absolute zero: The zero point on the Kelvin temperature scale, which represents the state at which there is no molecular vibration in a substance and hence no heat. Corresponds to −273.15 degrees on the Celsius scale.

Abyssal plain: Deep ocean floor where water depths range from 4,000 to 5,000 meters and life is very sparse.

Acid drainage: Acidic runoff from coal mines and related spoils that pollutes streams and groundwater.

Acid precipitation: Precipitation whose pH has been significantly lowered by air pollution from various oxides. A serious problem in southeastern Canada, northeastern United States, and northwestern Europe.

Active layer: The surface layer in a permafrost environment, which is characterized by freezing and thawing on an annual basis.

Acute toxicity: The occurrence of serious symptoms, including death, after a single exposure to a hazardous substance.

Adaptation: A change in an organism that brings it into better harmony with its environment; two types of adaptation are *acquired* and *genetic.*

Adiabatic lapse rate: The rate of temperature change in a rising or falling parcel of air due to compression and decompression; two rates are recognized—the dry rate (about 1°C/100 m) and the wet rate (about 0.6°C/100 m).

Adsorption: A chemical process by which ions are taken of by colloids.

Advanced treatment: The third level of sewage treatment; it involves reductions of phosphorus, suspended solids, organic compounds, advanced filtering, and application of a disinfectant. Also called *tertiary treatment.*

Aerial photography: The most widely used form of remote sensing imagery; records reflected sunlight to produce images of the Earth's surface. Photographic films used for remote sensing can produce panchromatic (black and white) or color images that include visible light and some wave-lengths of ultraviolet and infrared energy.

Aerosol: Any solid or liquid particle in the atmosphere from natural or human sources. Aerosols cause increased backscattering and reflection of solar radiation and provide condensation nuclei for precipitation droplets.

Agenda 21: The action plan from the 1992 United Nations Conference on Environment and Development (Earth Summit) to guide environmental protection and economic development in the twenty-first century.

Age structure: A demographic model used to illustrate the number of people in a country by age class.

Agroforestry: Trees planted with crops such as oats and potatoes in tree farms or plantations.

Albedo: The percentage of incident radiation reflected by a material. Usage in earth science is usually limited to shortwave radiation and landscape materials.

Allergen: Any biological, chemical, or physical substance that causes an allergic reaction. Plant pollen, dust, and smoke are common allergens.

Alluvium: Deposits of sediment laid down by streams and other runoff processes.

Almaz-1: Launched by the Soviet Union in 1991, this satellite was the first commercial *Earth resources satellite* to use remote sensing radar. It produced images in the S-band spectral region (10 cm) with ground resolution ranging from 10 to 30m during its ten-month operating lifetime.

Angiosperm: A flowering, seed-bearing plant; the angiosperms are the principal vascular plants on Earth.

Animal unit equivalent: A standard measure used in assigning different grazing animals to rangeland. For example, one cow is equivalent to five sheep.

Anion: A negatively charged ion.

Aquiclude: An impervious stratum or formation that impedes the movement of groundwater.

Aquifer: Any subsurface material that holds a relatively large quantity of usable groundwater and is able to transmit that water readily.

Azonal soil: A soil order under the traditional USDA soil classification scheme; soils without horizons; those that are usually found in geomorphically active environments such as sand dunes and river valleys.

Backscattering: That part of incoming solar radiation directed back into space as a result of diffusion by particles in the atmosphere; in a polluted atmosphere, backscattering is increased, resulting in lower rates of solar heating.

Base: In soil chemistry, a mineral that forms cations (e.g., magnesium and potassium).

Baseflow: The portion of streamflow contributed by groundwater; it is a steady flow that is slow to change even during rainless periods.

Bioaccumulation: The buildup of contaminants such as DDT in organisms as the contaminants are passed up food chains; also called *biomagnification.*

Biogeochemical cycle: The global system of materials that transfers elements among air, water, soil, and life; same as *nutrient cycle.*

Biogeography: A field of science shared by geography and biology that studies the dispersal and distribution of plants and animals.

Biological diversity: The number of species of plants, animals, and microorganisms in a prescribed geographic area, in an area of land or water. Also called *biodiversity.*

Biological oxygen demand: The amount of dissolved oxygen required by decomposers such as bacteria to break down the organic material in sewage and other wastes discharged into water. Also called *BOD.* High BODs may delete oxygen supplies for fish and other aquatic organisms.

Biomagnification: See *bioaccumulation.*

Biomass: The total weight of living organic matter per unit area of landscape; also, the total weight of the organic matter in an ecosystem.

Biomass fuels: Organic material harvested or gathered from active ecosystems that is used as an energy source for heating, cooking, and locomotion.

Biome: A large, biogeographical unit of land characterized by a particular combination of animals, vegetation, and climate; for example, grassland or tundra.

Black lung disease: A respiratory disorder contracted by coal miners from years of breathing coal dust in mines.

BOD: See *biological oxygen demand.*

Boreal forest: Subarctic conifer forests of North America and Eurasia; floristically homogeneous forests dominated by fir, spruce, and tamarack; in Russia, it is called taiga.

Breeder reactor: A nuclear fission reactor that produces more fissionable fuels than it uses; not currently available for commercial energy generation.

British thermal unit: A unit of energy used to measure heat; 1 BTU is equal to about 250 calories, the amount of heat required to raise the temperature of 1 pound of water by 1°F.

Bureau of Land Management: A federal agency responsible for managing vast tracts of public lands in the American West.

Caliche: An accumulation of calcium carbonate at or near the soil surface in an arid environment.

Calorie: A unit of energy; the amount of heat required to raise the temperature of water 1°C, from 14.5°C to 15.5°C.

Capillary water: Molecular water held around and between particles in the soil; the principal type of water utilized by plants.

Carbon dioxide: A minor atmospheric gas of major significance for its heat-absorbing capacity. Atmospheric CO_2, has been rising since the Industrial Revolution with growing world population and consumption of fossil fuels. If this trend continues, many scientists predict warming of the atmosphere.

Carcinogen: Any chemical or form of radiation that can cause cancer in humans and animals.

Carrying capacity: The number of organisms that can live in a long-term sustained balance with the environment at a reasonable quality of life. Often applies to a particular area of land, and may be defined as cultural or biological.

Cation exchange capacity: The total exchangeable cations that a soil can adsorb; the capacity increases with finer soil textures and organic content.

CFCs: See *chlorofluorocarbons.*

Chernobyl: The world's worst nuclear power plant accident, Soviet Union, 1986. It is estimated that as many as 8,000 people have died prematurely from the effects of Chernobyl.

Chlorofluorocarbons: Manufactured organic compounds made up of carbon, hydrogen, and fluorine atoms used in refrigerants, spray propellants, and some plastics. Lighter than air, this gas rises in the atmosphere and reacts with ozone in the stratosphere, depleting the ozone layer.

Chronic toxicity: The delayed occurrence of symptoms, usually after multiple exposures to a hazardous substance.

Chronospecies extinction: Extinction of a species whose genetic materials have been passed on to offspring species. See also *true extinction.*

Clade: A group of species sharing a common lineage; a species and all its descendant species.

Clean Water Act: The principal U.S. body of law on water pollution, originally titled the Water Pollution Control Act.

Clear-cutting: The logging practice of harvesting all trees in a forest stand. See also *selective cutting.*

Climate: The representative or general conditions of the atmosphere at a place on Earth; it is more than the average conditions of the atmosphere, for climate may also include extreme and infrequent conditions.

Climax community: A group of interrelated organisms living together in ecological equilibrium with the environment and therefore capable of maintaining long-term stability.

Coefficient of runoff: A number given to a type of ground surface representing the proportion of a rainfall converted to overland flow; it is a dimensionless number between 0 and 1.0 that varies inversely with the infiltration capacity; impervious surfaces have high coefficients of runoff.

Cold front: A contact between a cold air mass and a warm air mass in which the cold air is advancing on the warm air, driving the warm air upward.

Colloid: A small clay particle, less than 0.001 mm in diameter, that provides adsorption sites for ions. See also *cation exchange capacity.*

Combined sewer overflows: Municipal sewer systems that are linked to stormsewer systems so that when treatment plants become overloaded, raw sewage is discharged to natural waters via storm lines.

Commensalism: A form of symbiosis in which one species benefits and the other neither benefits nor is harmed.

Commercial agriculture: Farming for the purpose of selling crops, animals, and related products; often specialized around selected crops or livestock.

Community: A group of organisms that live together in an interdependent fashion.

Condensation: The physical process by which water changes from the vapor to the liquid phase.

Condensation nuclei: Very small particles of dust or salt suspended in the atmosphere around which condensation takes place to initiate the formation of a precipitation droplet.

Conservation-of-energy principle: The principle that energy in an isolated system can be neither destroyed nor created; thus, total energy in the system remains constant.

Conservation movement: A movement beginning in the 19th century based on the concept of sustained yield and multiple use of rural and wilderness lands. Also called progressive or scientific conservation.

Conservation tillage: Various techniques being adopted by farmers to reduce soil erosion and labor including reduced plowing and cultivation with heavier pesticide applications.

Consumerism: According to economists, the practice of utilizing economic goods; according to environmentalists, the practice of using up or squandering resources.

Consumptive water use: Water use in which the water is not returned to the environment in a liquid form; for example, irrigation leading to evaporation.

Continental shelf: The seaward-sloping margin of the continents under water to a depth of 200–300 m; the shoulders of the continents where the rate of biological activity is very high.

Contour plowing: The practice of plowing fields and planting crops along slopes, parallel to the elevation contour, to retard runoff and soil erosion.

Convectional precipitation: A type of precipitation resulting from the free ascent of unstable air; usually short-term, intensive rainfall that often produces thunderstorms.

Convergent precipitation: A type of precipitation that takes place when air moves into a low-pressure trough or topographic depression and escapes by moving upward: precipitation in the equatorial zone is at least partially convergent.

Coriolis effect: The deflection of moving airborne objects away from their line of flight; right deflection in the Northern Hemisphere; left in the Southern Hemisphere; this effect accounts for the long, curved path of winds around pressure cells.

Critical threshold: The point at which shear stress (driving force) equals shear strength (resisting force) and beyond which change, such as slope failure, rock rupture, plant damage, or soil erosion, is imminent.

Crop rotation: A farming practice in which exhausted fields are planted periodically with fertility-restoring crops such as clover.

Crude birth rate: Number of live births per 1,000 people in a country, continent, or other geographic area per year.

Crude death rate: Number of deaths per 1,000 people in a country, continent, or other geographic area per year.

Crude oil: A naturally occurring liquid hydrocarbon from which gasoline, fuel oil, kerosene, and other petroleum products are made.

Crust: The outermost zone of the lithosphere, which ranges from 8 to 65 km (5–40 miles) in thickness, and in which the ocean basins and mountain ranges are formed.

Culture: All things invented by humans, including language, law, land use, weapons, and tools, the ideas for which can be passed from one generation to the next.

Cyclone: A large, low-pressure cell characterized by convergent by airflow and internal instability; the two main classes of cyclones are midlatitude cyclones and tropical cyclones; in some parts of the United States, tornadoes are also called cyclones.

DDT: A chlorinated organic compound used as a pesticide; banned in the United States, Canada, and other developed countries, still widely used in some developing countries. See also *persistent chemical.*

Decomposers: Bacteria and fungi that break down organic debris in ecosystems. See also *detritivores.*

Deep-well injection: A hazardous waste disposal method in which waste is injected into rock at great depths.

Demographic transition: A population change model that describes population growth in countries as a result of economic development and improved health care.

Denitrification: The process by which bacteria free nitrogen in soil organic matter as it decomposes.

Density: A measure of the intensity of land use such as the area covered by facilities per acre of land.

Denudation: A term used to describe the erosion or wearing down of a land mass; also used to describe the process by which a site is stripped of its vegetative cover.

Deprived country: A very poor, undeveloped country such as Haiti.

Desertification: The transformation of vegetated landscapes into desert as a result of environmental degradation by humans and stress from drought.

Detritivores: Organisms such as buzzards and various worms, insects, and bacteria that feed on dead organic debris.

Development: Economic growth involving a change in land use to a more productive activity.

Dew point: The temperature at which air is saturated; 100 percent relative humidity.

Diffusion, agriculture: The spread of crop species and farming practices.

Diffusion, cultural: The spread of ideas and technology to other peoples and lands.

Diffusion, human: The geographic spread of humans over the earth.

Diffusion, radiation: The scattering of incoming solar radiation by molecules and particles in the atmosphere.

Digital data: Numeric data from remote sensors, digitizers, and scanners that can be processed by computers. Include GIS data from maps and images coded into series of digits (bits) consisting of 0s and Is and stored on magnetic disks or tapes.

Dioxin: A synthetic organic compound used in various industrial processes, but especially paper manufacturing; known to be a serious health risk, a cause of cancer.

Discharge: The rate of water flow in a stream channel; measured as the volume of water passing through a cross-section of a stream per unit of time, commonly expressed as cubic feet (or meters) per second. Also applies to *groundwater.*

Distal scale: The space farther away; at a distance from a feature, organism, or a location. See also *proximal scale.*

Disturbance: Physical disruption of the environment by humans, their machines, or animals.

Disturbance theory: A concept or model of ecological change in which the external (nonbiological) environment is the controlling force.

Drainage basin: The area that contributes runoff to a stream, river, or lake. Also called a *watershed.*

Drainage divide: The border of a drainage basin or watershed where runoff separates between adjacent areas.

Drainage net: A system of stream channels usually connected in a hierarchical fashion. Also called a *drainage network.*

Dust Bowl: The name given to the area of severe drought and wind erosion in the Great Plains during the 1930s.

Dust dome: Term given to the envelope of dirty, hazy air that often hangs over cities.

Dynamic equilibrium: A term used to describe the behavior of a system, such as a river network or an ecosystem, which is continually trending toward a state of equilibrium, but rarely reaches it.

Earth Observation System (EOS): The series of polar orbiting remote sensing satellites that will provide data for the *Mission to Planet Earth* Program. The first satellites of this series, Landsat 7 and Terra, were launched in 1999.

Earth resources satellites: Unmanned remote sensing satellites that produce relatively high resolution digital imagery from line scanners, radar, and other devices. See also *Landsat* and *SPOT, ERS-1, 2, Almaz-1,* and *Radarsat* for examples of Earth resources satellite programs.

Earth Summit: A United Nations meeting of world nations and nongovernmental organizations in 1992 to discuss environmental problems and formulate agreements for environmental protection and economic development.

Economic productivity: The amount of revenue a land use generates per acre or square kilometer of area.

Ecosystem: Groups of organisms linked together by energy flows in food chains; also, a community of organisms and their environment.

Ecotone: The transition zone between two habitats such as adjacent zones of different vegetation.

Edge habitats: Narrow zones on the margins of fields or along the border between different environments where the transition itself constitutes a distinct habitat. See also *ecotone.*

Efficiency: The effectiveness of plants in converting solar radiation into plant material through photosynthesis. For natural vegetation worldwide, efficiency is less than 1 percent.

Electromagnetic radiation: Radiant energy emitted from all objects in electromagnetic waves of varying wavelengths. It is the type of energy recorded by remote sensing systems in ultraviolet, visible, infrared, and microwave wavelength bands. Other bands of electromagnetic energy include x-rays, gamma rays, radio and TV waves, and electric current. See also *electromagnetic spectrum.*

Electromagnetic spectrum: The classification scheme used to describe the array of electromagnetic radiation; the various categories or bands of radiation are distinguished on the basis of wavelength or frequency. Gamma rays and x-rays are shortwave (or high frequency) radiation whereas radio and TV waves are longwave (or low frequency) radiation. Visible wavelengths, represented by the color spectrum, are relatively shortwave.

Electronic digitizing systems: Electronic digitizers are used to transform mapped features into digital data for computer mapping and GIS. *Coordinate digitizers* are used to manually trace the outline of map features using a sequence of lines connected by points. *Drum* or *laser scanning digitizers* systematically transform maps into digital data.

Emission standards: Air pollution limits established under the U.S. Clean Air Act for exhaust emissions from automobiles, power plants, and other facilities.

Endangered species: A species in imminent danger of elimination from all or a significant portion of its ranges. See also *threatened species.*

Endemic species: Species limited to a small range in a particular geographic location.

Energy: Generally, the capacity to do work; defined as any quantity that represents force times distance. Joules and calories are energy units commonly used in science.

Energy flux: The rate of energy flow into, from, or through a substance; also called radiant flux density and irradiance.

Energy pyramid: A model describing the energy structure of an ecosystem in which trophic levels are represented by tiers that grow smaller toward the apex.

Environmental accounting: An approach to economic accounting that factors in natural resources, changes in ecosystems, and other environmental conditions as part of economic performance.

Environmental impacts: Human activities that degrade the environment and reduce its potential to support life.

Environmental Summit Conference: See *Earth Summit.*

Environmentalism: A political or social movement with the express aim of defending the environment, conserving resources, and protecting natural phenomena and the quality of life.

Equatorial low pressure: The belt of low pressure near the equator that produces the cloudy, rainy conditions of the tropics.

Equatorial zone: The middle belt of latitude, extending 10°C or so north and south of the equator.

Equilibrium surface temperature: The overall surface temperature of the earth, 15°C (about 60°F), representing the earth's thermal balance.

ERS-1, 2: Radar-equipped earth-resources satellites operated by the European Space Agency that gather land and ocean imagery with 30-meter resolution.

Ethanol: Ethyl alcohol manufactured from grain, sugar beets, sugar cane, milk byproducts, or even garbage.

Eutrophication: Accelerated biological productivity in a water body as a result of the input of nutrients such as nitrogen and phosphorus from runoff and air pollution.

Evapotranspiration: The loss of water from the soil through evaporation and transpiration into the atmosphere.

Event: An episode of a process defined as some quantity of a variable such as a river discharge or wind velocity.

Exchange time: The time it takes for water in a lake, aquifer, or other water body to be completely replaced with new water.

Exotic rivers: A river such as the Nile or Colorado that flows through an arid region after gaining its flow elsewhere.

Extinction: In general terms, the end of a species on earth; see also *true extinction* and *chronospecies* (or *pseudo*) *extinction.*

Extinction event: A short-term or sudden increase in the rate of extinction.

Fallow: The practice of "resting" farmland for a period to allow it to recover from cropping.

Favelas: Shantytowns on the margins of, or well within, Brazilian cities. Similar squatter settlements occur in large urban areas in most developing countries.

Feedback: A return effect of a change; the consequences of a change have a feedback effect if they dampen or amplify the change or the causes of it.

Fertility rate: See *total fertility rate.*

Field capacity: The maximum amount of capillary water that can be held by a soil; it varies with soil texture and composition.

First-flush flow: Runoff produced from the first part of a rainstorm that flushes land-use surfaces of pollutants.

Flash flood: Sudden rush of water through a stream channel caused by intense runoff from a thunderstorm; common in arid and mountainous regions.

Food chain: A sequence of organisms in an ecosystem, each eating or decomposing its predecessor.

Food value: The total calories and the types of nutrients and vitamins in food crops.

Food web: A system of integrated food chains.

Forest mining: The practice of logging old-growth and/or virgin forests and then moving on to cut similar forests elsewhere.

Fossil fuels: Organic material in the form of oil, natural gas, and coal extracted from ancient deposits and burned as a source of energy.

Fragmentation: The spatial breakdown of ecosystems into smaller geographic units such as continuous forest into woodlots and fencerows.

Freon: See *chlorofluorocarbons.*

Front: In the atmosphere, the boundary between contrasting air masses or types of air; fronts are the main features of midlatitude cyclones. See also *frontal precipitation.*

Frontal precipitation: Precipitation produced by a midlatitude cyclone, principally its cold front and warm front. See also *cyclone.*

Frontier environment: Lightly settled geographic zones such as the deserts and tundra where the environment has posed serious limitations to settlement and land use.

Fuelwood: Various types of wood used for cooking and heating, mainly in developing countries.

Gaia concept: The idea that the biosphere is capable of regulating the global environment to maintain its suitability for life. A popular environmental concept on the role of life in shaping the Earth environment for life.

Gamma radiation: Very shortwave, ionizing radiation that reacts with and damages living tissue. See also *ionizing radiation.*

Gasification: A technological process by which coal is converted into gas for energy.

Gasohol: High octane fuel made up of 90 percent gasoline and 10 percent ethanol.

Gene splicing: Genetic engineering process in which the genetic makeup of a species is altered by extracting certain desirable genes from a specimen and inserting them into the cells of a target crop of livestock species. The resultant cells then carry this trait as part of their genetic code.

Genetic engineering: A technique involving gene splicing used in crop science to produce a superior crop or livestock variety. See also *gene splicing.*

Geochemical system: The global cycle of elements among the biosphere, lithosphere, hydrosphere, and atmosphere.

Geographic barrier: A physical feature such as a deep canyon, a large highway corridor or a continuous belt of farmland that limits the geographic distribution of an organism, a population, or an ecosystem.

Geographic Information System (GIS): A computer-based system that collects, stores, statistically analyzes, and displays geographic data, environmental features and their attributes. GIS references data by geographic location.

Geographic isolation: One of the two basic mechanisms of speciation; the splitting apart of a population by a geographical barrier such as a mountain range or deep canyon. See also *polyploidy.*

Geophysiology: The concept of Earth as a self-regulating organism. Also see *Gaia concept.*

Geosynchronous environmental satellites: Meteorology satellites launched into a geostationary orbit 38,500 km over the Earth's equator that provide atmospheric and other environmental data. They orbit in the same direction and speed as Earth's rotation and monitor approximately one-quarter of the Earth's surface from pole to pole. Such satellites are operated by the United States, India, the European Community, and Japan. See also *GOES* and *meteorology satellites.*

Geothermal heat: Energy from the Earth's interior. It is a very small quantity relative to solar energy, but in selected hot spots it is a major source of energy.

GIS: See *Geographic Information System.*

Glacial drift: A general term applied to all glacial deposits, a principal parent material for soils in the northern United States and most of Canada.

Global coordinate system: The network of east-west and for north–south reference lines (parallels and meridians) used to record locations on the Earth's surface.

Global warming: Atmospheric warming trend forecast by that scientists as a result of global increases in atmospheric carbon dioxide and several other gases. See also *carbon dioxide.*

GOES (Geostationary Operational Environmental Satellites): Meteorology satellites placed in a geosynchronous orbit operated by the U.S. National Oceanographic and Atmospheric Administration (NOAA).

Gravity water: Liquid water in the soil that moves by the force of gravity; groundwater is a collection of gravity water at depth in the soil and bedrock.

Greenhouse effect: The capacity of the atmosphere to act like a greenhouse, trapping and absorbing longwave radiation and heat. Principal greenhouse gases are water vapor and carbon dioxide. See also *global warming.*

Green revolution: A period beginning in the 1950s when technology greatly advanced the capacity of world agriculture to produce food. A period when agricultural output equaled or stayed slightly ahead of population demands.

Greenwich meridian: The zero meridian; also called the *Prime meridian.*

Groundwater: The mass of gravity water that occupies the subsoil and upper bedrock zone; the water occupying the based zone of saturation below the soil–water zone.

Groundwater vulnerability: The susceptibility of aquifers to pollution from land uses, spills, and other sources.

Gullying: Soil erosion characterized by the formation of narrow, steep-sided channels etched by rivulets or small streams of water. Gullying can be one of the most serious forms of soil erosion on cropland.

Habitat: The surrounding or local environment of an organism; the place from which it draws its resources; sometimes called the address of an organism.

Habitat corridors: Zones or ribbons of habitat and ecosystem complexes that can be traced through the landscape.

Hazardous waste: Toxic substances, such as heavy metals, solvents, and acids, that represent a threat to human health and the environment if handled or disposed of improperly.

Hazardous waste treatment: Various methods of processing hazardous waste to make it less harmful; for example, land application and incineration.

Heat balance: A concept or model of a heat system characterized by heat inflows, outflows, and storage. See also *equilibrium surface temperature.*

Heat island: The area or patch of relatively warm air that develops over urbanized areas.

Heat syndrome: A human health disorder caused by some combination of extreme heat, exposure to solar radiation, physical exertion, and loss of body fluids and salts; often pronounced in the heat islands of urban climates.

Heavy metals: High-density metals such as lead, arsenic, and zinc, which are toxic to humans and other organisms and as wastes pose serious disposal problems.

High-Resolution Visible scanner (HRV): See *SPOT.*

Homeostatis: The trend toward constancy or equilibrium within an organism.

Horizon: A layer in the soil that originates from the differentiation of particles and chemicals by moisture movement within the soil column; four major horizons are recognized in a standard soil profile: O, A, B, and C.

Hurricane: A large tropical cyclone characterized by convergent airflow, ascending air in the interior, fast winds, and heavy precipitation; the largest storm on earth.

Hydrocarbons: A wide variety of gases including methane and benzene emitted as pollutants from burning fossil fuels.

Hydroelectric power: Electrical power produced when flowing or falling water turns a turbine.

Hydrograph: A streamflow graph that shows the change in discharge over time, usually hours or days.

Hydrologic cycle: The planet's water system, described by the movement of water from the oceans to the atmosphere, to the continents, and back to the sea.

Hydrothermal energy: Hot water and steam from geothermal areas that are tapped and used to heat buildings and other facilities or to drive turbines for electrical power.

Illuviation: The process of accumulation of ions and colloids in a soil.

Impervious cover: A surface material that is impenetrable to surface water; concrete, asphalt, and building roofs.

Impoundment: A water body such as a lake, pond, or reservoir. May be natural or human-made.

Incineration: A method of hazardous waste treatment in which organic wastes such as those containing PBCs are burned at a high temperature.

Industrial smog: A form of air pollution common to industrial urban areas characterized by a mixture of sulfur dioxide, acid droplets, and particulates. Also see *photochemical smag.*

Infiltration: The process by which surface water from rainfall and melting snow enters the soil.

Infiltration capacity: The rate at which soil takes in water through the surface; measured in inches or centimeters per minute or hour.

Infrared radiation: Mainly longwave radiation of wavelengths between 0.7 and 100 micrometers.

Integrated landscape management: An approach to landscape planning and management that involves coordinating the efforts of businesses, governmental agencies, landowners, and others to conserve biodiversity.

Intensive subsistence agriculture: High-yield subsistence farming in densely populated areas such as China and India.

International Conference on Population and Development (ICPD): the third U.N. population conference held in Cairo, Egypt, in 1994. Its primary objective was to prepare a plan, the World Programme of Action, to stabilize world population over the next two decades.

Intercropping: Simultaneous production of several crops on the same field; planting of secondary crops such as fruit trees within and on the margins of fields as a part of intensive subsistence agriculture.

Intertropical convergence zone: The belt of convergent airflow and low pressure in the equatorial zone that is fed by the trade winds. Also called the *ITC zone.*

Inversion: See *thermal inversion.*

Ion: A minute particle of a dissolved mineral; usually an atom or group of atoms that are electrically charged.

Ionizing radiation: High-energy radiation emitted by radioisotopes; powerful enough to dislodge electrons from atoms to form ions that can damage living flesh.

Island biogeography: The study of biodiversity in relationship to the size of a prescribed area of habitat, such as an island, and distance from similar habitats.

Kilocalorie: One thousand calories.

Kilogram: A metric unit of mass (weight) equal to 2.208 pounds.

Land application: A method of hazardous waste treatment which waste is worked into biologically active soil where microorganisms break it down. Also called *land farming.*

Land cover: The materials such as vegetation and concrete that cover the ground; see also *land use.*

Land farming: See *land application.*

Landfill: A managed waste disposal site in which the waste is buried or covered with soil.

Landsat: The first international Earth resources program begun in 1972. Five polar orbiting Landsat satellites have produced multispectral digital imagery of the earth surface using two line scanning systems. The multispectral scanner (MSS) produces images with 80-meter resolution in two visible and two reflected infrared bands. The Thematic Mapper (TM) on Landsat 4 and 5 produces images with 30-meter ground resolution in six visible and reflected infrared bands and 120-meter resolution images in a thermal infrared band. The enhanced Thematic Mapper Plus (ETM+) on Landsat 7 produces an additional panchromatic band which generates images with 15 meter resolution.

Landscape: The composite of natural and human features that characterize the surface of the land at the base of the atmosphere; includes spatial, textural, compositional, and dynamic aspects of the land.

Land tenure: The type of ownership or rights to land; for example, in tenant farming, the land is owned by a landlord.

Land unit: A parcel, of land characterized by relatively homogeneous topography, soils, drainage, and other factors. The logical unit for land-use planning.

Land use: Any human activity that takes place on the land (e.g., agriculture and industry).

Land-use programs: Various approaches to land-use planning organized at different levels of government.

Lapse rate: The rate of temperature change through a substance; in the atmosphere, the ambient atmospheric lapse rate is the temperature change with elevation in the troposphere, which averages about −6.4°C per 1,000 m. See also *adiabatic lapse rate.*

Latent heat: The heat released or absorbed when a substance changes phase, as from liquid to gas. For water at 0°C, heat is absorbed or released at a rate of 600 calories per gram in the liquid to vapor phase change.

Latitude: System for referencing locations north and south of the equator; readings given in degrees north or south of the equator, which is 0 degrees latitude.

Leachate: The contaminated liquid that seeps from landfills.

Leaching: The removal of minerals in solution from a soil; the washing out of ions from one level to another in the soil.

Life cycle: The biological stages in the complete life of an organism.

Life form: The form of an individual organism or the form of the individual organs of an organism; also, the overall structure of the vegetative cover.

Line scanners: A widely used electronic imaging system in remote sensing that records reflected or emitted energy from the Earth in narrow strips either at right angles or parallel to the line of flight. Multispectral line scanners use in Earth resources satellites simultaneously produce images in several visible and infrared bands.

Lithosphere: The upper layer of the mantle; the unit in which the tectonic plates are defined; it is about 100 km thick and includes the crust; also, a general term used in reference to the entire solid Earth.

Littoral zone: The coastal zone of the ocean where wave action, tidal fluxes, and other processes mix the water.

Loess: Wind-deposited silt that covers large areas of the North American plains and prairies and many other areas of the world. A rich parent material for soils.

Longitude: System for referencing locations east and west of the Prime Meridian; readings given in degrees east or west of the Prime Meridian.

Love Canal: The most publicized environmental incident involving hazardous waste in the United States; it resulted in serious health problems for many people and the evacuation of 239 families.

Low-tech treatment: Sewage treatment using simple, nonmechanical methods with little or no chemical treatment. For communities it may involve soil filtering and wetland applications.

Macronutrients: Major elements utilized by living organisms such as nitrogen, phosphorus, and calcium.

Magnitude and frequency: The concept concerning the behavior of processes and the resultant changes they produce individually and collectively in the landscape; it involves which events render the greatest change and what kinds of change different-sized events render.

Malthusian theory: The argument that human population grows much faster than food supply, resulting inevitably in famine.

Malnutrition: Inadequate nutrition for good health; caused by a diet deficient in proper levels of protein, fat, minerals, vitamins, and other nutrients.

Marginal land: Land where the chances of successful crop farming are about 50–50 because of environmental limitations such as climate, slope, or drainage.

Megacities: Extremely large urbanized areas formed from the influx of huge numbers of people (such as Mexico City) or the geographical coalescence of neighboring cities with urban sprawl (such as U.S. eastern seaboard).

Meridians: A set of north–south running lines in the global coordinate system that are used to reference locations east and west. See also *longitude.*

Meteorology satellites (metsats): Unmanned satellites that generate visible and thermal infrared digital imagery as well as other measurements used primarily to monitor regional and global atmospheric conditions and provide data for weather forecasting. See also *GOES, NOAA,* and *Upper Atmosphere Research Satellite.*

Methane: A natural gas produced from decaying organic matter, animal digestive systems, and underground deposits. heat-absorbing gas that contributes to Earth's greenhouse effect.

Methanol: Methyl alcohol manufactured from biomass, natural gas, or coal.

Microendemic species: Species with very small ranges, as small as a group of trees or an island in a river.

Micrometer: The metric unit for one-millionth of a meter. It is a common measurement unit for wavelengths of electromagnetic energy in the visible and infrared bands. It is equivalent to the micron.

Micronutrients: Minor elements utilized by living organisms such as boron and copper.

Middle latitude: Generally, the zone between the pole and the equator in both hemispheres; usually given as 35° to 55° latitude.

Midnight dumping: Secretive and usually illegal dumping of hazardous waste in streams, ditches, vacant land, and other unprotected sites.

Migration: Movement of people to new geographic areas; it may take place within counties, between countries, or between continents.

Millibar: A unit of force (or pressure) equal to one-thousandth of a bar; normal atmospheric pressure at sea level is 1013.2 millibars (mb).

Minimization: See *source reduction.*

Mission to Planet Earth (MTPE): An international program proposed by NASA in 1992 to study global change using data collected by remote sensing satellites as well as by ground-based measurement systems. The ultimate goal of this program now named Earth Science Enterprise is to understand environmental processes well enough to predict environmental change, such as global warming. See *Earth Observation System.*

Moisture budget: The seasonal balance of water in the soil with inputs from rainfall and outputs through evapotranspiration and drainage processes.

Monoculture: An extreme form of ecological simplification in which one or two species of forestry, agricultural, or weed plants dominate an area. A common result of commercial farming and forestry.

Monsoon forest: A tropical forest found in the wet/dry monsoon climate mainly in South and Southeast Asia.

Montreal Protocol: An international meeting of nations held in Montreal in 1987 to discuss global environmental problems; among other things, agreements were made to build plans for limiting CFC and carbon dioxide emissions.

Multinucleated land use: A pattern of urban land use characterized by several or more nodes of concentrated activity within an urban area.

Multiple-use resource: A part of the environment such as a forest that can simultaneously be used for various land uses such as recreation, lumbering, and water supply.

Multispectral scanner (MSS): See *Landsat.*

Municipal solid waste: The refuse from households, businesses, and institutions; nonindustrial solid waste, mainly paper, plastic, and glass.

Mutagen: Any chemical or form of radiation that can induce genetic mutation in a living organism.

Mutualism: A form of biological symbiosis in which both species benefit.

National Aerial Photography Program (NAPP): The current U.S. government aerial photography program; acquires 1 : 40,000 scale (1 in. to 3333 ft.) panchromatic or color infrared photographs for federal and state agencies. Indexes are available from the U.S. Geological Survey.

National Ambient Air Quality Standards (NAAQS): Allowable air quality levels established under the U.S. Clean Air Act; primary and secondary standards for protection of human health and the environment.

National forest system: Forests on federal lands under the management of the U.S. Forest Service.

National grasslands: Prairie lands that are part of the U.S. *National forest system.*

National monuments: Facilities and special features of national significance managed by the U.S. National Park Service.

National recreation areas: Part of the system of lands under the U.S. National Park Service; mostly beaches, rivers, and reservoirs near urban areas.

National Wilderness Preservation System: The system of wilderness created under the *Wilderness Act of 1964.* Since then several hundred additional wilderness areas have been created on federal land.

Natural gas: Mainly methane mixed with smaller amounts of heavier hydrocarbons that liquify at normal atmospheric pressure.

Natural resource accounting: See *environmental accounting.*

Nature Conservancy: An international, nongovernmental organization that through purchases, arbitrations, and other means promotes sustainable approaches to development and protection of open lands and valuable ecosystems.

Neo-Malthusianism: Renewed and updated Malthusian theory arguing that population in some parts of the world is currently outgrowing available food supply leading to poverty, urban crowding, disease, and social unrest.

Net photosynthesis: The organic energy from photosynthesis remaining in a plant after respiration; the energy that goes into building new tissue.

Net primary production: The total organic matter added to an area by vegetation. Usually measured in grams per square meter per day or year.

New building sickness: A phenomenon characterized by various human health disorders associated with new buildings.

New Deal conservation: Government-sponsored conservation programs during the 1930s aimed at putting people to work planting trees and building roads, campgrounds, and other facilities.

Niche: A term used to define an organism's way of life or what it does in its habitat. See also *habitat.*

Nimbus: A U.S. experimental meteorology satellite program begun in 1964 and operated by NASA. It launched seven satellites to test a wide array of remote sensing instruments designed to acquire atmospheric and oceanographic data. One instrument, the Coastal Zone Color Scanner (CZCS), measured physical, biological, and chemical properties of the world's coastal waters.

Nitrogen fixation: The process by which certain bacteria growing on plant roots synthesize gaseous nitrogen and convert it into a solid form.

NOAA: The National Oceanographic and Atmospheric Administration; an agency of the federal government responsible for monitoring weather and related phenomena. NOAA operates a series of polar-orbiting satellites (metsats) that produce imagery and data for weather forecasting, hydrology, oceanography, vegetation studies, and climatology.

Nomadism: Mobile population or nonsedentary people usually engaged in herding.

Nonconsumptive water use: Water use in which the water is returned to the environment in a liquid form; for example, water used as a cooling medium in power plants.

Nonpoint source pollution: Water pollution generated by spatially-dispersed, usually nonspecific, sources such as agriculture. Most stormwater is nonpoint source pollution.

Nonrenewable resources: Resources such as oil and coal that are not naturally replenishable for human use.

Nonsedentary: A settlement type, such as nomadism, characterized by groups shifting about an area.

Nuclear energy: Energy released when atomic nuclei undergo a nuclear reaction.

Nuclear fission: A nuclear chain reaction involving the splitting apart of the nuclei of radioactive isotopes, usually uranium 235, when bombarded by neutrons. The energy released is used to heat water and generate electricity in power plants.

Nutrient cycle: The system of transfer of nutrients such as nitrogen through the organic, air, water, and soil environments. Also called *biogeochemical cycle.*

Nutrient retention: The capacity of soil to retain nutrients such as phosphorus and nitrogen; retention potential is usually high for finer grained soils with high organic contents.

Ocean dumping: A common method of waste disposal for garbage, sludge, hazardous materials, and spoils from harbor dredging. Despite restrictions in many developed countries, the practice is widespread in the world.

Oil shale: A source of nonliquid petroleum held in sedimentary rock such as shale and marlstone.

Old-growth forests: Uncut, primary (virgin) forest containing large trees and undisturbed ecosystems.

Open dumps: Traditional means of waste disposal involving little or no planning, management, and environmental protection.

Open lagoons: Dumping hazardous waste into unprotected soil pits; common around industrial areas in the United States as late as the 1970s; still common in some countries.

Open system: A system characterized by a through-flow of material and/or energy; a system in which energy or material is added and released over time.

Opportunistic species: Species such as weedy plants and various animals with a capacity to flourish where natural landscapes have been disturbed or eradicated.

Orogenic belt: Broad belts of mountains that lie along one or two sides of a continent.

Orographic precipitation: A type of precipitation that results when moist air is forced to rise when passing over a mountain range; most areas of exceptionally heavy rainfall are areas of orographic precipitation.

Outgassing: The production of gases such as water vapor and carbon dioxide from the Earth's exterior, as in volcanic activity.

Overland flow: Runoff from surfaces on which the intensity of precipitation or snowmelt exceeds the infiltration capacity; also called Horton overland flow, for hydrologist Robert E. Horton.

Oxidant: A type of chemical air pollutant associated with photochemical smog; ozone is a major oxidant.

Oxygen-demanding wastes: Organic material in sewage and other wastes discharged into water, which as they are degraded by chemical and biological processes, use up available oxygen in water. See also *biological oxygen demand.*

Ozone: One of the minor gases of the atmosphere that performs the important function of absorbing ultraviolet radiation in the stratosphere; in the lower atmosphere ozone is a pollutant—a pungent, irritating oxidant.

Ozone depletion: The process of reduction in natural atmospheric ozone due to air pollution by CFCs; process varies seasonally and geographically with greatest rates of depletion over the poles.

Palustrine wetland system: A class of wetlands defined by their inland location; the most abundant type of wetlands; swamps, marshes, and bogs.

Parallels: A set of east–west running lines in the global coordinate system labeled in degrees beginning at the equator.

Parasitism: A form of biological symbiosis in which one species is benefited and the other is harmed.

Parent material: The particulate material in which a soil forms; the two major classes of parent material are residual and transported.

Passive solar systems: See *solar heating system.*

Performance standards: Desired or mandated levels of operation, efficiency, and quality in environmental systems such as facility energy use, air pollution, and water pollution.

Permafrost: Ground in a permanently frozen state; most extensive in high latitudes of North America, Eurasia, and Antarctica.

Persistent organic chemical: Chemical such as DDT with a long half life; chemicals that take a long time to be broken down by the environment; also called conservative chemicals and POPs.

Pesticide: Any chemical used to inhibit or kill an organism people consider undesirable. Includes herbicides, insecticides, and fungicides.

Phosphorus: An element and macronutrient; a critical nutrient in agricultural fertilizers.

Photic zone: Surface layer of the ocean that is illuminated by sunlight and highly active biologically; about 100 m (330 ft) deep.

Photochemical smog: A complex mixture of secondary air pollutants produced by the reaction of nitrogen oxides and hydrocarbons in the presence of sunlight and primary pollutants such as particulates and carbon monoxide. A common form of urban air pollution where automobiles are the principal polluters and sunlight is abundant, e.g., Los Angeles.

Photovoltaic cells: Solar collector panel designed to produce electrical current directly from the solar panel.

Photosynthesis: The process by which green plants synthesize water, carbon dioxide, and the energy from absorbed light, and convert it into plant materials in the form of sugar and carbohydrates.

Physiologic density: A measure of population density based on the number of people in a country relative to the total area of cropland.

Phytoplankton: Tiny floating plants that live in the surface layer of the oceans and provide the energy base for aquatic ecosystems.

Pixel: A basic unit of ground resolution (or measurement) in a digital image. Rows and columns of pixels make up the digital image. A number is assigned to each pixel scaled to the quantitative value of the area represented.

Plant production: The rate of output of organic material by plants through photosynthesis; the total amount of organic matter added to the landscape over some period of time, usually measured in grams per square meter per day or year.

Plant productivity: Total organic matter added to an area by plants through photosynthesis. See also *plant production.*

Plate tectonics: A comprehensive geological theory that describes the processes by which the Earth's crust shifts slowly about, moving the continents and reshaping the ocean basins.

Pleistocene: An epoch in the geologic time scale dating from 1 to 2 million years ago, and according to some interpretations, extending through the present. A time of episodes of widespread glaciation.

Point source pollution: Water pollution emanating from a specific source such as a factory and released at a known discharge point.

Polar front: The zone or line of contact in the mid-latitudes between polar/arctic air and tropical air; it often coincides with the polar front jet stream.

Polar orbiting environmental satellites (POES): Metsats and Earth resources satellites that orbit the earth over both poles, with essentially north and south orbits these satellites can acquire imagery and other measurements of the entire Earth, over the course of a number of complete orbits. See also *NOAA*, *Landsat*, and *SPOT.*

Polluter pay laws: Laws that require former parties responsible for pollution to pay for cleanup of contaminated sites.

Pollution: Degradation of the environment as a result of contamination of some sort by chemicals, biological agents, sediment, radiation, or heat.

Pollution concentration: The amount of contaminant per unit volume of air, water, or soil; for example, milligrams of sediment per liter of water.

Pollution load: The total mass or volume of pollution discharged or received by air, water, or soil over an extended period such as a year.

Pollution plumes: Trains of polluted air emanating from factories, fires, or urban areas and driven by crosswinds. Plumes behave differently depending on atmospheric conditions.

Pollution sink: Any part of the environment that becomes a repository for pollutants; soil, groundwater, and lakes, for example.

Polyploidy: One of the two basic mechanisms of speciation; the development of mutations that interbreed with individuals within their population having the same gene traits. Less common than geographic isolation and limited mainly to plants. See also *geographic isolation.*

Pond: Generally a small lake, though there is no definitional rule on the size.

POP: see *Persistant organic chemical.*

Population density: The basic measure, called arithmetic or crude density, is the number of people living in a geographic area such as a country divided by the total area of land.

Porosity: The total volume of pore (void) space in a given volume of rock or soil; expressed as the percentage of void volume to the total volume of the soil or rock sample.

Poverty: Inability to meet one's basic needs for food, clothing, shelter, and other necessities. Definition varies among societies because of differences in the concept of basic needs.

Poverty cycle: A vicious cycle of human life characterized by the decline in available food, expansion of farming into marginal lands, degradation of the environment, abandonment of the land, and migration to cities.

Precipitation: The term used for all moisture—solid and liquid—that falls from the atmosphere.

Preservationist school: Part of the conservation movement that opposes encroachment on U.S. parklands for utilitarian purposes such as dams for water supply.

Pressure, atmospheric: The force applied to the Earth's surface by the weight of the atmosphere. Mean sea level pressure is 1013.2 millibars. See also *millibar.*

Pressure cell: Any body of air characterized by high or low pressure; at the global scale, pressure cells cover huge areas and generate the systems of prevailing winds.

Prevailing westerlies: See *westerlies.*

Primary air pollutant: Any pollutant added directly to the atmosphere by natural or human processes. See also *secondary air pollutant.*

Primary consumer: An organism that eats plants as its sole source of sustenance.

Primary treatment: The first level of sewage treatment in a treatment plant; it involves filtering, screening, and settling.

Principle of limiting factors: The biological principle that maximum obtainable rate of photosynthesis is limited by the one resource needed for plant growth that is in least supply.

Prior appropriation: A method of water allocation in which rights to water are assigned on a predetermined basis.

Productivity: In economics, a measure of output of goods and services; in ecology, the net output of organic matter by plants per unit area (m^2) of the Earth's surface. See also *plant production.*

Proximal scale: The space close to or immediately around a point of reference such as an organism.

Pseudo extinction: See *chronospecies extinction.*

Radar: Image forming radar systems designed for remote sensing. Radars send out energy beams that can penetrate the atmosphere, including clouds. Some of the energy striking the Earth's surface is reflected back to the radar antenna, which produces an electronic signal that is processed into imagery.

Radarsat: Canada's first Earth resources remote sensing satellite, launched in 1995. Radarsat 2 is scheduled for launch in 2003.

Radiation: The process or form by which radiant (electromagnetic) energy is transmitted through free space or transparent materials such as the atmosphere; the term used to describe electromagnetic energy, as in infrared radiation or shortwave radiation.

Radioactive waste: The contaminated byproducts of nuclear energy plants, military weapons installations, and various other sources. Waste is classed as high-level or low-level, depending on the magnitude of radiation output.

Radon: A radioactive gas produced from the Earth's crust that is linked to lung cancer in humans.

Rainforest: Forest formation of wet tropical and temperate maritime environments; dominated by a heavy cover of evergreen trees with abundant species of plants, animals, insects, and microorganisms.

Range: The geographic area occupied by a species, genus, or family of a plant, animal, or microorganism. Many different types of ranges are recognized including cosmopolitan, discontinuous, and endemic.

Recharge: The process by which groundwater or soil water is replenished.

Recycling: The practice of collecting and reusing waste materials; glass, metals, plastics, for example.

Reduction: The loss of area or total coverage of an ecosystem such as the loss of tropical rainforest to deforestation and land use; in waste management, the practice of reducing waste at its source.

Remote sensing: Methods for acquiring environmental data, usually from an aircraft or satellite. Cameras, line scanners, and radars are remote sensing systems used to gather primarily image data. Some other remote sensing instruments acquire nonpictorial data.

Renewable resources: Resources such as freshwater and managed forests that are replenishable.

Replacement fertility rate: The number of children a man and woman must have to replace themselves; about 2.1 in developed countries, but it is higher in some developing countries because of higher death rates among children.

Resource Conservation and Recovery Act (RCRA): A U.S. law providing legal definition of hazardous waste and guide-lines for waste handling, management, and storage.

Respiration: The internal cellular processes of a plant or animal by which energy is used for biological maintenance.

Rio de Janeiro Summit: See *Earth Summit.*

Riparian rights: A method of water allocation based on first come, first served for users bordering on a stream.

Rosy periwinkle: A flowering plant found on Madagascar from which cures for Hodgkin's disease and acute lymphcytic leukemia are derived.

Runoff: In the broadest sense, runoff refers to the flow of water from the land as both surface and subsurface discharge; the more restricted and common use, however, refers to runoff as surface discharge in the form of overland flow and channel flow.

SAS: See *soil absorption systems.*

Safe aquifer yield: The rate at which water can be pumped from an aquifer without significant decline in the aquifer.

Safe limit: The critical level of pollution beyond which change, degradation, illness, or death will result for humans, animals, or some aspect of the environment.

Salinization: Salt saturation of agricultural soils because of overirrigation and waterlogging in dry lands.

Saltwater intrusion: The invasion of saltwater into fresh groundwater because of overpumping of the fresh groundwater; a problem in coastal areas.

Sanitary landfill: A planned waste disposal site designed to secure waste and minimize its impacts on human health and safety.

Sanitary sewer system: The system of pipes used to collect sewage from homes and other buildings.

Scale: The size of features and the distances shown on a map relative to those on the Earth itself.

Scanner: See *line scanners.*

Scattering: See *backscattering.*

Seasat: The first scientific satellite designed primarily to monitor the oceans with remote sensing radar. Launched by NASA in 1978, it collected thousands of unique images that detected wave, current, and bottom configuration patterns.

Secondary air pollutant: Air pollutants formed from primary pollutants as a result of chemical processes in the atmosphere; for example, acid rain. See also *primary air pollutant.*

Secondary consumer: An animal that preys on primary consumers: carnivores representing the third trophic level in ecosystems.

Secondary forests: Forests that have replaced primary (original) forests that have been cut; includes natural second growth and tree plantations.

Secondary treatment: The second level of sewage treatment in a treatment plant; it involves aeration with additional filtering and settling.

Secure landfill: Any landfill for hazardous waste disposal that is lined, diked, and equipped with underdrains for leachate collection and monitoring wells for pollution detection.

Sedentary: Population that is geographically stable or permanent in a particular area; the opposite of nomadism.

Sedimentation: The accumulation of sediment in streams, wetlands, lakes, forests, and fields.

Sediment yield rate: The amount of sediment exported from watersheds and delivered to streams, reservoirs, and the ocean.

Selective cutting: The logging practice of harvesting only trees of certain species and sizes, leaving others in place. See also *clear-cutting.*

Sensible heat: Heat that raises the temperature of a substance such as air and thus can be sensed with a thermometer. In contrast to latent heat, it is sometimes called the heat of dry air.

Septic system: Specifically, a sewage system that relies on a septic tank to store and/or treat waste water; generally, an on-site (small-scale) sewage disposal system that depends on the soil to dispose of waste water.

Sewage treatment: See *treatment plant.*

Sharecropping: A form of tenant farming in which the crop produced is shared with the landowner by prior agreement.

Shield: The geological core area of the continents.

Shifting agriculture: An agricultural practice in tropical and equatorial areas characterized by the movement of farmers from plot to plot as soil becomes exhausted under cultivation. Also known as *slash-and-burn agriculture.*

Simplification: Reduction in the biodiversity of an ecosystem; usually accompanies replacement of native species by introduced species.

Sink: See *pollution sink.*

Slash-and-burn agriculture: See *shifting agriculture.*

Slope failure: Unstable soil or near-surface rock on hillsides that moves downslope under the force of gravity; e.g., landslide.

Smog: See *photochemical smog* and *industrial smog.*

Soil absorption systems: The term applied to sewage-disposal systems that rely on the soil to absorb waste water; usually site-scale systems, such as residential. See also *septic system.*

Soil-forming regimes: The bioclimatic conditions and related factors that combine to produce a soil with a distinctive set of traits.

Soil horizon: A layerlike zone or band in an active soil marked by a certain color, physical makeup, and chemical composition. See also *soil profile.*

Soil mantle: A traditional term used to describe the composite mass of soil material above the bedrock.

Soil orders: The major classes of soil in U.S. and other classification systems; (e.g., aridosol, spodosol, oxisol, and histosol).

Soil profile: The sequence of horizons, or layerlike zones, in a soil.

Soil texture: The cumulative sizes of particles in a soil sample; defined as the percentage by weight of sand, silt, and clay-sized particles.

Solar constant: The rate at which solar radiation is received on a surface (perpendicular to the radiation) at the edge of the atmosphere. Average strength is very close to 2 calories per square centimeter per minute.

Solar heating system: Solar energy collectors designed to provide heat for buildings; active systems use fans or pumps to circulate heat; passive systems work without mechanical assistance.

Solar radiation: The energy broadcast by the sun and received by the Earth and its atmosphere; mostly relatively shortwave radiation in the visible part of the electromagnetic spectrum. See also *solar constant.*

Solar thermal electric systems: Technologies using reflective solar collectors to generate high-temperature water to drive steam turbines and electric generators.

Solid waste: Any nongaseous and nonliquid debris discarded by cities, industry, agriculture, and mining. Nonhazardous waste; garbage and refuse of various types. See also *hazardous waste.*

Solum: The part of soil material capable of supporting life; the true soil according to the agronomist; the upper part of the soil mass, including the topsoil and soil horizons.

Source reduction: Various means of reducing hazardous waste at its source; for example, recovery and reprocessing systems. Also called *minimization.*

Speciation: The process by which new species originate in nature; see also *geographic isolation* and *polyploidy.*

Species: A group, or taxon, of individuals able to freely interbreed among themselves, but unable to breed with other groups; the smallest taxon generally recognized in biological classification.

Species diversity: See *biological diversity.*

Specific heat: A measure of the relative increase in the temperature of a substance with the absorption of energy.

SPOT (System Pour l' Observation de la Terre): The French *Earth resources satellite program* has launched four remote sensing satellites since 1986. Each satellite has two high-resolution visible (HRV) line scanners that operate either in a single wide, visible band with 10-meter ground resolution or in a three-band multispectral mode (green, red, and reflected infrared) with 20-meter ground resolution.

Squall line: The narrow zone of intensive turbulence, storm clouds, and rainfall along a cold front.

Staple crop: Basic food plants eaten by humans: rice, wheat, corn (maize), potatoes, and cassava or manioc.

Steppe: Semiarid grasslands such as the North American Great Plains.

Stormflow: The portion of streamflow that reaches the stream relatively quickly after a rainstorm, adding a surcharge of water to baseflow. Also called quickflow.

Storm surge: Elevated water surface in coastal areas caused by wind pressure associated with a hurricane.

Stratosphere: The division of the atmosphere above the troposphere, between 10 km and 50 km altitude, where zone is concentrated and most UV absorption takes place.

Stream order: The relative position, or rank, of a stream in a drainage network. Streams without tributaries, usually the small ones, are first-order, first-order streams with one or more tributaries are second-order, and so on.

Stress: A force acting on a body or substance or organism; in ecology, an environmental force of natural or human origin that can weaken or destabilize an organism, community, or ecosystem.

Strip cropping: The practice of planting strips of erosion-prone crops such as corn alternately with more soil protective crops such as oats or alfalfa.

Strip mining: The principal coal mining method in the United States; it involves stripping away surface rock and soil to extract coal.

Submarginal land: Land where the chances of successful crop farming are less than 50-50 because of environmental limitations.

Sub-Saharan Africa: The vast region of Africa south of the Sahara.

Subsistence farming: Farming to provide food primarily for the farmer's immediate family; a traditional form of farming. See also *intensive subsistence agriculture.*

Substitution: The replacement of one set of organisms for another in an ecosystem; for example, the replacement of native grasses by crops and weeds.

Subtropical high-pressure cells: Large cells of high pressure, centered at 25° to 30° latitude in both hemispheres, that are fed by air descending from aloft; these cells are the main cause of aridity in tropics and subtropics.

Subtropical zone: The zone of latitude near the tropics in both hemispheres; between 23.5° and 35°.

Succession theory: A widely held concept of vegetation change in which one community of plants and related organisms succeeds another, finally giving rise to a climax community. See also *community* and *climax community.*

Sun angle: The angle formed between the beam of incoming solar radiation and a level surface or plane at the Earth's surface.

Superfund: The U.S. act (Comprehensive Environmental Response Compensation and Liability Act) created primarily to clean up existing hazardous waste sites. Since 1980 relatively few sites have actually been cleaned up despite huge expenditures of public funds.

Surface deposits: Deposits of sediment left on the surface by streams, glaciers, wind, and other processes.

Sustainability: The capacity to build land-use systems that are environmentally balanced and enduring.

Sustainable development: Economic development designed to reduce environmental impacts and maintain the resource base over the long term.

Symbiosis: An intimate association between dissimilar species that may or may not benefit both species.

Synergism: The interaction of two or more factors with a net effect greater than the sum of the two acting separately.

Synfuel: Synthetic fuels; liquid or gaseous fuels manufactured from coal.

Synthetic crude oil: Petroleum produced from nonliquid sources such as tar sands and oil shale.

Synthetic organic compounds: Substances manufactured from organic chemicals (carbon-based) such as pesticides, cleaning compounds, and plastics that can cause various illnesses in humans and animals, including birth defects, cancers, and kidney disorders.

System: An interconnected set of objects or things; two or more components such as organisms, cities, or streams linked together in some fashion (e.g., energy systems, ecosystems, road systems).

Technology: The use of tools and machines in the extraction of resources, manufacturing of goods, and so on.

Tectonic plates: Large sections of the Earth's crust and lithosphere that shift slowing on the planet's surface.

Temperate forest: Forest of the midlatitudes including deciduous, evergreen, and mixed covers.

Temperature inversion: An atmospheric condition in which the cold air underlies warm air. Inversions are highly stable conditions, not conducive to atmospheric mixing, and are thus prone to pollution buildup where they overlie cities.

Tenant farming: Farming land without owning it; the land may be rented or the crop may be shared with the landowner for use of the land.

Tenure: See *land tenure.*

Teratogen: Any chemical that can cause damage to or spontaneous abortion of a developing fetus.

Tertiary consumer: A carnivore that preys on secondary as well as primary consumers; animals, such as birds of prey, that are near the ends of the food chains.

Tertiary recovery: Various means of "squeezing" additional oil from older oil fields with small and declining yields.

Tertiary treatment: Advanced sewage treatment following primary and secondary processes in a treatment plant.

Texture: See *soil texture.*

Thematic Mapper (TM): See *Landsat.*

Thermal gradient: The change in temperature over distance through a substance; usually expressed in degrees Celsius per centimeter or meter.

Thermal inversion: A stable thermal structure in the atmosphere in which warm air overlies cool air. Over cities and industrial areas, inversions cause buildup of air pollution.

Thermals: Drafts of upward-flowing air in a convectional storm.

Threatened species: A species with rapidly declining populations that is likely to become endangered in the foreseeable future. See also *endangered species.*

Threshold: The level of magnitude of a process at which sudden or rapid change is initiated, such as the magnitude of streamflow at which erosion begins.

Threshold of soil stability: The limit of maintenance of soil against erosion and failure where resistance is lowered by deforestation, agriculture, or other land uses.

Thunderstorm: A severe and often violent convectional storm characterized by strong winds, lightning and thunder, and occasionally tornadoes.

Tide: A large wave caused by bulges in the sea in response to the lunar and solar gravitational forces.

Tolerance: The range of stress or disturbance a living organism is able to withstand without damage or death.

Toposequence: The change in soil with topographic situation on a slope or other landform.

Topsoil: The uppermost layer of the soil, characterized by a high organic content; the organic-rich layer of active soil.

Tornado: A small, violent windstorm characterized by inward spiraling air of speeds reaching 300 miles per hour.

Total fertility rate: The average number of live children a woman will bear during her childbearing years (approximately age 15 to 44).

Toxic materials: Chemicals that can be harmful to humans and other organisms.

Toxicity: See *acute toxicity* and *chronic toxicity.*

Trade winds: The system of prevailing easterly winds that flow from the subtropical highs to the equatorial low in both hemispheres; also called the tropical easterlies.

Transpiration: The flow of water through the tissue of a plant and into the atmosphere via stomatal openings in the foliage.

Transported plant material: Soil formed in parent material comprised of deposits laid down by water, wind, glaciers, or other processes.

Treatment plant: A facility where municipal or industrial sewage is collected and processed to reduce its pollution load: three levels of treatment are performed in the most advanced plants: primary, secondary, and advanced or tertiary.

Tree farms: Planted forests made up of single species that replace wild forests. Also called tree plantations. See also *monoculture.*

Trophic level: The basic units or tiers of an energy pyramid that define the amount of energy associated with primary production followed by different levels of consumers.

Tropical forest: Any forest cover of the tropics including rainforest and tropical deciduous forest.

Tropical rainforest: Heavy forest cover of the wet tropics; biologically the richest forest in the world.

Tropical savanna: Landscape of the tropical wet/dry climate; grassland interspersed with trees and shrubs.

Tropical storm: A transient low-pressure cell in the tropics with strong winds but less than hurricane force; tropical storms may become hurricanes.

Tropics: Tropic of Capricorn, Tropic of Cancer; also the belt of latitude between these parallels.

Troposphere: The lowermost subdivision of the atmosphere with an average thickness of about 10 km; the layer that contains the bulk of the atmosphere's mass and is characterized by convectional mixing and decreasing temperature with altitude.

True extinction: The demise of a species in which its genetic lineage is not passed on to offspring (new) species.

Tsunami: A large and often destructive wave caused by tectonic activity such as faulting on the ocean floor.

Tundra: Landscape of polar and subpolar lands dominated by shrubs and ground plants and underlain by permafrost.

Typhoon: See *Hurricane.*

Ultraviolet radiation: Electromagnetic radiation of wavelengths shorter than visible, but longer than x-rays; toxic radiation absorbed by ozone in the stratosphere.

United Nations Conference on Environment and Development (UNCED): See *Earth Summit.*

Universal soil loss equation: A formula for estimating the soil loss from runoff on agricultural land; based on rainfall energy, soil, vegetative cover, slope, and crop management practices.

Upper Atmosphere Research Satellite (UARS): The advanced U.S. experimental meteorology satellite launched in 1991 designed to study the dynamics, chemistry, and energy balance of the upper atmosphere using a wide array of remote sensing instruments.

Uranium 235: The principal fuel used in fission nuclear reactors.

Urban climate: The climate in and around urban areas; it is usually somewhat warmer and foggier, with less solar radiation than the climate of the surrounding regions.

Urban sprawl: The spread of urban land uses far beyond the traditional limits of cities.

Urbanization: The term used to describe the process of urban development, including suburban residential and commercial development.

Variability: A measure of dependability or deviation from the average or mean.

Vascular plants: Plants in which cells are arranged into a pipelike system of conducting, or vascular, tissue; xylem and phloem are the two main types of vascular tissue.

Vegetation index image: Digital images acquired by the NOAA meteorology satellites that highlight vegetation growth. They are useful indicators of the condition or presence of natural or planted vegetation.

Vicariad: Endemic or microendemic species that are genetically very similar because they have evolved from a common population.

Visible radiation: Electromagnetic radiation at wavelengths between 0.4 and 0.7 micrometer; the radiation that comprises the bulk of energy emitted by the sun.

Vulnerable species: A species that is more prone to extinction than others because of its small range, specialized habitat, geographic situation, breeding habitats, or other factors. See also *endangered species* and *threatened species.*

Warm front: A contact between a cold air mass and a warm air mass in which the warm air is moving against the cold air, sliding upward along the contact.

Waste dumping: Various unregulated and unmanaged means of getting rid of waste, such as dumping in vacant land.

Waste exchanges: Organizations or arrangements among industrial installations designed to exchange and reprocess hazardous waste for its economic value.

Waste separation: The practice of segregating recyclable materials from other wastes and from one another.

Waterlogging: The saturation of soil due to overirrigation. Often associated with salinization in dry lands.

Water balance: The relative inflow and outflow of water in a lake, aquifer, or soil.

Water Pollution Control Act: See *Clean Water Act.*

Water pollution system: A simple three-part model of water pollution involving production of pollutants, removal from the production site, and delivery to a stream or water body.

Watershed: See *drainage basin.*

Water table: The upper boundary of the zone of groundwater; in fine-textured materials it is usually a transition zone rather than a boundary line. The configuration of the water table often approximates that of the overlying terrain.

Watt: A unit of power that is often used as an energy expression; equal to 1 joule (about 0.25 calorie) per second.

Weathering: The breakdown and decay of Earth materials, especially rock.

Wellhead protection: Management of groundwater pollution sources within the drainage area serving a well. See also *groundwater vulnerability.*

Westerlies: The prevailing west–east flow of air over land and water in the midlatitudes of both hemispheres; also called the Prevailing Westerlies.

Wetland: A term generally applied to an area where the ground is permanently wet or is wet most of the year. It is occupied by water-loving (or tolerant) vegetation such as cattails, mangrove, or cypress.

Wilderness Act of 1964: A federal law that established the National Wilderness Preservation System and placed millions of acres of national forest land in the NWPS.

Wilderness area: An area where the environment has not been seriously disturbed and altered by humans; open space with light human populations. See also *frontier environment.*

Wildlife preserve: An area set aside for wildlife in which the environment and animals are protected from land use, hunting, and related human activities.

Wind power: The power generated by wind; proportional to the cube of speed or velocity.

Work: A concept closely related to energy, work is the product of force and distance and is accomplished when the application of force yields movement of an object in the direction of the force.

World Programme of Action (WPOA): See *International Conference on Population and Development.*

Zenith: For any location on Earth, the point that is directly overhead to an observer. The zenith position of the sun is the one directly overhead.

Zonal circulation: A reference, to geographically extensive atmosphere circulation reaching across different zones of latitude.

Zonal soil: Soils with well-developed horizons that reflect the climate conditions of the region in which they are found.

Zonation: The levels or zones of vegetation and related organisms at different elevations on a mountain.

Zone of eluviation: The level, or zone, in a soil losing materials in the form of colloids and ions as a part of soil formation; the zone of mineral removal.

Zone of illuviation: The level, or zone, in a soil where colloids and ions accumulate; the zone of mineral accumulation.

Zone of saturation: See *groundwater.*

CREDITS

Photo Credits

Chapter 1 Figure 1.1 (a): Steve Kahn/FPG International. Figure 1.1 (b): John Eastcott/Yva Momatiuk/The Image Works. Figure 1.1 (c): Tony Savino/The Image Works. Figure 1.3 (a): Georg Gerster/Photo Researchers. Figure 1.3 (b): Yann Arthus-Bertrand/Corbis Images. Figure 1.3 (c): Roger Lemoyne/Liaison Agency, Inc. Figure 1.4: Georg Gerster/Photo Researchers.

Chapter 2 Figure 2.1 (a): Bill Bachman/Photo Researchers. Figure 2.1 (b): Grant Heilman Photography. Figure 2.13: Antonio Mari/Liaison Agency, Inc.

Chapter 3 Figure 3.2: Chris Steele-Perkin/Magnum Photos, Inc. Figure 3.3: Corbis Sygma. Figure 3.4: Dinodia/Omni-Photo Communications. Figure 3.8: Still Pictures/Peter Arnold, Inc.

Chapter 4 Figure 4.5: Gerry Ellis/ENP Images. Figure 4.6: Georg Gerster/Photo Researchers. Figure 4.11: © Alyeska Pipeline Service Company. Figure 4.12: Francois Ancellet/Rapho/Liaison Agency, Inc.

Chapter 5 Figure 5.2: Piero Pomponi/Liaison Agency, Inc.

Chapter 6 Figure 6.7 (a) & (b): Alex S. Maclean/Lanslides. Figure 6.7 (c): Charlie Ott/Photo Researchers. Figures 6.9 & 6.11: Courtesy NASA. Figure 6.12 (c): Tom Bean/DRK Photo. Figure 6.13 (a) & (b): Nina Marsh. Figure 6.14 (a): UNEP-Topham/The Image Works. Figure 6.14 (b): Bruce Gordon/Photo Researchers.

Chapter 7 Figure 7.10 (a): Roger Job/Liaison Agency, Inc.. Figure 7.10 (b): Filip Horvat/SABA. Figure 7.11: Alex S. MacLean/Landslides. Figure 7.13 (inset): Pascal Maitre/Gamma Liaison.

Chapter 8 Figure 8.8 (a): Ron Watts/Corbis Images. Figure 8.8 (b): Luiz C. Marigo/Peter Arnold, Inc. Figure 8.8 (c): Ed Kashi/Corbis Images. Figure 8.8 (d): Jim Sugar Photography/Corbis Images. Figure 8.9: Andrew Holbrooke/The Stock Market. Figure 8.12 (a): Lineair/Peter Arnold, Inc. Figure 8.12 (b): Howard Buffett/Grant Heilman Photography. Figure 8.13 (a): Roth Stein Library of Congress. Figure 8.13 (b): Mark Wagner/Stone. Figure 8.14 (b): Helene Caux/Liaison Agency, Inc. Figure 8.15 (a) & (b): Georg Gerster/Photo Researchers. Figure 8.15 (c): Kevin Flaming/Corbis Images.

Chapter 9 Figure 9.6 (a): Denneth Murray/Photo Researchers. Figure 9.6 (b): Grant Heilman Photography. Figure 9.6 (c): Courtesy National Coal Association. Figure 9.15 (a): Steve Dunwell/The Image Bank. Figure 9.15 (b): Stone. Figure 9.17: Calvin Larsen/Photo Researchers.

Chapter 10 Figure 10.12: Bartholomew/Liaison Agency, Inc.

Chapter 11 Figure 11.4 (a) & (b): Jim Mendenhall. Figure 11.6: Michael S. Yamashita/Corbis Images. Figure 11.13: Dillon Aerial. Figure 11.14 (a): Bobbie Kingsley/Photo Researchers. Figure 11.14 (b): Runk Schoenberger/Grant Heilman Photography.

Chapter 12 Figure 12.2: William Marsh. Figure 12.3: Charlie Ott/Photo Researchers. Figure 12.8 (a): B. Wisser/Liaison Agency, Inc. Figure 12.8 (b): Reuters/Corbis Images. Figure 12.12: Warren Winter/Sygma. Figure 12.13: A. Moldvay/Eriako Associates. Figure 12.19: Courtesy Cambridge University Collection of Air Photographs.

Chapter 13 Figure 13.2: Courtesy USGS. Figure 13.2 (inset): Andree Kaiser/Sipa Press. Figure 13.16: Robert P. Carr/Bruce Coleman, Inc.

Chapter 14 Figure 14.2: Courtesy City of New York Department of Sanitation. Figure 14.5: Corbis-Bettmann. Figure 14.5 (inset): Nathan Beck/Omni-Photo Communications. Figure 14.8: Courtesy United Nations. Figure 14.9: Michael Philippot/Sygma Photo News. Figure 14.12: Vince Streano/Corbis Images. Figure 14.13 (a): Corbis-Bettmann. Figure 14.13 (b): S. Dooley/Liaison Agency, Inc. Figure 14.13 (c): Phil Degginger/Bruce Coleman, Inc. Figure 14.14: Allen Green/Photo Researchers.

Chapter 15 Figure 15.3 (a): Frans Lanting/Minden Pictures, Inc. Figure 15.3 (b): Tom McHugh/Photo Researchers. Figure 15.14: Joyce Photographers/Photo Researchers. Figure 15.15: Georg Gerster/Photo Researchers.

Chapter 16 Figure 16.8: Courtesy NASA. Figure 16.10 (a) & (b): William Marsh. Figure 16.10 (c): Michael Horvy/Panos Pictures. Figure 16.12: Francois Gohier/Photo Researchers. Figure 16.13: Larry Lefever/Grant Heilman Photography. Figure 16.15: Courtesy USDA. Figure 16.16: Georg Gerster/Photo Researchers. Figure 16.18 (a): Courtesy NASA. Figure 16.18 (b): J. Lanley/Courtesy William Marsh.

Chapter 17 Figure 17.2 (a): Courtesy USDA. Figure 17.2 (b): Corbis-Bettmann. Figure 17.3: Courtesy United States Department of the Interior, National Park Service Photo. Figure 17.7 (a): John M. Burnley/Bruce Coleman, Inc. Figure 17.7 (b): Calvin Larsen/Photo Researchers. Figure 17.12 (a): Hubertus Kanus/Photo Researchers. Figure 17.12 (b): New Zealand Tourist Board. Figure 17.13 (a): Grant Heilman Photography. Figure 17.13 (b): Linde Waidhofer/Western Eye/Liaison Agency, Inc. Figure 17.13 (c): Runk/Schoenberger/Grant Heilman Photography. Figure 17.14: Landslides.

Chapter 18 Figure 18.2 (a): Roger Lemoyne/Liaison Agency, Inc. Figure 18.2 (b): Dermot Tatlow/Panos Pictures. Figure 18.2 (c): Noel Quidu/Liaison Agency, Inc. Figure 18.2 (d): Carleton Ray/Photo Researchers. Page 388: Lee Celano/Liaison Agency, Inc. Figure 18.5: William Marsh. Figure 18.7: Courtesy NOAA. Figure 18.8 (top): Courtesy Barbara Summey, NASA GSFC. Figure 18.8 (bottom left): Courtesy NASA/GSFC/MITI/ERSDAC/JAROS, and U.S./Japan ASTER Science Team. Figure 18.8 (bottom center): Courtesy NASA/GSFC/JPL, MISR Science Team. Figure 18.8 (bottom right): Courtesy Jacques Descloitres, MODIS, Land Science Team. Figure 18.9 (a): NASA/Science Source/Photo Researchers. Figure 18.9 (b) & (c): © CNES 1993/Spot Image Corporation. Figure 18.10: Courtesy NASA Goddard Space Flight Center. Figure 18.11: Courtesy Aerial Images, Inc. Figure 18.12: Courtesy Space Imaging. Figure 18.13: William Marsh. Figure 18.14: Courtesy NASA/JPL/NIMA. Figure 18.15: Courtesy Canadian Space Agency 1997. Received by the Candade Centre for Remote Sensing. Processed & distributed by RADARSAT International.

Line Illustrations

Chapter 1 Figure 1.2: United Nations. Figure 1.5: by W.M. Marsh. Figure 1.6: by W.M. Marsh. Figure 1.7: by W.M. Marsh and J. Grossa. Figure 1.8: by W.M. Marsh. Figure 1.9: by W.M. Marsh. Figure 1.10: by W.M. Marsh.

Chapter 2 Figures 2.2, 2.6, 2.7, 2.8: from Marsh, W.M., *Earthscape: A Physical Geography*, Wiley, 1987. Figure 2.3a: from Marsh, W.M., *Earthscape: A Physical Geography*, Wiley, 1987. Figure 2.3b: by W.M. Marsh. Figure 2.4: from W.M. Marsh, *Earthscape*, Wiley, 1989. Figure 2.5: from W. Marsh, *Earthscape*, Wiley, 1989. Figure 2.9: based on NASA imagery. Figure 2.10: by W.M. Marsh, 1995. Figure 2.11: from Roger Bilham, University of Colorado.

Chapter 3 Figures 3.1, 3.5: by W.M. Marsh, 1995. Figure 3.6: by J. Grossa. Figure 3.7: by W.M. Marsh.

Chapter 4 Figure 4.1: after National Geographic map. Figure 4.2: after Wolman, M.G. and Fournier, F.G.A. (eds), *Land Transformation by Agriculture*, SCOPE, Wiley, 1987. Figure 4.3: W.B. Greeley, "The Relation of Geography to Lumber Supply," *Economic Geography*, 1925. Figure 4.4: base map from DeBlij, H.J. and Muller, P.O., *Geography: Realms, Regions, and Concepts*, Wiley, 1994. Figures 4.7, 4.8, 4.9, 4.10: by W.M. Marsh, 1995. Figure 4.13: from Marsh, W.M., *Earthscape: A Physical Geography*, Wiley, 1987. Figure 4.14: by W.M. Marsh, 1995.

Chapter 5 Figure 5.1: adapted from Marsh, W.M., *Earthscape: A Physical Geography*, Wiley, 1987. Figures 5.3, 5.4, 5.5, 5.6: by W.M. Marsh, 1995. Figure 5.7: based on data from Budyko, M.I., *Climate and Life*, Academic Press, 1974. Figure 5.8: from Marsh, W.M., *Earthscape: A Physical Geography*, Wiley, 1987. Figures 5.9, 5.10, 5.11, 5.12, 5.13: by W.M. Marsh, 1995.

Chapter 6 Figures 6.1, 6.2, 6.3, 6.8, 6.15: by W.M. Marsh, 1995. Figures 6.3, 6.4, 6.10: adapted from Marsh, W.M., *Earthscape: A Physical Geography*, Wiley, 1987. Figure 6.6: W.M. Marsh, Earthscape, Wiley, 1987. Figure 6.10: from NASA imagery. Figure 6.16: based on data from Cottam, The Ecologist's Role in Problems of Pesticide Pollution, *BioScience:* 15, 1965; and Rudd, R.L., *Pesticides and the Living Landscape*, University of Wisconsin Press, 1964. Figure 6.17: after Marsh, W.M., *Landscape Planning: Environmental Applications*, Wiley, 1991.

Chapter 7 Figure 7.1: by J.M. Grossa, Jr., 1995. Figure 7.2: DeBlij, H.J. and Muller, P.O., *Geography: Realms, Regions, and Concepts*, Wiley, 1994. Figures 7.3, 7.5: data from Population Reference Bureau, World Population Data Sheet, 1994. Figure 7.4: by W.M. Marsh, 1995. Figure 7.7: by J.M. Grossa and W.M. Marsh, 1995. Figures 7.8 and 7.9: data from Jones, H., *Population Geography*, 2nd ed., London: P. Chapman Publishing Ltd., 1990. Figures 7.12 and 7.13rt: by W.M. Marsh, 1995. Figure 7.14: from Bongaarts, J., "Population Policy Options in the Developing World," *Science*, 263, 1994.

Chapter 8 Figure 8.1: by J.M. Grossa, Jr., and W.M. Marsh, 1995. Figures 8.2, 8.3, 8.4, 8.5, 8.7: based on various sources. Figure 8.10(a): data from *World Resources 1992–1993*, New York: Oxford Press, 1992. Figure 8.10(b): by W.M. Marsh, 1995. Figure 8.13: based on U.S. Geological Survey; (photo) Figure 8.14: from Micklin, P.P., "Desiccation of the Aral Sea: A Water Management Disaster in the Soviet Union," *Science*, 241, 1989. Figure 8.15a: by W.M. Marsh. Figure 8.15b: by W.M. Marsh. Figure 8.15c: by W.M. Marsh.

Chapter 9 Figures 9.1, 9.4, 9.8, 9.10, 9.11, 9.12: data from *Annual Energy Review*, 1991. Figure 9.2: adapted from Davis, G., "Energy for Planet Earth," *Scientific American*, 1990. Figure 9.3: by W.M. Marsh, 1995. Figures 9.5, 9.9, 9.13: *Annual Energy Review*, 1991, and other sources. Figure 9.7: by W.M. Marsh, 1995. Figure 9.14: data from *Annual Energy Review*, 1991, and Canada Department of Energy, 1995.

Chapter 10 Figures 10.1, 10.2, 10.3, 10.4, 10.9, 10.13: adapted from Marsh, W.M., *Earthscape: A Physical Geography*, Wiley, 1987. Figures 10.5, 10.6, 10.7, 10.8, 10.10(upper): by W.M. Marsh, 1995. Figures 10.10(lower), 10.12, 10.14: from NOAA.

Chapter 11 Figures 11.1, 11.4, 11.5, 11.7, 11.8, 11.9, 11.10: by W.M. Marsh, 1995. Figures 11.2, 11.3: data from *World Resources 1994–95*, New York: Oxford Press, 1994. Figure 11.11: after Likens, G.E. et.al. "Acid Rain," *Scientific American*, 1979, and U.S. EPA, 1987. Figure 11.18(b): data from U.S. Census.

Chapter 12 Figures 12.1, 12.5, 12.9, 12.11, 12.14, 12.16, 12.17, 12.20: by W.M. Marsh, 1995. Figures 12.2, 12.3: from Marsh, W.M., *Earthscape: A Physical Geography*, Wiley, 1987. Figures 12.15, 12.18: from U.S. Geological Survey.

Chapter 13 Figures 13.1, 13.3, 13.4, 13.9, 13.10, 13.11, 13.14, 13.15, 13.17: by W.M. Marsh, 1995. Figures 13.5, 13.7: from *Farmline*, U.S. Department of Agriculture. Figure 13.12: data from U.S. EPA and Environment Canada. Figure 13.13: data from *World Resources 1994–95*, New York: Oxford Press, 1994.

Chapter 14 Figure 14.1: various sources. Figures 14.3, 14.6, 14.7, 14.10, 14.14: by W.M. Marsh, 1995. Figure 14.11: from U.S. Geological Survey.

Chapter 15 Figures 15.1, 15.2, 15.6, 15.8, 15.9, 15.12, 15.16: by W.M. Marsh, 1995. Figures 15.4, 15.5, 15.7, 15.11, 15.18: from Marsh, W.M., *Earthscape: A Physical Geography*, Wiley, 1987. Figure 15.13 United Nations. Figure 15.17: National Research Council, *Alternative Agriculture*, National Academy Press, 1989.

Chapter 16 Figures 16.1, 16.2, 16.3, 16.9, 16.11, 16.14: by W.M. Marsh, 1995. Figure 16.5: adapted from Illies, J. *Introduction to Zoogeography*, London: MacMillan, 1974. Figure 16.6: partly from Collins, M. (ed.), *The Last Rainforect*, New York: Oxford University Press, 1990. Figure 16.7: from MacArthur, R.H. and E.D. Wilson, *The Theory of Island Biogeography*, Princeton, N.J.: Princeton University Press, 1967. Figure 16.12: by Frank R. Thompson, North Central Forest Experiment Station. Figure 16.13: from J.T. Kartesz, North Carolina Botanical Garden. Figure 16.17: from various sources.

Chapter 17 Figure 17.1: from Castillon, D.A., *Conservation of Natural Resources*, Dubuque, Iowa: W.C. Brown, 1992. Figures 17.4, 17.5: by W.M. Marsh, 1995. Figure 17.6: data from *World Resources 1994–1995*. Figure 17.8: data from Cutter, S.L., et al., *Exploitation, Conservation, and Preservation*, New York: Wiley, 1991. Figure 17.9: adapted from National Geographic map. Figure 17.11: adapted from Cutter, S.L., et al., *Exploitation, Conservation, Preservation*, New York: Wiley, 1991. Figure 17.15: data from *Wilderness Management* (?)

Chapter 18 Figures 18.3, 18.4, 18.17, 18.18, 18.19: by W.M. Marsh, 1995. Figure 18.6: from World Meteorological Organization, 1988. Figure 18.8: by J. Grossa. Figure 18.16: from Avery, T.E. and G.L. Berlin, *Fundamentals of Remote Sensing and Airphoto Interpretation*, 5th ed., New York: MacMillan, 1992.

INDEX